MATERIALS SURFACE PROCESSING BY DIRECTED ENERGY TECHNIQUES

Elsevier Internet Homepage - http://www.elsevier.com
Consult the Elsevier homepage for full catalogue information on all books, major reference works, journals, electronic products and services.

Elsevier Titles of Related Interest

Joining of Materials and Structures
Robert Messler
2004, 0-7506-7757-0

Fracture of Polymers, Composites and Adhesives II
J.G. Williams, A. Pavan, Bamber Blackman
2003, 0-08-044195-5

Adhesion Science and Engineering
Pocius
2002, 0-444-51140-7

Related Journals:

Elsevier publishes a wide-ranging portfolio of high quality research journals, encompassing the adhesives, composites, and polymer fields of materials science. A sample journal issue is available online by visiting the Elsevier web site (details at the top of this page). Leading titles include:

International Journal of Adhesion and Adhesives
Construction and Building Materials
Mechanics of Materials
Composites Science and Technology
Composites Part A: Applied Science and Manufacturing
Composites Part B: Engineering
Polymer
European Polymer Journal
Dental materials
NDT International
Reinforced Plastics
Diamond and Related Materials
Surface and Coatings Technology
Thin Solid Films
Vacuum

All journals are available online via ScienceDirect: www.sciencedirect.com

To contact the Publisher
Elsevier welcomes enquiries concerning publishing proposals: books, journal special issues, conference proceedings, etc. All formats and media can be considered. Should you have a publishing proposal you wish to discuss, please contact, without obligation, the publisher responsible for Elsevier's Material Science programme:

David Sleeman
Publishing Editor
Elsevier Ltd
The Boulevard, Langford Lane
Kidlington, Oxford
OX5 1GB, UK

Phone: +44 1865 843265
Fax: +44 1865 843987
E.mail: d.sleeman@elsevier.com

General enquiries, including placing orders, should be directed to Elsevier's Regional Sales Offices – please access the Elsevier homepage for full contact details (homepage details at the top of this page).

MATERIALS SURFACE PROCESSING BY DIRECTED ENERGY TECHNIQUES

EDITED BY

YVES PAULEAU
School of Electrochemical and Electrometallurgical Engineering
National Polytechnic Institute of Grenoble, Grenoble, France

Published in Association with

E·MRS
EUROPEAN
MATERIALS RESEARCH SOCIETY

2006

ELSEVIER

Amsterdam • Boston • Heidelberg • London • Oxford • New York
Paris • San Diego • San Francisco • Singapore • Sydney • Tokyo

Elsevier
The Boulevard, Langford Lane, Kidlington, Oxford, OX5 1GB, UK
84 Theobald's Road, London WC1X 8RR, UK
Radarweg 29, PO Box 211, 1000 AE Amsterdam, The Netherlands
525 B Street, Suite 1900, San Diego, CA 92101-4495, USA

First edition 2006-03-07

British Library Cataloguing in Publication Data
A catalogue record for this book is available from the British Library

Library of Congress Cataloging-in-Publication Data
A catalog record for this book is available from the Library of Congress

ISBN 13: 9-78-0-08-044496-3
ISBN 10: 0-08-044496-2

Typeset by Charon Tec Ltd, Chennai, India
www.charontec.com
Printed and bound in Great Britain

06 07 08 09 10 10 9 8 7 6 5 4 3 2 1

Contents

Preface

Multicomponent, nanostructured and functionally graded coatings or thin films may exhibit unique physical, mechanical, chemical properties ensuring remarkable degradation resistance where the surface protection of materials against wear, corrosion, friction is a key issue, in particular for mechanical assemblies operating in hostile environment (vacuum, extreme temperature, corrosive atmosphere). Nowadays, coating and thin-film technology is pervasive in a large variety of applications, including micromechanics, microelectronics, optics, etc.

In each field of applications, major scientific and technological advances depend on the ability to control the deposition and microstructure (at the atomic level) of coatings and thin films with thickness ranging from micrometers to tens of angströms. New deposition techniques using high-energy beams or particles (laser, ion, plasma processing) are required for the preparation and production of coatings and thin films at the industrial scale. A particular challenge for industry is to establish technological facilities for the production of coated parts with specific shapes and large sizes, and to develop alternative technologies replacing surface treatment techniques polluting the environment. The development of techniques and the achievement of good understanding of the basic physical-, chemical- and materials-related processes for the deposition and characterization of coatings, thin films and surface modifications are crucial points for successful applications and implementation of new advanced environmentally benign technologies.

A vast number of deposition techniques are available and in use today. Each technique has specific limitations involving compromises with respect to process specifications, substrate materials limitations, expected properties of coatings and thin films, and cost. Over the few past years, ion and laser beam materials processing technologies for surface modifications as well as plasma-based chemical and physical vapor deposition techniques experienced a very rapid growth. These techniques appropriate to solve a given surface engineering problem are highly differentiated. As a result, it becomes difficult to maintain a clear overlook and understanding in this very interdisciplinary field of research and applications.

This book provides a broad overview on modern coating and thin-film deposition techniques. The major purpose is to present and discuss various problems of physics and chemistry involved in the production, characterization and applications of coatings and thin films, which can be variously hard and wear resistant for cutting tools, corrosion and abrasion resistant for turbine blades, self-lubricant for ball bearings, etc. The current status of the science and technology related to coatings, thin films and surface modifications produced by directed energy techniques is assessed through a series of 20 chapters. For each chapter presented at a tutorial level, a balance is found between fundamental aspects and experimental results illustrating various models, mechanisms or theories. New trends and new results are also evoked to have an overlook about future developments and applications. The book is intended to assist scientists, engineers, and researchers in learning and updating various aspects of physics and chemistry involved in materials surface processing and deposition techniques.

I would like to express my gratitude to the authors with internationally recognized expertise from 15 different countries who contributed to the preparation of chapters published in this collective book. I am aware of the high personal commitment and time-consuming work having been necessary to elaborate such excellent chapters.

Yves Pauleau
Grenoble, France

List of Contributors

(Numbers in parenthesis indicate the pages on which the author's contribution begins.)

José M. Albella (345)
Instituto Ciencia de Materiales, CSIC
Cantoblanco, 28049 Madrid, Spain

James Arps (151)
Southwest Research Institute,
6220 Culebra Road
San Antonio, TX 78242, USA

Péter B. Barna (443)
Department of Thin Film Physics
Research Institute for Technical Physics and Materials Science of the Hungarian Academy of Sciences
H-1121 Budapest, Konkoly-Thege Miklós út 29–33, Hungary

Pavel Baroch (67)
Department of Physics, University of West Bohemia
Univerzitní 22, 306 14 Plzen,
Czech Republic

Gerhard Betz (1)
Institut für Allgemeine Physik,
Technische Universität Wien
Wiedner Hauptstr. 8–10/134, A-1040
Wien, Austria

José Castanho (537)
Faculty of Sciences and Technology of Coimbra University
ICEMS – Coimbra,
3030–201 Coimbra, Portugal

Jean-Pierre Celis (111)
Katholieke Universiteit Leuven,
Dept. MTM
Kasteelpark Arenberg 44,
B-3001 Leuven, Belgium

Costel Sorin Cojocaru (411)
Laboratoire de Physique des Interfaces et Couches Minces
UMR 7647 Bâtiment 406 École Polytechnique,
91128 Palaiseau Cedex
France

Goeffrey Dearnaley (151)
19826 Wittenburg Road
San Antonio, TX 78256, USA

Marie-Paule Delplancke-Ogletree (383)
Université Libre de Bruxelles,
Faculté des Sciences Appliquées
Chimie Industrielle
CP 165/63, 50 Avenue F.D. Roosevelt
1050 Bruxelles, Belgique

Jean-Paul Deville (411)
Institut de Physique et Chimie des Matériaux de Strasbourg (IPCMS)
CNRS, Groupe "Surfaces-Interfaces"
23 Rue du Loess, B.P. 43,
67034 Strasbourg Cedex 2, France

Christophe Donnet (573)
Member of the University Institute of France
Universite Jean Monnet de Saint-Etienne
Laboratoire Traitement du Signal et Instrumentation
UMR 5516
Bât. F–18 rue Professor Benoit Lauras
42000 Saint-Etienne
France

Wolfgang Ensinger (595)
Darmstadt University of Technology
Department of Materials and Geo Sciences, Chemical Analytics Group
Petersenstrasse 23, 64287 Darmstadt, Germany

Ali Erdemir (573)
Energy Technology Division,
Argonne National Laboratory
Argonne, IL 60439
USA

Pierre Fauchais (311)
Laboratoire Sciences des Procédés Céramiques et de Traitements de Surface,
SPCTS, UMR CNRS 6638, Université de Limoges,
123 Avenue Albert Thomas,
87060 Limoges Cedex, France

Anthony Fischer-Cripps (475)
CSIRO Division of Industrial Physics
Bradfield Rd, West Lindfield,
NSW 2087, Australia

Francesco Fuso (275)
Istituto Nazionale per la Fisica della Materia and Dipartimento di Fisica "Enrico Fermi"
Universita' di Pisa
Largo Bruno Pontecorvo 3,
56127 Pisa, Italy

Raúl Gago (345)
Centro de Microanálisis de Materiales,
Universidad Autónoma
Cantoblanco, 28049 Madrid, Spain

Norihiro Inagaki (659)
Faculty of Engineering,
Shizuoka University
3-5-1 Johoku, Hamamatsu, 432–8651
Japan

Yury Ivanov (205)
Institute of High-Current Electronics
Academichesky Avenue 2/3, 634021,
Tomsk, Russia

Ignacio Jiménez (345)
Instituto Ciencia y Tecnología de Polímeros, CSIC
Juan de la Cierva 3, 28006 Madrid, Spain

Cristina Louro (537)
Faculty of Sciences and Technology of Coimbra University
ICEMS – Coimbra,
3030–201 Coimbra, Portugal

Boguslaw Major (241)
Institute of Metallurgy and Materials Science
Polish Academy of Sciences
ul. Reymonta 25, 30–059 Cracow, Poland

Alexey Markov (205)
Institute of High-Current Electronics
Academichesky Avenue 2/3,
634021, Tomsk, Russia

Jindrich Musil (67)
Department of Physics,
University of West Bohemia
Univerzitní 22, 306 14 Plzen,
Czech Republic

Yves Pauleau (501)
National Polytechnic Institute of Grenoble
CNRS-LEMD, B.P. 166
25 Rue des Martyrs, 38042 Grenoble
Cedex 9, France

Balakrishnan Prakash (111)
Katholieke Universiteit Leuven,
Dept. MTM
Kasteelpark Arenberg 44,
B-3001 Leuven, Belgium

György Radnóczi (443)
Department of Thin Film Physics
Research Institute for Technical
Physics and Materials Science of the
Hungarian Academy of Sciences
H-1121 Budapest, Konkoly-Thege
Miklós u. 29–33, Hungary

Vladimir Rotshtein (205)
Tomsk State Pedagogical University
Komsomolsky Avenue.75, 634041,
Tomsk, Russia

Jørgen Schou (35)
Department of Optics and Plasma
Research
Risø National Laboratory
4000 Roskilde, Denmark

Armelle Vardelle (311)
Laboratoire Sciences des Procédés
Céramiques et de Traitements de
Surface
SPCTS, UMR CNRS 6638, Université de
Limoges
123 Avenue Albert Thomas,
87060 Limoges Cedex, France

Robert Vaßen (629)
Institute for Materials and Processes
in Energy Systems (IWV1)
Forschungszentrum Jülich GmbH, IWV1,
52425 Jülich, Germany

Teresa Vieira (537)
Faculty of Sciences and Technology
of Coimbra University
ICEMS – Coimbra,
3030–201 Coimbra, Portugal

Jaroslav Vlcek (67)
Department of Physics,
University of West Bohemia
Univerzithi 22, 306 14 Plzen,
Czech Republic

Chapter 1

Interaction of Ions and Electrons with Solid Surfaces

Gerhard Betz

1.1. Introduction

In this chapter the basic processes, if energetic ions or electrons enter a solid, like energy loss per unit length and range are discussed. The main difference between ions and electrons is the mass difference between them. Ions can transfer up to 100% of their energy and momentum to a target atom at equal mass, leading to energetic recoils, vacancies, interstitials and the development of a collision cascade, resulting in the emission target atoms, that is sputtering or surface erosion. On the other hand electrons are elastically reflected from the atoms, without loosing energy (except for bremsstrahlung at high energies) and will loose energy essentially due to inelastic processes, like excitation or ionization of bound or free electrons. No emission of target atoms, due to collision is observed.

1.2. Ion Bombardment of a Solid

If energetic ions or a beam of ions impinge on the surface of a solid target, a variety of different processes will occur. Usually, ions are used for bombardment as they can readily be accelerated to a defined energy and mass selected. As long as the equilibrium charge state of the penetrating particle is close to zero, no clear evidence has been found that bombardment by ions or otherwise identical neutral particles will lead to different results with respect to emitted atoms or clusters. This is definitely true for metals and for ion energies up to a few hundred keV. For semiconductors and insulators or very highly charged ions some differences have been reported by some authors. In the following we will usually use the term “ion”, but this means ion or neutral, except where specifically stated.

Materials Surface Processing by Directed Energy Techniques
Y. Pauleau (Editor)

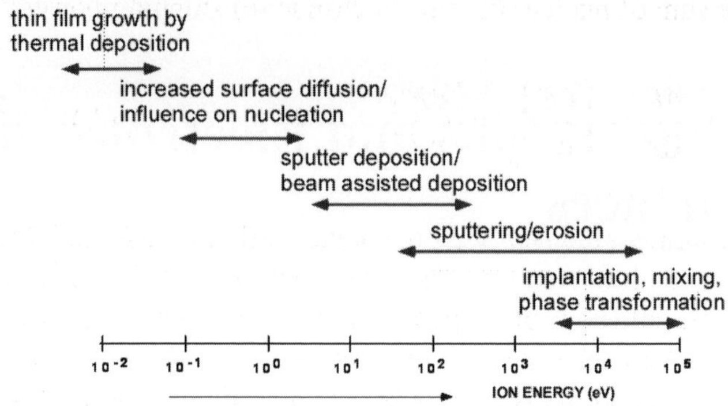

Figure 1.1: Different processes occurring under ion bombardment of a solid depending on ion energy.

Bombardment of a surface by ions gives rise to a variety of elastic and inelastic collision events leading to the ejection of a large variety of particles and radiation from the target, modifications of the structure of the target, like amorphization or phase changes, implantation or reflexion of the primary particles. The possible processes as a function of the energy of the primary ions are indicated in Fig. 1.1, and range from thin-film deposition at thermal energies, surface erosion (sputtering) mainly in the keV range, to ion implantation as widely used for doping of semiconductors, to mixing and phase changes.

These processes will be discussed starting at the lowest energies and up into the MeV range, where ion implantation is the dominant process. Bombardment at even higher energies (GeV), like observed in cosmic radiation penetrating a target, will not be discussed. Common to all these phenomena is the energy loss process of an energetic particle penetrating a solid, and this will be discussed before the individual processes are described.

For a more detailed information on this topic of ion bombardment of and implantation in a solid the reader is referred to the three volumes edited by Behrisch on sputtering [1–3], a book on fundamentals and applications of ion–solid interactions by Nastasi *et al.* [4] and on computer simulations of ion–solid interaction and sputtering the books from Eckstein [5] and Smith [6].

1.2.1. Background

If an energetic particle penetrates a target, its kinetic energy is dissipated via elastic (nuclear collisions) and inelastic (electronic excitation) processes. According to Lindhard *et al.* [7,8] the energy loss or stopping power per length unit in the target is given as the sum of nuclear (n) and electronic (e) stopping powers:

$$\frac{\mathrm{d}E}{\mathrm{d}x} = \left(\frac{\mathrm{d}E}{\mathrm{d}x}\right)_{\mathrm{n}} + \left(\frac{\mathrm{d}E}{\mathrm{d}x}\right)_{\mathrm{e}} = -N(S_{\mathrm{n}}(E) + S_{\mathrm{e}}(E)) \tag{1.1}$$

where N is the number density of atoms in the medium, $S_{\mathrm{n}}(E)$ and $S_{\mathrm{e}}(E)$ are the so-called nuclear and electronic stopping cross-sections. $S_{\mathrm{n}}(E)$ determines the energy loss due to elastic collisions of the penetrating ion with target atoms and $S_{\mathrm{e}}(E)$ determines the energy loss due to inelastic processes, like excitations of bound electrons of target atoms or bond breaking in a target consisting of molecules. This separation into elastic and inelastic processes can be made, as due to the large mass difference between ion and electron, excitation of an electron typically involves a small energy transfer and does not change the momentum of the ion. On the other hand the elastic collision of an ion with a target atom leads typically to a much higher energy and a large momentum transfer.

The nuclear stopping cross-section can be derived from appropriate screened Coulomb potentials [9,10], which were originally derived by Lindhard, Scharff and Schiott to calculate the range of ions in solids (LSS theory) [7–9]. Different interatomic potentials $V(r)$ have been proposed in the literature, like the Moliere potential [11], the krypton–carbon (Kr-C) potential [12], or the Ziegler–Biersack–Littmark (ZBL) or universal potential [13]. They all are of the form of a screened Coulomb potential:

$$V(r) = \frac{Z_1 Z_2 e^2}{r}\Phi(r/a) \tag{1.2}$$

where Φ is the so-called screening function, which screens the repulsive Coulomb forces of the nuclei, due to the partial shielding by the surrounding electron clouds. Only in the MeV range, at near head-on collisions the nuclei come so close to each other, that pure Rutherford scattering is observed. For the Kr-C potential Φ has the form [12]:

$$\Phi(x) = 0.190945\mathrm{e}^{-0.131825x} + 0.473674\mathrm{e}^{-0.63717x} + 0.335381\mathrm{e}^{-1.919249x} \tag{1.3a}$$

$$x = r/a_{\mathrm{L}} \quad a_{\mathrm{L}} = 0.8854a_0(Z_1^{2/3} + Z_2^{2/3})^{-1/2} \quad a_0 = 0.0529\,\mathrm{nm} \tag{1.3b}$$

where a_{L} is the Lindhard screening radius for the Thomas–Fermi interaction between atoms and a_0 is the Bohr radius. In first order the nuclear stopping cross-section

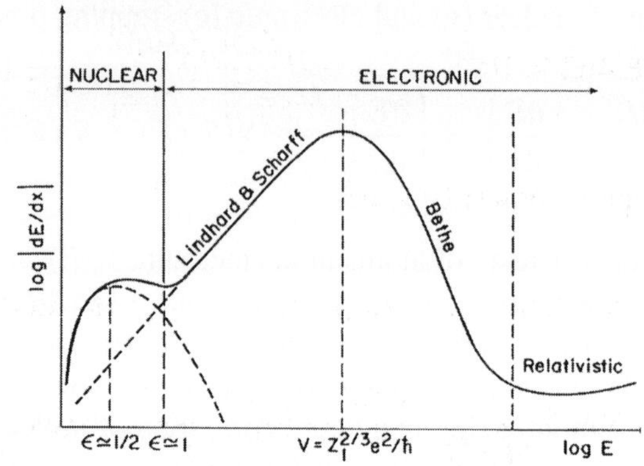

Figure 1.2: Stopping power of an ion as a function of the dimensionless energy ε. At low-velocities ($\varepsilon < 1$) nuclear stopping dominates for heavy and medium mass ions (keV range and below), and is competitive for light ions. At higher-velocities nuclear stopping decreases and electronic stopping becomes dominant (MeV range). Below the electronic stopping power maximum, the projectile is mainly neutral, above in the Bethe regime [16] it is preferably stripped. In the extreme relativistic regime the (electronic) stopping power increases again (after Ref. [17]).

increases linear with energy at low energies, reaches a plateau at intermediate energies, for example ~10–100 keV and then decreases at higher energies (see Fig. 1.2). In many cases a dimensionless energy ε instead of E and a dimensionless range ρ instead of x are used:

$$\varepsilon = \frac{M_2 E}{M_1 + M_2} \frac{a_L}{Z_1 Z_2 e^2} \qquad \rho = xNM_2\, 4\pi a_L^2 \frac{M_1}{(M_1 + M_2)^2}$$
$$\varepsilon = \frac{32.53 M_2 (Z_1^{2/3} + Z_1^{2/3})^{-1/2} E(\mathrm{keV})}{(M_1 + M_2) Z_1 Z_2} \tag{1.4}$$

where M_1 is the projectile mass, M_2 is the target atom mass and e is the electron charge. Also the nuclear stopping cross-section $S_n(E)$ can be expressed in a reduced form $s_n(\varepsilon)$ (see Fig. 1.2), independent on the ion target combination. The dimensionless reduced stopping cross-section $s_n(\varepsilon)$ is connected with $S_n(E)$ in units of eV cm^2 through:

$$S_n(E) = \frac{Z_1 Z_2 M_1}{M_1 + M_2} 4\pi e^2 a_L s_n(\varepsilon)$$

$$S_n(E) = 8.462 \times 10^{-15} \frac{Z_1 Z_2 M_1 s_n(\varepsilon)}{(M_1 + M_2)(Z_1^{2/3} + Z_1^{2/3})^{1/2}} \text{ (eV cm}^2\text{)} \tag{1.5a}$$

and the nuclear stopping power becomes:

$$\begin{aligned} \left(\frac{dE}{dx}\right)_n &= NS_n(E) \quad N \ldots \text{atoms cm}^{-3} \quad \rho \ldots \text{density(g cm}^{-3}\text{)} \\ \left(\frac{dE}{dx}\right)_n &= 509.6 \frac{\rho Z_1 Z_2 M_1 s_n(\varepsilon)}{M_2(M_1 + M_2)(Z_1^{2/3} + Z_1^{2/3})^{1/2}} \text{ (eV nm}^{-1}\text{)} \end{aligned} \tag{1.5b}$$

A simple analytical expression for $s_n(\varepsilon)$ based on the so-called Kr-C potential [13–15] was suggested by Wilson *et al.* [14]:

$$s_n(\varepsilon) = \frac{0.5 \ln(1 + \varepsilon)}{\varepsilon + 0.10718\varepsilon^{0.37544}} \tag{1.6a}$$

For the ZBL potential the corresponding analytical expression is of the form [13]:

$$\begin{aligned} s_n(\varepsilon) &= \frac{\ln(1 + 1.1383\varepsilon)}{2(\varepsilon + 0.01321\varepsilon^{0.21226} + 0.19593\varepsilon^{0.5})} && \text{for } \varepsilon \leqslant 30 \\ &= \frac{\ln(\varepsilon)}{2\varepsilon} && \text{for } \varepsilon \geqslant 30 \end{aligned} \tag{1.6b}$$

In the Eqs. (1.2)–(1.6) we have followed the notation of Lindhard *et al.* [7–9]. In other approaches to the screening function $\Phi(r/a)$ (Eqs. (1.2) and (1.3)) of the interatomic potential different screening lengths a have been defined. For a discussion as well as for other analytical expressions for $s_n(\varepsilon)$, see for example the book by Ziegler *et al.* [13].

Contrary to nuclear stopping, electronic stopping does not cause appreciable scattering of the penetrating particle, because of the small electron mass. At low projectile energies electronic stopping is proportional to the particle velocity. This so-called Lindhard–Scharff (LS) regime [8,9,13] extends up to projectile velocities v_1:

$$\begin{aligned} v_1 &\leq Z_1^{2/3} v_0 = Z_1^{2/3} \frac{e^2}{h} \quad v_0 = 2.18 \times 10^8 \text{ cm s}^{-1} \\ \text{or} \quad E(\text{keV}) &\leq 24.88 Z_1^{4/3} M_1 \end{aligned} \tag{1.7a}$$

which is well into the 100 keV range for medium mass projectiles. As a rule of thumb electronic and nuclear stopping are of the same order of magnitude for:

$$v_1 \approx 0.1 Z_1^{2/3} v_0 \quad \text{or} \quad E(\text{keV}) \approx 0.25 Z_1^{4/3} M_1 \tag{1.7b}$$

In the LS regime the electronic stopping cross-section can be approximated by [8]:

$$S_e(E) \approx 3.846 \times 10^{-15} Z_1^{1/6} \frac{Z_1 Z_2}{(Z_1^{2/3} + Z_2^{2/3})^{3/2}} \sqrt{\frac{E(\text{keV})}{M_1}} \; (\text{eV cm}^2) = KE^{1/2} \tag{1.8a}$$

and the electronic stopping power becomes similar to Eq. (1.5):

$$\left(\frac{dE}{dx}\right)_e = NS_e(E) \qquad N \ldots \text{atoms cm}^{-3} \qquad \rho \ldots \text{density(g cm}^{-3})$$

$$\left(\frac{dE}{dx}\right)_e = 231.6 \rho Z_1^{1/6} \frac{Z_1 Z_2}{(Z_1^{2/3} + Z_1^{2/3})^{3/2} M_2 \sqrt{M_1}} \sqrt{E(\text{keV})} (\text{eV nm}^{-1}) \tag{1.8b}$$

Also the electronic stopping cross-section is often expressed in a dimensionless reduced form $s_e(\varepsilon)$ (see Fig. 1.2) and is connected to $S_e(E)$ as $S_n(E)$ to $s_n(\varepsilon)$ (Eq. (1.5)) and becomes:

$$s_e(\varepsilon) = 0.0793 Z_1^{1/6} \frac{Z_1^{1/2} Z_2^{1/2} (M_1 + M_2)^{3/2}}{(Z_1^{2/3} + Z_1^{2/3})^{3/4} M_1^{3/2} M_2^{1/2}} \sqrt{\varepsilon} = k\varepsilon^{1/2} \tag{1.8c}$$

with k typically of the order of 0.1 to 0.2. At higher energies ($v \gg e^2/\eta$) electronic stopping decreases and is well described by the Bethe formula [16]. Already in the LS regime for energies $\varepsilon > 1$ (this corresponds to 117 keV for Ar on Ni), energy loss due to electronic excitations starts to dominate over nuclear stopping (Fig. 1.2) and becomes the dominant energy loss at higher energies.

The total path length (range) of the ion can be calculated by integration over the energy losses:

$$R(E) = \int dx = -\int_0^E \frac{dE}{[dE/dx]_n + [dE/dx]_e} \tag{1.9}$$

Table 1.1: Electronic (S_e) and nuclear (S_n) stopping cross-sections and stopping powers, average total path length for He, Ar and Xe ions bombarding a Cu target

Ion energy (keV)	ε	$s_n(\varepsilon)$	$S_n(E)$ (eV cm^2) $\times 10^{16}$	$(dE/dx)_n$ (eV nm^{-1})	$S_e(E)$ (eV cm^2) $\times 10^{16}$	$(dE/dx)_e$ (eV nm)	$s_e(\varepsilon)$	Path length (nm)
He–Cu								
0.1	0.016	0.205	17.9	15	10.8	9.1	0.183	7
1	0.16	0.347	30.4	26	34.2	28.9	0.390	29
10	1.59	0.277	24.3	21	108.0	91.3	1.230	133
100	15.90	0.087	7.6	7	342.0	289.0	3.900	574
Ar–Cu								
0.1	0.00095	0.054	227.0	191	24.6	21.0	0.006	1
1	0.095	0.168	710.0	598	78.0	66.0	0.018	3
10	0.095	0.326	1380.0	1162	240.0	208.0	0.058	11
100	0.95	0.316	1340.0	1130	780.0	659.0	0.184	65
Xe–Cu								
0.1	0.00014	0.018	322.0	272	28.0	23.6	0.001	1
1	0.0014	0.067	1220.0	1028	88.0	75.0	0.005	2
10	0.014	0.195	3570.0	3016	280.0	236.0	0.015	6
100	0.14	0.342	6270.0	5296	885.0	747.0	0.048	24

The LS regime of electronic stopping extends up to ion energies of 250 keV, 47,000 keV or 660 MeV for He, Ar and Xe, respectively. The factor k in Eq. (1.8) is 0.97, 0.19 or 0.13 for He, Ar or Xe, respectively.

Note that this is the average total path length of the ion and not the average depth it penetrates into the target, which is the projected range, that is the projection of $R(E)$ onto the target normal. Further discussion on the range will be given in Section 1.2.5 on ion implantation.

Using the above equations in combination with the Kr-C potential, one can get estimates for the different energy losses and ranges. Values for He, Ar and Xe ions bombarding a Cu target, using the equations given above, are listed in Table 1.1. Recoil atoms from collisions between the penetrating particle and atoms of the solid will, if sufficiently energetic, create secondary and higher-generation recoils and thus a collision cascade is produced, which can lead to sputtering radiation damage and mixing.

1.2.2. Thermal Deposition

At thermal energies the ions (atoms) are deposited and partly reflected at the surface, which leads to thin-film deposition as is typically the case in electron beam evaporation of neutral atoms followed by deposition on a substrate. Atoms (ions) will be deposited on the surface, no penetration occurs and depending on the target temperature diffusion processes take place, resulting in rearrangement of the randomly deposited atoms, and will lead to island formation or layer-by-layer growth.

1.2.3. Sputter- or Beam-Assisted Deposition

Deposition at hyperthermal energies and up to 1 eV, will increase the target temperature in the surroundings of the deposited particle for a few picoseconds and thus increase surface diffusion and therefore have a pronounced influence on nucleation. Particles will not penetrate below the first layer. Deposition is still the main process for energies up to a few 10 eV, penetration up to 1–2 monolayers can occur, but target atom emission is negligible. Nevertheless thin-film deposition at energies of a few 10 eV or using sputtered atoms and ions with kinetic energies of a few eV has important technologic applications. From molecular dynamics (MD) calculations [18] it was found that the optimum energy per atom for homo-epitaxial growth of Cu on Cu (at 100 K) is around 20 eV. Under thermal deposition, but otherwise identical conditions a rough surface, but no epitaxial growth occurred. With deposition energies of 20 eV epitaxial growth becomes actually independent of target temperature. The ideal energy of 20 eV per atom indicates, that at this energy atoms in the neighborhood of the impact point receive enough energy to allow sufficient diffusion during times of up to a millisecond to anneal the lattice. On the other hand no severe damage is done to the lattice and no mixing or sputtering occurs at these low energies.

1.2.4. Sputtering

Starting at energies of a few 10 eV the impinging ions have enough energy to penetrate the target and due to mainly elastic collision processes target atoms can be emitted. This leads to an erosion of the surface and is termed sputtering. Thus sputtering is the erosion of a target material under energetic atom or ion bombardment. It was observed for the first time in glow discharges in the middle of the 19th century [19,20]. Bombarding energies under which erosion is observed can vary from a few 10 eV up to MeV ions. Sputtering is the dominant process under ion impact, resulting in an effective surface erosion, in the energy range from 1 keV up to a few 10 keV. The key quantity of interest is the sputtering yield Y, defined as the number of emitted particles per incoming ion and a vast amount of measurements has been performed and the results have been compiled in tables [1,21].

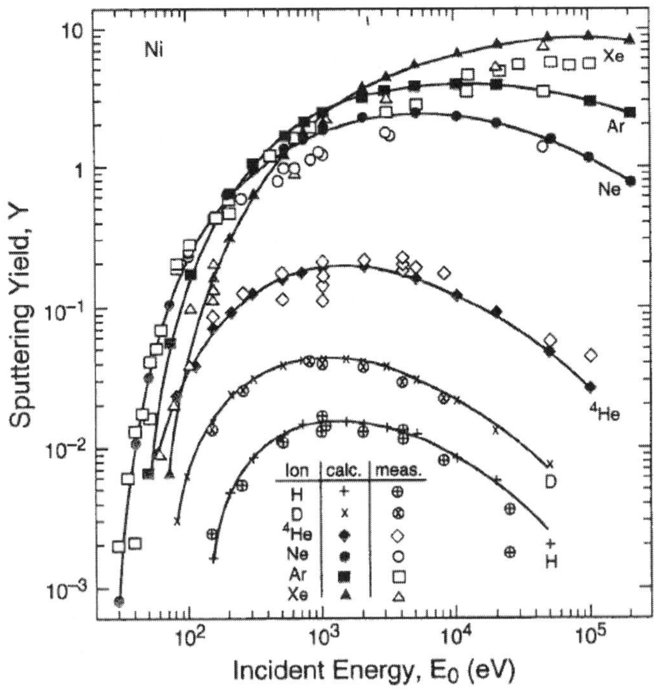

Figure 1.3: Comparison of TRIM calculations with experimental measurements of the sputtering yield *Y* vs. incident ion energy for Ni bombarded by H, D, He, Ne, Ar and Xe ions (after Ref. [22]).

Sputtering yields (Fig. 1.3) typically reach their maximum value at an ion energy between a few keV (for very light ions) and 100 keV. At energies of 1 MeV sputtering becomes negligible. Maximum sputtering yields reach values between 10 and 100.

During sputtering or surface erosion neutral, excited as well as ionized atoms, molecules and clusters are ejected, the bombarding particles are in part reflected, electrons as well as radiation can be emitted from the surface or from excited sputtered particles in front of the surface. This situation is shown schematically in Fig. 1.4. For clean metal surfaces and rare or inert gas ion bombardment the dominant fraction of emitted particles are neutral metal atoms and clusters as indicated in the figure. The composition of the sputtered particle flux can change drastically depending whether oxygen or other electro-negative atoms are adsorbed or incorporated in the surface from almost exclusively neutral atoms and clusters to a large contribution of ions and excited atoms. This effect has been the subject of many studies, partly because of its importance for quantitative surface analysis in secondary ion mass spectrometry (SIMS), but many questions are still not answered

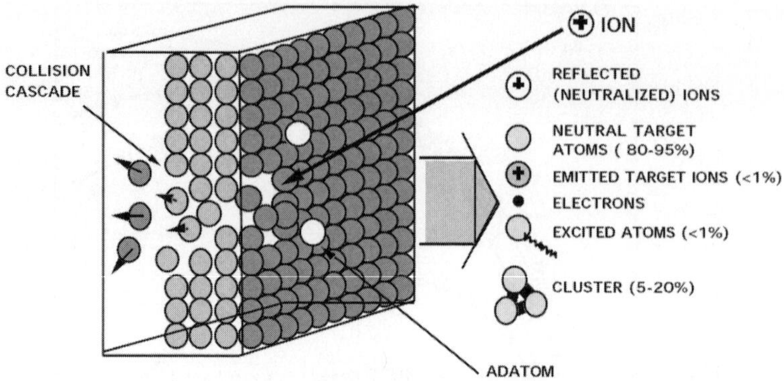

Figure 1.4: Schematic representation of particles, which are emitted under keV ion bombardment of a metal surface. The percent values given besides the emitted particles listed are approximate contributions to the sputtered flux for a clean metal surface.

and we are far from a quantitative understanding of the composition of the sputtered flux [23].

As discussed in Section 1.2.1, recoil atoms from collisions between the penetrating particle and atoms of the solid will, if sufficiently energetic, create secondary and higher-generation recoils and thus a collision cascade is produced. Sputtering occurs, if such moving atoms reach the surface and can overcome the surface-binding forces. For these processes leading to sputtering three regimes can be distinguished [17]: the single knock-on regime, the linear cascade regime and the spike regime. They are schematically shown in Fig. 1.5.

The single knock-on regime is characterized by only few recoil atoms, as is the case at low energies and/or for light ions (H, He at energies of at most a few keV). A recoil atom will not receive enough energy to produce further knock-ons and a collision cascade will not develop.

In the linear cascade regime the collisions between the incoming particle and the target atoms produce energetic recoils, which in turn can produce higher-generation recoils, resulting in a well-developed collision cascade. The density of moving atoms within the cascade volume is assumed to be small enough, so that collisions between moving atoms are rare. Then the collision cascade can be well described within the framework of transport theory. The energy range is keV up to a few hundred keV, except for heavy ions, where the spatial density of moving atoms becomes too high.

In the spike regime the density of moving atoms is so high, that essentially all atoms in the "spike volume" are moving and thus collisions are dominantly

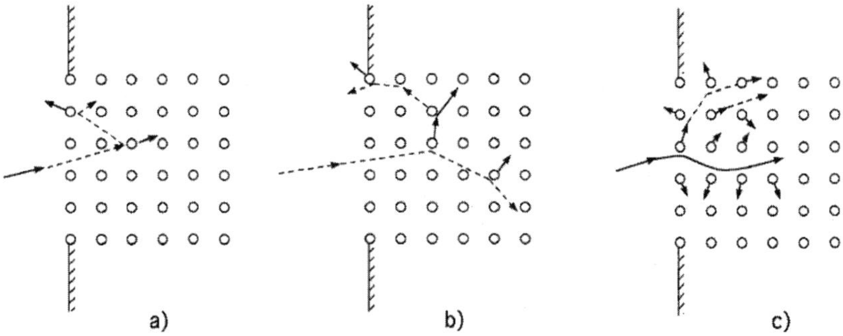

Figure 1.5: Three regimes of sputtering can be distinguished. (a) *The single knock-on regime*: Recoil atoms from ion target collisions receive enough energy to be emitted, but not enough atoms are set in motion to generate a collision cascade. (b) *The linear cascade regime*: Recoil atoms from ion target collisions receive enough energy to generate a collision cascade. The density of the moving atoms is sufficiently low, so that collisions between moving atoms can be neglected. (c) *The spike regime*: The density of atoms in the collision cascade is so high that most of the atoms within a certain volume (spike volume) are in motion (after Ref. [17]).

between moving atoms. This leads to non-linear effects. For a dilute collision cascade the superposition of the cascade will be linear, but this is no longer true in the spike regime. Experimentally this can be tested by molecule bombardment, as the molecule will dissociate immediately at the first collision, and superposition of two cascades at the same place and time occurs. Indeed, for heavy particles and energies above 100 keV non-linear effects could be demonstrated by measuring the sputtering yield [24,25].

In the linear cascade regime sputtering of metals and semiconductors in the keV energy range can be well described by Sigmund's linear collision cascade theory [17,26]. Within this concept the sputtering process can be broken down into three sub-processes, namely (a) slowing down and energy dissipation of the ion, (b) development of the collision cascade itself and (c) passage of atoms of the collision cascade through the surface, which then become sputtered particles.

Within this framework theoretical recoil (cascade) energy distributions inside the target ("internal spectra") are predicted to show an E^{-2} behavior. The experimental energy distribution ("external spectrum") of sputtered atoms evolves out of the internal distribution, but is critically influenced by the surface-binding forces during passage. This surface-binding energy is usually set equal to the heat of sublimation.

The theory of linear collision cascades has been developed for amorphous (random) targets involving the Boltzmann transport equations using appropriate cross-sections for elastic (nuclear) collisions as discussed in Section 1.2.1. If inelastic energy losses are included, they are usually dealt with as a continuous energy loss between collisions, not causing any momentum changes. The good agreement with experiments indicates that effects due to the regularity of the crystal lattice cancel out for a polycrystalline target for most quantities of interest. A basic assumption is the linearity of the cascade, which means that the cascade is so dilute that an already moving target atom has zero probability to collide with another moving target atom. Deviations from this concept become visible at energetic (100 keV) heavy ion bombardment, where the cascade becomes so dense that the assumption of linearity breaks down and the process belongs to the spike regime. Also the theory breaks down at low energies and especially for light ion bombardment, that is in the single knock-on regime, where the development of the collision cascade is incomplete and specific short collision sequences can be identified, which lead to sputtering.

The linear cascade theory predicts a cosine-like angular distribution for the sputtered flux. The energy distribution for the flux of sputtered atoms, which is often referenced as Sigmund–Thompson distribution [17,26,27] is predicted as:

$$\frac{\mathrm{d}\Phi}{\mathrm{d}E} \propto \frac{E}{(E + U_\mathrm{b})^3} \tag{1.10}$$

U_b is the heat of sublimation which is typically in the range from 2 to 8 eV. The theoretical shape of the energy distribution is independent on the primary ion energy, the angle of incidence and also on the emission direction of the sputtered atoms. The energy distribution exhibits a maximum at $U_\mathrm{b}/2$, which indicates that sputtered atoms have average energies of a few eV at most.

The Sigmund–Thompson energy distribution is obtained from the fact that the theoretical treatment of a random collision cascade results in an energy spectrum of the moving atoms in the cascade inside the target scaling with $1/E^2$ ("internal spectrum"), and under the assumption of a planar surface potential U_b for the surface barrier. A planar surface barrier implies that an atom loses an energy equal to the surface-binding energy U_b in passing through the surface, and is in addition also refracted toward the surface [26]. The value of the surface potential U_b or surface-binding energy, which is the energy needed to remove an atom from the surface, has been discussed frequently in the literature [28,29]. This quantity is also important with respect to the total sputtering yield, which should scale with $1/U_\mathrm{b}$. Typically U_b is taken equal to the heat of sublimation and there is clear evidence from all the experimental results obtained with metals, that the actual value as determined from energy distribution measurements, does not differ greatly from the heat of sublimation. Indeed the energy to remove an atom, which is emitted normally from a crystal with a relaxed perfect surface is typically 30–40% greater than the heat of sublimation [29], which is the average energy per atom to disassemble a crystal into individual atoms. However in sputtering many atoms will be removed from edges, steps, kink sites and ad-atom sites, with a lower binding energy, thus the average energy turns out to be close to the heat of sublimation.

So far the discussion of sputtering was for amorphous targets or polycrystalline targets assuming that single crystal effects will cancel out. Wehner [30,31] was the first, who observed that the emission of sputtered atoms from single crystals exhibits pronounced maxima into low-index (closed-packed) lattice directions (Wehner spots). Bombarding monocrystalline Ag, Cu, Ni, Fe, Ge and W targets by low-energy Hg ions ($E < 400$ eV) and collecting the sputtered particles on a flat collector, spot patterns characteristic for the crystal surfaces were observed. This strongly anisotropic emission has been confirmed over a wide range of ion energies (100–10^6 eV) and studied by many authors and is one of the most characteristic features of single crystal sputtering [32–34]. The spots are superimposed on the random cosine-like background as observed for polycrystalline targets. The ratio of spot to background intensity, however, is strongly ion energy and mass dependent. The fact, that these structures are observed at high fluences proves, that at least for metals the regular lattice structure of the target recovers fast enough, before the next particle hits near the previous impact point. For semiconductors (Si, Ge) a single crystal/amorphous phase transition under ion bombardment can be observed below a certain critical temperature, that is above this temperature the spots are persisting independent on fluence, but below amorphization due to ion bombardment takes place and the emission pattern is typical for a random (polycrystalline) target [35,36].

In 1957 Silsbee explained the spot patterns in terms of focusons [37], which are collision sequences along a straight row of uniformly spaced atoms as shown in Fig. 1.6. He showed, that using the hard-sphere approximation, such a collision sequence can operate along the <110> direction in face centered cubic (f.c.c.) lattices, whereby each consecutive atom moves at a smaller angle with the <110> direction than its predecessor, thus accomplishing momentum focusing. In the hard-sphere approximation a criterion for focusing can be derived from simple geometrical considerations: $D < 4R$; D is the distance between two neighboring atoms in the row and R is the hard-sphere atom radius. Thus a focuson can only occur along the most closely packed lattice directions. It was assumed, that atoms receive their energy in collisions several layers below the surface, causing ejection of surface atoms via collision sequences along closed-packed directions. As R is energy dependent it follows that above a given energy E_f (typically between 10 and 50 eV) a focusing collision sequence cannot exist. Thus only particles in the cascade recoil spectrum with energies below E_f can participate in correlated collision sequences.

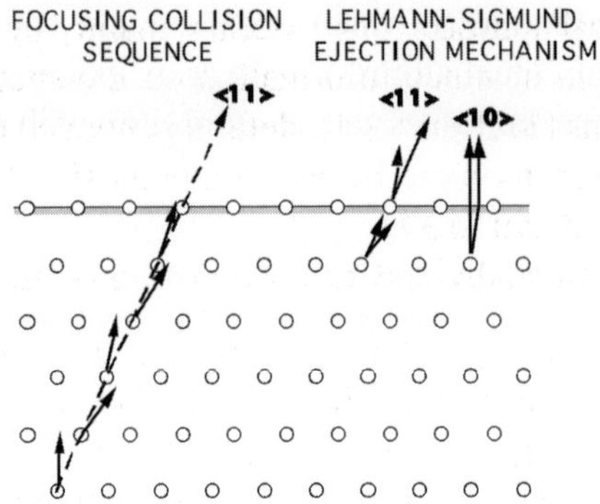

Figure 1.6: Schematic representation of a focusing collision sequence [37] (left side) and the surface ejection mechanism after Lehmann and Sigmund [38] (right side). The focusing collision sequence occurs along a low-index (closed-packed) lattice direction and the angle between the particle momentum and the <11> direction approaches zero in the sequence (momentum focusing). In the Lehmann–Sigmund mechanism preferential emission is due to the regularly ordered surface lattice. Preferential ejection will occur in the <11> direction due to nearly head-on collisions and in addition in the <11> direction due to the potential minimum of the surface-binding energy set up by the surface atoms.

About 10 years after the proposal to explain the directionality in particle emission by focusons, Lehmann and Sigmund [38] suggested a quite different, pure surface mechanism. In part this came from inconsistencies in explaining the experimental results with focusons at low ion energies and for special emission directions, where spots were observed, but could not be explained by focusons. This "surface mechanism" is also presented schematically in Fig. 1.6. The basic idea is that low-energy particles, which dominate the cascade, can only transfer sufficient energy to a surface atom in a near head-on collision to overcome the surface-binding energy (surface potential). Due to the lattice regularity this will also cause emission in closely packed directions. For second layer atoms penetration through the surface potential minimum between two surface atoms will cause directionality and momentum focusing in direction of the surface normal. For hexagonal closest packing (h.c.p.) crystals in the <2023> direction a spot is observed, which can only be explained by the surface model, but not by focusons. Most other emission anisotropies can be explained by both models (surface mechanism or "bulk" focusons) and they have been used concurrently to explain the observed spots. The amount focusons contribute to the emission spots is still an open question and seems to depend strongly on ion energy, mass and temperature [33].

Besides the theoretical approach to sputtering, mainly by the linear cascade theory and modifications to it, much information on the sputtering process, on the penetration depth of ions (Section 1.2.5), defects generated (Section 1.2.6) can and has been gained by computer simulations. Computer simulation codes and results gained are described in detail in Refs. [5,6].

The existing codes can be divided into two main groups:

(a) Codes based on the binary collision approximation (BCA) [6], which can be divided into lattice codes using a single crystal target structure and into Monte-Carlo (MC) codes using an amorphous target structure. BCA means that a collision occurs always only between two atoms: a moving atom and an atom at rest. An often used example for the first group is the MARLOWE code [39] and for the latter group the TRIM code [22].
(b) MD codes based on solving Newton's equations of motion as a function of time, which have been pioneered by Harrison for sputter simulations [40]. In these codes practically always a regular lattice structure is assumed.

While the BCA codes have the advantage of being fast and therefore good statistics can be easily achieved, their disadvantage is the break down of the BCA at low energies, where many body effects become dominant [6]. Thus the domain of such codes are higher energies (>1 keV). Typically such codes employ only a repulsive potential, thus the surface-binding energy is an input parameter in such codes and problems associated with cluster emission in sputtering are difficult to address. Probably the most widely employed BCA–MC code is the TRIM code [22]. It assumes an amorphous (random) target, but is mostly employed for polycrystalline targets. Sputtering yields, ion implantation and damage are often calculated with this code and compared to experimental results. An example for such a comparison of calculated (TRIM) and measured sputtering yields for Ni are shown in Fig. 1.3. An example for a TRIM ion implantation calculation is shown in Fig. 1.9.

MD codes on the other hand are very computer time intensive and reasonably good statistics can only be achieved at low energies (<1 keV). While earlier codes have mainly employed a combination of a repulsive potential (e.g. Born-Mayer) and an attractive Morse potential, in the last few years new many body potentials for the attractive part of the atom–atom interaction in a solid became available, like tight-binding potentials [29], EAM-potentials [41] and Tersoff potentials [42], and have been used in MD simulations of ion–solid interactions [43–45]. These new potentials are no longer pairwise additive and it has been found that such potentials can give better agreement with measured bulk and surface properties than the simple Morse potential, like surface relaxation and binding energies and bond lengths of small atom clusters. This latter point is especially important in calculations concerning the emission of neutral clusters under ion bombardment.

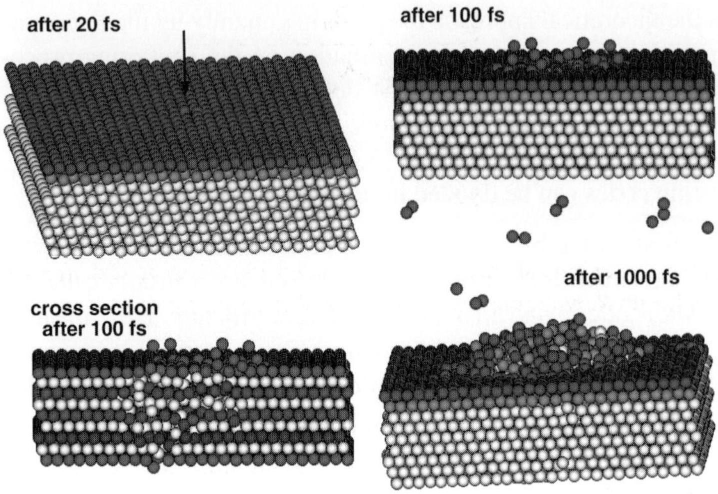

Figure 1.7: MD results for a 2 keV Ar ion bombarding a Cu (111) surface.

As an example a typical impact of a keV ion (2 keV Ar) on a single crystal metal target (Cu 111) and the time scale of such a process will be discussed. The development of the collision cascade and the changes at the surface of the bombarded crystal are shown in Fig. 1.7. Twenty femtoseconds after the particle has started 0.6 nm above the surface the ion has penetrated several layers deep into the crystal, but otherwise no damage except for one missing atom (which has moved inside the target due to a collision with the impinging ion) at the surface is visible. This atom became an energetic recoil and is after 100 fs down in layer 6. Analysis shows, that inside the crystal a number of energetic recoils have been created.

The cascade starts to develop and after 100 fs the cascade is dominated by atoms in the energy range between 1 and 10 eV. First layer atoms start to leave the target, that is, are being sputtered. The cross-section through the crystal reveals the size of the cascade and all the mixing processes taking place. The number of cascade atoms is increasing but their kinetic energy decreases with time and after 500 fs almost all cascade atoms have energies <1 eV, that is the cascade is rapidly cooling down. Most of the atoms which will be sputtered have already left or are leaving the surface. Sputtering has passed its maximum.

After 1 ps sputtering is over; all the cascade atoms have energies well below 1 eV, that is sputtering is no longer possible. The surface is still greatly disordered and atoms from the second and a few from the third layer have reached the surface. The typical cascade dimension is about $10 \times 10 \times 10$ lattice units large. It is still a very hot region, causing the surface to bulge out. This atomic protrusion will with time (ns–μs)

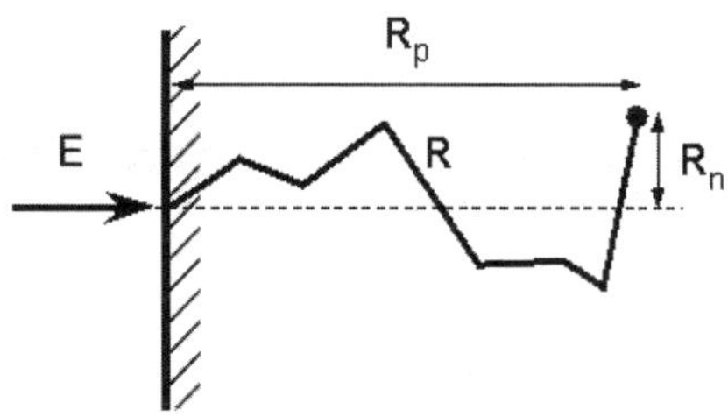

Figure 1.8: An ions of energy E enters a solid and comes to rest after a total path length R, with projected range R_{p} and transverse projected range R_{n}.

disappear but a large number of ad-atoms will remain at the surface. Typically the number of ad-atoms created is about five times larger than the number of sputtered atoms and has been studied using scanning tunneling microscopy [46,47].

We note that in this event all sputtered atoms were atoms from the surface layer and also that metal dimers (cluster) were emitted. Indeed this is a typical result, as even at higher energies 90–95% of all sputtered atoms are from the first layer, the rest mainly from the second layer. This is of great importance for the use of ion bombardment as a tool of surface analysis as in SIMS and for composition vs. depth profiling, where always sputtering is used to erode the surface and simultaneously the surface composition is recorded, for example by SIMS or Auger electron spectroscopy (AES).

1.2.5. Ion Implantation

As discussed in Section 1.2.1 the implanted ion looses energy by means of both nuclear and electronic interactions with the substrate atoms. The former interactions consist of individual elastic collisions between the ion and the target atom, whereas the electronic interactions can be viewed more as a continuous viscous drag phenomenon between the ion and the sea of electrons surrounding the target atom nuclei. Normally for energies considered here the nuclear contribution will dominate. Eq. (1.9) for the total range R of an ion was already derived in Section 1.2.1.

As already mentioned and shown in Fig. 1.8 the total range R is the average total path length of the ion and not the average depth it penetrates into the target, which is the projected range R_{p}, that is the projection of $R(E)$ onto the target normal (for normal ion incidence). Similarly a transverse projected range R_{n} can be defined.

As the stopping of an ion is a stochastic (random) process, the number of collisions, deflections and total path length vary from ion to ion. Probability (range) distributions $P(E, R)$ for the range as a function of depth R as well as $P_{\mathrm{p}}(E, R)$ and

$P_n(E, R)$ can be derived from the cross-sections given in Section 1.2.1. For example, $P_p(E, R)dR$ is the probability that an ion with initial energy E comes to rest after a projected path length between R and $R + dR$. Such distributions can be characterized by their moments and usually only the first 2 to 3 moments are used:

$$\begin{aligned} R_1(E) &= \int_0^\infty RP(E, R)\mathrm{d}R && \text{mean range} \\ R_2(E) &= \int_0^\infty R^2P(E, R)\mathrm{d}R && \text{straggling} \\ R_3(E) &= \int_0^\infty R^3P(E, R)\mathrm{d}R && \text{skewness} \end{aligned} \tag{1.11}$$

From these moments the concentration $c(R)$ of implanted ions as a function of depth can be reconstructed as:

$$\begin{aligned} c(R) &= \frac{1}{\sqrt{2\pi R_2}} \mathrm{e}^{-\xi^2/2} \left[1 - \frac{1}{6}\beta(3\xi - \xi^3)\right] \mathrm{d}R \\ \xi &= (R - R_1)/\sqrt{R_2} \qquad \beta = R_3^2/R_2^3 \end{aligned} \tag{1.12}$$

If R_3 is equal zero the distribution is Gaussian and the position of the peak of the distribution is equal to the mean range R_1, and R_2 determines the width of the symmetric peak. The skewness R_3 is a measure how asymmetric the distribution is.

Using the reduced energy ε as discussed in Section 1.2.1 and in addition a reduced range ρ (compare Eq. (1.4)) as a rule of thumb the following relation (neglecting electronic stopping) for the range of an ion in a target can be obtained independent of the ion target atom combination [7,8]:

$$\rho = 3.06\varepsilon \quad \text{with} \quad \rho = RNM_2 4\pi a_L^2 \frac{M_1}{(M_1 + M_2)^2} \tag{1.13a}$$

Eq. (1.13a) is equivalent to:

$$R(\mathrm{nm}) = \frac{13E(\mathrm{keV})}{\text{Density (g cm}^{-3})} \frac{1 + M_2/M_1}{Z_1^{2/3}} \tag{1.13b}$$

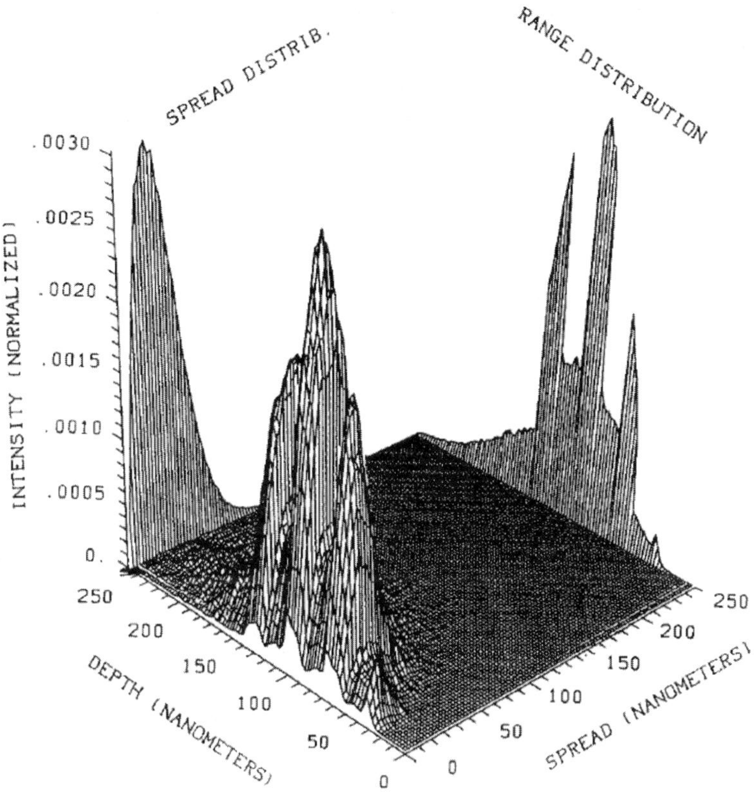

Figure 1.9: TRIM calculation for implantation of 400 keV Xe at normal incidence in a structure of eight layers on silicon: [Al[20 nm]/Mn(15 nm)]$_4$Si. The depth and spread (longitudinal and transverse projected range) correlation is shown in a three-dimensional plot. The depth (projected range) and the spread distributions are shown as shadows at the edges of the plane (from Eckstein [5]).

Projected ranges are as discussed above shorter than the ranges obtained by the above equations. Lindhard *et al.* [7,8] have derived the following rule of thumb:

$$R_p \approx \frac{R}{1 + M_2/3M_1} \tag{1.13c}$$

Detailed range distributions, their straggling as well as transverse range distribution can be easily obtained from MC computer codes, like the TRIM code discussed in Section 1.2.4. They also easily allow calculations for polyatomic targets. As an example Fig. 1.9 shows the results of a TRIM code calculation for 400 keV Xe ions into a multilayer structure on Si.

All considerations in this section were based on the assumption, that the stopping medium is disordered (i.e. amorphous). Results for polycrystalline targets agree well with these theoretical calculations, however for single crystals the orientation of the ion beam with respect to the crystallographic axes of the substrate can have a pronounced effect on the range distribution. The crystal orientation influence on ion penetration is called channeling. When an ion trajectory is in direction of a low-index crystallographic direction the crystal has wide channels, that is, is transparent. The potential of the rows of atoms surrounding such a channel (low-index direction) can steer the ion into the channel and results in very large ranges depending on target temperature. For a detailed discussion see for example Mayer *et al.* [48].

1.2.6. Radiation Damage, Phase Transformation and Ion Beam Mixing

As an ion slows down and comes to rest in a solid, it makes a number of collisions with target atoms. In these collisions sufficient energy can be transferred from the ion to displace an atom from its lattice site. Lattice atoms, which are displaced by incident ions are called primary knock-on atoms or PKAs. These PKAs can in turn displace other atoms, secondary knock-ons and so on – thus creating a cascade of atomic collisions or collision cascade. This leads to the formation of vacancies and interstitials in the region around the track of the impinging ion or collision cascade. If these defects do not anneal fast enough, before the next ion hits the same region the disordered regions will overlap and a heavily damaged layer or an amorphous region is formed. As discussed in Section 1.2.3 for many metals at room temperature annealing is fast enough, thus the crystal structure remains even after prolonged ion bombardment.

The minimum energy required to displace a lattice atom is the displacement energy E_d. If a lattice atom in a collision receives an energy $>E_d$ a vacancy and an interstitial (Frenckel defect) are formed (see Fig. 1.10). Typical minimum displacement energies for metals are between 20 and 40 eV [49]. The average number of displaced atoms in a cascade produced by a primary knock-on of energy E is the displacement damage function $N_d(E)$. A simple estimate was given by Kinchin and Pease [50], which gives for energies up to energies, where electronic stopping can be neglected (M_2keV, i.e. 64 keV for Cu):

$$N_d(E) = \begin{matrix} 0 \\ 1 \\ E/2E_d \end{matrix} \quad \text{for} \quad \begin{matrix} E < E_d \\ E_d < E < 2E_d \\ 2E_d < E \end{matrix} \tag{1.14}$$

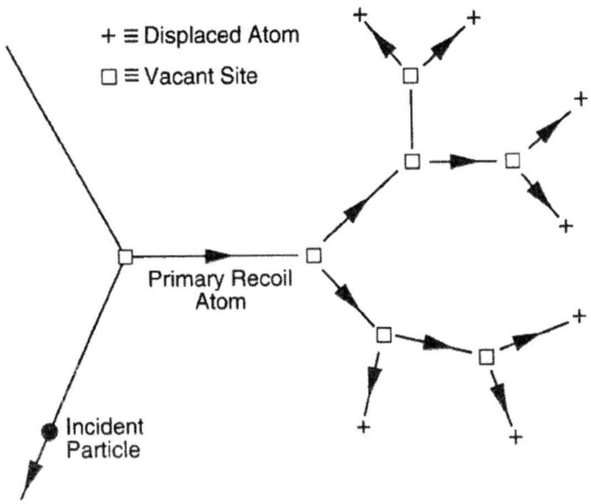

Figure 1.10: Schematic of the formation of a collision cascade and defect creation by a primary recoil atom (PKA) (from M.W. Thompson [52]).

Note that only the energy loss of an energetic ion/atom due to elastic nuclear collisions (nuclear stopping) will produce vacancies and interstitials, not the energy lost due to electronic collisions (electronic stopping). Therefore, if electronic stopping cannot be neglected, at first the part of the total energy of an ion or PKA due to nuclear stopping has to be calculated [51].

Eq. (1.14) gives the number of displaced atoms produced by a PKA. What however is needed is the number of displacements, taking into account the cumulative effects from the many PKAs of different energies that a primary ion creates on the course of its slowing down path [53–56]. A plot of the total number of atomic displacements produced by an ion of initial energy E_0 for different ions in Ag is shown in Fig. 1.11.

Another often used measure of irradiation damage is displacements per atom or dpa; 1 dpa means, that on the average, every atom in the irradiated volume has been displaced once form its original lattice site. For example, 50 keV Kr ions bombarding Al produce about 1000 displacements per ion and for an ion dose of $10^{14}\,\mathrm{Kr\,cm^{-2}}$ the Al target will experience a dpa of 0.45.

As outlined a material under ion bombardment undergoes significant atomic rearrangement. At an interface between two different materials this will lead to atomic intermixing and alloying can occur. The process is referred to as ion beam mixing [57,58]. Typically the substance to be alloyed with the host material is applied as a surface film by conventional means, often by deposition as a vapor. Accelerated ions such as Xe supply the energy needed to mix the substances of the film and the substrate. The excitation of atoms in the implanted region and the

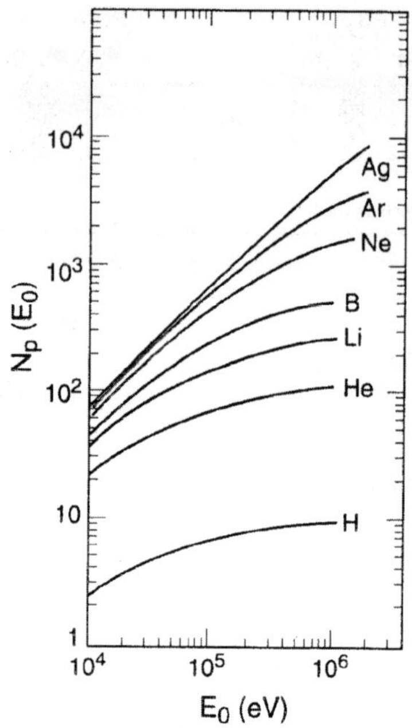

Figure 1.11: Total number of defects produced by different ions of energy E_0 in a Ag target. The displacement energy of Ag was taken as $E_d = 39\,eV$ (after Averback [54]).

defects left in the crystal structure by the bombardment also cause atoms to migrate. Atoms shift to fill vacancies in the lattice and to accommodate displaced atoms within the crystal structure. Extensive bombardment can thicken the mixed region until it approaches the surface. Post-annealing can produce further phase changes. An example is shown in Fig. 1.12. A Pd film is deposited on a Si target and afterwards bombarded with 40 keV Xe ions. Under ion bombardment the Pd metal is consumed forming Pd_2Si. With increasing bombardment dose at the interface a less Pd rich PdSi phase starts to grow consuming the Pd_2Si phase. Additional thermal post-annealing can produce further composition and phase changes. In the case of a Pd film finally a PdSi phase remains, for a Pt film different metastable phases like Pt_2Si_3 or Pt_4Si_9 can form.

Finally ion irradiation of metallic alloys or compounds can cause structural changes. Crystalline phases can become amorphous or can change to a metastable or another equilibrium phase. The driving force for such a transformation is provided by the energy deposited in the lattice by the impinging ion. A large amount of information has been accumulated on this subject for a wide variety of ion target combinations. An overview on this topic can be found in Refs. [59,60] and relevant chapters of this book.

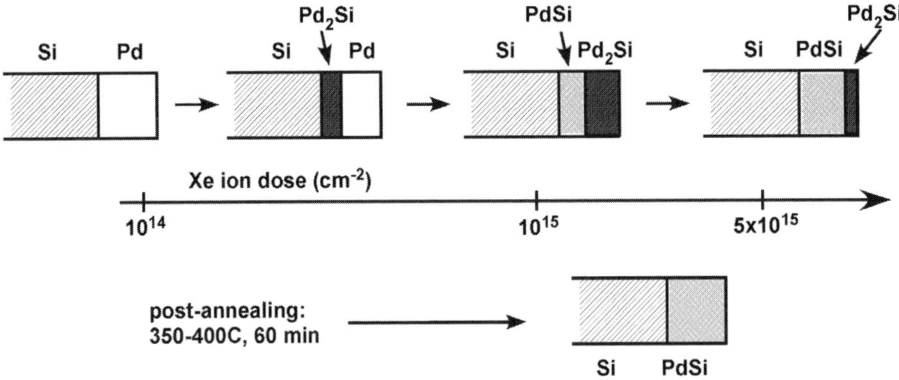

Figure 1.12: Ion beam mixing in a Pd film on Si due to irradiation with 40 keV Xe ions (after Mayer *et al.* [57]).

1.3. Electron Bombardment of a Solid

When an electron impinges on a target it penetrates the surface and undergoes complicated scattering processes. The penetration is associated with various phenomena, like generation of X-rays, Auger electrons, backscattered electrons, secondary electrons emitted from the surface, etc. Some of these phenomena are used as analytical tools to characterize the target and determine its composition near the surface, like electron probe microanalysis (EPMA), scanning electron microscopy (SEM) or Auger electron spectroscopy (AES). The main technique to describe the scattering processes of a penetrating electron and its slowing down is the MC method and a short discussion of the different scattering processes and the outline of an MC simulation will be given. In addition analytical approaches, like solving the Boltzmann transport equation exist [61], but are less suited to find answers to a specific question. In an MC calculation the electron penetration is approximated by a classical zigzag trajectory, each change of direction corresponding to a scattering event.

The two basic scattering processes are elastic scattering (with the atomic potential) and inelastic scattering through an electron–electron interaction. A similar separation of scattering processes into elastic (with the atom) and inelastic (electronic stopping) processes is usually also assumed for the ion–solid interaction as discussed in Section 1.2. Different to the ion target atom interaction the elastic electron target atom interaction is connected with almost no momentum transfer and energy loss due to the large mass difference between electron and atom. In principle, energetic electrons also radiate energy in form of photons, when they are strongly deflected by the nuclei of atoms during an elastic collision. This radiation

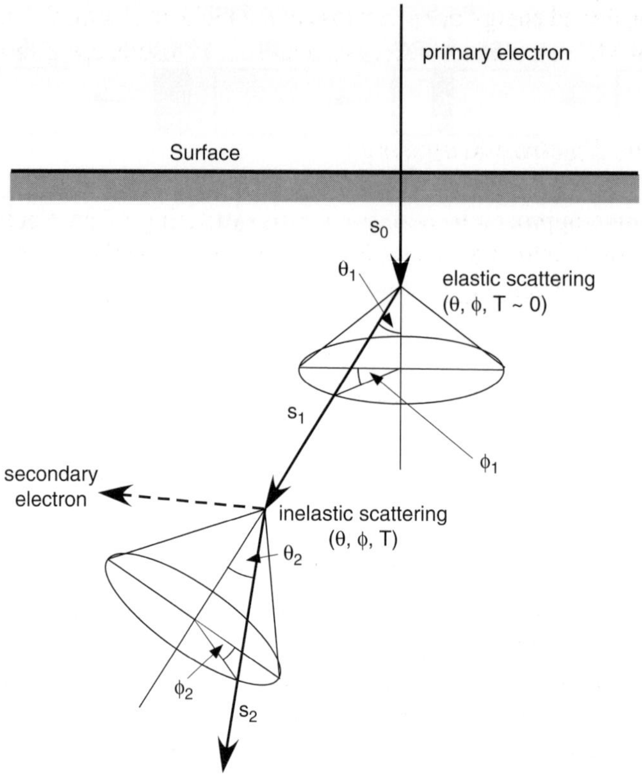

Figure 1.13: Schematic view of an electron trajectory in an MC simulation. T is the energy loss of the incident electron in an inelastic collision.

accompanying electron collisions is called bremsstrahlung; however, bremsstrahlung only becomes important for MeV electron energies and will not be discussed any further.

Inelastic electron–electrons scattering causes energy loss and change in momentum, while inelastic energy losses for ions (with electrons) give a small energy loss for each scattering process and nearly zero change in momentum. Thus for ion bombardment a continuous inelastic energy loss of the ion moving in a straight line is assumed in the MC TRIM code discussed in Section 1.2.4. This is not the case for inelastic electron scattering. The motion of an electron in an MC calculation is show schematically in Fig. 1.13.

At first the main input parameters for such an MC calculation will be discussed, which are mean free path for elastic and inelastic collisions, and the differential cross-sections both for elastic and inelastic collisions ($d\sigma(E, \theta)/d\theta$ or $d\sigma(E, T)/dT$

with T the transferred energy or energy loss of the incident electron). Then the basics of an electron MC calculation is discussed and some results are given.

1.3.1. Elastic Electron Scattering

The most simple approach to describe elastic scattering of an electron from the screened Coulomb field of a nucleus is the use of the Rutherford scattering formula with inclusion of a screening parameter β:

$$\frac{d\sigma}{d\Omega} = \frac{Z^2 e^4}{4E^2(1 - \cos\theta + 2\beta)} \tag{1.15}$$

Z is the atomic number of the target atom, E and θ are the kinetic energy and scattering angle of the electron, respectively. This approach is reasonable for keV electrons and light elements.

A better approach is to take into account the wave properties of electrons and calculate the differential cross-section based on quantum mechanics [62]. The incoming electron is represented by a plane wave which interacts with the Coulomb field of the nucleus producing a phase-shifted outgoing spherical wave in accordance with Huygen's principle. The total outgoing wave is then the superposition of incoming plane and outgoing spherical wave [63]. The differential cross-section $d\sigma_e(\theta)/d\Omega$ is the absolute square of the scattering amplitude $f(\theta)$ of the outgoing spherical wave. To obtain $f(\theta)$ the phase shift has to be evaluated by solving the Schrödinger equation for the given potential [64]. The marked difference between the Rutherford and the Mott cross-section is the appearance of lobes in the cross-section for certain directions as shown in Fig. 1.14. As can be seen the total scattering cross-section decreases with energy and increases with atomic number.

The average distance between successive elastic collisions, the so-called elastic mean free path (EMFP) λ_e, can be obtained by integrating the differential cross-section over the unit sphere (see Fig. 1.15):

$$\lambda_e^{-1} = N \int_{4\pi} \frac{d\sigma(\theta)}{d\Omega} d\Omega \qquad N \ldots \text{ atomic density} \tag{1.16}$$

Another important quantity is the so-called transport mean free path (TRMFP), which measures the momentum transfer along the initial direction:

$$\lambda_{tr}^{-1} = N \int_{4\pi} (1 - \cos\theta) \frac{d\sigma(\theta)}{d\Omega} d\Omega \qquad N \ldots \text{ atomic density} \tag{1.17}$$

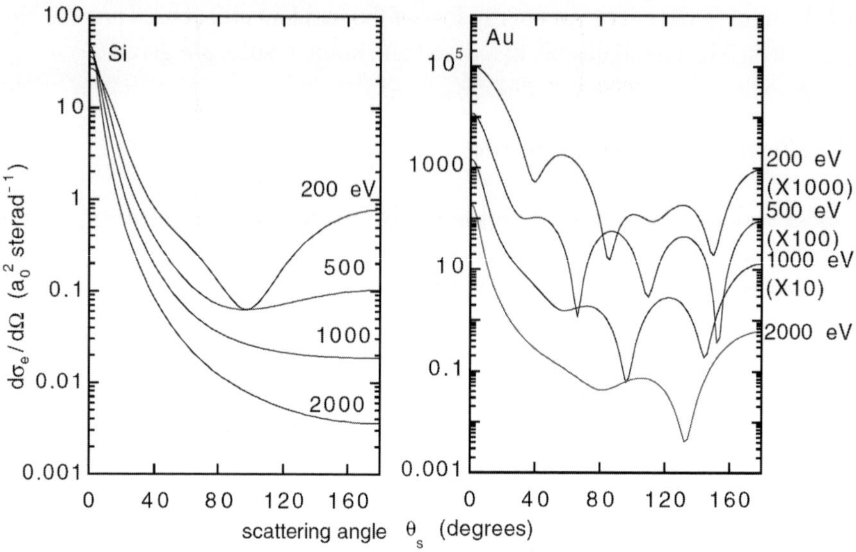

Figure 1.14: Differential cross-section for elastic scattering of electrons by Ag and Au atoms at different energies taking into account the wave properties of electrons (from Ref. [74]).

It can be shown that the TRMFP (see Fig. 1.15) is the typical distance a particle travels before it forgets its original direction owing to large angle deflections.

1.3.2. Inelastic Electron Scattering

For inelastic scattering of electrons (by electrons) the first expression for the stopping power of incident electrons was obtained by Bethe [65]:

$$\frac{\mathrm{d}E}{\mathrm{d}x} = -\frac{2\pi e^4 NZ}{E} \ln\left(\frac{1.166E}{I}\right) \qquad I = 9.76Z + 58.8Z^{-1.19}s \quad (1.18)$$

An empirical formula for the mean ionization energy I is also given in Eq. (1.18) [66]. The Bethe equation given above is valid for energies well above 1 keV, but for relativistic energies additional terms have to be included. Besides the limitation to high energies this continuous slowing down approximation does not allow to describe generation of secondary electrons.

Ritchie *et al.* [67] have shown that the stopping power described by Bethe's equation can be obtained by the summation of theoretical stopping powers for conduction

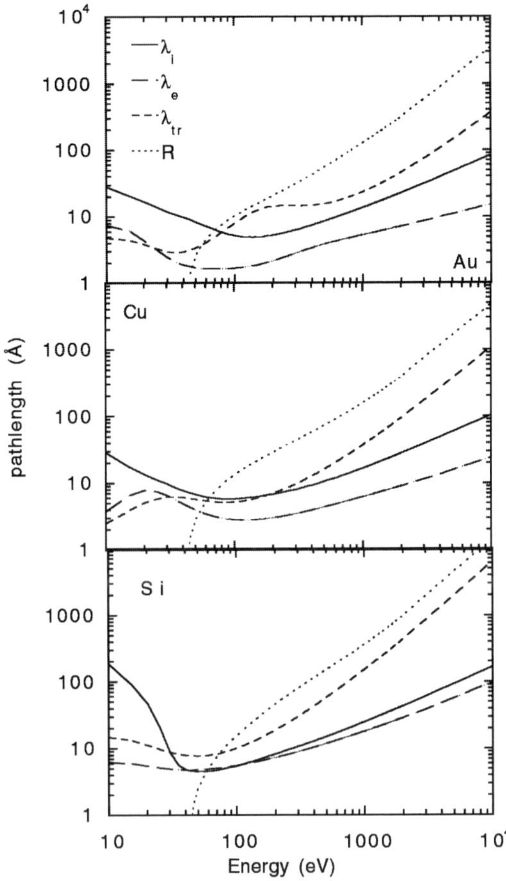

Figure 1.15: Inelastic (λ_i), elastic (λ_e) and transport (λ_{tr}) mean free paths and total path length R for electrons in Si, Cu and Au as a function of energy (from Ref. [74]).

electrons, plasmons and L-shell electronic excitations. For L-shell excitation a semi-quantum mechanical treatment by Gryzinski [68] yields a differential cross-section $d\sigma(E, T)/dT$ from which the angular deflection of the primary electron after an electron–electron interaction can be obtained from the classical binary collision model as $\sin^2\theta = T/E$. In some MC calculations the same excitation function was also used for conduction electrons [69]. For plasmon excitation a mean free path and an angular differential cross-section was obtained theoretically by Quinn [70], and Green and Leckey [71].

The theoretical formulation for inelastic scattering of an electron in a solid is well established in terms of the dielectric function [72–74]. Linear response theory describes the inelastic process as the work done by the probing electron in a polarization field in the medium, that is set up by an external perturbation. In the case of electron scattering inside a solid this polarization field is the dielectric response of the infinite medium to the probing electron itself. From this the differential cross-section is obtained as:

$$\frac{d^2\lambda^{-1}}{dT\,dq} = -\frac{1}{\pi a_0 E}\,\mathrm{Im}\left(\frac{-1}{\varepsilon(q,\omega)}\right)\frac{1}{q} \qquad a_0 \ldots \text{Bohr radius (0.0529 nm)} \tag{1.19}$$

with ηq the momentum transfer from and $T = \eta\omega$ the energy loss of an incident electron of kinetic energy E to the solid. The dielectric function $\varepsilon(0, \omega)$ for zero momentum transfer is available in compilations of optical data [75] and extrapolations to non-vanishing momentum transfers can be performed using linear response theory. Powell has proposed the following excitation function for inelastic scattering [76] by integrating Eq. (1.19):

$$\frac{d\lambda^{-1}}{dT} = -\frac{me^2}{2\pi h^2 E}\,\mathrm{Im}\left(\frac{-1}{\varepsilon(0,\omega)}\right)\ln\left(\frac{cE}{T}\right) \tag{1.20}$$

This equation defines the probability that an electron of kinetic energy $E = (\eta k)^2/2\,\mathrm{m}$ will loose the energy $T = \eta\omega$ per unit path length traveled in a solid. Thus, the differential inverse inelastic mean free path (DIIMFP) $d\lambda^{-1}/dT$ is equivalent to the differential inelastic scattering cross-section $d\sigma(E, T)$ with the scattering angle $\sin^2\theta = T/E$. The inelastic mean free path (IMFP) $\lambda_i(E)$ is defined as the mean distance between successive inelastic collisions regardless of the value of the energy loss. It is obtained by integrating $d\lambda^{-1}/dT$ over all possible energy losses T (see Fig. 1.15).

Finally the mean energy loss per unit path length $s\,dT(E)/ds$, which is termed stopping power, is obtained by integrating the DIIMFP over all energy losses T, and the total path length or electron range R (see Fig. 1.15) by a further integration:

$$\frac{dT}{ds} = \int_0^\infty T\frac{d\lambda^{-1}}{d(T)}\,dT \qquad R = \int_0^{E_0}\frac{dT}{[dT/ds]} \tag{1.21}$$

Results for the various quantities for Si, Cu and Au target are shown in Fig. 1.15.

1.3.3. MC Simulation

Various models for MC simulation have been developed. In the following a brief outline of the principles is given. For a detailed description as well as applications of the technique, mainly with respect to surface analysis methods like AES, the reader is referred to two review papers by Shimizu and Ding [73] and Werner [74].

In an MC calculation (see Fig. 1.13) one calculates a large number of stochastically (using random numbers) generated trajectories of electrons as representative for an electron beam of a given energy impinging on a target. Then averages are taken, like to determine the total path length, mean number of secondary electrons generated, etc. This implies that the path lengths between two collisions, scattering angles and energy losses are distributed according to the physical quantities describing the collision processes. This can be achieved using random numbers and is outlined for the case of the mean free path below.

One important quantity in an MC calculation is path length s between two successive scattering events. Assuming that the probability distribution $P(s)$ of the path length s obeys a Poisson stochastic process we have:

$$P(s) = \frac{1}{\lambda_{\mathrm{m}}} \mathrm{e}^{-s/\lambda_{\mathrm{m}}} \quad \text{equivalent to} \quad \lambda_{\mathrm{m}} = \int_0^{\infty} s\, P(s)\mathrm{d}s \tag{1.22}$$

where λ_{m} is the total mean free path, which is connected to the elastic (EMFP) and inelastic (IMFP) mean free paths λ_{e} and λ_{i}, respectively, by:

$$\frac{1}{\lambda_{\mathrm{m}}} = \frac{1}{\lambda_{\mathrm{e}}} + \frac{1}{\lambda_{\mathrm{i}}} \tag{1.23}$$

The determination of λ_{e} and λ_{i} was discussed in Sections 1.3.1 and 1.3.2. The actual value of s in a given situation in an MC computation is obtained with the help of a random number R between zero and one by:

$$R = \int_0^{s} P(s')\mathrm{d}s' = 1 - \mathrm{e}^{-s/\lambda_{\mathrm{m}}} \Leftrightarrow s = -\lambda_{\mathrm{m}} \ln(1 - R) \tag{1.24}$$

The correlation is such, that if s is determined N times from random numbers in this way the final distribution approaches $P(s)$ with $N \rightarrow \infty$. Similar to the decision, if the collision event after traveling the distance s is an elastic or inelastic collision it is

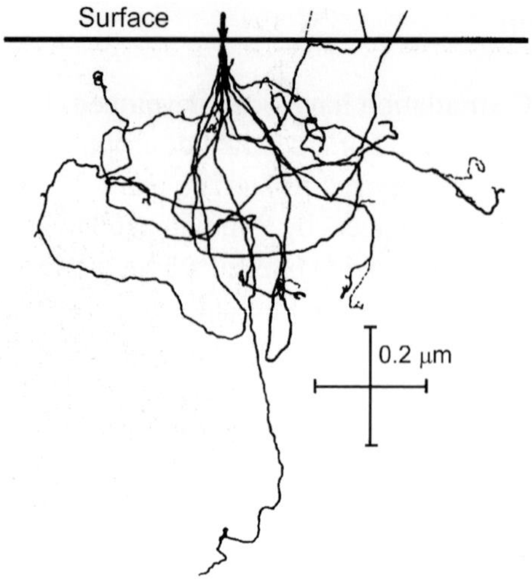

Figure 1.16: Trajectories of 10 keV electrons entering an Al target at normal incidence (from Ref. [77]).

determined by another random number R, in such a way that the event is elastic if $R < \lambda_m/\lambda_e$, otherwise inelastic.

Thus the steps in an MC calculation, using always random numbers for the actual values, are:

(a) Determination of the path length the electron travels until the next collision.
(b) Decision if the collision is elastic or inelastic.
(c) If the collision is elastic, the polar angle θ is obtained from the elastic cross-section $d\sigma(\theta, E)/d\Omega$ (e.g. using Eq. (1.15)), in a similar way as the actual path length s was obtain from the path length distribution using a random number. If the collision is inelastic the energy loss is obtained from the differential inelastic scattering cross-section $d\sigma(E, T)$ (e.g. using Eq. (1.20)). The energy of the electron is reduced by T and the polar scattering angle θ is obtained from $\sin^2\theta = T/E$. In both cases, the azimuthal angle ϕ can be assumed to be uniformly distributed over the interval $[0, 2\pi]$, that is $\phi = 2\pi R$ (R random number).
(d) The steps (a)–(c) are repeated until the electron energy is below a predefined minimum energy or the electron is backscattered and has left the target.

A few individual trajectories of 10 keV electrons entering an Al target at normal incidence are shown in Fig. 1.16. If the generation of secondary electrons is taken

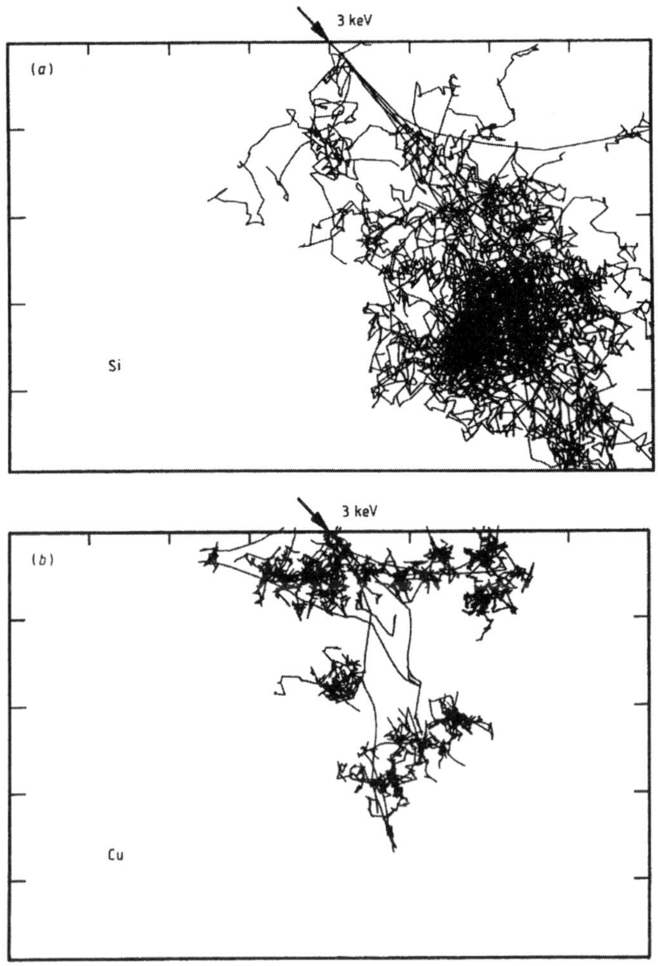

Figure 1.17: Trajectories for five 3 keV electrons entering at an angle of 45° a (a) Si and (b) Cu target. Also shown are the trajectories of the cascade electrons produced by the five primary electrons (from Ref. [78]).

into account, the initial positions and energies of these electrons can be obtained form step (c). In this case additional assumptions have to enter, like was the scattering with a shell electron or a Fermi sea electron, has a hole in an inner shell be produced, which can give rise to an Auger electron, etc. [74,78,79]. As an example for 3 keV electrons impinging on Cu and Si, respectively, the primary and secondary electron trajectories (electron cascade) are shown in Fig. 1.17.

References

[1] R. Behrisch (Ed.), *Sputtering by Particle Bombardment I, Topics in Applied Physics*, Vol. 47, Springer, Berlin, 1981.

[2] R. Behrisch (Ed.), *Sputtering by Particle Bombardment II, Topics in Applied Physics*, Vol. 52, Springer, Berlin, 1983.

[3] R. Behrisch and K. Wittmaack (Ed.), *Sputtering by Particle Bombardment III, Topics in Applied Physics*, Vol. 64, Springer, Berlin, 1991.

[4] M. Nastasi, J.W. Mayer and J.K. Hirvonen, *Ion–Solid Interactions: Fundamentals and Applications*, Cambridge University Press, Cambridge, 1996.

[5] W. Eckstein, *Computer Simulations of Ion–Solid Interactions, Springer Series in Materials Science*, Vol. 10, Springer, Berlin, 1991.

[6] R. Smith (Ed.), *Atomic and Ion Collisions in Solids and at Surfaces*, Cambridge University Press, Cambridge, 1997.

[7] J. Lindhard, V. Nielsen, M. Scharff and P.V. Thomsen, *Mat. Fys. Medd. Dan. Vid. Selsk.*, 33 (1963) 10.

[8] J. Lindhard, M. Scharff and H.E. Schiott, *Mat. Fys. Medd. Dan. Vid. Selsk.*, 33 (1963) 14.

[9] J. Lindhard, V. Nielsen and M. Scharff, *Mat. Fys. Medd. Dan. Vid. Selsk.*, 36 (1968) 10.

[10] K.B. Winterbon, P. Sigmund and J.B. Sanders, *Mat. Fys. Medd. Dan. Vid. Selsk.*, 37 (1970) 14.

[11] G. Moliere, *Z. Naturforsch. Teil A*, 2 (1947) 133.

[12] W.D. Wilson and C.L. Bisson, *Phys. Rev. B*, 3 (1971) 3984.

[13] J.F. Ziegler, J.P. Biersack and U. Littmark, *The Stopping and Range of Ions in Solids*, Pergamon Press, New York, 1985.

[14] W.D. Wilson, L.G. Haggmark and J.P. Biersack, *Phys. Rev. B*, 15 (1977) 2458.

[15] J.P. Biersack and J.F. Ziegler, *Nucl. Instrum. Method.*, 194 (1982) 93.

[16] H.A. Bethe, *Ann. Phys. (Leipzig)*, 5 (1930) 325.

[17] P. Sigmund, in *Sputtering by Particle Bombardment I*, Ed. R. Behrisch, *Topics in Applied Physics*, Vol. 47, Springer, Berlin, 1981, p. 9.

[18] G. Betz and W. Husinsky, *Nucl. Instrum. Method. B*, 193 (2002) 352.

[19] W.R. Grove, *Philos. Trans. Roy. Soc. London*, 142 (1853) 87.

[20] W.R. Grove, *Philos. Mag.*, 5 (1853) 203.

[21] Y. Yamamura and H. Tawara, *NIFS-DATA-23*, National Institute for Fusion Science, Nagoya, Japan, 1995.

[22] J.P. Biersack and W. Eckstein, *Appl. Phys. A*, 34 (1984) 73.

[23] G. Betz and W. Husinsky, *Nucl. Instrum. Method. B*, 32 (1988) 331.

[24] H.H. Andersen and H.L. Bay, *J. Appl. Phys.*, 45 (1974) 953.

[25] H.H. Andersen and H.L. Bay, *J. Appl. Phys.*, 46 (1975) 2416.

[26] P. Sigmund, *Phys. Rev.*, 184 (1969) 383.

[27] M.W. Thompson, *Philos. Mag.*, 18 (1968) 377.

[28] B.J. Garrison, N. Winograd, D. Lo, T.A. Tombrello, M.H. Shapiro and D.E. Harrison Jr., *Surf. Sci.*, 180 (1987) 129.

[29] H. Gades and H.M. Urbassek, *Nucl. Instrum. Method. B*, 69 (1992) 232.

[30] G.K. Wehner, *J. Appl. Phys.*, 26 (1955) 1056.
[31] G.K. Wehner, *Phys. Rev.*, 102 (1956) 690.
[32] W.O. Hofer and H. Gnaser, *Nucl. Instrum. Method. B*, 18 (1987) 605.
[33] W. Szymczak and K. Wittmaack, *Nucl. Instrum. Method. B*, 82 (1993) 220.
[34] W.O. Hofer, in *Sputtering by Particle Bombardment III*, Eds. R. Behrisch and K. Wittmaack, *Topics in Applied Physics*, Vol. 64, Springer, Berlin, 1911, p. 15.
[35] R.J. MacDonald, *Radiat. Eff.*, 3 (1970) 131.
[36] G. Holmen, *Radiat. Eff.*, 24 (1975) 7.
[37] R.H. Silsbee, *J. Appl. Phys.*, 28 (1975) 1246.
[38] C. Lehmann and P. Sigmund, *Phys. Status Solidi*, 16 (1966) 507.
[39] M.T. Robinson and I.M. Torrens, *Phys. Rev. B*, 9 (1974) 5008.
[40] D.E. Harrison Jr., *Crit. Rev. Solid State Mater. Sci.*, 14 (Suppl. 1) (1988).
[41] S.M. Foiles, M.I. Baskes and M.S. Daw, *Phys. Rev. B*, 33 (1986) 7983.
[42] J. Tersoff, *Phys. Rev. B*, 37 (1988) 6991.
[43] G. Betz, R. Kirchner, W. Husinsky, F. Rüdenauer and H.M. Urbassek, *Radiat. Eff. Defect. Solid.*, 130/131 (1994) 251.
[44] F. Karetta and H.M. Urbassek, *Appl. Phys. A*, 55 (1992) 374.
[45] R. Smith, D.E. Harrison Jr. and B.J. Garrison, *Nucl. Instrum. Method. B*, 18 (1987) 605.
[46] T. Michely and C. Teichert, *Phys. Rev. B*, 50 (1994) 11156.
[47] H. Gades and H.M. Urbassek, *Phys. Rev. B*, 50 (1994) 11167.
[48] J.W. Mayer, L. Eriksson and J.A. Davies, *Ion Implantation in Semiconductors*, Academic Press, New York, 1970.
[49] H.H. Andersen, *Appl. Phys.*, 18 (1979) 131.
[50] G.H. Kinchin and R.S. Pease, *Rep. Prog. Phys.*, 18 (1955) 1.
[51] M.J. Norgett, M.T. Robinson and I.M. Torrens, *Nucl. Eng. Des.*, 33 (1975) 50.
[52] M.W. Thompson, *Defects and Radiation Damage in Metals*, Cambridge University Press, Cambridge, 1969.
[53] R.S. Averback, R. Benedek and K.L. Merkle, *Phys. Rev. B*, 18 (1978) 4156.
[54] R.S. Averback, R. Benedek and K.L. Merkle, *J. Nucl. Mater.*, 69/70 (1978) 786.
[55] P. Sigmund, *Rev. Roum. Phys.*, 17 (1972) 823, 969, 1079.
[56] R.S. Averback, *Nucl. Instrum. Method. B*, 15 (1986) 675.
[57] J.W. Mayer, B.Y. Tsaur, S.S. Lau and L.S. Hung, *Nucl. Instrum. Method. B*, 182/183 (1981) 1.
[58] J.W. Mayer and S.S. Lau, in *Surface Modification and Alloying by Laser, Ion and Electron Beams*, Eds. J.M. Poate, G. Foti and D.C. Jacobson, Plenum Press, New York, 1983, p. 241.
[59] M. Nastasi and J.W. Mayer, *Mater. Sci. Rep.*, 6 (1991) 1.
[60] R. Kelly, in *Ion Bombardment Modification of Surfaces: Fundamentals and Applications*, Eds. O. Auciello and R. Kelly, Elsevier Science Publ. B.V., Amsterdam, 1984, p. 79.
[61] J. Schou, *Phys. Rev. B*, 22 (1980) 2141.
[62] N.F. Mott, *Proc. Roy. Soc. A*, 124 (1929) 425.

[63] L. Landau and E.M. Lifshitz, *Quantum Mechanics: Nonrelativistic Theory*, Pergamon Press, Oxford, 1977.
[64] A. Jablonski, *Surf. Interface Anal.*, 14 (1989) 659.
[65] H. Bethe, *Ann. Phys. (Leipzig)*, 5 (1930) 325.
[66] R.M. Sternheimer, S.M. Seltzer and M.J. Berger, *Phys. Rev. B*, 26 (1982) 6067.
[67] R.H. Ritchie, F.W. Garber, M.Y. Nakai and R.D. Birkhoff, *Adv. Radiat. Biol.*, 3 (1969) 1.
[68] M. Gryzinski, *Phys. Rev. A*, 138 (1965) 305, 322, 336.
[69] R. Shimizu, T. Ikuta, M. Kinoshita, T. Murayama, H. Nishizawa and T. Yamamota, *Jpn. J. Appl. Phys.*, 15 (1976) 967.
[70] J.J. Quinn, *Phys. Rev.*, 126 (1962) 1453.
[71] A.J. Green and R.C.G. Leckey, *J. Phys. D Appl. Phys.*, 9 (1976) 2123.
[72] D.R. Penn, *Phys. Rev. B*, 35 (1987) 482.
[73] R. Shimizu and Z.J. Ding, *Rep. Prog. Phys.*, 55 (1992) 487.
[74] W.S.M. Werner, *Surf. Interface Anal.*, 31 (2001) 141.
[75] B.L. Henke, E.M. Gullikson and J.C. Davies, *Atom. Data Nucl. Data Table.*, 54 (1993) 181.
[76] C.J. Powell, *Surf. Interface Anal.*, 7 (1985) 263.
[77] R. Shimizu, *Technol. Rep. Osaka Univ.*, 27 (1977) 69.
[78] Z.J. Ding, R. Shimizu, T. Sekine and T. Sakai, *Appl. Surf. Sci.*, 33/34 (1988) 99.
[79] Z.J. Ding and R. Shimizu, *Surf. Sci.*, 197 (1988) 539.

Chapter 2

Laser Beam–Solid Interactions: Fundamental Aspects

Jørgen Schou

2.1. Introduction

The interaction of laser beams with solids has played an important role in many disciplines, such as laser processing of solids, laser-induced mass spectrometry and film deposition by pulsed lasers [1–7]. Until 1990 the preferred lasers in materials science were the carbon dioxide laser with the wavelength at 10.6 μm and the Nd:YAG laser at 1064 nm in the infrared (IR) regime. The field of laser–solid interactions has undergone a tremendous development with new intense and reliable lasers in the ultraviolet (UV) regime and lasers with ultra-short pulses, but many problems are still not resolved. The interaction of photons with matter depends critically on the wavelength, since the absorption of light in a solid or a gas in the IR regime deviates strongly from that in the UV. A second complicating issue is the gradually increasing absorption during nanosecond or even shorter pulses, due to the enhanced number of free electrons generated earlier during the pulse.

A frequently used term for material removal by intense laser light is "ablation" which originates from the Latin "ablatum", *taken away*. Laser ablation is typically used in a broad sense to denote any laser-induced material removal, including the removal of volatile products from chemical etching. This terminology will be used in the present chapter as well.

The field of laser–solid interactions has been promoted by the widespread use of pulsed lasers to deposit thin films on practically all types of substrates. The technique *pulsed laser deposition* (PLD) is now used on a worldwide basis to produce films, in particular of materials and combinations of materials which cannot or can be produced only with great difficulties by other methods [1,8–13]. Recently, high-quality films with a variety of special properties have been produced by PLD [14–16]. The primary reason for the advancement of PLD is that target material of complicated stoichiometry can be transferred to a suitable substrate during the film

Materials Surface Processing by Directed Energy Techniques
Y. Pauleau (Editor)

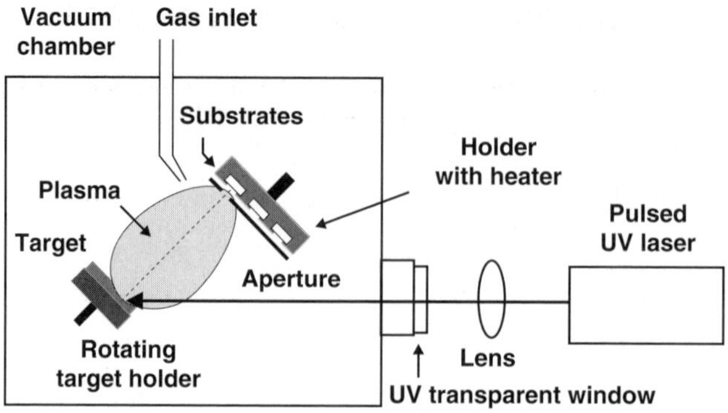

Figure 2.1: A standard PLD setup with a UV laser that strikes the target at an oblique angle of incidence such that the plume material can be collected along the normal of the target surface. The deposition is largest in this direction (from Ref. [17]).

production process without changing the stoichiometry, but the technique is also advantageous in other respects (see also Chapter 8). In summary, films can be produced by PLD, in such a way that:

- The stoichiometry of the target is preserved by the material transfer to the substrate.
- The film deposition can take place in vacuum, inert or reactive gases (e.g. oxygen).
- Atoms and ions in the plume possess kinetic and internal excitation energy which is available for enhanced sticking, surface mobility and film nucleation.
- Target change is comparatively simple, that is "sandwich" films of different materials can easily be produced in a multitarget system chamber.
- Highly perfect interfaces in multilayer systems can be achieved by controlling the laser repetition rate and the number of pulses.

A typical setup for a PLD system is shown in Fig. 2.1 [17]. A UV laser with nanosecond pulses is directed onto a target in a chamber in such a way that material is ablated. This "explosive removal" of material expands like a flow in a direction perpendicular to the target surface and is collected on a substrate in a holder system. The accumulation of target material on the substrate from a large number of laser pulses leads to the gradual formation of a film. Even though this film deposition process appears to be simple, and PLD has been used for more than two decades, many features are still not fully understood.

Film production by PLD is primarily based on the use of nanosecond laser pulses in the UV regime. Thus the main emphasis of the present treatment will be on the interaction of UV light with solids and the gas plume which is formed in front of the solid. Typical wavelengths for PLD are obtained from excimer lasers

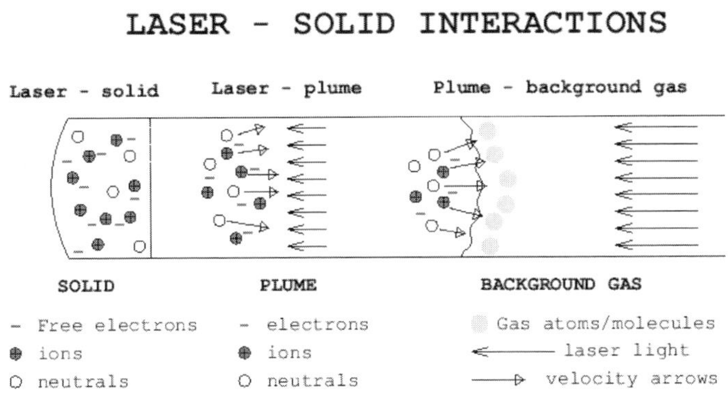

Figure 2.2: A schematic view of the processes that take place during ablation by a nanosecond laser pulse. From left to right: light absorption in the solid (stage 1), ejection of the ablated material in a plasma plume and the interaction of the light with the plume (stages 2–3) and the expansion in a background gas (stage 4).

at 248 and 193 nm and by tripled Nd:YAG lasers at 355 nm, but also laser light of other wavelengths (10.6 μm and 1064 nm) will occasionally be included in the discussion. The main emphasis is on nanosecond pulses, but aspects of femtosecond and picosecond laser irradiation will be discussed as well.

The interaction between an incoming laser beam and a solid is complicated, since the absorption process and the subsequent material ejection are characterized by different processes which usually even occur simultaneously. A schematic survey of the ablation processes is shown in Fig. 2.2:

1. The laser light strikes the solid and interacts with the electrons in the solid. After a period of tens of picosecond the electrons and the atoms equilibrate, which leads to a strong heating of the irradiated volume (stage 1).
2. Material from the heated volume is ejected and continuously absorbing energy from the laser, resulting in formation of an expanding one-dimensional plasma plume (stage 2).
3. After the termination of the laser pulse the plume expands adiabatically in three dimensions. If the expansion takes place in vacuum, the plume atoms will eventually flow away with a constant velocity (stage 3).
4. If the expansion occurs in a background gas, the high plume pressure initially drives the expansion as if it were in vacuum. After several microsecond the plume propagation is completely determined by the interaction of the plume atoms with the atoms and molecules of the ambient gas (stage 4).

Each of these stages is discussed in the following sections.

2.2. Light Absorption in a Solid: Stage 1

The reflection of light on the surface of a solid plays an important role for the efficiency of the light for ablation processes. The reflectivity depends on the wavelength, angle of incidence and polarization. UV light has a low reflectivity for many metals (0.5–0.9) and other materials in contrast to IR light [1]. This has made UV lasers attractive for materials processing and for PLD.

The incident light is absorbed by electronic transitions in the solid. In a metal or semiconductor the light produces excited electrons which subsequently interact with the atoms. In an insulator the light is absorbed by interband transitions or transitions from impurity levels for photon energies which exceed the energy of the band gap or the impurity level. Insulating materials are transparent for photons of energy smaller than that of the band gap. With increasing intensity other absorption processes (e.g. multiphoton and avalanche processes) play an increasing role [18]. At sufficiently high intensities "seed" electrons from impurities or multiphoton processes act as "absorbers" by inverse bremsstrahlung (IB) processes (see Section 2.3) for the avalanche processes which eventually can lead to an optical breakdown [1,4,19].

A typical intensity profile can be found from the exponential attenuation (Lambert–Beer's law) of the incident laser light with an intensity I_0 (energy/(time × area)):

$$I(z) = (1 - R)I_0 \exp(-\alpha z) \tag{2.1}$$

where R is the reflectivity and α, the absorption coefficient. The attenuation length $1/\alpha$ is typically around 10 nm for most metals. A complicating feature is that the reflectivity can change during a nanosecond laser pulse due to the surface roughness which can be produced already in the first part of a pulse [20]. Also the heating of a solid by the laser light can induce changes in the absorption coefficient [1].

2.2.1. *Heating of Target*

The temperature of the solid increases fast as a result of the impact of the laser pulse. Temperatures up to several thousands kelvin are easily achieved with nanosecond pulses used for PLD, which means that the heating rate of the absorbing volume can exceed values of $10^{12}\,\mathrm{K\,s^{-1}}$. Except for the "hardest" material the temperature of the target surface reaches the melting and boiling point in a fraction of a nanosecond. For ultra-short laser pulses this happens even faster, since the atoms and electrons in a metal are in equilibrium due to the strong electron–photon coupling already after 10–20 ps [1,21].

The conduction of heat induced by a laser pulse is governed by the diffusion equation for a stationary target with a mass density $\rho(T)$, a thermal conductivity $\kappa(T)$ and the specific heat (at constant pressure) $c_p(T)$. For the one-dimensional case the equation is:

$$\rho(T)c_p(T)[T(z,t)/t] - \nabla[\kappa(T)\,\nabla T(z,t)] = Q(z,t) \tag{2.2}$$

where the source term is:

$$Q(z,t) = (1 - R)I_0(t)\alpha \exp(-\alpha z) \tag{2.3}$$

During a typical laser pulse for PLD the temperature of the surface rapidly passes the melting point and increases to a value which may be far above the boiling point (at standard conditions) [1,10,22–27].

For conducting materials the energy, which is deposited by the laser light within the absorption layer $z = 1/\alpha$, propagates more deeply into the bulk. The critical parameter is the thermal diffusivity:

$$D = \kappa/(\rho c_p) \tag{2.4}$$

A typical length scale for thermal processing is the heat diffusion length:

$$l_{th} = 2(D\tau_L)^{1/2} \tag{2.5}$$

where τ_L is the pulse length of the laser. For 6-ns long pulse the thermal diffusion length in silver, the best conductor of all, is $l_{th} \approx 1\,\mu m$ which is several orders of magnitude larger than the optical absorption length $1/\alpha \approx 20\,nm$ at the wavelength 250 nm. Matthias *et al.* [28] have demonstrated that the heat can be considered to be dissipated within a volume that is determined by the laser beam spot and l_{th}. Surface melting and evaporation occur at a much higher laser fluence for silver than for nickel due to the much higher thermal diffusivity in silver than in nickel. The heat diffusion equation has been solved in great detail in Bäuerle [1] and von Allmen and Blatter [4].

Essentially it means that the spatial deposition of the laser light is determined by the largest value of $1/\alpha$ or l_{th}. Except for very short pulses the heat in metals is distributed throughout the entire "thermal volume" with the boundary at the depth l_{th}. For insulating materials the limiting depth of the "thermal volume" is the absorption depth $1/\alpha$, and the solid will be heated only to this depth, independent of the laser pulse duration. In the absence of any significant heat conduction the surface temperature increases much faster than that of a metal. For multicomponent solids, a small thermal diffusion depth is a necessary condition for stoichiometric ablation [10].

2.2.2. The Initial Material Ejection from the Solid

As discussed above the simplest estimate for the ablation yield Y (ablated atoms/pulse) induced by a laser fluence:

$$F = \int I_0(t)\,\mathrm{d}t \tag{2.6}$$

during a laser pulse of duration τ_L onto an area A:

$$Y = FA/U_0 \tag{2.7}$$

given by the cohesive energy per atom U_0, is usually much too high, but Eq. (2.7) is instructive as a starting point for a discussion. U_0 is typically between 1 and 2 eV atom^{-1} for the most volatile metals and 7 and 10 eV atom^{-1} for the most refractory metals. Unfortunately, U_0 is not a well-defined quantity for many solids (e.g. multicomponent targets) for which one may use the enthalpy difference (per atom) between the solid at room temperature and the evaporating gas [1,4]. The ablation yield from a selection of metals with different cohesive energy at a laser fluence of 2 J cm^{-2} is shown in Fig. 2.3 [29]. One notes that the ablation yield decreases significantly with increasing cohesive energy, but also that the yield from elements with a similar cohesive energy may differ from one to another.

In Eq. (2.7) not only the laser light reflection at the surface has been disregarded, but also the energy loss by heat conduction from the surface has not been taken into account. At high laser fluence the target surface becomes shielded by the plume of ablating particles as well. Usually, the experimentally obtained yield is at least one order of magnitude smaller than that given by Eq. (2.7), because of the loss of energy from the target surface to the bulk by heat conduction. Willmott and Huber [10] reported that a 60-g rod of titanium had a temperature of 60 K over room temperature after irradiation with 50,000 laser shots of 100 mJ each.

The ablation rate $j(T_s)$ from a solid or liquid surface (number of atoms/(unit area × time)) of atoms with mass m is controlled by the surface temperature T_s according to the Hertz–Knudsen equation for evaporation [1]:

$$j_s = s\,\frac{p_s(T_s) - p_\infty}{(2\pi m k_B T_s)^{1/2}} \tag{2.8}$$

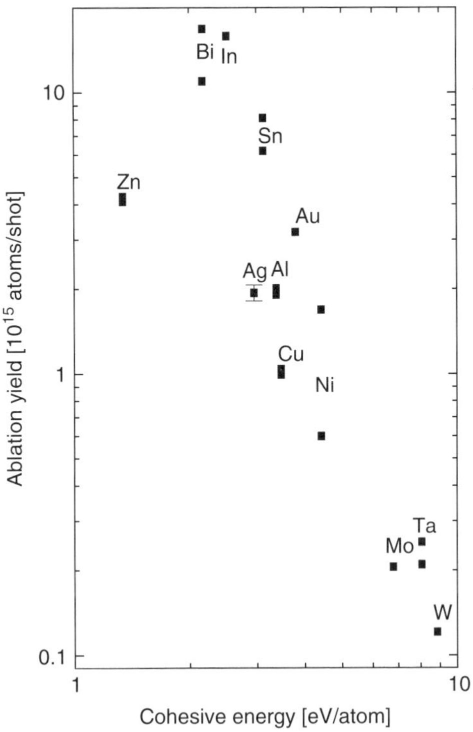

Figure 2.3: Total ablation yield per pulse from a variety of metals measured by weight loss [29]. The cohesive energy is from Kittel [30]. Laser wavelength: 355 nm, pulse length: 6 ns; fluence: 2.0 J cm^{-2} and beam spot area: 0.04 cm^2.

where p_s is the pressure at the surface and p_∞, the pressure at infinity ($\approx$0 in vacuum). s is a factor that accounts for the recondensation at the surface (sticking coefficient), and k_B is a Boltzmann's constant. The pressure at the surface p_s is determined by the Clausius–Clapeyron relation:

$$\begin{aligned} p_s(T_s) &= p_0 \exp(-\Delta H_b/k_B T_s) \cong p_0 \exp(-U_0/k_B T_s) \\ &= p_b \exp(-U_0/k_B T_s + \Delta H_b/k_B T_b) \end{aligned} \tag{2.9}$$

where we have approximated the cohesive energy per atom by the largest contribution, the enthalpy of evaporation ΔH_b per atom. p_0 is determined by the boiling pressure at the temperature T_b (e.g. for the standard pressure p_b = 1 bar, T_b is the normal boiling point).

The first step of modeling laser ablation for a solid with the atomic density n is usually to determine the surface temperature T_s from the heat diffusion equation (2.2) with the absorbed laser energy from Eq. (2.1). As the second step the ablation rate j_s and the surface recession velocity $v_s = j_s/n$ are calculated. Since the surface moves inwards because the outermost layers evaporate, the heat diffusion equation (2.2) needs be modified to include a moving surface with a new depth coordinate $\xi = z - z_0(t)$, where $z_0(t)$ is the instantaneous position of the surface [22,26,27, 31,32]:

$$\rho(T)c_p(T)\frac{\partial T}{\partial t} = \frac{\partial}{\partial \xi}\left(\kappa(T)\frac{\partial T}{\partial \xi}\right) + \rho(T)c_p(T)v_s\frac{\partial T}{\partial \xi} + Q(\xi,t) \qquad (2.10)$$

A slightly different approach based on an energy-balance equation has been used by Amoruso [24] for computing the absolute ablation rate. The rate for a 6-ns pulse

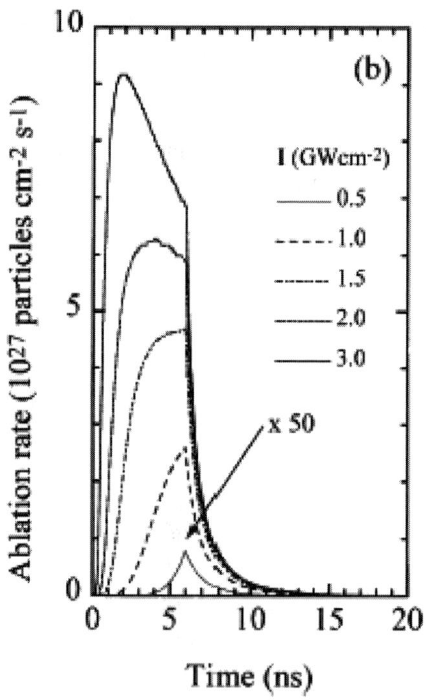

Figure 2.4: Modeling of the time evolution of the ablation rate from aluminum from Amoruso [24]. The calculations are made for a 6-ns pulse (where the maximum is seen for low intensities) at a wavelength of 350 nm; 1 GW cm^{-2} corresponds to a fluence of 6 J cm^{-2} (courtesy of Springer Verlag).

on aluminum for intensities from 0.5 to 3 GW cm^{-2} (fluence range 3–18 J cm^{-2}) is shown in Fig. 2.4. For the lowest intensities the rate increases monotonically because of the increasing surface temperature, until the pulse terminates at $t = 6$ ns. Even after the laser pulse the surface temperature is so high ($T_s > 2000$ K) that the ablation continues for several nanoseconds. At higher intensities the ablation rate reaches a maximum already during the laser pulse due to the plume shielding of the incident laser beam (see Section 2.3). The figure also demonstrates that the rate and total yield (integral of the rate) increase strongly with intensity or fluence.

2.3. One-Dimensional Material Ejection during the Laser Pulse: Stage 2

The initial heating of the solid leads to a strong evaporative ejection of material, which primarily consists of atoms, but also ions and electrons can be emitted. Since the heating of the solid is extremely fast, surface temperatures close to the thermodynamically critical temperature can be reached. With increasing temperature the material ejection can change from evaporation and boiling to explosive boiling (phase explosion), by which also nano- and microparticles can be ejected [21,23,25,27]. As discussed below, the laser light is partly absorbed by this plume, and partly transmitted through the plume to the (strongly heated) solid. The plume is heated by various absorption processes and starts the expansion away from the target. A simplified picture is that the propagation of the plume is driven in a dynamical equilibrium, in such a way that the absorption of laser light provides the kinetic energy for the expansion.

After the termination of the laser pulse a layer of a dense gas of a thickness up to 100 μm is propagating away from the surface. For typical PLD conditions with a fluence from 1 to 10 J cm^{-2} the ablation yield ranges from 10^{15} to 10^{16} atoms pulse^{-1} [10,24,33]. A comprehensive case for aluminum irradiated at 9 J cm^{-2} at 350 nm was modeled by Amoruso [24]. The partly ionized gas layer achieves a temperature of 2.5×10^4 K and a density of 0.005 of that of the solid. In terms of the ideal gas law this gives a pressure of 1 kbar which drives the expansion of the plume into vacuum or into a background gas. These magnitudes agree with values assumed or computed by others [34,35]. Already after 1 ns the gas parameters are close to these final values, and it is this extreme matter that interacts with the incident laser light.

2.3.1. *Absorption of Light in the Initial Plume*

It has been known for decades that the incoming laser light can be absorbed partly or completely by the plume. For IR light an opaque plasma can be formed already at moderate densities [1,11,36,37]. At plasma densities above the critical density:

$$n_{cr} = \frac{4\pi^2\varepsilon_0 c^2 m_e}{e^2\lambda^2} = 1.115 \times 10^{21}/\lambda^2 \ (\mathrm{cm}^{-3}\ [\lambda \text{ in } \mu\mathrm{m}]) \tag{2.11}$$

the light is strongly reflected and does not penetrate the plasma plume. For carbon dioxide lasers (at 10.6 μm) it means that already a modest plume density around $10^{19}\,\mathrm{cm}^{-3}$ prevents the laser light to pass through the plume and reach the target. For UV lasers the critical electron density is of the same magnitude as that of the free electrons in a metal. In the example in Section 2.3, the plume density of 3×10^{20} Al-atoms cm^{-3} is much below the critical density $n_{cr} = 9 \times 10^{21}\,\mathrm{Al\,cm}^{-3}$ for laser light at 350 nm (where we even have assumed that all atoms are singly ionized). Even for densities far below the critical value there is a considerable absorption of light in the plume [1,23,24,26,27,31,38,39]. A simple way of handling this effect of the laser beam has been to include an attenuation of the laser intensity by the area density of the mass in the plume with the same tabulated values as if the laser light passed a solid [22,31,38]. Then, the intensity I of the light that enters into the solid is:

$$I = I_0(1 - R)\exp[-\alpha z_p(n_p/n)] \tag{2.12}$$

where α is the attenuation coefficient of the solid, and z_p and n_p are the thickness and the density of the plume. The results from Fähler and Krebs [38] show that the temperature of an iron surface exhibits a maximum before the laser pulse terminates already at a fluence around $4\,\mathrm{J\,cm}^{-2}$ because of the increasing absorption in the ablated material similar to the results shown in Fig. 2.4. In this case the ablation depth is about 9 nm which is a significant fraction of the attenuation depth $1/\alpha \cong$ 16 nm at the wavelength 248 nm, and about 0.6 of the incoming intensity is absorbed at the end of the pulse. While this plume allows the laser beam to reach the target surface, this is not the case for the example in Section 2.3, where the optical properties of the aluminum solid would lead to a unrealistic reduction of the incoming laser intensity in the plume by a factor 0.001.

The laser light is absorbed in the plume by IB processes and direct single-photon excitation processes. At the very low electron densities and high neutral atoms densities occurring in the initial evaporating flow, the electrons will absorb photons in free–free transitions via collisions with neutral atoms. This process is much less efficient than free–free transitions involving ions, but is an important channel for production of free electrons for a high concentration of neutrals. Once a sufficient number of free electrons have been produced, absorption by free–free transitions via ions becomes the dominant process. The absorption coefficient α_{IB} can be expressed [36,40] as:

$$\alpha_{IB} = 1.369 \times 10^{-23} \left(\frac{Z^3 n_i^2 \lambda^3}{T_e^{1/2}} \right) \left[1 - \exp\left(-\frac{h\nu}{k_B T_e} \right) \right] \ (\text{cm}^{-1}) \qquad (2.13)$$

where Z is the charge state of the ion, for example $Z = 1$ for singly ionized atoms, n_i, the ion density (cm^{-3}), λ, the wavelength (cm), T_e, the electron temperature (K), h, Planck's constant, ν, the laser frequency and c, the velocity of light. Since the ratio of the laser energy $h\nu$ to the plasma temperature $k_B T_e$ enters into the exponential as

$$\frac{h\nu}{k_B T_e} = \frac{hc}{\lambda k_B T_e} \qquad (2.14)$$

the dynamic behavior of the system changes for variations in the wavelength and the plasma energy, which makes it difficult to evaluate the absorption for laser pulses with photon energies of the same order of magnitude as the plasma temperature. For all realistic plasma temperatures the absorption increases drastically with increasing wavelength; for example, carbon dioxide laser light at 10.6 μm has an absorption coefficient which is several orders of magnitude larger than that of IR light at 1064 nm and that for UV lasers. For IR lasers and for a plasma of high temperatures the absorption will exhibit a $T_e^{-1/2}$ dependence, while it shows a weaker dependence $T_e^{-3/2}$ for UV light.

Initially, there are very few free electrons in the plume, and the IB process described above does not give any significant contribution. A minor contribution arises from photoemission and thermo-ionic emission, but as shown by Dreyfus [41] multiphoton ionization is important for the production of "seed" electrons in the UV regime. Amoruso *et al.* [42] have pointed out that also IB by neutral atoms is important. The cross-section for such a process is one to two orders of magnitude smaller than that for absorption by free electrons and ions, but since the number of neutrals can be several orders of magnitude larger than that of ions, this contribution can be important as well. Once a sufficient number of free electrons have been produced, the absorption of photons from electrons as in Eq. (2.13) becomes increasingly more efficient due to the factor n_i^2 of the plume, and the ionization of the plume proceeds via electron impact collisions. This may lead to a fast breakdown of the plume. For UV laser irradiation photoionization of electron-excited species is important as well [24,26,39], and the laser light may even partly delete the population of excited atoms [24].

2.3.2. Ionization in the Plume

For a gas of temperature T_p the number of ions in dynamic equilibrium can be described by the Saha equation. In this equilibrium the rate of generation of ions is equal to the recombination rate. For low temperatures the density of singly ionized ions n_i can be expressed by the total density of the plume, $n_p = n_e + n_o = n_i + n_o$, where n_o is the density of neutral atoms and U_i, the energy of the (lowest) ionization level [1,10]:

$$n_i = [c_{Saha} T_p^{3/2} n_p \exp(-U_i / k_B T_p)]^{1/2} \tag{2.15}$$

where $c_{Saha} = 2.4 \times 10^{15}$, when n_i, n_p are in cm^{-3}, T_p in K and U_i in eV [10].

Willmott and Huber [10] have pointed out that even for gas temperatures of 5000 K obtained by laser heating of a solid it is difficult to achieve any ionized fraction of more 0.01. This behavior is clearly determined by the exponential, since U_i is typically several eV, and any initial ionization can only happen as a result of strong non-uniformities in the electron density.

Under normal laser irradiation with PLD conditions, the Saha equation is only of limited value. The seed electrons are produced by low-level impurities, multi-photon ionization or electron impact ionization. In fact, in most of the literature the ionization breakdown of the matter is caused by a dynamic balance between these production mechanisms and recombination, which is strongly dependent on laser pulse length and wavelength.

2.3.3. The Knudsen Layer

The high-density gas layer which is produced by a laser pulse under PLD conditions has been treated comprehensively in the past. All ablated particles move away from the surface, and the flow of particles is typically described as a half-range Maxwellian distribution (only velocities $v_z > 0$ are possible). After a few collisions the distribution of particles can be fitted by a shifted Maxwellian distribution with an average flow velocity u and a temperature T_z [43]:

$$F(z, v_z, t) = \left(\frac{m}{2\pi k_B T_z} \right)^{1/2} \exp\left(-\frac{m(v_z - u)^2}{2k_B T_z} \right) \tag{2.16}$$

which means that at this point the flow has equilibrated in the flow direction by collisions. Strictly speaking T_z is the variance of the Maxwellian distribution around the flow velocity, and a proper temperature only exists if not only the velocity, but also the temperature in all directions coincide ($T_z = T_x = T_y$). Nevertheless, such an expression as Eq. (2.16) has often been used to characterize ablation flows [34,43,44].

Sibold and Urbassek [43] found from Monte Carlo simulations that the thickness z_K of the Knudsen layer is approximately determined by the mean free path λ:

$$z_K \cong 19\lambda = 19(\sqrt{2}n_p\sigma)^{-1} \tag{2.17}$$

where σ is the collision cross-section (typically a few times $10^{-16}\,cm^2$) in the ablated layer, and n_p, the density in the plume. For the example given in the beginning of this section, $n_p = 3 \times 10^{20}$ Al-atoms cm^{-3}, and with a cross-section of $\sigma = 5 \times 10^{-16}\,cm^2$, one arrives at a thickness of the Knudsen layer $z_K \cong 0.9\,\mu m$, which is considerably less than the thickness ($\cong$100 μm) of the full ablated layer. At this thickness the temperature T_z in the flow direction coincides with the lateral temperatures ($T_x = T_y$). The Mach number increases monotonically from the surface and is equal to unity at z_K, so that velocity of the flow is subsonic in the Knudsen layer and supersonic in the outer layer. Analytical treatments typically involve assumptions of the form of the distribution of the particles backscattered from the surface [43]. After the end of the ablation process the Knudsen layer tends to become destroyed because the supply of atoms from the surface ceases.

2.3.4. Acceleration of Plume Particles by Space Charge Effects

A surprising feature of laser ablation is the existence of very fast particles in the ablation plume. Time-of-flight (TOF) spectra show kinetic energies which exceed 100 eV even at comparatively low fluences [45]. In addition to the hydrodynamic processes, a few authors have tried to estimate the acceleration of ions due to ambipolar effects [10,46–48].

2.4. Three-Dimensional Expansion of the Plume in Vacuum: Stage 3

After the termination of the laser pulse the initial gas layer has an extension of about 100 μm. From this point there is little further transfer of energy or mass to the ablation plume. Therefore, plume propagation can be considered as an adiabatic expansion. The three-dimensional expansion of the plume is largely governed by the shape of the plume at the end of the laser pulse. Basically, the dynamics is determined by the pressure gradients in this initial plume and the expansion will take place in all directions into the hemisphere.

A model of plume expansion must account for three features:

1. The strongly peaked plume in a direction perpendicular to the surface.
2. Broadening of the angular distribution with decreasing laser spot size.
3. The flip-over effect.

The broadening of the angular distribution has been systematically studied in a number of cases [8,49–52]. In a typical experiment the fluence is kept constant, while the size of the beam spot varies. It gives the initially counterintuitive result that the angular distribution of the ablated atoms becomes narrower as the spot size increases. The flip-over effect occurs when the plume of a non-circular spot evolves in such a way that the fastest component is in the lateral direction where the initial plume was most narrow. For an elliptical or rectangular spot it means that the plume "turns" in such a way during expansion that the major axis of the expanded plume lies 90° to that of the initial beam spot. Also this effect has been studied recently [33,52–54]. Even though the three features are listed separately, they are closely connected and all determined by the pressure gradients in the initial plume.

*2.4.1. Anisimov's Expansion Model**

The adiabatic expansion has been treated by Anisimov *et al.* [55,56] and Singh and Narayan [34]. The treatments are to some extent similar, but exhibit differences as well. Singh and Narayan include the formation of a Gaussian plume during the laser pulse without specifying the mechanism of particle ejection from the solid. After the laser pulse is terminated, the plume expands adiabatically with the additional assumptions that there are no spatial variations in the plume temperature (i.e. $\nabla T = 0$). As pointed out by Anisimov *et al.* this approximation is undoubtedly questionable. The major drawback of the model by Singh and Narayan [34] is that no explicit solution of the gas-dynamic equations is given. The results from their calculations of the angular and energy distribution of the ablated atoms are for a multicomponent system (YBaCuO target) and cannot easily be generalized to other materials.

* In Sections 2.4 and 2.5 the Z-axis points away from the target surface (as in Fig. 2.5).

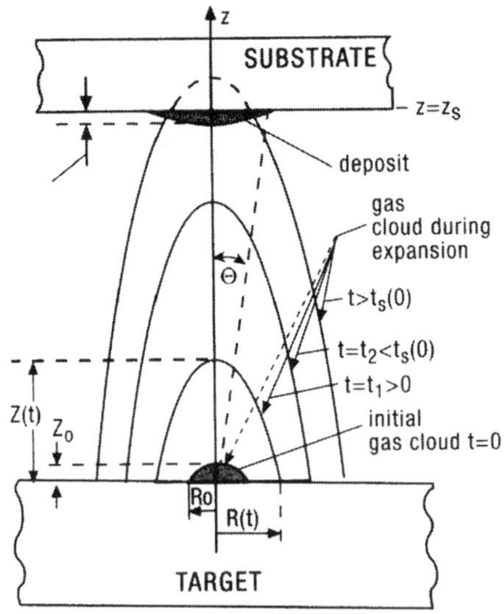

Figure 2.5: Geometry of the plume expansion in Anisimov's model [55]. The initial gas cloud (at time $t \approx 0$) lies as a thin "bubble" on the target. The figure shows a rotationally symmetric expansion ($X(t) = Y(t)$) of an ellipsoidal plume. A substrate for collecting ablated material has been included as well.

Anisimov's model is based on Lie group transformation theory which simplifies the solution of the gas-dynamic equations. The expanding plume has an semi-ellipsoidal shape with a front determined by the axes $X(t)$, $Y(t)$ and $Z(t)$ at a time t, where the Z-axis is normal to the surface (see Fig. 2.5). The characteristic properties of the expansion, such as the density and the pressure (see below), are constant on ellipsoidal surfaces, that is $x^2/X(t)^2 + y^2/Y(t)^2 + z^2/Z(t)^2 = \text{constant}$.

The complete hydrodynamic motion of all particles in the plume is determined by self-similarity from the motion of the ellipsoidal plume front with the axes ($X(t)$, $Y(t)$, $Z(t)$) in which all fluid elements of the plume are embedded (see Fig. 2.5). The velocity v of any fluid particle at (x, y, z) is then determined by the relative position to the front:

$$v_x = \frac{x(\mathrm{d}X/\mathrm{d}t)}{X(t)} \qquad v_y = \frac{y(\mathrm{d}Y/\mathrm{d}t)}{Y(t)} \qquad v_z = \frac{z(\mathrm{d}Z/\mathrm{d}t)}{Z(t)} \tag{2.18}$$

which was also assumed by Singh and Narayan [34].

In Anisimov's theory the initial conditions for this semi-ellipsoidal gas bubble with the axes X_0, Y_0 and Z_0, which starts the expansion after the termination of the laser pulse at $t = \tau_L \approx 0$, are:

$$X(0) = X_0 \qquad Y(0) = Y_0 \qquad Z(0) = Z_0 \tag{2.19}$$

and

$$\frac{dX}{dt}(0) = 0 \qquad \frac{dY}{dt}(0) = 0 \qquad \frac{dZ}{dt}(0) = 0 \tag{2.20}$$

The underlying assumption is that the kinetic energy of the initial plume is much smaller than the thermal energy. For a given value of the adiabatic constant $\gamma = c_p/c_v$, the mass density, pressure and temperature profiles for the adiabatic expansion have the form:

$$\rho(x, y, z) = \frac{M_p}{I_1(\gamma)XYZ}\left(1 - \frac{x^2}{X^2} - \frac{y^2}{Y^2} - \frac{z^2}{Z^2}\right)^{1/(1-\gamma)} \tag{2.21}$$

$$p(x, y, z) = \frac{E_p}{I_2(\gamma)XYZ}\left(\frac{X_0Y_0Z_0}{XYZ}\right)^{\gamma-1}\left(1 - \frac{x^2}{X^2} - \frac{y^2}{Y^2} - \frac{z^2}{Z^2}\right)^{\gamma/(1-\gamma)} \tag{2.22}$$

and

$$T(x, y, z) \sim \rho(x, y, z)^{\gamma-1} \tag{2.23}$$

M_p is the plume mass and the E_p, the initial energy of the plume. The expressions for $I_1(\gamma)$ and $I_2(\gamma)$ are given in Anisimov *et al.* [55,56]. The sharp edges of the pressure and density profiles at the front at which the density is equal to zero mean that the temperature at the edge approaches zero as well. This is in contrast to Singh and Narayan's model, in which the temperature T_p of the plume is considered constant.

With the profiles from Eqs. (2.21)–(2.22) and the velocity determined from Eq. (2.20), Anisimov *et al.* [55] as well as Singh and Narayan [34] arrive at the differential equations which control the expansion of the plume:

$$X(t)\frac{d^2X}{dt^2} = Y(t)\frac{d^2Y}{dt^2} = Z(t)\frac{d^2Z}{dt^2} = A\left(\frac{X_0Y_0Z_0}{XYZ}\right)^{\gamma-1} = \left(\frac{V_0}{V}\right)^{\gamma-1} \tag{2.24}$$

where Anisimov's model gives:

$$A = (5\gamma - 3)E_p/M_p \tag{2.25}$$

and Singh and Narayan's model:

$$A = k_B T_p/\mathrm{m} \tag{2.26}$$

Intuitively, the form of the equations can be understood from simple arguments, as also described by Zel'dovich and Raizer [40]. The expansion is determined by the pressure gradients in the initial plume. Since the dimensions of the laser beam spot usually is much larger than the thickness of the initial plume, the pressure gradient $\partial p/\partial r$ is much larger along the surface normal than parallel to the surface. The gradient is thus driving the expansion away from the surface much faster than in lateral direction. The force per mass unit, that is $\mathrm{d}^2X/\mathrm{d}t^2$ is given by the ratio of the pressure gradient to the mass density $(1/\rho)\partial p/\partial r$. This quantity is of the order of magnitude of $(1/\rho)p/X_0$, where p and ρ are averaged over the plume mass at the time t. The pressure p is determined from the ratio of the thermal energy of the gas to its volume $p \sim E_{\mathrm{heat}}/V$, which falls off as $V^{1-\gamma}/V = V^{-\gamma}$ $(=(X_0)^{-3\gamma}$ for a spherical plume). Since the density falls of like V^{-1}, the ratio p/ρ approaches $V^{-(\gamma-1)}$, which gives the expression of the right-hand side of Eq. (2.24). The coordinate $X(t)$ on the left-hand side of the equation originates from the gradient of the pressure.

The system of equations (2.24)–(2.26) is coupled so that the propagation of the front in all directions depends on the instantaneous volume $V \sim XYZ$. There is one immediate conclusion from Eq. (2.24). Since $\gamma > 1$, the acceleration of the front will approach zero with increasing expansion of the plume. It means that the plume expands with constant velocity in the asymptotic limit.

The expansion features (1)–(3) can be explained by Eq. (2.24). As discussed above, the strong forward peaking follows from fact that the acceleration of the front in the Z-direction $\mathrm{d}^2Z/\mathrm{d}t^2$ is proportional to Z^{-1}. Since the initial plume usually is relatively thin, the acceleration is much greater away from the target than in lateral direction. The effect of the beam spot size follows from a similar argument. If the initial lateral axes are large, the acceleration is reduced in these directions, and the plume will be strongly forward directed. Finally, also the flip-over effect can be explained from the structure of the equations. If X_0 is less than Y_0, the plume accelerates faster in the X-direction.

In general, the equations have to be solved numerically. However, Anisimov *et al.* [55,56] have defined a number of reduced variables, which can be used for the calculation of the plume shape at a later time. An extensive set of curves and tabulations are given in Refs. [55,56].

A simple expression can be found for the asymptotic front velocities from this model [57,58]. Let us assume a spherical expansion ($X_0 = Y_0 = Z_0 = R_0$), which means that the propagation can be expressed by the radius R of the front. The total plume energy is a sum of the kinetic energy (first term) and the heat in the plume (second term):

$$E_p = \frac{3(\gamma - 1)}{10\gamma - 6} M_p \left(\frac{dR}{dt}\right)^2 + E_p \left(\frac{R_0}{R}\right)^{3(\gamma-1)} \quad (2.27)$$

Since the second term vanishes for large plume radii, the plume energy is eventually converted to kinetic energy. The asymptotic velocity dR/dt of the front is then given by:

$$\left(\frac{dR}{dt}\right) = \left[\frac{2(5\gamma - 3)}{3(\gamma - 1)}\right]^{1/2} \left(\frac{E_p}{M_p}\right)^{1/2} \quad (2.28)$$

For a one-component plume it means that the asymptotic velocity is proportional to $m^{-0.5}$ during the expansion. However, this result is usually not confirmed by experiments because plume expansion in the stage 2 takes place as a non-adiabatic process during the laser pulse (see in Section 2.3.1) and because species of different mass interact during the initial expansion [34].

2.4.2. *The Angular Distribution of Ablated Species in Vacuum*

In the literature the angular distribution of the ablated species $F(\theta)$ is often expressed as:

$$F(\theta) = F(0)\cos^n\theta + B\cos\theta \quad (2.29)$$

with a strong peak in forward direction superimposed on a $\cos\theta$ – background, which in many cases is not included ($B = 0$) [59]. Typically, the value of n for an expansion in vacuum is in the range of 5–20, but even higher values have been reported [60]. This approximation is purely empirical, and in many cases it only works for small angles, so that one needs a sum of $\cos^n\theta$ – terms rather than a simple expression as Eq. (2.29).

Anisimov *et al.* [55] arrive at the following expression for the angular distribution of species $F(\theta)$ collected on a planar substrate in the distance z_s:

$$F(\theta) = \frac{F(0)}{(1 + k_z^2 \tan^2\theta)^{3/2}} \tag{2.30}$$

where the value of the area density $F(0)$ in normal direction $\theta = 0$ is:

$$F(0) = \frac{Yk_z^2}{2\pi z_s^2} \tag{2.31}$$

and $k_z = Z_{inf}/X_{inf}$ is the ratio of the limiting value of the position of the cloud front along the *Z*-axis and the value of the front in horizontal direction along the *X*-axis (and *Y* is the total number of ablated atoms). A large value of k_z corresponds to a strongly peaked angular distribution, while the trivial case of $k_z = 1$ gives an isotropic distribution. A typical value of k_z is 2.5–3.5 for irradiation with a UV laser under conditions which are typical for PLD [33]. Also expressions for two-dimensional distributions are available [56]. For small values of θ the distribution can be approximated by $\cos^n \theta$ with $n = 3k^2$. One should bear in mind that the formula does not include self-sputtering or incomplete sticking [8,60]. Self-sputtering tends to reduce the deposition on a substrate along the target normal, where the most energetic ablated particles occur. This means that the angular distribution will be less peaked in forward direction if self-sputtering occurs. Self-sputtering takes primarily place at high fluence and for volatile materials [61]. Occasionally, incomplete sticking and self-sputtering have been incorporated in deposition models [62,63].

For the angular distribution on a hemisphere [64,65], the expression from Eq. (2.30) has to be multiplied by a geometric factor $(1 + \tan^2 \theta)^{3/2}$ [60]:

$$F(\theta) = F(0)\left(\frac{1 + \tan^2\theta}{1 + k_z^2 \tan^2\theta}\right)^{3/2} \tag{2.32}$$

This distribution is used for the deposition of ablated atoms on a circular stripe in the *X*–*Z*-plane [33,66]. Actually, this formula is often used to describe a deposition profile with k_z as a fitting parameter. The expression is a good approximation also for the ion component of the ablated flux with a realistic value of k_z [65].

How well Eq. (2.32) approximates the angular distribution for different values of k_z can be seen in Fig. 2.6. The distribution of silver atoms for an elliptical beam

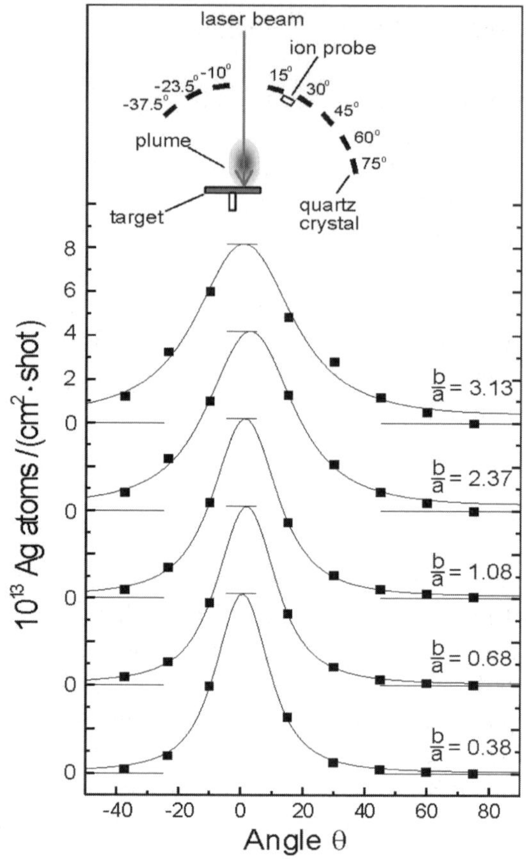

Figure 2.6: The flip-over effect measured for silver in vacuum [33]. The angular distribution of the plume in the *X*–*Z*-plane has been determined from the deposition on the quartz crystal microbalances. The geometry of the setup is shown in the upper part of the figure. The elliptical beam spots had different aspect ratios (b/a), but same area. The data points have been fitted with Eq. (2.32), solid line. Laser wavelength: 355 nm; pulse length: 6 ns; fluence: $2.0\,\mathrm{J\,cm^{-2}}$ and beam spot area: $0.04\,\mathrm{cm^2}$.

spot with different aspect ratios b/a shows the flip-over effect in a convincing way. The flip-over effect has been observed for laser beam spots of more complicated geometry as well [54].

The dependence on the laser fluence does not emerge from Anisimov's model. However, there seems to be clear indication of a more strongly peaked distribution in forward direction for high fluence [51].

2.4.3. The Value of the Adiabatic Constant γ

The value for a monatomic gas at room temperature is $\gamma = 5/3$, but the value decreases for polyatomic and ionized species. For plume particles γ has turned out to be considerably lower. Toftmann *et al.* [33] have determined an experimental value of $\gamma = 1.42 \pm 0.07$ for a plasma plume under typical PLD conditions from a detailed study of the flip-over effect. The ion component was studied by Hansen *et al.* [64], who found a value of $\gamma = 1.2 \pm 0.1$, which is consistent with the value of $\gamma = 1.24$ suggested for a plasma plume in this regime by Zel'dovich and Raizer [40]. The value of γ for a multicomponent plume of lead zirconia titanate (PZT) reported by Tyunina and Leppävuori turned out to be very small ($\gamma = 1.1$–1.2) and also difficult to explain [52].

2.4.4. Charge States

Spectroscopic and Langmuir probe measurements have revealed that there is a considerable fraction of ions in the plume at UV laser fluences relevant for PLD. At moderately high fluences ($>3\,\mathrm{J\,cm^{-2}}$) a fraction of more than 0.5 of the plume particles are ionized and the fraction increases with fluence [61,65,67]. Thestrup *et al.* [65] have reported that the ion fraction is somewhat smaller for volatile materials. At UV laser fluences below $2\,\mathrm{J\,cm^{-2}}$ the ion fraction primarily consists of singly charged ions [68]. The high fraction of ions in the plume generated by UV laser light is also a clear indication that photoionization processes are important in the initial plume formation [24,26,39]. For IR laser irradiation at $\gamma = 1064\,\mathrm{nm}$ the fraction of ions from a metal target is a few percent even at comparatively high fluences in the range 20–80 $\mathrm{J\,cm^{-2}}$ [69].

2.4.5. Plume Expansion from Femtosecond Laser Pulses

The formation of a plume from femtosecond pulses is different from the plume generation at nanosecond pulses. There is no significant interaction between the laser light and the plume which is generated tens of picosecond after the laser irradiation. However, the propagation of the plume takes place on a microsecond time scale similar to plumes generated by nanosecond pulses.

Since the initial plume after the termination of the laser pulse is an ultra-thin layer on top of the irradiated solid (atoms with a velocity of $10\,\mathrm{km\,s^{-1}}$ move 1 nm in 100 fs) one would in principle expect an extremely narrow angular distribution in forward direction on the basis of Anisimov's model. Such a calculation has been performed by Komashko *et al.* [70], but the results disagree with all existing measurements. One problem is that the surface undergoes roughening which tends to smear out all monodirectional distributions. This can only be avoided by using single shots on a virgin surface, but in that case the usual preconditioning of the surface by laser cleaning cannot be carried out and the plume will contain impurities as well. In any case, the existing angular distributions from femtosecond laser irradiation are somewhat narrower than the corresponding ones from nanosecond pulses [71,72].

2.5. Plume Expansion in Background Gases: Stage 4

The expansion of a laser ablation plume in a background gas involves additional interaction processes between the plume particles and the background gas atoms and molecules compared with stage 3, the plume expansion in free space. A full treatment of the ablation process and the subsequent plume dynamics has turned out to be extremely complex due to the many different processes that take place during the expansion [57]. Any modeling based on cross-section of collisions between plume atoms and background gas atoms or molecules will involve poorly known quantities; for example, cross-sections for the impact of atoms with a kinetic energy range 10–100 eV on gas or other plume atoms are hardly known. In some cases combinations of different computational models have led to satisfactory agreement with experimental results [73].

The dynamics of the plume depends critically on the pressure of the ambient gas. Since the pressure of the plume is much higher than that of a typical background gas, the plume will expand rapidly, until the driving pressure has decreased considerably. Three regimes appear for the plume dynamics with increasing pressure of the ambient gas. For the lowest pressure up to 0.01 mbar, the behavior of the plume is essentially *vacuum-like*, then there is a *transition* regime with a division of the plume into two propagating components, one moving fast as in the vacuum-like regime and one more slowly. Finally, at high pressure the plume is slowed down in the *diffusion-like* regime as a result of shock wave formation. Eventually, the plume stops and the particles become thermalized [57,66,74]. In a simple picture the expanding plume acts on the surrounding gas as a piston which is a shock-compressed gas blanket. In the *transition* regime the fast component of the plume penetrates the background gas, while the slow component pushes the "piston" away from the target. At a high background pressure, in the *diffusion-like* regime, all plume particles are slowed down by pushing this compressed gas layer further away from the target. This process releases an internal as well as an external shock wave [57]. Also, Itina *et al.* [73] identify three pressure regimes based on similar concepts observed from the vicinity of the target rather than from an observation point in some distance as Amoruso *et al.* [66].

The knowledge of plume slowing down and shock wave formation has advanced tremendously due to the many recent spectroscopic studies with fast imaging techniques [75–77].

2.5.1. Film Deposition in a Background Gas

The majority of all film depositions by PLD takes place in an ambient gas with a pressure up to 0.5 mbar. The gases can be reactive, as oxygen or nitrogen, but also the inert gas, argon, is frequently used. Basically, the background gas serves to:

1. maintain the chemical equilibrium of a volatile component in the growing film;
2. induce gas chemical reactions between the plume atoms and the gas molecules during the transfer of target material to a substrate (e.g. nitrides in a nitrogen background gas);
3. act as a moderator for the kinetic energy of the incident particle flow on the growing film.

The last point is important because the incident atoms should have sufficient kinetic energy to induce surface reactions, bond breaking and enhance the mobility of the atoms on the surface. However, the impact of high-energy species that induce lattice displacements in deeper layers must be reduced as much as possible. The optimum kinetic energy, which fulfills both requirements, is not known precisely, but is suggested to be below 20–30 eV [8,10].

2.5.2. From the Vacuum-Like to the Transition Regime: Shock Wave Formation and Plume Splitting

According to Zel'dovich and Raizer [40] the shock wave formation becomes important, when the mass of the displaced ambient gas is comparable to the mass of the plume. For an ambient gas of mass density ρ_g the distance R_{SW}, at which shock wave formation for a hemispherical expansion starts, can be approximated by:

$$(2/3)\,\pi R_{SW}^3 \rho_g \approx M_p \tag{2.33}$$

This approximation demonstrates that for a given target (i.e. the same plume mass M_p), the shock wave production starts much closer to the target for a background gas with a high pressure than for one with low pressure. Similarly, the shock production also starts closer to the target for a heavier gas than for a light one.

Since the pressure p of an ideal gas, where n_g and m_g are the density and atomic/molecular mass of the background gas, is:

$$p = n_g k_B T = (\rho_g / m_g) k_B T \tag{2.34}$$

we obtain:

$$p \approx \frac{3 M_p k_B T}{2\pi R_{SW}^3} \times \frac{1}{m_g} \tag{2.35}$$

This relationship tells us that for a fixed distance ($\sim R_{SW}$) the pressure at which shock wave production becomes important is inversely proportional to the atomic/molecular mass m_g of the background gas. However, the formula cannot be used to give accurate predictions, since Eq. (2.32) essentially is an estimate.

The shock wave production does not appear abruptly, but is the result of an increasing hydrodynamic interaction of the plume atoms with the atoms and molecules in the background gas. At a certain distance the fraction of ablated particles which penetrates the surrounding ambient gas as a freely expanding plume is decreasing with increasing gas pressure. This decrease is accompanied by a large enhancement of the slowly propagating component of the plume and leads to the so-called *plume splitting*. This phenomenon was first observed by Geohegan [75] and Geohegan and Puretzky [78,79] for the ion component of the plume and later explained by Wood *et al.* [80,81] by a hydrodynamic model which includes multiple scattering. Plume splitting of ions from a one-component metallic target, as well as from a multicomponent target, has turned out to be a general feature which occurs within a relatively narrow pressure interval [66,76,77,82]. Also neutrals show this behavior in TOF distributions from one-component metals [77,83] as well as from a multicomponent oxide target [84].

Typical TOF signals in this transition regime for silver ions in four different background gases are shown in Fig. 2.7 [85]. There is a clear plume splitting for silver ions in xenon, argon and oxygen, whereas the structure is less pronounced for the very light helium gas. The pressure, at which the plume splitting occurs, is for all gases except helium in the 10^{-2}mbar range (where it also occurs for Geohegan and Puretzky [79]) and increases clearly with decreasing background gas mass, even though the figure only gives a qualitative view. The second peak in Fig. 2.7 is the ion component which gradually is slowed down because of shock

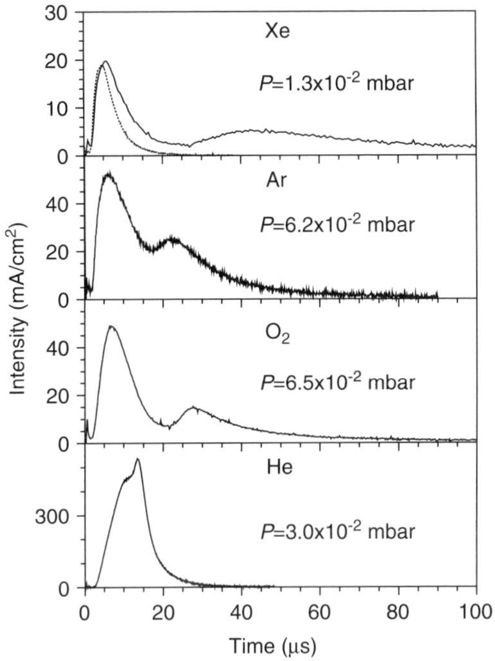

Figure 2.7: The transition regime: TOF spectra of ions from a silver plasma plume in different background gases. The spectra have been measured with the Langmuir probe shown in Fig. 2.6 in a distance of 75 mm from target. All spectra (except that for helium) show a distinct splitting of the plume into a fast and slow component. Laser fluence: 2.5 J cm^{-2} (otherwise experimental conditions as in Fig. 2.6). The dashed curve in the upper panel shows the ion signal recorded in vacuum (the vacuum-like regime) with a different scale (from Amoruso *et al.* [85]).

wave production and finally becomes confined. Therefore, the flow of plume particles through the background gas to a substrate or a Langmuir probe decreases with increasing pressure. Toftmann *et al.* [86] have shown that attenuation of the integrated ion signal obtained with a probe can be described by a dependence on the atomic/molecular mass of the background gas, $p \sim m_g^{-1.2}$ similar to that in Eq. (2.35).

The two ion components in the TOF distribution of silver atoms in argon can be separated by a fitting procedure so that the average time of arrival at a Langmuir probe can be determined [77] (Fig. 2.8). Initially, only the first peak is present. The velocity of the ions is constant, $z \sim t$, and equal to the free (e.g. vacuum) expansion velocity up to a fairly large pressure. This component is exponentially reduced (not seen in this figure) and the signal vanishes at about 0.2 mbar. The slow component (i.e. the second peak) is slowed because of the shock wave production as discussed

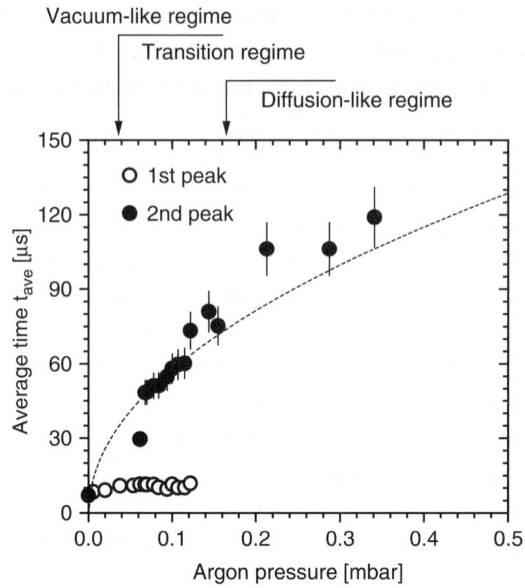

Figure 2.8: Time development of the two ion peaks seen in Fig. 2.7 for a silver plume in an argon gas. The average time (found from the integral of the peaks in Fig. 2.7) is plotted as a function of the pressure. The dotted line is a fit to the point blast wave model (Eq. (2.36)). The three pressure regimes are separated by arrows. Experimental conditions as in Fig. 2.7 (from Toftmann *et al.* [86]).

above. The average flight time is approximated by a point blast wave [40,57,58], which appears to fit the data well up to 0.2 mbar:

$$z = \alpha t^{0.4} \tag{2.36}$$

where α is a constant.

The three regimes are indicated in Fig. 2.8 as well. In the vacuum-like regime only the fast peak is seen, while in the second regime, the transition regime, the second, slow peak gradually increases relatively to the first peak, and in the third regime, the diffusion-like regime, the third peak is dominant. As mentioned, this is also the regime in which the shock wave dynamics largely controls the expansion.

2.5.3. *Shock Wave and Diffusion: The Third Regime*

As seen in Fig. 2.8 the slow component follows a point blast model well up to a relatively high pressure. Then the plume atoms are slowed by thermalization [57] and move in a diffusion-like way, until they reach the collecting system [58,66].

The shock wave dynamics can be characterized by a contact surface which separates the slow part of the plume; for example the ions as well as the slow neutrals in the second peak from the background gas. The ambient gas is compressed and pushed away as a compressed shell around the plume with a contact pressure which is considerably higher than the instantaneous internal plume pressure and the ambient pressure. Both an external as well as an internal shock wave are launched by the expansion. During the expansion the internal shock wave eventually reaches the center of the plume and a significant part of the plume energy becomes thermal rather than kinetic. In the later expansion phase the kinetic energy is mainly carried by the mass of the external shock wave, but eventually the plume stops because of the finite pressure of the ambient gas and the energy of plume and of the shock waves becomes converted to thermal energy [57].

During the expansion at the late stage the plume atoms diffuse into the shock wave shell. This mixing leads to strong emission of light in the shock wave region [58,75,83,87]. After the stopping of the plume, the plume atoms eventually diffuse away from the confined volume and become thermalized [58,66,74]. The many collisions with the ambient gas molecules or atoms reduce the energy of the plume atoms to a value below the thermalization limit (<0.1 eV). In particular, this diffusion-like behavior is dominant at distances far away from the target at relatively high pressures, where the plume comes to rest close to the target. The dynamics of the diffusion becomes much more complex on a millisecond scale, since the configuration and the parameters depend on the particular vacuum system (e.g. pumping speed and adsorption dynamics on the surfaces). The behavior of the ablation flow on this time scale has been treated by Amoruso *et al.* [66] on the basis of a three-dimensional spherical solution of the diffusion equation.

Actually, the shock wave behavior may start already during the laser pulse in stage 2. This expansion has been studied by Zhang *et al.* numerically [35] and analytically by Zhang and Gogos [88].

2.5.4. The Angular Distribution in Background Gases

Knowledge of the angular distribution of the ablated particles in a background gas is obviously important for the production of film of uniform thickness. Saenger [60] has compiled experimental results, but only few results are available compared with the distribution in vacuum. Itina *et al.* [89] have made computer simulations for a few cases.

Even though formula (2.32) was derived for the expansion of a plume in vacuum, it has also turned out to be a good approximation for the angular distribution in a background gas with a suitable choice of the fitting constant k_z [66,77,90,91].

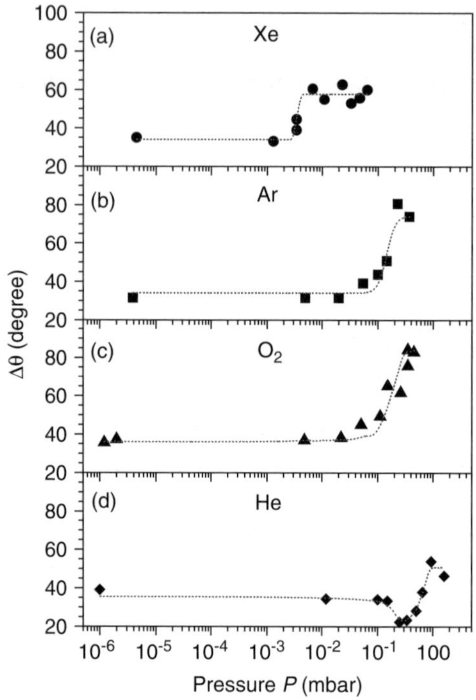

Figure 2.9: The full width at half maximum $\Delta\theta$ of the angular distribution of a silver plume (ions as well as neutrals) in different background gases (measured by deposition as in Fig. 2.6). The distributions broaden with pressure except for helium, where plume narrowing is observed in a small pressure regime. Experimental conditions as in Fig. 2.7 (from Amoruso *et al.* [85]).

With an increasing number of collisions between plume atoms and background gas atoms/molecules the angular distribution broadens (k_z decreases), and finally the distribution may reach a saturation level. The full width $\Delta\theta$ at half maximum is shown for a silver ablation plume in four different gases (Fig. 2.9) [85]. Only the data for a silver plume in xenon shows a distinct saturation, and the possible saturation level seems to vary from one gas to another. For a helium background gas a sharpening even occurs in a narrow pressure range above 10^{-1} mbar [92]. However, there is a clear trend of broadening at a smaller pressure with increasing mass of the background gas atoms/molecules. The broadening of the angular distribution starts when the background pressure reaches the transition regime and the saturation occurs at pressure values in the diffusion-like regime [66].

Acknowledgments

The author of this chapter, Jørgen Schou, thanks Dr. Salvatore Amoruso and Prof. James G. Lunney for comments to the manuscript.

References

[1] D. Bäuerle, *Laser Processing and Chemistry*, 3rd edition, Springer, Berlin, 2000.
[2] J.C. Miller and R.F. Haglund (Eds.), *Laser Ablation and Desorption, Experimental Methods in the Physical Sciences*, Vol. 30, Academic Press, New York, 1998.
[3] A. Vertes, R. Gijbels and F. Adams (Eds.), in *Laser Ionization Mass Analysis, Chemical Analysis Series 124*, John Wiley, New York, 1993.
[4] M. von Allmen and A. Blatter, *Laser–Beam Interactions with Materials: Physical Principles and Applications*, Springer, Berlin, 1995.
[5] S.I. Anisimov and B.S. Luk'yanchuk, *Physics-Uspekhi*, 45 (2002) 293.
[6] S. Gergiou and F. Hillenkamp (Eds.), *Chem. Rev.*, 103 (2003) 317.
[7] T. Lippert (Ed.), *Polymers and Light, Adv. Polymer Sci.*, 168 (2004) 51.
[8] D.H. Lowndes, in *Laser Ablation and Desorption*, Eds. J.C. Miller and R.F. Haglund, *Experimental Methods in the Physical Sciences*, Vol. 30, Academic Press, New York, 1998, p. 475.
[9] D.B. Chrisey and G.K. Hubler (Eds.), *Pulsed Laser Deposition of Thin Films*, Wiley, New York, 1994.
[10] P.R. Willmott and J.R. Huber, *Rev. Mod. Phys.* 72 (2000) 315.
[11] A. Gorbunoff, *Laser-Assisted Synthesis of Nanostructured Materials*, Fortschritt-Berichte; Reihe 9, VDI 357,VDI Verlag, Düsseldorf, 2002.
[12] G.K. Hubler (Ed.), *Mater. Res. Bull.*, XVII (2) (1992).
[13] M.N.R. Ashfold, F. Claeyssens, G.M. Fuge and S.J. Henley, *Chem. Soc. Rev.*, 33 (2004) 23.
[14] P.R. Willmott, P. Manoravi and K. Holliday, *Appl. Phys. A*, 70 (2000) 425.
[15] C. Darshan, S.B. Kundaliya, S.E. Ogale, S. Lofland, C.J. Dhar, S.R. Metting, Z. Shinde, B. Ma, K.V. Varughese, L. Ramanujachary, Salamanca-Riba and T. Venkatesan, *Nature Mat.*, 3 (2004) 709.
[16] V. Moshnyaga, B. Damaschke, O. Shapoval, A. Belenchuk, J. Faupel, O.I. Lebedev, J. Verbeeck, G. van Tendeloo, M. Mücksch, V. Tsurkan, R. Tidecks and K. Samwer, *Nature Mat.*, 2 (2003) 247.
[17] B. Thestrup, J. Schou, A. Nordskov and N.B. Larsen, *Appl. Surf. Sci.*, 142 (1999) 248.
[18] M. Reichling, in *Laser Ablation and Desorption*, Eds. J.C. Miller and R.F. Haglund, *Experimental Methods in the Physical Sciences*, Vol. 30, Academic Press, New York, 1998, 573.
[19] A. Vogel and V. Venugopalan, *Chem. Rev.*, 103 (2003) 577.
[20] B. Toftmann, J. Schou and N.B. Larsen, *Appl. Phys. A*, 69 (1999) S811.

[21] D.S. Ivanov and L.V. Zhigilei, *Phys. Rev. B*, 68 (2003) 064114.
[22] W. Svendsen, O. Ellegaard and J. Schou, *Appl. Phys. A*, 63 (1996) 247.
[23] K.H. Song and X. Xu, *Appl. Surf. Sci.*, 127–129 (1998) 111.
[24] S. Amoruso, *Appl. Phys. A*, 69 (1999) 323.
[25] A. Miotello and R. Kelly, *Appl. Phys. A*, 69 (1999) S67.
[26] J. Neamtu, I.N. Mihailescu, C. Ristoscu and J. Hermann, *J. Appl. Phys.*, 86 (1999) 6096.
[27] N.M. Bulgakova and A.V. Bulgakov, *Appl. Phys. A*, 73 (2001) 199.
[28] E. Matthias, M. Reichling, J. Siegel, O.W. Käding, S. Petzoldt, H. Skurk, P. Bizenberger and E. Neske, *Appl. Phys. A*, 58 (1994) 129.
[29] J. Schou, B. Toftmann and S. Amoruso, *SPIE Proceedings, Thirteenth International School on Quantum Electronics and Laser Physics and Applications*, Bourgas, Bulgaria, September 2005.
[30] C. Kittel, *Introduction to Solid State Physics*, 7th edition, Wiley, New York, 1996.
[31] V.N. Tokarev, J.G. Lunney, W. Marine and M. Sentis, *J. Appl. Phys.*, 78 (1995) 1241.
[32] A. Miotello and R. Kelly, *Appl. Phys. Lett.*, 67 (1995) 3535.
[33] B. Toftmann, J. Schou and J.G. Lunney, *Phys. Rev. B*, 67 (2003) 104101.
[34] R.K. Singh and J. Narayan, *Phys. Rev. B*, 41 (1990) 8843.
[35] Z. Zhang, Z.-X. Han and G.S. Dulikravich, *J. Appl. Phys.*, 90 (2001) 5889.
[36] C.R. Phipps and R.W. Dreyfus, in *Laser Ionization Mass Analysis*, Eds. A. Vertes, R. Gijbels and F. Adams, *Chemical Analysis Series 124*, Wiley, New York, 1993, p. 369.
[37] P.P. Pronko, Z. Zhang and P.A. VanRompay, *Appl. Surf. Sci.*, 208–209 (2003) 492.
[38] S. Fähler and H.-U. Krebs, *Appl. Surf. Sci.*, 96–98 (1996) 61.
[39] J.G. Lunney and R. Jordan, *Appl. Surf. Sci.*, 127–129 (1998) 941.
[40] Ya. B. Zel'dovich and Yu. P. Raizer, *Physics of Shock Waves and High-Temperature Hydrodynamic Phenomena*, Dover, Minneola, New York, 2001.
[41] R.W. Dreyfus, *J. Appl. Phys.*, 69 (1991) 1721.
[42] S. Amoruso, M. Armenante, V. Berardi, R. Bruzzese, N. Spinelli, *Appl. Phys. A*, 65 (1997) 265.
[43] D. Sibold and H.M. Urbassek, *Phys. Rev. A*, 43 (1991) 6722.
[44] J.C.S. Kools, T.S. Baller, S.T. De Zwart and J. Dieleman, *J. Appl. Phys.*, 71 (1992) 4547.
[45] T.N. Hansen, J. Schou and J.G. Lunney, *Appl. Phys. Lett.*, 72 (1998) 1829.
[46] P.R. Willmott, R. Timm and J.R. Huber, *J. Appl. Phys.*, 82 (1997) 2082.
[47] S. Amoruso, X. Wang, C. Altucci, C. De Lisio, M. Armenante, R. Bruzzese, N. Spinelli and R. Velotta, *Appl. Surf. Sci.*, 186 (2002) 358.
[48] T. Nedelea and H.M. Urbassek, *Phys. Plasmas*, 9 (2002) 3209.
[49] D. Sibold and H.M. Urbassek, *J. Appl. Phys.*, 73 (1993) 8544.
[50] M. Eyett and D. Bäuerle, *Appl. Phys. Lett.*, 51 (1987) 2054.
[51] I. Weaver and C.L.S. Lewis, *J. Appl. Phys.*, 79 (1996) 7216.
[52] M. Tyunina and S. Leppävuori, *J. Appl. Phys.*, 87 (2000) 8132.
[53] A. Mele, A. Giardini Guidoni, R. Kelly, A. Miotello, S. Orlando, R. Teghil and C. Flamini, *Nucl. Instr. Meth. B*, 116 (1996) 257.

[54] R. Kelly, A. Miotello, A. Mele and A. Giardini Guidoni, in *Laser Ablation and Desorption*, Eds. J.C. Miller and R.F. Haglund, *Experimental Methods in the Physical Sciences*, Vol. 30, Academic Press, New York, 1998, p. 225.
[55] S.I. Anisimov, D. Bäuerle and B.S. Luk'yanchuk, *Phys. Rev. B*, 48 (1993) 12076.
[56] S.I. Anisimov, B.S. Luk'yanchuk and A. Luches, *Appl. Surf. Sci.*, 96–98 (1996) 24.
[57] N. Arnold, J. Gruber and J. Heitz, *Appl. Phys. A*, 69 (1999) S87.
[58] S. Amoruso, R. Bruzzese, N. Spinelli, R. Velotta, M. Vitiello and X. Wang, *Phys. Rev. B*, 67 (2003) 224503.
[59] E. Buttini, A. Thum-Jäger and K. Rohr, *J. Phys. D*, 31 (1998) 2165.
[60] K.L. Saenger, in *Pulsed Laser Deposition of Thin Films*, Eds. D.B. Chrisey and G.K. Hubler, Wiley, New York, 1994, p. 199.
[61] S. Fähler, K. Sturm and H.-U. Krebs, *Appl. Phys. Lett.*, 75 (1999) 3766.
[62] M. Tyunina, K. Sreenivas, C. Bjormander, J. Wittborn and K.V. Rao, *Appl. Surf. Sci.*, 96–98 (1996) 831.
[63] M. Tyunina, J. Levoska and S. Leppävuori, *J. Appl. Phys.*, 86 (1999) 2901.
[64] T.N. Hansen, B. Toftmann, J. Schou and J.G. Lunney, *Appl. Phys. A*, 69 (1999) S601.
[65] B. Thestrup, B. Toftmann, J. Schou, B. Doggett and J.G. Lunney, *Appl. Surf. Sci.*, 197–198 (2002) 175.
[66] S. Amoruso, B. Toftmann and J. Schou, *Phys. Rev. E*, 69 (2004) 056403.
[67] F.J. Gordillo-Vázquez and J. Gonzalo, *J. Appl. Phys.*, 94 (2003) 7427.
[68] S. Amoruso, R. Bruzzese, N. Spinelli and R. Velotta, *J. Phys. B*, 32 (1999) R 131.
[69] A. Thum-Jäger and K. Rohr, *J. Phys. D*, 32 (1999) 2827.
[70] A.M. Komashko, M.D. Feit, A.M. Rubenchik, in *SPIE*, Vol. 3935, Laser Plasma Generation and Diagnostics, 2000, p. 97.
[71] J. Perrière, E. Millon, W. Seiler, C. Boulmer-Leborgne, V. Cracuin, O. Albert, J.C. Loulergue and J. Etchepare, *J. Appl. Phys.*, 91 (2002) 690.
[72] O. Albert, S. Roger, Y. Glinec, J.C. Loulergue, J. Etchepare, C. Boulmer-Leborgne, J. Perrière and E. Millon, *Appl. Phys. A*, 76 (2003) 319.
[73] T.E. Itina, J. Hermann, P. Delaporte and M. Sentis, *Phys. Rev. E*, 66 (2002) 066406.
[74] T.E. Itina, W. Marine and M. Autric, *J. Appl. Phys.*, 82 (1997) 3536.
[75] D.B. Geohegan, *Thin Solid Films*, 220 (1992) 138.
[76] S.S. Harilal, C.V. Bindhu, M.S. Tillack, F. Najmabadi and A.C. Gaeris, *J. Phys. D. Appl. Phys.*, 35 (2002) 2935.
[77] S. Amoruso, B. Toftmann, J. Schou, R. Velotta, X. Wang, *Thin Solid Films*, 453–454 (2004) 562.
[78] D.B. Geohegan and A.A. Puretzky, *Appl. Phys. Lett.*, 67 (1995) 197.
[79] D.B. Geohegan and A.A. Puretzky, *Appl. Surf. Sci.*, 96–98 (1996) 131.
[80] R.F. Wood, K.R. Chen, J.N. Leboeuf, A.A. Puretzky and D.B. Geohegan, *Phys. Rev. Lett.*, 79 (1997) 1571.
[81] R.F. Wood, J.N. Leboeuf, D.B. Geohegan, A.A. Puretzky and K.R. Chen, *Phys. Rev. B*, 58 (1998) 1533.
[82] B. Thestrup, B. Toftmann, J. Schou, B. Doggett and J.G. Lunney, *Appl. Surf. Sci.*, 208–209 (2003) 33.

[83] S.S. Harilal, C.V. Bindhu, M.S. Tillack, F. Najmabadi and A.C. Gaeris, *J. Appl. Phys.*, 93 (2003) 2380.
[84] S. Amoruso, R. Bruzzese, R. Velotta, N. Spinelli, M. Vitiello and X. Wang, *Appl. Surf. Sci.*, July, 2005.
[85] S. Amoruso, B. Toftmann and J. Schou, *Appl. Surf. Sci.*, July, 2005.
[86] B. Toftmann, S. Amoruso, J. Schou and J.G. Lunney (2005, to be published).
[87] P.E. Dyer, A. Issa and P.H. Key, *Appl. Phys. Lett.*, 57 (1990) 186.
[88] Z. Zhang and G. Gogos, *Phys. Rev. B*, 69 (2004) 235403.
[89] T.E. Itina, A.A. Katassonov, W. Marine and M. Autric, *J. Appl. Phys.*, 83 (1998) 6050.
[90] T. Szörényi and J.M. Ballesteros, *Appl. Surf. Sci.*, 109–110 (1997) 327.
[91] M. Tyunina, J. Wittborn, C. Björmander and K.V. Rao, *J. Vac. Sci. Technol. A*, 16 (1998) 2381.
[92] S. Amoruso, B. Toftmann and J. Schou, *Appl. Phys. A*, 79 (2004) 1311.

Chapter 3

Magnetron Discharges for Thin Films Plasma Processing

Jindrich Musil, Jaroslav Vlcek and Pavel Baroch

3.1. Introduction

The sputtering and evaporation of solid materials are fundamental physical processes currently used in the physical vapor deposition (PVD) of thin films. Up to the mid-1970s, however, the evaporation dominated over the sputtering in PVD technologies. This was mainly due to a very low sputter deposition rate a_{Ds} ($\ll \mu m\, min^{-1}$) of the film and relatively high pressures p (>1 Pa) needed to sustain the sputtering discharge. The breakthrough arrived in 1974 when the planar magnetron was discovered by Chapin [1]. From this moment, a very strong development of the sputtering method started; see excellent review and breakthrough papers [2–35]. As we will see further, every important progress in magnetron technology is, however, vitally dependent not only on a deep understanding of the physical basis of magnetron discharges and significant improvements of existing sputtering systems but mainly on the development of new ones operating under new physical conditions.

3.2. Milestones in Sputtering

3.2.1. Sputtering Sources

Main milestones in the development of sputtering systems are the following:

1. Diode sputtering system.
2. Conventional magnetron (CM).
3. Unbalanced magnetron (UM) [9,10,36,37].
4. Low-pressure magnetron (LPM) [22,24,26,36,38,39].
5. Magnetron with enhanced ionization (MEI) using an electron beam, rf or microwave discharge [24,37,40,41].

Materials Surface Processing by Direct Energy Techniques
Y. Pauleau (Editor)

6. Ionized magnetron (IM) [42–46].
7. High-power, high-rate magnetron (HPM) [21].
8. Closed field magnetron sputtering system (CFMSS) [16,31].
9. Dual magnetron [31,47,48].

Principles of the function of individual magnetrons are shortly described below.

3.2.1.1. Diode Sputtering Source

In diode sputtering the substrate is placed on an anode and is fully immersed in plasma (see Fig. 3.1(a)). Therefore, the substrate is exposed to a flux of electrons incident on it. This means that in the diode sputtering system the films cannot be formed by an ion plating process (IPP), which makes possible to control the microstructure of the growing film by the energy of bombarding ions. The diode discharge is sustained at high pressures (>1 Pa) and high discharge voltages ranging from about 1 to 5 kV. The ionization of sputtering gas is low and so a_D is also very low ($\ll \mu m\ min^{-1}$). The deposition rate a_{Ds} of sputtered films is considerably lower than that of evaporated films a_{De}, that is, $a_{De} \gg a_{Ds}$. This is the main reason why the diode sputtering has not been utilized in the industrial fabrication of thin films.

3.2.1.2. Conventional Magnetron

In magnetron sputtering a magnetic field is used to sustain the discharge in a close vicinity of the sputtered cathode (target). The magnetic circuit placed behind the sputtered cathode forms above its surface a tunnel of semitoroidal magnetic field **B**. In this closed **B** field tunnel of **B** field lines the plasma is confined and due to a drift of electrons along the tunnel axis in crossed **E** and **B** fields the sputtered gas is very efficiently ionized. This sputtering system is called the CM. The discharge of the CM is distributed in a very closed vicinity of the sputtered target, where the

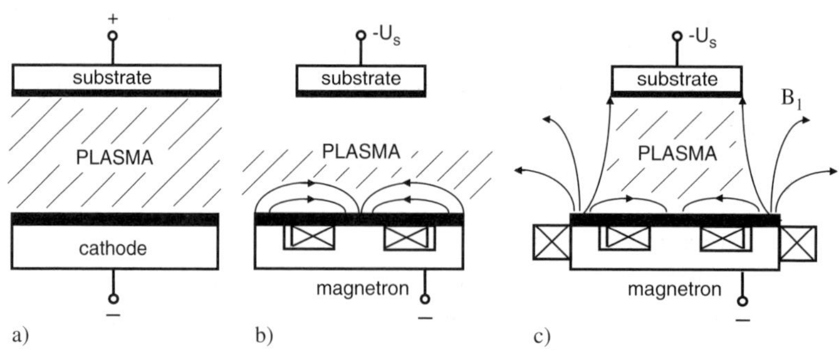

Figure 3.1: Comparison of (a) diode sputtering system with (b) CM and (c) UM.

magnetic field **B** is strongest (see Fig. 3.1(b)). It enables to place the substrate in the low-density plasma or even completely outside the discharge. This makes it possible to deposit films also on heat sensitive substrates. A high-density plasma near the target surface results in a decrease of the plasma impedance and so in: (1) the decrease of the discharge voltage U_d down to approximately 500 V and (2) the dramatic increase of a_D up to several tenths of micrometers per minute. This value of a_D is already sufficiently high to be competitive with a_D of the evaporated films. The high value of a_D, simple design and so also a very reliable magnetron were main reasons for an extremely rapid transfer of the magnetron technology into industry.

3.2.1.3. Unbalanced Magnetron

Due to the very efficient plasma confinement near the magnetron target the substrate lies in the low-density plasma. Therefore, from this plasma only small ion currents I_s can be extracted and the substrate ion current density i_s is small, very often considerably lower than 1 mA cm^{-2}. Low values of i_s are, however, insufficient (i) to control the microstructure of growing films and (ii) to produce dense, compact films. This can be achieved with an UM, which is equipped with an external magnet added to the magnetic circuit of CM [9]. The external magnet has opposite polarity to that of CM and produces an unbalanced magnetic field $\mathbf{B}_1$. This new configuration of **B** field above the sputtered target makes it possible to expand the discharge from the target and to transport the plasma to the substrate (see Fig. 3.1(c)). This makes it possible to extract from the plasma large I_s to the substrate and large values of $i_s > 1$ mA cm^{-2} can be easily achieved. Under these conditions, the ratio of the ion and coating-material fluxes, incident on the substrate, can be easily greater than 1. This means the UM is fully equivalent to the ion plating systems based on the evaporation.

The main advantage of UM equipped with two electromagnets is the possibility (i) to control continuously the **B** field distribution above the target and (ii) to keep the magnetron discharge voltage U_d constant during the whole lifetime of the target; during sputtering the magnitude of **B** on the target surface increases due to thinning of target thickness and this results in the improvement of plasma confinement and subsequent decrease of U_d. The UM with permanent magnets has a much simpler construction compared to that of the UM with electromagnets – the outer magnet is stronger than the inner magnet – but its discharge voltage U_d cannot be controlled.

3.2.1.4. Low-pressure Magnetron

The main problem in sputtering at low pressures is to eliminate losses of charged particles from the discharge. This can be achieved by an improvement of the plasma confinement and/or by an additional ionization of the sputtering gas. Principles of both methods are schematically shown in Fig. 3.2. However, the most efficient way

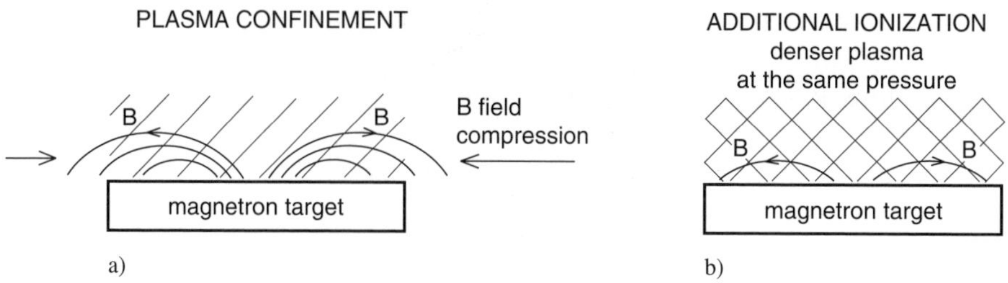

Figure 3.2: The basic principles of low-pressure sputtering discharge based on (a) the improvement of plasma confinement and (b) the additional ionization of magnetron discharge.

to decrease the sputtering gas pressure is to prevent the magnetic-field lines **B** to go beyond the edge of sputtered target (see Fig. 3.2). This can be done by an optimizing of the magnetic-field distribution above the target [39]. The low-pressure UMs can operate down to 10^{-2} Pa with large magnetron currents I_d ranging from 0.1 A to several amperes. These magnetrons make it possible to realize a collisionless, line-of-sight deposition, that is, new technological processes, for instance, the metalization of trenches in the submicrometer microelectronics.

The plasma confinement can be controlled by the shape of the magnetic-field lines **B** in the sputter deposition device. Typical representatives of such sputtering systems are the magnetron with multipolar magnetic field plasma confinement [36] and two magnetically coupled magnetrons with electric mirrors [49,50], see Fig. 3.3. In the first sputtering system (Fig. 3.3(a)) the magnetic field **B** is closed in the plane perpendicular to the surface of sputtered target and so the plasma confinement is strongly enhanced.

Two last sputtering systems (Figs. 3.3(b) and(c)) strongly reduce the losses of charged particles from the discharge volume due to the oscillating movement of electrons along **B** lines between both cathodes. Moreover, this oscillation motion of electrons results in a higher ionization of the sputtering gas. All these three sputtering systems can be operated at pressures lower than 0.1 Pa.

3.2.1.5. Magnetron with Additional Ionization

The losses of charged particles from the discharge can be compensated by an additional ionization of the sputtering gas (see Fig. 3.4).

This can be achieved, for instance, either using a hot cathode electron emission [51] (Fig. 3.4(a)) or using a hollow cathode arc electron source [52] (Fig. 3.4(b)). In both systems a considerable increase of i_s at the substrate, placed in a standard substrate-to-target distance $d_{s\text{-}t} = 50$ mm, was achieved compared with the sputtering system equipped with the CM only.

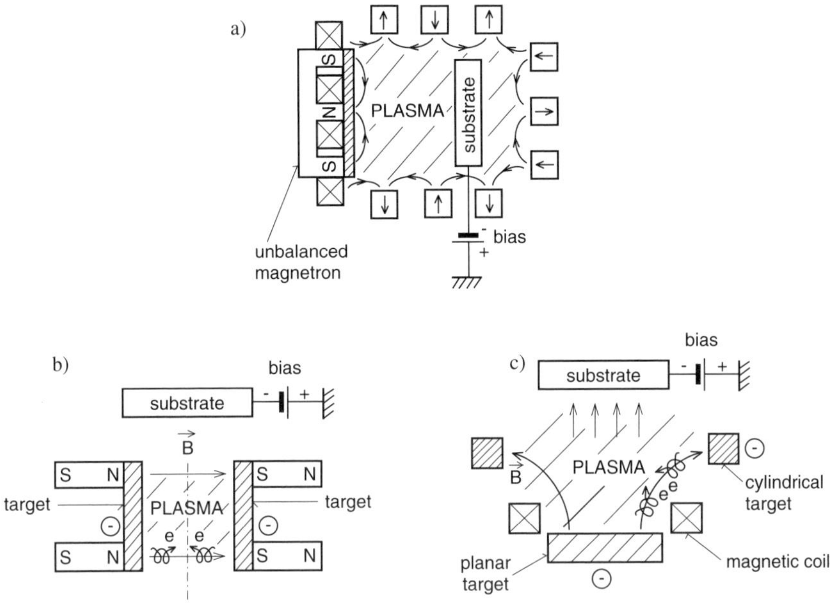

Figure 3.3: Schematic illustration of (a) magnetron with multipolar magnetic field plasma confinement, (b) and (c) two magnetrons with electric mirrors.

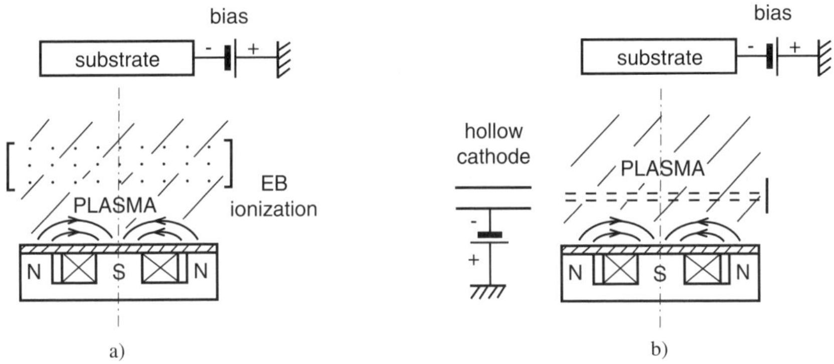

Figure 3.4: Magnetron with additional gas ionization (a) CM with the hot cathode electron beam and (b) CM with the hollow cathode arc electron source.

3.2.1.6. Ionized Magnetron

In magnetron discharges mainly Ar^+ ions are presented. The ionization of sputtered atoms is very low, approximately 1% or less. The ionization of sputtered atoms is, however, very desirable. The ionized sputtered atoms can be used in many applications, particularly for a homogeneous deposition of the films into narrow deep trenches in the microelectronics [20]. Therefore, the IM was developed [43–46]. The principle of the IM is shown in Fig. 3.5. A several turns coil with a diameter greater than that of the sputtered target, placed between the magnetron and the substrate, is supplied from an rf power supply. When the rf power is on, an additional discharge in a space between the magnetron and the substrate is generated and the sputtered atoms penetrating through it are ionized. The ionization of sputtered atoms, however, strongly depends on the sputtering gas pressure p_{Ar}. It is low at $p_{Ar} \leqslant 0.1$ Pa, but increases with increasing p_{Ar} and reaches a maximum at $p_{Ar} \geqslant$ several Pa. This means that the IM can effectively ionize sputtered atoms only at relatively high values of p_{Ar}. It is a certain limitation, which can be avoided when a high-power sputtering is used. A strong ionization of the sputtered atoms at low ($\leqslant$0.1 Pa) values of p_{Ar} can be, however, achieved only in the case when sputtered magnetron targets are made of materials with a high ($\geqslant$1) sputtering yield γ and dc HPM is used.

3.2.1.7. *High-power, High-deposition Rate Magnetron*

A typical averaged target power density $W_t = U_d I_d / S$ of the magnetrons described above and those usually used in industrial sputtering machines are relatively low and ranges from several W cm^{-2} to approximately 10 W cm^{-2} (calculated over the whole surface of sputtered target); here S is the total area of sputtered target. Recently, the attention has begun to be focused on the high-rate sputtering due to its technological potential. There are three main reasons why to develop this type of sputtering: (1) to increase substantially the deposition rate a_D of the sputtered film and so (i) to shorten its formation time, what makes the industrial production cheaper and (ii) to develop an alternative technology, which could replace fast, but ecology strongly damaging, galvanic-coating processes; (2) to ionize the sputtered

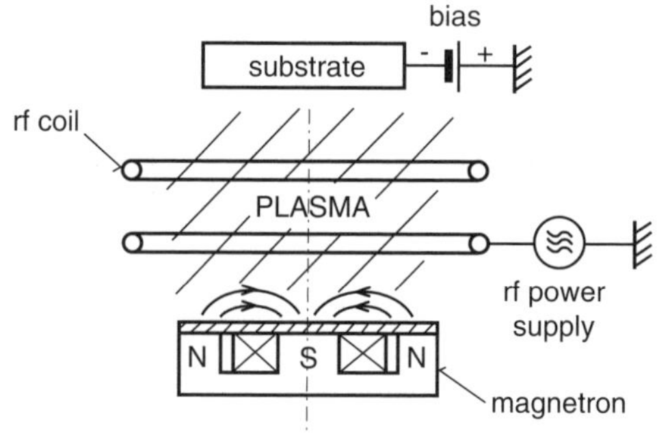

Figure 3.5: Schematic principle of IM.

material and (3) to eliminate the inert sputtering gas from the deposition process. A high-rate sputtering in the absence of inert gas is called *a self-sputtering*.

Argon is usually used as a sputtering gas. According to the magnitude of sputtering gas pressure p_{Ar}, the magnetron sputtering can be divided into four groups:

1. Conventional sputtering at $p_{Ar} \geqslant 0.1$ Pa.
2. Low-pressure sputtering at $p_{Ar} \leqslant 0.1$ Pa.
3. High-rate sputtering at $p_{Ar} > p_0 \approx 10^{-3}$ Pa.
4. Self-sputtering at $p_{Ar} = 0$.

Here p_0 is the base pressure in the deposition chamber. This classification is based on measurements of the extinction pressure p_{ex} of the magnetron discharge, generated by the UM with Cu target of 100 mm in diameter, as a function of the discharge current I_d (see Fig. 3.6). The increase of p_{ex} with increasing I_d in the region A–B is caused by a higher density of charged particles needed to carry increasing I_d. A decrease of p_{ex} in the region B–C is due to the ionization of sputtered Cu atoms. These copper ions Cu^+ substitute for the argon ions Ar^+, and the magnetron discharge is sustained in the mixture of Ar + Cu vapor. The sharp decrease of p_{ex} in the B–C region is caused by the strong increase of the ionization of Cu vapor at high values of the target power density $W_t \approx 50\,W\,cm^{-2}$. At $W_t > 50\,W\,cm^{-2}$ (region C–D) the ionization of Cu atoms is so efficient that the magnetron discharge is sustained in an ionized Cu vapor only. More details are given in the Ref. [21].

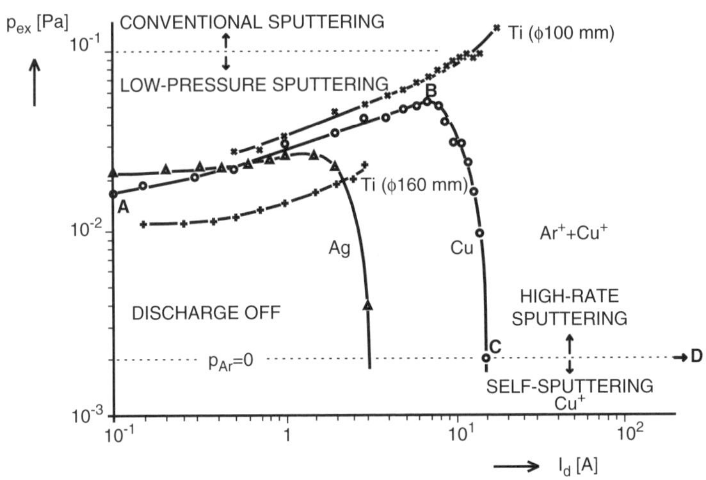

Figure 3.6: The extinction pressure p_{ex} of UM discharge as a function of discharge current I_d for UM equipped with cathodes made of Cu, Ag and Ti [21].

The generation of the magnetron discharge is strongly dependent on (i) the material from which the sputtered target is made and (ii) the efficiency of plasma confinement. These factors are of a great importance in low-pressure sputtering and self-sputtering [22].

From Fig. 3.6 it can be seen that W_t necessary for the self-sputtering is considerably lower for Ag than that for Cu. This is due to a lower sputtering yield γ of Cu compared to that of Ag; $\gamma_{Cu} = 2.33$ and $\gamma_{Ag} = 2.91$. The measured minimum values of W_t ensuring the self-sputtering of several target materials are given in Table 3.1 [22].

There are many materials, which cannot be self-sputtered due their low sputtering yield ($\gamma < 1$). A typical example is Ti with $\gamma_{Ti} = 0.57$ (see Fig. 3.6). The extinction pressure p_{ex} continuously increases with increasing I_d up to 18 A ($P_d = 9.3$ kW, i.e. $W_t = 120$ W cm^{-2}).

In addition, it is necessary to note that the magnetron discharge can be sustained at lower values of p_{Ar} if the magnetron with a larger target, which ensures a more efficient ionization of the sputtering gas, is used; compare the curves for Ti target of 100 and 160 mm in diameter in Fig. 3.6. No self-sputtering of materials with $\gamma < 1$ can be realized using dc magnetron with the solid target.

For both processes, the high-rate sputtering and self-sputtering, a maximum value of the deposition rate a_D is limited by the target cooling. Therefore, a design of the magnetron is of crucial importance. Using dc HPM sputtering thin films can be deposited with high values of a_D achieving up to several μm min^{-1} on the substrate placed in the substrate-to-target distance $d_{s\text{-}t} = 100$ mm. Also, it is worthwhile to note that at $W_t \geqslant 100$ W cm^{-2} a partial melting of the target along a line of the maximum sputtering (around a centerline in the erosion zone) occurs. This fact strongly influences the growth mechanism of the film and its physical properties because it is formed from both the sputtered and evaporated atoms. More details are given in Ref. [21].

Table 3.1: Conditions for the self-sputtering of several target materials in planar round UMs [22]

Material	Target (mm)	U_d (V)	I_d (A)	W_t (W cm^{-2})	p_{Ar} (Pa)
Cu	100, 124	585–720	7.5–16	57–100	$>5 \times 10^{-4}$
Ag	100	760–820	2.9–5	29–50	$>5 \times 10^{-4}$
Pb	100	500–700	0.4–0.7	2.1–6.5*	$>2 \times 10^{-3}$
Cd	100	850–1250	0.8–1.4	11–22*	$>6 \times 10^{-3}$
Brass	100, 124	615–820	4.9–15	33–100	$>5 \times 10^{-4}$
Al bronze	124	750–820	11.5–16	77–120	$>5 \times 10^{-4}$

* With Ar gas support.

3.2.1.8. CFMSS

A perfect plasma confinement of the magnetron discharge is a basic assumption for the optimum operation of advanced sputtering systems. The optimum operation can be achieved in the case when one or several magnetrons operate in a closed magnetic field **B** configuration, see Figs. 3.3(a) and 3.7, respectively. In the second case, the closed magnetic field **B** can be formed only if even number of the magnetrons is used. The substrates, for example, cutting tools, mechanical parts, etc., are placed at a substrate holder, which is immersed inside the plasma and rotates. The direct contact of substrates with the plasma enables to extract high ($>1\,\mathrm{mA\,cm^{-2}}$) current ion densities to the substrates if they are negatively biased. Such a process is called *the ion plating process*. This process enables to control the mechanism of the film growth and so also to control its final physical properties. At present, many sputtering systems are equipped with two or four magnetrons in the closed field configuration.

3.2.1.9. Main Characteristics of Magnetrons

Main features and limits of the magnetrons described above are given in Table 3.2.

3.2.2. Sputtering Process

The basic task in sputtering of thin films is to produce the film with prescribed physical and functional properties in a fully reproducible way. Therefore, we need the sputtering source to be operated under different physical conditions, particularly at different values of U_d, I_d, $p_T = p_{Ar} + p_{RG}$, U_s, i_s and T_s, that is, (i) at low or

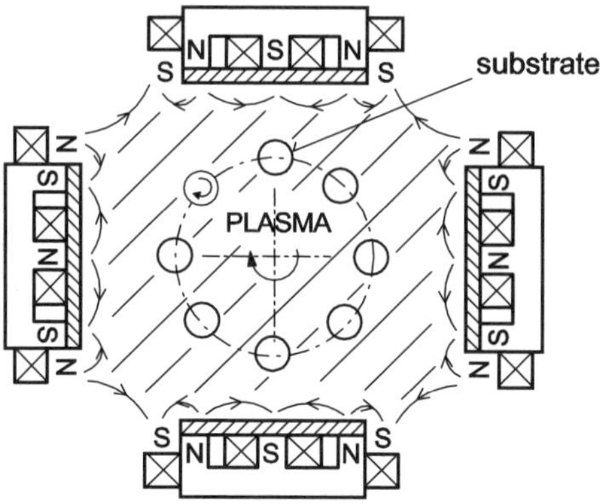

Figure 3.7: Schematic diagram of the sputtering machine with four magnetrons in a closed **B** field configuration.

Table 3.2: Main characteristics and limits of sputtering sources and systems

System	Typical features	Main limits
Diode	$U_d = 1$–$5\,kV$	Low $a_{Ds} \ll a_{De}$
	$U_s > 0, p \gg 1\,Pa$	No substrate bias can be applied
CM	$a_{Ds} < a_{De}$	$i_s < 1\,mA\,cm^{-2}$
	$U_d \approx 500\,V$	Substrate outside plasma or immersed in a weak plasma
	$p > 0.1\,Pa, \lambda \leqslant d_{s\text{-}t}$	
UM	$i_s > 1\,mA\,cm^{-2}$	Substrate in contact with plasma
MEI	$p < 0.1\,Pa, \lambda \gg d_{s\text{-}t}$	Low a_{Ds}
	Collisionless process	Complex system
LPM	$p \geqslant 0.01\,Pa, \lambda \gg d_{s\text{-}t}$	Low a_{Ds}
IM	Ionized sputtered atoms Me^+	$p \geqslant 5\,Pa$
HPM	$W_t \leqslant 150\,W\,cm^{-2}$, $a_D > 1\,\mu m\,min^{-1}$	Target cooling
CFMSS	Plasma confinement in closed Magnetic field **B**	

a_{Ds} and a_{De}: deposition rate for sputtering and evaporation, respectively; i_s: substrate ion current density; U_s: substrate bias; p: sputtering gas pressure; λ: mean free path.

high values of a_D, (ii) in collision ($\lambda < d_{s\text{-}t}$) or collisionless ($\lambda > d_{s\text{-}t}$) medium, (iii) in pure argon or with assistance of a reactive gas (p_{RG}) and (iv) under weak or strong ion bombardment of the growing film.

According to process parameters used, the sputtering process can be classified as follows:

1. Non-reactive dc or rf sputtering.
2. Reactive sputtering.
3. Argon ions assisted sputtering.
4. Low-pressure sputtering.
5. Pulsed dc or rf sputtering.
6. High-rate sputtering.
7. Self-sputtering.
8. Sputtered material ions assisted sputtering.
9. Pulsed bias sputtering.
10. Low-energy bias sputtering.

The processes given in points 1–4 are well developed and now are currently used in the industrial production of coatings. The other processes are mostly in the stage of very intense development and are expected to be introduced into the industry in a very near future. The last processes open new very exciting possibilities for the production of new advanced films and for the development of new advanced technological processes of their production. For instance, the high-rate sputtering is a hot candidate to replace the widely used and the ecology very damaging galvanic process of the production of protective coatings. Some of the processes given above are described in details below.

3.2.2.1. Reactive Magnetron Sputtering

When a metallic target of the magnetron is sputtered in a mixture of argon and reactive gas (RG), the compounds formed from the atoms of sputtered target and the atoms of RG, such as nitrides, oxides, carbides, etc. and their combinations, can be easily formed. This process is called *the reactive magnetron sputtering*. The reactive magnetron sputtering process can be, according to the amount of RG used in the film formation, divided into three modes: (1) metallic, (2) transition and (3) reactive. A typical characteristic of the reactive magnetron sputtering is the low-deposition rate of the compound $a_{\mathrm{D\,com}}$ produced in the reactive mode compared to that of the pure metallic or alloyed film $a_{\mathrm{D\,Me}}$ produced in the metallic mode (see Fig. 3.8(a)). The ratio $a_{\mathrm{D\,Me}}/a_{\mathrm{D\,com}}$ is relatively small (3–4) for nitrides but achieves high (10–15) values for oxides.

The decrease in a_{D} of films sputtered in the reactive mode is due to a reaction of the RG with the surface of sputtered target and its conversion to a compound, for instance, Ti changes into TiN or TiO_2 when nitrogen or oxygen is used as RG. This phenomenon results in: (1) the decrease of sputtering yield ($\gamma_{\mathrm{Me}} > \gamma_{\mathrm{com}}$, e.g. $\gamma_{\mathrm{Ti}} > \gamma_{\mathrm{TiN}}$), (2) the change of magnetron voltage U_{d} and (3) the rise of a hysteresis effect, see Fig. 3.8(b). The decrease in a_{D} strongly depends, particularly on the material of sputtered target and the kind of RG. In addition, when the reaction product is an electrical insulator two further problems occur:

1. The non-sputtered surfaces of target (very significant in planar magnetrons) are covered by dielectric layers, which are charged up and when this charge achieves a threshold value causes an arcing.
2. The anode of the magnetron disappears also due to its covering by a dielectric layer. These are very serious problems in the dc reactive magnetron sputtering. Therefore, the hysteresis effect and the methods allowing:
   - to increase a_{D} of oxides;
   - to avoid arcing in the dc reactive sputtering of electrically insulating films, for example, TiO_2, are in more detailed discussed below.

3.2.2.2. Hysteresis Effect and Ways of Its Elimination

The hysteresis effect arises in consequence of two competitive processes: (1) the sputtering of the target surface and (2) the covering of its surface by reaction products.

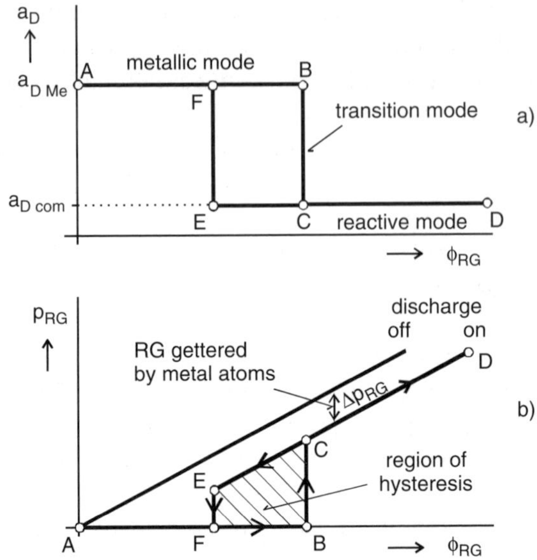

Figure 3.8: Schematic illustration of (a) deposition rate a_D and (b) partial pressure of RG p_{RG} of sputtered films as a function of flow rate of RG ϕ_{RG}.

Fig. 3.8 illustrates the dependence of RG partial pressure p_{RG} as a function of the flow rate ϕ_{RG} of RG entering into the deposition chamber without and with the discharge, that is, with the magnetron discharge off and on, respectively. The hysteresis occurs in the presence of the discharge only.

At low values of ϕ_{RG} (interval A–B) all RG is gettered by the sputtered metal. In the point B the flow rate ϕ_{RG} (B) of RG into the chamber is equal to the gettering rate of sputtered metal. Every small increase in ϕ_{RG} results in a jump (1) increase of p_{RG} in the deposition chamber and (2) decrease of the film deposition rate a_D. A further increase of ϕ_{RG} (interval C–D) results in a linear increase of p_{RG} and almost constant value of a_D, which is typical for reactive mode of sputtering. The decrease of ϕ_{RG} from D to E is accompanied by the decrease of p_{RG}, but a return to the metallic mode (interval E–C) is delayed. This is because p_{RG} remains high until the compound layer on the surface of sputtered target is fully removed and metal is again exposed to be sputtered. As a result, the consumption of RG increases and p_{RG} decreases to a background level. This way a closed hysteresis loop is formed.

The jump increase of p_{RG} from B to C (Fig. 3.8) is undesirable phenomenon because (1) prevents to form $MeRG_x$ films with the stoichiometry x corresponding to p_{RG} in the interval B–C and (2) is responsible for the unstable sputtering at ϕ_{RG} corresponding to the transition from B to C and its close vicinity. Therefore, a considerable effort has been devoted to find means to eliminate this jump. First, it was suggested by Maniv *et al.* [53] to introduce a baffle (grid) between the substrate and the target with Ar inlet into the baffle and RG inlet into the system on the substrate side. However, this modified magnetron system called "baffled" magnetron suffers from three principal drawbacks: (1) frequent cleaning of the baffle grid, (2) reduction of the metal flux to the substrate through grid and (3) reduction of the ion bombardment of growing film if the baffle is grounded. Despite that the last problem may be solved by either placing a positively biased electrode in the substrate region to pull out the plasma from the baffle, or by biasing the substrate. This solution is not, however, suitable for the industrial production of coatings. For this purpose, simple systems with no elements between the sputtered target and the substrate, which can either fully eliminate the hysteresis effect or ensure the stable sputtering in the transition mode, are needed. These requirements meet the sputtering systems with a pumping speed S larger than a critical pumping speed S_c [54,55], sputtering systems controlled by plasma emission monitoring (PEM) [56,57] or gas pulsing [58,59], dual magnetrons [47,48], magnetrons with a full target erosion [60–62], magnetrons equipped with sub-stoichiometric ceramic conductive targets [63], and recently developed advanced two targets sputtering systems with closed magnetic field **B** and full target erosion.

In case the elimination of the hysteresis effect is realized by increase of pumping speed of pumping system, a condition necessary for the elimination of the hysteresis effect can be derived from an equilibrium state between the gettering of RG by the sputtered material and the removal of RG from the deposition chamber by the pumping system [54]:

$$\phi_{RG} = p_{RG}S_{RG} + {}^{s}\phi_{RG} \tag{3.1}$$

where S_{RG} is the pumping speed of pumping system for the RG and ${}^{s}\phi_{RG}$ is the flow rate of RG gettered by the sputtered material. The hysteresis effect occurs if the system is unstable. It was found that the solution of Eq. (3.1) is stable under the following condition:

$$S_{RG} > S_c = (-\partial^{s}\phi_{RG}/\partial p_{RG})_{max} \tag{3.2}$$

where S_c is *the critical pumping speed* of the pumping system. This means that the hysteresis effect can be eliminated if S_{RG} of the pumping system is greater than a critical value of S_c defined by Eq. (3.2). A possibility to remove the hysteresis effect is schematically shown in Fig. 3.9. Here, the dependencies ${}^{s}\phi_{RG} = f(p_{RG})$ and $\phi_{RG} = f(p_{RG})$ for $S_{RG} < S_c$ and $S_{RG} > S_c$ are given. In the case when $S_{RG} > S_c$

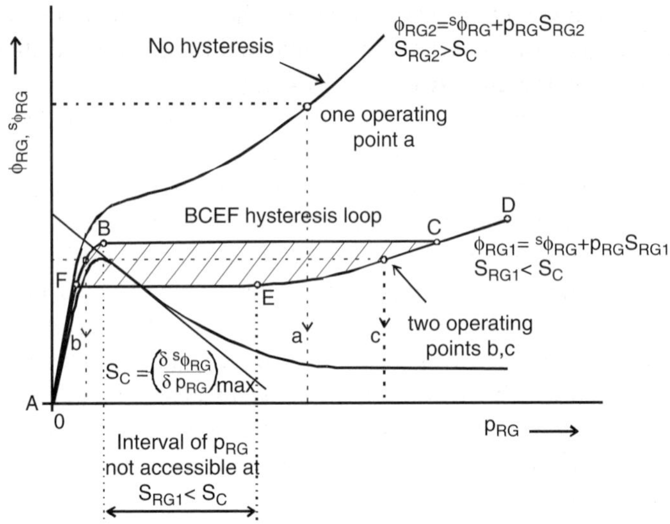

Figure 3.9: Schematic illustration of ϕ_{RG} as a function of p_{RG} in sputtering system with ($S_{\mathrm{RG}\,1} < S_{\mathrm{c}}$) and without ($S_{\mathrm{RG}\,2} > S_{\mathrm{c}}$) the hysteresis effect. The dependence of flow of RG gettered by sputtered target material ${}^{\mathrm{s}}\phi_{\mathrm{RG}}$ as a function of p_{RG} is also given.

the hysteresis effect can be eliminated even despite the fact the derivative $\partial^{\mathrm{s}}\phi_{\mathrm{RG}}/\partial p_{\mathrm{RG}}$ is negative because the dependence $\phi_{\mathrm{RG}} = f(p_{\mathrm{RG}})$ is an unambiguous function. The validity of this conclusion was verified experimentally for the case of reactive dc magnetron sputtering of titanium nitride films [55].

A main problem with increasing of the pumping speed S_{RG} lies in adding considerable cost at the building stage of new machines. Despite this fact, a new generation of high-speed diffusion-pumped coaters with $6500\,\mathrm{l\,s^{-1}}$ diffusion pumps was put into operation in the mid of 1990s by the Von Ardenne Coating Technology (formerly BOC Coating Technology) and recently, the advanced turbomolecular coaters with four $1000–2000\,\mathrm{l\,s^{-1}}$ turbomolecular pumps surrounding each magnetron position (Von Ardenne Anlagentechnik “780” mm) were developed [33]. This example demonstrates a clear trend to enhance the pumping speed of new coaters.

3.2.2.3. Arcing in Reactive Sputtering of Insulating Films

An arcing rises in the dc reactive magnetron sputtering of insulating films. It occurs in consequence of a charging of insulating layers formed on the uneroded areas of sputtered target. There are two ways, which allow to suppress or eliminate the arcing: (1) to remove the uneroded areas on the sputtered target and so to avoid the formation of insulating areas or (2) to remove the accumulated charge from the insulated surfaces on the uneroded areas [64–66].

The existence of uneroded areas is an inherent property of the planar magnetrons. They are formed everywhere outside the magnetron racetrack, strongly decrease the efficiency of target utilization (TU) and thus they decide on the target lifetime. In principle, there are two ways to eliminate the uneroded areas: (1) to scan a magnetic circuit (magnets) under the planar target or (2) to use a rotated cylindrical target. During the scanning of magnetic circuit under the target or the rotation of cylindrical target above the stationary magnetic circuit, the magnetron discharge "sweeps" the whole surface of sputtered target. Therefore, in every moment a certain part of the target is without the discharge and reacts with the RG. This means that even these magnetrons, which achieve almost full target erosion, are covered by a thin compound layer ($MeRG_x$), whose thickness depends on the speed of scanning or rotation, that is, on a reaction time of the RG with the sputtered metal. This thin ($\geqslant$10 nm) compound layer on the target surface results in a change of the mode of sputtering from the metallic to the compound and so in (i) the rise of the hysteresis effect, which was already experimentally demonstrated for the C-MAG with full target erosion [17] and (ii) the arcing.

The hysteresis effect and arcing can be avoided only in the case when the whole surface of the sputtered target is uninterruptedly exposed to the magnetron discharge and the compound layer is continuously removed. To our knowledge, only two sputtering sources meet such a requirement: (1) the rectangular magnetron with many racetracks (i) oriented perpendicularly to a larger side of the target, (ii) covering the whole target surface along its length [62] and (iii) continuously moving along its longer axis, that is, the magnetron with an endless one-way direction movement of many magnetic circuits, producing many separated racetracks, under the target [63] and (2) the modified facing target sputtering systems.

At present, the arcing in dc reactive sputtering of insulating films can be reliably and successfully eliminated if either the dc pulse dual magnetron or the conductive, sub-stoichiometric ceramic target, for instance, TiO_{2-x}, is used.

The principle of the dc pulse dual magnetron sputtering is displayed in Fig. 3.10. The voltage polarity of magnetron cathodes periodically changes from negative to positive. When the cathode voltage U_d is negative the target material is sputtered. On the contrary, when U_d is positive the charge accumulated at the surface of insulating layer formed on the uneroded areas of target is discharged as a result of the electron bombardment.

Such a process is called bipolar pulsed magnetron sputtering. The charge accumulation can results in an arcing. The arcing can be eliminated if the positive charge accumulated on the uneroded areas is discharged before reaching a critical

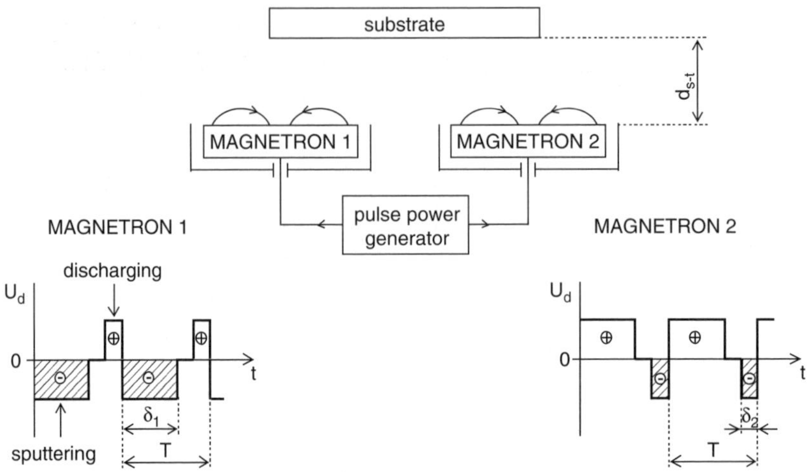

Figure 3.10: Schematic diagram of symmetric bipolar dc pulsed dual magnetron sputtering.

value at which the electric field across the insulating layer overpasses its dielectric strength. This can be reached only in the case when the frequency of discharging f_r is higher than a critical frequency f_{cr}. The value of the critical frequency f_{cr} can be estimated from the following formula [30]:

$$f_{cr} = (\varepsilon_r \varepsilon_0 E_B / J_i)^{-1} \qquad (3.3)$$

where E_B, ε_r, ε_0 and J_i are the dielectric strength of insulating layer, the dielectric constant of this layer, the permittivity of free space and the discharge current density during the negative voltage pulse. Eq. (3.3) shows that f_{cr} is a function of the material to be sputtered. Therefore, whereas no arcing is achieved for the reactive sputtering of TiO_2 at a pulsing frequency of 30 kHz, it takes a frequency between 50 and 70 kHz for all arcing to disappear for Al_2O_3 [28].

3.2.2.4. High-power, Pulsed Magnetron Sputtering

Magnetrons operated at high values of the target power density $W_t = I_d U_d / S_t > 100\,\mathrm{W\,cm^{-2}}$ are called *the high-power magnetrons* (HPM). A maximum power of the HPMs is limited by the cooling of magnetron target. In the case of dc magnetron sputtering of Cu, which exhibits high values of (i) the sputtering yield $\gamma_{Cu} = 2.33$ atom/ion and (ii) the thermal conductivity, the maximum values of $W_{t\,max}$ achieve up to approximately $150\,\mathrm{W\,cm^{-2}}$ [21].

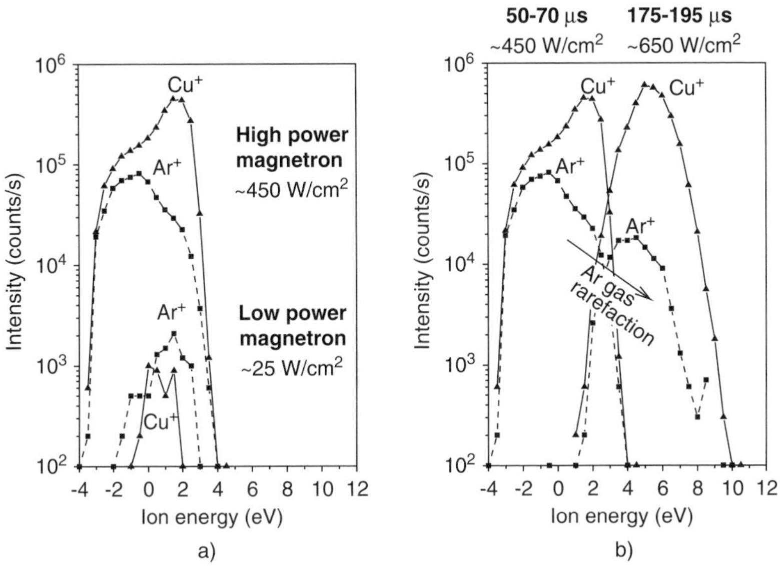

Figure 3.11: Energy distribution of Cu^+ and Ar^+ ions measured in front of substrate ($d_{s\text{-}t} = 100\,\text{mm}$) in the pulsed discharge of UM with Cu target ($\phi = 100\,\text{mm}$) at $f_r = 1\,\text{kHz}$, $\tau = 200\,\mu\text{s}$, $\tau/T = 0.2$, $p_{Ar} = 1\,\text{Pa}$ in (a) interval 50–70 μs after the pulse beginning at average pulse current $I_{da} = 5\,\text{A}$ ($\sim 25\,\text{W cm}^{-2}$) and $I_{da} = 50\,\text{A}$ ($\sim 450\,\text{W cm}^{-2}$) and (b) two intervals 50–70 μs ($\sim 450\,\text{W cm}^{-2}$) and 175–195 μs ($\sim 650\,\text{W cm}^{-2}$) at $I_{da} = 50\,\text{A}$.

The values of $W_{t\,max}$ can be, however, considerably increased if a pulsed magnetron sputtering is used. The selection of a duty cycle, that is, the ratio τ/T, makes it possible to control W_{tp} in the pulse while keeping (i) the average target loading $W_{av} = W_{tp}(\tau/T)$ and (ii) the target cooling constant; here τ is the length of pulse and T is the period of pulses. W_{tp} increases with decreasing τ/T and easily reach high values of approximately $1000\,\text{W cm}^{-2}$ and more, that is, W_{pt} can be considerably higher than $W_{t\,max}$ achieved in dc HPM sputtering. Here, it is necessary to note that the increase of W_{tp}, however, results in a decrease of the deposition rate a_D due to the shortening of duty cycle τ/T. This means that the pulsed magnetron sputtering using the HPM with very high values of W_{tp} does not allow the industrial production of coatings with high-deposition rates a_D.

Very important feature of the high-power, pulsed magnetron (HPPM) discharge is a production of ions of sputtered metal Me^+. However, to produce Me^+ ions the HPPM must operate at high values of W_{tp}. It results in an intensive sputtering of the target and the rarefaction of argon gas (see Fig. 3.11). The last effect is caused by

Ar gas heating in collisions due to energy transfer from the copper atoms leaving the target with a kinetic energy of several eV. In this process the temperature T of Ar gas increases and this increase in T is compensated by the decrease of Ar atoms density n approximately according to the equation $p = nkT$, where p is the gas pressure. Under these conditions an efficient ionization of Cu atoms in the bulk plasma takes place due to its low ionization energy (7.72 eV) and Cu^+ ions prevail over Ar^+ ions in the plasma (see Fig. 3.11(a) and (b)). Here, the effect of the target power density on Cu^+ ions production and Ar gas rarefaction is illustrated. This experiment clearly shows that in the pulsed HPM sputtering discharge the flux of Me^+ ions can be more than 10 times higher than that of Ar^+ ions and progressively increases with increasing time t from the pulse beginning. At $I_{da} = 50$ A ($\sim$450 W cm^{-2}) Cu^+ ions dominate over Ar^+ ions; 82% and 96% Cu^+ ions of the total amount of ions in the discharge is produced in the intervals 50–70 μs and 175–195 μs (the end of pulses), respectively, (see Fig. 3.11(b)).

The increase of number of Cu^+ ions is caused by the increase of pulse discharge current I_d with increasing t from the pulse beginning and so by the increase of W_{tp} at a constant value of U_d during the whole length of pulse (see Fig. 3.12). The energy

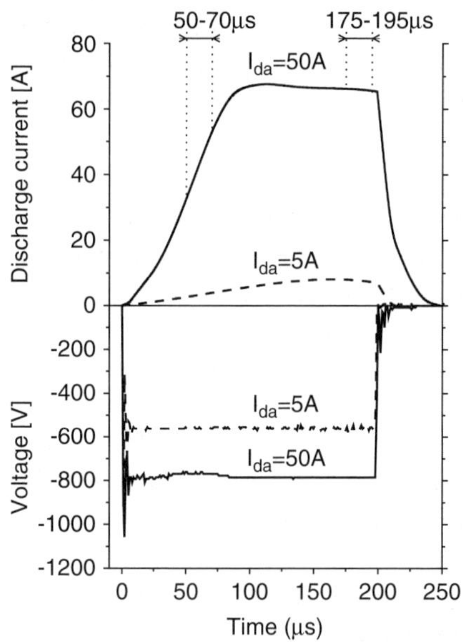

Figure 3.12: Time development of U_d and I_d in pulsed UM discharge sustained at pre-set average pulse current $I_{da} = 50$ A, $f_r = 1$ kHz, $\tau = 200$ μs and Ar pressure $p_{Ar} = 1$ Pa.

distribution of positive ions in front of grounded substrate was measured with a time- and energy-resolved mass spectrometer (EQP 300 Hiden Analytical) placed perpendicularly with respect to the target-substrate axis. More detailed analysis is given in Ref. [35].

The HPPM sputtering is very advanced technology. It is characterized by three main advantages: (1) very efficient ionization of atoms sputtered from the target; number of Me^+ ions in the discharge achieves up to approximately 95% of all ions (only ~5% of Ar^+ ions at the end of each pulse), (2) very high deposition rate a_D of films achieving up to several $\mu m\,min^{-1}$ on substrates placed at the substrate-to-target distance $d_{s\text{-}t} = 100\,mm$ and (3) no formation of macroparticles. This process opens new possibilities in thin films processing. It is a hot candidate to replace (1) the cathodic arc evaporation producing macroparticles and (2) the wet, ecologically strongly unfriendly galvanic-coating processes.

3.3. General Properties of Magnetron Discharge

3.3.1. I–V Characteristics of Magnetrons

The magnetron operates correctly only in the case when its (*I–V*) characteristics are the current ones, that is, $U_d = f(I_d) \approx$ constant. Such a mode of operation is called "the current mode". In the ideal case U_d remains almost constant with increasing I_d, and their shape strongly depends on the magnitude of (i) parallel component of the magnetic field $\mathbf{B}_{par}$ on the surface of sputtered target and (ii) sputtering gas pressure p. The effect of $\mathbf{B}_{par}$ and p on a shape of the *I–V* characteristics of the CM equipped with an electromagnet is illustrated in Fig. 3.13. $\mathbf{B}_{par}$ on the target surface is controlled by the current I_m in the coil of electromagnet and increases from 330 to 540 G with I_m increasing from 1.5 to 2.5 A.

From Fig. 3.13 it is seen that the magnetron operates correctly at $\mathbf{B}_{par} = 540\,G$ and $p_{Ar} \geqslant 0.5\,Pa$. A decrease in $\mathbf{B}_{par}$ results in the increase of U_d with increasing I_d and this increase in U_d is the higher the smaller is p_{Ar}. However, the magnetron discharge starts to be unstable at p_{Ar} below 0.5 Pa due to high losses of electrons escaping from it. Imperfect magnetic plasma confinement is the reason why the magnetron cannot be operated at $p_{Ar} < 0.5\,Pa$. Also, a maximum magnetron current $I_{d\,max}$ decreases with decreasing both $\mathbf{B}_{par}$ and p_{Ar}. The value of $I_{d\,max}$ is limited by a maximum voltage U_0 of the power supply, which usually is 1 kV. Also, it is necessary to note that during sputtering the thickness of sputtered target decreases and in consequence of it $\mathbf{B}_{par}$ on its surface increases. This results in the decrease of U_d and so also in decrease of a_D. This fact results in changes of the properties of the film when it is sputtered under the same deposition conditions (I_d, U_s, i_s, T_s, p_{RG}, p_T, $d_{s\text{-}t}$) in different

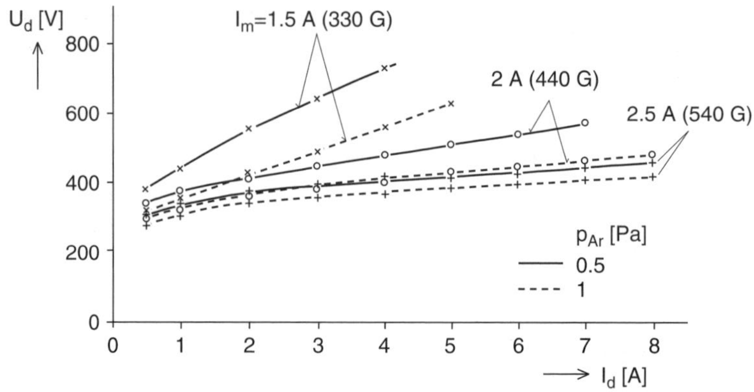

Figure 3.13: *I–V* characteristics of CM equipped with electromagnet and Cu target of 100 mm in diameter measured at three values of current I_m = 1.5, 2 and 2.5 A in coil of electromagnet and two pressures p_{Ar} = 0.5 and 1 Pa.

time *t* from the beginning of sputtering of the new target, that is, in different time *t* during the lifetime of target.

The minimum operating pressure of every magnetron is given by its construction design and particularly by (1) the magnitude of $\mathbf{B}_{par}$ on the surface of sputtered target, (2) the distribution of magnetic field **B** lines on sputtered target and eventually (3) the additional means, such as additional EB, rf or microwave discharges (Fig. 3.4), which enhance the ionization of the sputtered gas and so compensate losses of charged particles from the magnetron discharge. This means that an increase of the magnitude of $\mathbf{B}_{par}$ alone is not sufficient to decrease the minimum operating pressure p_{min} of the magnetron. A good plasma confinement of the magnetron discharge and efficient ionization of the sputtering gas are crucial parameters needed to achieve the minimum value of p_{min} of the magnetron. For instance, Fig. 3.14 shows a development of the *I–V* characteristics of the UM, which has a considerably better plasma confinement compared to the CM, with decreasing argon pressure. This figure clearly shows that the UM, compared to the CM (Fig. 3.13), can be operated at considerably lower values of p_{Ar}, down to 0.075 Pa. Here, it is necessary to note that even at the lowest operating pressure p_{min} = 0.075 Pa the UM can sustain the high discharge current I_d = 1 A. This is a clear evidence for the good plasma confinement of this UM.

3.3.1.1. Effect of Length of Magnetron Racetrack on Minimum Operating Pressure

The length of magnetron racetrack, that is, the length of a closed line along a center of the erosion zone on the sputtered target, strongly influences p_{min}. The longer

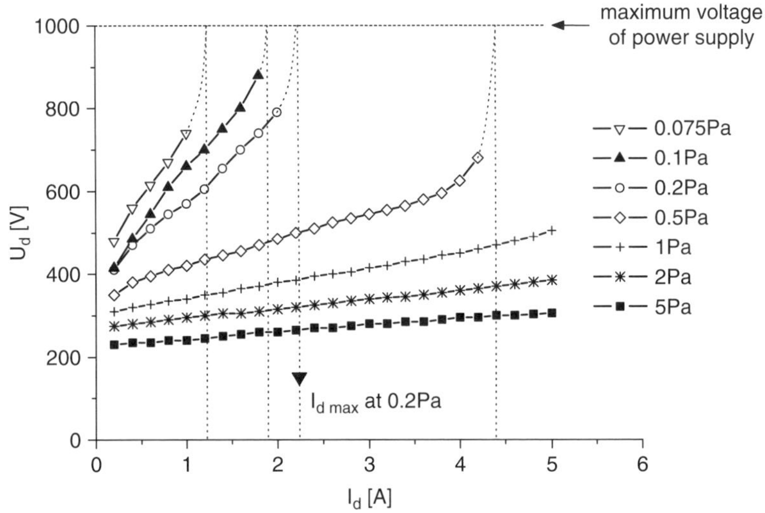

Figure 3.14: *I–V* characteristics of UM equipped with two electromagnets and target made of VT6 (6 at.% Al, 4 at.% V and balance Ti) of diameter 100 mm measured at different values of argon pressures.

is the racetrack, the higher is the ionization of sputtered gas and so the lower values of p_{min} can be achieved (see Fig. 3.15). This is due to the fact that electrons traveling along a longer line in the center of the closed tunnel of magnetic-field lines **B** above the erosion zone of sputtered target can perform more collisions before leaving the magnetically confined plasma. This results in a more efficient ionization of the sputtering gas and covering of losses of electrons leaving the discharge down to lower values p_{min}.

Therefore, there is no great problem to operate long rectangular magnetrons at low values of p_{min} of about 10^{-2} Pa or less. However, the operation of round magnetrons with small ($\leqslant$50 mm) targets is a serious problem. To operate small round magnetrons with small diameters, high values of $\mathbf{B}_{par} \approx 1000$ G or even higher are needed. Also, the better plasma confinement is needed if small magnetrons are operated at lower values p_{min}. One of possible solutions is to operate these small magnetrons in a configuration called a dual magnetron, see Section 3.3.3.

3.3.1.2. Effect of Sputtering Gas Pre-ionization on Minimum Operating Pressure

A very efficient way to decrease the minimum operating pressure of the magnetron is the pre-ionization of sputtering gas. For this purpose very suitable is rf or microwave discharge generated in a magnetic field [40,41,67]. These discharges,

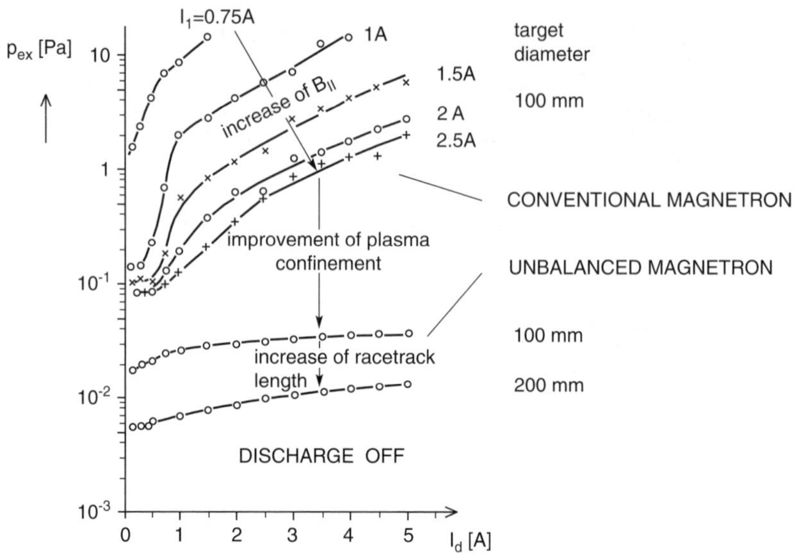

Figure 3.15: Effect of magnitude and shape of **B** field, and length of the magnetron racetrack on extinction pressure p_{ex}, at which magnetron discharge extinguishes.

due to a good magnetic plasma confinement, can be easily sustained at low pressures down to 0.01 Pa. Using a helix coil, the rf or microwave pre-ionization discharge can be generated just in front of the magnetron target (see Fig. 3.16(a)). The rf (13.56 MHz) and microwave (2.4 GHz) power is introduced into the vacuum chamber via the helix coil surrounded by eight SmCo magnets, azimuthally spaced by 45°. All magnets have the same polarity and are coupled to the external pole of the magnetron (see Fig. 3.16(b)). The UM is used and its magnetic field is produced by two electromagnets. The pre-ionization discharge is a distributed electron cyclotron resonance one.

The *I–V* characteristics of the UM with a Cu target 100 mm in diameter operated with and without the microwave discharge are given in Fig. 3.17. Dc magnetron discharge is self-sustained, that is, no microwave discharge is necessary ($P_\mu = 0$), at $p_{Ar} \geqslant 0.025$ Pa. The *I–V* characteristic of this discharge exhibits a purer current mode of operation. However, in the case when the magnetron discharge is enhanced by the additional microwave discharge the *I–V* characteristic is a combination of two modes: (1) the voltage mode for $U_d < 300$ V and (2) the current mode for $U_d > 300$ V (see Fig. 3.17(a)). In addition, the existence of the voltage mode enables to sustain the sputtering discharge at low values of U_d continuously

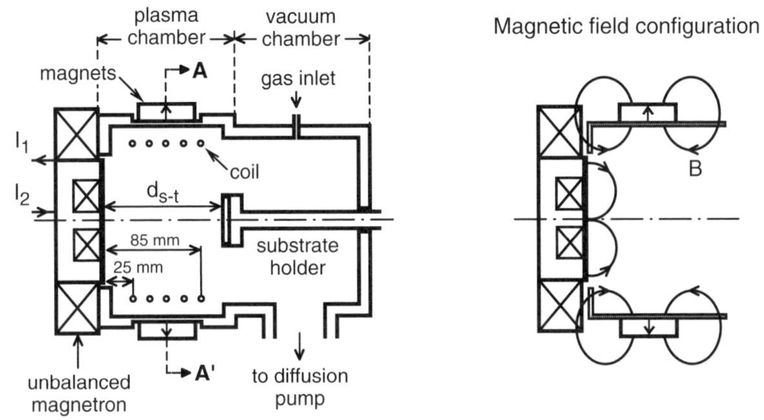

Figure 3.16: Schematic of planar magnetron whose discharge is enhanced or pre-ionized using rf or microwave discharge generated in external magnetic field [40,41].

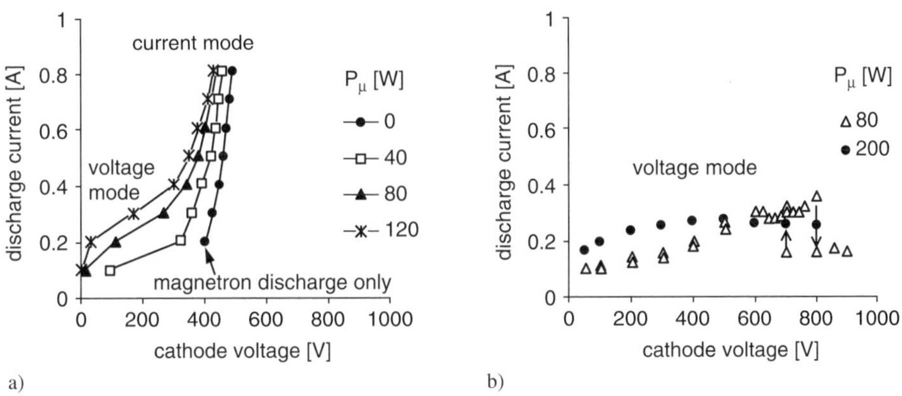

Figure 3.17: *I–V* characteristics of the discharge of round magnetron with Cu target 100 mm in diameter enhanced by the microwave discharge at argon pressures (a) 0.025 Pa and (b) 0.01 Pa [41].

decreasing down to zero. At low pressures $p \leqslant 0.025$ Pa, when the magnetron discharge needs a pre-ionization to be generated the *I–V* characteristics exhibit a pure voltage mode (see Fig. 3.17(b)). A switching between two plasma modes of the magnetron operation is seen for a lower value of the microwave power $P_\mu = 80$ W. The current of the magnetron discharge I_d is, however, low and saturates with

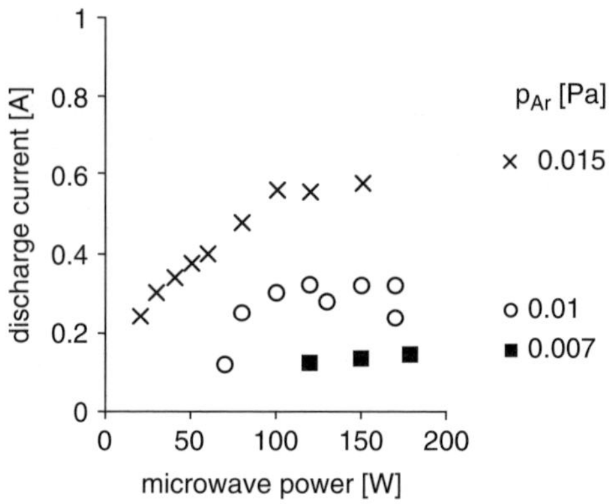

Figure 3.18: I_d of the discharge of the round magnetron with Cu target 100 mm in diameter enhanced by the microwave discharge as a function of microwave power P_μ and three values of $p_{Ar} = 0.007$, 0.01 and 0.015 Pa.

increasing both the cathode voltage U_d and the microwave power P_μ. With the pre-ionization of argon the magnetron with target 100 mm in diameter can be operated down to 0.007 Pa.

The magnetron discharge current I_d decreases with decreasing argon pressure p_{Ar} and the microwave power P_μ needed for pre-ionization increases with decreasing p_{Ar} [41], see Fig. 3.18.

3.3.2. Magnetron Discharges Sustained at Low Discharge Voltages

Present magnetrons suffer from one fundamental limit – a strong increase of the magnetron discharge voltage U_d with decreasing sputtering gas pressure and interruption of the magnetron discharge when the maximum voltage of the power supply is achieved (see Fig. 3.14). Therefore, recently two new types of magnetrons: (1) the magnetron with a grooved target [26] (Fig. 3.19(a)) and (2) the magnetron with discharge enhanced by one or several additional hollow cathode discharges [68,69] (Fig. 3.19(b)) – were developed with the aim to decrease the magnetron discharge voltage U_d. The *I–V* characteristics of these magnetrons demonstrate that a modification of the shape of sputtered target is, besides the perfect plasma confinement, also a very efficient way to enhance the magnetron discharge, what allows the magnetron to be operated at considerably lower values of U_d compared to those of the standard magnetrons.

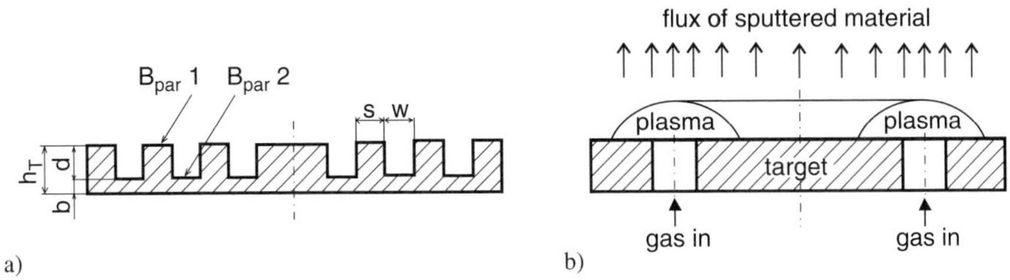

Figure 3.19: (a) Grooved target and (b) target with holes in which additional hollow cathode discharges can be sustained.

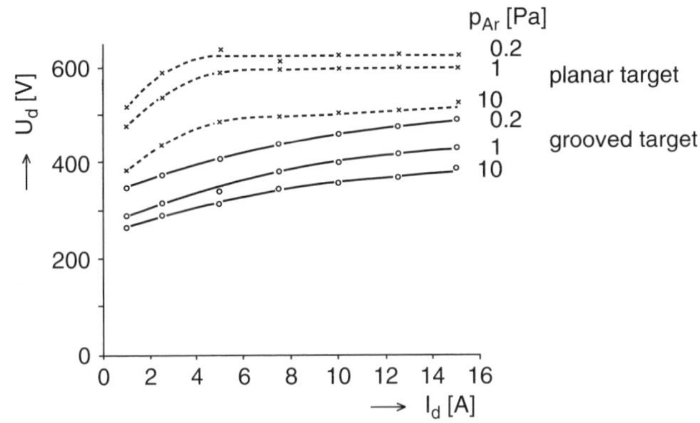

Figure 3.20: Comparison of *I–V* characteristics of the UM with full planar and grooved target 100 mm in diameter.

3.3.2.1. Magnetron with Grooved Target

The principle of the magnetron with a grooved target is based on the fact that $\mathbf{B}_{par\ 2}$ at the bottom of grooves is considerably higher than $\mathbf{B}_{par\ 1}$ on the surface of target outside the grooves. The magnetic field $\mathbf{B}_{par\ 2}$, which is much greater than $\mathbf{B}_{par\ 1}$, not only decides about a more favorable ignition of the sputtering discharge but also results in (i) the dramatic decrease of U_d needed to sustain a given I_d compared to the full planar target, see Fig. 3.20, and (ii) the unchangeability of *I–V* characteristics of the sputtering discharge during the whole lifetime of sputtered target.

Experiments, described in detail in Ref. [69], indicate that grooves with optimized geometrical dimensions can operate as hollow cathodes and the discharges sustained inside grooves enhance the sputtering discharge by the generation of new charged particles. A replacement of the charged particles leaving the sputtering discharge, in consequence of an imperfect plasma confinement, by newly generated ones is a basic process, which enables a low-pressure operation of the magnetron. These facts are main and unique features of the magnetron with grooved target.

3.3.2.2. Magnetron with Gas Injection Through Holes in Target

The magnetron with gas injection through holes in the target operates on the same principle as a gas-jet. The gas is fed into the holes, which are usually (i) drilled perpendicularly to the target surface (Fig. 3.19(b)) and (ii) distributed equidistantly along a closed erosion zone. The holes fulfill two main functions: (1) inject the gas directly inside the magnetron discharge and (2) form hollow cathodes, in which (i) the hollow cathode discharges are generated and (ii) the sputtering gas is ionized and/or activated. The second function requires optimizing the diameter d of the holes.

The injection of ionized and/or activated gas inside the magnetron discharge strongly changes its *I–V* characteristics (see Fig. 3.21). In this figures the *I–V* characteristics of UM with NdFeB magnets equipped with (1) the full target operated with *Ar introduced into the deposition chamber*, that is, from outside of the magnetron discharge and (2) the target with four holes of diameter 2 mm, machined equidistantly in the erosion zone, and operated with *Ar injected inside the magnetron discharge* through holes in the target are compared. This comparison shows that the magnetron operated with the gas injection inside the discharge easily operates at (1) low values of p down to 0.025 Pa, (2) high values of I_d up to 5 A (a physical limit for $I_{d\,max}$ needs to be determined) and (3) relatively low values of U_d of approximately 400 V. This is a great improvement because the UM with a full target (100 mm in diameter) operates at low pressures $p < 0.1$ Pa at (i) $U_d > 500$ V

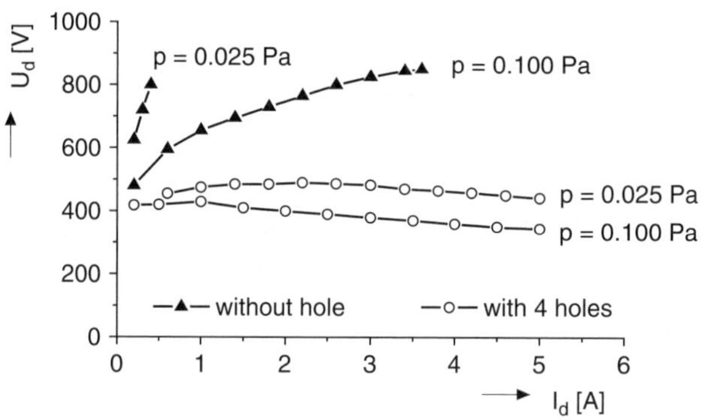

Figure 3.21: Comparison of *I–V* characteristics of two UMs (1) UM with full target and Ar supplied to the chamber and (2) UM with target with holes and Ar injection through holes inside the discharge. Both targets: Cu (ϕ 100 mm, 6 mm thick).

and (ii) low values of I_d only. $I_{d\,max}$ is the lower, the lower is $p_{Ar;}$; at $p_{Ar} = 0.025$ Pa only small values of $I_{d\,max} \leqslant 0.5$ A can be achieved (see Fig. 3.21).

Also, it is necessary to note that the decrease in U_d has three fundamental consequences: (1) the decrease in the deposition rate a_D of sputtered films, (2) the decrease of the energy of fast neutral particles at low (<0.1 Pa) pressures, that is, in the regime of collisionless sputtering) and (3) the increase of the ion substrate current I_s extracted from a highly ionized sputtering gas.

3.3.3. Discharge of Dual Magnetron

A discharge in front of the dual magnetron cathodes strongly depends on the polarity of magnets inside both magnetrons. There are two possible arrangements: (1) both magnetrons have the same polarity of magnets, so-called "mirror **B** field" (Fig. 3.22(b)) or (2) magnetrons mutually differ in the polarity of magnets, so-called "closed **B** field" (Fig. 3.22(a)). In the first case, discharges of both magnetrons are mutually repelled. On the contrary, in the second case, the discharges are mutually coupled and the plasma is well confined between the cathodes of both magnetrons.

This fact is clearly demonstrated by the photos of discharge of the dual magnetron with different polarities of magnets in each magnetron (see Fig. 3.23). The dual magnetron consists of two magnetrons with targets 50 mm in diameter tilted to the vertical axis at an angle 20°.

A distribution of the discharge in front of the cathodes of dual magnetron exactly corresponds to basic principles of a magnetic confinement of the plasma now currently used in fusion machines. In addition, in the dual magnetron with a closed **B** field the magnetic plasma confinement is enhanced by an electrostatic confinement. Electrons are repelled from the magnetron cathodes and oscillate between them. This phenomenon strongly enhances the sputtering discharge. An intensive

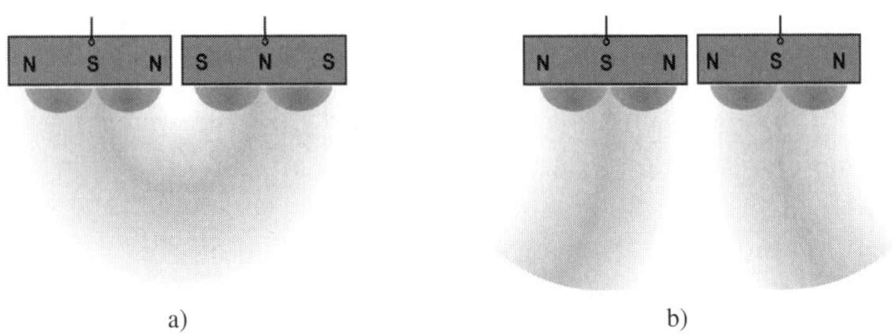

a) b)

Figure 3.22: Schematic of discharges and polarity of magnets in the dual magnetron operated with (a) closed **B** field and (b) mirror **B** field.

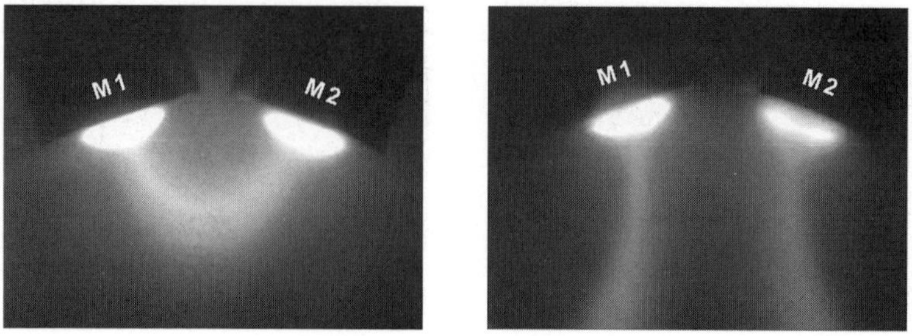

a) b)

Figure 3.23: Photos of the discharge of the dual magnetron operated in a bipolar mode at $f_r = 100\,\text{kHz}$, $I_{da\,1,2} = 0.5\,\text{A}$ in nitrogen at $p_{N2} = 0.5\,\text{Pa}$. (a) Closed **B** field and (b) mirror **B** field.

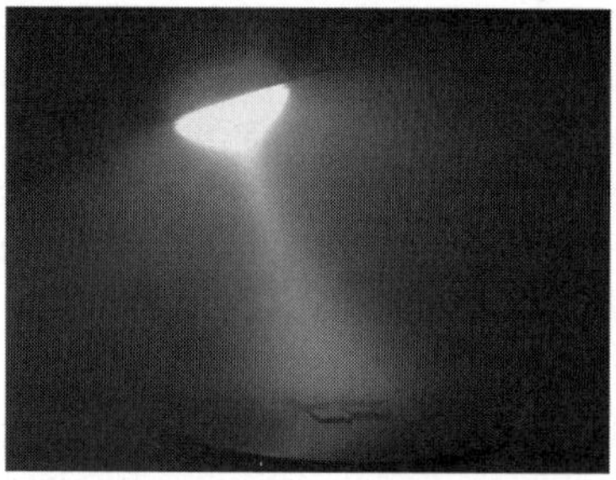

Figure 3.24: Photograph of the discharge of single magnetron tilted at angle 20° with respect to the vertical axis and operated in an asymmetric bipolar mode $f_r = 100\,\text{kHz}$, $I_{da} = 0.5\,\text{A}$ in argon at $p_{Ar} = 0.5\,\text{Pa}$.

discharge in the dual magnetron with closed **B** field is clearly seen in Fig. 3.23(a). In the case of a mirror **B** field configuration, the magnetron discharge is bended from the magnetron axis, which is perpendicular to the cathode (target) surface (see Fig. 3.23(b)). The magnetron discharge can be expanded along the axis only in two cases: (1) when one magnetron is removed from the sputtering device, that is, the dual magnetron is converted into the single magnetron, see Fig. 3.24 or (2) if the magnetic **B** field of the inactive magnetron is generated by electromagnets and currents flowing in the electromagnets are switched off.

The magnetic field **B** between magnetrons strongly influences the geometry of the dual magnetron discharge. Therefore, the geometry of the discharge does not change when one of the two magnetrons is switched off (see Fig. 3.25). In this case, the intensity of discharge is weakened only.

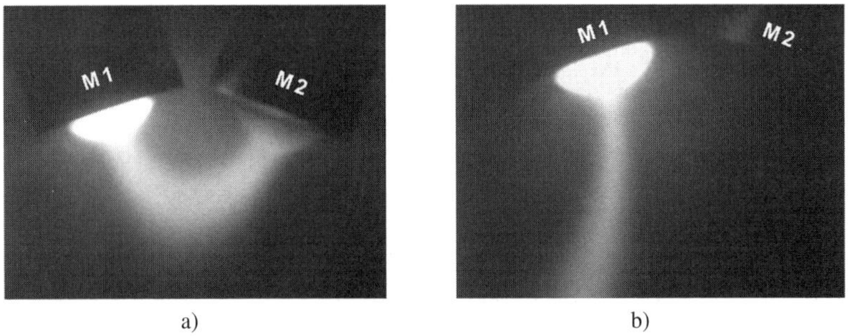

Figure 3.25: Photos of a discharge of the dual magnetron, in which one magnetron on right side is switched off. Asymmetric bipolar mode of operation at $f_r = 100\,\text{kHz}$, $I_{da} = 0.5\,\text{A}$ in oxygen at $p_{O2} = 0.5\,\text{Pa}$. (a) Closed **B** field and (b) mirror **B** field.

3.4. Role of Energy in Formation of Sputtered Films

The energy E delivered to the film during its growth by (i) bombarding ions (E_i) and (ii) condensing atoms (E_{ca}) is a key parameter, which decides on the mechanism of its growth, elemental and phase composition, microstructure and structure and so on its final physical properties. What energy E is delivered, it depends on the design of magnetron or sputtering system and their I–V characteristics, particularly on the efficiency of ionization of (i) inert sputtering gas, (ii) RG in the case of dc sputtering of compounds and (iii) atoms of sputtered material, the discharge voltage U_d and current I_d and total pressure of sputtering gas $p_T = p_{Ar} + p_{RG}$.

In collisionless plasma the energy E_i is defined by the following formula:

$$E_i[\text{J/cm}^{-3}] = e(V_p - V_s)\nu_i/\nu_{ca} \approx U_s(i_s/a_D) \quad \text{at } T_s = \text{constant} \tag{3.4}$$

where V_p and V_s are the plasma and substrate potential, respectively, ν_i and ν_{ca} are the flux of ions and condensing atoms, respectively, U_s is the substrate bias and a_D is the deposition rate. From this formula it is seen that E_i depends not only on U_s but also on i_s and a_D. The magnitude of i_s strongly depends on the design of magnetron.

3.4.1. Ion Current Density at Substrate in Sputtering Magnetron Discharge

The process in which the growing film is bombarded by ions is called *the ion plating process* (IPP). In this process the substrate is negatively biased. Typical I–V

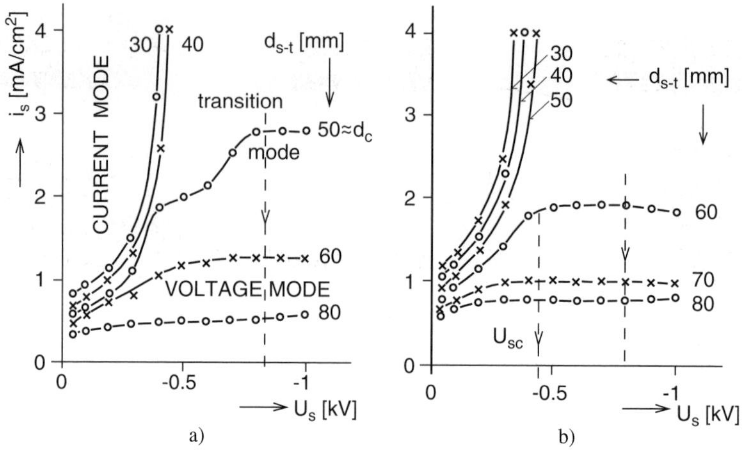

Figure 3.26: Development of substrate ion current density i_s in the discharge of CM, equipped with (i) Cu target (ϕ = 100 mm) and (ii) electromagnet, with increasing U_s at p_{Ar} = 0.5 Pa, I_m = 2 A, different values of $d_{s\text{-}t}$ and (a) P_d = 1 kW and (b) P_d = 1.5 kW. I_m is the current in the coil of electromagnet.

characteristics of the substrate immersed in the plasma of CM are displayed in Fig. 3.26. Three modes of the IPP can be realized: (1) voltage mode at $d_{s\text{-}t} > d_c$, (2) transition mode at $d_{s\text{-}t} \approx d_c$ and (3) current mode at $d_{s\text{-}t} < d_c$; here d_c is the critical substrate-to-target distance.

The critical distance d_c decreases with increasing power $P_d = I_d U_d$ delivered to the magnetron discharge, that is, with expanding of the discharge toward the substrate. For IPP the voltage mode is used. In this mode of operation, first i_s increases with increasing U_s up to U_{sc} and for $|U_s| > |U_{sc}|$ the value of i_s saturates. According to Eq. (3.4) and under the assumption that a_D is approximately constant this means that E_i can be controlled by U_s. This makes it possible to investigate the effect of U_s on the properties of sputtered film at a constant value of i_s.

The increase in U_s can, however, result in a resputtering of the growing film if i_s is sufficiently large and exceeds 1 mA cm^{-2}. Therefore, the sputtering is usually realized at low negative values of $|U_s| < |100\,V|$ and high ion fluxes, that is, high values of i_s. An extraction of high ion currents I_s, corresponding to $i_s > 1$ mA cm^{-2}, from the CM to the substrate placed at $d_{s\text{-}t} \approx$ 100 mm is impossible (see Fig. 3.26). To achieve $i_s > 1$ mA cm^{-2}, the magnetrons with (i) plasma expanding from the sputtered target toward the substrate, that is, UMs, and/or (ii) enhanced plasma ionization, for instance, the magnetron with extended magnet and target of the "dish" type (Fig. 3.27(c)), must be used. This statement was confirmed experimentally (see Fig. 3.27). In this figure the dependencies $i_s = f(U_s)$ for the UM with (a) two electromagnets,

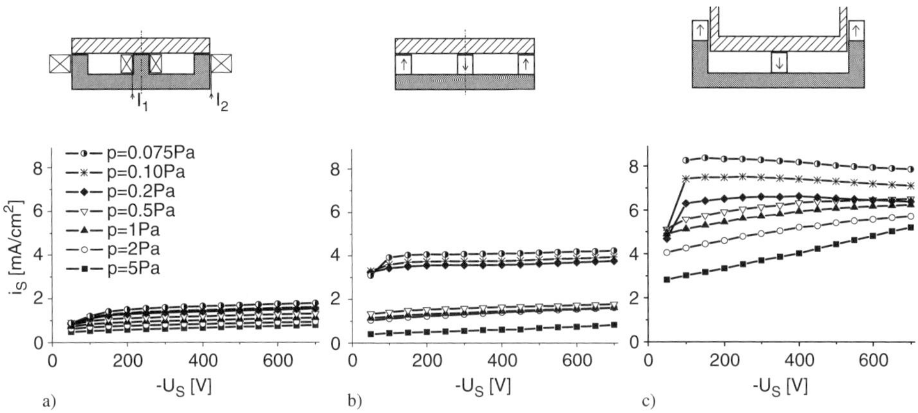

Figure 3.27: Comparison of $i_s = f(U_s)$ for the UM with 8 mm thick Cu target (ϕ = 100 mm) and equipped with (a) two electromagnets (I_1 = 1.6 A, I_2 = 0.5 A), (b) NdFeB magnets and (c) NdFeB magnets (periphery magnet is extended) and "dished" target.

(b) strong NdFeB magnets and (c) external NdFeB ring magnet extended above the sputtered surface of the dished target are given. The increase of the magnitude of $\mathbf{B}_{par\ max}$ on the surface of sputtered target (compare Fig. 3.27(a) and (b)) and the optimization of $\mathbf{B}_{par}$ distribution above the target combined with the additional electrostatic plasma confinement in the "dish" target (Fig. 3.27(c)) result in the enhancement of magnetron discharge and in consequence of this enhancement also in the increase of i_s. High values of $i_s \geqslant 5\,mA\,cm^{-2}$ can be achieved at $p_{Ar} \leqslant 1$ Pa on the substrates placed at $d_{s\text{-}t}$ = 100 mm. These experimental data clearly show that the operational characteristics of the magnetron strongly depend on its design.

3.4.2. *Ion Current Density on Substrate in Cleaning Magnetron Discharge*

The magnetrons, which create a closed discharge above the target at low values of $U_d \leqslant 200$ V and high values of $I_d > 1$ A, are called *the cleaning magnetrons*. Due to low U_d they exhibit a low ($\leqslant$100 nm min^{-1}) a_D and due to high I_d they enable to extract high ion currents I_s to the substrate. The deposition rate a_D of such magnetrons can be further strongly decreased if the magnetron is equipped with (i) the oxide target (e.g. TiO, TiO_2, etc.) or (ii) easily oxidized metal target (e.g. Ti, Al, etc.) and the sputtering is carried out in the oxide mode. In such a case, a_D can be further decreased even below 10 nm min^{-1} (see Fig. 3.28).

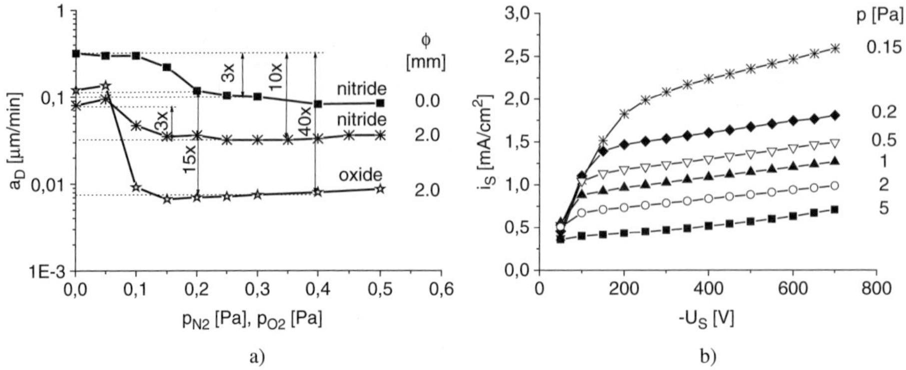

Figure 3.28: (a) a_D of Ti(Al,V)N_x and Ti(Al,V)O_x films magnetron sputtered with gas injection into discharge through four holes (Ø2 mm) in the target at $I_d = 2$ A, $T_s = 300$°C vs. p_{N2} and p_{O2}, respectively and (b) i_s vs. U_s at $d_{s\text{-}t} = 100$ mm, $p_{O2} = p_{Ar}$ and $p = p_{O2} + p_{Ar}$.

From this figure it is seen that high (>1 mA cm^{-2}) values of i_s can be easily achieved at negatively biased substrate placed at $d_{s\text{-}t} = 100$ mm. The value of i_s can be controlled by (i) the kind and pressure of sputtering gas and (ii) the magnetron discharge current I_d. For instance, on the substrate located at $d_{s\text{-}t} = 100$ mm the substrate ion current density $i_s = 2.5$ mA cm^{-2} can be achieved at $I_d = 2$ A, $U_s = -700$ V and $p = 0.15$ Pa. This value can be further increased if $I_d > 2$ A is used. Under such conditions no film is deposited and the substrate surface is efficiently sputtered. Such a magnetron discharge is very suitable for the cleaning and activation of surfaces prior to the film deposition or for the modification of substrate surfaces only.

3.4.3. *Effect of Deposition Rate* a_D *on Energy* E_i *Delivered to the Growing Film*

The energy E_i delivered to the growing film by bombarding ions decreases with increasing a_D (see Eq. (3.4)). In general, the deposition rate a_D increases with increasing (i) magnetron discharge current I_d and (ii) energy of ions incident on the sputtered target, that is, with increasing U_d (see Fig. 3.29). This means that also a_D is limited by the magnetron design, particularly by its cooling and the perfection of its plasma confinement, that is, by the discharge voltage U_d.

The deposition rate a_D changes also in dc reactive sputtering of compounds (e.g. nitrides, oxides, etc.). The value of a_D decreases with increasing partial pressure of RG at a constant value of I_d due to the target poisoning (see Fig. 3.30). In

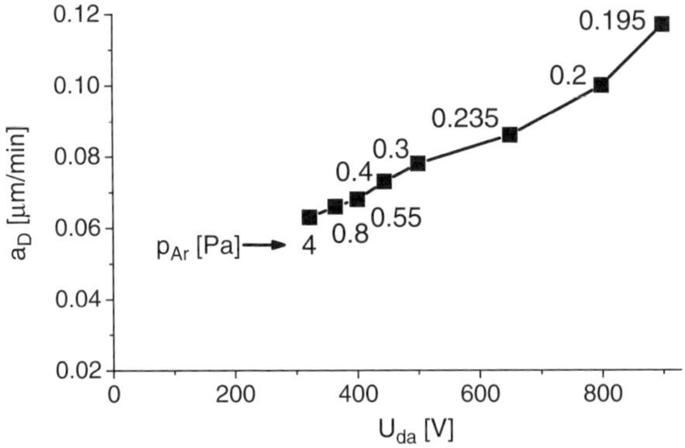

Figure 3.29: Effect of discharge voltage on a_D of Cu films dc pulsed sputtered in argon using UM at $f_r = 100\,\mathrm{kHz}$, $\tau/T = 0.5$, $I_{da} = 1\,\mathrm{A}$, $d_{s\text{-}t} = 100\,\mathrm{mm}$, Cu target ($\phi = 100\,\mathrm{mm}$).

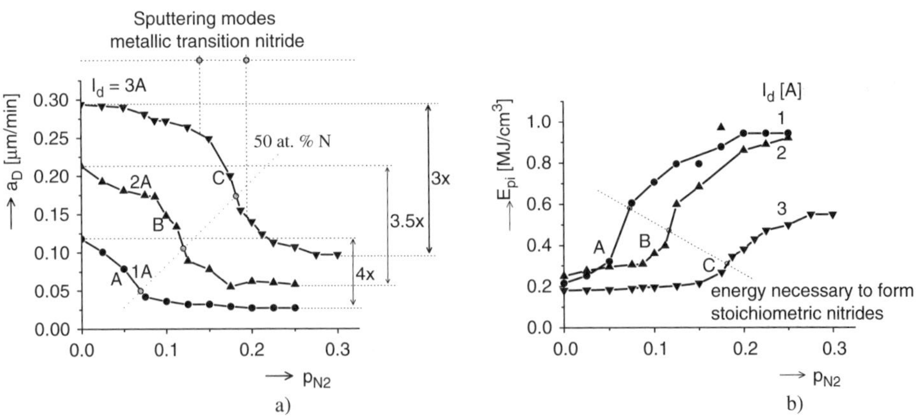

Figure 3.30: (a) Deposition rate a_D of Ti(Fe)N_x films and (b) energy E_{pi} delivered to them during their growth by bombarding ions as a function of p_{N2}. The films were sputtered at (i) $I_d = 1\,\mathrm{A}$, $i_s = 0.5\,\mathrm{mA\,cm^{-2}}$, (ii) $I_d = 2\,\mathrm{A}$, $i_s = 1\,\mathrm{mA\,cm^{-2}}$, (iii) $I_d = 3\,\mathrm{A}$, $i_s = 1\,\mathrm{mA\,cm^{-2}}$ and $U_s = -100\,\mathrm{V}$, $T_s = 300°\mathrm{C}$, $d_{s\text{-}t} = 60\,\mathrm{mm}$ and $p_T = 0.5\,\mathrm{Pa}$.

agreement with Eq. (3.4), a large decrease in a_D with increasing p_{N2} results in the large (3–4x) increase of the energy E_i delivered to the growing film by bombarding ions despite that all other parameters with the exception of p_{N2} are kept constant. This increase in E_i has dramatic consequences and is responsible for unexpected, very strong changes of the film structure and so its final physical properties. For details see the Ref. [70].

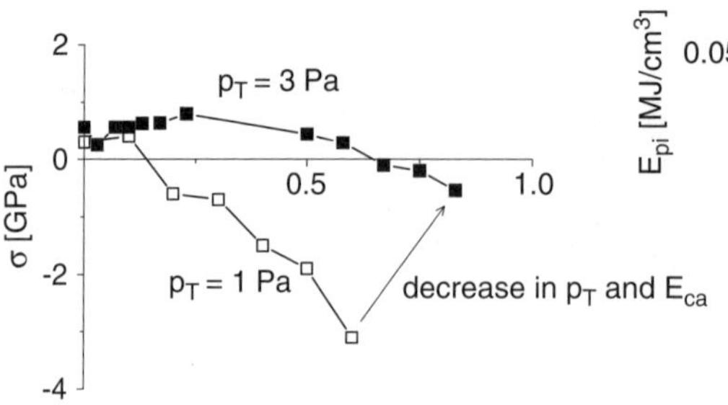

Figure 3.31: Effect of E_{ca} on macrostress σ in Mo-Al-N films reactively sputtered under a weak ion bombardment [71].

3.4.4. Energy Delivered by Fast Condensing Atoms

The energy E_{ca} delivered to the growing film by fast condensing atoms dominates over that of E_i in a collisionless plasma when the mean free path of atoms $\lambda_{mfp} > d_{s\text{-}t}$ and the ion bombardment is weak; here λ_{mfp} (cm) $= 0.4/p$ (Pa). This fact is illustrated by the experiment results of which are displayed in Fig. 3.31.

This figure shows a development of (a) the growth macrostress σ and (b) the energy E_i in the Mo-Al-N film with a low ($\leqslant$10 at.%) Al content, reactively sputtered under a weak ion bombardment ($U_s = -20$ V, $i_s = 1$ mA cm^{-2}) and two values of $p_T = p_{Ar} + p_{N2} = 1$ and 3 Pa, with increasing ration p_{N2}/p_T. A small difference in E_{pi} (Fig. 3.31(b)) cannot explain the large difference in σ and indicates that the large decrease of σ in the films sputtered at $p_T = 3$ Pa is due to the decrease in E_{ca} when p_T is increased. These experimental data clearly show that (1) the energy E_{ca} can be used to control the mechanism of the growth of insulating films or films sputtered on insulating substrates, where dc bias does not operate and (2) the magnetrons operating at low $p_T < 0.1$ Pa are very perspective.

3.5. Advanced Sputtering Sources and Systems

3.5.1. Magnetron with Extended Anode

The magnetron with a floating cathode and an extended anode is displayed in Fig. 3.32.

The anode is a metallic ring located around the substrate holder. Between the cathode and the chamber walls a resistance R is inserted. The voltage U_{ag} between the anode and the ground decreases with increasing R up to U_{fl} at $R = \infty$ (see Fig. 3.33).

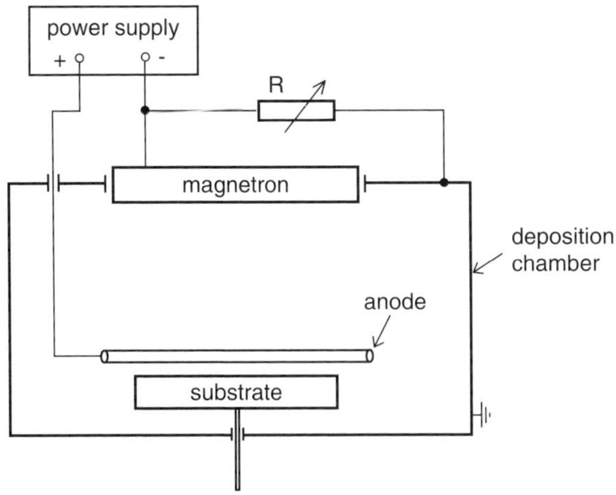

Figure 3.32: Schematic diagram of the magnetron with extended anode.

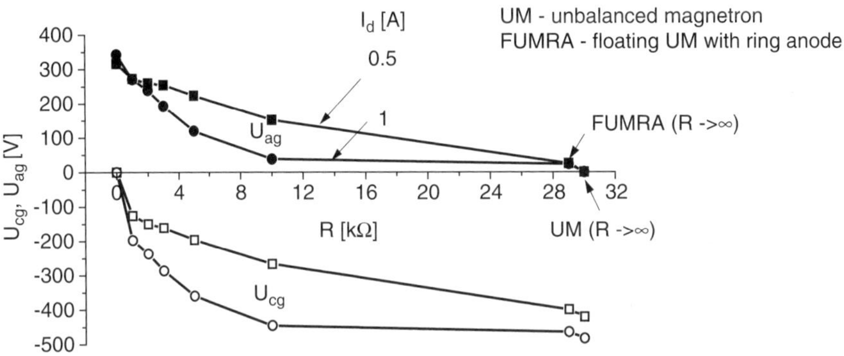

Figure 3.33: Dependencies of anode U_{ag} and cathode U_{cg} voltages, measured with respect to ground, for the magnetron with extended ring anode operated at $I_d = 0.5$ and 1 A, $p_{Ar} = 0.5$ Pa and $d_{s\text{-}t} = 100$ mm.

On the contrary, the decrease in R results in increase of (1) U_{ag} and (2) sputtering of chamber walls (see Fig. 3.34(c)). Therefore, the resistance R has to be sufficiently high to avoid sputtering of chamber walls.

The geometrical arrangement of this magnetron and its electrical connection to the power supply result in two principal advantages: (1) the plasma is expanded from the sputtered target and easily fills up a space between it and the substrate holder and (2) the plasma potential U_p, measured with respect to the ground, can be increased to a positive higher value compared to that of CM ($U_p \approx 10$–20 V).

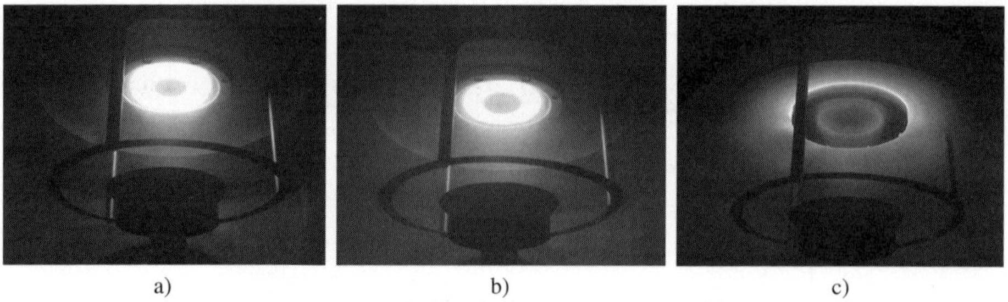

Figure 3.34: Photos of the discharge of (a) CM and magnetron with extended anode electrically connected to the grounded chamber through, (b) $R = 10\,k\Omega$ and (c) $R = 0\,\Omega$ (the discharge on chamber wall is seen).

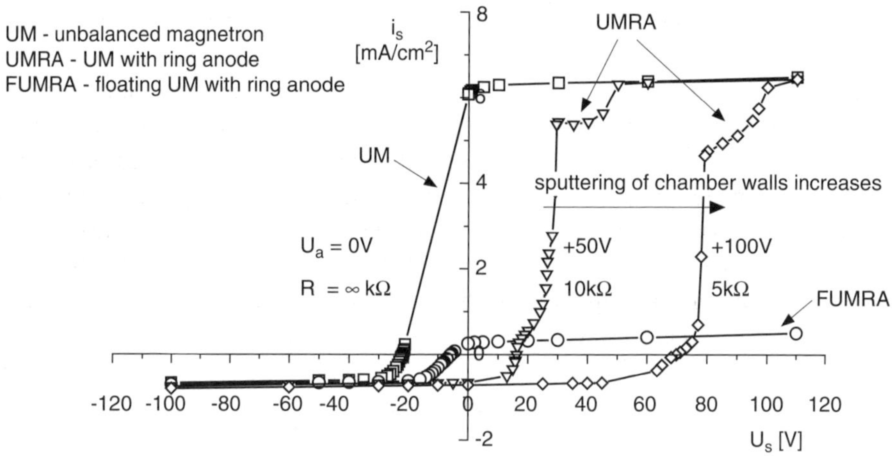

Figure 3.35: Comparison of *I–V* characteristics of the substrate holder placed in the discharge of CM and magnetron with extended anode. Process parameters: $I_d = 0.5$ A, $p_{Ar} = 0.5$ Pa.

I–V characteristics of the substrate holder (Ø = 100 mm) located at the distance $d_{s\text{-}t} = 100$ mm in the discharge of CM and the magnetron with extended anode are given in Fig. 3.35.

Three issues can be drawn from this experiment: (1) while a strong electron current I_e is flowing on a conductive grounded substrate in the discharge of CM, the same substrate is bombarded by ions in the discharge of magnetron with extended anode, (2) I_e can be strongly reduced in the case when the magnetron anode is floating and (3) the increase of U_p with decreasing R results in no change of the difference $U_p - U_{fl}$, which remains almost constant with varying R. The last fact means that there is no increase in the ion bombardment of floating substrates. Moreover, the film is contaminated by the material sputtered from chamber walls due to their sputtering, which increases with increasing U_p, that is, decreasing R. The magnetron with extended anode can be, however, used for the efficient ion bombardment of the film growing on grounded substrates. The voltage between the anode and ground decreases with increasing R and so its value has to be optimized.

3.5.2. Rectangular Magnetron with Full Target Erosion

At present, there is an urgent need to increase the TU of rectangular magnetrons. Magnetrons currently produced exhibit a low TU of approximately 20–30%. This value of TU can be increased up to 50–60% in the case when the magnetic field of magnetron is optimized. This is a limit because the magnetron discharge forms a closed loop and so always a certain part of the target surface inside it is not sputtered. In principle, there are two ways which allow to achieve a full target erosion: (1) to design the magnetron with the target without a central part, which is not sputtered or (2) to scan the magnetic circuit, which forms a closed magnetic field of the magnetron, under the target and so to expose every point of it to the sputtering discharge.

The first method resulted in realization of several magnetrons, for instance, "S-gun" magnetron [72], INSET™ target magnetron [73], interpole target magnetron [74,75], toroidal plasma type magnetron [76] and Torus type magnetron [77]. The INSET and interpole targets were used for rectangular magnetrons. The second method can be easily realized in the case of round planar magnetron when a simple excentric rotation of the magnetic circuit is used [78]. For the rectangular magnetron with full target erosion a translation movement of the closed magnetron discharge is, however, needed. This solution was described by Garrett in the US patent [79].

The magnetic circuit can be moved in two directions: (i) perpendicularly to the long side of target and (ii) along the long side of target. For long rectangular magnetrons with one closed magnetron discharge the perpendicular movement is relatively complicated. Therefore, a new magnetron with several mutually separated closed magnetron discharges (racetracks), which are generated simultaneously and swept along the long side of target, was designed [62], see Fig. 3.36. In this way the non-sputtered regions inside individual racetracks were completely eliminated.

A successful operation of this magnetron, however, requires to avoid the interaction between neighboring discharges. In the case when neighboring magnetic circuits have the same polarity of magnets and are located too close each to other, the touching parts of separate neighboring discharges disappear due to the opposite electron drifts in the touching parts of discharges. In consequence of this phenomenon separate discharges on the target convert into one closed discharge (see Fig. 3.37(a)).

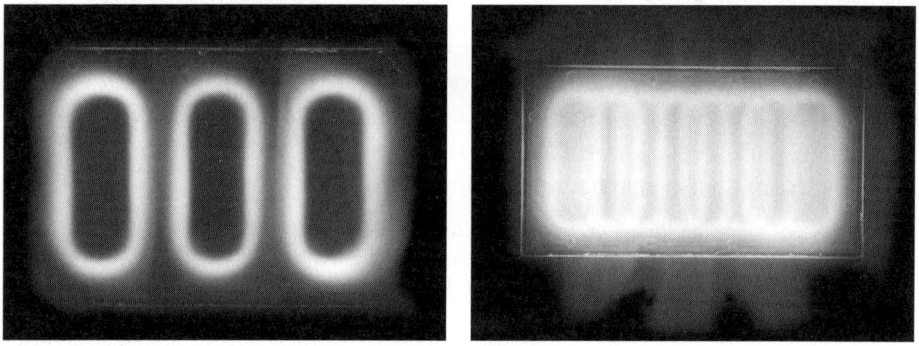

a) b)

Figure 3.36: Photos of (a) three closed magnetron discharges simultaneously formed on the surface of Cu target generated by three stationary magnetic circuits and (b) magnetron discharges during scanning of the magnetic circuits generated in argon at $I_d = 0.5$ A and $p = 2.2$ and 0.5 Pa, respectively.

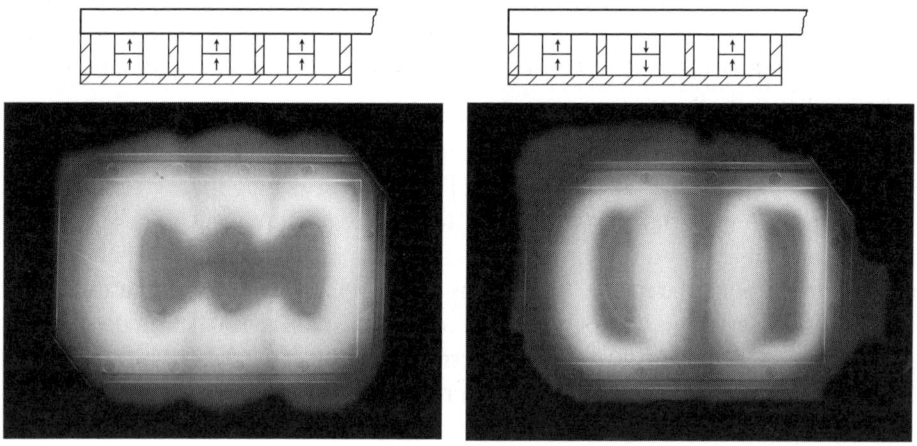

a) b)

Figure 3.37: Photos of interacting magnetron discharges produced in argon at $p = 0.8$ Pa and $I_d = 0.5$ A by three magnetic circuits with (a) the same and (b) alternating polarity of magnets.

On the contrary, in the case when discharges are produced by magnetic circuits with alternating polarity of magnets there is the same direction of electron drifts in adjacent parts of discharges (Fig. 3.37(b)). This results in an enhancement of side-closed discharges at the expense of central one, which is interrupted and disappears. Therefore, the distance between neighboring magnetic circuits must be correctly designed. For more details see the Ref. [62].

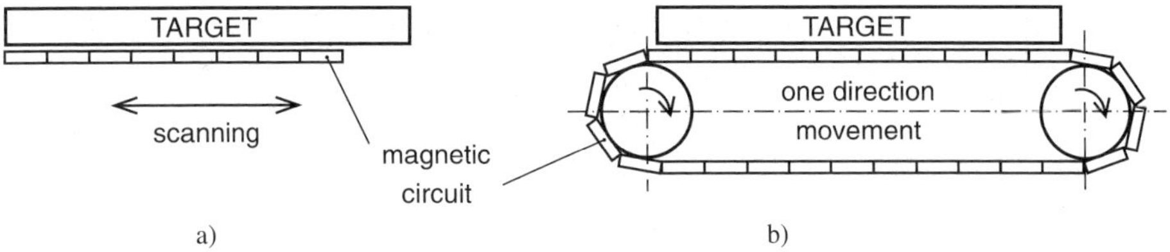

Figure 3.38: Schematic diagram of the movement of magnetic circuits under the target (a) scanning movement and (b) one direction continuous "endless" movement.

The main drawback of the rectangular magnetron with many closed discharges scanning on the surface of target is the overlapping of neighboring discharges in certain regions along its long axis. This overlapping results in an inhomogeneous (i) sputtering of the target and (ii) thickness of sputtered film along the long side of rectangular magnetron. This serious drawback can be fully avoided if the scanning of the magnetic circuits under the target is replaced by a continuous "endless" movement of all magnetic circuits *in one direction* (see Fig. 3.38(b)). This magnetron, which ensures a full and homogeneous erosion of the rectangular target, was already manufactured and now is successfully used in laboratory and production sputtering machines.

3.5.3. Cluster of Magnetrons

Recently, it was recognized that nanoscale multilayer and multicomponent films exhibit unique and enhanced properties compared to the coarse-grained single-layer films. For the production of such films a cluster of several magnetrons is usually used. Targets of individual magnetrons are made of different elements, alloys or compounds, for example TiAl, TiO, etc. Such sputtering systems make it possible to form multicomponent films. In addition, the amount of individual elements in the film can be controlled by the power delivered to the individual magnetrons. Also, multilayer films can be produced by alternating switching on and off of individual magnetrons.

Usually, the cluster magnetron system is composed of three or four small magnetrons with targets of diameter from 30 to 50 mm. Such sputtering systems are successfully used, for instance, for the production of high-temperature superconductive films (e.g. YbaCuO) or multilayer magnetic films for high-density memories in microelectronics. Also, the nanocomposite films are very perspective for the formation of new advanced optical coatings. For production of large area coatings a cluster of rectangular magnetrons were developed (see Fig. 3.39). While in the

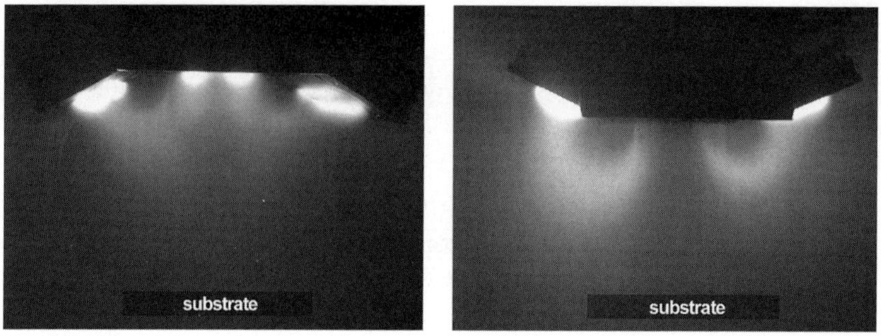

Figure 3.39: Photos of discharges of three rectangular magnetrons with closed magnetic field in argon at $p_{Ar} = 0.5$ Pa and $I_d = 1$ A. (a) All magnetrons are switched on and (b) the central magnetron is switched off.

production of single multiphase films all magnetrons operate simultaneously (Fig. 3.39(a)), in the production of multilayers individual magnetrons are alternatively switched on and off (Fig. 3.39(b)). The targets of individual magnetrons can be mutually tilted (compare Figs. 3.39(a) and (b)). This enables to control the width of a sputtered lane below the sputtering system. A mutual coupling between individual magnetrons is also very important because in the case of dual magnetron operation it is possible, for instance, to perform a high-rate dc reactive sputtering of oxides. New advanced sputtering systems comprising the cluster of rectangular magnetrons are expected to be developed soon.

3.6. Conclusions

At present, the magnetron technology is well developed and successfully transferred into the industrial use. In spite of this fact, new requirements for its improvement and modification continuously occur. Therefore, new magnetrons and sputtering systems have been continuously developed. In this connection, it is necessary to note that the development of new advanced magnetrons and particularly those operating under new physical conditions is possible only on the basis of the deep knowledge of physical limits of existing magnetrons. We believe that:

(i) the perfect plasma confinement,
(ii) increase of the ionization of both the atoms of inert sputtering gas and the sputtered atoms,
(iii) increase of the target power loading up to the evaporation of target material, that is, the development of HPPMs, operating with simultaneous sputtering

and partial evaporation of target material, magnetrons sputtering from molten targets, etc.,

(iv) dissociation and activation of RG,

(v) control of the energy of condensing and bombarding ions over a wide range of sputtering gas pressure, and

(vi) sputtering systems based on cluster of several magnetrons with mutually coupled discharges will be leading directions in the development of new advanced magnetron technology in the near future.

Acknowledgments

This work was supported in part by the Ministry of Education of the Czech Republic under Project No. MSM 235200002.

References

[1] J.S. Chapin, *US Patent Application*, 438–482, 1974.

[2] J.L. Vossen and J.J. Cuomo, in *Glow Discharge Sputter Deposition*, Eds. J.L. Vossen and W. Kern, *Thin Film Processes*, Academic Press, Inc., London, 1978, p. 12.

[3] J.A. Thornton and A.S. Penfold, in *Cylindrical Magnetron Sputtering*, Eds. J.L. Vossen and W. Kern, *Thin Film Processes*, Academic Press, Inc., London, 1978, p. 76.

[4] D.B. Fraser, in *The Sputter and S-gun Magnetrons*, Eds. J.L. Vossen and W. Kern, *Thin Film Processes*, Academic Press, Inc., London, 1978, p. 115.

[5] R.K. Waits, in *Planar Magnetron Sputtering*, Eds. J.L. Vossen and W. Kern, *Thin Film Processes*, Academic Press, Inc., London, 1978, p. 131.

[6] B. Chapman, Sputtering (Chapter 6), in *Glow Discharge Processes; Sputtering and Plasma Etching*, John Wiley and Sons, New York, 1980, p. 177.

[7] J.A. Thornton, in *Coating Deposition by Sputtering*, Eds. R.F. Bunshah, *et al.*, *Deposition Technologies for Films and Coatings*, Noyes Publications, Park Ridge, New Jersey, 1980, p. 170.

[8] A.R. Nyaiesh, *Thin Solid Films*, 86 (1981) 267.

[9] B. Window and N. Savvides, *J. Vac. Sci. Technol.*, A4 (2) (1986) 196.

[10] B. Window and N. Savvides, *J. Vac. Sci. Technol.*, A4 (3) (1986) 453.

[11] S. Schiller, U. Heisig, Ch. Korndorfer, G. Beister, J. Reschke, K. Steinfelder and J. Strumpfel, *Surf. Coat. Technol.*, 33 (1987) 405.

[12] S.M. Rossnagel, *Thin Solid Films*, 171 (1989) 125.

[13] S.M. Rossnagel, in *Magnetron Plasma Deposition Processes*, Eds. S.M. Rossnagel, J.J. Cuomo and W.D. Westwood, *Handbook of Plasma Processing Technology*, Noyes Publications, Park Ridge, New Jersey, 1990, p. 160.

[14] W.D. Westwood, in *Reactive Sputter Deposition*, Eds. S.M. Rossnagel, J.J. Cuomo and W.D. Westwood, *Handbook of Plasma Processing Technology*, Noyes Publications, Park Ridge, New Jersey, 1990, pp. 160, 233.
[15] J. Musil and S. Kadlec, *Vacuum* 40 (1990) 435.
[16] W.D. Munz, F.J.M. Hauzer, D. Schulze, B. Buil, *Surf. Coat. Technol.*, 49 (1991) 161.
[17] A. Belkind, W. Gerristead, Z. Orban, D. Dow, J. Felts and R. Laid, *Surf. Coat. Technol.*, 49 (1991) 155.
[18] B.P. Hichwa, Low temperature, high rate sputter deposition of metal oxides and transparent conductors, *Proceedings of the 2nd international Symposium on ISSP-93*, Tokyo, 1993, p. 61.
[19] D.P. Palicki and A. Matthews, Recent developments in magnetron sputtering systems, *Finishing*, November (1993) 36.
[20] S.M. Rossnagel, *Thin Solid Films*, 263 (1995) 1.
[21] J. Musil, A. Rajský, A.J. Bell, J. Matouš, M. Cepera and J. Zeman, *J. Vac. Sci. Technol.*, A14 (1996) 2187.
[22] S. Kadlec and J. Musil, *Vacuum*, 47 (1996) 307.
[23] B. Window, *Surf. Coat. Technol.*, 81(1996) 92.
[24] J. Musil, Basic properties of low-pressure plasma, *Proceedings of the International School of Physics* "Enrico Fermi", Course CXXXV, Eds. A. Paoletti and A. Tucciarone, IOS Press, Amsterdam, 1997, p. 2636.
[25] J.M. Schneider and W.D. Sproul, in *Reactive Pulsed dc Magnetron Sputtering and Control, Handbook of Thin Film Process Technology*, IOP Publishing Ltd, 1998, A5.1:1–A5.1:12.
[26] J. Musil, *Vacuum*, 50 (1998) 363.
[27] J. Musil, *Surf. Coat. Technol.*, 100–101 (1998) 280.
[28] W.D. Sproul, *Vacuum*, 51 (4) (1998) 641.
[29] J.M. Schneider, S. Rodhe, W.D. Sproul and A. Matthews, *J. Phys. D Appl. Phys.*, 33 (2000) 173.
[30] I. Safi, *Surf. Coat. Technol.*, 127 (2000) 203.
[31] P.J. Kelly and R.D. Arnell, *Vacuum*, 56 (3) (2000) 159.
[32] J.O. Brien and P.J. Kelly, *Surf. Coat. Technol.*, 142–144 (2001) 621.
[33] S.J. Nadel, P. Greene, J. Rietzel and J. Strumpfel, *Thin Solid Films*, 442 (2003) 11.
[34] A. Matthews, *J. Vac. Sci. Technol.*, A21 (3) (2003) 224.
[35] J. Vlcek, A.D. Pajdarova and J. Musil, *Cont. Plasma Phys.*, 44 (5–6) (2004) 437.
[36] S. Kadlec, J. Musil and W.D. Munz, *J. Vac. Sci. Technol.*, A8 (1990) 1318.
[37] J. Musil, S. Kadlec and W.-D. Munz, *J. Vac. Sci. Technol.*, A9 (3) (1991) 1171.
[38] J. Musil, K. Rusnák, V. Jezek and J. Vlcek, *Vacuum*, 46 (1995) 341.
[39] S. Kadlec and J. Musil, *J. Vac. Sci. Technol.*, A13 (1995) 389.
[40] J. Musil and M. Mišina, *Czech. J. Phys.*, 46 (1996) 353.
[41] J. Musil, M. Misina and D. Hovorka, *J. Vac. Sci. Technol.*, A15 (4) (1997) 1999.
[42] M. Yamashita, *J. Vac. Sci. Technol.*, A7 (1989) 151.
[43] S.M. Rossnagel and J. Hopwood, *Appl. Phys. Lett.*, 63 (24) (1993) 3285.
[44] S.M. Rossnagel and J. Hopwood, *J. Vac. Sci. Technol.*, B12 (1) (1994) 449.

[45] Y. Setsuhara, M. Kamai, S. Miyake and J. Musil, *Jpn. J. Appl. Phys.*, 36 (1997) 4568.
[46] C. Christou and Z.H. Barber, *J. Vac. Sci. Technol.*, A18 (6) (2000) 2897.
[47] G. Brauer, J. Szczyrbowski and G. Techner, *J. Non-Cryst. Solids*, 218 (1997) 19.
[48] V. Kirchoff, T. Kopte, T. Winkler, M. Schulze and P. Wiedemuth, *Surf. Coat. Technol.*, 98 (1998) 82.
[49] M. Matsuoka, Y. Hoshi and M. Naoe, *J. Appl. Phys.*, 60 (1986) 2096.
[50] M. Matsuoka and K. Ono, *Appl. Phys.*, Letter 53 (1988) 2025.
[51] R. Adachi and K. Takeshita, *J. Vac. Sci. Technol.*, 20 (1982) 98.
[52] J.J. Cuomo and S.M. Rossnagel, *J. Vac. Sci. Technol.*, A4 (1986) 393.
[53] S. Maniv, C.J. Miner and W.D. Westwood, *J. Vac. Sci. Technol.*, 18 (2) (1981) 195; A1 (3) (1983) 1370.
[54] S. Kadlec, J. Musil and J. Vyskocil, *J. Phys. D Appl. Phys.*, 19 (1986) 187.
[55] S. Kadlec, J. Musil and J. Vyskocil, *Vacuum*, 37 (1987) 729.
[56] S. Schiller, U. Heisig, K. Steinfelder, J. Strumpfel, R. Voigt, R. Fendler and G. Teschner, *Thin Solid Films*, 96 (1982) 235.
[57] A.G. Spencer and R.P. Howson, *Thin Solid Films*, 186 (1990) 129.
[58] R.P. Howson, N. Danson and I. Safi, *Thin Solid Films*, 351 (1999) 32.
[59] N. Martin, A.R. Bally, P. Hones, R. Sanjines and F. Levy, *Thin Solid Films*, 377–378 (2000) 550.
[60] J.J. Hofmann, C-MAG, *Proceedings of the 2nd Annual Technical Conference*, Society of Vacuum Coaters, 1989, p. 297.
[61] U. Heister, J. Krempel-Hesse, J. Szczyrbowski and G. Brauer, New developments in the field of MF-sputtering with dual magnetron to obtain higher productivity for large area coatings, *Proceedings of the 41st Annual Technical Conference*, Society of Vacuum Coaters, 1998, p. 187.
[62] J. Musil, *J. Vac. Sci. Technol.*, A17 (2) (1999) 555.
[63] H. Ohsaki, Y. Tachibana, A. Mitsui and Y. Hayashi, *Thin Solid Films*, 392 (2001) 169.
[64] R.L. Cormia, T.A. Trumbly and S. Andersen, *US Patent No.* 4,046,659.
[65] S. Schiller, K. Goedicke, J. Reschke, V. Kirchhoff, S. Schneider and F. Milde, *Surf. Coat. Technol.*, 61 (1993) 331.
[66] M.E. Graham and W.D. Sproul, *37th Annual Technical Conference*, Society of Vacuum Coaters, 1994, p. 275.
[67] J. Musil and M.Mišina, Czech. *J. Phys.*, 50 (2000) 785.
[68] J.W. Bradley, D.M. Willett and Y.A. Gonzalvo, *J. Vac. Sci. Technol.*, A17 (1) (1999) 3333.
[69] J. Musil, P. Baroch, H. Poláková, J. Vlcek, K.H. Nam and J.G. Han, *Surf. Coat. Technol.*, 148 (2001) 296.
[70] J. Musil, H. Poláková, J. Šuna and J. Vlcek, *Surf. Coat. Technol.*, 177–178 (2004) 289.
[71] J. Musil and J. Šuna, The role of energy in formation of sputtered nanocomposite films, *International Conference on New Frontiers of Process Science and Engineering in Advanced Materials*, Kyoto, Japan, 2004, Paper No. IA-109.
[72] J.A. Thornton, *Met. Finish.*, 77 (1997) 45.
[73] W.H. Class, *US Patent No.* 4 198 283, MRC INSET™ target.

[74] R. Kukla, J. Kieser and M. Mayer, *IEEE Trans. Magn.*, 23 (1987) 137.
[75] H. Aichert, J. Kieser and R. Kukla, *US Patent No.* 4 572 776, 25.2.1986.
[76] T. Takahashi and M. Naoe, *IEEE Trans. Magn.*, 25(1989) 4174.
[77] Bulletin Kurt J. Lesker Company, Magnetron sputter deposition sources, Torus®, 412-233-4200, 1995.
[78] Bulletin of Sloan Technology Corporation, Dynamic Planar Magnetron Sputtering Source, Orbitor™, T-133A-2M-782, *US Patent No.* 3 878 085.
[79] C.B. Garrett, *US Patent No.* 4 444 643, April 24, 1984.

Chapter 4

Surface Modification of Materials by Plasma Immersion Ion Implantation

Jean-Pierre Celis and Balakrishnan Prakash

4.1. Introduction

Modification of surfaces is an attractive way to achieve the desired property at the surface or in the near surface region of any bulk material. Surface modifications can be done by alloying/mixing elements at the surface of bulk materials or by an adherent overlayer on bulk materials. Plasma immersion ion implantation (PIII) is one among the surface modification techniques and was developed by Conrad and his co-workers at the University of Wisconsin in 1991 [1]. PIII is an emerging technology for the surface modification of semiconductors, metals, and insulators. Materials to be treated are immersed in a plasma at a given potential. Intricate/complex shapes can be treated rather uniformly with this technique. Indeed, this technology offers a substantially uniform ion bombardment of components, removing the line-of-sight restrictions of conventional ion beam implanters, as well as providing a more simple way to treat large surface areas. As there is no ion beam rastering as in conventional ion implantation, the treatment time can be reduced with PIII. This could support the introduction of PIII into manufacturing processes in a competitive manner compared to conventional ion implantation. The physics and technology of the PIII process are discussed in Section 4.2. In Section 4.3, the potential of this technique for depositing low friction and wear resistant layers is discussed. In addition, the nitriding of stainless steel and the formation of intermixed layers are discussed. Applications of PIII in the fields of microelectronics and medicine are discussed in Sections 4.4 and 4.5.

4.2. PIII and its Classification

In PIII, materials to be treated are immersed in a plasma-containing ions of the species to be implanted. The PIII system does not use a separate ion source, extracting and accelerating coils or deflection plates as in conventional raster beam

Materials Surface Processing by Directed Energy Techniques
Y. Pauleau (Editor)

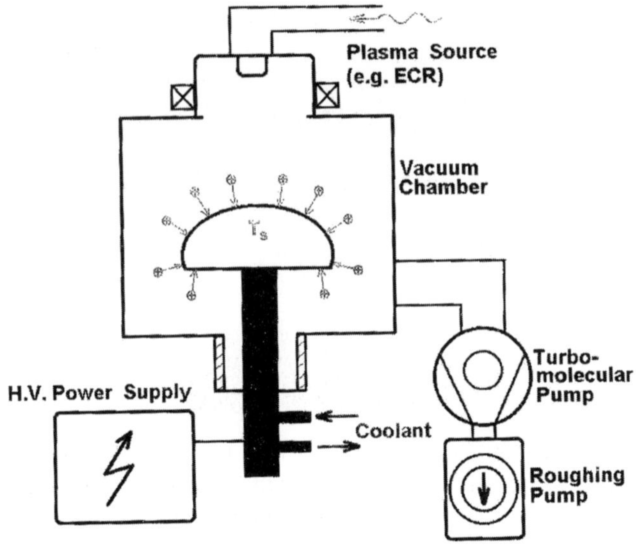

Figure 4.1: Schematic of the GPIII (reproduced with permission from Gunzel).

ion implanters. Usually samples to be coated are subjected to negative potential pulses. When a negative potential is applied, electrons start moving away from the sample and the potential drops around the sample. Positive ions of the species present in the plasma accelerate normal toward the sample surface, and get implanted.

The PIII technique can be classified broadly based on the kind of source used for the production of the plasma. Plasma of species to be implanted can be produced from a gas or a solid material. The broad classification can thus be: (1) gaseous PIII (GPIII) and (2) metal PIII (MPIII). Metal plasma can be produced by the cathodic arc principle or by sputtering. Hence, MPIII can be sub-classified as: (a) cathodic arc MPIII and (b) sputter-assisted MPIII.

4.2.1. Gaseous PIII

GPIII is a PIII technique in which a gas is used as a source for producing the plasma. The schematic of the GPIII is shown in Fig. 4.1. The workpiece to be treated is electrically insulated and placed inside a vacuum chamber. By combining rotary and turbo-molecular pumps, an ultimate pressure is created inside the chamber.

This is followed by the flow of the working gas inside the chamber. Plasma of the gas can be produced with a simple filament discharge, radio frequency (RF) or a microwave excitation. The plasma source ion implantation (PSII) process as

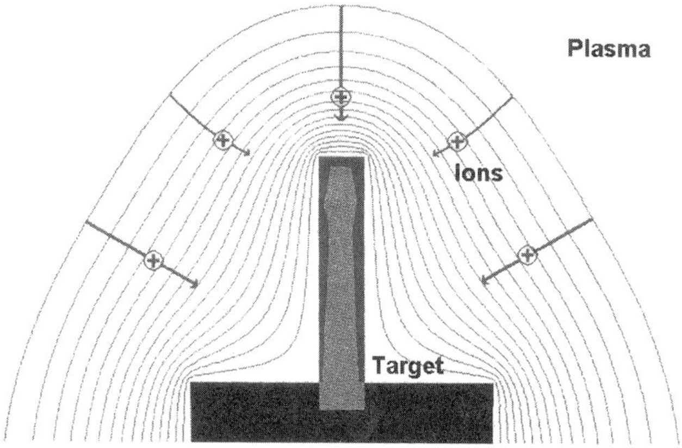

Figure 4.2: Schematics of the plasma sheath around the workpiece (normal lines to the potential gradient denote the direction of ions toward the workpiece) (reproduced with permission from Gunzel).

developed by Conrad uses a simple filament discharge to ionize the gas in the process chamber [2].

The workpiece can be DC or negative pulse biased. The applied potential drops around the workpiece. The region around the workpiece where the applied potential drops, is called the plasma sheath. The schematics of the plasma sheath is shown in Fig. 4.2. This figure illustrates the gradient in applied potential around a workpiece. When a DC bias is applied to the workpiece, the thickness, d, of the steady-state Langmuir sheath can be calculated from the following equation [3]:

$$j_s = 4/9\, \varepsilon_0\, (2e/m_i)^{1/2}\, (-U)^{3/2}/d^2 \tag{4.1}$$

with j_s the ion current density, U the applied bias voltage, m_i the mass of the implanted species, ε_0 the free-space permittivity, and e the electronic charge.

Before reaching a steady-state, three time scales govern the response of the sheath to the applied voltage pulse [4]. Once the potential is applied, a sheath forms around the electrode. Thickness of this sheath is a function of the applied potential, the electrode radius, and the plasma density. This sheath expands and reaches the steady-state Langmuir sheath.

Conrad discussed the response of the sheath to an applied voltage, and calculated the sheath thickness and potential profiles of the ion–matrix sheath forms on cylindrical, planar, and spherical electrodes [5]. The sheath response is schematically shown in Fig. 4.3. When a negative pulsed potential is applied on the electrode, electrons near the electrode repel away. This repulsion takes place at a time

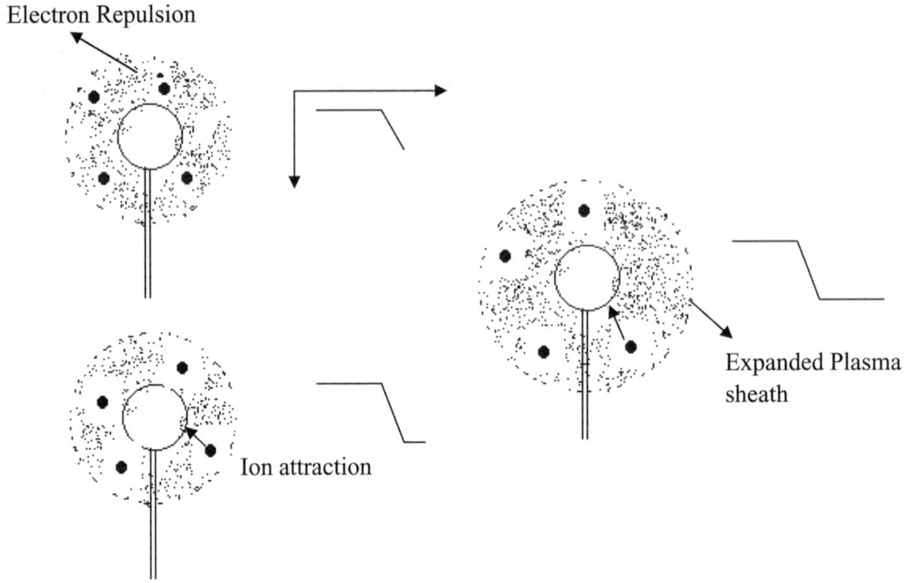

Figure 4.3: Response of the sheath to a voltage pulse applied to the workpiece.

scale related to the inverse of the electron plasma frequency, ω_{pe}^{-1}. During this time, the ions remain stationary around the electrode and give rise to a potential around the electrode. When the time becomes longer than the inverse ion plasma frequency, $\omega_{pi}{}^{-1}$, ions around the electrode start moving toward the electrode. On further increase of the time to values much larger than the inverse of ion frequency, the ions decrease in the sheath and this results in an electron decrease required to maintain the charge neutrality. The ion decrease inside the sheath is compensated by a flow of ions originating from the pre-sheath. That pre-sheath is a region between the ion–matrix sheath and the bulk plasma, where ions move at a Bohm or ion-acoustic velocity. The flow of ions from the pre-sheath gives rise to an increase of the sheath thickness. Ultimately, the sheath thickness reaches conditions corresponding to a steady-state child Langmuir sheath.

When a DC bias is applied, it can give rise to mono or multi-energetic ions depending on the mean free path, λ, of ions inside the chamber. This mean free path is the distance between two adjacent collisions of ions. When the mean free path λ is equal or greater than the plasma sheath thickness, d, the probability for the collision inside the plasma sheath becomes less, and only the existence of collisionless ions can be seen inside the sheath. This results in a bombardment of the workpiece by ions possessing a single energy and gives rise to a Gaussian distribution of implanted species inside the workpiece. When the mean free path is smaller than the plasma sheath thickness, ions accelerated from the plasma sheath edge, undergo a continuous collisions. This results in ions with a broad energy distribution inside the plasma sheath. The bombardment of a workpiece by such ions gives rise to a uniform distribution of implanted species inside the workpiece.

A pulsed negative bias to the electrode can be applied for many reasons [6]. It avoids sustained high-voltage arcing and avoids the contact of an expanding sheath with the chamber wall. A pulsed negative bias results also in a spatial and depth uniformity of implanted species. This can also result in a low-energy bombardment and give rise to a dense layer. Finally a pulsed negative bias allows both deposition and implantation to be carried out simultaneously.

4.2.2. *Metal PIII*

MPIII is a technique in which solid material is used as a source for producing the plasma. Plasma of metal can be produced by a cathodic process or by sputtering. The plasma produced by any of these processes surrounds the workpiece. On applying a negative pulsed bias to the workpiece, ions are accelerated toward the workpiece, and modify the surfaces. In the cathodic arc process a very high current is applied to the solid metal (cathode) producing a plasma of metal from its surface without any liquid phase. The Lawrence Berkeley Laboratory of the University of California has operated with more than 50 different metallic elements and a range of alloys and compounds [7]. In the sputter-assisted MPIII, metal ions are produced from a solid surface by bombarding it with inert gas ions.

4.2.2.1. *Cathodic Arc/Vacuum Arc MPIII*

A vacuum arc metal plasma is obtained by creating an arc discharge in between two electrodes. When the current applied between the two electrodes is more than 1 kA, cathode spots are formed on the surface of the cathode [8]. Such cathode spots are micron sized and move rapidly over the cathode. These spots undergo a transition from solid phase to plasma via liquid and dense, equilibrium non-ideal plasma phases. This pressure of the plasma at the cathode spots is extremely high (10^5 mbar), this high pressure makes this plasma to expand out of the cathode at a very high velocity of about $10^3\,\mathrm{m\,s^{-1}}$.

The ion-acoustic velocity, v_s, can be calculated from:

$$v_s = (2KT/m_i)^{1/2} \tag{4.2}$$

with K the Boltzman constant, T the electron temperature, and m_i the mass of the ion.

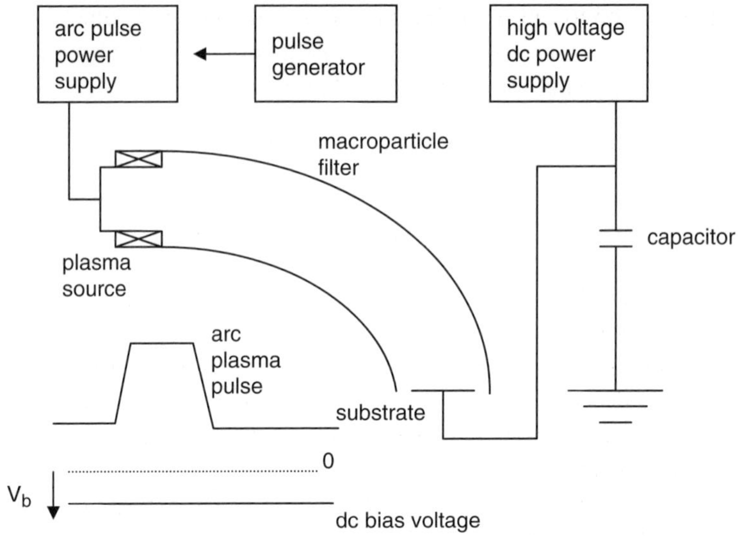

Figure 4.4: Schematics of the cathodic arc-assisted MPIII [9].

The final velocity of the ions (v_i) is of the order of $10^4\,m\,s^{-1}$, and is independent of the mass of the ions. The calculated ion velocity ($10^4\,m\,s^{-1}$) is less than the final ion velocity ($10^3\,m\,s^{-1}$). The difference in the pressure between cathode spot and vacuum chamber, results in an ejection of the plasma at a high velocity from the cathode spot. This is an interesting feature of this cathodic arc or vacuum arc MPIII. The MPIII is shown schematically in Fig. 4.4.

The arc established between cathode and anode, produces a local heating of the cathode surface. Such cathode spots produce liquid droplets of cathode material in addition to the plasma. The size of these droplets ranges from 0.1 to 10 μm. These liquid droplets/macroparticles can however degrade the property of the deposited film/implanted layer. Such macroparticles cannot be eliminated with magnetic particle filters. Macroparticles move straight as a result of their inertia, whereas plasma can be guided with magnetic fields. Ions in the plasma are coupled to electrons by electric field. By applying a magnetic field in a direction perpendicular to the motion of electrons, a force, F, acts on the electrons in a direction perpendicular to the magnetic field, B, and to the motion of the electrons.

The force acting on the electrons is:

$$F = q\,(v \times B) \tag{4.3}$$

with q charge, v velocity, and B magnetic flux density.

This must be equal to the centrifugal force, F_c, acting on the electron:

$$F_c = m_e v^2/r \tag{4.4}$$

with m_e mass of electron, v velocity, and r radius of curvature.

Hence:

$$r_e = m_e v/qB \approx (8kT_e m_e/\pi)^{1/2}/qB \tag{4.5}$$

Plasma can be guided through a magnetic duct by varying the magnetic flux density B. The removal of macroparticles is not efficient by using 45°, 60°, or 90° curved ducts. The use of S-shaped magnetic filter increases the transport efficiency of the plasma and decreases the transport of macroparticle to a large extent. When a metal plasma surrounds the workpiece, metal condenses on the surface of the workpiece and forms a film. The sticking coefficient is unity even without an external substrate bias. This is an interesting feature in MPIII. By varying the bias voltage, the layers deposited become modified. When the workpiece is biased to a constant DC, pure implantation profiles can be obtained without any deposited films. The bias voltage should be selected in such a way that the surface sputtering due to the energetic ion bombardment should produces implanted layers without a surface film. When the substrate is pulsed biased, ions get implanted in the substrate during the high-voltage pulse and deposited when the substrate bias is zero. During the high-voltage pulse, ions implant in the substrate and also produce a recoil implantation of the deposited metal layer. Both deposition and implantation result in an adherent modified layer onto the substrate. Variation of the bias voltage, variation of the pulse length of the arc pulse, and phasing of the bias voltage pulse with respect to the arc pulse, can be tailored to get the desired surface modified layer. Apart from metallic films, a wide range of compound films and ceramics including oxides and nitrides can be deposited by flowing gases in the chamber during implantation and deposition processes [10–24].

4.2.2.2. Sputter-assisted MPIII

The source for metal atoms in this technique is a solid material. Atoms from the solid surface can be removed by sputtering. Sputtering is a physical phenomenon in which atoms are expelled from the solid target by an energetic bombardment with inert atoms. Atoms produced by sputtering are ionized, and this plasma surrounds the workpiece to be coated/implanted. The sputter-assisted MPIII is shown schematically in Fig. 4.5. The target is fixed to the magnetron and the workpiece is connected to the power supply. Combined rotary and turbo-molecular pumps create a vacuum inside the chamber. Inert Ar ions are allowed inside the chamber as a tool to sputter. RF/electron cyclotron resonance (ECR) coils can be used to create

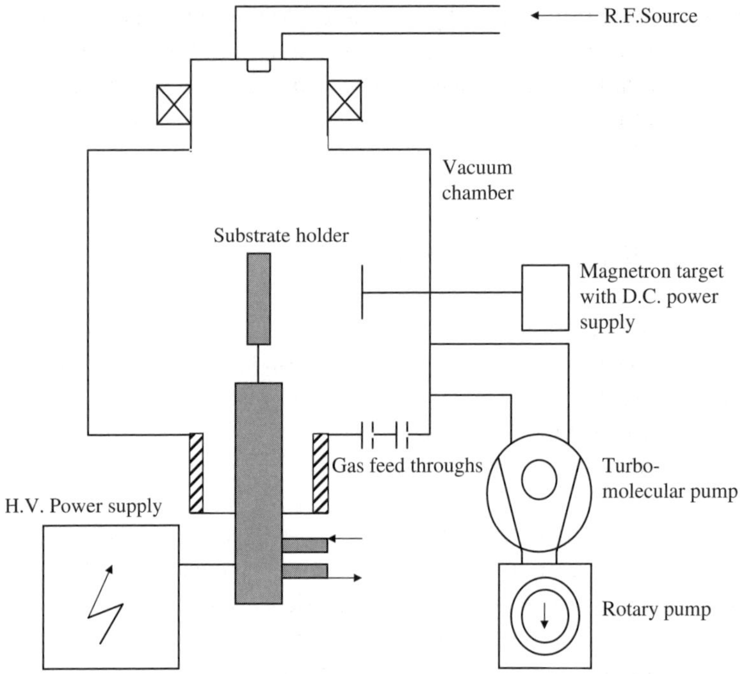

Figure 4.5: Schematics of the sputter-assisted MPIII [25].

or to increase the ionization efficiency in the plasma. DC voltage applied to the magnetron target creates a DC discharge and the magnetic field nearer to the target confines the electrons and increases the density of the sputter argon ions nearer to the target. These argon ions accelerate toward the target as a result of the negative potential applied to the target and bombard it. This energetic bombardment by inert argon ions sputters atoms from the target. These atoms get ionized by the electrons confined in the magnetic field and also by the RF coil. On applying a bias voltage to the substrate, ions originating from the target material accelerate, acquire a very high energy, and modify the surface of the substrate.

4.3. Low Friction PIII-Treated Materials

Tribology is a branch of science and engineering dealing with bodies in relative motion under contact. Broadly, this branch studies friction, wear, and lubrication. Such phenomena come across in various part of our life. The surface modification of materials plays a key role in getting the desired tribologic property. Metal working equipments, aircraft engines, turbines, and automobiles are few among the many products available after surface modification [26]. PIII technique is unique in comparison to other physical vapor deposition (PVD) techniques in treating intricate shapes.

4.3.1. Titanium Diboride Coatings by PIII

Ceramics are of a large interest for tribologic applications. Among these materials, metal borides are very promising as hard coatings on low-cost engineering materials. Titanium diboride (TiB_2), Ti-B-X, and transition metal-based composite coatings are attractive due to their high hardness, high melting point, and their unique functional properties like low friction, low wear, and high corrosion resistance [27–30]. TiB_2 is a transition metal-based refractory ceramic with a hexagonal structure and a metallic chemical bonding character. TiB_2 coatings can be deposited by a variety of techniques like electroplating, laser deposition, pulsed electrode surfacing, sputtering, ion-beam-assisted deposition (IBAD), PIII, and chemical vapor deposition (CVD) [31–36]. The compressive internal stresses in TiB_2 coatings deposited by magnetron sputtering lead frequently to a spontaneous coating failure at a coating thickness above about 4 μm [37]. That lack of adhesion can be by-passed by a PIII processing in which both implantation and deposition steps are carried out simultaneously or subsequently. So, for example, Treglio *et al.* applied a high-voltage bias to the substrate during vacuum arc PSII to reduce compressive internal stresses [38]. They succeeded in depositing TiB_2 at a thickness of more than 10 μm. Ti-B-based coatings deposited by a two steps PMIII assisted by sputtering was reported by Prakash *et al.* [25]. High-speed steel (HSS) substrates made of Vanadis 23 with a Rockwell hardness HRC 60 to 65 were used. The PIII deposition treatments were carried out in two steps consisting namely of a phase 1 where implantation was carried out followed by a phase 2 where a combined deposition and implantation occurred. In phase 1, a voltage of 15 kV was applied to the substrate for 10 min. In this phase a sputter current of 0.1 A was applied on the TiB_2 target in presence of pure argon gas. In phase 2, a voltage in the range 0 to 2 kV was applied to the substrate. A mixture of 80% Ar and 20% H_2 was used on depositing Ti-B. As-deposited Ti-B coatings were found to be amorphous. The coefficient of friction for these coatings sliding against corundum at temperatures between room temperature (RT) and 500°C is shown in Fig. 4.6.

The coefficient of friction recorded with as-deposited PIII Ti-B increases from 0.15 to 0.65 during the running-in phase, and remains then constant on further testing at RT. At 100°C, the coefficient of friction remains constant at 0.8 from

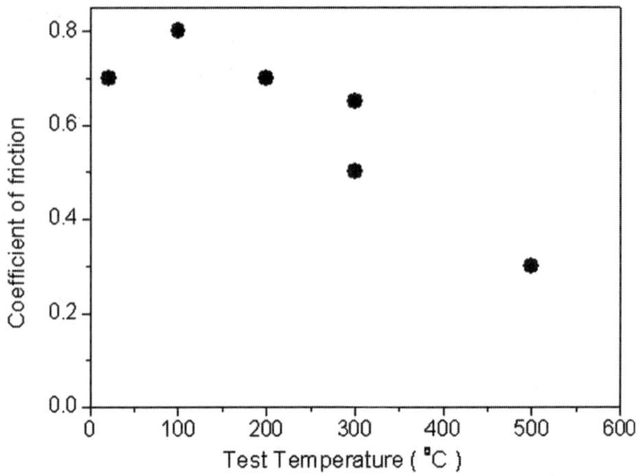

Figure 4.6: Variation of the coefficient of friction with test temperature [39].

almost the beginning of the fretting tests till the end. At 300°C, the coefficient of friction fluctuates between 0.7 and 0.5 throughout the whole fretting tests. At 500°C, the coefficient of friction increases up to 0.65 during the first 2500 cycles and starts to decrease down to 0.3 at cycle 5000. The coefficient of friction remains then at that value of 0.3 on further testing up to 10^4 fretting cycles. Raman spectra which reveals the presence of B_2O_3 in the wear track of the sample tested at 500°C, are shown in Fig. 4.7.

SEM of the fretting wear track on a Ti-B sample tested at 500°C revealed the presence of a smooth boron oxide layer with a very few numbers of delamination compared to the samples tested at lower temperatures (Fig. 4.8). This boron oxide layer can act as a lubricant at and above its melting point of 500°C. This is in good agreement with the work reported on the high-temperature frictional behavior of B_2O_3 by Peterson *et al.* [40].

4.3.2. Ti-B-C Coatings by PIII

Ti-B-C coatings deposited by magnetron-assisted PIII showed a low friction under fretting contact with corundum at RT [25]. In the first phase of the PIII process, the magnetron was switched off and pure methane was used as working gas. In the second phase, a gas mixture of 50% Ar and 50% CH_4 was used. The resulting Ti-B-C coatings were found to consist of hard TiB_2 and lubricating diamond-like carbon (DLC) phases. By mixing the DLC phase with TiB_2 in Ti-B-C coatings, the

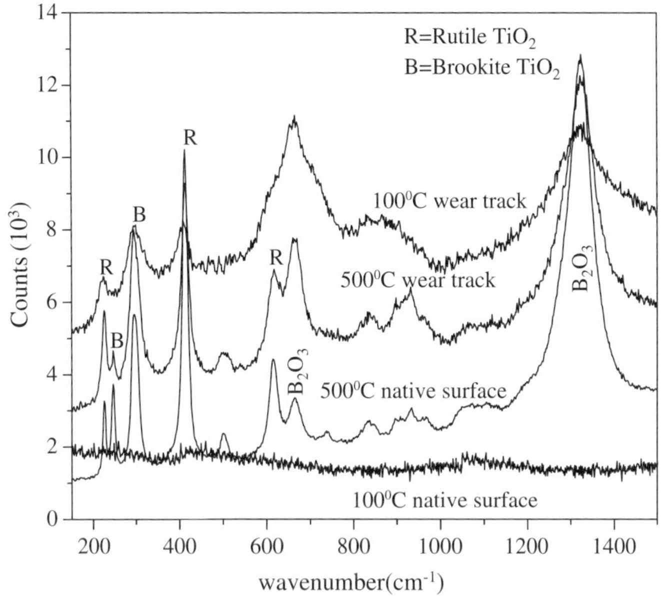

Figure 4.7: Raman spectra of PIII Ti-B coatings after fretting tests performed at different temperatures [39].

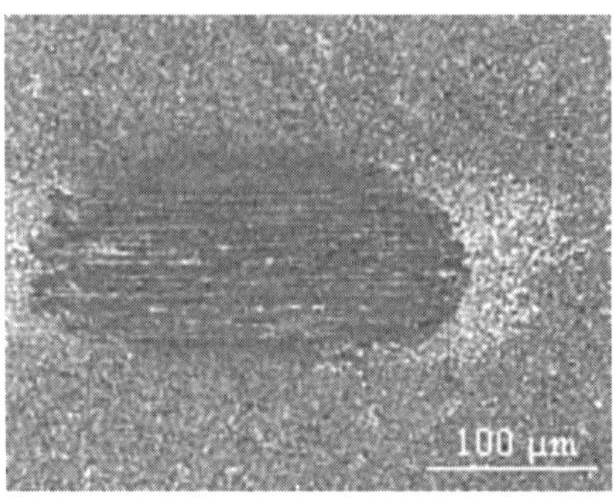

Figure 4.8: SEM micrograph of a wear track on PIII Ti-B after a fretting test performed at 500°C [39].

coefficient of friction on sliding against alumina dropped to 0.25 at RT and remained constant at this value for nearly 15,000 cycles. Variation in the coefficient of friction with fretting cycles on Ti-B-C coatings against corundum counterbody is shown in Fig. 4.9.

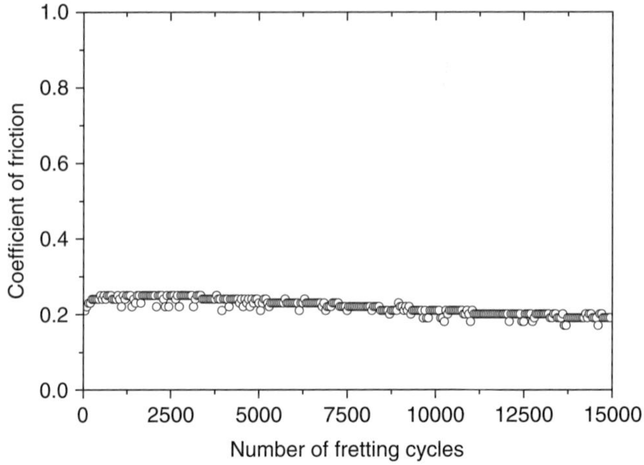

Figure 4.9: Evolution of the coefficient of friction of PIII Ti-B-C coating during fretting [25].

4.4. Wear Resistant Coatings by PIII

Loss of material due to a relative motion of contacting materials is defined as wear. Generally hard coatings are very resistant to wear. The surface modification of Ti6Al4V, Al alloys, polymers, and the deposition of hard coatings like TiN, TiB_2, Ti-B-X, DLC can be done by PIII.

4.4.1. Treatment of Ti6Al4V Alloy

Ti6Al4V is commonly used as a material for manufacturing surgical tools. The wear resistance of Ti6Al4V alloys is improved by fourfold in comparison to untreated alloy after a PIII treatment with nitrogen. A PIII treatment of this alloy at a temperature of 550°C and above, results in an improvement of the wear resistance [41]. This is thought to be associated with the formation of a greater volume of nitride and a deeper hardened case. The total wear volume of untreated, ion-implanted, plasma-nitrided, and PIII Ti6Al4V disks after 1000 revolutions in pin-on-disk wear tests performed at 2 N load, is shown in Fig. 4.10.

This result is also supported by the work carried out by Wang *et al.* They found that the wear of Ti6Al4V decreases with increasing implantation dose [42]. Ueda *et al.* studied the tribologic properties of Ti6Al4V alloy after nitrogen implantation by DC glow discharge plasma [43]. Nitrogen-implanted Ti alloys showed a better

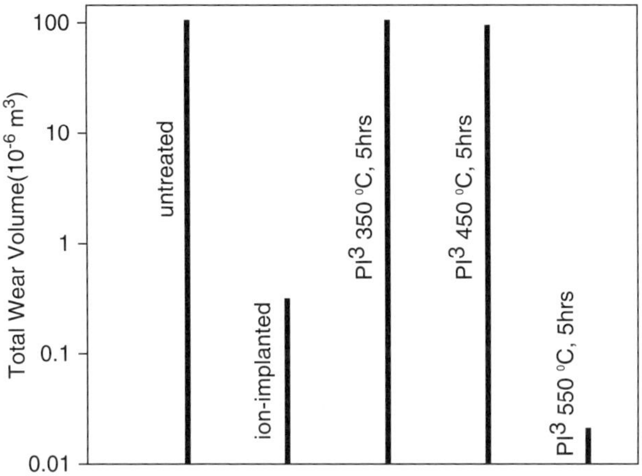

Figure 4.10: Wear resistance of ion-implanted and PIII-treated Ti6Al4V alloy [41].

wear resistance than unimplanted ones. The implantation temperature in this case was below 200°C. The increased wear resistance in this case was attributed to the formation of a thin titanium oxinitride layer. Peak implantation depth is an important concern when tribologic applications are concerned [44]. Yttrium can be implanted before the implantation of nitrogen to increase the peak implantation depth. Wang *et al.* implanted yttrium and nitrogen ions at voltages of 20 and 30 kV, respectively [44]. In the samples pre-implanted with yttrium for 30 min, the peak depth of implantation increased from 50 to 100 nm. Methane PIII can also be used to implant carbon in Ti-6Al-4V at a bias voltage of 30 keV. The wear resistance of such treated Ti alloy was found to be superior to that of the untreated one [45].

4.4.2. Surface Modification of Al Alloys by PIII

The surface modification of aluminum and aluminum alloys can be achieved by PIII to improve friction and wear properties. DLC coatings were deposited on oxidized Al alloy by PIII to improve the wear resistance, frictional behavior, and load carrying capability [46]. The wear rate of Al, alumina, and PIII DLC/alumina are shown in the Fig. 4.11. The wear rate of alumina layer was found to be undisturbed with PIII DLC overlayer [46].

Richter *et al.* compared the wear behavior of plasma nitrided aluminum alloys and PIII-treated Al alloys [47]. A treatment temperature above 400°C is necessary to form an AlN layer on the aluminum alloys. The aluminum alloy were treated by PIII at a bias voltage of 40 kV for 6 h maintaining the treatment temperature at

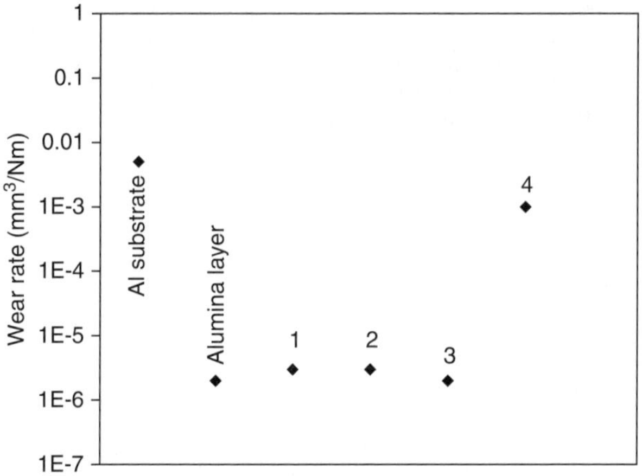

Figure 4.11: Wear rate of Al, alumina, and DLC/alumina (data points 1, 2, 3, 4 denote the wear rate of DLC layer deposited under different set of conditions over alumina) [46].

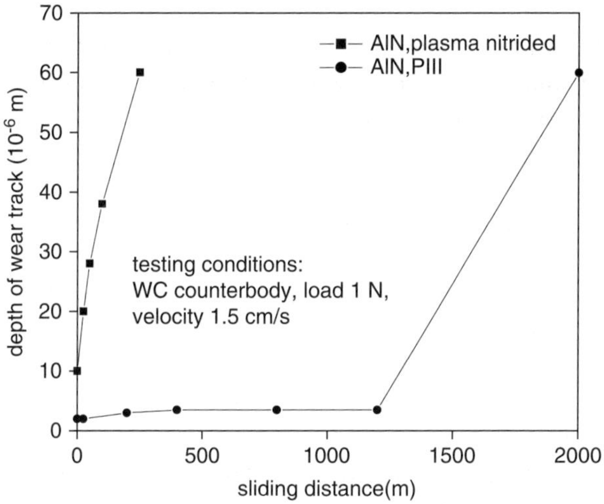

Figure 4.12: Evolution of depth of wear track with sliding distance for plasma nitrided and nitrogen PIII-treated aluminum [47].

500°C. Fig. 4.12 shows the evolution of wear depth in nitrided Al alloy with sliding distance when tested against WC balls. The PIII process resulted in substantially better wear properties in comparison to plasma nitrided Al alloys.

Similar studies were carried out by Zhan *et al.* who found that the wear resistance of Al alloys increases with nitrogen implantation dose [48]. The presence of

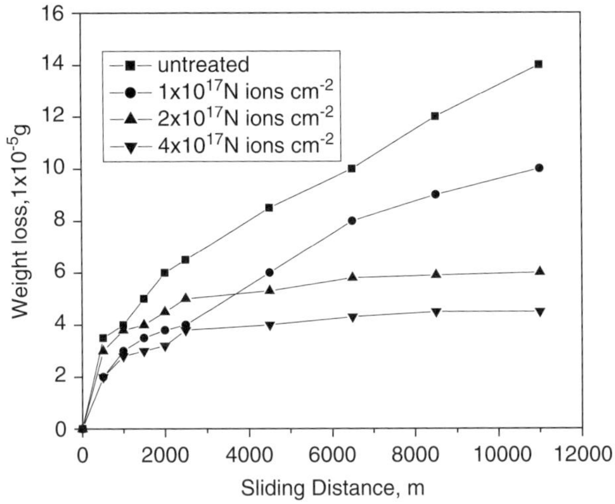

Figure 4.13: Effect of PIII nitrogen implantation dose on the weight loss of polyethylene [52].

fine AlN precipitates and supersaturated solid solution of nitrogen increased the wear resistance by dispersion strengthening of the matrix. Data on the implantation temperature were not reported in this work.

4.4.3. PIII Treatment of Polymers

The surface modification of insulating polymers was reported in the literature [49–51]. Ultra high-molecular-weight polyethylene (UHMWPE) is a material used in total joint replacement prostheses. Due to its low wear resistance, this material encounters pre-mature failure during service. PIII can improve the wear resistance of this polymeric material by nitrogen implantation.

Shi *et al.* improved the wear resistance of UHMWPE by nitrogen ion implantation [52]. Evolution of the weight loss with sliding distance for untreated and PIII-treated UHMWPE is shown in Fig. 4.13.

PIII-treated samples have a better wear resistance than untreated ones. At increasing an ion fluence, the wear resistance increases. The bombardment of ions could cause a displacement of the target atoms or disturbance in the electron cloud of the target atom. The latter one induces cross-linking between lamellar crystallites. The replacement of the weak bonds between the lamellas by covalent bonds in the ion bombarded/PIII-treated samples could be a possible explanation reason for the increase in surface hardness and wear resistance in comparison to untreated samples. The increase in wear resistance with increasing ion fluence could be the result of an increased degree of cross-linking.

4.4.4. Deposition of Hard Coatings

The demand of high-precision machining requires cutting tools with a good wear resistance and very sharp cutting edges. Deposition of hard coatings by PVD and CVD techniques result in lack of good adhesion between coating and substrate, and blunting of the sharp edges of the cutting tools. TiN coatings were deposited by cathodic arc-assisted PIII on different substrates at different substrate bias and pulse duration [53]. Bias voltage applied varied from 0 to 2.5 kV. The structure and properties of the TiN film did not vary with the kind of substrate material. The texture of the deposited film was found to become stronger with increasing bias voltage. TiN coatings with a thickness of 0.8 μm deposited over cemented carbide gun drills were tested for their cutting performance. The experimental results revealed that TiN coatings deposited by this cathodic arc-assisted PIII improved the lifetime of gun drills by a factor of 2.5.

Ti-B and Ti-B-C coatings deposited by sputter-assisted PIII showed that the wear rate is independent of the substrate bias voltage applied in the range from 0 to 2 kV [25]. The wear rate can be expressed as a wear volume per unit dissipated energy dissipated in overcoming friction during fretting tests [54]. That wear rate is shown in Fig. 4.14.

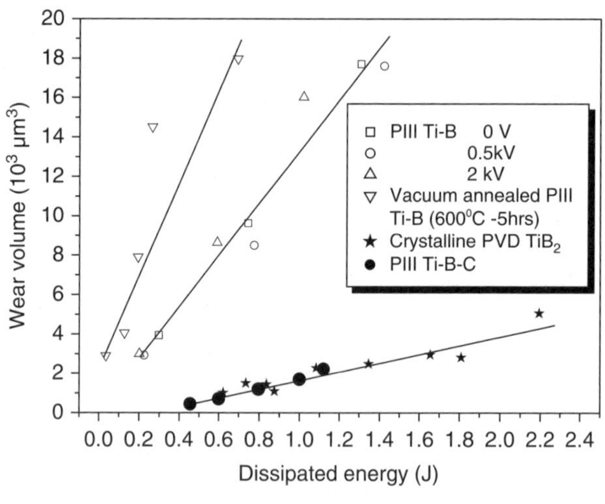

Figure 4.14: Wear rate of PIII Ti-B and Ti-B-C coatings [25].

In fretting tests against a corundum counterbody, Ti-B coatings showed a better wear resistance than TiN coatings. In the case of Ti-B-C coating, the presence of hard TiB_2 in the lubricating DLC phase increased the resistance to wear and decreased the coefficient of friction in comparison to magnetron sputtered TiN, TiB_2, and PIII Ti-B.

4.5. Nitriding of Steels by PIII

Steels have a relatively low hardness and a poor wear resistance in dry sliding. Nitriding of steel improves the hardness, wear resistance, and the load bearing capacity [55].

When steel is exposed to nitrogen at a temperature above 400°C, chromium in the steel forms chromium nitrides. The decrease in the amount of chromium in the solid solution decreases the ability of the steel to form a passive layer and become prone to corrosion [56]. So, treatment temperature is an important concern in nitriding [57]. Ion beam nitriding, gas nitriding, plasma nitriding, conventional beam line, and PIII nitriding are various nitriding techniques available [58]. When compared with other nitriding processes, beam line and PIII techniques offer a superior nitrided layers with higher amounts of supersaturated nitrogen and deeper penetration depth [58].

The reason could be linked to the removal of the oxide barrier from the surface and the implantation of nitrogen at elevated temperature promoting the faster diffusion of nitrogen into the steels (see Fig. 4.15).

In stainless steels treated at 400°C, the presence of f.c.c. N solid solution of austenite was observed [59]. At temperatures above 450°C, the presence of CrN was observed and this could be caused by an enhanced diffusion of substitutional Cr. Blawert *et al.* studied the behavior of austenitic stainless steel (AISI 321) and austenitic–ferritic stainless steel (AISI 318) after nitrogen implantation [59]. The X-ray diffraction (XRD) pattern of the steels after treatment at 400°C and 500°C are shown in the Fig. 4.16. At 400°C, austenite in austenitic steels transformed into expanded austenite, whereas in austenitic–ferritic steels all the ferrite transformed into expanded austenite. Phases detected in the austenitic steel after treatment at 500°C look similar to the one treated at 400°C. Presence of CrN was not detected at this temperature by XRD. In the case of austenitic–ferritic steel, ferrite, and CrN were detected at the treatment temperature of 500°C. The presence of austenite stabilizing elements in this steel transformed the metastable austenite into ferrite and CrN. The presence of CrN precipitates along with ferrite resulted in a better wear resistance than at 400°C but with a reduced corrosion resistance. Studies carried out by Samandi *et al.* on these similar materials showed similar results [60,61].

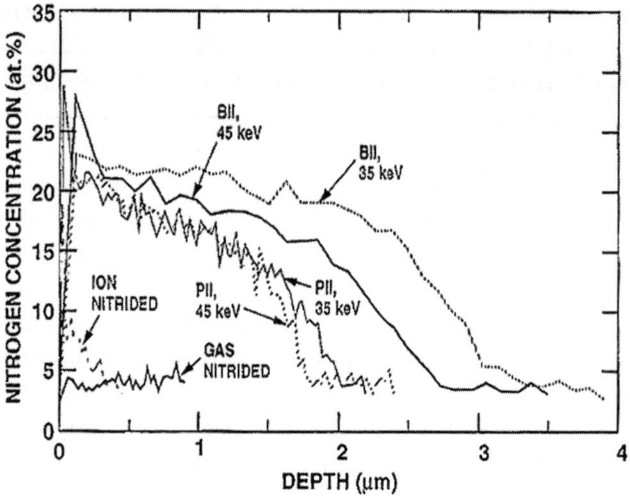

Figure 4.15: Atomic concentration of nitrogen with depth in nitrided steel samples treated by different processes [58].

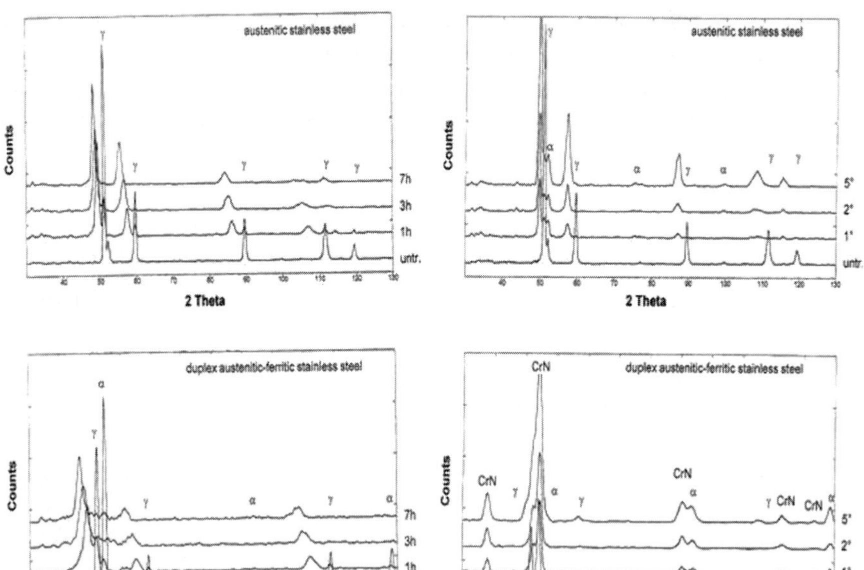

Figure 4.16: XRD pattern of austenitic and austenitic–ferritic stainless steel nitrided at different temperatures and exposure time [60].

Lattice parameters measured by them at different temperature showed the increase in the lattice parameter of the austenite till 450°C (Fig. 4.17).

The presence of iron nitride was detected in the surface layer after a treatment at 520°C, whereas in plasma nitriding the presence of iron nitride layers was observed at all treatment temperatures till 500°C. This limits the diffusion of nitrogen and the concentration of nitrogen in the modified layer. Collins *et al.* carried out TEM and XRD analyses on austenitic stainless steel after implanting with nitrogen by PIII [56]. They implanted nitrogen between 350°C and 520°C at treatment times between 0.5 and 5 h. The implanted nitrogen dose varied with treatment temperature and hence the treatment time was selected to have the same dose in all the samples irrespective of temperature. XRD analyses confirmed results observed by other researchers [60]. Investigation with TEM showed the presence of two zones in the modified layer. The presence of diffuse rings and few dim spots in the microbeam electron diffraction (MED) pattern confirms the presence of an amorphous structure in the outermost layer of the samples treated at 450°C. The observed dim spots and diffuse rings correspond to the d-spacing of CrN and ferrite. The MED pattern obtained in a second layer confirms the presence of austenite, and the thickness of this layer was nearly 2 μm. In the sample treated at 520°C, the presence of a thin (0.4–0.5 μm) nanocrystalline layer was noticed in the outermost layer. MED pattern confirmed the presence of CrN, ferrite, iron nitride, and hexogonal phases. The underlying layer consisted of lamellar precipitates of CrN and ferrite. The thickness of this layer is 7.5–8.5 μm. Atomic force microscopy (AFM) and magnetic force microscopy (MFM) studies by Fewell

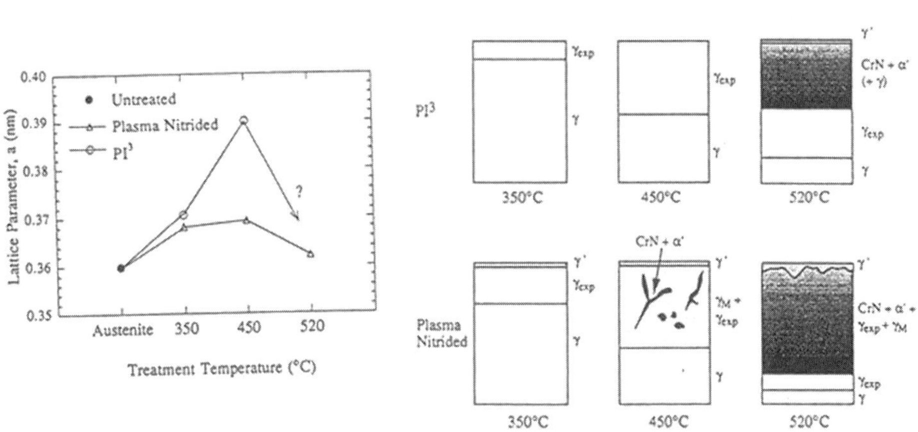

Figure 4.17: The variation in the lattice parameter of austenite with treatment temperature and the distribution of phases in the modified layer [61].

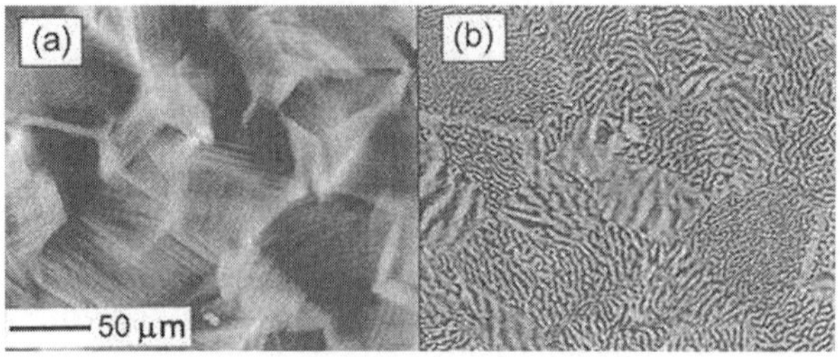

Figure 4.18: Surface topography of the PIII nitrided austenitic stainless steel samples with (a) AFM and (b) MFM [62].

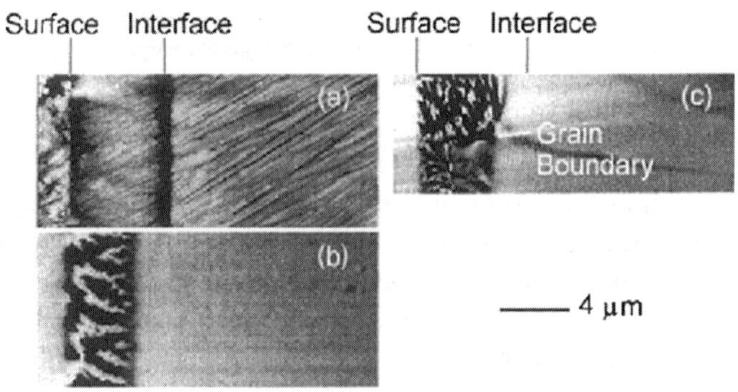

Figure 4.19: Cross-sectional view of the PIII nitrided sample with (a) AFM and (b) and (c) with MFM [62].

et al. explored the topographic and magnetic nature of the nitrogen enriched layer in PIII-treated austenitic steel [62]. Fig. 4.18 shows the AFM and MFM images of the surface of a nitrided sample. Topography observed with AFM reveals the grains and slipbands within the individual grains. MFM of the same region is shown in Fig. 4.18(b).

The light and dark bands in the image are characteristic for the presence of a gradient in the magnetic force. The density and form of magnetic domain variation are related to the orientation of different grains. If the presence of CrN is considerable in the top surface layer, then the MFM image only shows a small island of magnetic domains. The cross-sectional image of nitrogen PIII austenitic steel is shown in Fig. 4.19 [62]. The AFM image shows the nitrogen-implanted layer, the interface, and the steel substrate. Magnetic domains are visible in cross-sectional MFM images. The magnetic domains extend from the outer surface till 80% of the total

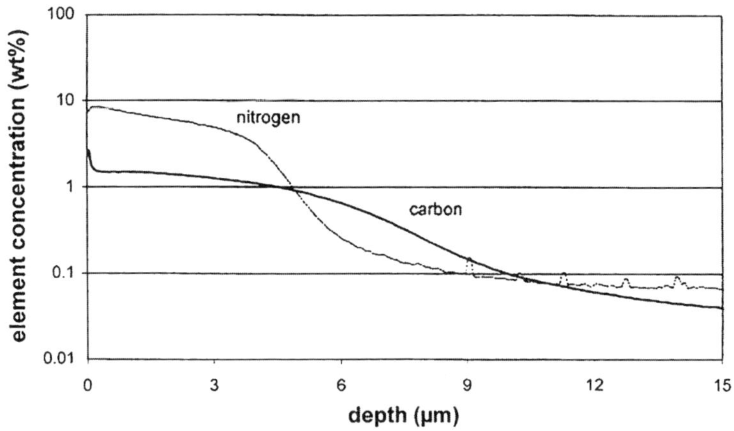

Figure 4.20: Elemental depth profile of carbon and nitrogen implanted in austenitic stainless steel [63].

thickness of the implanted layer. This confirms the ferromagnetic nature of the top layer, and the paramagnetism of the underlying layer.

The dependence of the Curie temperature on the nitrogen concentration or change in the dislocation density at the interface could be a possible reason for the change in the magnetic properties with depth of the nitrided layer. Similar studies were carried out by Blawert *et al.* on nitrogen and carbon expanded austenite [63]. They found ferromagnetism only in the outer layer with the thickness half of the total thickness of the nitrogen-implanted layer. Expanded austenite can be produced either by nitrogen or carbon implantation by PIII. Elemental concentration observed in the nitrogen-implanted austenitic steel is more than the carbon implanted (see Fig. 4.20).

This could be because of a higher interaction of nitrogen with substitutional chromium. The more the interaction, implanted element occupy the position around the substitutional Cr in the interstitials, hence more the elemental concentration and lesser the diffusion. Higher amounts of nitrogen inside the steel resulted in a relatively higher lattice expansion and defect density in comparison to carbon implanted. The lattice expansion and defect density contributed to a higher hardness and a better wear resistance of nitrogen-implanted steel in comparison to carbon-implanted one. Hardness and wear test results are shown in Fig. 4.21. However, the corrosion resistance of nitrogen-implanted austenitic stainless steel decreases. The increase in defect density and lattice expansion, dissolved the Cr from the solid solution and formed precipitates of Cr, which ultimately reduced the capacity to form the passive layer and hence decreased the corrosion resistance.

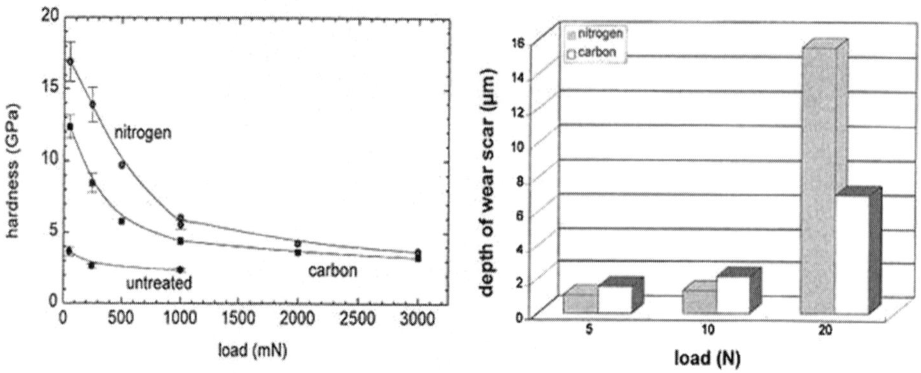

Figure 4.21: Hardness and wear test results of carbon and nitrogen expanded austenite [63].

4.6. Formation of Intermixed Layer

The formation of intermixed layer is an important tool to improve the adhesion between coating and substrate. Intermixed layers has also been successfully applied in multi quantum well (MQW) materials to get tunable band gap in selected areas of the same substrate [64].

Gunzel *et al.* studied TiN layers deposited over HSS by cathodic arc-assisted PIII [65]. Cross-sectional analysis by TEM is shown in Fig. 4.22. The right corner is the substrate material. Above the substrate, two layers are visible. The top layer is amorphous TiN with a thickness of 40 nm. The layer below this is the intermixed layer with a thickness of approximately 10 nm. An improvement in the adhesion of the coating with the substrate was achieved by this intermixed layer.

Similar observation was done by Prakash *et al.* in the Ti-B-C coating deposited over HSS by sputter-assisted PIII [25]. The deposition was carried out in two phases. In phase 1, the magnetron was switched off and pure methane was used as working gas. The substrate was given a negative potential of 15 kV. Treatment was carried out for 10 min. Temperature raised to 500°C at the end of phase 1. In phase 2, a gas mixture of 50% Ar and 50% CH_4 was used. Magnetron TiB_2 target was given a voltage of 620 V whereas the substrate was biased at a negative potential of 2 kV. Cross-sectional view of the Ti-B-C coating is shown in Fig. 4.23.

The left-hand side is the HSS substrate. An interlayer enriched with carbon is visible in between the Ti-B-C coating and the substrate. The presence of an intermixed layer in this Ti-B-C coating improved the adhesion and wear behavior at high temperature.

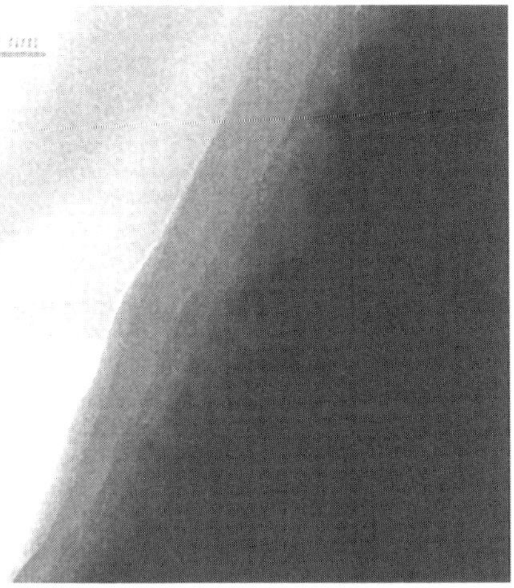

Figure 4.22: Cross-sectional TEM of PIII-treated TiN over steel substrate [65].

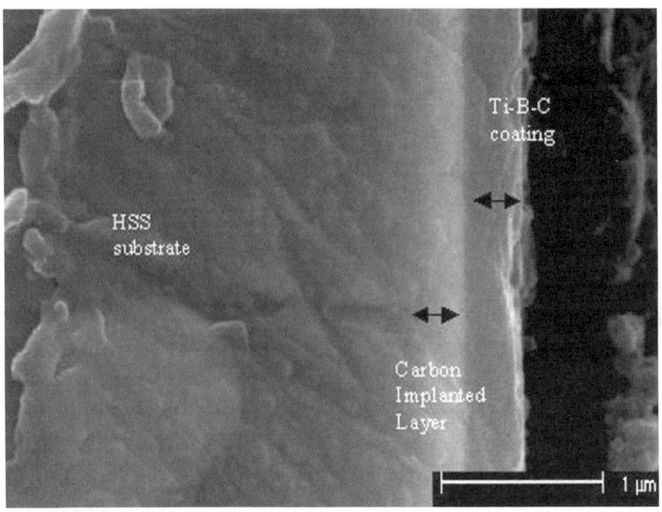

Figure 4.23: Cross-sectional view of PIII Ti-B-C coating deposited over steel substrate [25].

Disordering or shift in the band gap of MQW materials is of research interest in the field of photoelectronics. Disordering can be produced by photoabsorption, ion implantation, etc. [66,67]. PIII can also induce QW disordering. MQW intermixing using the PIII has the advantage of treating the large and intricate shaped

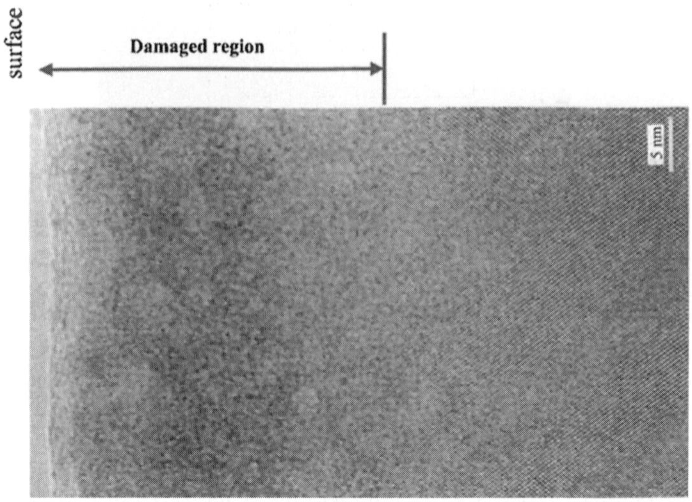

Figure 4.24: Cross-sectional view of the Ar-implanted InGaAsP [64].

samples. Ho *et al.* studied the band gap shift in MQW materials after implanting with Ar ions and thermal annealing [64]. Bombardment of MQW materials with Ar ions creates point defects in the material. Cross-sectional TEM of the InGaAsP after Ar ion implantation is shown in Fig. 4.24. PIII was performed at an implantation energy of 20 kV with Ar ion dose ranging from 10^{13} to $10^{16}\,cm^{-2}$. Annealing was carried out for 30–90 min at 650°C. Few samples were also implanted with different energies keeping the dose as constant. The region with lot of point defects on the surface diffuses toward the MQW region, whereas the host atoms from the MQW region move toward the surface. This results in intermixing. The precise control on the intermixing can be done by varying the ion dose and implantation energy. Similar intermixing results were also reported by Paquette *et al.* [68].

4.7. PIII in Microelectronics

PIII is an attractive surface modification tool in the microelectronics industries. The primary feature which makes this attractive is the implantation dose can be achieved in a time independent of the wafer area. This lowers the production cost as compared to the conventional raster beam ion implantation. PIII is used in fabricating silicon on insulators (SOI), doping the trench walls, fabrication of ultra shallow junctions.

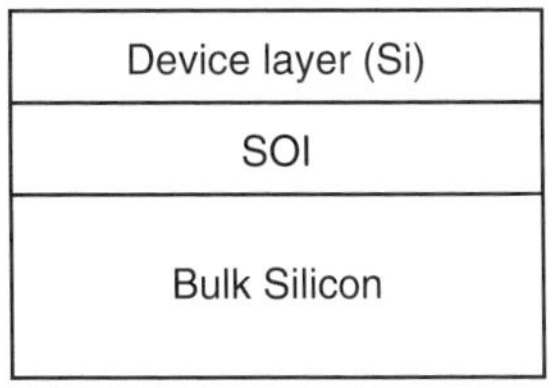

Figure 4.25: Schematic of the SOI.

4.7.1. SOI Fabrication

SOI technologies offer many inherent advantages in microelectronics, especially in complementary metal oxide semiconductors (CMOS) integrated circuits [69]. The major advantage of building the device over the SOI wafers are:

1. The operating speed increases by 20–30% in comparison to the devices on bulk silicon. The devices fabricated over the bulk silicon operate at relatively low switching speed because of the presence of the large volume of semiconductor material underlying the devices and hence more charge is needed to turn on and off.
2. Higher device packing density can be achieved.
3. Minimizes the current leakage.
4. Lower the power consumption.

The advantages mentioned above may lead the SOI technology to fabricate devices to operate in gigahertz frequency and with battery or solar cell power. The schematic of the SOI is shown in Fig. 4.25. The top layer is the device layer and the layer below this is an insulating SiO_2. There are few methods available now to fabricate SOI [69,70]. The broad classification of the methods are implantation and bonding. Separation by ion implantation of oxygen (SIMOX) is a technique coming under the category of implantation. Fabricating SOI with SIMOX is expensive because of the expensive implanters and long time required to implant the high dose of oxygen required to form the buried oxide. However all the methods for fabricating SOI are very expensive. Separation by plasma implantation of oxygen (SPIMOX) is a promising tool for the fabrication of SOI. Implantation time is independent of wafer diameter and hence it can reduce the cost of SOI wafers (see Fig. 4.26) [71].

4.7.2. Separation by Plasma Implantation of Oxygen

SPIMOX is an implantation process which uses PIII technique to fabricate SOI wafers. With the PIII technique oxygen ions are implanted in the silicon wafers. During the implantation the temperature of the silicon wafer is maintained at a

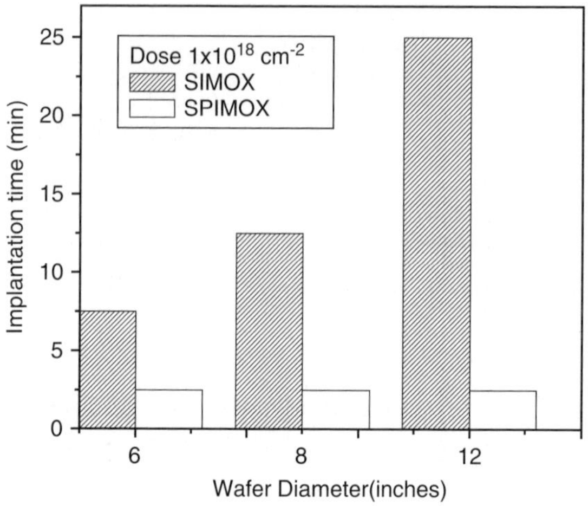

Figure 4.26: Dependence of implantation time with wafer diameter in SIMOX and SPIMOX [71].

temperature of 600°C to maintain the crystallinity and to avoid the amorphization [72]. The implantation of right dose of the monoenergetic oxygen ions results in the formation of SiO_2 precipitates at the implantation depth. After implantation, the wafer should be capped with the oxide and nitride layers to protect the top silicon device layer from pitting during annealing in the next step. The implanted wafers are then annealed at 1300°C for 6 h to form a continuous buried oxide and SOI structure. It is very important to know the implantation parameters, implantation constraints, and the operating regions in the SPIMOX process to form the desired dimension of the SOI structure [71].

An arbitrary value cannot be given to the pressure and implantation time because of the constraints in the SPIMOX process. There are few operating conditions to be satisfied for getting the desired SOI and for the implantation process to happen. Hence, this put forward few constraints to know the region of operation in the pressure and time scale. There are four implantation constraints to be considered to get the desired dimension of SOI and the implantation process to happen. In every constraint a relation connecting the pressure and implantation time is derived and finally the region of operation is defined in the plotted graph. The constraints taken into account allowed us to know the region of operation in the pressure and time scale. The allowed region of operation is shown as the shaded region (see Fig. 4.27).

The silicon wafers implanted with oxygen at a temperature of 600°C were annealed at 1200°C to form SOI [72]. XTEM micrograph of the buried oxide layer formed after annealing is shown in Fig. 4.28. The wafer implanted with

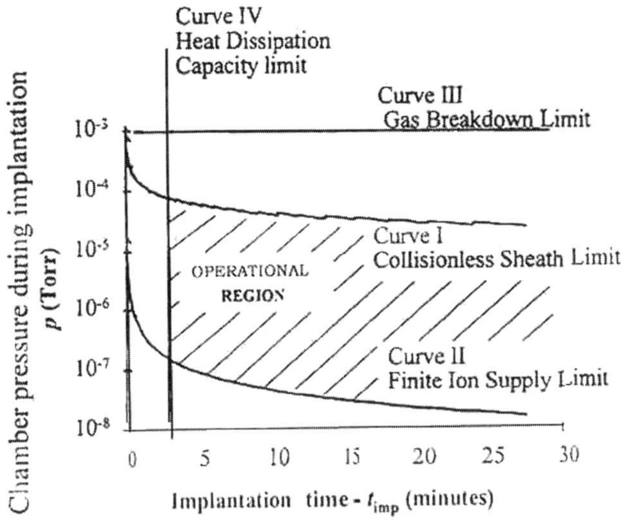

Figure 4.27: Operational phase space for a PIII unit to form SOI [71].

$1 \times 10^{17} cm^{-2}$ of oxygen showed precipitates of SiO_2 after annealing. With increase in the dose to $3 \times 10^{17} cm^{-2}$ of oxygen, a continuous buried oxide layer formed. If the plasma contains both the O^+ and the O_2^+ ions, then the implantation of these ions in the silicon wafer give rise to two continuous oxide layers. Such dual oxide structures may find application in the three dimensional devices.

4.7.3. Trench Wall Doping

The size of dynamic random access memory (DRAM) devices are scaled down, to increase the density and speed of DRAM chips [73]. So increasing the area of trench capacitors became an important aspect in the ultra large-scale integration (ULSI) processing. A deep trench capacitor used as charge storage element in DRAM consist of a thin node film and capacitor electrode, and the n-type region in the p-type Si substrate surrounding the trenches is the another capacitor electrode. Fig. 4.29 shows the SEM micrograph of an array of trenches of 6 μm deep and 0.175 μm wide in the DRAM cell [74].

These trenches were implanted by the PIII process using AsH_3 plasma with a density of $10^{10} cm^{-3}$. The DRAM cells were biased at 7 kV. The conformal doping of the trenches is possible with the PIII technique. With conventional implantation, doping of the sidewalls can be done by multiple implantation with various tilt and

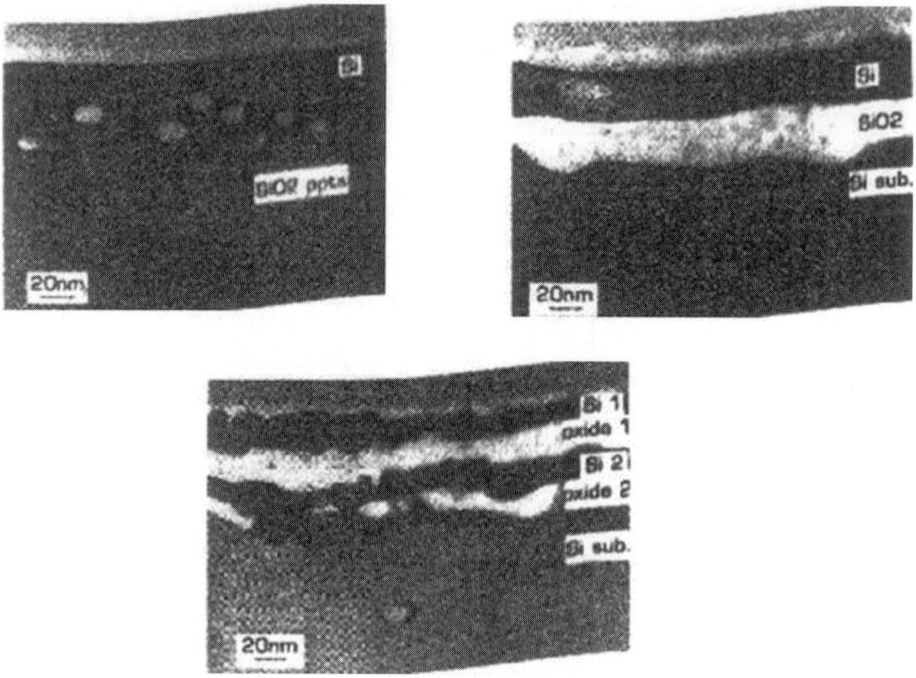

Figure 4.28: XTEM micrograph formed after annealing the implanted silicon with (a) 1×10^{17} O_2^+ (b) 3×10^{17} O_2^+ (c) 1×10^{17} O^+ and O_2^+ [72].

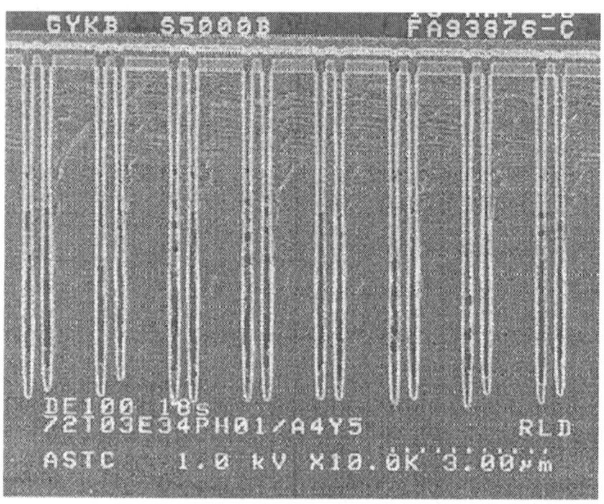

Figure 4.29: SEM micrograph of deep trench array [74].

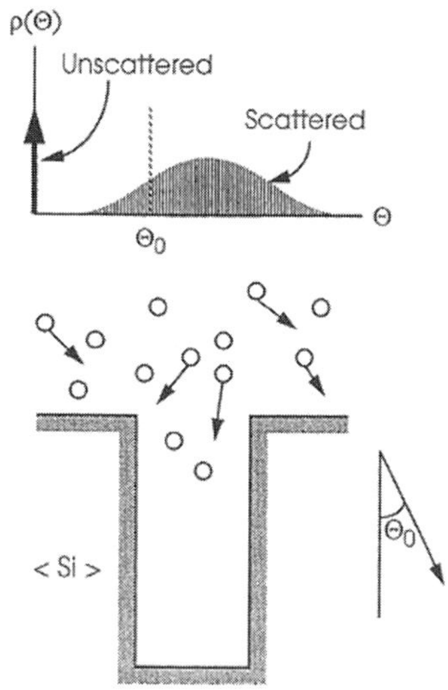

Figure 4.30: Angular distribution of ions entering the trench [75].

rotation. Even then the conformal doping of the trenches with high aspect ratio is not possible with this implantation technique. In the PIII, ions in the plasma sheath move in different directions toward the trench. This angular distribution of the ions is dependent on the direction in which the ions are entering the sheath and the collision of the ions with the neutrals in the plasma sheath [75]. The angular distribution of the ions is shown in Fig. 4.30. The angular divergence of the ions represents the scattered ions. The aspect ratio of the trenches should be in a way to accept the ions with broad angular divergence and hence the uniform doping around the walls in the trench. A trench with the width of 0.45 μm and a depth of 2.8 μm in the silicon wafer was implanted with boron by Mizuno *et al.* [76]. This has the aspect ratio of 6. Higher aspect ratio trenches have been doped with elements by this PIII technique [74]. The trench showed in the previous Fig. 4.29 is with an aspect ratio of 35. The conformal doping of the trench depends also on the aspect ratio of the trench and the implantation energy.

Sano *et al.* carried out deposition and implantation by the PIII in the trenches with the width and depth of 16 mm [77]. For the trench placed parallel to the direction of the ion emission, the thickness of the deposited and implanted layer at different walls is shown in Fig. 4.31. The ratio of the thickness of the deposited layer between different walls is nearly 10, whereas the ratio of the thickness of the implanted layer is only 2.5. The ratio of the thickness of the implanted layer is found to be lesser than the ratio of deposited layer both in the parallel and perpendicular trench.

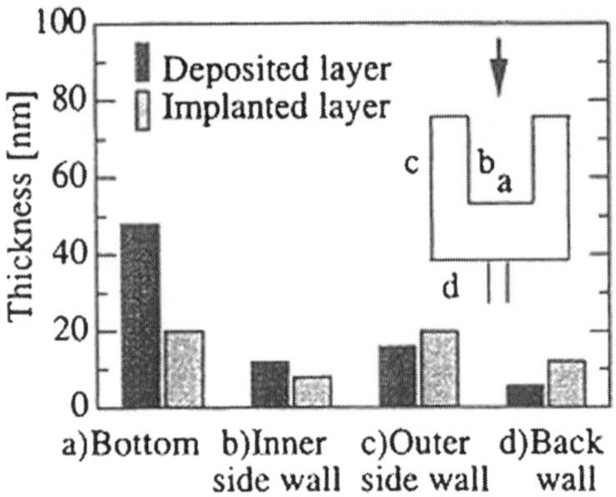

Figure 4.31: Thickness of the deposited and implanted layer on different walls in the trench [77].

4.7.4. Formation of Ultra Shallow Junctions

In CMOS technology the fabrication of ultra shallow junction is required. The extension of the source and drain in the lateral and vertical directions should be shallow. Shallow junctions are formed by dopant implantation and followed by an annealing. The 150 nm junctions are relatively easy to process [78]. Reducing the junction depth to 60 nm is possible with rapid thermal annealing. Junction of 10–50 nm can be formed using P and As and junctions of 60 nm have been shown with B dopant [79]. Direct BF_3 doping result in junction depth greater than 100 nm, because of rapid diffusion of boron during the annealing step.

Qian *et al.* formed sub-100 nm p+/n junctions with BF_3 doping and by pre-amorphization with SiF_4 doping [80]. Preamorphization retards the diffusion of boron during the dopant activation annealing. Fig. 4.32 shows the variation in the B dopant concentration with depth. The concentration of boron decreased from the surface till the depth of 80 nm and then maintained at a constant value. This sample was preamorphized with SiF_4 at a DC bias voltage of 4 kV for 10 s followed by BF_3 doping at 2 kV. The dopant activation annealing was carried out at 1060°C for 1 s. Two-step rapid annealing can reduce the junction depth compared to the single step annealing [81]. The junction depth of PIII diodes formed by different groups with a background concentration of $10^{18}\,cm^{-3}$ is shown in Fig. 4.33 [75]. The junction depth increased with increase in the bias voltage. The deviation in the junction depth value could be mostly by the difference in the annealing cycles. So, it is recommended to work with lower bias voltage to form ultra shallow junctions.

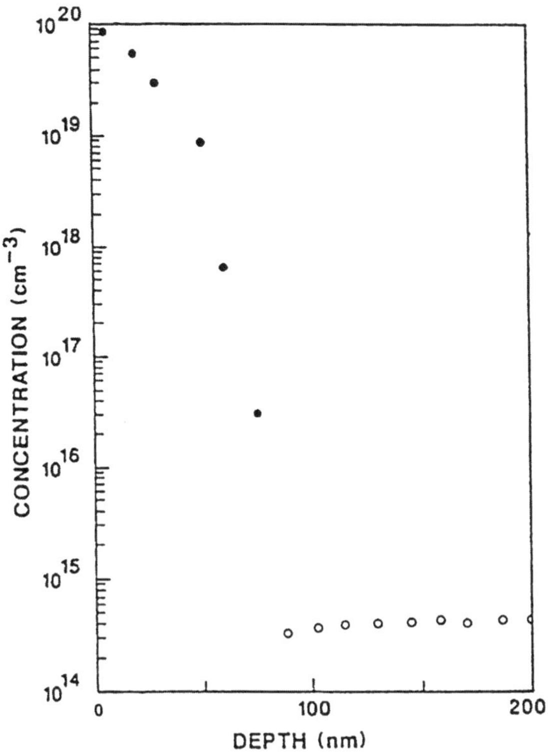

Figure 4.32: Variation in the boron dopant concentration with depth [80].

4.8. PIII in Bio-medical Applications

Bio-integration is the ideal outcome expected of an artificial implant. The phenomenon that occurs at the interface between the implant and the host tissue/blood do not induce any deleterious effects such as formation of unusual tissues or blood coagulation. Hence, it is very important to design biomaterials with best surface properties. Recently DLC films and Ti-O films has been proposed for use in blood-contacting devices such as rotary blood pumps, cardiovascular stents, and artificial heart valves. PIII-treated DLC, Ti, and Ti-O were studied for their bio-compatibility by several researchers [82–89].

4.8.1. Deposition of DLC by PIII

Yang *et al.* studied the effect of bias voltage and annealing temperature on the blood compatibility of DLC films [82]. Hydrogenated amorphous carbon films were

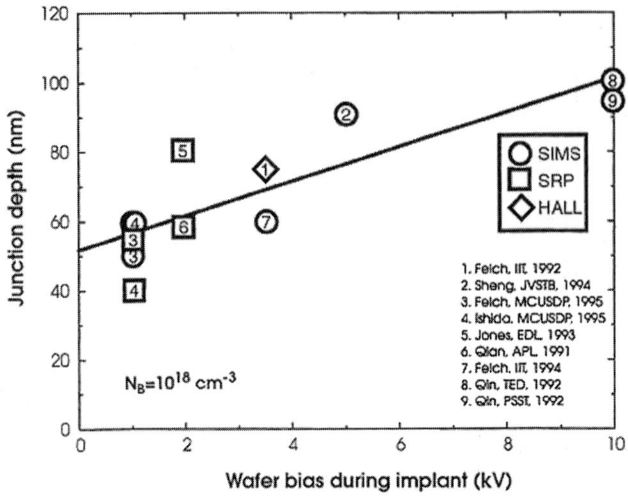

Figure 4.33: Shallow junction depth formed by different groups with different bias voltage to the wafer [75].

fabricated over silicon substrate by PIII deposition. Substrate bias voltage varied in the range −75 to −900 V. Acetylene and argon was used as working gas. To examine the interaction of blood with the material, platelet adhesion experiments were carried out by incubating the a-C:H films in the platelet-rich plasma (PRP) for 15 min at 37°C. The total number of platelets and the percentage of unactivated platelets on the film were counted after the incubation period (Fig. 4.34). DLC film showed good blood compatibility as like low-temperature isotropic (LTI) carbon and better than stainless steel. Number of platelet adhering on the film increased with increase in the bias voltage, whereas the percentage of unactivated platelets decreased. Poor blood compatibility at higher bias voltage resulted as a result of graphitization.

Similar results was noticed by the same researchers with the a-C:H films annealed at different temperatures [83].

Samples annealed above 400°C showed graphitization and poor blood compatibility. The first step in the process of blood clotting is the adsorption of protein in the blood. If the adsorbed protein is denatured, blood platelets adhere on the surface and result in the clotting of blood. Denaturation of the adsorbed protein depends on the transfer of charges to the material and this is related to the electronic property of the material [84,85]. When a material possess wider band gap value than the adsorbed protein, protein denaturation will be inhibited. Graphitization result in the band gap value less than the adsorbed protein and hence the protein is easily denatured and hence the poor blood compatibility. The blood platelet adhesion and

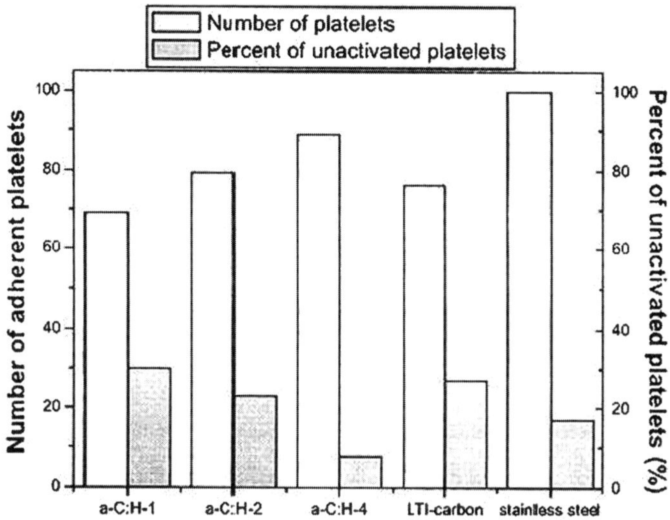

Figure 4.34: Number of platelets adhered on amorphous carbon films deposited at different bias voltage (a-C:H-1, a-C:H-2, a-C:H-3 are in the order of increasing bias voltage) [82].

activation on the surface of implant can be accessed by its shape and size. Fig. 4.35 shows the SEM micrograph of the blood platelets on the surface of a-C:H film. Film deposited at low bias voltage showed the disk shaped platelet with 2–3 μm size. Slight pseudopodium was observed. Morphology exhibiting heavily developed pseudopodia, platelets with size larger than 5 μm and more depressed shape indicates the activated blood platelets. Such appearance was observed with carbon films deposited at higher bias voltage and in stainless steel.

4.8.2. Deposition of Ti-O Film by PIII

Yang *et al.* studied the blood compatibility of Ti-O thin films fabricated by PIII and showed it is superior than the LTI carbon [88]. Ti-O films were deposited over Ti substrates by PIII. The implanted disks were inserted in dog's body for 30 days and then removed to study its blood compatibility. Fig. 4.36 shows the amount of fibrinogen adsorbed on the surface of Ti-O and LTI carbon. With increase in the incubation time, the amount of fibrinogen adsorbed also increased in LTI carbon. In the case of Ti-O films, the amount of fibrinogen adsorbed is less and maintained at a steady level with increase in the incubation time. SEM micrograph of the materials after 30 days of implantation is shown in Fig. 4.37. Only a few blood platelets

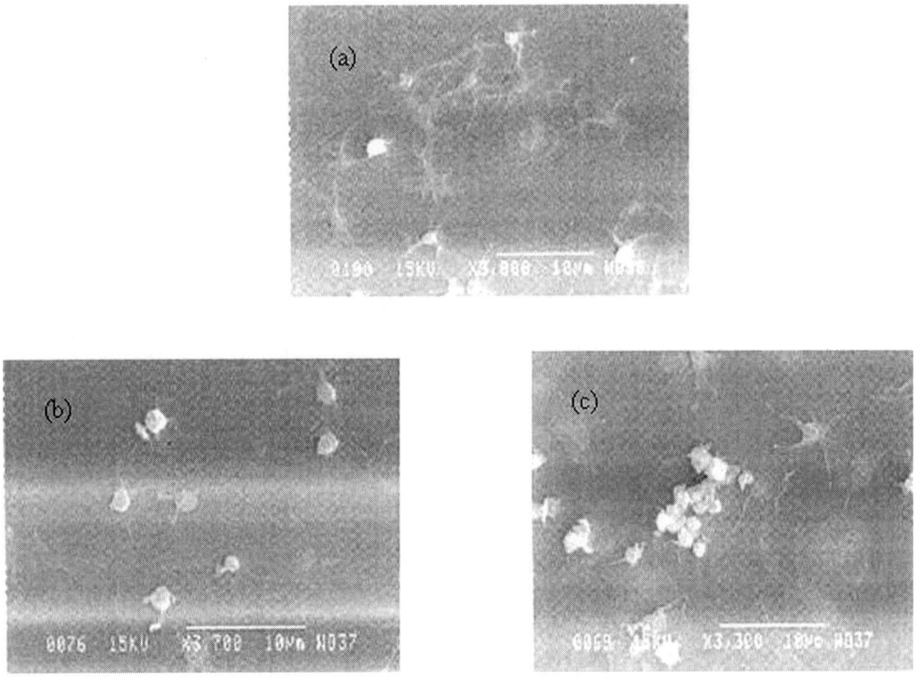

Figure 4.35: SEM micrograph of blood platelets adhering on (a) stainless steel (b) a-C:H-1 and (c) a-C:H-2 [82].

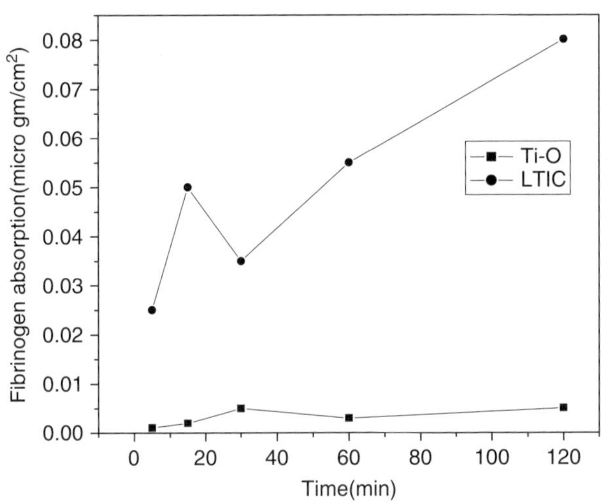

Figure 4.36: Amount of fibrinogen absorbed on the surface of Ti-O and LTI carbon [88].

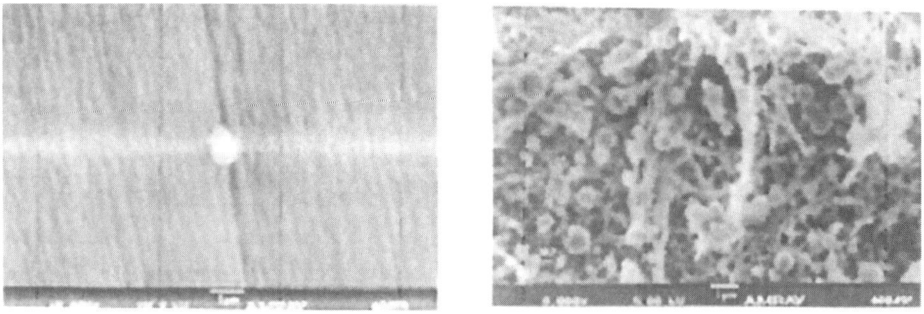

Figure 4.37: Amount of blood platelets and proteins absorbed on the surface of Ti-O and LTI carbon after 30 days in dog's body [90].

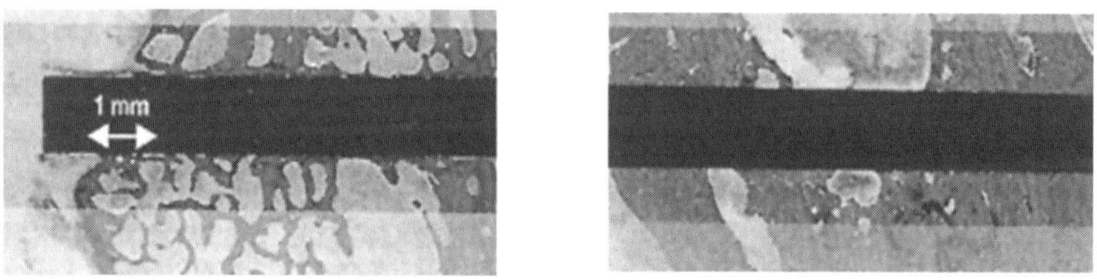

Figure 4.38: Cross-sectional view of the pure Ti and PIII-treated Ti after the removal from the rat body [91].

were observed in Ti-O film, whereas in the case of LTI carbon, thick multilayered fibrin formation enhanced the platelet adhesion and activation. Mechanical and structural studies were also carried out on this PIII Ti-O films [90], and they showed good performance.

4.8.3. PIII Treatment of Ti

Compatibility of tissue and bone with Ti before and after oxygen PIII was studied by Mandal *et al.* [91]. A thick oxide layer over Ti can be produced by PIII. This passive oxide layer forms a layer of hydroxyl group in the aqueous solution and improves the bio-compatibility. Untreated and treated Ti and anodized Ti were implanted in the proximal femur of rats for 3 months and then they were evaluated for its bio-compatibility. Fig. 4.38 shows the cross-sectional view of the implanted samples in the rat body. In comparison to pure Ti, pure Ti after PIII treatment showed the regrowth of the bone toward the implant and increased the fraction of the direct bone contact.

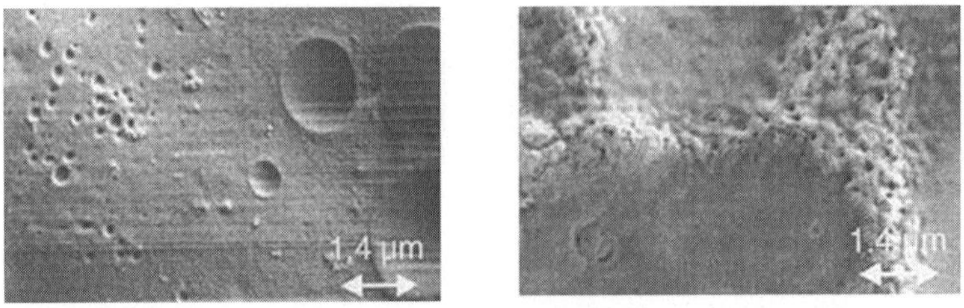

Figure 4.39: SEM topography of the anodized Ti and PIII-treated anodized Ti after the removal from the rat body [91].

A more detailed investigation of the implanted samples was performed after their removal from the test animals. Fig. 4.39 shows the topography of the anodized Ti and anodized Ti treated with PIII after removal. A smooth surface with very few craters were observed in anodized Ti, whereas very fine features were observed in the anodized Ti after PIII treatment. This confirms the regrowth of bone and tissue in the PIII-treated samples in a significant way than the untreated one.

References

[1] J.R. Conrad, J.L. Radtke, R.A. Dodd, F.J. Worzala and N.C. Tran, *J. Appl. Phys.*, 62 (1987) 4591.
[2] J.R. Conrad, *Mat. Sci. Eng.*, A116 (1989) 197.
[3] P.K. Chu, S. Qin, C. Chan, N.W. Cheung and P.K. Ko, *IEEE Trans. Plasma Sci.*, 26 (1) (1998) 79.
[4] J.R. Conrad, J.L. Radtke, R.A. Dodd, F.J. Worzala and N.C. Tran, *J. Appl. Phys.*, 62 (1987) 4591.
[5] J.R. Conrad, *J. Appl. Phys.*, 62 (3) (1987) 777.
[6] A. Anders, *Surf. Coat. Technol.*, 93 (1997) 158.
[7] I.G. Brown and X. Godechot, *IEEE Trans. Plasma Sci.*, PS-19 (1991) 713.
[8] A. Anders, S. Anders, B. Juttner, W. Botticher, H. Luck and G. Schroder, *IEEE Trans. Plasma Sci.*, 20 (1992) 466.
[9] T. Sorda, S. Meassick and C. Chan, *Appl. Phys. Lett.*, 60 (1992)1076.
[10] A. Anders, S. Anders, I.G. Brown and K.M. Yu, *Nucl. Instrum. Meth. Phys. Res.*, B102 (1995) 132.
[11] A. Anders, S. Anders, I.G. Brown and P. Chow, *Mat. Res. Soc. Symp. Proc.*, 314 (1993) 205.
[12] R.A. MacGill, S. Anders, A. Anders, R.A. Castro, M.R. Dickinson, K.M. yu and I.G. Brown, *Surf. Coat. Technol.*, 78 (1996) 168.
[13] J. Robertson, *Prog. Solid State Chem.*, 21(1991) 199.
[14] I.I. Aksenov and V.E. Strelnitskij, *Surf. Coat. Technol.*, 47 (1991) 98.

[15] B.F. Coll, P. Sathrum, R. Aharonov and M.A. Tamor, *Thin Solid Films*, 209 (1992) 165.
[16] J.J. Cuomo, D.L. Pappas, J. Bruley, J.P. Doyle and K.K. Saenger, *J. Appl. Phys.*, 70 (1991) 1706.
[17] R. Lossy, D.L. Pappas, R.A. Roy, J.J. Cuomo and A.H. Sura, *Appl. Phys. Lett.*, 61 (1992) 171.
[18] E.G. Gerstner, D.R. McKenzie, M.K. Puchert, P.Y. Timbell and J. Zou, *J. Vac. Sci. Technol.*, A13 (1995) 406.
[19] D.R. Mckenzie, *Diamond. Relat. Mater.*, 1 (1991) 51.
[20] S. Anders, A. Anders, I.G. Brown, B. Wei, K. Komvopoulos, J.W. Ager and K.M. Yu, *Surf. Coat. Technol.*, 6869 (1994) 388.
[21] K. Komvopoulos, B. Wei, S. Anders, A. Anders, I.G. Brown and C.S. Bhatia, *J. Appl. Phys.*, 76 (1994) 1656.
[22] B. Bhushan, B.K. Gupta, R. Sundaram, S. Dey, S. Anders, A. Anders and I.G. Brown, *IEEE Trans. Magn.*, 31 (1995) 2976.
[23] J.W. Ager, S. Anders, A. Anders and I.G. Brown, *Appl. Phys. Lett.*, 66 (1995) 3444.
[24] S. Anders, A. Anders, J.W. Ager, Z. Wang, G.M. Pharr, T.Y. Tsui, I.G. Brown and C.S. Bhatia, *Mat. Res. Soc. Proc.*, 383 (1995) 453.
[25] B. Prakash, E. Richter, H. Pattyn and J.P. Celis, *Surf. Coat. Technol.*, 173 (2–3) (2003) 150.
[26] K. Miyoshi, *Solid Lubrication Fundamentals and Applications*, Marcel Dekker Publisher, 2001.
[27] R.G. Munro, *J. Res. Natl. Inst. Stand. Technol.*, 105 (5) (2000) 709.
[28] M. Berger, L. Karlsson, M. Larson and S. Hogmark, *Thin Solid Films*, 401 (2001) 179.
[29] R. Wiedemann, V. Weihnacht and H. Oettel, *Surf. Coat. Technol.*, 116–119 (1999) 302.
[30] T.P. Mollart, M. Baker, J. Haupt, A. Steiner, P. Hammer and W. Gissler, *Surf. Coat. Technol.*, 74–75 (1995) 491.
[31] J. Elders, P.A. Quist, B. Rooswijk, J.D.W. van Voorst and J.V. Nieuwkoop, *Surf. Coat. Technol.*, 45 (1991) 105.
[32] E. Kelesoglu, C. Mittere, M.K. Kazmanli and M. Urgen, *Surf. Coat. Technol.*, 116–119 (1999) 133.
[33] A. Agarwal and N.B. Dahotre, *Mater. Charact.*, 42 (1999) 31.
[34] Wieslaw, Zdaniewski, J. Wu, C.G. Subhash and K. Oxorn, *J. Mater. Res.*, 6 (5) (1991) 1066.
[35] P.V. Ananthapadmanabhan, K.P. Sreekumar, P.V. Ravindran and N. Venkatramani, *J. Mater. Sci.*, 28 (1993) 1655.
[36] J.R. Treglio, S. Trujilo and A.J. Perry, *Surf. Coat. Technol.*, 61 (1993) 315.
[37] M. Berger, L. Karlsson, M. Larson and S. Hogmark, *Thin Solid Films*, 401 (2001) 179.
[38] J.R. Treglio, S. Trujilo and A.J. Perry, *Surf. Coat. Technol.*, 61 (1993) 315.
[39] B. Prakash and J.P. Celis, *PSE 2004 Conference Proceedings*, Garmisch, Germany.
[40] M.B. Peterson, S.F. Murray and J.J. Florek, *ASLE Trans.*, 2 (1960) 225.
[41] S.M. Johns, T. Bell, M. Samandi and G.A. Collins, *Surf. Coat. Technol.*, 85 (1996) 7.
[42] S.Y. Wang, P.K. Chu, B.Y. Tang, X.C. Zeng, Y.B. Chen and X.F. Wang, *Surf. Coat. Technol.*, 93 (1997) 309.

[43] M. Ueda, M.M. Silva, C. Otani, H. Reuther, M. Yatsuzuka, C.M. Lepienski and L.A. Berni, *Surf. Coat. Technol.*, 169–170 (2003) 408.

[44] L.P. Wang, B.Y. Tang, X.F. Wang, S.Y. Wang and K.C. Paul, *J. Rare Earth.*, 19 (1) (2001) 40.

[45] J. Chen, J.R. Conrad and R.A. Dodd, *J. Mater. Process. Tech.*, 49 (1995) 115.

[46] X. Nie, A. Wilson, A. Leyland and A. Matthews, *Surf. Coat. Technol.*, 121 (2000) 506.

[47] E. Richter, R. Gunzel, S. Parasacandola, T. Telbizova, O. Kruse and W. Moller, *Surf. Coat. Technol.*, 128–129 (2000) 21.

[48] Z. Zhan, X. Ma, L. Feng, Y. Sun and L. Xia, *Wear*, 220 (1998) 161.

[49] S. Li and A.H. Burstein, *J. Bone Joint Surg.*, 76A (1994) 1080.

[50] B. Wroblewski, P. Siney, P. Dowson and S. Collins, *J. Bone Joint Surg.*, 78B (1996) 280.

[51] N. Chang, A. Bellare, R.E. Cohen and M. Spector, *Wear*, 241 (2000) 109.

[52] W. Shi, X.Y. Li and H. Dong, *Wear*, 250 (2001) 544.

[53] R. Gunzel, N. Shevshenko and W. Matz, *Surf. Coat. Technol.*, 142 (2001) 978.

[54] H. Mohrbacher, J.P. Celis and J.R. Roos, *Tribol. Int.*, 28 (5) (1995) 269.

[55] M. Ueda, G.F. Gomes, E. Abramof and H. Reuther, *Nucl. Instrum. Meth. B*, 206 (2003) 749.

[56] S. Rossi, Y. Massiani, E. Bertassi, F. Torregrosa and L. Fedrizzi, *Thin Solid Films*, 416 (2002) 160.

[57] G.A. Collins, R. Hutchings, K.T. Short, J. Tendys, X. Li and M. Samandi, *Surf. Coat. Technol.*, 74 (1995) 417.

[58] R. Wei, J.J. Vajo, J.N. Matossian, P.J. Wilbur, J.A. Davis, D.L. Williamson and G.A. Collins, *Surf. Coat. Technol.*, 83 (1996) 235.

[59] C. Blawert, A. Weisheit, B.L. Mordike and F.M. Knoop, *Surf. Coat. Technol.*, 85 (1996) 15.

[60] M. Samandi, B.A. Shedden, D.I. Smith, G.A. Collins, R. Hutchings and J. Tendys, *Surf. Coat. Technol.*, 59 (1993) 261.

[61] M. Samandi, B.A. Shedden, T. Bell, G.A. Collins, R. Hutchings and J. Tendys, *J. Vac. Sci. Technol. B*, 12 (2) (1994) 935.

[62] M.P. Fewell, D.R.G. Mitchell, J.M. Priest, K.T. Short and G.A. Collins, *Surf. Coat. Technol.*, 131 (2000) 300.

[63] C. Blawert, H. Kalvelage, B.L. Mordike, G.A. Collins, K.T. Short, Y. Jiraskova and O. Schneeweiss, *Surf. Coat. Technol.*, 136 (2001) 181.

[64] H.P. Ho, K.C. Lo, P.K. Chu, K.S. Chan, J. Li, D.T.K. Kwok, E.Y.B. Pun and J.H. Marsh, *Nucl. Instrum. Meth. B*, 173 (2001) 304.

[65] R. Gunzel, N. Shevshenko, W. Matz, A. Mucklich and J.P. Celis, *Surf. Coat. Technol.*, 142–144 (2001) 978.

[66] T. Wolf, C.L. Shieh, R. Engelmann, K. Alavi and J. Mantz, *Appl. Phys. Lett.*, 55 (1989) 1412.

[67] P.J. Poole, S. Charbonneau, M. Dion, G.C. Aers, M. Buchanan, R.D. Goldberg and I.V. Mitchell, *IEEE Photon. Technol. Lett.*, 8 (1996) 16.

[68] M. Paquette, J. Beauvais, J. Beerens, P.J. Poole, S. Charbonneau, C.J. Miner and C. Blaauw, *Appl. Phys. Lett.*, 71 (1997) 3749.

[69] F.J. Henley and M.I. Current, *Semiconductor Fabtech*, 12th edition, 2000, p. 201.
[70] C.H. Yun, A.B. Wengrow and N.W. Cheung, *Electrochem. Soc. Proc.*, 99-3 (1998) 125.
[71] S.S.K. Iyer, X. Lu, J. Liu, J. Min, Z. Fan, P.K. Chu and C. Hu, *IEEE Trans. Plasma Sci.*, 25 (5) (1997) 1128.
[72] J. Min, P.K. Chu, Y.C. Cheng, J.B. Liu, S.S.K. Iyer and N.W. Cheung, *Surf. Coat. Technol.*, 85 (1996) 60.
[73] N.C.C. Lu, *IEEE Circuit Dev. Mag.* (1989) 27.
[74] P.K. Chu, *Plasma Phys. Contr. Fusion*, 45 (2003) 555.
[75] E.C. Jones, B.P. Linder and N.W. Cheung, *Jpn. J. Appl. Phys.*, 35 (1996) 1027.
[76] B. Mizuno, I. Nakayama, N. Aoi, M. Kubota and T. Komeda, *Appl. Phys. Lett.*, 53 (1988) 2059.
[77] M. Sano, T. Teramoto, K. Yukimura and T. Maruyama, *Surf. Coat. Technol.*, 136 (2001) 168.
[78] P. Singer, *Semicond. Int.*, 18 (8) (1995) 58.
[79] B.G. Park, J. Bokor, H.S. Luftman, C.S. Rafferty and M.R. Pinto, *IEEE Trans. Electron Device Lett.*, 13 (1992) 507.
[80] W. Ensinger, *Nucl. Instrum. Meth. B*, 120 (1996) 270.
[81] S.B. Felch, D. Chapek, S.M. Malek, P. Maillot, E. Ishida and C.W. Magee, *Proc. Meas. Char. Ultra-Shall. Doping Profiles Semicond.*, 3 (1995) 25.1.
[82] P. Yang, N. Huang, Y.X. Leng, J.Y. Chen, R.K.Y. Fu, S.C.H. Kwok, Y. Leng and P.K. Chu, *Biomaterials*, 24 (2003) 2821.
[83] P. Yang, S.C.H. Kwok, R.K.Y. Fu, Y.X. Leng, J. Wang, G.J. Wan, N. Huang, Y. Leng and P.K. Chu, *Surf. Coat. Technol.*, 177–178 (2004) 747.
[84] L. Feng and D.J. Andrade, *J. Colloid Interf. Sci.*, 166 (1994) 419.
[85] N. Huang, P. Yang, X. Cheng, Y. Leng, X. Zheng and G. Cai, *Biomaterials*, 19 (1998) 771.
[86] J.Y. Chen, L.P. Wang, K.Y. Fu, N. Huang, Y. Leng, Y.X. Leng, P. Yang, J. Wang, G.J. Wan, H. Sun, X.B. Tian and P.K. Chu, *Surf. Coat. Technol.*, 156 (2002) 289.
[87] P. Yang, S.C.H. Kwok, P.K. Chu, Y.X. Leng, J.Y. Chen, J. Wang and N. Huang, *Nucl. Instrum. Meth. B*, 206 (2003) 721.
[88] P. Yang, N. Huang, Y.X. Leng, J.Y. Chen, H. Sun, J. Wang, F. Chen and P.K. Chu, *Surf. Coat. Technol.*, 156 (2002) 284.
[89] Y.X. Leng, J.Y. Chen, Z.M. Zeng, X.B. Tian, P. Yang, N. Huang, Z.R. Zhou and P.K. Chu, *Thin Solid Films*, 377–378 (2000) 573.
[90] Y.X. Leng, N. Huang, P. Yang, J.Y. Chen, H. Sun, J. Wang, G.J. Wan, X.B. Tian, R.K.Y. Fu, L.P. Wang and P.K. Chu, *Surf. Coat. Technol.*, 156 (2002) 295.
[91] S. Mandl, R. Sader, G. Thorwarth, D. Krause, H.F. Zeilhofer, H.H. Horch and B. Rauschenbach, *Nucl. Instrum. Meth. B*, 206 (2003) 517.

Chapter 5

Ion Surface Treatment of Materials

Goeffrey Dearnaley and James Arps

5.1. Introduction

5.1.1. A Brief History

The modification of materials by ion beams was initiated about 40 years ago in the early 1960s, as a spin-off from nuclear science. For good reasons, the first objective was to alter the electrical properties of semiconducting materials by the implantation of active dopants such as boron into silicon. It was recognized that energetic ion bombardment is accompanied by radiation damage which in covalently bonded crystals must be removed by thermal annealing. Nuclear particle accelerators and isotope separators were adapted for this new technology, with immediate success. Device yields were improved by one, and sometimes two orders of magnitude in comparison with thermo-chemical diffusion of dopants. A decade later, ion implantation became the preferred method for doping electronic microcircuits, and a multibillion dollar industry has been built on this, nowadays utilizing ion energies from a few eV to several MeV.

This success led a few workers, in the 1970s, to explore the value of ion implantation in other technologically important materials, such as steels and titanium alloys. It was found that small amounts of implanted reactive elements, such as yttrium will strongly reduce thermal oxidation of stainless steels, without impairing their bulk properties. Others made use of the versatility of ion implantation to prepare large varieties of specimens for studies in aqueous corrosion, atmospheric corrosion and oxidation. Such work continues.

But the most successful applications of ion beam surface treatments, in commercial terms, were for the reduction of wear, especially in expensive molding tools. For this purpose, easily generated beams of nitrogen ions, at 50–200 keV energy, were remarkably successful in providing tool life extensions of 4 times to as much as 20 times, without heating. This is despite the relatively shallow penetration of the ions, to depths of only a fraction of a micron.

Materials Surface Processing by Directed Energy Techniques
Y. Pauleau (Editor)

Over the years, ion implantation has been employed to improve the surface properties also of polymers, ceramics, and cemented tungsten carbide composites. In China, it has even been used to bring about mutations in seeds! In comparison with other surface treatment processes, it is remarkably versatile and controllable. In the semiconductor industry, it is a method of very high precision, and for critical applications in medicine, aeronautics and the nuclear industry the integrity of the process is a powerful advantage.

5.1.2. Distinctions from PBII

In the previous chapter (Section 4.2) there was described a method for carrying out ion implantation by electrically pulsing the workpieces immersed in a plasma of the requisite gas or vapor. The maximum ion energy is limited by the tendency for arcing to occur, and the counter-current of secondary electrons released from the target consumes a great deal of the electrical power, and at high potentials becomes a source of hazardous X-rays.

Directed ion beams are extracted from the plasma within an ion source, and are subsequently accelerated to the required energy. They may be broad or collimated. Bo Torp [1] has described the benefits of scanning such a collimated ion beam over specific, functional surfaces of a tool rather than distributing the beam to parts not prone to wear. The X-ray problem is minimal with a directed beam accelerator, and is confined to the area of the ion source.

Another distinction, and one of high importance in the semiconductor industry, is the fact that an accelerated ion beam can be mass analyzed and separated in an applied magnetic field in order to select a specific species for implantation. This is important in research generally on materials modification in any comparison of the effects of introducing different elements (e.g. to study corrosion resistance). We shall see how this is done in Section 5.1.4.

In other cases, especially when very broad area beams are required, it may not be necessary to apply mass analysis. A good example is in the widespread use of nitrogen ion beams, generated from a plasma of nitrogen gas, or in the use of an argon ion beam for sputter cleaning.

The most powerful ion sources (e.g. of the bucket source design) can produce extracted beams of up to 100 mA or so, and even more is feasible from large ion guns developed for fusion research. But at an accelerating potential of 100 kV a 100 mA beam has a power of 10 kW. Even if this is distributed over a wide area this is often as much power as can be handled: the conductive cooling of workpieces in vacuum is difficult, and overheating is undesirable in a process that offers low-temperature treatments and no thermal distortion of components or tools.

PBII offers certain other advantages, as have been discussed in Section 4.2. Large effective ion beam currents are easily achieved and the ions impinge radially from all directions. Where it is required to treat all the surfaces of a workpiece this is an obvious advantage because it avoids the need for workpiece manipulation. Large process rolls provide a very good example of expensive tools that need overall and uniform treatment.

In the case of small items, say a few centimeters across, another factor comes into effect. In PBII, one must arrange that the items are spaced apart at a distance sufficient to allow an ion sheath to develop around each one. With a large-area directed beam the workpieces can be packed much more closely together, bearing in mind only shadowing, and larger batches are feasible.

PBII is sometimes used in a hybrid mode in which workpiece heating is advantageous. This is when ion implantation, during each pulse, is supplemented by thermal diffusion taking place continuously from the plasma. A good example is that of plasma nitriding combined with PBII for the hardening of austenitic steels.

The relative economics of PBII and directed beam ion implantation depend upon the particular application, and cannot be generalized. There is clearly room for both the technologies in the future, in industry and in research.

5.1.3. Commonly Used Ion Sources

For many years the most widely used ion source for ion implantation was that developed by Dearnaley *et al.* [2] at the Harwell Laboratory (UK). This was a more advanced, compact version of sources used at Oak Ridge National Laboratory to separate uranium isotopes during WWII.

The principle of all ion sources is similar. It is necessary to ionize the feed gas or vapor using electrons, and for this to be efficient electron energies of 100–200 eV are required. Hence it is necessary to operate at a reduced pressure to lessen energy dissipation in close collisions.

In the Freeman source, electrons are emitted thermionically from a linear cathode, and caused to spiral through the gas in an applied magnetic field, so increasing the ionization probability. A schematic diagram of this source is shown in Fig. 5.1; in other versions feed materials are generated from a heated crucible or by sputtering from a suitable electrode.

For widely used ion beams of gaseous species such as nitrogen or argon, the sophisticated versatility of the Freeman source is not required, and it may also be desirable to extract ions from over a larger area of the source plasma, to generate a broad beam. Then a type of ion source developed for the injection of energetic ions into plasmas, in order to heat them in a fusion reactor, is a suitable choice. A. Holmes

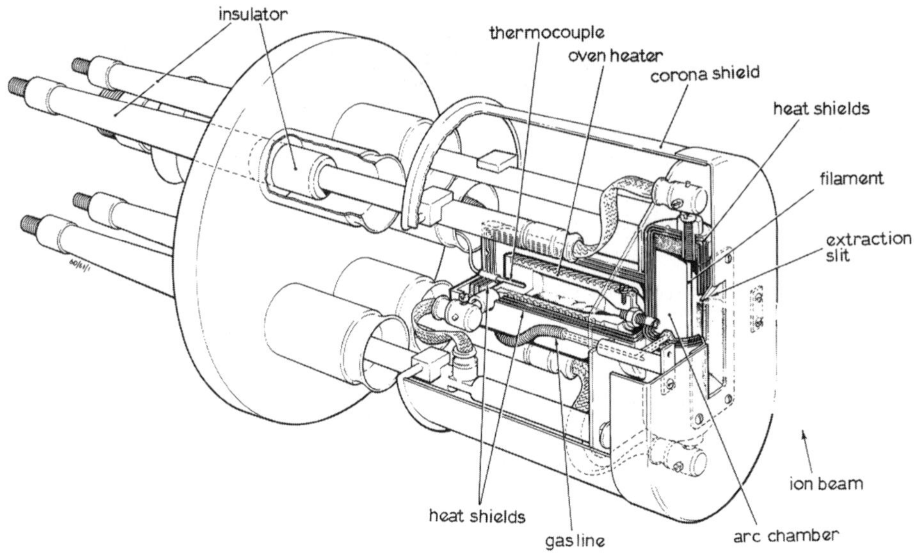

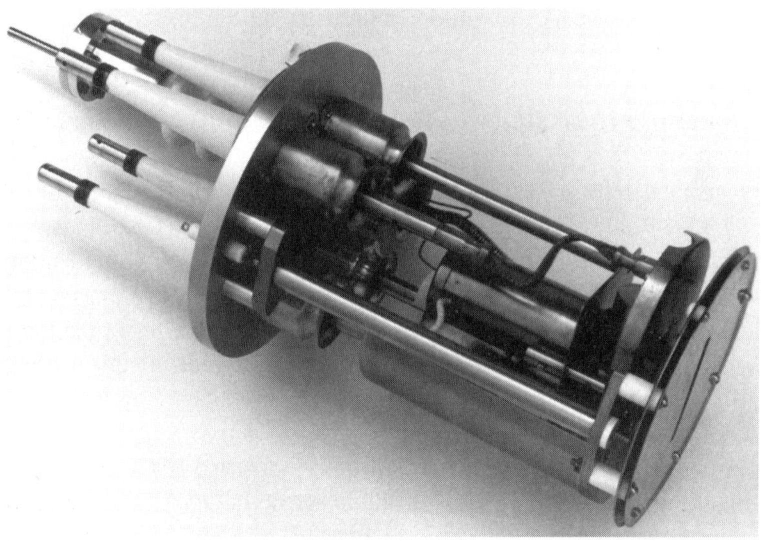

Figure 5.1: The Freeman ion source, the most widely used and versatile ion source for implantation purposes, was developed originally for isotope separation.

and P.D. Goode adapted such a "bucket" ion source for nitrogen ion implantation, with a large number of circular extraction apertures in a hexagonal array, designed so that the overlap of the multi-beams would produce a uniform current density. A diagram of this source is shown in Fig. 5.2. Once again, the electrons are generated from

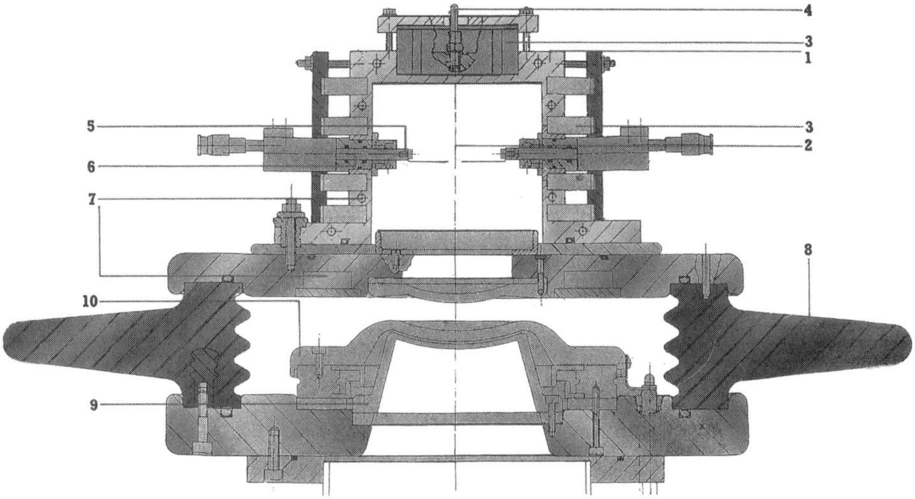

Figure 5.2: A plasma bucket ion source, mostly used for gaseous ions. The model shown can deliver up to 100 mA of nitrogen, mostly N_2^+, at or extraction voltage of 100 kV. The beam divergence angle is 40; enabling broad area treatment.

hot filaments and a magnetic field, confining the plasma, is generated by an array of powerful rare earth magnets on the wall of the copper chamber. Large versions of this source could readily produce ion beams of 1-A intensity. The use of thermionic filaments in ion sources limits the time between maintenance, for filament replacement. The bucket source above uses four interchangeable hairpin filaments, but even so it must be periodically dismantled for replacement. There are cases, in industry, where a much longer source lifetime without maintenance may be required, even at the expense of reduced beam intensity. The answer lies in the cold cathode source, and one that has been particularly successful and simple in construction is the saddle-field or twin-anode source developed originally by J. Franks. Within a cylindrical cathodes are arranged two parallel anode rods about 1 cm apart. A potential of about 8 kV applied to these rods causes electrons to oscillate in paths at right angles to the anodes, back and forth, thus enhancing the ionization efficiency. A linear beam, extracted through a slot can be accelerated by the potential on the surrounding, coaxial cylinder. Source life is in the region of 10,000 h. Though the beam intensities are

Figure 5.3: A long (75 cm) twin-anode ion source. The structure is very simple, with two anode rods mounted on insulators, spaced about 8 mm apart. The beam emerges from a slot opposite the gap between the rods. The feed gas pipe is seen on the right. These sources operate only with gaseous materials.

limited to 5–10 mA, it is easy to provide many such inexpensive sources to obtain multiple beams, for better coverage. The small cylindrical source can also be inserted along the axis if a tubular workpiece, and rotated within. The saddle field ion source is shown in Fig. 5.3.

Bo Torp and others have demonstrated the value of a "pencil" ion beam, 1 cm or so in diameter, that can be scanned electromagnetically over the working surfaces of tools or components. None of the sources above are suitable of this type. Torp, Nielsen and their co-workers adapted a cold and hot reflex discharge ion source (CHORDIS) for this application. The design uses a linear multicusp magnetic field to confine the electron paths, and there are reflector electrodes at the two ends. One version of this source is shown in Fig. 5.4, and it offers the following alternative configurations:

- A "cold" version of gases such as nitrogen or argon.
- A heated version for vaporization of volatiles such as metal chlorides.
- The design as shown with a sputter target inside the discharge chamber.

Beam currents after magnetic analysis are typically a few mA.

5.1.4. General Principles

The conceptual principle of ion implantation is extremely simple: chosen ions are electrically accelerated to high velocities and injected, or implanted, into the workpiece in order to provide improved performance (electrical, tribologic, corrosion, etc.)

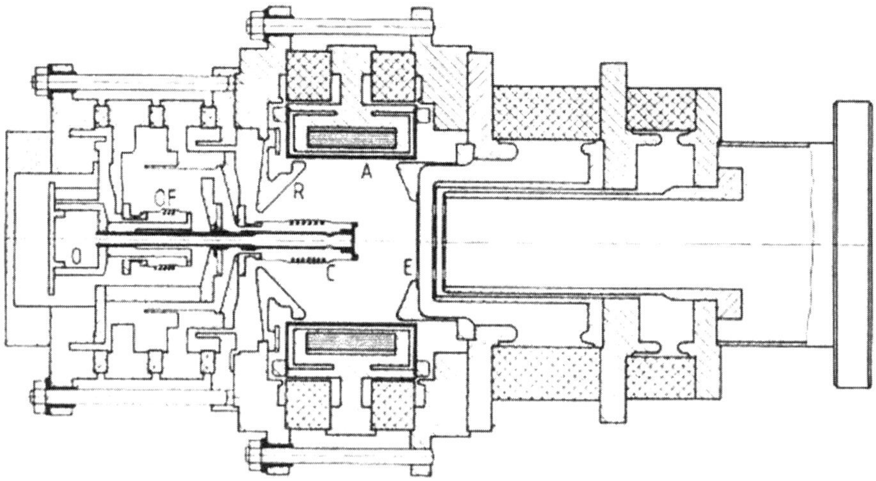

OF: oven filaments
R: reflector on cathode side
C: cathode filament
A: hot running anode shield
E: extraction system

Figure 5.4: One version of the CHORDIS ion source, showing the oven and cathode chamber on the left, the discharge chamber in the center, and the ion extraction system on the right. Another version incorporates a sputter electrode. Adapted from Torp *et al.* [1].

In practice, this calls for a number of features that must be designed into a system. Besides the ion source (Section 5.1.3) a high-voltage system is required, and these are generally of the solid-state switched mode design, with inductive protection against electrical surges.

For mass analysis an electromagnet with a circular cross-section is needed and in order that ions can pass efficiently through the separator careful design of the ion optics is required, for example, by using the code TRANSPORT developed by K.L. Brown. A slit positioned at the point of maximum mass dispersion enables the required ion species to be selected.

The equipment must be evacuated to avoid scattering of the ion beam. For rapid pump-down a combination of a Roots blower and oil diffusion pumps is a typical choice: a pump-down time of about 20 min is generally achievable, with an overnight base pressure around 10^{-6} mbar.

The work-chamber is an important feature in a versatile industrial implanter. It needs to be large enough to accommodate sizeable tools, or batches of smaller components. Correspondingly, a large door is needed. Because of the directed nature of the ion beam, workpiece manipulation (by rotation or transversal) must be provided, and different designs of rotatable tables or drums have been devised. Unlike the case in PBII, there is no complication in the form of a need to have the workpiece electrically insulated, for high-voltage pulse application. This greatly simplifies the mechanical system and the provision, where necessary, of water cooling.

Finally a reliable control system is a key component, providing control of the vacuum system, beam current and positioning, workpiece temperature, etc. in the form that is easily observable by the operator.

There are basically two types of industrial ion implanters. The simpler one comprises a large-area ion source, generally of the bucket variety, operated on a gaseous feedstock such as nitrogen. The entire output is directed towards the work chamber without mass analysis. Fig. 5.5 illustrates such a system embodying a 2.6-m diameter

Figure 5.5: Large industrial ion implanter (the Blue Tank machine at the Harwell, UK Laboratory). The ion source at this time was the twin-anode unit shown in Fig. 5.3: this was later replaced by a plasma bucket source (Fig. 5.2).

work-chamber with a tiltable rotatable table. Workpieces up to 500 kg in weight have been implanted, with ion beam currents up to 50 mA at 90 keV.

The second type incorporates magnetic analysis of the beam, and is much more versatile in regard to the ion species. A good example to the Triion Model 1100 implanter, with a CHORDIS source (seen in Section 5.1.3) and a 180° electromagnet. Additional electromagnets enable the beam to be steered towards critical areas of the workpieces, reducing processing costs by avoiding wasteful use of the beam.

5.2. Ion Implantation for Improved Wear Resistance

One of the earliest demonstrations of the application of ion surface treatments of materials other than semiconductors was the significant improvement in wear resistance of nitrogen ion-implanted steel, by Dearnaley and Hartley [3]. Pin-on-disc laboratory tests established that doses of around 3×10^{17} ions cm^{-2} are required, and this was soon followed by successful tests on small tools, mainly for cutting or stamping operations. This work commenced in Britain but was soon taken up in the USA, where the emphasis was upon military applications.

Later, the technology was extended to polymers, in which substantial increases in surface hardness were achieved, and also to ceramics such as alumina. The mechanisms are different in each case, as will be discussed in the following three sections. Virtually any solid material, including a composite, is amenable to property improvement by ion beam treatment.

5.2.1. Applications to Metal and Alloys

In each of these next sections we shall discuss first the mechanisms by which the desired tribologic effects are achieved, and then go on to cite some industrial applications.

5.2.1.1. Mechanisms

Based upon known physical metallurgy, Dearnaley and Hartley were able to propose in 1976 that nitrogen implanted into steels acts by pinning mobile dislocations – the cause of plasticity in metals. It can act in two ways. Dissolved nitrogen, atomically dispersed, can decorate dislocations and pin them according to the model proposed by A.H. Cottrell. In addition, precipitates of second-phase nitrides, such as CrN, or iron nitrides, serve to pin dislocations. There is an optimum size of precipitate that creates the maximum amount of local lattice strain, and ion implantation offers a highly controllable means for arriving at this optimum, while thermochemical nitriding at elevated temperatures does not. For this reason, nitrogen ion implantation can beneficially be applied after a bright nitriding process, to obtain the greatest surface hardness and wear resistance.

Hardness increases by factors of 2–3 have been registered using sensitive instruments, and it is this, together with the uniformity of the microstructure that gives a performance which can extend tool life dramatically, despite the shallowness of the ion penetration. In steel, a 100-keV nitrogen ion has a range less than one-third of a micron.

All steels benefit from nitrogen ion implantation but the effects are particularly noticeable in the alloy tool steels M2 and D2. In H13 steel the maximum benefits require post implantation heat treatment to 500°C. In steels such as 52,100 in which a large amount of strain-hardening is already in effect, the benefits provided by additional dislocation pinning are understandably less noticeable.

Two groups in the USA have studied the wear resistance of steels implanted with titanium and carbon with the idea forming a dispersion of TiC particulates. This is indeed very effective, but only if the doses of both species are high, around 5×10^{17} ions cm^{-2}. This would therefore be a costly process.

A much more economic use of metallic ions is by solid solution strengthening in which individual atoms in pairs act instead of precipitates containing perhaps 10^5 atoms. R. Fleischer in the 1960s showed that it is best achieved by an oversized and an undersized atom forming a substitutional-interstitial pair. Good choices in ferrous alloys are yttrium, which is 40% greater in radius that iron, coupled with nitrogen. Yttrium is too large to be dissolved in iron by conventional alloying, and Fleischer suggested that "an energetic beam" might be used instead. Dearnaley *et al.* did just that, and moreover showed, by isotopic tracer techniques, that the nitrogen relocates to decorate the immobile yttrium sites. The result is that even at relatively low doses around 10^{16} ions cm^{-2} there is great strengthening. In one pin-on-disc test in stainless steel a wear rate reduction of almost 1000 times was determined, with an accompanying low friction coefficient which is probably due to the maintenance of a surface oxide.

Other ion combinations also appear to have been successful. Langguth and Ryssel [4] found that aluminum and oxygen when implanted into steel gave 100-fold reduction in wear in a laboratory test. It seems likely that there is pairing between the atoms of these highly reactive species.

Kluge and colleagues [5] compared the effectiveness of nitrogen, boron, carbon, silver, tin and lead ions implanted at high doses into various steels. Of these, only nitrogen, boron and carbon gave significant wear reductions and nitrogen was particularly effective in steels with a high chromium content. This is consistent with the formation of chromium nitride precipitates.

In other alloys results are comparable. Thus in the important titanium alloy Ti–6Al–4V nitrogen ion implantation produces a fine dispersion of titanium nitride precipitates with a strong increase in wear resistance. Carbon and boron ions are also effective.

Aluminum alloys are soft and very prone to wear. Nitrogen ion implantation again produces strengthening by the formation of AlN precipitates a few nano meter in size. Nitriding of aluminum is made difficult by the presence of a coherent oxide coating, but this is easily penetrated by an energetic ion beam.

5.2.1.2. Applications to Metals

The two commonest modes of wear in metals are abrasive and adhesive wear, sometimes exacerbated by corrosion. Adhesive wear arises due to localized welding together of asperities, followed by tearing and it is very common in titanium and aluminum alloys. Such wear gives rise to work-hardened or oxidized wear debris that induces secondary wear by an abrasive mechanism. A typical example of abrasive wear is in the injection molding of filled polymers containing fine abrasive particles (e.g. of pigment) or of glass fiber.

Both types of wear are reduced by surface hardening. According to Archard's law the wear rate is inversely proportional to the hardness, but in the case of abrasive wear Moore [6] showed that the wear rate falls exponentially, and by orders of magnitude, if the surface hardness is greater than that of any of the abrading particles. The concepts are exemplified in the following case studies of ion-implanted components.

There are now many successful case studies in the literature and even more that have not been publicly reported because of company confidentiality. Space here allows us only to give examples of typical results in the various categories.

Knives of D2 steel are used to cut aluminum sheet and there is wear due to the abrasiveness of aluminum oxide. Straede and Mikkelsen [7] reported that nitrogen ion implantation increased tool life 10-fold, from about 80,000 to 800,000 cuts before resharpening. Fig. 5.6 shows a milling tool of high speed steel that gave 2.5 times normal life after nitrogen ion implantation.

Punches and dies used to stamp out transformer laminations from high-silicon iron sheet gave six times the life following nitrogen ion implantation, and Straede [8] found similar benefits in a punch used to cut out rubber discs containing the abrasive pigment, TiO_2. The best results appear to be obtained in the punching of abrasive materials.

Drills are very widely used, but they are cheap, throwaway tools that in general do not justify the use of ion implantation. Taps and reamers are more favorable, with cutting edges on the flanks, and have higher value. Delves [9] found a fivefold increase in life of taps used in filled phenolic resin, and Lempert reported a sixfold increase in life of steel reamers.

Metal-forming tools, such as those used for the manufacture of clutch housings in automobiles, have been treated by ion implantation as an alternative to hard chrome

Figure 5.6: A high-speed steel milling cutter that was given a life extension of 2.5 times as a result of ion implantation. Such tools can be stacked side-by-side for efficient treatment.

plating. Unplated steel tools failed after producing only a few parts, but nitrogen ion implantation over the "nose" area of the tools raised this number to about 2000. The dual treatment with yttrium plus nitrogen raised this figure to 23,000 parts. Straede and Mikkelsen found that nitrogen ion implantation of D2 steel drawing dies for can manufacturing enabled them to last 6 months instead of 3 weeks.

It is in the area of plastics injection molding and extrusion that ion implantation has had its main commercial success, chiefly in Europe. Such tools are expensive because they are machined to high precision from pre-hardened tool steels (such as P20). Polymer resins are frequently reinforced with glass fibers and inorganic fillers, and pigmented with abrasive powders such as titanium dioxide. The resulting fine abrasive wear reduces tool life to a few weeks and the refurbishing of molding tools is an expensive process. This presents an ideal situation for nitrogen ion implantation, which can be carried out at temperatures that cause no thermal distortion or tool softening.

Fig. 5.7 shows a tool used for producing coffee pots from a high-temperature resistant polymer, polyphenylene sulfide (PPS), containing a high filler content. Nitrogen ion implantation extended the tool life by about 10 times for a cost which was a small fraction of the tool cost. Fig. 5.8 shows a glassfiber-reinforced polyester

Figure 5.7: A steel tool used for the molding of coffee percolators, from filled PPS resin, being prepared for nitrogen ion implantation.

coil former used in the motors of fans. The precision of the mold used to make this is important. Nitrogen ion implantation in this case increased the life by an impressive 22 times. Finally, a small but costly mold is shown in Fig. 5.9 used to mold glassfiber-reinforced printer wheels: once again a 10-fold increase in life was achieved. This is a very good example of a tool in which the ion beam treatment may be confined easily to the area of the alphanumeric impressions. In Denmark, Straede and Mikkelsen report that chromium ion implantation into steel molds used at LEGO System Ltd. to mold PVC blocks increased the life by 12–13 times (3.7 million shots rather than the usual 300,000). In this case the problem is due to corrosive wear induced by hydrochloric acid released from the resin and hence the use of chromium ions.

Longer tool life with ion implantation

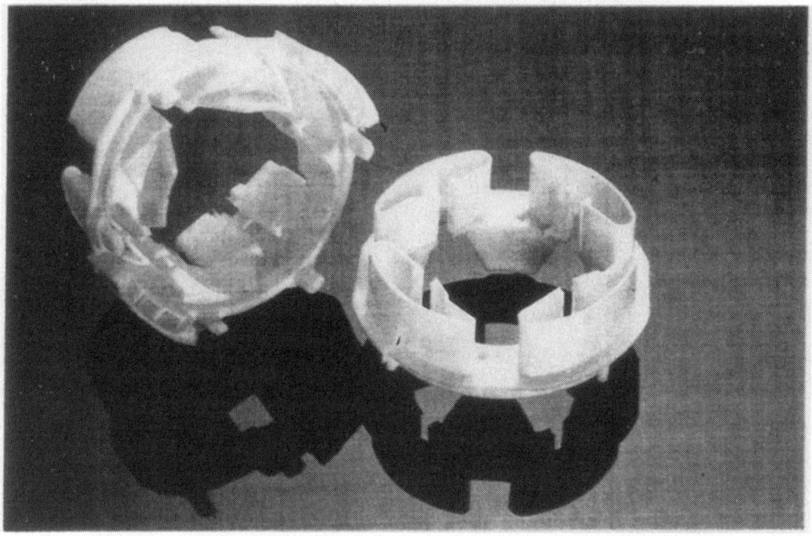

Build up of flash in the shallow angle of this moulding was retarded by ion implantation.

THE benefits of ion implantation in prolonging the active life of tooling continue to spread. Xpelair reports a 22-fold increase in the life of moulds used for insulation mouldings, following treatment by Tech-Ni-Plant.

The two-impression tooling, which costs around £20,000, is used with 15 per cent glass-reinforced polyester and was prone to wear which led eventually to flash, which then interfered with work-flow during assembly and caused broken coil windings. The normal practice when this became a problem was to have the tooling refurbished, with hard-chrome welded to the worn faces. Since ion implantation treatment, the tooling has produced 684,000 moulding without problems.

British Plastics & Rubber July/August 1989

Figure 5.8: Build-up of flash in the shallow angle of this molding was retarded by ion implantation.

Nozzles, screws, barrels and extrusion tools have also benefited from ion implantation, as reported by Delves and Dearnaley [9].

Aluminum alloy molds are increasingly made for shorter production runs because the machining cost is far lower. However, their life is short for reinforced polymers, but Ozate *et al.* [10] in Spain-implanted nitrogen to increase tool life by over two orders of magnitude.

Note that in almost all of these cases it is wasteful to apply ion implantation to all surfaces of the tools, and only to the working areas. Consequently directed beam implantation is more economic than PBII. A further advantage is that very

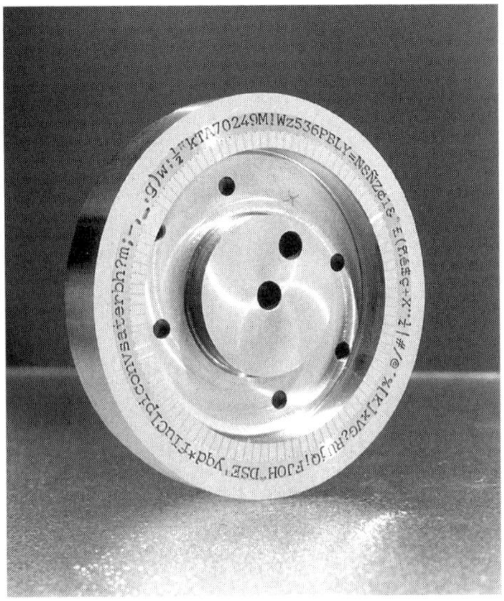

Figure 5.9: A small molding tool (10-cm diameter) used for the production of glassfiber-reinforced pherolic resin printer wheels, successfully treated by ion implantation of nitrogen.

heavy molds do not need to be mounted on electrical insulators. Finally, there is a finite risk in PBII of an arc: in an expensive tool this arc pitting would be disastrous. In summary, the nature of the abrasive wear process and the high cost of tooling make this an excellent area for application of ion beam implantation.

5.2.1.3. Applications to Cemented Carbides

There is a very large market for composite metal-cutting tools made from tungsten carbide grains cemented together with cobalt. The cobalt content can range from about 4% to 13%. These tools are relatively small and costly and so are economically attractive for treatment by ion beams.

Larsen-Basse [11], and Hong and Gurland [12] have studied the wear and failure mechanisms in cemented carbides and it is evident that these mainly originate within the softer binder phase, due to abrasion, extrusion or fatigue, allowing carbide grains to be pulled out. Eric Almond [13] commented that "the possibility of producing improved properties in WC-Co hard metals by precipitation hardening of the binder phase through alloying additions does not appear to have been investigated in detail." Dearnaley [14] has concluded that this is just what is achieved by nitrogen ion implantation, and it is important to recognize that the binder contains a few percent of dissolved tungsten, which has much greater affinity for nitrogen that does cobalt.

Thus it is understandable that the best results have been obtained as a result of ion implantation of nitrogen at around 400°C, allowing migration of the nitrogen to tungsten sites. Secondary ion mass spectrometry (SIMS) analysis of implanted materials has shown evidence for WN molecules in the emitted spectrum.

5.2.1.4. Case Studies in Implanted Carbides

Tests on implanted tools for very harsh conditions in rock drilling were not successful. This is when the failure mode is by cleavage of the brittle carbide phase rather that by wear or plastic deformation of the binders.

However, circular tungsten carbide knives used for cutting flat gaskets from carbon-filled neoprene tubing showed a life increase of about 11 times (from 3000 to 33,000 cuts, with a further savings in down-time for tool changing. It is an interesting reflection on ion implantation that this latter saving had to be negotiated with the labour unions because workers had previously been paid an agreed amount for tool changing.)

Guzman *et al.* [15] measured life increases of about 3.5 for carbide wire-drawing dies, and in other cases (for finishing dies) factors of up to six times life were observed.

Bo Torp has reported especially good results using nitrogen ion implantation to treat cemented carbide tools used for cutting wood, with life increases ranging from 5 to as much as 50 times (50 times). It is perhaps significant that, to increase the toughness of such tools, they are made from high-cobalt composites.

In Russia Vesnovsky *et al.* [16] investigated ion-implanted tungsten carbide and mixed WC-TiC carbide tools for machining application. Life extensions ranged up to 20 times. This group also concluded that the benefits came about by strengthening of the cobalt binder and also by the creation of compressive stresses in the surface of the carbide grains.

5.2.2. Applications to Ceramics

The interatomic bonding and material properties of ceramics are very different from those in metals. Ceramics are well known to be very strong in compression, but the can readily fail under tensile stress due to brittle fracture. Consequently, the effects of ion bombardment and implantation in ceramic materials are very different from those in metals. Nevertheless, some improvements are achievable, as long as the mechanisms at work are properly understood.

Good accounts of the effects of ion implantation on the mechanical properties of ceramics have been published by Burnett and Page [17], and Bolse and Peteves [18]. It is shown that one must distinguish between the “low-dose regime” in which the material may be damaged but is still crystalline from the "high-dose regime" in which the irradiated material has become amorphous. The critical level of energy deposition for amorphization of alumina is about 44 kJ mm^{-3} (3 $\times$ 10^{23} keV cm^{-3}) but it is far lower in covalently bonded substances such as silicon nitride or silicon carbide. At low doses, ion implantation produces an increase in hardness, sometimes by as much as 50%, but higher doses cause progressive softening. Implantation at low temperatures (e.g. 100 K in sapphire) enhance these effects. At elevated temperatures (e.g. 500 K), there may be progressive hardening with increasing ion dose.

It is concluded that the introduction of radiation damage leads to an increase in compressive stress up to the point at which amorphization occurs, following which there is extensive stress relief and surface swelling. At high temperatures at which some annealing of the disorder occurs, it is feasible to introduce enough foreign elements such as yttrium or nickel into the alumina to provide solid solution hardening, as in a metal.

Bolse and Peteves conclude that the improvement in ceramic fracture strength usually observed in the low-fluence regime can be without doubt attributed to the additional amount of tensile stresses needed to overcome the compression due to the radiation damage. In practice, ceramic species contain numerous surface flaws and microcracks that raise the local stresses under tension. High dose ion implantation, by causing swelling, is able to reduce the effective size of critical surface flaws and so improve fracture strength despite the accompanying softening of the materials. It follows that the ion range used should be chosen to match the depth of the critical flaws. There have been few published accounts of the practical application of ion beam treatment of ceramic components, despite the fact that these will generally be of higher value than the steel equivalents, and the ion doses required for treatment are lower.

A patent has been awarded to Southwest Research Institute for the treatment of ceramic orthopedic components by ion beams. Such items, of alumina or zirconia, give low wear rates but have occasionally been prone to fracture, with serious consequences. Ion bombardment by closing up surface microcracks can improve the fracture resistance and fracture toughness of these brittle materials.

The strengthening of ceramics appears to be a promising area for the application of ion beams. The technology is simpler that in the case of PBII because of the dielectric properties of ceramics.

5.2.3. Applications to Polymers

Ion bombardment brings about radical changes in the composition and structure of polymeric materials by mechanisms very different from those in metals and ceramics.

Once again, it is necessary to understand these mechanisms if one is to be able to make use of them.

5.2.3.1. Mechanisms

Ion irradiation readily ruptures the carbon–hydrogen bond present in most polymers and the mobile hydrogen atoms are able to escape into the vacuum. Thus the C:H ratio progressively increases with dose, to as much as 10:1. Residual hydrogen may be physically trapped at regions of low electron density. There is extensive formation of new C–C bonds and adjacent polymer chains become cross-linked, a process that adds considerable strength and resistance to deformation of the polymer. As in pyrolytic carbons formed by furnace treatment of polymers, excessive cross-linkage can lead to embrittlement. Another well-recognized mechanism that occurs is chain scission, in which atomic displacement can rupture the polymer chains, with loss of strength.

Besides the mechanical property changes described, there are powerful electrical modifications in which with increasing ion dose the electronic conductivity may be changed by 10 orders of magnitude, and in a controllable way.

Some polymers, such as polytetrafluorethylene (PTFE) contain no hydrogen. The behavior under ion bombardment is different, and it is found that volatile molecular compounds of carbon and fluorine are emitted without extensive graphitization. The benefits are therefore limited.

Polymers differ in regard to the amount of ion bombardment they can tolerate without mechanical degradation, and it is perhaps not surprising that those able to resist high temperature, such as Kapton are able to withstand remarkably high doses of ions.

From numerous studies that have been made of the effects of using different ion beams, it has been possible to deduce that the energy deposited in ionizing or electron-exciting processes is more effective for hardening than atomic collision, or elastic processes. The latter tend to induce chain scission. Thus light ions such as protons, boron ions and nitrogen ions have proved to be some of the most successful for surface hardening. The hardness of a number of polymers, including epoxy resins, has thereby been brought up to that of a mild steel (i.e. about 200 kg mm^{-3}). This is sufficient to resist many kinds of mild abrasion.

In the study of dual treatment, comprising metal ion implantation and metal ion deposition in polyethylene terephthalate (PET) Yuguang *et al.* [19 a, b] reported that implanted tungsten ions, at 40 keV and a dose of 10^{17} ions cm^{-2} will increase the hardness of the polymer by a factor of 8–4 GPa. Adhesion of the deposited metal coating was improved by prior ion implantation of the same species. The authors suggest that the hardness increase is due to the formation of tungsten carbide.

Venkatesan [20] irradiated several polymers with 2 MeV helium or argon ions and determined their electrical resistively over a range of temperatures. At doses approaching 10^{17} ions cm^{-2} resistivities below 10^{-2} ohm cm^{-1} were achieved and the temperature variation indicated charge transport by hopping between conducting islands within an insulating matrix. In an organo-silicon (silylene) polymer the resistively remained very high, and the author noted that these films became very hard and could not be scratched by sharp steel or quartz tools. It probably remains true, as the author remarked, that this is a promising and multidisciplinary field for research.

This challenge was taken up by Lee *et al.* [21] who implanted a variety of species into Kapton, Teflon, Tefzel (a partially fluorinated polymer) and Mylar (PET). The ion beams at energies from 0.5 to 5 MeV were applied singly or with dual or triple beam simultaneously, from a suite of three Van de Graaff accelerators. Measurements were made of the hardness and wear resistance of the treated polymers against nylon or steel balls.

The greatest increase in microhardness occurred in Kapton (polypyromellitimide), an aromatic chosen compound, when implanted with boron and carbon or with boron, carbon and nitrogen beams, the hardness increased by about, 25 times to 10–11 GPa, little change took place, however, in Teflon, A good correlation was found between the hardness and the degree of carbon enrichment.

Ion-implanted Kapton showed remarkable wear resistance, attributed to extensive cross-linking, rather than to formation of BN or BC within the surface. Rather, a B–H adsorption line was observed by FTIR spectroscopy. The treatment described, using a multiplicity of Van de Graaff accelerators is not, however, a practicable industrial process.

5.2.3.2. Polymer Applications

Again, since they are electrically insulating materials, polymer components are more readily treated by directed ion beams than by PBII.

Bhattacharya [22] has published details of work on improving the scratch resistance of eyewear made from polycarbonate (PC) or CR-20 (allyldiglycolcarbonate). 2.5 MeV oxygen ions were implanted at low doses, from 5×10^{13} to 3×10^{14} ions cm^{-2}. The color of the polymers changed through light yellow to brown, and to black with increasing dose. With an acceptable transmittance of almost 40%, it was possible to improve the abrasion resistance of PC or CR-20 lenses in standardized tests conducted by Bausch & Lomb Co. by an order of magnitude, as measured by a hazemeter and with a dose of only 10^{-14} ions cm^{-2}. The method is particularly suitable for sunglasses which transmit 15–20% of visible light, because it allows scope for the incorporation of a variety of desirable colors prior to ion implantation.

Dearnaley participated in a development relating to printing by the intaglio method. Steel cylinders were coated with a precision coating of epoxy resin which was subjected to excimer laser ablation to create small pits designed to hold ink. Paper, however, contains abrasive particulates and filler, and produces wear of a polymer surface. Dearnaley proposed proton irradiation at 100 keV, and the variety of angles of incidence of the beam over a cylindrical surface produced a treatment that was sufficiently uniform in depth, the ion penetration being about 1 μm. The abrasion resistance and hardness were increased substantially to about 200 VHN. The project was, however, terminated for non-technical reasons.

These few examples serve to illustrate the potential for ion beam surface treatment of polymers, a process that may be combined with subsequent coating, for example with diamond-like carbon (DLC).

The electrical property changes brought about by ion bombardment have also been applied, for example, to eliminate static charge build-up on polymer film used to wrap components in spacecraft. It is recognized that a surface resistivity of about 1 M ohm m^{-2} is appropriate for this purpose, and Dearnaley *et al.* achieved this in Kapton film by bombardment with nitrogen ions at 90 keV energy. There was no mechanical degradation (e.g. by crinkling of the film) which developed a silvery appearance.

5.3. Friction and Surface Energy Modification

There are many important applications in which the provision of a low static or dynamic friction coefficient is more valuable than protection against wear, although good wear resistance may also ensue because of the reduction in shear forces applied to the component. This may be especially the case in ceramics, which tend to have high friction coefficients and low strength in tension. In other instances, a low surface energy may be desirable, or a modified tendency for wetting.

One approach is to use soft materials that can shear easily, or even melt at the hot spots associated with load bearing asperities. Aware that tin appears to behave as such a soft lubricant when tin-plate is used for can manufacture, Hartley [23] implanted tin ions into steel to a dose of 2×10^{16} ions cm^{-2} and did observe reduced friction. This result was confirmed in later work by Baumvol *et al.* [24].

In another example the mechanism is very different. Oxides generally behave as lubricants on metal surfaces as long as they can be retained (i.e. do not wear off). Yttrium is known to improve the adherence of the oxide on stainless steels, which consist mainly of chromium oxide or oxyhydroxide. Yttrium and rare earths produce strengthening in such oxides by segregating to grain boundaries, and by grain refinement. Dearnaley *et al.* [25] implanted yttrium and nitrogen into austenitic stainless steel and observed major reductions (by about a factor of 3) in sliding friction in a pin-on-disc test. Periodically, the friction coefficient would rise and correspondingly the wear rate increased, but then the low friction condition was reestablished. The authors concluded that when the protective oxide is worn away, the friction rises, the temperature at the asperities will then rise and reoxidation of the steel appears to take place so that the friction coefficient jumps between two stable states in a manner governed by the load and sliding speed.

Boron oxide is known to act as an effective solid lubricant at high temperatures. High doses of boron implanted into wear-resistant steels may be expected to provide the same low friction at 500°C as was observed by Erdemir and Bindal [26] in borided steel. Boron also reduces the wear rate, by hardening mechanisms.

A totally different approach makes use of the catalytic activity of nickel for inducing carbon deposition from hydrocarbon atmospheres at temperatures above about 750°C. Lankford and Kossowsky [27] found that ion beam-assisted coatings of nickel and titanium on ceramics such as silicon nitride would provide a low friction coefficient at 800°C. Ion-implanted nickel and titanium may be expected to be effective also in causing the continuous deposition of carbon or to the surface at high temperatures. Titanium dioxide acts as a good support for the nickel or nickel oxide catalyst, enhancing its activity. In the case of titanium alloys only the nickel would need to be implanted.

Following Hartley's work in steels, Watkins [28] implanted tin ions into the titanium alloy Ti–6Al–4V and measured friction and wear in the pin-on-disc test. There was a significant reduction in friction coefficient, accompanied by low wear. Titanium alloys have poor frictional properties due to metal-to-metal adhesion and welding, but modifications to the oxide on the Ti–Sn surface alloy may prevent this as long as the load is not too high.

In chromium and its alloys, chromium oxide Cr_2O_3 is known to act as a solid lubricant, as long as the wear rate allows it to be retained. In a patented process, Dearnaley has proposed the use of ion-implanted CO^+ or CO_2^+ ions in order to form both chromium oxide and carbide to a greater depth than the native oxide. The carbide strengthens the alloy while the oxide improves lubrication.

An interesting application for ion implantation for friction and wear reductions in the textile industry has been reported by Lohmann and van Valkenhoef [29]. Hard chrome-plated coatings are often used in the manmade fiber industry on thread guides and tools for drawing, heating and winding the fiber. Some fibers are pigmented with TiO_2, for example, in Diolen™ and the abrasive wear rate can be high. Fifteen different ion species were implanted into chromium at a dose of 4×10^{17} ions cm^{-2} at 150 keV energy. Titanium together with nitrogen gave friction coefficient reductions by factors of 2–3, and, more importantly, a more stable friction by factors up to 20-fold. Correspondingly, the wear rate fell by a factor of

24 times for the titanium plus nitrogen implant. Lead ions yielded an even lower wear reduction of 54 times. These treatments may be beneficial in other applications of chromium plating, reducing the need for re-plating and thus providing environmental benefits.

5.4. Modification of Corrosion and Oxidation

Corrosion, like wear, accounts for huge losses each year in the developed nations. As I write, corrosion has thinned a steam pipe wall at a Japanese power plant in Mihama and the rupture killed four workers and has resulted in very costly consequences.

Corrosion, like wear, is a complex phenomenon involving chemical reactions, ionic diffusion, mechanical stresses and interfacial behaviors. Its study is an ongoing research field. Protection against corrosion is a multi-billion dollar per year industry.

Ion implantation allows the controllable modification of metal surfaces in a very versatile manner, without the need for new alloy development or any change in the bulk properties of the material. Instead, it places the protection where it is needed. We shall see, however, that the degree and duration of this protection depends strongly on the nature of the corrosion or oxidative environment.

5.4.1. Aqueous Corrosion

In an aqueous corrosion situation, all parts of the immersed component are electrically connected through the pathway provided by the water, or other electrolyte. As a consequence, if some areas have little or no protection a galvanic couple will be established and the localized corrosion or pitting at these areas may be severe. The problem with ion implantation is that it is a very shallow modification, typically to a small fraction of a micron. Particles of dust on the surface of the workpiece and exposed inclusions may be larger that this, with the result that these areas are shadowed or not effectively treated: they become sites for galvanic corrosion. For this reason, the practical applications of ion implantation (or PBII) to aqueous corrosion have been very limited.

However, as a tool for studying the aqueous corrosion behavior of a wide variety of surface alloys, ion implantation has been invaluable. Moreover, because of the non-equilibrium nature of the implantation process, it has been possible to prepare alloy systems that are not amenable to conventional metallurgy, for example, Al–Mo and to investigate their corrosion resistance.

The most useful test procedures invoke the potentiodynamic method in which the specimen is forced cyclically into the anodic and cathodic regimes. A criterion of interest is the so-called pitting potential, at which rapid localized corrosion occurs. These sites act as powerful stress-raisers if the component is subject to stress (e.g. in the case of a bearing).

Some of the most interesting tests involve the ion implantation of species that cannot be incorporated by convention alloying, such as tantalum into steel. Ashworth *et al.* [30] and his group were able to demonstrate that Ta is more effective than Cr in preventing corrosion of iron (Fig. 5.10). Sometimes, oversized or undersized implanted species will disturb the microstructure to the extent that it becomes amorphous. Then the grain boundaries that represent the most vulnerable regions for corrosion are no longer present. It is well established that amorphous or glassy metals resist corrosion very well. However, they are difficult and expensive to fabricate, and their bulk properties are inferior in terms of toughness. Ion implantation of the surface therefore provides an excellent alternative.

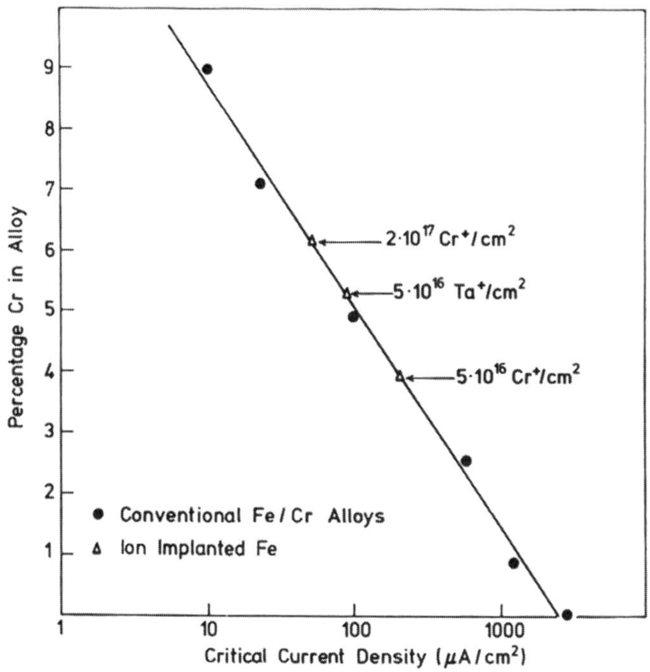

Figure 5.10: The corrosion rate of iron implanted with Cr^+ or Ta^+ ions, compared with that of conventional iron–chromium alloys, showing that tantalum is about four times as effective as chromium in reducing the critical current density (adapted from Ashworth *et al.* [30]).

An example, well studied by Cooney and Potter [31] is that of high dose phosphorus implanted into steel. Doses over 3×10^{19} ions cm^{-2} at 175 keV energy render the steel amorphous and the corrosion resistance was shown to increase by up to two orders of magnitude. This study was in Type 304 stainless steel.

In an extension of Ashworth *et al.*'s work, Natishan *et al.* [32] implanted Si, Zr, Nb, Cr, Mo, Zn or Al into aluminum and investigated the pitting potential. All of the species, except zinc, increased passivity, but Zn reduced the resistance to pitting. Aluminum self-ions had no effect, showing that chemical effects are dominant.

A different approach was taken by Hirvonen *et al.* [33] in the case of AISI 420 stainless steel.

Nitrogen, chromium, or chromium plus nitrogen were ion implanted, at doses up to 5×10^{17} ions cm^{-2}. The ion energies of 50 keV for N^+ and 230 keV for Cr^+ were chosen to provide similar ion ranges. Potentiometric measurements in IN sulfuric acid solution showed that nitrogen and chromium implanted separately had little effect on corrosion, but the implantation of both species shifted the corrosion potential in the noble direction by about 500 mV. and the corrosion current density fell by four orders of magnitude, so that the active peak almost vanished. After polarization, the specimen was analyzed by SIMS and evidence was seen for CrN and interestingly, a peak in sulfur concentration at the surface.

5.4.2. Atmospheric Corrosion and Tarnishing

This is a form of aqueous corrosion combined with room-temperature oxidation and sulfidation and with which we are all familiar. In this case, the metal surface is not connected via an electrolyte, and so the prospects for a shallow surface treatment are much better. Generally there is no wear combined with tarnishing and the effects are chiefly cosmetic. Occasionally, there are instances in which atmospheric corrosion is utilized in order to achieve a cosmetic appearance, as in Corten® steel.

The choice of elements for ion implantation to reduce tarnishing is generally similar to those which convey resistance to aqueous corrosion. Thus, in copper, Dearnaley showed that implanted chromium has a long-lasting effect, maintaining a bright surface. Copper implanted with aluminum ions obtains a pleasing golden appearance that also resists tarnishing.

As in the case of aqueous corrosion, surfaces rendered amorphous by ion implantation resist atmospheric corrosion well.

5.4.2.1. Applications

The practical application of ion beams to the corrosion of metals has usually been hampered because of the large area required treatment. Unlike in a tool, processing cannot be restricted to a particular area. However, in the 1970s a bold attempted was begun to achieve the large-scale ion implantation of carbon steel sheet, used for automobile bodies.

The process evaluated, at the Swinden Research Laboratory of British Steel Corporation, was the high dose implantation of phosphorus. Prolonged, standardized exposure tests on implanted steel sheets showed that as time went on (over a year) the steel became more and more "stainless." Plans and cost-benefit estimates were drawn up for a large plant to treat steel in a reel-to-reel system. The power consumption would have been several megawatts (as it is indeed in the sputter coating of architectural glass). Ion sources would have been similar to the large Calutrons operated at Oak Ridge to separate uranium isotopes in WWII.

This ambitious, though feasible, project was terminated by the closure of the corporation's Swinden Laboratory, in London, and dispersal of the staff. It remains a viable technical prospect, possibly for PBII.

The atmospheric corrosion of depleted uranium poses problems in certain military applications, especially in air with a high relative humidity (RH). Anthony [34] studied electropolished uranium specimens ion implanted with seven selected species. It was found that molybdenum, at a dose of 3×10^{16} ions cm^{-2} at 350 keV energy, improving oxidation and corrosion resistance by about two orders of magnitude, in air with 75% RH, at 80°C.

A particularly severe form of atmospheric corrosion can take place in a marine environment, in which humidity is combined with the presence of chloride ions, which induce pitting corrosion in steels and aluminum. As mentioned above, such pits strongly intensify local stresses under load and mechanical failure or fatigue can easily ensue.

This is a situation which arises when aeroengine main bearings, made of M50 high-strength steel, are stored on board an aircraft carrier. Pitting corrosion can render bearings useless in as little as 3 months. A comprehensive account of a project in which ion implantation was studied as a means for overcoming this problem was published by Grabowski *et al.* [35]. From a range of ion species, chromium and chromium plus phosphorus emerged as the best choices. A dose of 1.5×10^{17} Cr^+ ions cm^{-2} plus 5×10^{16} P^+ ions cm^{-2} renders the surface of the steel amorphous and reduces the critical current density by about a factor of 50. Equipment for the manipulation of bearings in an ion beam to ensure uniform treatment was designed, and a cost estimate (1985) for the process suggested about 0.7 cm^{-2}, which evidently would be acceptable.

5.4.3. Thermal Oxidation

In technologic applications to high temperature chemical plant, nuclear reactors, aeroengines and power generation equipment thermal oxidation is of major importance. This is normally achieved by coatings, but under stress and with thermal cycling these may decohere exposing the substrate. Ion-implanted layers are free from this risk.

Once again, it is important to understand the mechanisms that operate during thermal oxidations, and ion implantation has provided a highly versatile tool in such research. Insoluble, non-equilibrium species can be introduced, such as over sized atoms, and the redistribution of the implanted layer can be followed in detail by microanalysis. Since there is no electrical connectivity across the surface, as via an electrolyte, there is no risk of galvanic corrosion. As a result, very long-lasting benefits have been observed in ion-implanted alloys subjected to temperatures as high as 1100°C. In fact the redistribution of implanted atoms (e.g. to grain boundaries and to the surface), or to the oxide–metal interface, during such exposures can sometimes enhance the effectiveness of the implanted species.

Over a wide range of temperatures the transport processes that are responsible for thermal oxidation do not occur homogeneously throughout the oxide scale, but predominantly at alloy grain boundaries or dislocations. To be effective, ion-implanted atoms must be able to segregate to these fault planes or lines and there modify the oxide microstructure so as to impede the movement of ions, whether it be oxygen in-diffusion or cation out-diffusion.

In several technologically important alloys the oxide scale provides good protection as long as it stays in place. Oxide and metal have different thermal coefficients of expansion, however, and during thermal cycling this can impose great stresses on the oxide, causing it to decohere or spall. Not only does this weaken the substrate by attrition, but the abrasive spalled oxide may cause significant problems.

The first step, in corrosion, is the absorption of reactant molecules on the exposed surface. Sometimes the implantation of a catalyst such as platinum (see Section 5.7.1) will modify the process, especially if water vapor is involved, or (in some cases) CO.

5.4.3.1. Specific Mechanisms

We shall here consider mechanisms of oxidation and protection that are specific to important alloys. This will throw light on how, sometimes, a small amount of selected ion-implanted element can provide long-lasting protection.

5.4.3.1.1. Chromia-Forming Stainless Steels

High-chrome steels, with over 18% chromium, resist high-temperature oxidation by development of a scale largely composed of chromium oxide, Cr_2O_3. This grows parabolically with time by the out-diffusion of chromium cations. According to Smeggil [36], the adherence of this scale to the substrate is sometimes jeopardized by the segregation of a trace impurity, sulfur, to the oxide–metal interface. Both these mechanisms can be controlled by the ion implantation of yttrium, or certain other reactive elements, including the rare earths. Yttrium is an oversized atom and tends to segregate to oxide grain boundaries, modifying their microstructure to reduce cation mobility. It segregates also to the metal–oxide interface where it combines preferentially with sulfur to provide a more cohesive compound. Oxidation rates and oxide spallation are both reduced.

5.4.3.1.2. Titanium Alloys

Titanium, by contrast, oxidizes by the inward diffusion of oxygen or oxygen anions and therefore the oxide develops at the surface of the metal. Benjamin and Dearnaley [37] implanted 28 different species into polished titanium and found that barium was particularly effective in reducing oxidation. It was proposed that this is due to the formation of precipitates of barium titanate ($BaTiO_3$) within the oxide. Lucke *et al.* [38] subsequently examined similar specimens by selected-area electron diffraction, and determined that $BaTiO_3$ is indeed formed, and exists in a topotaxial relationship to the cubic sub-oxide, TiO, present at the oxide–metal interface. It is inferred that the barium titanate acts as a diffusion barrier to oxygen ions. A similar effect is known to occur in aluminum doped with NiO, the spinel $NiAl_2O_4$ precipitating coherently with NiO.

5.4.3.1.3. Zirconium Alloys

Despite the chemical similarity of Zr and Ti, their oxidation behavior is markedly different, due to the differences in oxide microstructure and defect character. Extensive studies by Bentini *et al.* [39] established that all the ion-implanted elements that reduce oxidation of Zr at 450°C have ionic sizes between 80% and 100% of the value for the Zr^{4+} ion. It was conjectured that incorporation of these smaller cations into the oxide lattice facilitate dislocation climb, which is the likeliest mechanism of plastic flow in oxides. Some experimental measurements of plasticity support this model. Such flow will reduce the growth stress induced by oxygen in-diffusion and a consequent increase in volume. Thus ion-implanted molybdenum can reduce the oxidation rate by almost 40%.

5.4.3.1.4. Alumina-forming Ferritic Steels

Certain iron–chromium–aluminum–yttrium alloys owe this high-temperature performance to the ability to develop a scale which is largely the very coherent oxide, Al_2O_3. This requires an Al content above about 5 wt.%, and unfortunately this reduces the ductility of the steel. Bernabai *et al.* [40] therefore investigated an alloy with only 1.45% Al content, and ion-implanted additional aluminum into the surface. The oxidation rate at 1100°C was thereby reduced by a factor of 140. Analysis revealed the formation of a 3-μm-thick Al_2O_3 scale which must have formed by migration of Al atoms form the bulk, combining with oxygen. The protective barrier radically altered the oxidation mechanism.

5.4.3.2. Applications to Thermal Oxidation

In this section three representative examples will be given of the application of the principles outlined above to technologic problems of sufficient importance to justify the relatively high cost of ion implantation.

5.4.3.2.1. Nuclear Fuel Cladding

The cladding or casing for the nuclear fuel in the British Advanced Gas-Cooled Reactor (AGR) is made from a high-chrome austenitic stainless steel, Fe–20Cr–25Ni–Nb. Thermal oxidation at 900–1000°C leads to the formation of a protective Cr_2O_3 scale which, however, is subject to spallation during thermal cycling of the reactor. Not only does this cause loss of protection, but also the abrasive oxide spall, now highly radioactive, can collect at locations around the coolant loop, with hazardous consequences.

In a decade-long research program Bennett *et al.* [41] investigated ion implantation and found that yttrium, at a dose of 10^{16} ions cm^{-2} will reduce oxidation by a factor of 4 but most importantly, the time taken to reach the point at which stress causes oxide spallation to occur was much increased.

Design studies showed that a single ion implantation facility, with a Freeman ion source would be sufficient to treat all the fuel elements required for UK nuclear power production each year. The project did not go ahead, not for technical reasons, but because the government decided to curtail the nuclear power program for environmental and other political reasons.

5.4.3.2.2. Aeroengines

The nickel-based superalloy Inconel is widely used as a heat-resisting material in aeroengines but it is subject to oxidative wear at 700–900°C. Wright, at Rolls-Royce Ltd., studied yttrium-implanted Inconel in a button-on-plate wear test at high temperatures, and found a substantial reduction in wear. The improved adherence of oxide scale may have also been contributing factor. The ion dose was 10^{16} ions cm^{-8}. It is also reported that implanted components performed well in an engine test over a 1000-h period.

5.4.3.2.3. Power Generation Plant

High temperatures and corrosive atmosphere are encountered in plant used for electrical power generation from oil or coal. This example is from an oil-fired power station at Fawley, UK and concerns the burner orifices through which fuel oil is injected to produce a flame that heats the boiler tubes. This power station has

128 burners each generating 17 MW of electrical power. The rapidly injected oil has entrained within its particles of sand, from its place of origin, and these together with oxidation and sulfidation lead to oxidative wear of the 6-mm diameter orifice. If this wears (usually in an asymmetric manner) control of the flame direction is lost and repair of the boiler tubes is costly. The normal life of such a steel burner is about 2500 h, and it is taken out of service when the orifice diameter has increased by about 100 μm.

Dearnaley and Peplow [42] investigated several ion-implanted species to reduce the problem. Nitrogen ions were ineffective. Cerium ions, while successfully reducing sulfidation, gave rise to increased oxidation and the scale was red, with Fe_3O_4 rather than black with FeS. Titanium and zirconium at a dose of 10^{17} ions cm^{-2} at 300 keV energy were then assessed. Ion-implanted burners provided over 8000 h of operation, representing a reduction in the erosion rate by a factor exceeding 12 times. There are other potential applications for the treatment in combustion equipment.

Note that once again, this is an application in which a small area focused ion beam is preferable to implant just the periphery of the orifice. Broad beams, or PBII, would be inappropriate. The equipment shown in Fig. 5.11 would, however, be well suited for this application.

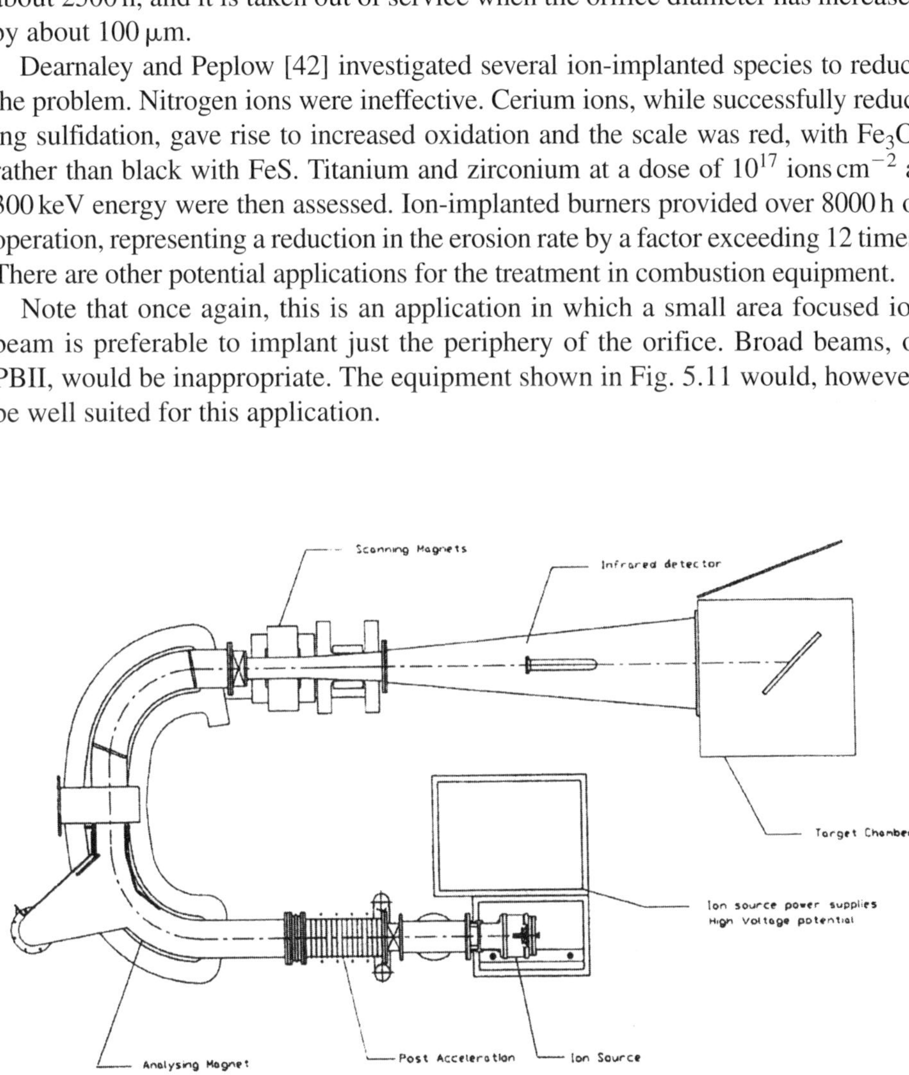

Figure 5.11: The Triion Model 100 ion implanter. The overall dimensions are 4.5 m × 3.6 m × 2.6 m high. Adapted from Torp *et al.* [1]. The ion source is the CHORDIS type (Fig. 5.4).

5.5. Fatigue Life Enhancement

Cyclically applied stresses of sufficient magnitude may cause a metallic component to fail by a sequence of steps involving dislocation migration, slip, crack initiation and crack propagation. It is a significant cause of failure in many high performance systems such as aircraft, rotating machinery, press tools and bearings.

Hard coatings are not the solution because they may crack and so cause high local stresses to develop that propagate rapidly. Soft coatings are easily worn away. Therefore, about 25 years ago, interest developed in the use of ion implantation to improve fatigue life. The principles behind this are as follows. Fatigue crack initiation and slip are processes that occur at the outer surface of the material. They involve dislocation movement and pileup, and so the ion implantation of a fine-dispersion of precipitates can prevent these mechanisms. Furthermore, in many situations (e.g. in a rotating bend test) the peak stresses are at the surface, so that here again a surface treatment is appropriate. Ion bombardment can introduce large compressive stresses into the surface and this is beneficial in countering applied tensile stresses.

5.5.1. Steels

The early investigations were made on the fatigue life of nitrogen-implanted steel, by Hirvonen. Doses of around 3×10^{17} N^+ ions cm^{-2} were implanted into Type 1018 steel specimens for a rotating bend test. Dramatic increases of one to two orders of magnitude in fatigue life were reported, the best results being achieved after a thermal anneal at 100–150°C for several hours. Similar work was carried out by Hartley [43].

A more common form of fatigue is that which occurs in bearings: an example involving pitting corrosion was cited in the previous section. Generally, rolling contact fatigue (RCF) in bearings is not exacerbated by corrosion. White and Dearnaley [44] used a standardized four-ball test to investigate the performance of steel bearings of Type EN31 steel implanted with about 3×10^{17} nitrogen ions cm^{-2} under lubricated conditions. A significant improvement in RCF performance was observed, and presented in the form of Weibull plots of the failure times.

Likewise, in simple rotating bend tests of cylindrical specimens Hu *et al.* [45] reported dramatic increases in fatigue life of nitrogen ion implanted 1018 steel, especially if the specimens were aged for 6 h at 100°C following a dose of 2×10^{17} N^+ ions cm^{-2}. Lo Russo *et al.* [46] also reported increases in the number of cycle to failure of over 50 times, but their results were dependent upon ion dose and dose rate. The optimum dose in a low alloy steel was about 2×10^{17} N^+ ions cm^{-2}, similar to that used by Hu *et al.*

These results suggest that the performance is dependent upon the growth of ion nitride precipitates such as $Fe_{16}N_2$ and Fe_2N by migration and aggregation of implanted nitrogen atoms, sometimes in association with iron vacancies. It is well established that there is an optimum size of precipitate that produces the maximum amount of local lattice strain, and is thereby most effective in hampering dislocation movement. In turn, this reduces the tendency to form persistent slip bands resulting in a smoother, more homogeneous surface.

In low-cycle fatigue, however, in which applied stresses are much greater other factors became more important. Jata and Starke [47] suggest that the introduction of a high surface compressive stress may be beneficial. This is borne out in unpublished work at Rolls-Royce Ltd. on cerium implanted Waspaloy, a nickel-based superalloy. Remarkable improvements in low-cycle fatigue life were observed in test specimens 25-mm thick. Oversized atoms like cerium in this case, or aluminum in copper, discussed by Jata and Starke, appear to be effective in strain controlled or low-cycle fatigue.

5.5.2. Titanium Alloys

Titanium alloys are favored for rotating machinery because of their lightness, but they are subject to fatigue failure, which can have catastrophic results in, for example, an aircraft turbine compressor.

Vardiman *et al.* [48] ion-implanted nitrogen and carbon into the alloy Ti–6Al–4V at 75 keV and to a dose $2 \times 10^{17}\,cm^{-2}$. The rotating bend specimens were then thermally annealed at 500–400°C, respectively to facilitate the growth of second-phase precipitates. In high-cycle fatigue tests the carbon implant was particularly successful extending fatigue life by at least a factor of 3. Similar results have been reported for nitrogen implanted Ti–24V alloy, but it was not found to be an effective treatment for strain controlled low-cycle fatigue.

5.5.2.1. Fretting Fatigue in Titanium Alloys

Fretting is a type of surface degradation that results from relatively small amplitude reciprocating movement, or vibration, under load. If cyclic stresses are imposed, then the process is called fretting fatigue and it can dramatically reduce the life of components. An example of this is in turbine blades which are secured to the disk by dovetailed roots. These blades are subjected to severe vibration.

Fretting wear and fretting fatigue are both associated with the local production of oxidized wear debris, which by remaining in place goes on to induce secondary abrasive wear. The native oxide or the metal spalls off under stress: the metal then reoxidizes and the cycle continues.

Following their prior success with carbon and nitrogen ion implantation into Ti–6Al–4V to extend fatigue life, Chakravarty *et al.* [49] went on to investigate these treatments for fretting fatigue in a loaded two bridge system. The results were disappointing in that there was no significant improvement: the customary treatment by shot peening to impart a surface compressive stress was much more effective.

This example shows that hardening the surface of the alloy is not the way to reduce fretting fatigue. The wear debris, containing TiC or TiN will be even harder than TiO_2 and hence more likely to abrade the oxide. Neither of these species reduces the oxidation rate or renders the oxide more adherent.

Syers [50] therefore took a different approach. Having shown that ion-implanted barium will reduce oxidation of titanium, they chose to implant Ba^+ to a dose of 2×10^{16} ions cm^{-2}. In some cases the alloy, Ti–6Al–4V, was first shot peened to an Almen intensity in the range 8–12. Pads of the treated materials were tested at Rolls-Royce Ltd. in an Amsler Vibrophore machine at a contact pressure of 93 Mpa. The tests showed that out of 40 different surface treatments, this ion-implantation procedure was equal second, being exceeded only by shot peening followed by D-gun detonation coating with tungsten carbide. There was no optimization of the barium ion dose, and therefore this proved to be a very successful test in comparison with well-established processes. There is the advantage that components would not require final dimensional corrections or special surface preparation. It is interesting to note that, by comparison with barium, ion-implanted cesium, strontium and ytterbium were less effective in enhancing fretting fatigue life, and in the same ranking order as their ability to reduce the oxidation rate. In keeping with this observation is the fact that the least effective treatment (out of 40) was the Tiduran™ process, in which the alloy is deliberately oxidized.

5.6. Related Techniques

5.6.1. Ion Beam Mixing

Ion beam mixing provides an alternative to ion implantation as a means for introducing a chosen additive. It is to be distinguished from ion beam-assisted deposition (IBAD) which aims to produce a coating (see Section 10.4).

There are several mechanisms that can be involved in the mixing of a thin superficial layer with its substrate under ion bombardment. These are:

- recoil implantation;
- cascade mixing;
- radiation-enhanced diffusion.

The first of these requires a direct hit between an incident ion and a coating atom, transferring kinetic energy. Such head-on collisions are rather rare: Nelson calculated the under 100 keV neon ion bombardment of a thin boron coating on silicon, only 2% of the boron recoils would have an energy greater than 1 keV. Mixing by recoil implantation therefore occurs over a distance of a few nm.

A more efficient process of mixing across an interface is provided when relatively heavy, energetic ions such as Xe^+ are incident. These will generally displace, by collisions, every atom along their path and secondary displacements result from the recoils. Coating atoms become intermingled by a random-walk diffusion-like mechanism and it is possible to arrive at an effective diffusion coefficient. This should not obscure the fact that this is a purely collisional process that occurs very rapidly and does not involve thermal activation. Further information is given in a review article by Dearnaley [51].

5.6.1.1. Radiation Enhanced Diffusion

Under appropriate conditions the redistribution of coating atoms by their interaction with point defects, generated in large numbers during ion bombardment, can provide a far more efficient means of mixing, and one that lends itself more readily to practical applications. Impurity-defect complexes may be relatively mobile and other thermally activated diffusion processes can occur in the presence of a concentration gradient of the defect population. Such diffusion takes place over a time-scale that is very long compared with that of purely collisional relocation. The diffusion length is determined by the density of sinks, such as dislocations or grain boundaries, at which defects are trapped. Under some conditions the diffusion may extend beyond the collision cascade or damage profile of the incident ions.

If the binding energy between a solute atom and a vacancy (or an interstitial) is large, then the complex will migrate as an entity in the same direction as the flux of point defects. If the binding is weak, however, the flux of vacancies in one direction is accompanied by a reverse current of both solute and solvent atoms. Marwick has studied these effects in irradiated dilute alloys, showing how silicon in nickel is transported out of the damage profile induced by irradiation with 75 keV nickel ion at 500°C. Dearnaley quotes unpublished work by Watkins in which various metallic ions implanted into titanium at temperatures above 500°C became transported beyond the ion range, to create a bimodal depth distribution. Watkins and Dearnaley attribute the deeper peak to vacancy-assisted migration to sinks and liken it to the manner in which seaweed is transported to form a line at the high water mark.

Another effect occurs in the weak binding situation, and is described by the non-equilibrium thermodynamics discussed by Howard and Lidiard. If the solute, or coating atom has a greater probability (jump frequency) of jumping into a neighboring vacancy site than do the host atoms, then the impurity will migrate up the concentration gradient of the said vacancies. This is called the inverse Kirkendall effect, because it is the opposite of the flow of vacancies generated by interdiffusion of two species. It has been discussed in more detail by Dearnaley and by Marwick [52] who proposed an expression for the solute flux, J_s as:

$$J_s C_v^e = n\, C_s (D_s - D_l)\, \nabla C_v - n\, D_s\, C_v\, \nabla C_s$$

where C_v^e is the equilibrium vacancy concentration, C_v is the local vacancy concentration under irradiation, C_s is the solute concentration, D_s is the thermal diffusion coefficient of the solute, and a D_l is the thermal self-diffusion coefficient of the host atoms. The first term on the right indicates the solute flow induced by a concentration gradient of vacancies, and a difference between D_s and D_l. The second term denotes the normal radiation-enhanced diffusion of solute driven by its concentration gradient. Marwick's formula does not take into account the presence of divacancies nor the role of interstitials but it conveys a clear model of a mechanism of mixing that can be remarkably efficient.

Thus Dearnaley [51] has reported the bombardment-diffusion of thin metal coatings on gold. A film of cobalt, about 40-nm thick, was irradiated with 400 keV nitrogen ions at 400 keV energy at temperatures of 350°C and above. Analysis showed that concentrations of up to 30 at.% of cobalt were introduced to depths around 300 nm. About one coating atom per ion was transported. This concentration of cobalt well exceeds the equilibrium solubility limit, but the ability to mix insoluble species using ion beams was further demonstrated with a coating of iridium on gold irradiated at 600°C. The combination was chosen because neither element has any significant equilibrium solubility in the other, and yet efficient mixing took place. The high non-equilibrium vacancy concentration appears to act like another constituent in the alloy, and to stabilize the incorporation of oversized solute atom.

5.6.1.2. Applications of Ion Beam Mixing

The earliest applications of radiation-enhanced diffusion were in semiconductor technology, for the formation of conducting interconnects of metal silicides. Platinum films or silicon can be mixed under ion bombardment to yield platinum silicide. Cobalt silicide is more conductive, with an electrical resistivity of 9×10^{-6} ohm cm.

In metals, Watkins and Dearnaley modified the titanium alloy Ti–6Al–4V by bombardment-diffused tin, deposited to a thickness of about 70 nm. A dose of nitrogen ions of $4 \times 10^{16}\,\mathrm{cm}^{-2}$ transported all the tin, at a temperature around 500°C, at which vacancies are mobile in titanium. Analysis showed tin at depths up to about

0.5 μm. The authors explain these results as follows. Tin is oversized in titanium and consequently is likely to have a strong binding to vacancies. Diffusion of tin-vacancy complexes at 500°C is evidently able to take place well beyond the range of 200 keV nitrogen ions. Tin atoms create around them several sites that are favorable for the trapping of nitrogen: it is used in Zircaloy to avoid the deleterious effects of nitrogen in solid solution. Hence the final distribution may contain Sn-N substitutional-interstitial pairs that would be highly effective obstacles to dislocation movement, reducing both wear and fatigue. In pin-on-disc wear tests there was dramatic reduction in friction and wear between titanium pins and the treated discs (Fig. 5.12).

In another series of experiments aimed at practical applications, Galerie irradiated 140 nm thick coatings of silicon on iron with 300 keV argon ions to a dose of $2 \times 10^{17}\,cm^{-2}$ over the temperature range of 350–510°C. The radiation enhanced diffusion coefficient was constant over this range at $10^{-14}\,cm^2\,s^{-1}$ which is orders of magnitude greater than the equilibrium thermal diffusion rates. The transport of silicon was very efficient, with about 3 Si atoms per ion.

In thermal oxidation in dry O_2 at 600°C the rate of oxidation was 45 times less than that of iron itself: the oxidation kinetics remained parabolic, and it was deduced that sufficient silicon has been introduced to form a coherent film of SiO_2 at the metal–oxide interface.

Thus ion beam mixing can provide a viable method for introducing chosen species into the surface of metals and alloys, without the limitations imposed by equilibrium solid solubility. Ion beam mixing has also been applied for the preparation of bimetallic catalysts, the subject of Section 5.7.1 below.

5.6.2. Ion Beam Texturing

Ion beams, generally of easily generated argon ions at around 40 keV energy, have successfully been applied to the polishing or micro-texturing of solid surfaces. Paradoxically, the process can be used to smooth surfaces or by contrast to impart a roughened topology which is accompanied by a large increase in surface area, changes in reflectivity, etc.

Both of these applications are a result of sputtering, the kinetic removal of surface atoms, and in some cases, molecules. It has long been known that the sputtering coefficient, *S*, the number of sputtered atoms per incident ion, depends upon the angle in incidence of the beam to the surface. *S* increases roughly as the inverse cosine of this angle due to the variation in the rate of collisional energy loss as a function of depth normal to the surface. Hence it is easy to see that the rate of removal of micro-protrusions or asperities will be greater than that from a plane surface

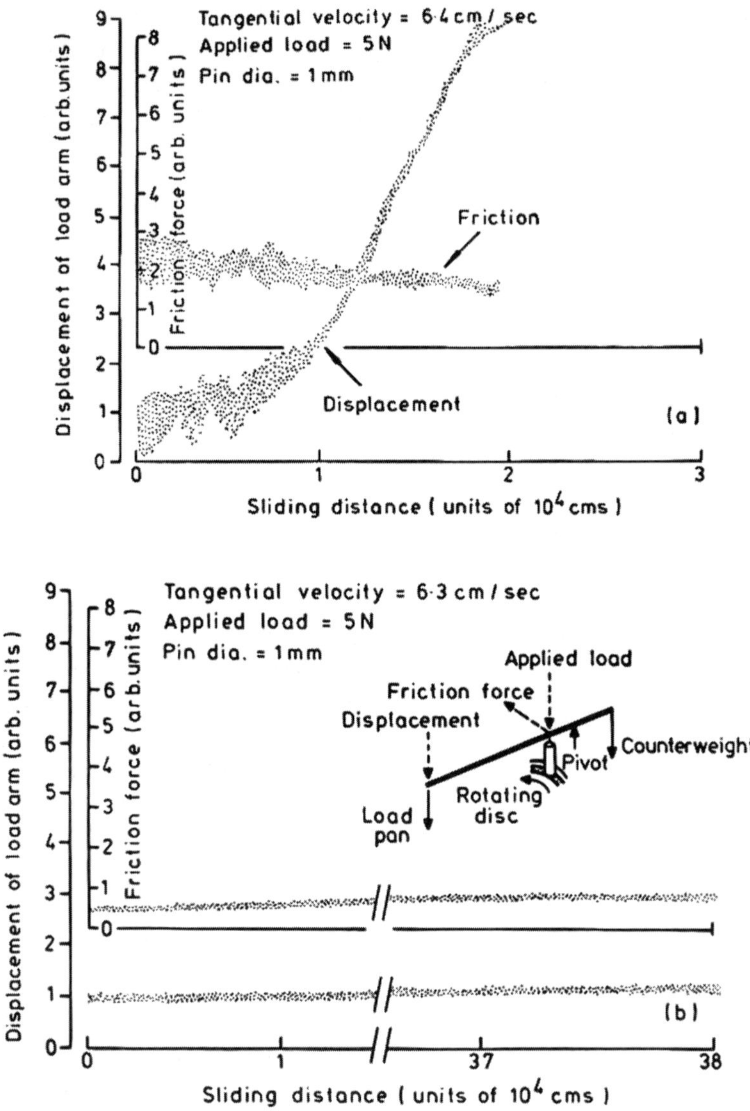

Figure 5.12: The results of tests of friction and wear between titanium pins and discs; (a) without treatment; (b) after ion beam mixing of a 70 nm deposited layer of tin, using 4×10^{17} nitrogen ions cm^{-2} at 450°C. The displacement of the load arm is a measure of the wear of the couple. Note the greatly extended time-scale in the lower figure. (after Dearnaley and Watkins, unpublished).

exposed to a beam at normal incidence. This produces a measurable smoothing of an already relatively smooth surface such as that of a steel bearing. Such effects have been observed by Dearnaley and by Hubler, though not published. It is a consequence of relatively short periods of ion bombardment, but is a useful means for removing protruding asperities that are prone to wear, yet leaving re-entrant grooves or pits that are valuable as a means for retaining lubricant.

In order to ensure that sputtering removes asperities uniformly it is preferable to rotate the workpiece about the normal to its surface. This procedure has successfully been applied to glass. Schroeder *et al.* [53] measured the fraction of light scattered from glass surfaces and found that ion beam polishing readily equaled the very best mechanically polished surfaces. The fraction of light scattered could be as low as 0.06%. However, if the argon ion energy used is too high, implanted argon will form micro-bubbles and blisters that scatter incident light. Schroeder *et al.* established that an argon ion energy of 30 keV will avoid this problem.

In the case of a perfectly homogeneous, non-crystalline, unstrained surface it may be expected that sputter erosion will occur uniformly, and indeed this is the basis for SIMS and Auger electron spectroscopy (AES) analysis as a function of depth. However, real solid surfaces are not so homogeneous, and contain a number of intrinsic and extrinsic defects which modify the sputtering rate.

Extrinsic defects include impurity inclusions, local variations in native oxide thickness and possible surface contamination. Intrinsic defects include grain boundaries, dislocations and defects, which aggregate during heavy ion bombardment. The progressive interaction of all these defects with sputter-induced profiles is highly complex: a good introduction to this subject has been given by Carter *et al.* [54].

Carbide inclusions and localized thickening of the oxide on a metal surface will generally diminish ion sputtering, causing the development of a conical protrusion upon which sometimes a protective cap can be observed. Whitton [55] has shown that such cones, generally with planar crystallographic facets, can also occur at ridges along grain boundaries or at any elevated discontinuities. When pits, analogous to etch pits, form the cones will often occur within them, and it has been suggested that material redeposition within the pits is a contributory factor.

Localized regions of strain associated with dislocations that intersect the surface are a cause of pit formation, and these sometimes occur in rows or "trains". The underlying mechanism is that surface binding energies are reduced at dislocations. The situation is complicated by the fact that energetic ion bombardment will induce defects that aggregate into a dislocation network and the known repulsion between neighboring dislocations produces local variations of surface binding energy. These effects can be minimized if the ion energy is kept low although Navin©ek [56] has observed dense arrays of pits on tungsten sputtered by 4-keV Ar^+ ions.

Other structures frequently observed include periodic ripples and associated repetitive surface microstructures which can lie parallel or normal to an ion beam incident at an angle to the surface. Nelson and Mazey [57] have suggested that such effects arise from the presence of stable dislocation structures.

More complex structures develop in the presence of a reactive gas in the process known as reactive ion etching. Though this is generally carried out in a plasma, similar effects can be observed during low-energy-directed ion bombardment. A recent example by Manohara [58] illustrates the potential usefulness of the method. A so-called deep-reaction ion etching (DRIE) process consists of cycles in which a surface such as silicon is alternately exposed to passivating and etching steps, sometimes referred to as the Bosch process. Silicon is first subjected to reactive ion etching in a mixture of oxygen and carbon tetrafluoride which creates a random distribution of fluorine-based compounds on the surface. Then, ion etching was performed in sulfur hexafluoride, SF_6, interspersed with passivation steps in octaflurobutane, C_4F_8. The outcome is a dense array of silicon nanotips a few microns in height with extremely sharp tips and facetted sides. Manohara suggests that these may have applications as nanoscale probes in biomedical studies.

The rich variety of parameters, including ion energy, incident angle, ion dose and dose rate, combined with the presence (or not) of reactive gases creates a possibility for producing many kinds of surface texture on solid surfaces, for potential application in catalysis, electron emission, microbiology and optics, many of which have still to be realized.

5.6.3. Intense Ion Beam Surface Treatment

Novel mechanisms come into play when a material is subjected to very intense pulses of energetic ions. In some respects these resemble the effects of intense laser irradiation, but there are differences in the manner in which the radiation interacts with a surface: ion beams undergo very little reflection, for example, and therefore provide a more predictable absorption of energy. They also offer the means for alteration of surface composition.

This research commenced, and has largely continued at well-equipped nuclear research laboratories such as the Research Institute for Nuclear Physics in Tomsk, and Los Almos National Laboratory, because of the specialized nature of the equipment required. Various acronyms have been applied to name the process, including HPPIB, PPIB and IBEST among others: the terminology has not yet stabilized.

Remnev *et al.* [59] have described in outline two systems in use in Tomsk, based upon magnetically isolated vacuum diodes as ion sources. One machine bearing the acronym MUK operates up to 150 keV with a pulse duration of 20–20 ns and is capable of generating beams of metallic ions such as Al, Mg, Fe and W in multiply-charged states. The beam intensity at the target can reach 1–10 A cm^{-2}. The second unit, called TEMP generates ions of carbon or hydrogen at up to 300 keV and it is claimed that the intensity, during 50-ns pulse, can reach 40–200 A cm^{-2}. In early work on steels [60] it was observed that the surface hardness was almost doubled, but the smooth surfaces developed many small craters.

Shulov *et al.* [61] applied carbon and hydrogen ion implantation to titanium and nickel-based alloys that are used in gas turbine engines. They claimed an increase in fatigue strength by 30–200%, improvement by two times to three times in oxidation resistance and by six times to eight times in salt corrosion testing. In a related paper [62] Shulov *et al.* showed that 300 keV ions would modify properties of titanium alloys to a depth of up to 100 μm. In this study up to 100 intense pulses were delivered, with fluxes of 20–220 A cm^{-2}. Korotaev *et al.* [63] further studied crater formation in irradiated metals, and described dislocation and micro-twin rotation defects, microcracks, dislocation networks and amorphization.

Davis *et al.* [64] at Los Alamos reported an accelerator called CHAMP operating at 150-200 keV, with 15,000 A cm^{-2} beam intensities during 1 μs pulses at up to 1-Hz repetition frequency. Applications were anticipated for film deposition, ion beam mixing, surface cleaning and for improving corrosion and wear resistance.

Gavrilov and Oks [65] discussed various types of ion source required for intense pulse ion beam production, giving reasons why cold cathode sources are preferred since they provide a longer life. The type of source varies depending upon the ion species required.

Han *et al.* [66] showed that the depth of the modified layer in steel far exceeds the ion range. They too observed cratering on polished steel surfaces, the craters typically being about 1-μm deep and some 10 μm in diameter.

Valyaev *et al.* [67] correlated the dislocation density produced in α-Fe by intense ion bombardment with the microhardness, and Akamatsu *et al.* [68] also working with steel (a high-speed cutting tool steel 5KH51) reported a dramatic reduction (170 ×) in grain size in the surface austenitic phase. Their work was done with 160 keV hydrogen ions at a current density of 500 A cm^{-2} during a pulse of 65-ns duration.

Applications of these effects in semiconductors, metals and ceramics, with beam pulses of $10^5 \times 10^8$ W cm^{-2} were reviewed by Piekoszewski *et al.* [69]. In ceramics improved properties were attributed to surface melting, but the effects are not always beneficial, as was reported by Shulov *et al.* [70]. In their work the fatigue strength of a VT18U alloy decreased from 380 to 240 MPa after pulsed beam treatment, while titanium alloys treated by HPPIB are more subject to pitting corrosion in saline solutions. Zhang *et al.* [71] showed that some results depend critically on the power density of the ion beam. Working with a γ^1- based superalloy they found that 25 and 37.5 MW cm^{-2} gave a significant reduction in the oxidation rate, but 50 MW cm^2 impaired the oxidation resistance. Zhang *et al.* Showed that the normal-layered scale rich in Ni was dramatically changed to one rich in Mo and Al, in a two-layered scale. This suggests that the modified alloy is more able to produce a scale containing elements with a higher affinity for oxygen than nickel has.

In summary, this body of work shows that the effects of intense pulsed ion beam treatment are relatively complex. They involve the thermal consequences of extremely large power densities, for brief periods, which are similar to those seen in pulsed laser treatment, but also there are effects due to atomic displacements by collisional processes. These defects are able to migrate at the high temperatures that briefly exist, causing damage beyond the ion range. Many of the bombardment-induced defects agglomerate into dense dislocation networks, as is the case in materials irradiated in nuclear reactors. This is probably the cause of the observed hardening and wear resistance.

It should be noted that HPPIB is not a process that brings about major compositional changes in the target. The Russian TEMP machine, producing peak currents of 200 A cm^{-2} has short pulses of 50-ns duration. One hundred such pulses deliver an ion dose of about 6×10^{15} ions cm^2 (and less if the ions are multiply-charged), which is not a high dose: the power density-related effects are much more important. Melting and very rapid cooling give rise to the fine grain structure or in other cases amorphous condition that is better able to resist corrosion. Cratering effects and other rugosities on the scale of a micron in depth are deleterious in materials used for bearings and they may, as in Shulov's work degrade fatigue life by providing stress-rising pits.

The most recent work in the field indicates that care must be taken to choose the optimum power density that is not too high. Useful benefits can be obtained. The cost effectiveness of the process has yet to be evaluated, but treatment times are relatively short.

5.7. Other Applications

5.7.1. The Use of Ion Beams in Catalysis

Heterogeneous catalysis takes place on the surface of solids, and this is just the zone that is amenable to modification by ion implantation. We have seen how it is feasible to produce ultra-fine and uniform dispersions of nano-precipitates, and also to produce metastable solid solutions of immiscible constituents. Materials can be rendered amorphous, and the presence of defects will alter the local environment of surface atoms in ways that can affect their catalytic behavior. For example, atoms may be present in unusual oxidation states. In many cases, sputtering serves to increase the effective area of the surface, to advantage in catalysis.

For these reasons, there has been a long history of studies of ion-implanted catalytic systems, both for gas-phase reactions and for electrocatalytic reactions in various solutions. More studies have been carried out in the later area because it allows the measurement and control of electrode potentials.

5.7.1.1. Gas-Phase Catalyst Studies

There have been relatively few investigations of the behavior of ion-implanted surfaces in gas phase reactions, the reason being that monitoring of such reactions is more difficult than it is in the case of electrocatalysis. Furthermore, because the density of the reactants is low it is generally necessary to use a very large area of catalytically active surface, although, as we shall see, this is not always the case.

In an early study, Wolf [72] investigated the effects of argon bombardment on the catalytic activity of various metals for the ethylene hydrogenation reaction. The activity of nickel foil increased by a factor of 100, and that of platinum foil by a factor of 10. Since argon has no catalytic activity these effects must be due to the physical consequences of ion bombardment. An increase in surface area takes place due to sputter erosion but especially in the case of nickel, there must also be a contribution to the enhanced activity due to the presence of atomic defects and the creation of surface atoms in metastable states, and perhaps unusual oxidation states.

Adopting a different approach, Rabette *et al.* [73] implanted platinum ions into single crystal alumina, magnesia and silica. Oxides such as alumina are good support media for metal catalysts, and indeed good activity was observed for Pt^+ implanted alumina, and less so in magnesia, but only after a high-temperature anneal up to 1200 k in argon. The authors propose that this treatment brings about diffusion of platinum to the surface where it forms metallic clusters. In an attempt to reduce dependence upon precious metal catalysts in automobile exhaust systems, for the conversion of CO to CO_2, etc. Brook and Dearnaley [74] implanted 20-μm diameter spheres of alumina (made by a sol-gel process) with copper and chromium, in a 1:2 ratio, in order to induce formation of copper chromite, $CuCr_2O_4$ in the surface layers. The ion implantation, at 40 keV energy, was carried out with the beam from an isotope separator, deflected electrostatically into a downward direction to impinge on the alumina spheres in a vessel that was vibrated electromagnetically. A fluidized bed was so produced, bringing about a uniform treatment of the alumina surface. The 1:2 ratio of implanted ions was confirmed by Rutherford Backscattering spectrometry (RBS) and the surface dose was estimated to be about 10^{15} ions cm^{-2}. Copper chromite is a known catalyst for CO conversion. The material was tested in the exhaust of the Honda 1.5-l gasoline-powered engine. It was found that the implanted spheres gave good catalytic activity as temperatures above 300°C and at

400°C were comparable to platinum. Poisoning of the catalyst did occur after several hours, due to constituents of the fuel.

This work exemplifies the relative ease with which a catalyst can be prepared for study: it is well suited for modern combinatorial methods for selecting an optimum catalytic combination, in binary or tertiary systems.

5.7.1.2. Electrocatalytic Studies

The principles and mechanisms involved here have been reviewed in an excellent article by Poplavskii *et al.* [75] The effects are dominated by the catalytic activity of implanted atoms: the physical effects that accompany ion bombardment have little influence. Cyclic potentiodynamic curves and related electrochemical methods are valuable tools. These authors tabulate no fewer than 44 different experiments that have been carried out, with the species, ion energies and doses. In most of the cases, an increase in catalytic activity was observed, sometimes to a very significant degree. Most studies have been of hydrogen or oxygen redox reactions, but a few relate to chlorine evolution, or to electrocatalytic conversion of organic compounds such as formic acid.

In a series of studies, Kasten and Wolf [76] implanted gold or platinum ions into iron and obtained two to three orders of magnitude increase in the activity of the hydrogen evolution reaction. This was about one order of magnitude greater than that of smooth platinum, and though some of this may be due to an increase in the effective area of the surface, some appears to be due to the existence of platinum in a highly active state.

Problems understandably arise when the catalyst is buried deeply, by ion implantation at energies of 100 keV and above. Greness *et al.* [77] encountered this in studying the modification of catalytic properties of tungsten oxide by implantation of 200 keV Pt^+ ions. A high electrocatalytic activity for the hydrogen evolution reaction was obtained only after potentiodynamic cycling which brought about the anodic dissolution of the surface, exposing buried catalyst.

Sometimes, additional problems arise during such dissolution of near-surface alloys. Thus Appleton *et al.* [78] implanted platinum into titanium at 100 to 270 keV energy. During potentiodynamic tests the catalytic activity increased, as Pt was exposed, but then fell again as the platinum atoms transformed into an "inactive" state. This is possibly due to their migration to form larger clusters, but the presence of adventitious carbon in the samples (possibly derived from pump oil) may have had an effect. For these reasons, Poplavskii *et al.* advise against the use of high-implantation energies. Ion implantation introduces a homogeneous dispersion of the implant material, in general. It is known that the highest activity is achieved if there are nano-scale clusters, and this can be the result of post-implantation thermal annealing. Thus Okada *et al.* found that heat treatment of gold-implanted titanium would increase the catalytic activity for hydrogen evaluation and the stability of the electrode properties [79].

In carbon, however, Wolf [80] found that 80 keV ion-implanted gold or platinum did not migrate to the surface during a 900°C anneal. Ion beam mixing of a thin sputtered layer of platinum did achieve good activity, but it is important to avoid loss of catalytic material by sputtering, in this process carbon is probably a poor choice of substrate for ion beam mixing for, as was seen in Section 5.6.1, diffusional mixing requires that there be a significant solute diffusion coefficient, D_s to be enhanced by a vacancy flux.

5.7.1.3. Applications

The authors are not aware of any practical applications that have so far been made as a result of the many studies, but it must be recognized that ion implantation has been a very versatile and controllable means for preparing catalytic surface for a better understanding of mechanisms.

Here we will suggest a few feasible application areas that take into account the line-of-sight nature of directed beam implantation and its cost. In the area of gas-phase reactions there are applications for small area, specially developed catalyst for the detection of contaminants, such as methane or CO in the environment. Appropriate ions would be implanted into a chosen support material, which might consist of alumina, titania or carbon, with the aim of achieving higher sensitivity as a result of an optimal distribution of surface nanoclusters, possibly incorporating more than one constituent (e.g. Pt–Ru). In the area of electrocatalysis an important application is in the preparation of electrodes for fuel cells. There is a need to conserve precious metals if their use becomes widespread, and as Poplavskii *et al.* pointed out, ion implantation involves only a "small consumption of the implanted activating element", they also remark that the incorporation can be carried out in a single operation, unlike many alternative methods.

There has been intense activity in the field of fuel cells, and for the anodes of cells utilizing methanol or impure hydrogen, one of the favored materials is a hydrogen tungsten bronze, with the general formula H_xWO_3 ($0.3 < X < 0.5$). These are metallic conductors and their catalytic activity is raised by a surface distribution of platinum, or platinum–ruthenium, as discussed by Tseung *et al.* [81] Their suggestion is to make use of the "hydrogen spillover" mechanism in which some of the catalytic steps involved in the oxidation of hydrogen spill over to what is described as an active support. The oxidation of methanol involves many steps, the first of which is a dehydrogenation reaction. Tseung *et al.* demonstrate that Pt/WO_3 can be significantly more active than platinum. The standard method of preparation involves the thermal decomposition of chloroplatinic acid, mix with tungsten oxide and carbon, to produce active powders. In view of the success of

Greness *et al.* (see above) with ion-implanted platinum in tungsten oxide, an alternative would be to prepare an electrode of hydrogen tungsten bronze and to implant this with Pt+ ions, or alternatively to carry out ion beam mixing of deposited platinum. One would choose a lower implantation energy than the 200 keV used in Greness' experiments. In this way an extremely low loading of noble metal would be required.

5.7.2. Superconductors

Both high and low T_c superconductors are finding increasingly wide application especially in instrumentation and magneto-optical devices. In this work it is important to be able to control the superconducting transition temperature, sometimes within a well-defined region. Ion implantation has provided a valuable means for achieving this, and in this section we shall give a number of examples of how it has been used.

Firstly, ion implantation has been applied in the fabrication of superconducting thin films. Bordes *et al.* [82] irradiated [001] strontium titanate substrate with 400 keV Ne^+ ions at a variety of doses, prior to deposition of $Y\,Ba_2\,Cu_3\,O_7$ (YBCO). A ion dose of $10^{15}\,cm^{-2}$ gave a significant improvement in epitaxy and an increase in T_c by comparison with a non-implanted control. This may have been brought about by a slight and controllable change in the lattice parameter of the titanate.

More recently, Peng *et al.* [83] in a patent application has described the fabrication of magnesium diboride (MgB_2) superconducting thin films and devices by ion implantation of boron ions.

In most of the published work-use has been made of the radiation damage or disorder produced by ion bombardment in order to reduce T_c controllable. Young *et al.* [84] describe how, for use in sensitive transition edge sensors (TES) and bolometers low T_c values around 10–200 mK are desirable in order to achieve the minimum Johnson noise. Thus aluminum has a T_c of about 1.4 K but this can be reduced by ion implantation of Mn^+ ions. In molybdenum films Fe^+ ions are effective in reducing T_c in good agreement with the Abrikosov–Gor'kov pair breaking model [85]. In tungsten and titanium certain transition element ions such as Fe or Mn will reduce T_c. This recent work continues.

Kuhn *et al.* [86] implanted O^+ ion at 180 keV into YBCO films and found that the superconductor could be patterned by a dose of 5×10^{14} ions cm^{-2}. The work of Hollkott *et al.* [87] showed that it is possible to achieve sub-100-nm lithography by suitable masks in the fabrication of Josephson junctions using 200-keV O^+ ions in YBCO. All the junctions could be well described by the resistively shunted junction (RSJ) model. La Graff *et al.* [88] chose silicon ions for the multilayer processing of

YBCO circuits, and in this case a dose of 10^{15}–10^{16} ions cm^{-2} was optimal. Barkow *et al.* [89] approached the problem of creating a homogeneous depth distribution over which T_c was suppressed in YBCO using multiple ion energies for oxygen, neon, nitrogen, argon and aluminum ions. The homogeneity was further improved by thermal annealing at up to 150°C. The structures produced acted as superconducting–normal–superconducting or S–N–S junctions which have a much longer barriers than conventional Josephson junctions. SQUIDS made by oxygen implantation were fabricated reproducibly and were successfully applied in measurement systems.

In Barkow *et al.*'s work the masks were a photolithographically patterned film of polymethylmethacylate (PMMA) but this is not suitable when deeper ion penetration is required for high-aspect ratio devices. Consequently, Blamire *et al.* deposited masks of gold and irradiated YBCO with protons at 30–350 keV energy. This produced high-quality Josephson junctions displaying clear AC/DC Josephson effects. The authors claim that there is good potential for high-density integration of the said devices [90].

In summary, ion implantation has been shown to provide a versatile and controllable method for tailoring and patterning superconducting properties for a variety of small-scale devices, mostly based on the Josephson junction. The ion doses required are relatively low.

5.7.3. Radiotherapy

The precision and versatility of ion implantation are valuable in its applications to radiotherapy, the usefulness of which is well recognized in medicine as a means for the selective destruction of cells.

The first of these is in angioplasty, which is a widely used and important procedure for expanding narrow arteries resulting from atherosclerosis. This is done either by expanding a small balloon or by the introduction of a metallic mesh tube, which is then expanded: this device is known as a stent. It is an all too common problem that after this so-called stenosis there is cellular growth near the end of the stent that causes reduction of the diameter of the artery, a process known as restenosis. There are several methods for countering this, for example, by powerful drugs that are bonded to the stent. Another, described by Bottcher *et al.* in 1994 [91] and Fischell *et al.* [92] makes use of low-dose beta emission from isotopes such as phosphorus 32 or yttrium 90, the effectiveness of which extends beyond the physical dimensions of the stent itself to hinder cell growth or scar tissue formation.

Ensinger *et al.* [93] have described how the ion implantation of phosphorus 32 to a dose of 5×10^{16} ions cm^{-2} at an energy of 60 keV has been used to provide the short-term beta irradiation required. Stents in this case were of AISI 316L stainless steel with a length of 15 mm. A stent has a complex geometry and thus directed ions will impinge over a wide range of incident angles. Ensinger *et al.* have computed the projected ranges of P^+ ions at these different angles, and related this to the "washout" of radio phosphorus determined following a 23-min period of ultrasonic vibration in 0.9 sodium chloride solution, a test designed to evaluate the risk of dispersion of the implanted ions as a result of corrosion in body fluids. As the authors remark, implanted phosphorus is likely to reduce corrosion in steel, as was described in Section 5.4.2, according to the work of Cooney *et al.* [94]. Only some 0.5% of the implanted phosphorus washed out during the test, a figure that is acceptably low.

In the same publication, Ensinger *et al.* described the use of ion beam mixed coating of the radioisotope yttrium 90 deposited by electron beam evaporation, and by electrodeposition. See Section 10.4 for more information on this technique. However the medical profession may be more concerned about a coating process in which, during flexure of the stent, the said radioactive coating may decohere than about the ion implantation method in which the radioactive isotope is embedded within the metallic matrix of the stent.

Fortin *et al.* [95] describe how plasma immersion ion implantation has been used to implant phosphorus 32 into stents. In their work a coaxial plasma reactor was developed: see Section 4.2.

However recent clinical tests have revealed that radioactive stents are not always successful. Golombek *et al.* [96] describe an edge effect known as "edge restenosis" in which narrowing of the artery occurs mainly at the proximal edge of the stent. Early work by Carter and Fischell [97] had deposited about 1 μCi of phosphorus 32 into Palmaz–Schatz stents. Later, Wardeh *et al.* [98] increased the activity to 6–12 μCi but nevertheless observed edge restenosis in 44% of patients, and concluded that the procedure is "clinically non-applicable". The conclusion was assessed critically by Arab, Bode and Hehrlein in an article entitled "The radioactive stent – Any chance of a resurrection?" [99]. Meanwhile good success has been reported for drug-eluted stents. It is not impossible that future designs of stents may overcome the problem of edge restenosis, but for the present the use of radioisotopes is in abeyance.

In another application of ion-implanted radiophosphorus Raymond *et al.* [100] have described the results of a pilot study on the endovascular treatment of intracranial aneurysms. This procedure is carried out with coils of platinum wire, ranging from 2 to 14 mm in diameter, inserted into the aneurysm. The beta radiation from P-32 reduces the risk of recanalization [101]. Total activities from 2 to 80 μCI were used in this study of 41 patients. The goal was to reduce the incidence of recurrences from about 20% to below 10%. Some reduction in the number of reccurrences was observed during a 6-month followup period, by about 30% (11 instead of an expected 16), and no complications from the beta radiation were observed over 10 months. However the authors estimate that it will require a larger randomized study on at least 500 patients to verify whether the ion-implanted coils can diminish recurrences of aneurysms.

The third area of application of ion implantation is in brachytherapy which is a procedure by which radioactive "seeds" are surgically introduced into the body to irradiate localized cancers, as in the case of prostate cancer. In this case a favored isotope is iodine-125, the half-life of which is about 60 days, which is a convenient period for distribution and use of the seeds. The method of preparation is a two-step process. Non-radioactive xenon-124 is ion implanted into the seed cores, and these items can very safely stored and inventoried. As needed these cores are sent to a nuclear reactor in order that some of the xenon-124 is transmuted into iodine-125 by neutron capture: the neutron capture cross-section is about 100 barns, which is favorable, but Xe-124 is a rare isotope with a natural abundance of only 0.096%. Thus efficient use of the Xe^+ ion beam must be ensured in order that this process is economically viable. It appears the US company Implant Sciences Corporation has achieved this.

It is likely that other applications of ion implantation of short to medium half-life radioactive species will develop. The list of potential radioactive isotopes includes iodine-125, palladium-103, cesium-131 and ytterbium-169 besides those mentioned. Suitable precursors for thermal neutron activation include rhenium-185, rhenium-187 and tungsten-186.

The iodine-125 brachytherapy process based on ion implantation is patented by Armini and Burkes according to US Patent 6,010,445 "Radioactive Medical Device and Process," granted in January 2000 [102], and it is noted in the patent that the ion-implanted xenon will aid the enhancement of the X-ray opacity to improve the visibility of the implantable medical device. Ion-implanted heavy elements may have other such applications in regard to their radiopacity.

The advantages of ion implantation lie in the fact that the radioactive material is embedded below the surface of the artifact, and the dose and therefore the activity can be controlled to a higher degree than in electrochemical or solution-based techniques.

5.7.4. Diffusion Barriers

It is frequently necessary to reduce the influx or permeation of certain gases, usually of low molecular weight, such as hydrogen, oxygen or water vapor. Of these, oxygen and water vapor are troublesome causes of spoilage in packaged food and beverages. Hydrogen, on the other hand, is more of a problem in metals due to embrittlement, which may be associated with grain boundaries which provide regions of lower electron density. Hydrogen may be released during the corrosion of reactive metals such as titanium, and penetrate to the interior. Means for substantially reducing the ingress or permeation of the said gases in metals and polymers are therefore of great technologic interest.

Various mechanisms involved in the ion beam modification of hydrogen adsorption by metals have been reviewed by Sugizaki *et al.* [103]. These include the electrocatalytic activity of implanted noble metal ions such as platinum by promoting the hydrogen evolution reaction at the surface. Hydrogen ingress may also be inhibited by implantation of species such as titanium that have a strong chemical affinity for hydrogen. The result is that hydrogen atoms are trapped near the surface and do not easily migrate into the bulk.

Thus Zamanadeh *et al.* [104] showed that platinum ions implanted into a thin iron membrane will significantly reduce permeation of hydrogen. In this study the permeation rate was measured as a function of the charging current density.

Ensinger and Wolf [105] demonstrated the reduced ingress of hydrogen into platinum-implanted tantalum by measuring the ductility of the metal in a bending test. An order of magnitude improvement was reported. Since native oxide also has a protective effect, as a diffusion barrier, Ensinger and Wolf recommend platinum ion implantation into the oxide layer to obtain the optimum effect.

Schwenk *et al.* [106] studied Ti^+ -implanted nickel and found substantial amounts of gettering of hydrogen in the surface layers, to a depth of about 0.1 μm. Hydrogen permeation into the bulk is suppressed. Myers *et al.* [107] found similar effects in yttrium-implanted iron.

Dearnaley *et al.* [25], in unpublished work examined the effect of platinum ion implantation on the distribution of deuterium introduced into titanium by heating in the gas. Subsequent Pt+ implantation brought about a reduction in near-surface deuterium, indicating enhanced out-diffusion of the dissolved gas.

Ferber and Wolf [108] showed that nitrogen ion implantation into titanium to a high dose of 5×10^{17} ions cm^{-2} will reduce hydrogen permeation by providing an effective diffusion barrier of TiN. Miotello *et al.* [109] deposited silicon carbide, SiC onto stainless steel as a hydrogen diffusion barrier, and went on to show that nitrogen implantation into the coating to a dose of 5×10^{17} ions cm^{-2} will significantly reduce the hydrogen penetration to a level of about 10% of that in uncoated steel.

5.7.4.1. Diffusion Barriers in Polymers

Polymers such as polyethyleneterephthalate (PET) are widely used as packaging films for food stuffs or for beverage containers. Thin films are subject to permeation by oxygen and water molecules and over a period will bring about degradation of the product. One method for suppressing this diffusion is by evaporative coating with aluminum which has a very low permeation rate. However, there is the disadvantage that the food is no longer visible, to the dissatisfaction of customers.

Studies have been made using ion implantation in order to densify the surface layers by cross-linking and by ejection of hydrogen following desorption of C–H bonds. The hydrogen released easily diffuses away. Sakudo *et al.* [110] implanted 10 keV nitrogen ions into PET to a dose of 6×10^{16} ions cm^{-2} (actually by PBII, but similar results can be expected for directed beams). The permeability of oxygen through 100-μm-thick films was reduced by a factor of 6 compared with untreated films, and of CO_2 by a factor of 10. Allowing for the fact that the modified layer is only about 0.1-μm thick, the reduction in permeability within the ion-implanted layer must be of the order of 10^4 times. Arps and Dearnaley, in unpublished work, implanted similar doses of nitrogen ions into films of nylon 66 and measured the permeability to water vapor. Again, a factor of 10-fold reduction was observed.

Yoshida *et al.* [111] deposited DLC on to PET film, together with bombardment by 15 keV argon ions. The oxygen transmission rate fell by 150 times, which is comparable with that achievable by aluminum coating. However the opaque DLC coating has the same disadvantage that the packaged product can no longer be seen.

In using ion implantation to reduce gas permeability in polymers, care must be taken not to use so high a dose that the material becomes brittle and subject to cracking, because then the permeability rises again. There has not been sufficient investigation of the optimum ion dose to reduce gas permeability in polymers.

5.8. Conclusions

We have seen that, although ion implantation of metals, ceramics and polymers has not had the major impact achieved by the technique in semiconductor device fabrication, it nevertheless has been shown to bring benefits that far exceed the cost of treatment. One example is in the improvement of the wear resistance of expensive plastic molding tools, such as the one illustrated in Fig. 5.7. Such work has provided steady business for a number of small companies, mainly in Europe. In the medical field, too, there have been successes.

The chief advantages of ion implantation lie in its versatility, controllability and thus reproducibility. It has also proved to be an excellent tool for research, for example in corrosion science, enabling the rapid preparation of test samples. In this, and other fields, the lack of the normal constraints imposed on alloy formation or solid solubility allows novel systems to be evaluated.

In many cases plasma immersion ion implantation offers lower processing costs, but this is by no means always the case as authors such as Bo Torp have stressed.

It may be advantageous to localize treatment to specific areas of a tool or component, such as the one illustrated in Fig. 5.9, by means of a directed, steerable ion beam. There is clearly room for both types of treatment.

It may be too early to judge whether intense ion beam surface treatment will be commercially successful: surface cratering and other problems still exist and the equipment is still unfamiliar to many.

Novel, often unexpected, applications for this technology and related plasma-based methods. They are both firmly established among the many processes for surface treatment of materials of all kinds.

Acknowledgment

The author wishes to thank Dr. James H. Arps of Southwest Research Institute® for many valuable suggestions and comments during the compilation of this chapter.

References

[1] B. Torp, B.R. Nielsen, D.M. Rück, H. Emig, P. Spadtke and B.H. Wolf, *Rev. Sci. Instrum.*, 61 (1990) 595.
[2] G. Dearnaley, J.H. Freeman, R.S. Nelson and J.H. Stephen, *Ion Implantation*, North-Holland, Amsterdam, 1973.
[3] G. Dearnaley and N.E.W. Hartley, *Proceedings of the 4th Conference on Scientific and Industrial Applications of Small Accelerators*, Denton, TX (IEEE, NY) 1976, p. 20.
[4] K. Langguth and H. Ryssel, *Tribol. Int.*, 26 (1993) 121.
[5] A. Kluge, K. Langguth, R. Ochsner, K. Kobs and H. Ryssel, *Mater. Sci. Eng.*, A115 (1989) 261.
[6] M.A. Moore, *Mater. Eng. Appl.*, 1 (1978) 97.
[7] C.A. Straede and N.J. Mikkelsen, *Surf Coat. Technol.*, 84 (1996) 567.
[8] C.A. Straede, *Wear*, 130 (1989) 113.
[9] B.G. Delves and G. Dearnaley, *Plastics Rubber Process*. Appl., 3 (1983) 267.
[10] J.I. Ozate, F. Alonso, I. Braceras, A.L. Saenz and R.J. Rodriguez, *Surf. Coat. Technol.*, 103 (1998) 185.
[11] J. Larsen-Basse, in *Science of Hard Materials*, Eds. R. Viswanadham, D.J. Rowcliffe and J. Gurland Plenum Press, NY, 1981, p. 797.
[12] J. Hong and J. Gurland, *Loc. Cit.*, p. 649.
[13] E.A. Almond, *Loc. Cit.*, p. 517.
[14] G. Dearnaley, *Loc. Cit.*, p. 467.
[15] L. Guzman, L. Giaghi, F. Giacomozzi, E. Voltolini, A. Peacock, G. Dearnaley and P. Gardner, *Mater. Sci. Eng.*, A116 (1981) 183.
[16] O.K. Vesnovsky, *Surf Coat. Technol.*, 52 (1992) 297.

[17] P.J. Burnett and T.F. Page, *J. Mater. Sci.*, 19 (1984) 3524.
[18] W. Bolse and S.D. Peteves, *Nucl. Instrum. Method. Phys. Res.*, B68 (1992) 331.
[19] (a) W. Yuguang, Vacuum, 69 (2003) 461; (b) W. Yuguang, Z. Tonghe, L. Andong and Z. Gu, *Surf. Coat. Technol.*, 157 (2002) 262.
[20] T. Venkatesan, *Nucl. Instrum. Method. Phys. Res.*, B7/8 (1985) 461.
[21] E.H. Lee, M.B. Lewis, P.J. Blau and L.K. Mansur, *J. Mater. Res.*, 6 (1991) 610.
[22] R.S. Bhattacharya, *Surf. Coat. Technol.*, 103–104 (1998) 151.
[23] N.E.W. Hartley, Wear, 34 (1975) 427.
[24] I.J.R. Baumvol, R.E.J. Watkins, G. Longworth and G. Dearnaley, *Proceedings of the Conference. on Low Energy Ion Beams*, IOP Conference Series.
[25] G. Dearnaley, P.D. Goode, F.J. Minter, A.T. Peacock and C.N. Waddell, *J. Vac. Sci. Technol.*, A3 (1985) 2684.
[26] A. Erdemir and C. Bindal, *Surf. Coat. Tecnol.*, 76–77 (1995) 443.
[27] W. Wei, J. Lankford and R. Kossowsky, *Mater. Sci. Eng.*, 90 (1987) 307.
[28] R.E.J. Watkins see G. Dearnaley and P. D. Goode, *Nucl. Instrum. Mech.*, 189 (1981) 117 (unpublished).
[29] W. Lohmann and J.G.P van Volkenhoef, *Mater. Sci. Eng.*, A116 (1989) 177.
[30] V. Ashworth, W.A. Grant, R.P.M. Procter and E.J. Wright, *Corros. Sci.*, 18 (1978) 681.
[31] E.C. Cooney and D.I. Potter, *Corros. Sci.*, 34 (1993) 2007.
[32] P.M. Natishan, E. McCafferty and G.K. Hubler, in *Environmental Degradation of Ion and Laser Beam Treated Surfaces*, Eds. G.S. Was and K.S. Grabowski, The Minerals, Metals & Materials Society, 1989, p. 167.
[33] J.P. Hirvonen, D. Rueck, S. Yan, A. Mahiout, P. Torri and J. Likonen, *Surf. Coat. Technol.*, 74–75 (1995) 760.
[34] S.R. Anthony, AWRE Report No. O 10/79 (1979) (unpublished).
[35] K.S. Grabowski, G.K. Hubler, E.T. Hodge, R.A. Jeffries, C.R. Clayton, Y.F. Wang and G. Kuhlman, in *Naval Research Laboratory Report 5898*, Ed. F.A. Smidt, NRL, Washington, DC, 1986, p. 191.
[36] J.G. Smeggil, A.W. Funkenbusch and N.S. Bornstein, *Met. Trans.*, 174 (1986) 923.
[37] J.D. Benjamin and G. Dearnaley, *Inst. Phys. Conf. Series*, 28 (1976) 141.
[38] W.H. Lucke, J.A. Sprague and J.K. Hirvonen, *Nucl. Ionbeam Method.*, 149 (1978) 295.
[39] G.G. Bentini, M. Berti, A. Carnera, G. Della Mea, A.V. Drigo, S.Lo Russo, P. Mazzoldi and G. Dearnaley, *Corros. Sci.*, 20 (1980) 27.
[40] U. Bernabai, M. Cavallini, G. Bombara, G. Dearnaley and M.A. Wilkins, *Corros. Sci.*, 20 (1980) 19.
[41] M.J. Bennett, G. Dearnaley, M.R. Houlton, and R.W.M. Hawes in *Ion Implantation into Metals*, Ed. V. Ashworth, Pergamon Press, Oxford, 1982, p. 264.
[42] G. Dearnaley and D. Peplow (unpublished work for the UK Central Electricity Generating Board). See also G. Dearnaley, *J. Met.* (1982) 18.
[43] N.E.W. Hartley, *Inst. Phys. Conf. Series.*, 28 (1976) 210.
[44] G. White and G. Dearnaley, *Wear*, 64 (1980) 327.

[45] W.W. Hu, C.R. Clayton, H. Herman and J.K. Hirvonen, *Scr. Metall.*, 12 (1978) 697.
[46] S. Lo Russo, P. Mazzoldi, I. Scotoni, C. Tosello, and S. Tosto, *Appl. Phys. Lett.*, 36 (1980) 822.
[47] K.V. Jata and E.A. Starke, *Defect Diffus. Forum*, 61 (1988) 37.
[48] R.E. Vardiman and R.A. Kant, *J. Appl. Phys.*, 53 (1982) 690.
[49] S. Chakravarty, R.G. Andrews, P.C. Patnaik and A.K. Koul, *J. Met.* (1995) 31.
[50] G. Syers, Materials Laboratory Report LLR 23941/FS/80, Rolls-Royce Ltd., 1980 (unpublished).
[51] G. Dearnaley, *Radiat. Effects*, 63 (1982) 25.
[52] A.D. Marwick, *J. Phys. F.*, 8 (1978) 1849.
[53] J.B. Schroeder, H.D. Dieselman and J.W. Douglass, *Appl. Optics.*, 10 (1971) 295.
[54] G. Carter, B. Navinsek and J.L. Whitton, in *Sputtering by Particle Bombardment II*, Ed. R. Behrisch, Springer-Verlag. Berlin, 1983, pp. 231–269.
[55] J.L. Whitton, T. Tanovic and J.S. Williams, *Appl. Surf. Sci.*, 1 (1978) 408.
[56] B. Navinsek, *Prog. Surf. Sci.*, 7 (1976) 49.
[57] R.S. Nelson and D.J. Mazey, In *Ion Surface Interactions*, Eds. R. Behrich and W. Heiland, Gordon & Breach, London, 1973, p. 199.
[58] H. Manohara, *NASA Tech. Briefs*, 28 (11) (2004) 62.
[59] G.E. Remnev, I.F. Isakov, M.S. Opekounov, G.J. Kotlyarevsky, V.S. Lopatin, V.M. Matvienko, A.V. Potyomkin, V.A. Tarbokov, V.L. Kutuzov and M.Y. Ovsyannikov, *Surf. Coat. Technol.*, 96 (1997) 103.
[60] D.J. Rej, H.A. Davis, M.A. Nastasi, J.C. Olson, E.J. Peterson, R.D. Reiswig, K.C. Walter, R.W. Stinnett, G.E. Remnev and V.K. Struts, *Nucl. Instrum. Method. Phys. Res.*, B127–B128 (1997) 987.
[61] V.A. Shulov, N.A. Nochovnaya, G.E. Remnev, R. Pellerin and P. Monge-Cadet, *Surf. Coat. Technol.*, 99 (1998) 74.
[62] V.A. Shulov, N.A. Nochovnaya and G.E. Remnev, *Mater Sci. Eng.*, A213 (1998) 290.
[63] A.D. Korotaev, S.V. Ovchinnikov, Y.I. Pochivalov, A.N. Tyumentsev, D.A. Shchipakin, M.V. Tretjak, I.F. Isakov and G.E. Remnev, *Surf. Coat. Technol.*, 105 (1998) 84.
[64] H.A. Davis, B.P. Wood, C.P. Manson, L.J. Bitteker, M.A. Nastasi, D.J. Rej, W.J. Waganaar, K.C. Walter, D.M. Coates and H.M. Schleinitz, *Mater. Chem. Phys.*, 54 (1998) 213.
[65] N.V. Gavrilov and E.M. Oks, *Nucl. Instrum. Method. Phys. Res.*, A139 (2000) 31.
[66] B.X. Han, S.Yan, X.Y. Le, W.J. Ahao, G.E. Remnev, M.S. Opekounov, I.F. Isakov and I.I. Grushin, *Surf. Coat. Technol.*, 1280–129 (2000) 387.
[67] A.N. Valyaev, V.S. Ladysev, D.R. Mendygaliev, A.D. Pogrebnjak, A.A. Valyaev and N.A. Pogrebnjak, *Nucl. Instrum. Meth. Phys. Res.*, B171 (2000) 481.
[68] H. Akamatsu, T. Ikeda, K. Azuma, E. Fujiwara and M. Yatsuzuka, *Surf. Coat. Technol.*, 136 (2001) 269.
[69] J. Piekoszewski, Z. Werner and W. Szymczyk, *Vacuum*, 63 (2001) 475.
[70] V.A. Shulov, N.A. Nochovnaya and G.E. Remnev, *Surf. Coat. Technol.*, 158–159 (2002) 488.
[71] H. Zhang, T. Wang, C. Wang, B. Han, S. Yan, W. Zhao and Y. Han, *Nucl. Instrum. Method. Phys. Res.*, B197 (2002) 83.

[72] G.K. Wolf, *Radiat. Effects*, 48 (1980) 237.
[73] P. Rabette, A.R. Gonzales-Elipe, M.M. Bedi, O. Delafosse, M. Che, A.U. Tench and A. Masson, *Surf. Sci.*, 106 (1981) 484.
[74] R. Brook and G. Dearnaley, Unpublished.
[75] V.V. Poplavskii, G.I. Novikov and E.V. Ratnikov, *Zhernal Prikladnoi Khimii*, 59 (1986) 995.
[76] H. Kasten and G.K. Wolf, *Electrochim. Acta.*, 25 (1980) 23.
[77] M. Greness, M.W. Thompson and R.W. Cahn, *J. Appl. Electrochem.*, 4 (1974) 211.
[78] B.R. Appleton, E.J. Kelly, C.W. White, N.G. Thompson and B.D. Lichter, *Nucl. Instrum. Method.*, 182–183 (1981) 991.
[79] T. Okada, *Electrochim. Acta.*, 27 (1972) 1273.
[80] G.K. Wolf, *Nucl. Instrum. Method.*, 182–183 (1981) 875.
[81] A.C.C. Tseung, P.K. Shen and K.Y. Chen, *J. Power Sources*, 61 (1996) 223.
[82] N. Bordes, S.N. Basu, A.D. Rollett and M. Nastasi, *Nucl. Instrum. Res.*, B59–60 (1991) 1399.
[83] N. Peng, C. Jeynes, R. Webb and G. Shao, PCT Application (May 2004) "Fabrication of Mg B_2superconducting Thin Films and Electronic Devices by Ion Implantation."
[84] B.A. Young, J.R. Williams, S.W. Deiker, S.T. Ruggiero and B. Cabrera, *Nucl. Instrum. Method. Phys. Res.*, A520 (2004) 307.
[85] A. Abrikosov and L.P. Gor'kov, *Sov. Phys. JEPT.*, 12 (1961) 1243.
[86] M. Kuhn, B. Schey, R. Kalrmann, W. Biegel, B. Stritzker, J. Eisenmenger and P. Liederer, *Physica C Supercond.*, 294 (1998) 1–6.
[87] J. Hollkott, F. Kahlmann, C. Jaekel, S. Hu, B. Spangenberg and H. Kurz, *Microelectron.* Energy, 41–42 (1998) 403.
[88] J.R. LaGraff, G.Z. Pan and K.N. Tu, *Physica C Supercond.*, 338 (2000) 269.
[89] U. Barkow, D. Menzel and S.S. Tinchev, *Physica C Supercond.*, 370 (2002) 246.
[90] M.G. Blamire, D.J. Kang, G. Burnell, N.H. Peng, R. Webb, C. Jeynes, H.H. Yun, S.H. Moon, and B. Oh, *Vacuum*, 19 (2002) 11–15.
[91] H.D. Bottcher, B. Schopol, D. Liermann, J. Kollath and I.A. Adamietz, *Int. J. Jadial. Oncol. Biol. Phys.*, 29 (1994) 183.
[92] T.A. Fischell, A.J. Carter and J.R. Laird, *Am. J. Cardiol.*, 78 (1996) 45.
[93] W. Ensinger, P. Vater, S. Heise, A. Moeslang. And K. Schloesser, *Surf. Coat. Techol.*
[94] E.C. Cooney, N.L. Lee, G.B. Fisher and D.I. Potter, *Mater. Sci. Eng.*, A116 (1989) 27.
[95] M.A. Fortin, B. Stansfield, R. Paynter, F. Marion, A. Sarassion and B. Terreault (to be published).
[96] M.A. Golombek, S. Heise, K. Schloesser, B. Schuesser and H. Schweickert, *Nucl. Instrum. Method. Phys. Res.*, B206 (2003) 495.
[97] A.J. Carter and T.A. Fischell, *Int. J. Radiat. Oncol. Biol. Phys.*, 41 (1998) 127.
[98] A.J. Wardeh, A.H.M. Knook, I.P. Kay, M. Sabate *et al.*, *Euro. Heart J.*, 22 (2001) 669.
[99] A. Arab, C. Bode and C. Hehrlein, *Euro. Heart J.*, 22 (2001) 1245.
[100] J. Raymond, D. Roy, P. Leblanc, S. Roorda, C. Janichi, L. Normandeau, F. Morel, G. Gevry, J.P. Bahary, M. Chagnon, F. Guilbert and A. Weill, *Stroke*, 34 (2003) 2801; 34 (2003) 1035.

[101] A. Molyneux, R. Kerr, I. Stratton, P. Sandercock, M. Clarke, J. Shrimpton and R. Holman, *Lancet*, 360 (2002) 1267.
[102] A.J. Armini and S.N. Bunker *U. S. Patent*, 6, 101, 445 (January, 2000).
[103] Y. Sugizaki, T. Yasunaga and H. Satoh, *Surf. Coat. Technol.*, 65 (1994) 40.
[104] M. Zamanadeh, H. Allam, H.W. Pickering and G.K. Hubler, *J. Electrochem. Soc.*, 127 (1980) 1688.
[105] W. Ensinger and G.K. Wolf, *Surf. Coat. Technol.*, 51 (1992) 41.
[106] H. Schwenk, H. Baumann, K. Bethge, S. Blumner, T. Lenz, R. Mohr and F. Rauch, in *Proceedings of the 3rd International Conference on Hydrogen and Materials*, Ed. A. Pierre, 1982 p. 485.
[107] S.M. Myers, S.T. Picraux and R.E. Stolz, *Appl. Phys. Lett.*, 37 (1980) 168.
[108] H. Ferber and G.K. Wolf, *Mater. Sci. Eng.*, 90 (1987) 213.
[109] A. Miotello, R. Kelly and N. Laidani, *Surf. Coat. Technol.*, 65 (1994) 45.
[110] N. Sakudo, H. Endo, R. Yoneda, Y. Ohmura and N. Ikenaga, to be published.
[111] M. Yoshida, T. Tanaka, S. Watanabe, M. Shinohara, J.-W. Lee and T. Takagi, *Surf. Coat. Technol.*, 174/175 (2003) 1033.

Chapter 6

Surface Treatment of Materials with Low-Energy, High-Current Electron Beams

Vladimir Rotshtein, Yury Ivanov and Alexey Markov

6.1. Introduction

During the last decade Proskurovsky and Ozur [1–3] at the Institute of High Current Electronics (Tomsk) have developed and built sources of wide-aperture ($\geqslant 50\,cm^2$) low-energy (up to $\sim 40\,keV$), high-current (up to $\sim 50\,kA$) electron beams (LEHCEBs) of microsecond duration, intended for surface treatment of materials. At the present time, these sources are superior in performance to other known sources of pulsed electron beams, pulsed lasers, sources of high-power ion beams, and pulsed plasma generators used for surface treatment. Important advantages of the LEHCEB sources are their reliability, high electrical efficiency, X-ray safety, and low cost. The promise of LEHCEBs as candidates for surface modification of materials is due to the following circumstances. The electron energy is absorbed within a thin ($\sim 1\,\mu m$) surface layer, and this, in combination with the short pulse duration, allows one to realize pulsed melting and smoothing of the surface at rather low energy densities ($2–5\,J\,cm^{-2}$). As a result of the heat transfer into underlying layers of the material, the resolidification of the molten layer occurs at a cooling rate of $\sim 10^9\,K\,s^{-1}$, which is high enough for nonequilibrium microstructures, including amorphous and nanocrystalline ones, to form. One can easily realize controllable evaporation of the near-surface layers by varying the energy density and the number of pulses, and this, in particular, makes it possible to efficiently clean the surface of oxides and other contaminants. Finally, pulsed melting of film–substrate systems allows one to produce nonequilibrium surface alloys. Thus, LEHCEB sources offer a unique tool for investigations in materials science and for developing new vacuum technologies for surface treatment of materials.

In this chapter, we consider the methods and results of simulations of the non-stationary fields of temperatures and stresses formed in metal targets in the LEHCEB-affected zone and describe the characteristics and mechanisms of evolution of the

Materials Surface Processing by Directed Energy Techniques
Y. Pauleau (Editor)

microstructure of pure metals, film–substrate systems, and alloys under pulsed electron melting. Examples of using LEHCEBs for improving surface-sensitive characteristics of alloys and hard metals are also given.

6.2. Simulation of the Temperature and Stress Fields in Metals Irradiated with Pulsed Electron Beams

6.2.1. *Temperature Fields*

The structure-phase transformations that occur in a metallic target during and after irradiation, and the microstructure and properties of the irradiated target material are determined by the temperature field induced by the pulsed electron beam. To measure the temperature of a target during its irradiation is a rather complicated problem; therefore, the temperature field is routinely determined by calculations. The problem of finding a temperature field over a wide range of power densities (up to $10^{13}\,\mathrm{W\,m^{-2}}$) is reduced to the solution of an inhomogeneous nonstationary heat equation [4,5]. Assuming that the target is homogeneous, energy is released uniformly over the beam cross section, and the beam diameter D is much greater than the dimension of the region heated due to heat conduction within the time of observation of the process, t, that is,

$$D \gg \sqrt{at} \tag{6.1}$$

where a is the heat diffusivity, we can restrict ourselves to the solution of a one-dimensional heat equation of the form:

$$\rho[c(T) + q_{\mathrm{m}}\delta(T - T_{\mathrm{m}})]\frac{\partial T(x,t)}{\partial t} = \frac{\partial}{\partial x}\left(k(T)\frac{\partial T(x,t)}{\partial x}\right) + L_{\mathrm{V}}(x,t) \tag{6.2}$$

where ρ, c, q_{m}, T_{m}, and k are, respectively, the mass density, specific heat, latent heat of melting, melting temperature, and thermal conductivity of the material; δ is the Dirac delta function, and L_{V} is the heat source function, which determines the amount of energy released in the target in a unit volume per unit time.

The heat source function can be written as follows:

$$L_{\mathrm{V}}(x,t) = \frac{j(t)E_0(t)}{r(t)e} f(x/r,t) \tag{6.3}$$

where j is the electron current density, E_0 is the initial energy of the electrons, e is the electron charge, r is the extrapolated range of electrons of given energy E, and $f(x, t)$ is the depth-normalized function of electron energy losses related to the function of specific electron energy losses as follows:

$$f(x/r) = -\frac{r}{(1 - E_\eta)E_0}\frac{\mathrm{d}E}{\mathrm{d}x}(x/r) \tag{6.4}$$

Here, E_η is the energy carried away by reflected electrons.

The energy of the electrons and the electron current density are determined from the oscillograms of the accelerating voltage and of the current flowing through the target. The function $f(x, t)$ and the range r can be found by the well-known semi-empirical formulas [6,7]:

$$f(x/r) = 1.4\exp[-(2x/r - 0.66)^2] \tag{6.5}$$

$$r = C(E_0/e)^{3/2}/\rho \tag{6.6}$$

where C is a constant equal to $10^{-17/2}\,\mathrm{kg\,m^{-2}\,V^{-(3/2)}}$.

Eq. (6.2) should be complemented with boundary-value conditions. Generally used for these conditions are the initial condition:

$$T(x, 0) = G(x) \tag{6.7}$$

where $G(x)$ is the depth distribution of the temperature in the target at the initial time, and the boundary conditions of the form:

$$k(T)\left.\frac{\partial T(x,t)}{\partial x}\right|_{x=x_\mathrm{e}(t)} = \rho v_\mathrm{e}(T)\left(q_\mathrm{e} - \frac{RT}{2M_\mathrm{r}}\right) \tag{6.8}$$

and

$$k(T)\left.\frac{\partial T(x,t)}{\partial x}\right|_{x=l} = 0 \tag{6.9}$$

Here, q_e and M_r are, respectively, the evaporation heat and molar mass of the target material; R is the universal gas constant; l is the thickness of the target; and x_e is the coordinate of the target surface that moves in a fixed coordinate system as a result of evaporation, which is written as:

$$x_e(t) = \int_0^t v_e(t)\,dt \tag{6.10}$$

where the velocity of motion of the evaporation front, v_e, can be determined by the formula:

$$v_e(T) = \left(\frac{3}{4\pi}\right)^{1/3} v_s \exp\left(-\frac{q_e M_r}{RT}\right) \tag{6.11}$$

Here, v_s is the velocity of sound in the target.

Note that the quantity q_e depends on temperature as follows:

$$q_e(T) = q_0 \tanh\left(\frac{q_e T_c}{q_0 T}\right) \tag{6.12}$$

where q_0 is the evaporation heat at absolute zero and T_c is the critical temperature.

Thus, the heat Eq. (6.2) with the boundary-value conditions (6.7)–(6.9) and the auxiliary Eqs. (6.3), (6.5), (6.6), and (6.10)–(6.12) constitute the system of equations that is used to calculate the temperature field in a target irradiated with an intense pulsed electron beam. The process of melting is simulated by the method of effective heat capacity with the help of the second bracketed term on the left of Eq. (6.2). The process of evaporation is taken into account in terms of an approximate model of thermal destruction with the help of the boundary condition on the irradiated surface (6.8). The boundary condition (6.9) reflects the fact that the back side of the target is heat insulated. This system of equations can be solved by numerical methods.

Fig. 6.1(a) gives calculated values of the highest temperature achievable on the surface of various metal targets (Z being the nuclear charge) irradiated with LEHCEBs of energy density 6 and 9 J cm^{-2}. It can be seen that these values vary from 0.75 to $4T_m$, where T_m is the melting temperature. Fig. 6.1(b) gives, for the same metals, values of the energy density necessary to melt a layer of thickness 3 μm. These values fall in the range from 3 (for Al) to 15 (for W) J cm^{-2}. The lifetime of the melt varies from 1.4 to 3.6 μs, that is, it slightly differs from the pulse duration τ ($\tau = 2.1$ μs). The cooling rate at the resolidification front and the velocity of motion of the melt–solid interface are on the average ~10^9 K s^{-1} and 1–5 m s^{-1}, respectively.

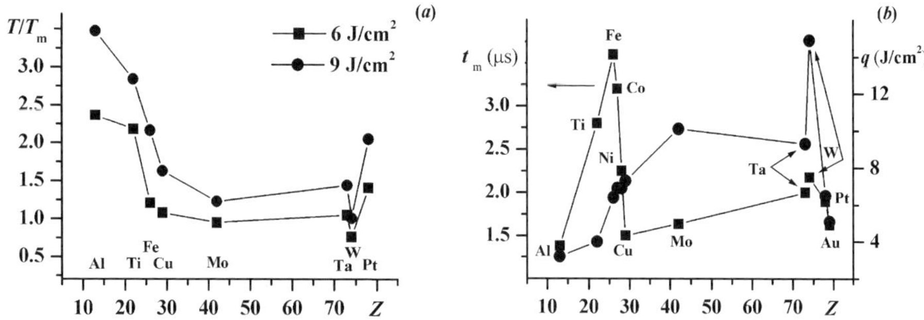

Figure 6.1: (a) Maximum surface temperature for metal targets irradiated with different energy densities; (b) the energy density q corresponding to a 3 μm melt depth and the lifetime of this melt, t_{m}. Pulse duration $\tau = 2.1$ μs.

6.2.2. *Stress Fields*

Alongside with nonstationary temperature fields, another essential factor of the action of pulsed electron beams on solids is the generation of fields of mechanical stresses and strains, which can operate not only within but also beyond the heat-affected zone (HAZ). Generally, the attention of researchers is focused on the σ_{11} stress wave generated in the irradiated target, which, in fact, is the normal component of the stress tensor that is applied to the planes parallel to the irradiated surface and acts perpendicularly to that [2,8]. If the material does not undergo phase transformations and its viscosity and plasticity can be neglected, the calculation of the longitudinal stress wave is reduced to the solution of a wave equation of the form:

$$v_{\mathrm{s}}^2 \frac{\partial^2 \sigma_{11}}{\partial x^2} - \frac{\partial^2 \sigma_{11}}{\partial t^2} = (3\lambda + 2\mu)\alpha_{\mathrm{t}} \frac{\partial^2 T}{\partial t^2} \tag{6.13}$$

where λ and μ are Lame constants, and α_{t} is the temperature coefficient of linear expansion.

The boundary-value conditions for this equation are as follows:

- The initial conditions

$$\begin{aligned} \sigma_{11}(0, x) &= 0 \\ \left.\frac{\partial \sigma_{11}(x, t)}{\partial t}\right|_{t=0} &= 0 \end{aligned} \tag{6.14}$$

- The boundary conditions

$$\sigma_{11}(0,t) = \sigma_{11}(l,t) = 0 \tag{6.15}$$

Eq. (6.13) with the boundary-value conditions (6.14) and (6.15) together with the above equations for the temperature field, constitute the system of equations of uncoupled thermoelasticity. In this case, the connectedness of the system of equations implies that not only variations in temperature field give rise to strain and stress fields (this can be seen from the right part of the wave equation where the driving force is described by the term proportional to the second time derivative of temperature), but also variations in strain field vary the temperature field. Mathematically, the connectedness is expressed by one more term proportional to the strain rate which appears in the heat equation. However, it can be shown that in our case the connectedness is not essential and can be neglected [9].

An important parameter, which describes the ability of a system to relax acoustically during irradiation and determines the stress waveform, is the ratio of the pulse duration to the acoustic relaxation time:

$$w = \frac{\tau}{r/v_s} = \frac{\tau v_s}{r} \tag{6.16}$$

With $w \leqslant 1$ a bipolar stress wave consisting of a compression pulse followed by a tensile pulse is generated in a target [10]. If a target is subject to the action of a microsecond LEHCEB, such that $w \sim 10^3 >> 1$, and with a nonmonotonic time dependence of the power density, which leads of a change of sign of the driving force in the right part of Eq. (6.13), an alternating stress wave is generated in which, generally speaking, an arbitrary number of stress pulses of different polarity follow one after another [11].

Fig. 6.2 shows stress wave profiles in an iron target of thickness 3.6 mm at various points of time. It can be seen that initially a compression pulse of amplitude $\sim 10^5$ Pa is formed and then propagates from the surface into the target bulk. The amplitude of the compressive stresses decreases in time, and the stress wave reverses its polarity for the first time: a tensile pulse is formed behind the compression pulse. As the irradiation is completed, several events of reversal of the stress wave polarity occur, and the formation of this wave is in fact over. Notwithstanding the small amplitude, such a stress wave can lead to changes in microstructure of the material beyond the HAZ, which seems to be related to the high-frequency alternating character of the loading [2].

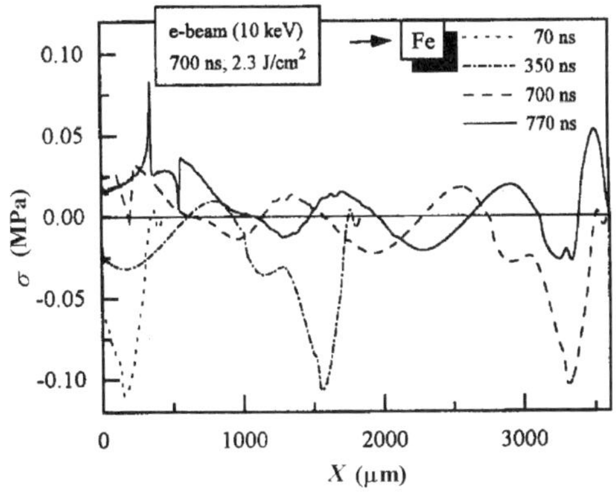

Figure 6.2: Stress wave profiles in an iron target at different points in time.

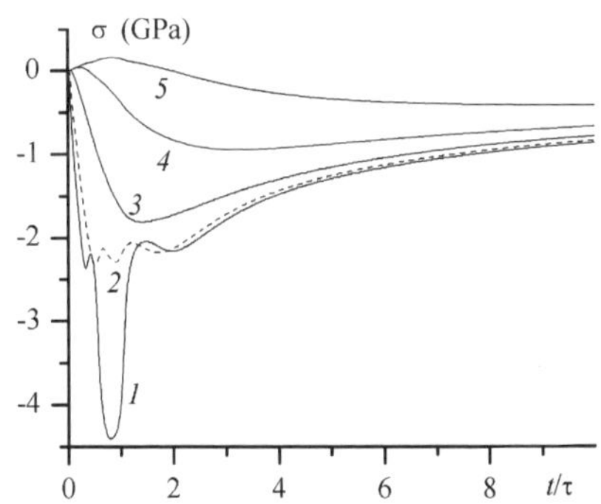

Figure 6.3: Time dependence of the quasistatic stresses at the axes of a 0.4 mm thick iron plate irradiated in the premelting mode (2.5 J cm^{-2}, 2 μs) at 0 (1), 2 (2), 5 (3), 10 (4), and 20 μm (5) from the surface.

In addition to the normal component of the stress tensor, σ_{11}, considered above, it is necessary to take into account two other normal components, which sometimes are called quasistatic. By virtue of the symmetry of the problem, they are generally equal each other ($\sigma_{22} = \sigma_{33} = \sigma$). The simplest case of calculation of these components is that of an irradiated plate, since there occurs a flat stressed state, and five of the nine components of the stress tensor are equal to zero [12]. The quasistatic components of the stress tensor are determined by the temperature field in the target and by the function that describes the bending of the target, which is calculated by solving a biharmonic Eq. [12].

Fig. 6.3 shows the evolution of quasistatic stresses in a Fe plate at different distances from the irradiated surface. These stresses are compressive, and they substantially exceed the yield strength. Comparing Figs. 6.2 and 6.3, one can see that the magnitude of the quasistatic stresses is more than four orders greater than the amplitude of the stress wave, and these stresses, of course, play an important role in the modification of the material, both within and beyond the HAZ [13].

6.3. Evolution of the Microstructure of Pure Metals

We now consider the evolution of the microstructure of pure metals treated with LEHCEBs by the example of pure iron irradiated in modes preceding to melting and in modes of uniform surface melting. It has been shown [14] that pulsed heating of pure ($\sim 10^{-3}$ at.% C) recrystallized iron near the melting threshold ($\sim$20 keV, 0.8 μs, and 2–2.5 J cm^{-2}) leads to plastic deformation of the surface layer (Fig. 6.4(a)) with the formation of an extended ($\sim$100 μm) nonuniformly hardened zone with a microhardness maximum located at a depth of $\sim$20 μm from the irradiated surface. The microhardness (Vickers) profile $H_{\mu}(x)$ follows the dislocation density depth distribution $\rho(x)$ (see Fig. 6.5), which is measured using transmission electron microscopy (TEM).

In the surface layer of thickness up to $\sim$5 μm, plastic deformation is accompanied by primary recrystallization with recrystallization nuclei (2–5 μm) appearing at grain boundaries (Fig. 6.4(b)). From calculations of temperature and stress fields, it follows that the nonmonotonic depth variations in the material substructure and microhardness are related to the plastic deformation caused by compressing (in the surface plane) quasistatic thermoelastic stresses and to the recrystallization processes caused by the heat transfer into the material bulk. The stress wave, because of its low amplitude (see Fig. 6.2), has a minor effect on the material structure and properties.

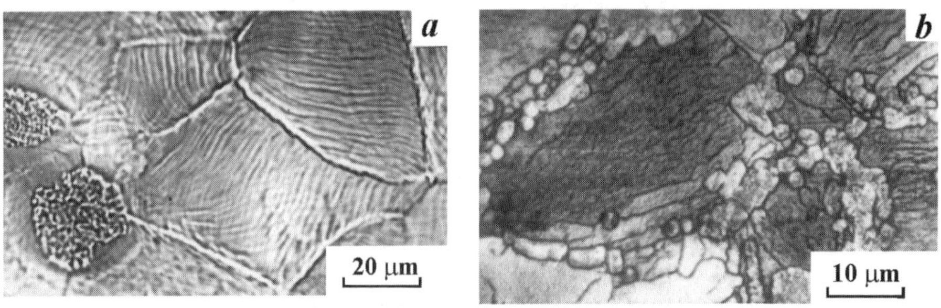

Figure 6.4: Microstructure of an iron surface irradiated near the melting threshold (2.2 J cm^{-2}, 0.8 μs, $N = 5$) (a) before and (b) after chemical etching. Transcrystalline slip traces are seen in (a). The presence of local melting regions of size $\sim$30 μm in (a) testifies to the premelting mode of irradiation.

According to calculations, the formation of a recrystallization nucleus and its growth to the characteristic size $d \sim 5\,\mu m$ occurs within $\sim 5 \cdot 10^{-6}$ s. Hence, the average rate of growth of a recrystallization nucleus (grain) is $\sim 1\,m\,s^{-1}$, which is four orders of magnitude greater than the stationary value and is comparable to the velocity of motion of the resolidification front under pulsed melting. Estimates show that at $T \sim 750$ K the diffusion range of iron atoms for a time of $\sim 5 \cdot 10^{-6}$ s is a few nanometers, which is too small for grains of micrometer size to form by the diffusion mechanism. It can be assumed that under the conditions of pulsed heating accompanied by intense plastic deformation under the action of transient quasistatic compressive stresses, the mechanism of fast dynamic recrystallization based on the feasibility of the martensitic mechanism of nucleation [15] is realized.

To determine experimentally the stresses in surface layers that operate during irradiation, a method based on microplastic stress–strain tests of thin (0.4 mm) plates subjected to bending was used [16]. The character of the bending of thin samples has confirmed that compressive stresses operated in the surface plane during pulsed heating. Estimates have shown that upon single irradiation the resulting stresses are as high as $\sim$400 MPa. These stresses are approximately equal to the dynamic yield strength for Fe, but they are about an order of magnitude lower than the calculated stresses (Fig. 6.3). This discrepancy is related to the fact that the simulation of the stress fields (see Section 6.2.2) did not took into account the plasticity of iron and the transformation stresses associated with the $\alpha \leftrightarrow \gamma$ allotropic transformation. According to calculations, this transformation occurs in a surface layer of thickness $\sim 2\,\mu m$. Since the specific volume of γ-Fe (f.c.c.) is smaller than that of α-Fe (b.c.c.) by $\sim 1\%$, this layer will be compressed during the lifetime of the γ-Fe ($\sim 10^{-6}$ s), and this will result in the appearance of tensile stresses. These stresses

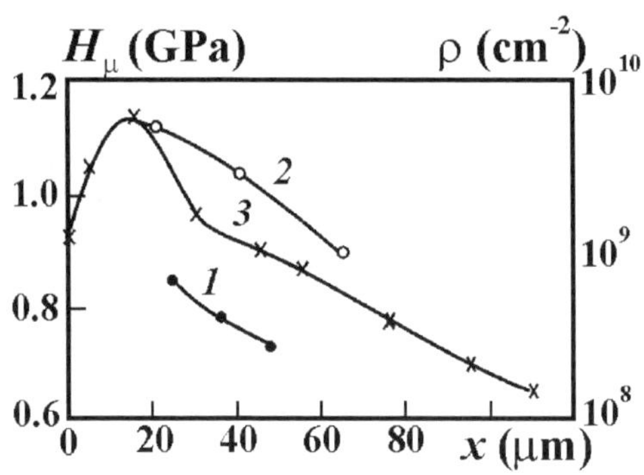

Figure 6.5: Depth distribution of dislocation density (1, 2) and microhardness (3) for iron after irradiation (2.2 J cm^{-2}, 0.8 μs): pulse number: $N = 5$ (1, 3); $N = 300$ (2). The Vickers diamond pyramid load was 0.2 N.

should substantially reduce the thermal compressive stresses, and this agrees with experimental data.

Once the pulse is completed, the near-surface layer is cooled and compressed; therefore, residual tensile stresses are induced. Measurements have shown that for the number of pulses $N = 1$–50 these stresses range between 40 and 80 MPa; that is, they are about an order of magnitude lower than the compressive stresses operating under irradiation. Experiments on samples electropolished layer by layer on the irradiated side have shown that the residual stresses have a maximum at a depth of ~5 μm and are effective within the HAZ of thickness of ~20 μm. The existence of a maximum agrees with the data on recrystallization in the surface layer of ~5 μm. The monotonic decrease in residual stresses beyond the recrystallized layer correlates with the $H_{\mu}(x)$ and $\rho(x)$ profiles (see Fig. 6.5).

As the melting threshold is achieved, the dominant effect on the microstructure of the surface layer is exerted by the processes of rapid resolidification and plastic deformation in dynamic stress fields. It follows from calculations ($\tau = 0.8$ μs) that as the energy density is increased in the range $E_s = 2.3$–$5.2\,\mathrm{J\,cm^{-2}}$, the melt thickness and lifetime increase in the ranges 0.7–2.5 μm and 0.5–3 μs, respectively. With this increase in melt thickness the melt cooling rate decreases from $\sim 10^{10}$ to $\sim 10^{9}\,\mathrm{K\,s^{-1}}$ and the resolidification front velocity decreases from ~5 to $\sim 2\,\mathrm{m\,s^{-1}}$.

Analysis of the spectra of positron lifetimes (slow positron annihilation spectroscopy) in the surface layer of recrystallized iron ($\sim 10^{-2}$% C) has shown that as a result of quenching from melt of vacancy-type defects, mainly, vacancy clasters (divacancies and trivacancies) are formed [17]. Fig. 6.6 presents the S parameter of the Doppler broadening of the annihilation peak as a function of positron energy for samples irradiated with various E_s values (the S parameter is defined as the ratio of central area to the total area of the annihilation peak).

The increase in the S parameter directly correlated with the increase in the vacancy-like defect concentration. Thus, the plots in Fig. 6.6 correspond to the vacancy concentration profiles. It can be seen that quenching from melt increases the defect concentration against the original state. Besides, as follows from Fig. 6.6, the dependence of the defect concentration on E_s at depths $x \geqslant 500$ nm is represented by a curve with a maximum at $E_s = 4.2\,\mathrm{J\,cm^{-2}}$. This dependence can be explained as follows. The concentration of quenched vacancies is determined by their nucleation at the resolidification front and the recombination at the dislocations formed under the action of thermal stresses. As E_s is increased and correspondingly increases the melt thickness and lifetime, the vacancy concentration decreases. The dependence of the dislocation density on E_s has the same character. This is associated with the decrease in temperature gradients with an increase in the thickness of the energy absorption zone and, hence, with the reduction of the quasistatic stresses. For low E_s ($2.3\,\mathrm{J\,cm^{-2}}$), a significant fraction of nonequilibrium vacancies may recombine at dislocations since

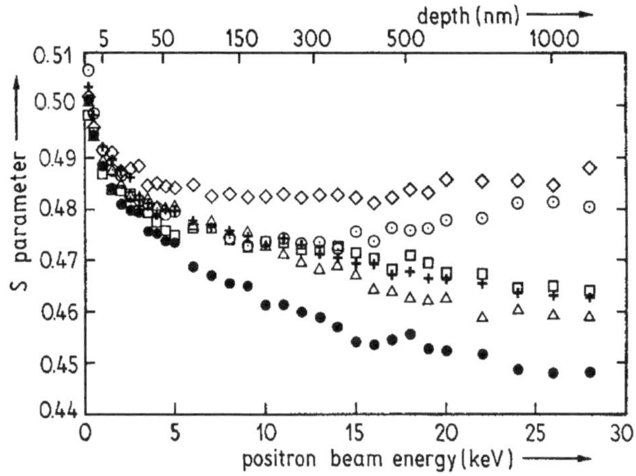

Figure 6.6: Depth distribution of the S parameter for iron before and after pulsed melting: untreated sample (●); 2.3 J cm^{-2} (△, +); 4.2 J cm^{-2} (◇,⊙); 5.2 J cm^{-2} (□).

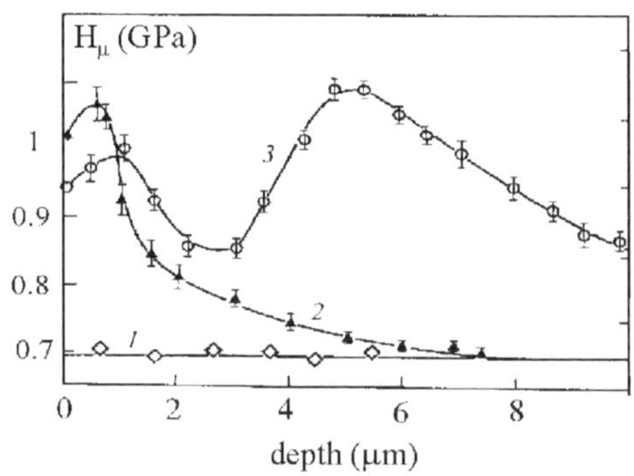

Figure 6.7: Microhardness depth profiles for iron: (1) untreated, (2) 3.3 J cm^{-2}, (3) 5.2 J cm^{-2}. The Vickers diamond pyramid load was 0.002 N.

the dislocation density is high. As E_s is increased, the probability of recombination falls because of the decrease in dislocation density. Thus, there exists a critical energy density at which the vacancy concentration is a maximum.

The $H_\mu(x)$ depth profiles are shown in Fig. 6.7. An increase in microhardness is accompanied by an increase in the wear resistance of the surface layer [18]. As follows from calculations (see Fig. 6.8), the near-surface hardened layer is formed on quenching from the liquid state. The increased microhardness in this layer is due to strain hardening under the action of quasistatic stresses formed at the stage of cooling after the completion of resolidification. The disperse carbides formed upon

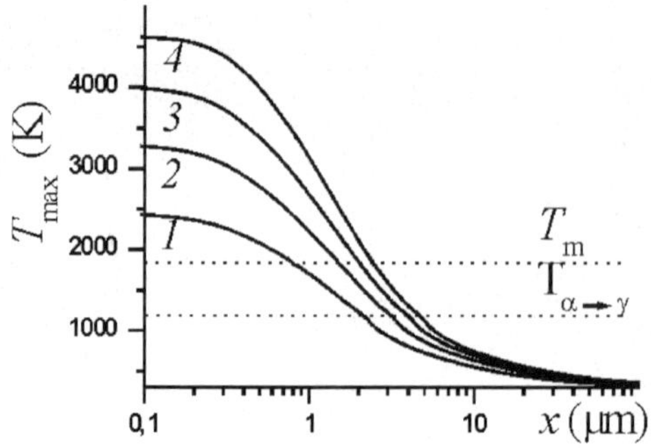

Figure 6.8: Maximum temperature vs. depth for iron: (1) 2.3 J cm^{-2}; (2) 3.3 J cm^{-2}; (3) 4.2 J cm^{-2}; (4) 5.2 J cm^{-2}.

resolidification due to the presence of carbon in the residual atmosphere of vacuum chamber and the existence of carbon-containing films on the surface can make some contribution to hardening.

The formation of the second microhardness maximum at a depth of ~5 μm for 5.2 J cm^{-2} might be related to the fact that, according to the calculations (see Fig. 6.8), the peak temperature at this depth reaches the point of allotropic $\alpha \leftrightarrow \gamma$ transformation (1083 K). Therefore, at $x \leqslant 5$ μm the compressive stresses resulting from pulsed heating will be compensated by transformation tensile stresses. The decrease in compressive stresses will reduce microhardness. At a depth $x \sim 5$ μm, the transformation stresses disappear. This should result in an increase in compressive stresses and, accordingly, in the formation of the second microhardness peak. The stress wave formed in this irradiation mode is small in amplitude, and its influence can be neglected.

6.4. Surface Cratering

Irradiation of metals and alloys with pulsed (10^{-8}–10^{-6} s) electron, ion, and laser beams leads to surface cratering. This phenomenon was first revealed for stainless steel irradiated with nanosecond laser beams [19]. First experiments performed on samples made of oxygen-free copper (~10^{-3} wt.% O) irradiated with an electron beam (20 ns, 20 keV, ~0.3 J cm^{-2}) have shown [20] that microcraters occur at a temperature of ~800°C (i.e., below the melting threshold for copper). Subsequent experiments with the use of LEHCEBs of microsecond duration have shown that microcraters are formed under irradiation of all commercial metals and alloys, preferentially, as the melting threshold is attained for the matrix [20,21] (Fig. 6.9).

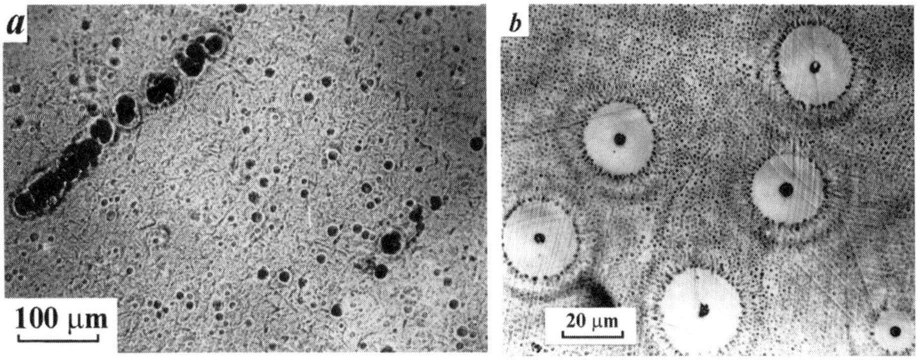

Figure 6.9: Optical micrographs of the surface of alloys irradiated with LEHCEB: (a) steel 304 (2.5 μs, $4\,\mathrm{J\,cm^{-2}}$, $N = 1$); (b) precipitation hardened α-brass (2.5 μs, $4\,\mathrm{J\,cm^{-2}}$, $N = 5$).

For the same irradiation conditions, the sizes and density of microcraters depend on the purity and phase composition of the material. For example, vacuum annealing of oxygen-free copper abruptly reduces the probability of cratering, and this is related to the removal of impurities from the surface layers.

The typical topography and the presence of "nuclei" at the center suggest that the most probable reason for cratering is the metal local overheat and melting in microvolumes at the sites of localization of impurities possessing a lower thermal conductivity and/or a lower melting temperature in comparison with the matrix or capable of creating eutectics with the matrix. For instance, in copper these impurities are insoluble Pb and Bi, which form eutectics at 326°C and 270°C, respectively, as well as oxygen and sulfur present in the Cu_2O and Cu_2S second phases that enter into eutectics. Note that the thermal conductivity of these phases is more than two orders of magnitude lower than that of copper. Direct evidence of the dominant role of second phase particles is that microcraters are absent on monocrystals of pure copper (99.999%) irradiated with an ion (mixture of C, H, and O ions) beam in surface melting modes (300 keV, 400 ns, 2 and $5\,\mathrm{J\,cm^{-2}}$, $N = 1$–10) [22].

The increase in energy density and the switching to the evaporation mode result in removal of crater-initiating impurities from the surface layer owing to which the density of microcraters abruptly decreases. Multiple irradiation in evaporation modes allows one to practically completely avoid cratering in some cases. It should also be noted that at a pulse duration $\tau \geqslant 10$ μs the probability of cratering decreases as well. This is due to the fact that the increase in the melt thickness and its lifetime have the result that second phase particles have time to be completely dissolved in the matrix [20].

Cratering leads to an increase in the surface roughness and to the formation of local regions with highly nonuniform microstructure in the near-surface layer. In the case of alloys of reduced plasticity, microcracks may appear near microcraters. The reason for this is high local tensile stresses resulting from the difference in coefficients of thermal expansion between the second phase particles and the matrix. Thus, cratering is an Achilles' heel of the surface treatment with LEHCEBs. It can result in deterioration of the wear resistance (increase in friction coefficient), corrosion resistance, and fatigue strength. Examples illustrating the possibility of inhibition of cratering will be considered below.

6.5. Pulsed Melting and Mixing of Film–Substrate Systems

The pulsed melting of the film–substrate systems is of considerable interest since it allows one to produce metastable phases, amorphous ones included, such that the thickness of the modified layer may be about an order of magnitude greater than that achievable on high-dose ion implantation [23]. From the practical point of view, LEHCEBs of duration $\sim 10^{-6}$ s have the advantage over nanosecond laser and electron beams [23] that they offer the possibility to produce surface alloys of greater thickness with a lower level of stresses. Investigations were performed for two-layer (Ta–Fe) [24] and multilayer (Al–Si and Al–C) systems [21]. These systems are characterized by a limited solvability of the components in the solid state, and the system Al–C is inconsistent in the liquid state. Also, a Cu/steel 316 system, which is of some practical interest, was investigated [25].

6.5.1. Ta–Fe

On irradiation of the Ta–Fe system (0.8–1.5 μs, 1.5–6 J cm^{-2}), as shown by calculations, the presence of a refractory Ta film (100 nm) increases severalfold the thickness and lifetime of the molten layer of the substrate (Fe) and decreases the velocity of the resolidification front as compared to pure Fe. With the Rutherford Backscattering spectrometry (RBS) and Auger electron spectroscopy (AES) it has been shown that the liquid-phase mixing has the result that the thickness of the Ta-alloyed layer is two or three times greater than that of the film. Increasing beam energy density favors the mixing and increases the thickness of the mixed layer. The effective coefficient of liquid-phase diffusion of Ta into Fe is $\sim 5 \times 10^{-5}\,cm^2\,s^{-1}$. With TEM, it has been established that the structure formed as a result of rapid quenching from melt has a complex phase composition and is inhomogeneous in depth. In the close vicinity of the surface (at a depth of

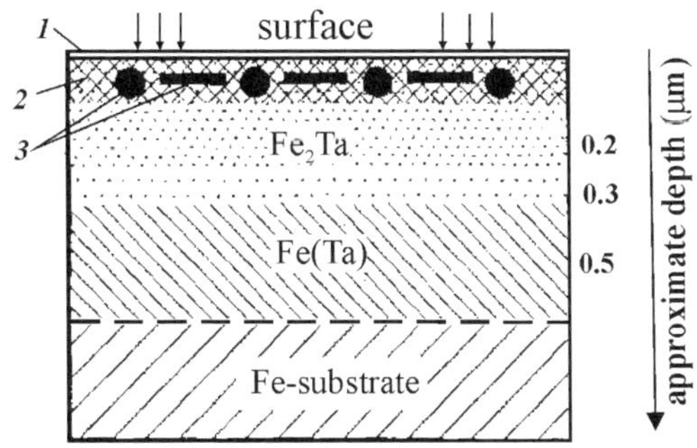

Figure 6.10: Schematic of the structure of the surface layer of a Ta–Fe system after pulsed melting: (1) oxide film; (2) Ta–Fe amorphous phase, and (3) Ta.

~0.2 μm), particles (⩽0.3 μm) of undissolved Ta, the Fe–Ta amorphous phase, and dispersed segregates of the intermetallic compound Fe_2Ta are observed. The deeper layers consist predominantly of Fe_2Ta particles and Fe(Ta) solid solution grains with a high dislocation density. The cross-section of this structure is shown in Fig. 6.10. The layer-by-layer disposition of the structures is due to the decrease in Ta concentration with depth, resulting from the diffusion of Ta into the liquid phase, and to the difference in the cooling rates at the surface and in the deeper layers of the melt. In the process of thermal cycling realized under multiple pulsed surface melting, the presence of the refractory Ta film has the result that in the near-surface layer a microstructure is formed with the dislocation density and long-range stresses elevated as compared to pure iron [24].

6.5.2. Multilayer Al–Si and Al–C Systems

In the original state, the multilayer (Al/Si/Al/Si/Al/Si/Al) system deposited on an Al substrate consists of four polycrystalline Al layers (~350 nm) separated with amorphous silicon interlayers (~50 nm). According to AES data, carbon (up to 5 at.%) is present at the Al/Si interfaces. On irradiation in the mode of melting of the Al layers below the melting threshold for Si (2.5 μs, 2.3–3.1 J cm^{-2}), the multilayered texture is retained; however, the concentration of Si in the Al layers increases abruptly due to the diffusion of Si atoms into the liquid Al. The rapid resolidification of the Si-enriched Al layers from melt results in the formation of nanosized (~30 nm) Al crystallites combined in complexes of size ~100 nm (Fig. 6.11). At the Al crystallite boundaries, nanosized particles of the second phase ($Al_4Si_2C_5$) are segregated. In the amorphous Si–Al–C interlayers, partial crystallization with the formation of nanosized Si particles takes place. As the melting

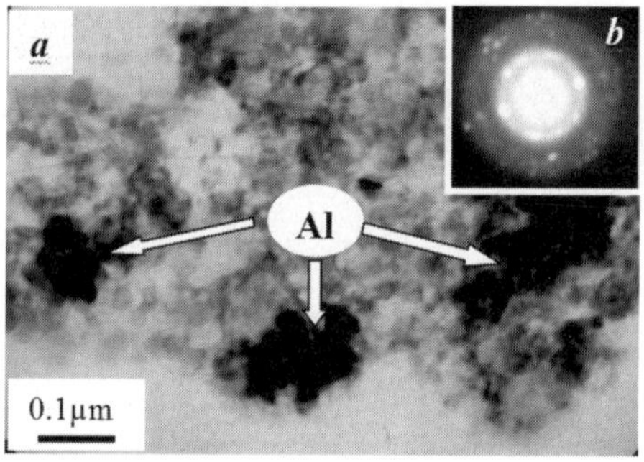

Figure 6.11: Microstructure of a multilayer (Al/Si/Al/Si/Al/Si/Al) system on an Al substrate after pulsed melting of Al: (a) bright-field image, (b) diffraction pattern (2.5 μs, 2.3 J cm^{-2}, $N = 1$).

threshold for Si is achieved ($E_s \geqslant 4$ J cm^{-2}), the efficiency of mixing decreases abruptly because of the ablation of the coating [21].

The original (Al/C/Al/C) system consists of alternating two polycrystalline Al layers (~700 and ~500 nm) and two amorphous carbon layers (~200 and ~100 nm) deposited on an Al substrate. On irradiation in the mode of melting of the Al layers (2.5 μs, 1.8 J cm^{-2}), the following structures are formed: coarse (up to ~20 μm) molten-off Al particles; regions consisting of submicrometer Al grains with nanosized (2–3 nm) particles of cubic-lattice carbon segregated along the grain boundaries; regions consisting of 50–100 nm cells of amorphous carbon with nanosized (3–5 nm) Al segregates (major type structures, Fig. 6.12), and, finally, amorphous carbon particles (a few micrometers) formed on destruction of the original carbon layers. The formation of these structures points to the stratification of Al and C on pulsed melting under the given conditions. As distinct from the previous system, the second phases (Al_4C_3) are absent, which may be related to the rather low temperature (<1700°C) and short lifetime (~10^{-6} s) of the Al melt [21].

6.5.3. Cu/Steel 316

Copper films were deposited on substrates made of austenitic stainless steel (SS) 316 by sputtering of a Cu target in the Ar plasma of a microwave discharge. To inhibit cratering, which leads to local delamination of the film on irradiation, the substrates were pre-irradiated with 8–10 J cm^{-2} and $N = 30$. The Cu (512 nm)/steel 316 samples were irradiated with $\tau = 2.5$ μs, $E_s = 2$–8.3 J cm^{-2}, and $N = 1$–5 [25].

According to calculations, the surface melting of Cu films is attained at 2–2.5 J cm^{-2}, which is in good agreement with experiment results. With increasing E_s

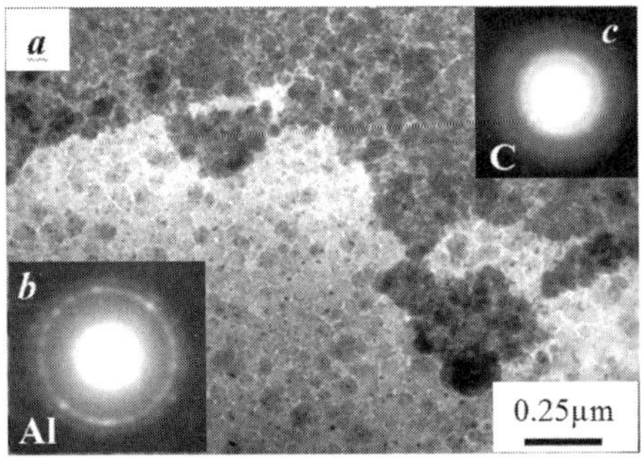

Figure 6.12: Microstructure of a multilayer (Al/C/Al/C) system on an Al substrate after pulsed melting of Al: (a) bright-field image, (b, c) diffraction patterns for bottom and top parts of (a), respectively (2.5 μs, 1.8 J cm^{-2}, $N = 1$).

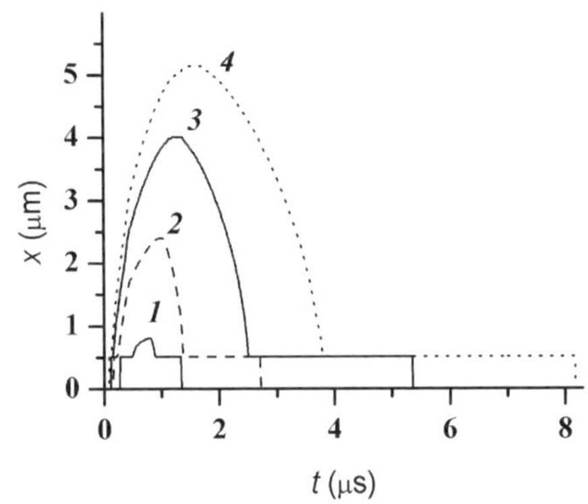

Figure 6.13: Coordinate of the melt–solid interface vs. time in a Cu/SS 316 system irradiated with 2.8 (1), 4.3 (2), 6.3 (3) and 8.4 J cm^{-2} (4).

the total thickness of the molten layer increases in the range 0.8–5 μm, and the lifetime of the molten substrate increases in the range 0.5–4 μs (Fig. 6.13). As a result, the velocity of the resolidification front decreases from 8 to 4 m s^{-1} and the cooling rate of the substrate decreases from 6.4×10^8 to 2×10^8 K s^{-1}. After resolidification of the substrate, the Cu film remains in the liquid state for 0.4–4.4 μs, depending on E_s (horizontal lines in Fig. 6.13), and then it solidifies as well. The comparatively long lifetime of Cu in the liquid state is due to the low thermal conductivity of SS 316.

According to AES data, the as-deposited Cu film contains impurities of C and O, which are concentrated, mainly, in the near-surface (~150 nm) layer. Fig. 6.14 shows element concentration profiles for Cu/SS 316 samples after irradiation. For $N = 1$ (Fig. 6.14(a)), the O and C impurities are removed and a transitory layer is formed at

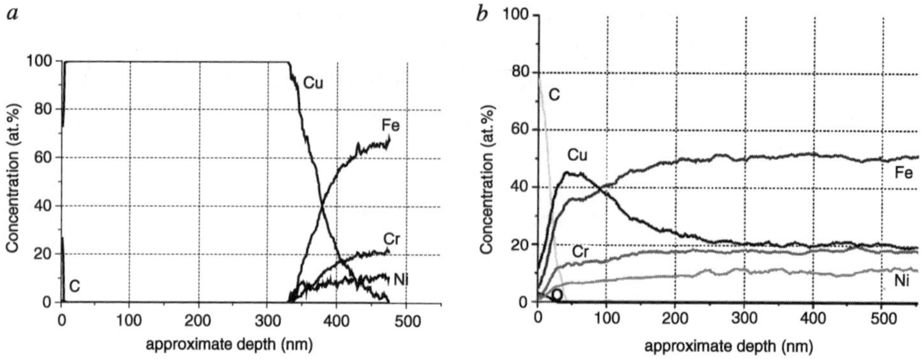

Figure 6.14: AES element profiles in a Cu/SS 316 system irradiated with (a) 6.3 J cm^{-2}, $N = 1$ and (b) 5 ± 1 J cm^{-2}, $N = 5$.

the film–substrate interface, whose thickness (120–170 nm) weakly depends on E_s. Assuming, in accordance with calculations (see Fig. 6.13), that the characteristic time during which the film and the substrate exist simultaneously in the liquid state, $t_m \sim 10^{-6}$ s, and putting the liquid-phase diffusivity, $D \sim 5 \cdot 10^{-5}$ $cm^2 s^{-1}$, we obtain the thickness of the diffusion layer $d \sim (2Dt_m)^{1/2} \sim 100$ nm. It follows that this layer was formed as a result of liquid-phase mixing of the film–substrate system.

Since increasing E_s increases the thickness and lifetime of the molten substrate, the thickness of the diffusion layer should also increase. The weak E_s dependence of the thickness of the mixed layer for $N = 1$ can be related to the fact that an increase in thickness and lifetime of the molten substrate is accompanied by a slowdown of the process of its resolidification and by a decrease in the rate of quenching from the liquid state. In this connection, it is highly probable that the Cu atoms dissolved in the liquid substrate will be pushed out from the growing crystal into the near-surface layers, restricting the concentration of Cu in the solid solution and, hence, the thickness of the mixed layer. In contrast to this, an increase in number of pulses abruptly increases the thickness of the diffusion layer, and with $N = 5$ a surface alloy containing ~20 at.% Cu is formed (Fig. 6.14(b)). Assuming that copper is uniformly distributed in depth, we obtain that the thickness of the Cu-alloyed layer is 2–2.5 μm.

Experiments have shown that for single pulsed melting in the surface layer of thickness 0.5–1 μm, including the Cu film and the diffusion layer, the nanohardness and the wear resistance nonmonotonicly vary with energy density, reaching a maximum in the range 4.3–6.3 J cm^{-2}. The improvement of these properties observed in [25] can be related to the hardening of this layer due to its fast quenching from the liquid state.

To produce thicker alloyed layers, it is convenient to deposit the coating by the electroplating. Experiments on a model Ni/Cu system have shown that this method allows one to create alloyed surface layers of thickness a few micrometers [26]. If an auxiliary target (e.g., a grid) is placed in front of the irradiated surface, it is possible to execute evaporation of the target, transfer of the target material in the form of vapors or ions to the surface, and pulsed melting of the surface layer [1]. The content of alloying elements and the thickness of the alloyed layer can be gradually increased by increasing the number of irradiation cycles. Investigations performed on Cu–Fe, Al–Fe, Ti–Fe, and other systems have shown [26] that the concentration of the alloying element in the near-surface layer is substantially greater than its equilibrium concentration. The thickness of the layers produced is several micrometers, which is an order of magnitude greater than that typical of high-dose ion implantation.

6.6. Changes in Microstructure and Properties of Alloys Subjected to Pulsed Melting

6.6.1. Carbon Steels

Examination of Steel 45 (Fe–0.45 wt.% C) irradiated with a low-energy (~20 keV) electron beam of pulse duration $\tau \geqslant 10\,\mu s$ and energy density $E_s \geqslant 10\,J\,cm^{-2}$ sufficient to melt a surface layer of thickness $\geqslant 5\,\mu m$, has shown that, irrespective of the original state, a quenched structure (martensite and residual austenite) with enhanced microhardness and wear resistance is formed in this layer. Metallographically, this structure shows up as a weakly etchable ("white") layer. The highest degree of hardening is achieved on irradiation of pre-quenched steels [20,21].

As the pulse duration is shortened to ~1 μs (LEHCEB), the martensitic transformation induced by quenching from the melt and, hence, efficient hardening of the surface layer are realized if only the energy density is such that the melt thickness is over the critical one equal to ~5 μm. This is associated with the existence of a minimum critical grain size for the γ-phase (~0.5 μm) necessary for the nucleation of martensitic crystals [27]. This criterion is in good agreement with the data on the effect of the grain size on the martensite start temperature, M_s, that are available in the literature and allows an interpretation of both our results and the data on pulsed laser treatment of steels [20].

On irradiation of carbon steels with ferrite–pearlitic structure in initial melting modes ($\tau \sim 1\,\mu s$, $E_s = 2.5\,J\,cm^{-2}$), such that the thickness of the melt is appreciably lower than the critical one, the original structure is practically retained.

Pulsed melting of pre-quenched steel 45 in a similar irradiation mode leads to dramatic microstructural changes in the HAZ. In the layer of thickness ~100 nm adjacent to the surface, a structure is formed which consists of α-phase and γ-phase grains with an average grain size of ~30 nm (Fig. 6.15(a)). The nanocrystalline layer goes into a

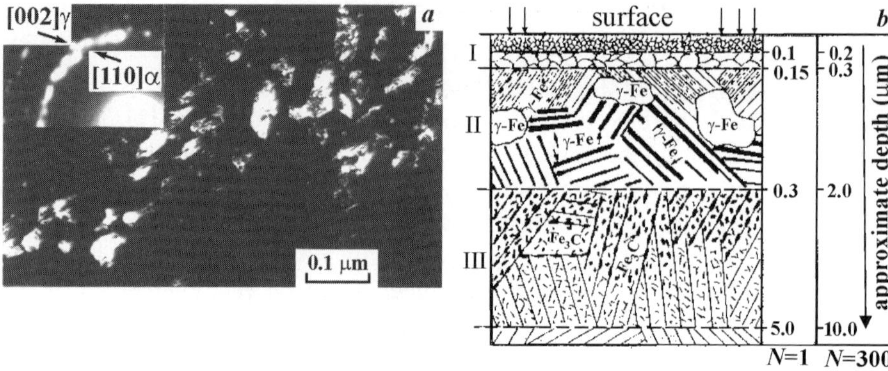

Figure 6.15: (a) TEM dark-field image (reflex of [110] α-Fe) and a diffraction pattern of the near-surface layer of pre-quenched steel 45 irradiated with an LEHCEB (2.2 J cm^{-2}, 0.8 μs, $N = 300$); (b) schematic of the structure of the HAZ.

coarser-grain layer (with an average grain size of ~200 nm) consisting of α-phase only. This layer is followed by structures characteristic of martensite-quenched steels (Fig. 6.15(b)). The formation of nanostructures correlates with the fact that the rate of quenching from melt and the velocity of the resolidification front reach their maxima (~10^{10} Kc^{-1} and ~5 ms^{-1}, respectively) at the surface [28].

6.6.2. High-Speed Steel

The structure-phase transformations in the near-surface layers (up to ~1 μm) of a pre-quenched high-speed steel (HSS) S6-5-2 (Fe–0.87 C–6.4 W–5.0 Mo–4.3 Cr–1.9 V, wt.%) irradiated with LEHCEBs (2.5 μs, 3–18 J cm^{-2}, $N = 5$) are described in [29]. According to X-ray diffraction (XRD), electron dispersion X ray (EDX) and TEM data, the original structure consists of crystals of lamellar α-martensite and globular particles of type Fe_3M_3C carbide, where M = Mo + W. The threshold for surface melting is 3.2 J cm^{-2}. An increase in energy density results in gradual dissolving of original carbide particles in the near-surface layer (Fig. 6.16). As a result of super-fast quenching from melt, a structure consisting of α-martensite and residual austenite (γ-phase) is formed; the γ-phase increases almost linearly with energy density and reaches ~90% at 15.6 J cm^{-2} (Fig. 6.17).

The formation of a near-surface layer with a high content of austenite might be associated with the liquid-phase dissolving of the initial carbide particles and the stabilization of the γ-phase due to the increase in carbon concentration in the solid solution. Moreover, the stabilization of austenite is favored by the thermal stresses [30] appearing in the surface layers under irradiation. Actually, from XRD data it

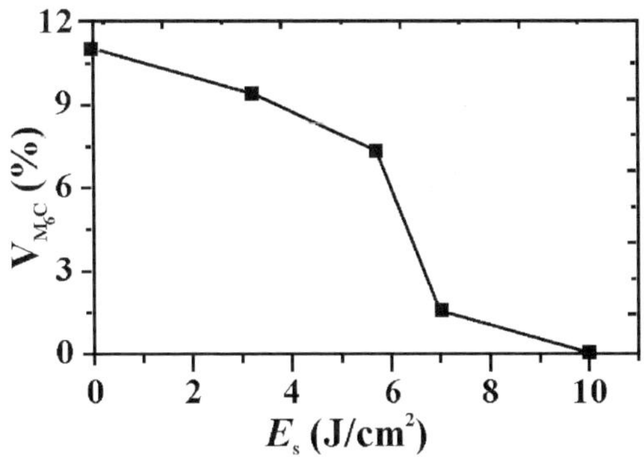

Figure 6.16: Volume content of the carbide particles observed in SEM images at the surface of HSS S6-5-2 vs. energy density. At higher energy densities the particles are fully dissolved. The description as M_6C results from XRD data.

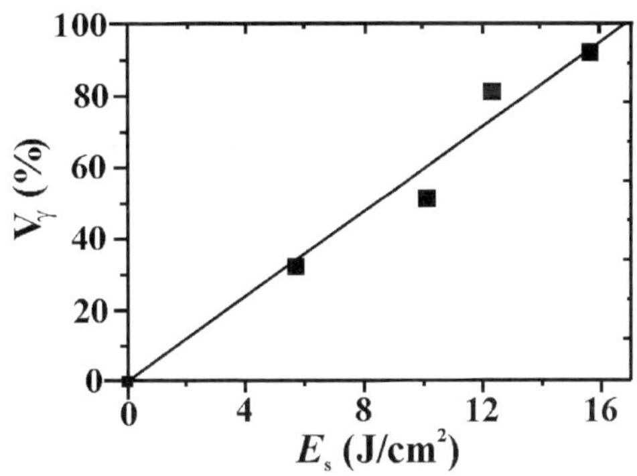

Figure 6.17: Volume content of the γ-phase in the near-surface (0.77 μm) layer of HSS S6-5-2 vs. energy density, estimated from XRD data

follows that after irradiation with 7 J cm^{-2}, the lines of the α- and γ-phases are shifted toward the smaller angles. This suggests that residual compressing stresses are induced in the plane of the irradiated surface. The magnitude of these stresses for the γ-phase, estimated from the shift of the γ-phase main lines, is $\sim$1 GPa.

According to TEM data obtained for 7 J cm^{-2}, which corresponds to an abrupt fall in the content of carbide particles (see Fig. 6.16), the phase composition in the near-surface (up to $\sim$1 μm) layer quenched from melt is the same as in original state: martensite crystals, globular particles of M_6C carbide, and nanosized M_3C particles. Around comparatively coarse particles of type M_6C carbide, a transitory spherical layer of submicron thickness is generally formed which has the α-phase lattice (Fig. 6.18). This layer results from the contact liquid-phase dissolving of an

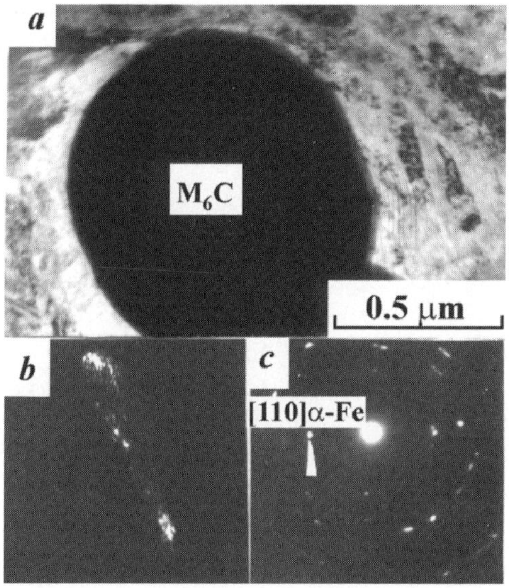

Figure 6.18: TEM images of a globular carbide particle at the near-surface layer of HSS S6-5-2 irradiated with 7 J cm^{-2} ($N = 5$): (a, b) bright-field and dark-field images, respectively; (c) diffraction pattern for (b).

original particle at the interface between the particle and the α-matrix. Obviously, this effect should increase the adhesive strength at the interface.

There is some difference in the phase composition of the irradiated steel, determined by the XRD and TEM methods. According to XRD, for 7 J cm^{-2} the near-surface layer of thickness 0.77 μm contains ~50% γ-phase (see Fig. 6.17), while according to TEM, the γ-phase is absent in this layer and, instead, plate martensite oriented in the surface plane is detected. The reason for this contradiction is the following. As mentioned above, the high level of residual thermal stresses is a factor favorable to the stabilization of the γ-phase in the near-surface layer quenched from melt. During the preparation of a foil for TEM, the residual stresses relax through elastic bending of the foil, resulting in the $\gamma \rightarrow \alpha$ martensitic transformation [30].

To elucidate the effect of pulsed melting on the wear resistance of HSS, drills of diameter 8 mm were tested in the original state and after irradiation with 7 J cm^{-2} on which an abrupt increase in the content of M_6C carbide particles was observed. The testing results have shown that irradiation reduces the wear of the cutting edge by a factor of ~1.7 compared to as-received drills. The enhancement of wear resistance is associated with the following changes in the near-surface layer microstructure: fixing of comparatively coarse original carbide particles in the matrix, formation of new disperse M_3C carbide particles, an increase in content of the metastable

γ-phase and its provable transformation into martensite during cutting, and formation of residual compressive stresses in the γ-phase.

After irradiation in intense melting modes (15–18 J cm^{-2}), according to TEM, complete dissolving of globular M_6C particles takes place in the near-surface layer, which agrees with scanning electron microscopy (SEM) and XRD data. Rapid resolidification of the melt leads to the formation of a structure consisting of cellular regions of size ~5 μm (Fig. 6.19(a–d)) alternate with regions of martensitic–austenitic structure (Fig. 6.19(e–g)). The cellular-structure regions occupy ~70% of the surface area. Analysis of the electron diffraction patterns (Fig. 6.19(c) and (d)) has shown that the cellular structure contains regions consisting of γ-phase (~70%) and α-phase cells. In addition, the α-phase is observed in the form of thin plates (Fig. 6.19(b)) which seem to be located near the irradiated surface. In the regions with martensitic–austenitic structure, the γ-phase content makes up 60%. From the TEM images for 18 J cm^{-2} it follows that the total γ-phase content in the near-surface layer is ~70%, while according to XRD data, it approaches ~100%. The lower values of the γ-phase volume content determined by the TEM method may be explained, as mentioned above, to the relaxation of elastic stresses in the near-surface layer in the process of foil preparation and to the formation of martensite.

In most cases, the cells are irregular in shape; their average size is $d \sim 130$ nm. The cells are separated by ~10 nm thick second-phase interlayers. It can be supposed that the interlayers are multicomponent high-alloy phases of nonstoichiometric composition.

Layer-by-layer TEM examination has shown that the nucleation of the cellular structure occurred as a result of the liquid-state dissolving of globular M_6C carbide particles (Fig. 6.19(h)). The structure model of the near-surface layer of thickness up to ~1 μm formed on irradiation in intense melting modes (15–18 J cm^{-2}) is shown schematically in Fig. 6.20. It consist of a weakly misoriented nanosized $(\gamma + \alpha)$ cellular structure (see Fig. 6.19(a) and (c)), a martensitic-austenitic structure (see Fig. 6.19 (e) and (f)), and a weakly misoriented nanosized γ-cell structure surrounding a partially dissolved M_6C globule (see Fig. 6.19(h), (k), and (j)).

Experiments have shown that cellular structures of this type are also formed on pulsed melting of a carbon steel (Fe–0.7 wt.% C), tool steels of the types 01 and 440A, and a nodular (ductile) iron [2,31]. In turn, this effect is not detected on irradiation of stainless steels of the types 304 and 15-5PH and Hadfield steel (Fe–1.1 C–13 Mn, wt.%). Analysis of the conditions under which cellular structures are formed has shown that these structures are observed only in alloys which contain rather large ($d \geqslant 0.5$–1 μm) globular carbide particles. In this case, it is necessary that the lifetime of the melt be long enough for efficient liquid-phase dissolving of carbide globules, providing a certain critical carbon concentration in the melt. For LEHCEBs, this lifetime is $t_m \sim 4$–10 μs, and the cell size is $d \sim 70$–150 nm.

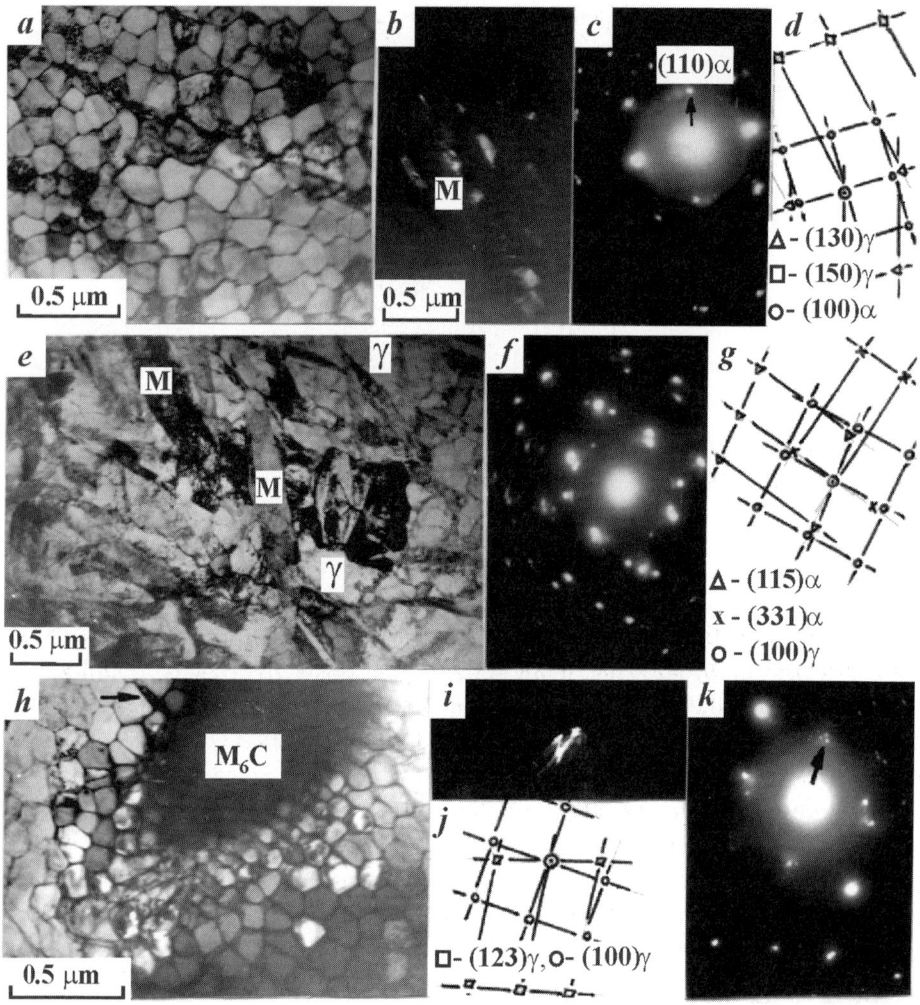

Figure 6.19: TEM images of the near-surface layer of HSS S6-5-2 irradiated with $18\,\mathrm{J\,cm^{-2}}$ and five pulses: (a, e, h) bright-field images; (b, i) dark-field images of (a, h), respectively; (c, f, k) diffraction patterns of (a, e, h), respectively; (d, g, j) identification schemes of diffraction patterns (c, f, k), respectively. The dark-field image (i) corresponds to the second-phase sublayer shown in image (h) with an arrow. The arrows in (c) and (k) show the reflections corresponding to dark-field images (b) and (i), respectively.

Cellular structures similar to those described above were observed earlier in Pb–Sn, Pb–Sb, Al–Cu, and other crystals at conventional velocity of crystal growth ($v \sim 10^{-5}\,\mathrm{m\,s^{-1}}$) [32] and after nanosecond laser melting of ion-implanted layers in silicon ($v \sim 3$–$4\,\mathrm{m\,s^{-1}}$) [23,33,34]. In ion-implanted silicon, a cellular structure

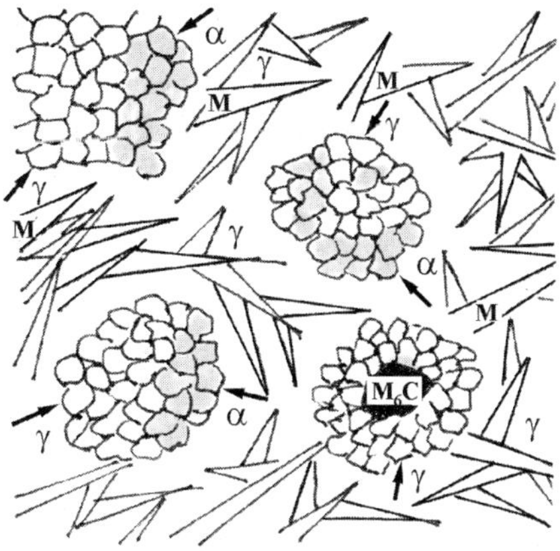

Figure 6.20: Schematic of the microstructure of the near-surface layer of HSS irradiated with LEHCEB (18 J cm^{-2}). The location of the α- and γ-phases, martensite (M), and carbide (M_6C) is indicated.

is formed if the partition coefficient $k' = C_s/C_l < 1$, where C_s and C_l being the dopant concentrations in the solid and in the liquid phase, respectively, at the rapidly moving interface, and the dopant concentration in the implanted layer is over some critical value. The characteristic size of the silicon cells separated by segregates of the impurity is ~100 nm [23,33,34].

According to the theory originally developed by Mullins and Secerka [35] and proposed by Cullis *et al.* [33], as applied to the nanosecond laser annealing of ion-implanted silicon, the formation of a cellular structure is due to the perturbation of the planar solid–liquid interface. This is associated with a large degree of constitutional supercooling of the melt when the destabilizing effect of the solute redistribution outweighs the stabilization provided by temperature gradients. With $k' < 1$, during resolidification the dopant will be pulled off the growing crystal into the melt and move into the near-surface layers. As a result, its concentration will peak at the surface. During the epitaxial resolidification of the near-surface layer enriched with the impurity, the latter is segregated from the supersaturated solid solution. This leads to the formation of cells separated by interlayers of the second phase. The cell size is determined by the diffusion distance as D/v, where D is the coefficient of diffusion of the impurity into the melt. In this case, the expected value is $D/v \sim 10^{-4}$ cm/100 = 10 nm. The deviation from experimental data can be associated with the fact that at high beam energy densities (16–18 J cm^{-2}), at which a cellular structure is formed, the actual value of D can be much greater than 10^{-4} cm^2 s^{-1} because of significant overheating of the melt.

6.6.3. Austenitic Stainless Steels

It has been demonstrated that multiple repeated ($N = 20$–50) LEHCEB irradiation of stainless steels in modes of surface melting and evaporation substantially enhances the corrosion resistance [2,36] and electric strength of the vacuum insulation [37]. The improved performance is associated with the removal of impurities from the near-surface layer and with the smoothing of the surface due to melting [1–3]. To elucidate the origins of the mentioned effects, comparative studies have been performed on the influence of LEHCEB irradiation (20–30 keV, 2.5 μs, 2–12 J cm^{-2}) on the surface morphology, characteristics of cratering, chemical composition, and microstructure of type AISI 304L and 316L austenitic stainless steels [38]. Samples were made of rod stock (steels 304L and 316L; the irradiated surface was normal to the stock axis) and plate stock (steel 304L; the irradiated surface was parallel to the stock plane).

It has been established that the main reason for cratering is the local overheating followed by explosive ejection of the material at the sites of localization of the second phases (Fig. 6.21(a)).

According to TEM data, the seats of microcraters might be associated with submicrometer low-melting-point FeS_2 particles (Fig. 6.22). It has been demonstrated that the intensity of cratering for plate stock samples is substantially lower than for rod stock ones. Multiply repeated pulsed melting of steels 304L (plate) and 316L (rod) almost completely suppresses cratering (Fig. 6.21(b)) and reduces the surface roughness as compared to the original (after cutting) surface condition.

With the use of AES it has been established that the surface smoothing is accompanied by the removal of O, C, and N impurities from the near-surface layer (~50 nm). According to TEM data, as a result of rapid quenching from the melt, a single-phase (γ-phase) microstructure is formed in the near-surface layer of thickness

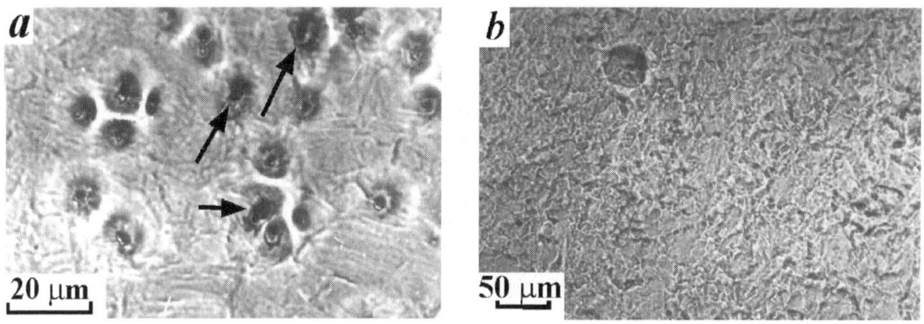

Figure 6.21: Optical micrographs of the surface of SS 304L samples irradiated with LEHCEB: (a) rod (~3 J cm^{-2}, $N = 1$); (b) plate (8 J cm^{-2}, $N = 50$). Microcrater nucleation centers are shown with arrows.

~0.5 μm. The grain size of this microstructure is 0.2–0.6 μm (Fig. 6.23), which is almost two orders of magnitude lower than that of the mentioned steels in the original state.

The formation of a single-phase homogeneous submicrometer structure with a low content of impurities at intergrain boundaries offers the possibility to considerably enhance the corrosion resistance and electric strength of a vacuum insulation. Fig. 6.24 presents the results of electric strength (first breakdown voltage) tests for vacuum gaps with electrodes of diameter 8 cm (steel 316L, rod) in the original state and after multiply repeated pulsed melting [3]. It can be seen that after irradiation the electric strength increased by a factor of ~1.8, which is in qualitative agreement with

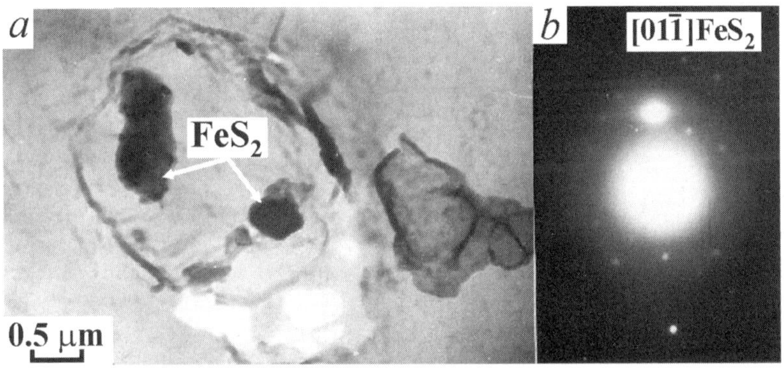

Figure 6.22: (a) TEM bright-field image of the near-surface layer of SS 304L (rod) in the original state and (b) diffraction pattern. The image was obtained by the carbon replica technique after revealing the grain structure.

Figure 6.23: TEM bright-field image of the near-surface layer of SS 304L (plate) after irradiation (8 J cm^{-2}, $N = 30$).

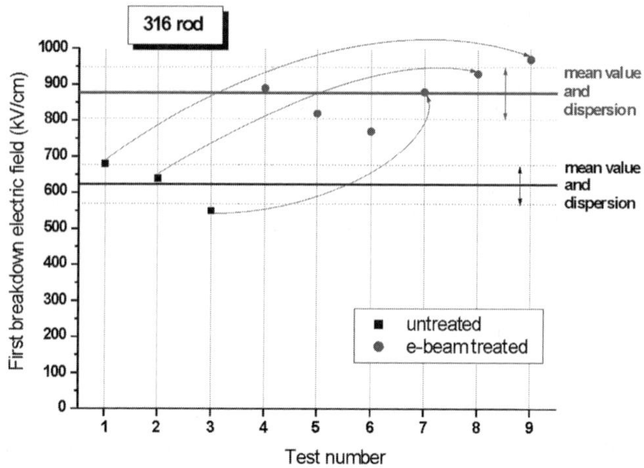

Figure 6.24: First breakdown electric field for vacuum gaps with electrodes of 8 cm diameter made of steel 316L (rod) in the original state and after LEHCEB irradiation ($10\,J\,cm^{-2}$, $N = 30$).

microstructural changes. This approach was also used to increase the pulse duration of microwaves generated by a high-power relativistic backward wave oscillator [39].

6.6.4. Aluminum Alloys

Investigations were performed on samples made of Al2024 (Al–4.3 Cu–1.5 Mg, wt.%) and Al6061 (Al–0.6 Si–0.7 Fe, wt.%) alloys irradiated with five pulses in two different modes: 0.8 μs, $5\,J\,cm^{-2}$ and 2.5 μs, $25\,J\,cm^{-2}$. For the Al2024 alloy irradiated in the first mode (melt thickness ~7 μm), complete dissolving of coarse (~1 μm) particles takes place in the near-surface layer of thickness up to ~0.4 μm and the second-phase segregates become smaller as compared to the original state. For the Al6061 alloy, a similar dispersion of second-phase particles is observed except for the coarse inclusions.

After irradiation of the Al2024 alloy in a more intense mode (with the melt thickness ~25 μm), a cellular structure is formed as a result of the dissolving of second-phase coarse particles followed by the decay of the supersaturated solid solution. This structure consists of nonmisoriented matrix-phase grains of size 200–300 nm with nanosized $CuAl_2$ segregates coherently bound with the matrix and localized along the grain boundaries. The reasons for the formation of cellular structures were discussed in Section 6.6.2. Microcracks are also observed which can be associated with the shrinkage of the molten surface layer on resolidification. For the Al6061 alloy irradiated under similar conditions, complete dissolving of all

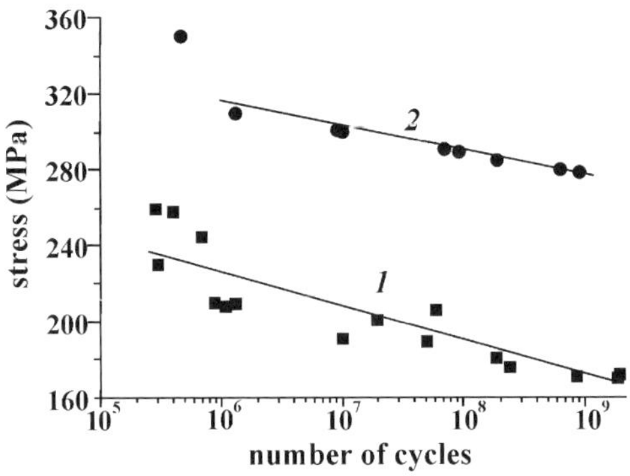

Figure 6.25: (1) Fatigue S–N curves for samples made of VT8M alloy in the original state (2) and after irradiation (2.5 μs, 2.5 J cm^{-2}, $N = 40$) and subsequent vacuum annealing (500–550°C, 10 h).

original second-phase particles (nanosized segregates included) is observed. In this case, an Al_2O_3 film uniform in thickness (~30 nm) is formed on the surface, while for the original samples the thickness of such a film varies in the range from 5 to 40 nm. These two effects substantially enhance the corrosion resistance of the material, which is evidenced by corrosion resistance measurements [2,40].

6.6.5. Titanium Alloys

Experiments were performed on samples made of BT8M (Ti–5.8 Al–3.7 Mo, wt.%) and BT18Y (Ti–6.3 Al–4.5 Mo, wt.%) alloys subjected to preliminary annealing at 920°C and 940°C and aging at 550 and 600°C, respectively. It has been found [2,3,41,42] that pulsed melting makes it possible to clean the surface of oxygen and carbon impurities, to increase the aluminum content in the near-surface layer to 20%, to make the component distribution in this layer more uniform, and to reduce the surface roughness to 0.1 μm. However, as this takes place, the phase composition changes to some extent, tensile weak residual stresses are induced in the near-surface layer, and microcraters appear on the surface. When irradiating these materials in optimum modes, cratering can be suppressed, and subsequent annealing makes it possible to restore the original phase composition and to enhance substantially the operating properties of the materials, namely, to increase the fatigue strength by more than 20%, the fatigue life more than tenfold (see Fig. 6.25), the resistance to dust erosion at low loads more than twice, and the tensile strength by up to 8%, with the plasticity being substantially improved. At the same time, the surface microhardness and the heat resistance are retained on the original level.

The enhancement of the fatigue properties of the materials is related to the smoothing of the surface due to its melting, to the cleaning of the surface of impurities, and to the increase in aluminum content in the near-surface layer. Fractography has shown that under static loading, pulsed melting changes the fracture mechanism from intercrystalline, or quasi-ductile, to ductile. Under cyclic loading, the fracture of original samples is initiated at the surface of sharp edges, while subsurface nucleation of cracks is characteristic of irradiated samples. This circumstance just determines the improvement of the fatigue properties of irradiated samples.

6.6.6. Metallic Biomaterials

The main metallic biomaterials are titanium and its alloys, Co-based alloys, and stainless steels. Experiments [43] were performed to investigate the effect of pulsed melting on the surface morphology and properties of metal dentures prepared by the lost wax casting method. The need for these investigations was dictated by the fact that the conventional methods of cleaning and smoothing of the denture surface, including grinding and mechanical polishing, are labor-consuming and low-productive. Most of the experiments were performed with samples and dentures made of dental materials: commercially pure (cp) Ti, Ti–6 Al–4 V and 61 Co–31 Cr–6 Mo alloys. For cp Ti and the Ti alloy, the surface was chemically etched prior to irradiation to remove the oxide layer (80–100 μm) formed on casting. For the Co–Cr–Mo alloy, in some experiments, the original (on casting) surface was coated with Ti by the PVD method. Electron-beam treatment ($7\,\mathrm{J\,cm^{-2}}$, $N \leqslant 30$) was performed on a specially designed automated electron-beam setup based on an LEHCEB source. The setup was furnished with a manipulator that allowed 12 dentures to be treated in one vacuum cycle.

It has been established that for cp Ti the presence of residues of the oxide layer, not completely removed by chemical etching, lead to microcracking of the irradiated surface. Obviously, this is associated with the high brittleness of TiO_2. If the oxide layer is completely removed, microcracks do not appear on the irradiated surface. For optimal treatments, the surface roughness R_a decreases to 0.27 μm (against 1.62 μm in the original state) and the reflection coefficient almost doubles. Measurements of the anode polarization characteristics in the 1 N HCl solution have shown that after irradiation the passivation current density decreased to one-third of its original value.

Similar experiments performed on Ti–6 Al–4 V samples have shown that after multiple pulsed melting the passivation current density decreases by about two order of magnitude as compared to the original state [44]. The increase in corrosion resistance well correlates with the smoothing of the surface and with its cleaning of oxygen and carbon impurities, mentioned above.

As for the Co–Cr–Mo alloy treated with an LEHCEB, microcraters formed on the surface, and, in contrast to Ti alloys and stainless steels, cratering could not be suppressed even by multiple repeated pulsed melting. However, if a Ti film was previously deposited on the surface, irradiation in a proper mode made it possible to noticeably reduce the microcrater density and to obtain rather good surface condition (Fig. 6.26) [43]. The effects observed can be explained as follows. The as-cast Co–Cr–Mo–C alloy contains ~5% of comparatively large carbide particles of type $M_{23}C_6$, where M = Cr + Mo [45]. The melting point of the particles is approximately 1500 K, which is 150 K lower than that of the matrix [45,46]. Furthermore, the thermal conductivity of the carbides is significantly lower than that of the matrix. Hence, the overheating of carbides under irradiation leads to intense cratering.

In an irradiated Co–Cr–Mo alloy pre-coated with Ti film, a liquid phase containing both the substrate and the film elements is formed. When rapid resolidification of the melt takes place, titanium carbide particles are mainly formed. This is due to the fact that the affinity of Ti to carbon is much greater than that to Cr and Mo [47]. Titanium carbide particles are more thermally stable and much smaller compared to Cr and Mo particles [47]. These factors might be responsible for the dramatic decrease of cratering on the irradiated Co–Cr–Mo alloy pre-coated with Ti.

The results obtained, in view of the large cross section of the beam and the high productivity of the electron-beam setup, can be considered as a basis for the development of a new technology for surface modification of metallic biomaterials. On the other hand, these methods may appear to be highly promising for improving corrosion resistance and biocompatibility of biomedical implants (artificial hip and knee joints, dentures, etc.). In addition, the data reported in Section 6.6.5 suggest that there exists a possibility to increase the fatigue strength of implants made of Ti alloys.

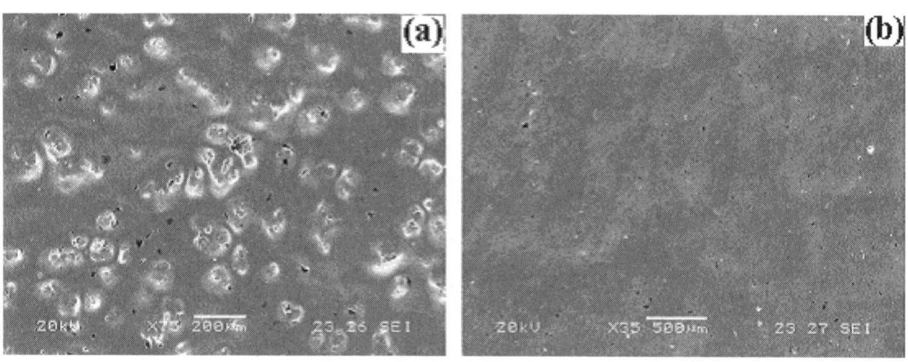

Figure 6.26: SEM images of the surface of a Co–Cr–Mo alloy irradiated with an LEHCEB: (a) without Ti coating, (b) with Ti pre-coating.

6.7. Surface Modification of Hard Metals

It has been shown [2,48] that irradiation of cemented carbide inserts in surface melting modes (2.5 μs, 1–7 J cm^{-2}) increases about threefold the wear resistance of the inserts at high cutting speeds. TEM examination of irradiated T15K6 (WC–15 TiC–6 Co, wt.%) inserts has shown that in the near-surface layer ($\leqslant$1 μm) the following microstructural changes are observed: a subgrain (0.1–0.2 μm) structure is formed in the binding phase, a nanosized (~10 nm) particles of new carbide phases are segregated in the near-boundary regions, and the α-WC $\rightarrow$ WC (f.c.c.) transformation occurs in part. In the process of cutting, intense plastic deformation of the material takes place, which is accompanied by annealing of defects in the near-surface contact zone. In unirradiated tools, the relaxation of stresses leads to the formation of a relatively depth-homogeneous dislocation structure with a low dislocation density in the cobalt binder and to the formation of micropores at the interphase boundaries. In irradiated inserts, just below the wearing surface layer, a substructure is formed which is characterized by a high dislocation density in the cobalt and carbide (WC) phases. This substructure contains microtwins in the carbide phase (Fig. 6.27(a)) and nanosized second carbide segregates in the binder and at the interphase boundaries (Fig. 6.27(b)). It follows that the main reason for the enhancement of the wear resistance of the inserts is the dispersion and grain-boundary hardening of the cobalt binder under pulsed melting. Due to this, in the process of cutting, a dislocation structure stable to high thermomechanical loads is continuously being formed beneath the wearing surface. The stability of this structure is due to the fact that dislocations are fixed by the nanosized segregates formed in the process of cutting.

An investigation of the action of an LEHCEB on the microstructure, microhardness, and wear resistance of the WC–30% Hadfield steel hard metal is described

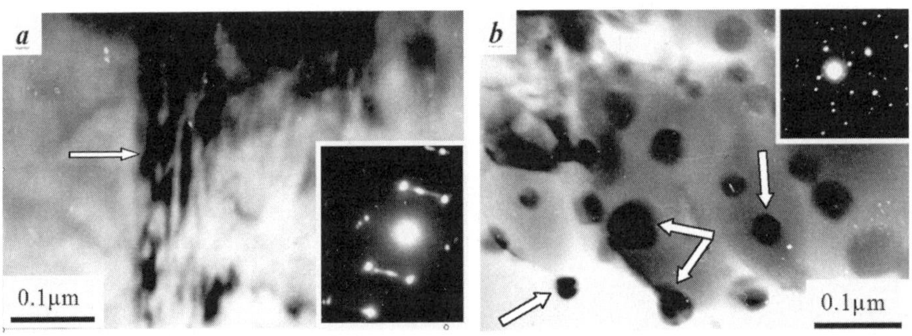

Figure 6.27: Microstructures of the sublayer of the cutting edge of a pre-irradiated T15K6 carbide insert. The cutting time is 120 s. The arrows indicate (a) microtwins and (b) nanosized carbide segregates.

elsewhere [26,49]. The binding phase had the γ-lattice (f.c.c.), which is characterized, as distinct from cobalt, by an elevated structural instability in deformation. For irradiation with $E_s \leqslant 30\,\mathrm{J\,cm^{-2}}$, irrespective of the number of pulses, the microhardness in the surface layer of thickness up to $\sim$100 μm is observed to increase by a factor of $\sim$1.5 as compared to the original state. Wear tests have shown that the surface hardening is accompanied by a decrease in coefficient of friction by about a half (Fig. 6.28) and by a substantial increase in wear resistance as compared to the original state. It has been demonstrated that pulsed melting has the result that there occurs $\gamma \rightarrow \alpha$ martensitic transformation in the binding phase and the formation of nanosized and submicrometer particles of W_2C, $M_{12}C$, and $M_{23}C_6$ carbides. Similar structure-phase transformations are observed on the wear surface of original samples upon steady-state friction, which is characterized by a minimum wear. It follows that these transformations are responsible for the hardening and enhanced wear resistance of the given material on irradiation.

The effect of the irradiation with an LEHCEB (2.5 μs, 3–5 $\mathrm{J\,cm^{-2}}$, five pulses) on the microstructure and wear resistance of TiN coatings deposited by the physical vapor deposition (PVD) on carbide (WC–TiC–TaC–Co) inserts is described elsewhere [50]. The lower level of E_s corresponded to initial melting of the coating. The experiments have shown that after irradiation, the coating was cracked because of the high intensity of tensile stresses formed at the stage of cooling. Notwithstanding this, the coating retained its good adhesion to the substrate. The stoichiometry and

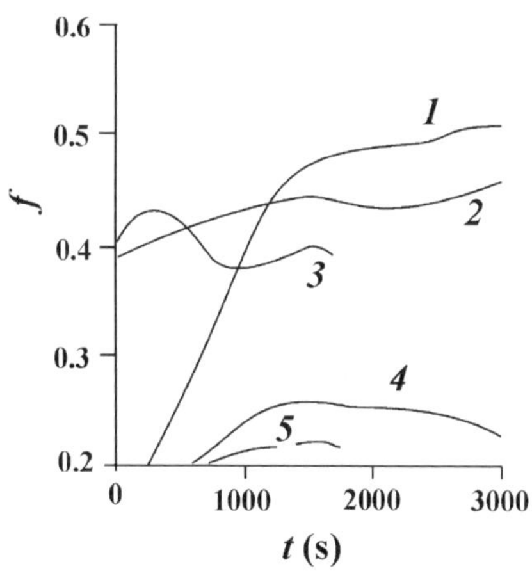

Figure 6.28: Friction coefficient vs. time for a pin-on-disk system. The pins were made of WC–30% Hadfield steel hard metal: (1, 2, 3) in the original state; (4, 5) after irradiation (20 $\mathrm{J\,cm^{-2}}$, 10 pulses). The disk was made of tool steel (HRC 52). Dry friction; normal load 175 H; slide speed ($\mathrm{m\,s^{-1}}$): 0.65 (1), 1.4 (2, 4), and 2.8 (3, 5).

phase composition of the coating, except the surface itself, remained unchanged. However, the residual compressive stresses dropped to zero and the concentration of vacancies noticeably decreased, indicating that pulsed annealing of defects took place. These substructural changes made it possible (with $3\,J\,cm^{-2}$) to double the wear resistance of the inserts in steel cutting.

6.8. Conclusion

The characteristics and mechanisms of the modification of the microstructure and properties of the surface layers of pure metals, alloys, and hard metals irradiated with LEHCEBs have been discussed. The results presented above suggest that LEHCEB sources are candidates for the development of new technologies for surface treatment of materials. For some applications, such as medical ones, the surface alloying based on liquid-phase mixing of film–substrate systems might be of considerable interest.

References

[1] D.I. Proskurovsky, V.P. Rotshtein and G.E. Ozur, *Surf. Coat. Technol.*, 96 (1) (1997) 115.
[2] D.I. Proskurovsky, V.P. Rotshtein, G.E. Ozur, A.B. Markov, D.S. Nazarov, V.A. Shulov, Yu.F. Ivanov and R.G. Buchheit, *J. Vac. Sci. Tech.*, 16 (1998) 2480.
[3] G.E. Ozur, D.I. Proskurovsky, V.P. Rotshtein and A.B. Markov, *Laser. Part. Beam.*, 21 (2003) 157.
[4] S.I. Anisimov, Ya.A. Imas, G.S. Romanov and Yu.V. Khodyko, *Action of High-Power Irradiation on Metals*, Nauka, Moscow, 1970 (in Russian).
[5] A.B. Markov and V.P. Rotshtein, *Nucl. Instrum. Meth. Phys. Res. B.*, 132 (1997) 79.
[6] Z. Shiller, U. Gaizich and Z. Pantser, *Electron-Beam Technology*, Energiya, Moscow, 1980 (in Russian).
[7] V.F. Kovalenko, *Electron. SVCh*, 1 (1972) 3 (in Russian).
[8] A.B. Markov, Yu.F. Ivanov, D.I. Proskurovsky and V.P. Rotshtein, *Mater. Manuf. Process.*, 14 (1999) 205.
[9] G.A. Bleikher, V.P. Krivobokov and O.V. Paschenko, *Heat and Mass Transfer In Solid Under the Action of High-Power Charged Particle Beams*, Nauka, Novosibirsk, 1999 (in Russian).
[10] R.B. Oswald, F.B. McLean, D.R. Schallhorn and L.D. Buxton, *J. Appl. Phys.*, 42 (1971) 3463.
[11] A.B. Markov, *Proceedings of 6th International Conference on Modification of Materials with Particle Beams and Plasma Flows*, Tomsk, Russia, 2002, p. 675.
[12] B.A. Boley and J. Weiner, *Theory of Thermal Stresses*, Wiley, New York, 1960.

[13] A.L. Bardenshtein, L.S. Bushnev, E.F. Dudarev, A.B. Markov and V.P. Rotshtein, *Proceedings of 5th International Conference on Modification of Materials with Particle Beams and Plasma Flows*, Tomsk, Russia, Vol. 3, 2000 p. 43.

[14] E. Dudarev, L. Kornienko, C. Lykov, A. Markov, G. Pochivalova, V. Rotshtein and T. Chubenko, *Izv. Vyssh. Uchebn. Zaved. Fiz.*, 5 (1993) 42 (in Russian).

[15] F. Haessner, *Recrystallization of Metallic Materials*, Springer-Verlag, Stuttgart, 1978.

[16] E. Dudarev, G. Pochivalova, D. Proskurovsky, V. Rotshtein and A. Markov, *Izv. Vyssh. Uchebn. Zaved. Fiz.*, 3 (1996) 126 (in Russian).

[17] A. Zecca, R. Brusa, M. Duarte Naia, J. Paridaens, A. Pogrebnjak, A. Markov, G. Ozur, D. Proskurovsky and V. Rotshtein, *Phys. Lett. A.*, 175, 6 (1993) 433.

[18] A. Pogrebnyak, R. Oshner, A. Zecca, V. Rotshtein and A. Mihalyov, *Fizika i Himiya Obrabotki Materialov.*, 1 (1996) 29 (in Russian).

[19] K. Vogel and P. Backland, *J. Appl. Phys.*, 36 (12) (1965) 3697.

[20] V. Rotshtein, DSc Thesis, Institute of High Current Electronics, Tomsk, 1995.

[21] D.I. Proskurovsky, V. P. Rotshtein, G.E. Ozur, Yu. F. Ivanov and A.B. Markov, *Surf. Coat. Technol.*, 125 (2000) 49.

[22] B. Wood, A. Perry, L. Bitteker and W. Waganaar, *Surf. Coat. Technol.*, 108–109 (1998) 171.

[23] J. Poate, G. Remini and D. Jacobson (Eds.), *Surface Modification and Alloying by Laser, Ion, and Electron Beams*, Plenum Press, New York (London), 1983.

[24] Yu. Ivanov, Yu. Kryuchkov, A. Markov, D. Nazarov, G. Ozur, A. Pogrebnjak, D. Proskurovsky and V. Rotshtein, *Poverhnost'*. 10–11 (1994) 95 (in Russian).

[25] V.P. Rotshtein, A.B. Markov, Yu.F. Ivanov, K.V. Karlik, B.V. Uglov, A.K. Kuleshov, M.V. Novitskaya, S.N. Dub, M.V.Y. Pauleau, F. Thièry and I.A. Shulepov, *Proceedings of 7th International Conference on Modification of Materials with Particle Beams and Plasma Flows*, Tomsk, Russia, 2004, p. 258.

[26] V.P. Rotshtein, D.I. Proskurovsky, G.E. Ozur, Yu.F. Ivanov and A.B. Markov, *Surf. Coat. Technol.*, 180–181(2004) 377.

[27] Yu. Ivanov, M. Kashchenko, A. Markov and V. Rotshtein, *Zh. Tekhn. Fiz.*, 65, 3 (1995) 98 (in Russian).

[28] Yu.F. Ivanov, V.I. Itin, S.V. Lykov, A.B. Markov, V.P. Rotshtein, A.A. Tukhvatullin and N.P. Dikii, *Fiz. Met. Metalloved.*, 75 (1993) 103 (in Russian).

[29] Yu. Ivanov, W. Matz, V. Rotshtein, R. Guenzel and N. Shevchenko, *Surf. Coat. Technol.*, 150 (2002) 188.

[30] V.M. Schastlivtsev, D.A. Mirzaev and I.L. Yakovleva, *Structure of Thermally Treated Steels*, *Metallurgia*, Moscow, 1994 (in Russian).

[31] Yu.F. Ivanov, V.P. Rotshtein and A.B. Markov, *Proceedings of 5th International Conference on Modification of Materials with Particle Beams and Plasma Flows*, Tomsk, Russia, 2000, p. 58.

[32] M. Flemings, *Solidification processing*, McGraw-Hill Book Company, New York, 1974.

[33] A. Cullis, D. Hurle, H. Webber, N. Chew, J. Poate, P. Baeri and G. Foti, *Appl. Phys. Lett.*, 38 (8) (1981) 642.

[34] J. Narayan, *J. Appl. Phys.*, 52 (3) (1981) 1289.

[35] W. Mullins and R. Secerka, *J. Appl. Phys.*, 35 (1964) 444.
[36] I.M. Goncharenko, V.I. Itin, S.V. Isichenko, S.V. Lykov, A.B. Markov, O.I. Nalesnik, G.E. Ozur, D.I. Proskurovsky and V.P. Rotshtein, *Metal Protection*, 29 (1993) 932 (in Russsian).
[37] A.V. Batrakov, A.B. Markov, G.E. Ozur, D.I. Proskurovsky and V.P. Rotshtein, *IEEE Trans. Dielect. Electr. Insul.*, 2 (1995) 237.
[38] V.P. Rotshtein, Yu.F. Ivanov, D.I Proskurovsky, K.V. Karlik, I.A. Shulepov and A.B. Markov, *Surf. Coat. Technol.*, 180–181 (2004) 382.
[39] A.V. Batrakov, K.V. Karlik, S.A. Kitsanov, A.I. Klimov, I.N. Konovalov, S.D. Korovin, G.A. Metsyats, G.E. Ozur, I.V. Pegel, S.D. Polevin and D.I. Proskurovsky, *Tech. Phys. Let.*, 27, 150.
[40] D.S. Nazarov, A.B. Markov, G.E. Ozur, D.I. Proskurovsky, I.M. Goncharenko, V.P. Rotshtein, L.S. Bushnev, R.G. Buchheit, E.F. Dudarev and G.P. Pochivalova, *Proceedings of 5th International Conference on Electron Beam Technologies*, Varna, Bulgaria, 1997, p. 209.
[41] N.A. Nochovnaya., V.A. Shulov, V.P. Rotshtein, A.B. Markov, D.S. Nazarov, G.E. Ozur and D.I. Proskurovsky, *Proceedings of 5th International Conference on Electron Beam Technologies*, Varna, Bulgaria, 1997, p. 215.
[42] N.A. Nochovnaya, V.A. Shulov, D.S. Nazarov, G.E. Ozur, D.I. Proskurovsky, V.P. Rotshtein and I.G. Karpova, *Fizika i Himiya Obrabotki Materialov.*, 1 (1998) 27 (in Russian).
[43] P. Raharjo, H. Wada, Y. Nomura, G.E. Ozur, D.I. Proskurovsky, V.P. Rotshtein and K. Uemura, *Proceedings of 6th International Conference on Modification of Materials with Particle Beams and Plasma Flows*, Tomsk, Russia, 2002, p. 679.
[44] P. Raharjo, K. Uemura, A. Okada and Y. Uno, *Proceedings of 7th International Conference on Modification of Materials with Particle Beams and Plasma Flows*, Tomsk, Russia, 2004, p. 263.
[45] M. Caudillo, M. Herrera-Trejo, M.R. Castro, E. Ramirez, C.R. Gonzalez and J.I. Juarez, *J. Biomed. Mater. Res.*, 59 (2) (2001) 378.
[46] T. Kilner, R.M. Pilliar, G.C. Weatherly and C. Allibert, *J. Biomed. Mater. Res.*, 16 (1982) 63.
[47] H.J. Goldschmidt, *Interstitial Alloys*, Butterwords, London, 1967.
[48] Yu. Ivanov, V. Rotshtein, D. Proskurovsky, P. Orlov, K. Polestchenko, G. Ozur and I. Goncharenko, *Surf. Coat. Technol.*, 125 (2000) 251.
[49] S. Gnyusov, S. Tarasov, Yu. Ivanov and V. Rotshtein, *Wear*, 257 (2004) 97.
[50] A. Perry, J. Matossian, S. Bull, D. Proskurovsky, P. Rice-Evans, T. Page, D. Geist, J. Taylor, J. Vajo, R. Doty, V. Rotshtein and A. Markov, *Metall. Mater. Trans A*, 30A (1999) 2391.

Chapter 7

Laser Processing for Surface Modification by Remelting and Alloying of Metallic Systems

Boguslaw Major

7.1. Introduction

The number of applications of lasers in technology has been growing exponentially. In modern materials engineering, lasers are used for cutting, welding, surface heating and shock hardening, drilling, surface melting and alloying as well as for ablation of materials and deposition to produce surface coating, and even for non-contact bending [1–5]. The applications range from low power density processes of surface heating without melting – like: transformation hardening and bending – and spread over processes involving surface melting and even ablation, requiring high power densities to overcome latent heat effects and significant heat conduction losses. Melting processes also include those, where a material is mixed into the melt pool – as in surface alloying and particle injection – or fused with the thin surface melt, as in cladding. The origin of microstructure modification by laser melting or laser alloying is attributed to the specific solidification fine structure, often comprising high concentration of alloying elements.

Laser–surface-treatment processing, consisting of remelting and alloying, is a field of considerable interest at present, because it seems to offer a chance to reduce the need for expensive materials or to improve components with idealized surfaces and bulk properties. It can also be used in repairing elements of tools working in severe conditions that loose their technologic properties. Recently, the hybrid treatment consisting of laser processing combined with turning or rolling in one operation has been applied in technologic processes [6]. The combination of these processes makes it possible for very hard materials to be machined by turning, as in the case of ceramic materials, hard tool alloys and high-alloy steels, or to improve surface roughness, when the process of material rolling occurs just after laser remelting or alloying with overlapping laser tracks.

Materials Surface Processing by Directed Energy Techniques
Y. Pauleau (Editor)

7.2. Overview of Laser Processing by Remelting

Rapid solidification process (RSP) is the application of high cooling rates or high undercooling to obtain high solidification rate ($u > 1\,\mathrm{cm\,s^{-1}}$) [7–10]. RSP is the basis for microstructure selection in laser modification of material surface layer. Immediately after the first high-power ruby laser became available, the interaction of intense laser beams with solid surfaces became the focus of scientific interest. Thorough understanding of the peculiarities of laser–surface interactions relies on the knowledge of the intrinsic properties of both lasers and surfaces [5]. Moreover, the solidification microstructure, with the accompanying solute segregation profile, largely, and often definitely, controls the properties and the quantity of the final product in the form of laser-modified surface layers [11]. Thus, from the practical viewpoint of technologic needs, it is important to know the processing conditions which will produce a microstructure that provides a material with the desired properties. The development of specific microstructure depends on the solidification conditions and the shape of the solid–liquid (S/L) interface, the transformation or even reaction processes which occur in the solid during the process of cooling down to the room temperature. Nucleation and growth of the solid phase is closely correlated with the applied laser processing, cooling rate and type of material [8,9]. Different metastable phases of various morphology (dendrite, eutectic, quasi-crystal) can be formed. Thus, knowledge of a phase diagram is important as it allows one to predict the microstructure formation. The process of rapid solidification can appear in two cases: (i) undercooling of liquid by its slow cooling down in the absence of heterogeneous nucleation or by its rapid cooling, as in powder fabrication by atomization or (ii) at fast movement of temperature field which appears in the process of surface melting or welding with the concentrated heat sources like laser or electron beams [7]. As properties of the laser-modified surface are related to the formed microstructure, it is of great importance to understand the nucleation and growth of micro-scale crystals during solidification in the layer subjected to laser melting.

7.2.1. Solidification Microstructures

Microstructures belong to the field at the center of materials science and engineering. They are the strategic link between materials processing and materials behavior. Therefore, microstructure control is essential for any processing activity. Solidification is one of the most important processing routes for many materials, especially metals and alloys. Recently, RSP and microstructure formation have focused a lot of interest due to the formation of refined microstructure that can be applied in industry [5,8–11]. In RSP, the growth rate of the S/L interface becomes sufficiently

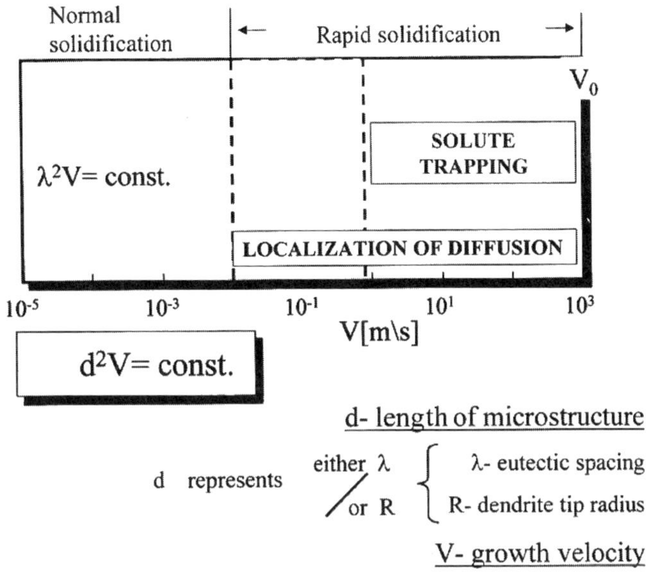

Figure 7.1: The range of solidification velocities encountered in normal and RSP, and characteristic phenomena appearing in metals under conditions of rapid solidification.

high for certain limits to be reached, which results in the modification of the global transformation behavior of materials. These limits are of threefold nature (Fig. 7.1):

(i) Localization of diffusion with respect to the size of microstructure.
(ii) Localization of diffusion with respect to the width of the interface.
(iii) Collision limit of atom attachment.

There have been developed many RSP, such as splat quenching, melt spinning, planar flow casting, laser or electron beam re-solidification, atomization, bulk undercooling, etc. However, they can be grouped into three categories: (a) laser–surface treatment, welding; (b) planar flow casting, melt spinning and (c) bulk undercooling, atomization [8] (Fig. 7.2).

The most important variable in RSP is the interface growth rate V and the temperature gradient in the liquid ahead of the S/L interface G. Further important variables are: the cooling rate $\dot{T} = dT/dt$ and the G/V ratio, which controls planar interface stability at low rates (constitutional undercooling). The most important differences between the process types with respect to the above-mentioned solidification conditions consist in the variation of velocity (V), temperature gradient (G), cooling rate ($\dot{T}$) and G/V ratio in respect to the direction "z" which is perpendicular to the surface [8].

Rapid movement of high-energy heat source is found in laser–surface treatment. The laser beam moves over the specimen surface with velocity V_{beam}. The solidification rate ranges from zero at the S/L interface to reach its maximum on the surface. The temperature gradient is the highest at the bottom of the trace and decreases toward the surface. The absolute value of the cooling rate increases strongly while approaching the surface (Fig. 7.2). Electron or laser beams re-solidification and electron beam or laser welding belongs to this group of processes.

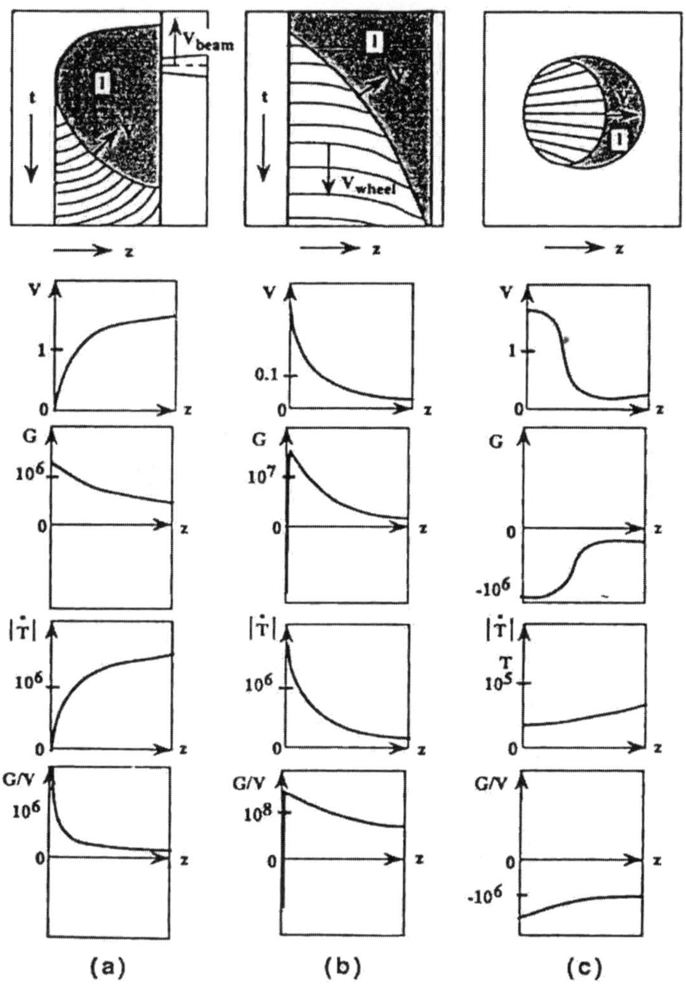

Figure 7.2: Characteristic solidification conditions of three main RSP: (a) laser–surface treatment or welding; (b) planar flow casting, melt spinning and (c) atomization.

Depending on the solidification conditions, various microstructures are formed, namely planar, cells and dendrites. All these morphologies may be obtained in single-phase solidification or in multi-phase (cutectic) solidification (Fig. 7.3) [7].

Moving from the lower left to the upper right along the lines at 45° leads to a refinement of the structure without changing the morphology (G/V = constant). Crossing these lines by passing from the lower right to the upper left leads to changes in morphology (from planar, to cellular, to dendritic growth), and the scale of the microstructure remains essentially the same. The gray bands define the region over which one structure changes into another.

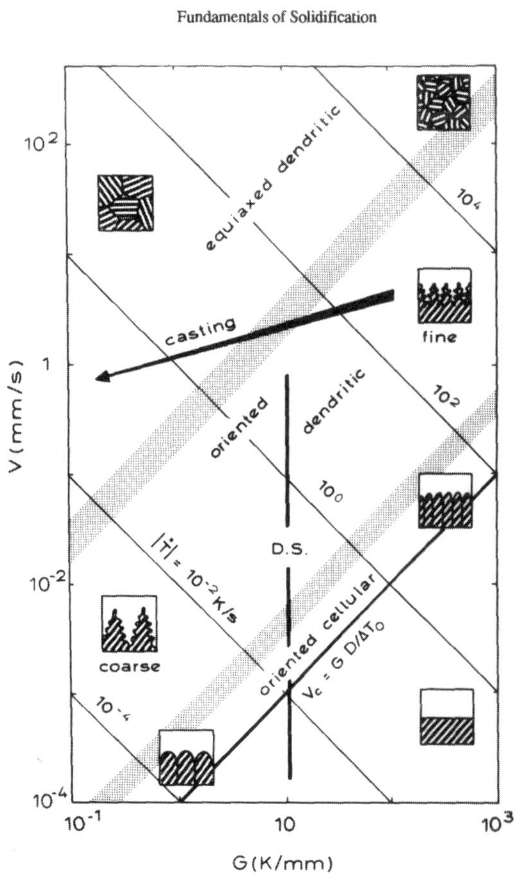

Figure 7.3: Schematic summary of single-phase solidification morphologies.

7.2.2. Laser–Surface Modification

Large and rapid local temperature changes in the surface layer subjected to laser processing by melting lead to the development of microstructures unattainable by conventional thermomechanical treatment. Between the bottom and the top of a laser trace, the solidification rate varies from zero to maximum, which leads to the appearance of various microstructures, even to amorphous phases. Laser–surface-treatment techniques are used to obtain improvements in hardness, corrosion or wear resistance.

Laser–surface melting yields fine microstructures, with little penetration into the remelted substrate. Alloying also gives fine microstructures and allows many metals and non-metals – like carbides, borides, etc. – to be alloyed with different substrates. Cladding also enables one metal to be joined with another, but in this case, there should be no mixing between the cladding layer and substrate. Laser alloying can be realized by laser melting of the substrate previously covered with alloying material which can be placed in the melt zone by: electroplating, vacuum evaporation, pre-placed powder coating, thin-foil application, ion implantation, diffusion, powder blowing or reactive gas shroud. The basic schemes of laser processing by remelting are presented in Fig. 7.4 and the scheme of longitudinal cross-section of laser track is shown in Fig. 7.5 [9].

Laser melting can be used to create supersaturated and highly alloyed materials with novel structures. Most of the beneficial effects of laser treatment can be attributed to specific types of solidification fine structure. The high-melt-pool temperature, resulting from a high power density of laser beam, enables the dissolution of even thermodynamically stable intermetallic phases and the formation of metastable phases due to the high cooling rates [8,9].

7.2.3. Structures Formed at Rapid Rates of Solidification

The microstructure formed in RSP is characterized by its phases and growth morphology. As far as solidification parameters are concerned, there can appear plane, cellular or dendritic solidification front of one phase or multi-phase eutectic morphology. The growth of microstructure is determined by the movement of laser beam and thermal conductivity of the treated material. Each phase of transformation needs a driving force. Solidification finds the force in undercooling of S/L interphase. The undercooling and equilibrium temperature determine the conditions of crystallite growth in a given laser-processing technique. The criterion of microstructure selection at directional solidification with nucleation can be formulated as follows: “For a given solidification rate, this phase is formed for which the

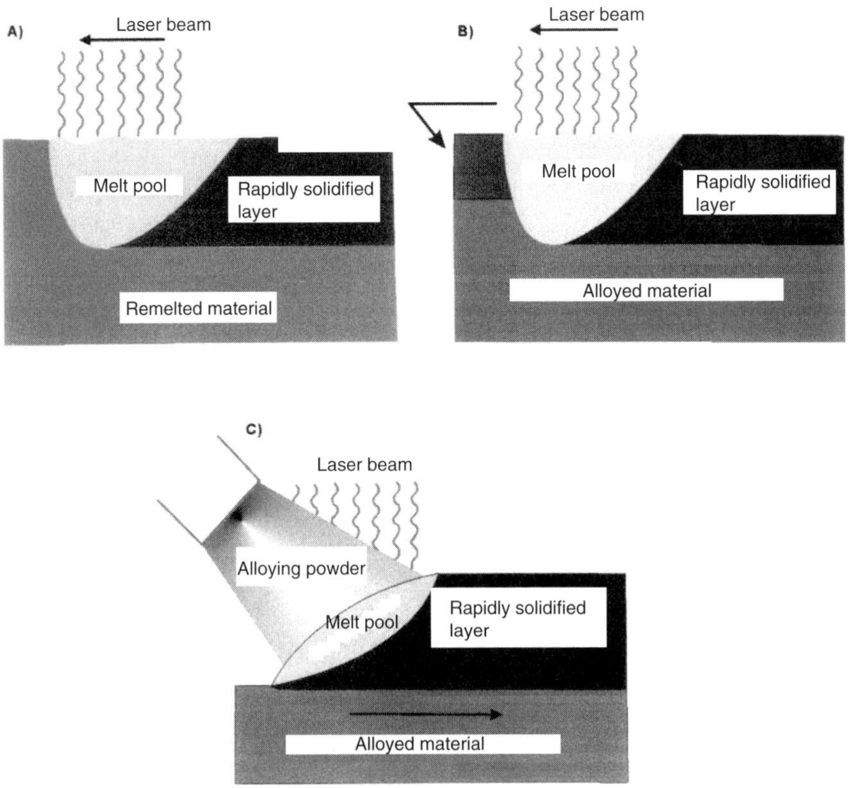

Figure 7.4: Schemes of laser treatment by remelting: (a) laser melting without alloying; (b) laser alloying by remelting of alloying layer and (c) laser alloying by remelting by powder blowing.

liquidus temperature is the highest" [9]. For the criterion to be applied, one should know the temperature of growth in relation to the function of growth rate for all structure components. The hierarchy of solidification processes in respect to the increasing solidification rate and undercooling are as follows [7,12]:

- *Full diffusional equilibrium*
  - No chemical potential gradient (phase compositions are uniform).
  - No temperature gradients.
  - Lever rule applies.
- *Local interfacial equilibrium*
  - Phase diagram gives compositions and temperatures only at S/L interface.
  - Corrections made for interface curvature (Gibbs–Thomson effect).

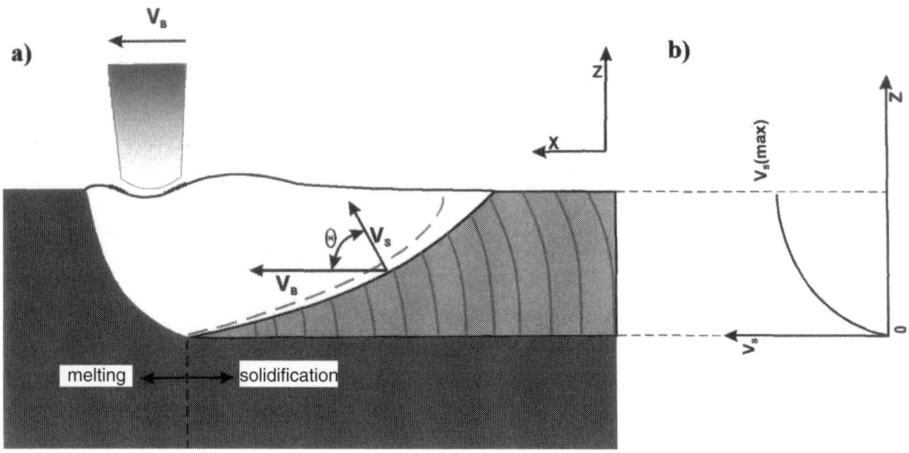

Figure 7.5: Scheme of longitudinal cross-section of laser track: (a) local solidification rate $V_s = V_B \cos T$ and (b) variation of solidification rate with respect to axis z.

- *Metastable local interface equilibrium*
  - Stable phase cannot nucleate or grow sufficiently fast.
  - Metastable phase diagram (true thermodynamic phase diagram) describes the interface conditions.
- *Interfacial non-equilibrium*
  - Phase diagram fails at the interface.
  - Chemical potentials are not equal at the interface.
  - Free energy functions of phases still lead to criteria which predict impossible reactions.

Stability of solidification front is also observed at high solidification rates. It is possible to show that for a given alloy and for the positive temperature gradient, the plane solidification front is formed when the growth rate is sufficiently high. The transition from the plane to cellular, dendritic and again cellular and plane solidification front occurs when the growth rate increases at small temperature gradient. The plot of temperature gradient vs. solidification rate and solidification morphology showing a range of stable interface morphology is presented in Fig. 7.6 [7].

When a positive temperature gradient, G, is imposed on an alloy of given composition there is, at low growth rates ($V = V_c$), a transition from a planar to a cellular morphology due to constitutional undercooling. A reverse transition from cells to a planar front is observed at high rates ($V > V_a$). The latter is essentially independent of the temperature gradient. Above a certain temperature gradient, G_a, the planar form is always stable.

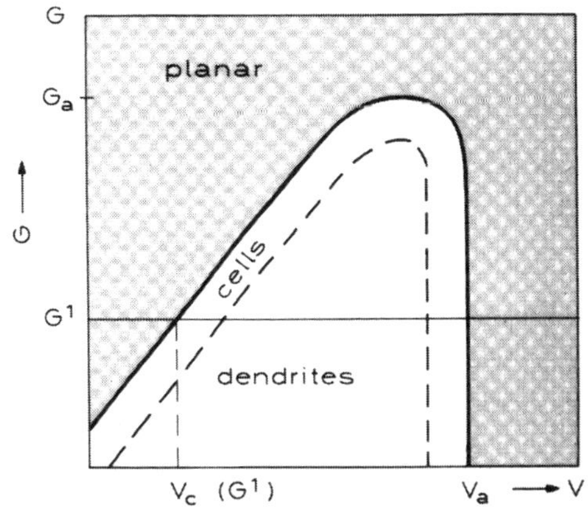

Figure 7.6: The range of stable (planar) interface morphology.

7.2.4. Constitutional Undercooling

During the solidification of an alloy, there is a substantial change in the concentration ahead of the interface. This change will affect the local equilibrium solidification temperature T_l of the liquid, which is related to the composition by the following formula:

$$T_l(C_0) - T_l = m(C_0 - C_l) \tag{7.1}$$

where $T_l(C_0)$ is the liquidus temperature corresponding to the initial alloy composition. This relationship is shown in Fig. 7.7 and indicates that the concentration boundary layer can be converted, using the phase diagram, into a liquidus temperature boundary layer. The liquidus temperature increases with increasing z, when the value of k is less than unity, because the value of m is then negative. It represents the local equilibrium temperature for the solidification of a corresponding volume of melted material. In order to investigate stability, it is also necessary to determine the temperature, T_q, imposed by the heat flux. Both temperatures must be equal at the interface. In the steady-state growth of a planar interface, this will correspond to the solidus temperature for the composition, C_0, as shown in Fig. 7.7. Depending on the temperature gradient,

$$G = (\mathrm{d}T_q/\mathrm{d}z)_{z=0} \tag{7.2}$$

in the liquid at the S/L interface (which is imposed by the extreme heat flux) there may, or may not, exist a zone of constitutional undercooling. This zone is defined

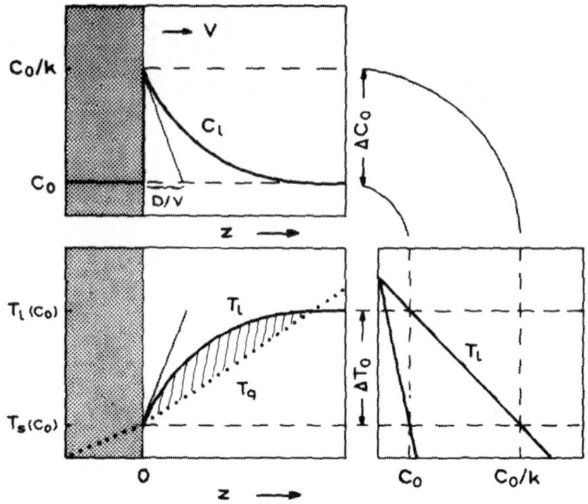

Figure 7.7: Constitutional undercooling in alloy solidification.

to be that volume of melt ahead of the interface within the actual temperature T_q is lower than the local equilibrium solidification temperature T_l. The melt in this zone is thus undercooled, that is in a metastable state.

Solidification proceeds as either a stable planar front or as an unstable front leading to dendrites or cells and, in consequence, to microstructure variation in remelted layer. We can see, that the process depends on the occurrence of constitutional undercooling. As it was shown, the constitutional undercooling is caused by the thermal gradient being less steep than the melting point gradient, which is the result of partition effects taking place at the solidification front, giving rise to composition variation in this region (Fig. 7.7) [7].

As the liquid concentration, C_l, decreases with distance, z, the liquidus temperature, T_l (i.e. the melting point), of the alloy will increase as indicated by the phase diagram. This means that if small volumes of liquid at various distances ahead of the S/L interface were extracted by some means and solidified, their equilibrium freezing point would vary with position in the manner described by the heavy curve in the lower left-hand diagram. However, each volume element finds itself at a temperature, T_q, which is imposed by the temperature gradient arising from the heat flow occurring in the casting. Since, at the S/L interface ($z = 0$), T_q must be less than or equal to T_s in order to drive the atomic addition mechanism, there may exist a volume of liquid which is undercooled when the gradient of T_q is less than the gradient of T_l. This (cross-hatched) region is called the zone of "constitutional undercooling". There exists a driving force for the development of perturbations in this volume.

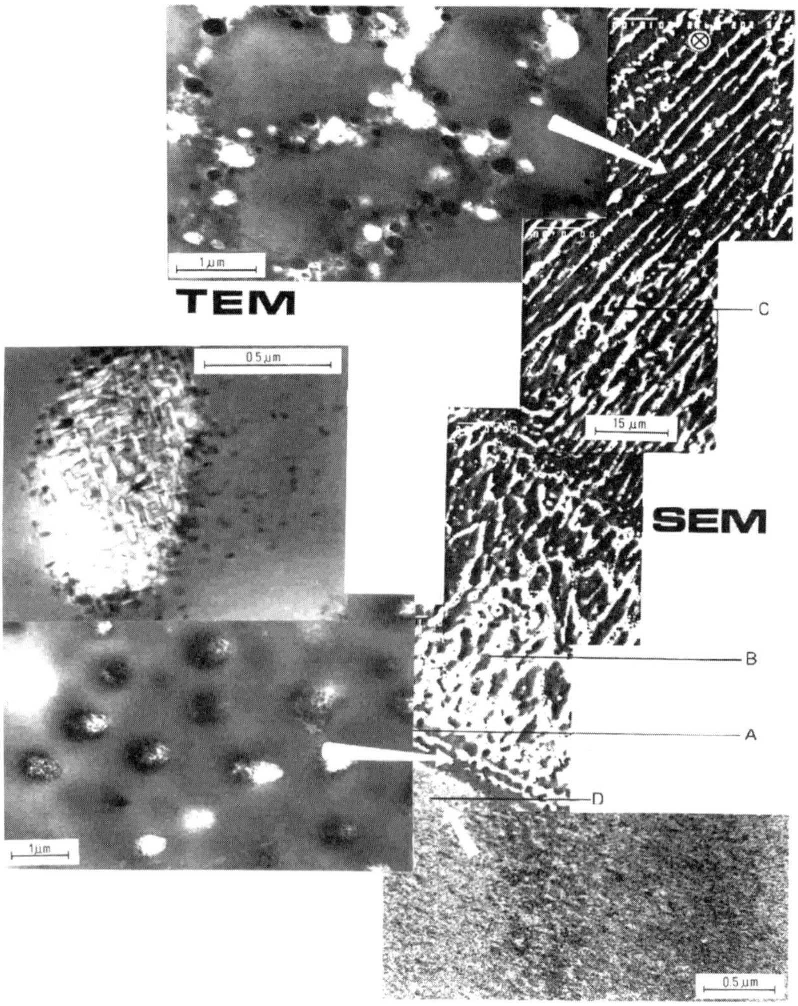

Figure 7.8: SEM micrographs of a cross-section through the solidified pool and TEM details of foils cut parallel to the specimen surface from the layer at various depths below the surface: (A) section with frozen droplets, (B) and (C) sections perpendicular to dendrite axes and (D) unmelted substrate.

Morphologic instability of the S/L interface caused by the constitutional undercooling was observed in the solidification process in laser-melted surfaces made of Al–30wt.%Zn alloys [13] (Fig. 7.8).

Details of solidification process produced during epitaxial regrowth of partially melted grains and revealing fluctuation of steady and unsteady solidification front

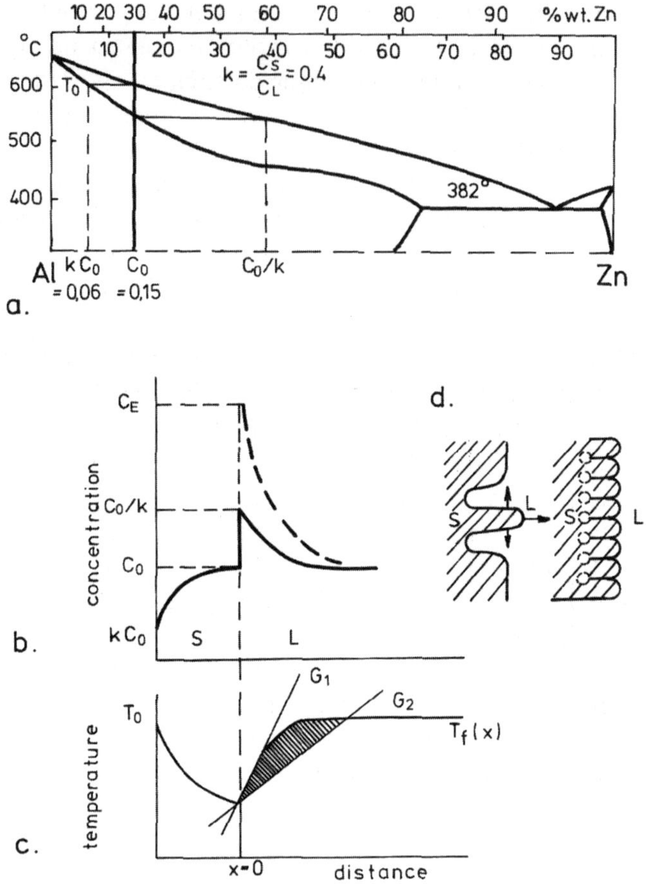

Figure 7.9: (a) Part of the phase diagram of Al–Zn illustrating the partition coefficient $k = C_S/C_L$ and (b) schematic drawing of the solute profile along S/L phases; (c) freezing temperature profile showing the development of a constitutionally undercooled liquid zone at the S/L interface, when the temperature gradient falls from G_1 to G_2 and (d) schematic development of a cellular interface due to the rejection of solute atoms in directions both normal to the average S/L interface from its tip and lateral from its side.

were identified on the basis of scanning electron microscopic (SEM) micrographs of the cross-section of the solidified pool and confirmed with transmission electron microscopic (TEM) microstructures of the respective areas shown in Fig. 7.8.

The difference in contrast between the undissolved part of grains with the average mole fraction $C_0 = 0.15$ of Zn (Al–30wt.%Zn alloy) in the investigated alloy and the regrowing grain at the onset of solidification indicates that the growth starts (at liquidus temperature of 610°C, Fig. 7.9(a)) with the mole fraction $k_0C_0 = 0.06$ of Zn; where $k_0 = 0.4$ is the equilibrium partition coefficient (Fig. 7.9(a)).

Excess Zn atoms rejected from the solidification grains into the melt at the locally planar S/L interface will built up there a solute-rich layer, due to limited diffusion rate and convection within the melt (Fig. 7.9(b)). Further growth thus proceeds at the temperature decreasing steadily below T_0 at the advancing interface. A gradually increasing amount of Zn in the melt close to the S/L interface and its rapid drop to C_0 ahead of it (especially due to the rising growth rate of grains) will lead to a constitutional undercooling of the melt within the Zn-enriched layer (Fig. 7.9(c)). This can promote a rapid, but still planar non-equilibrium growth of grains, leading to the appearance of a solute-enriched band within the growing grain already at the distance of about 2 μm from the deepest S/L interface, which is apparent at the bottom of SEM micrograph in Fig. 7.8. The appearance of this band is accompanied by the rise of the temperature at the S/L interface almost to T_0 again, so that the equilibrium planar growth can be restored again. For higher Zn content (30wt.%Zn), the local inhomogeneities in the solute distribution within the liquid layer close to the advanced S/L interface may lead to local temporary corrugations (Fig. 7.9(d)). This is associated with the trapping of solute-enriched droplets in the solidifying layer (Fig. 7.8, zone A). Such droplets solidify later, during the drop of temperature of the solute-rich band, reaching the eutectic composition C_E (Fig. 7.9(b)). Even β precipitates can be formed inside the solidifying droplets of the melt at the eutectic temperature of 382°C. Such plate-like precipitates can be resolved within the solidified droplets in the lower TEM micrograph in Fig. 7.8 (bottom-left). The consecutive formation of two solute-enriched bands during this planar growth are apparent in Fig. 7.8. The presented results clearly indicate that the undercooling of the liquid layer close to the planar S/L interface does not necessarily lead to its geometrical instability, associated with the development of dendritic growth claimed by Tiller *et al.* [14]. Only further grain growth with higher velocities is associated with persistent corrugations of the planar S/L interface. The formation of wavy solute-rich layer is apparent in Fig. 7.8, at the distance slightly above 6 μm. The rejection of the solute atoms aside and perpendicularly to the protruding tips of the corrugated S/L interface permits these tips to advance rapidly at T_0 (Fig. 7.9(d)). This leads to the establishment of irregular (Fig. 7.8, zone B) and later of arrayed cells elongated in the direction close to the temperature gradient, which is apparent from SEM micrograph in Fig. 7.8 (zone C). The rapid advance of dendrite tips at T_0 is followed by their slow thickening at lower temperatures, which is accompanied by solute enrichment of the melt within the interdendritic grooves. The final freezing proceeds there at the eutectic temperature and is associated with the precipitation of β particles. This is evident in

TEM details of columnar structure (cut perpendicularly to cell axes) situated in the upper part of Fig. 7.8 (left-top).

7.3. Laser–Surface Melting and Alloying

Laser–surface-treatment techniques are used to obtain improvements in hardness, corrosion or wear resistance. Rapid solidification after laser melting yields fine microstructures, little thermal penetration into the substrate, while alloying performed simultaneously with laser melting yields fine microstructures and allows many metals or non-metals to be alloyed with different substrates. Laser melting is performed in a short time, in which only a small part of the absorbed heat energy is transferred into material. This leads to formation of a high temperature gradient between the melt surface layer and the bulk material. Intensive mixing is observed during laser melting due to the convection movements caused by temperature changes between the surface and the bottom of the melt pool as well as by intensive blowing of shield gas (Fig. 7.10) [15].

Rapid solidification occurs because of a high temperature gradient, leading to fine-grained columnar or dendritic structures elongated along the heat flow. The internal microstructure of the formed crystals is related to the solidification rate, chemical composition, phase transformation and the course of cooling in the solid state (Figs. 7.3 and 7.6).

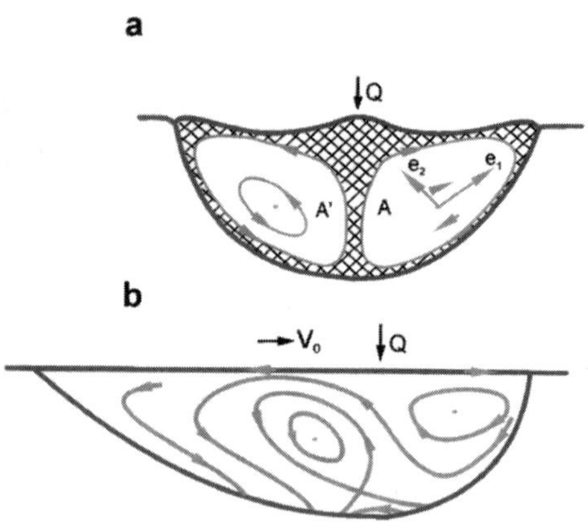

Figure 7.10: Scheme of convection movements in the melt pool: (a) cross-section perpendicular to the movement of laser beam and (b) cross-section longitudinal to the movement of laser beam; where Q is the laser beam and V_0 scan rate; points A and A′ indicate area where liquid velocity is bound to zero; e_1 and e_2 axis of elliptical co-ordinate system.

7.3.1. Experimental Arrangement and Lasers Used

The experimental arrangement used in laser melting or alloying is similar to that applied in transformational hardening, except for the use of focused beam in the latter case. The relevant scheme is presented in Fig. 7.11.

Continuous or pulsed lasers are used in the technologies of material-surface treatment. In most cases, CO_2 (10,600 nm) or Nd:YAG (1064 nm) lasers and recently diode lasers (800–900 nm) of high power are used for laser melting or alloying.

The surface to be melted is shrouded by an inert atmosphere and the main characteristics of the process are [1]: (i) moderate to rapid solidification rates producing fine, near homogeneous structures; (ii) little thermal penetration, resulting in little distortion and the possibility of operating near thermally sensitive materials; (iii) surface finishing of around 25 μm is fairly easy to obtain and can be significantly reduced by post-processing; (iv) process flexibility due to software control and (v) possibilities in automation.

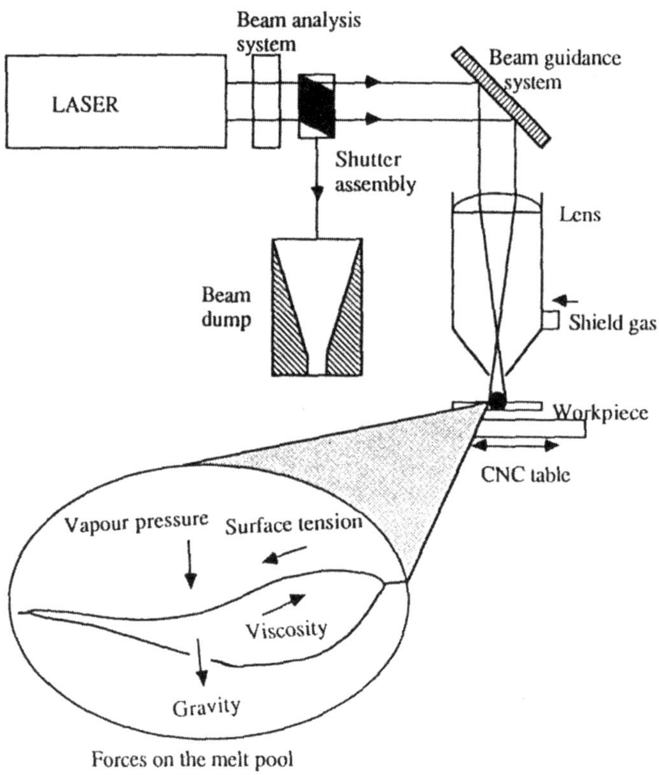

Figure 7.11: The scheme of experimental arrangement for laser heat treatment.

In order to understand the process of laser melting, it is necessary to quantitatively evaluate: (i) the thermal gradient and solidification rates which define the solidification process; (ii) the thermal gradient and temperature impact time related to stress generation and diffusion or dissolution effects and (iii) the conditions maintaining the material in melted form, and surface-tension effects and bubbles escape time in particular.

7.3.2. Aluminum Alloys

Laser treatment of aluminum and its alloys is somewhat difficult due to their low absorption of laser radiation [16]. In laser-treated Al–Si alloys (between 4 wt.% and 20 wt.%Si), the hypoeutectic alloys showed aluminum dendrites surrounded by flake-like eutectic regions. In the eutectic regions, a transition occurred from the eutectic type of solidification (at low laser scan velocity) to the dendritic type of solidification (at high scan velocity). Lamellar eutectic type of solidification was mainly found [17] in hypereutectic alloys. The analysis of microhardness distribution along the cross-section of laser track in Al–12wt.%Si alloy showed approximately twofold increase with respect to the parent material. A similar increase has been observed for this material laser alloyed by means of SiC and Ni [18,19]. Examinations of rapidly solidified Al–Fe alloys subjected to laser–surface melting revealed different types of microstructures as to the growth rate; that is, at low growth rates, cellular/dendritic structures and at high growth rates, a banded structure consisting of a succession of light and dark bands which lay approximately parallel to the S/L interface [20]. SEM examinations of Al–6wt.%Fe and Al–6wt.%Fe–3wt.%Ni alloys showed the micro-dendritic cells elongated along the temperature gradient. The large precipitates of the Al_3Fe phase, existing in the as-cast structure, have been considerably refined by laser melting. During TEM examinations, three types of microstructures were revealed, comprising the α-Al and Al_3Fe phases in the cellular, lamellar and needle forms. Laser melting caused an approximately threefold increase in hardness HV, compared to the parent material [18]. Examinations of Al–Zn alloys [13] revealed that extensive segregation of solute atoms took place, even during rapid solidification of laser-melted surface layer. This turns out to be advantageous for resolving details of solidification processes in these alloys presented in this Section 7.2.4 (Figs. 7.8 and 7.9). Examinations of microstructure using light microscopy (LM) and SEM of Al–5wt.%Zn–3wt.%Mg–1wt.%Cu alloy showed that a layer about 50 μm thick was formed with cellular fine-grained structure after laser melting (Fig. 7.12) [18]. The distribution of Zn was uniform within this layer. The formation of precipitates characteristic for rapid solidification with tenfold symmetry axis has also been observed in this zone [21]. An increase in hardness HV of about

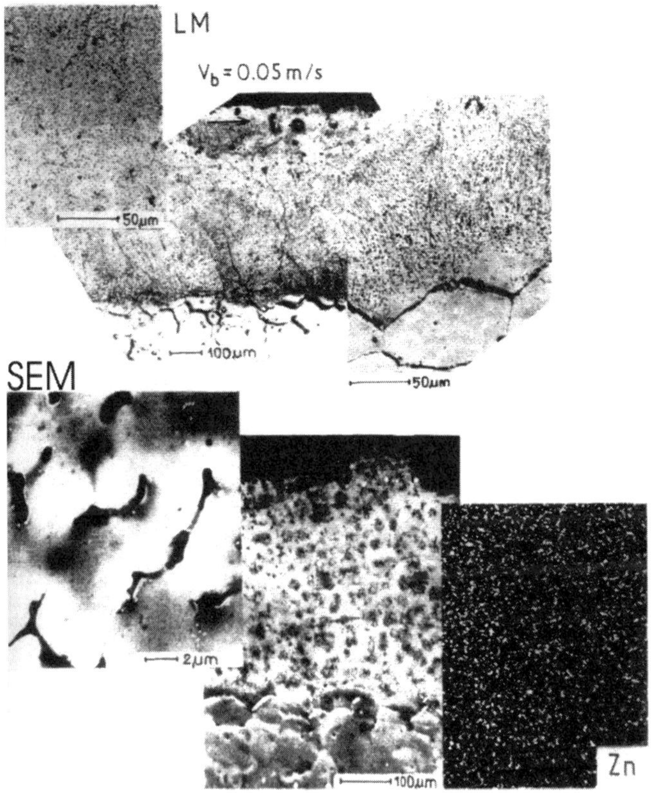

Figure 7.12: Micrographs (LM and SEM) of Al–5wt.%Zn–3wt.%Mg–1wt.%Cu alloy with the distribution of Zn; the alloy subjected to ageing at 150°C for 3 h after laser treatment.

20% has been observed in heat-affected zone (HAZ), while for the laser remelted one, the observed increase was of about 10%, however, cracks were also observed in the remelted zone formed during solidification.

7.3.3. Steels

The most beneficial effects of laser treatment of steels can be attributed to specific solidification structures, mostly of fine martensite in the case of laser melting [22–25]. Different alloying materials are used to improve the properties of various grades of steels, for instance chromium, nickel, tantalum as well as non-metallic materials, like carbides or borides [18,19,26–28].

7.3.3.1. Carbon Steels

The application of continuous wave (cw) CO_2 laser to melt C45 revealed the formation of fine martensite microstructure. In the area of overlapping laser track, the martensitic microstructure was even finer [18,19]. Laser alloying of carbon steels C45 with chromium [29,30] or carbides B_4C and borides VB_2, CrB [19,27,28] was used to improve in this way surface properties of bulk material (Fig. 7.13) [19].

Laser alloying of C45 steel by blowing CrB powder into the melt pool resulted in the formation of high amounts of residual austenite due to increased content of chromium. Irregular eutectic mixture containing $\alpha'/M_3(C,B)/M_2B$-type was observed at about eutectic composition, while the primary borides of the M_2B-type were formed for the hypereutectic compositions.

The blocky type of borides appeared for the C45 steel laser alloyed using VB_2. The M_3B_2 phase containing about 50 at.%B was identified in hypereutectic alloys, where the M_2B phase of about 39 at.%B was also observed for the higher concentrations of V in laser tracks that approached 14wt.%V and 6wt.%B (24 at.%B) [28].

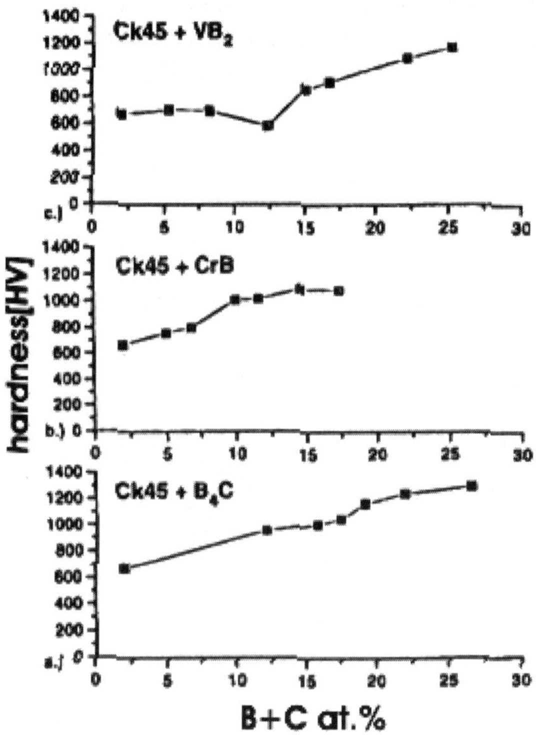

Figure 7.13: Hardness HV3 of the VB_2, CrB and B_4C laser-alloyed C45 steel.

Introduction of B_4C resulted in the disappearance of cellular dendritic microstructure. A fine eutectic with the composition of about 12.5 at.% C was observed, while the primary precipitates of the $Fe_{23}(C, B)_6$ phase were formed at the hypereutectic composition [27].

A satisfactory correlation was observed between the results of structure examinations and the phase diagram Fe–C–B calculated with the Thermo-Calc program [27]. Laser alloying of C45 by means of borides: CrB, VB_2 and carbide B_4C increased the hardness HV to about 1200 HV (Fig. 7.13).

In experiments, cw CO_2 laser was used for alloying low (0.17wt.%C) as well as high (1.13wt.%C) carbon steels by applying: chromium, SiC and tantalum in the form of a paste, consisting of alloying powder and binder, which was sprayed on the modified surface. Then, the surface was subjected to laser melting [30,31]. Organic (varnish) and non-organic (water glass) substances were used as the binder. The microstructure was observed to vary, depending on operation conditions, power density, scanning rate and thickness of the melt zone. The application of organic binder to powder-alloying material, like varnish, allowed to increase *C* content in the modified zone, which increased the hardness, compared to that obtained by the application of non-organic binder (water glass). Burning of varnish led to carbonizing of the laser-treated zone. Cellular microstructure with carbide precipitates along cell grains was observed (Fig. 7.14) [30].

7.3.3.2. Constructional Steels

The use of cw CO_2 laser to melt valve steel (X45CrSi9-3 ISO) resulted in the formation of a lath-type martensite. Such martensitic microstructure was also observed in C50 (ISO) steel. The hardness of material increased about 2–3 times in respect

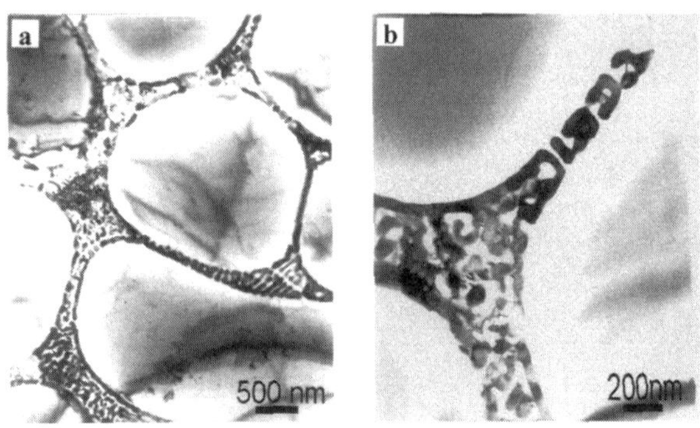

Figure 7.14: TEM micrographs presenting (a) eutectic mixtures, formed in the area enriched with chromium after laser melting of low-carbon steel, containing organic binder of chromium powder and (b) higher magnification of the area with cellular structure and ledeburitic network along the cell boarders.

to the material in annealed state. It is worth nothing, that subsequent laser melting of the surface, previously covered with nickel, resulted in the formation of austenite, causing a deep decrease in the hardness of this covered layer, while the hardness beneath the austenite layer returned to the value characteristic of martensite [18,19]. A high value of HV obtained in the HAZ, similar to that observed in this alloy after remelting or alloying with WC, could be caused by a diffusion of alloying elements into this region, which allowed them to reach the optimum combination and prevent austenite retention.

C50 steel was laser remelted and alloyed with WC and Cr by means of cw CO_2 laser [18,19]. The formation of martensite structure was observed in C50 steel after laser remelting. This was correlated with an approximately twofold increase of HV. A high rate of solidification and the specific characteristics of the laser beam caused inhomogeneity in the layer that was laser alloyed with WC within a martensite structure. A cellular structure was observed in the region of higher concentration of W. Retained austenite was observed in X-ray examinations X-ray diffraction (XRD).

7.3.3.3. High-Speed Steels

High-speed steels (HSS) belong to the group of ledeburitic tool steels and they are predominantly used as material for tools of complex shapes, because HSS offer a good combination of a range of properties, namely strength, toughness, tempering resistance, etc., and they can be easily machined to obtain complex shapes. HSS are characterized by superior toughness, but their wear resistance is usually lower than that of other tool materials. Improved wear resistance can be expected from HSS with higher amount of primary and secondary carbides [22–24]. This requires HSS with higher amount of carbon- and carbide-forming elements, but the addition of these elements is restricted. High contents of alloying elements combined with low cooling rates results in coarse and brittle networks of various carbides in the solidification structures, which deteriorate hot workability and toughness [32]. Increased contents of an alloying element can be obtained by the application of technologies that enable high cooling rates during solidification, for example powder metallurgy (PM) or surface-melting techniques. Some of PM-HSS grades contain an increased amount of alloying element, but there exist some limitations, too [32]. Surface-melting techniques, like laser or electron beam melting, offer new possibilities for the modification of the structure in near-surface regions [9–11,33,34]. The accompanying rapid solidification modifies the microstructure by structure refinement and by the supersaturation of crystals. In this way, the content of strong-carbide-forming elements can be increased significantly by laser alloying. One can even obtain higher contents than in PM-HSS [18,19,35,36]. All commercial HSS grades are based on alloying with carbon- and strong-carbide-forming elements. Laser alloying additionally offers the possibility of introducing other types of hard phases, like borides [28]. Carbon and boron are known to form intermetallic compounds with most metallic elements, but the solubility of boron in most metals is usually significantly lower than that of carbon. The structure of alloys containing boron and carbon is complex, because not only borides and carbides form binary systems, but also ternary carbo-borides may precipitate from the melt. Especially boron is known to form numerous stable and metastable compounds with almost all metallic elements. The resulting mixture of different phases in the alloyed tracks makes it difficult to provide a complete metallographic and crystallographic characterization of the phases. Usually, a large number of partially overlapping peaks appear in the XRD pattern of laser-alloyed melt tracks. The microstructure of the laser-melted HSS of HS6-5-2 (ISO) grade will be taken as a reference for the discussion of the effect of element additions through laser alloying. Moreover, the application of 2.5-kW diode laser for melting will be presented for HS6-5-2, HS2-10-1-8 and HS10-2-5-8 [35,36].

The typical chemical composition of the substrate material HSS used in experiments is presented below in wt.%:

Substrate	% C	% Cr	% W	% Mo	% V	% Co
ISO HS6-5-2 (conventional metallurgy)	0.9	4.10	6.30	5.00	1.90	–
ISO HS2-10-1-8 (conventional metallurgy)	1.10	3.9	1.4	9.2	1.2	7.8
ISO HS10-2-5-8 (powder metallurgy)	1.6	4.8	10.5	2.0	5.0	8.0

The melting of HS6-5-2 grade with 2.5 kW cw CO_2 laser-produced laser-melted tracks with a width and depth of about 2 mm and 1 mm, respectively, and resulted in a fine solidification microstructure, consisting of metallic dendrites surrounded by a carbide eutectic mix [26]. The spacing of dendrite arms in the laser-melted HSS fitted within the range of about 3–5 μm, compared to about 80–120 μm in conventional HSS. As a result of the supersaturation of metallic dendrites, a high amount of retained austenite – up to 50 vol% – was present after cooling to room temperature, which is responsible for the relatively low hardness of about 650–750 HV3 after laser melting. The volume fraction of the eutectic was about 5%, which means about a half of the eutectic in conventional HS6-5-2. XRD studies showed M_6C, MC and M_2C to be the main carbides in the eutectic mixture. Laser melting with 2.5-kW cw diode laser covered the area that was 6 mm wide and 0.5 mm deep (Fig. 7.15) [35].

The structure refinement due to rapid solidification was observed in the laser-melted zone of HS2-10-1-8, HS6-5-2 and HS10-2-5-8. Carbide phases with chemical composition different from the parent material were also observed to form (Fig. 7.16) [35,36].

The contribution of diode laser modification to the microstructure and properties of HSS obtained by conventional and PM was studied [35,36]. After laser melting,

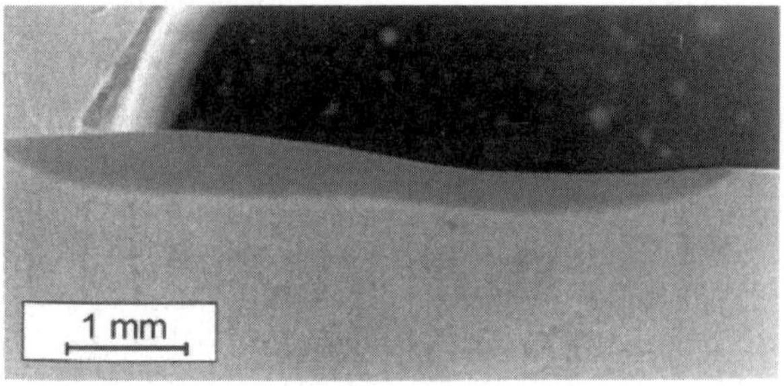

Figure 7.15: Microstructure of the cross-section of HS6-5-2 surface layer melted with a diode laser.

materials were subjected to tempering at 520°C/3 × 2 h. It was observed that laser melting and subsequent tempering increased hardness and improved wear resistance in the conditions of friction cause by static load. However, the resistance to dynamic loads was diminished [35–37].

Alloying of HS6-5-2 with 2.5-kW cw CO_2 laser was performed for vanadium carbide (VC), which is a thermodynamic stable intermetallic compound [18,26]. Vanadium belongs to typical alloying additions to HSS and therefore its content in common HSS is limited to about 3 wt.%, because higher vanadium content leads to the formation of carbides in the carbon-containing melt, prior to solidification followed by a separation due to density differences. On adding VC into the laser-alloying process, melt tracks were achieved with vanadium contents up to about 50 wt.% [18,26]. Blowing of the VC powder with shielding argon gas into the melt pool was applied. The microstructure of laser-alloyed HS6-5-2 changed with increasing VC addition as follows: VC additions cause an increasing amount of eutectic in the interdendritic space. The morphology of the eutectic also changed, from a feathery to spherical shape. A new type of eutectic was created at more than 4 wt.%V. It contains eutectic cells beside metallic dendrites. When the value of about 8 wt.%V is reached, the structure gets fully eutectic. When V content exceeds 8 wt.%, solidification starts with the precipitation of monocarbides. For comparison, an isopleth along the path HS6-5-2 + VC was calculated with Thermo-Calc program, and a good correspondence with the microstructure was observed. VC addition increased the amount of carbides and improved the hardness from about 700 HV3 in the remelted HS6-5-2 to about 1040 HV3 in the highest-alloyed melt tracks [38].

Microstructure evolution was studied in HS6-5-2 alloyed with Mo_2C by means of a 2.5-kW cw CO_2 laser [38]. Mo_2C powder was blown into the melt pool, with argon shielding. Solidification process was discussed on the basis of microstructure

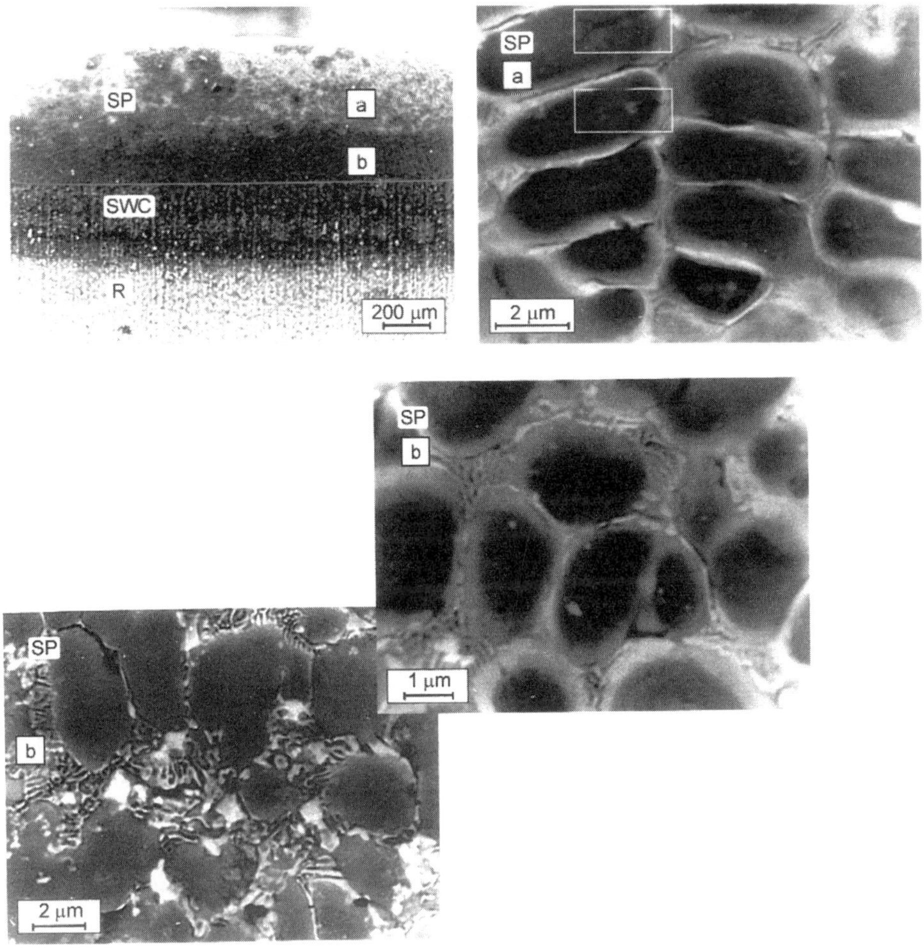

Figure 7.16: SEM micrographs of the cross-section of HS6-5-2 surface layer melted with a diode laser: SP – laser-melted zone; SWC – HAZ; R – substrate and enlargement of the regions marked with "a" and "b", where the carbide eutectic along the cell boundaries (small cracking inside the cells) and eutectics with carbides located in the interdendritic areas could be seen.

examinations for a wide range of addition: from 5 to 54wt.%Mo. An increase of the alloying addition (Mo_2C) caused the metallic dendritic component to diminish. At the solidification rates occurring during laser melting, the austenite was observed and carbide segregated round this phase. Formation of the primary M_6C carbides was observed for hypereutectic compositions. Thermodynamic calculations of the constitutional changes were performed with Thermo-Calc program. Experimental results confirmed that high solidification rates led to early solidification of M_2C phase. Exclusion of the M_6C stable phase from calculations resulted in shifting of eutectic composition to about 25wt.%Mo. Thus, the shape of the Mo_2C phase became more probable.

Boron is not used as alloying element in common HSS, although boron forms stable and hard intermetallic phases. One of the reasons is that even very low boron additions lead to an eutectic with iron at about 1150°C, which is lower than the conventional hardening temperatures of HSS. Moreover, boron additions impair the ductility and toughness. In spite of these disadvantages, boron seems to be an interesting alloying element, especially in the search of an increased amount of hard and wear-resistant particles. Boron is known as an element that forms a variety of intermetallic phases with metals, which makes the alloy development extremely difficult [28]. Enhancement of the alloy development by means of thermodynamic calculations is almost impossible in case of boride alloying because the thermodynamic functions of the various phases are insufficiently known. It is therefore necessary to start with the thermodynamic modeling of simple systems. Such models were calculated for the ternary system Fe–B–C [27]. Experimental investigations within this system also revealed a good correlation between the microstructure of laser-alloyed tracks and calculated constitution. Laser alloying was done by moving the laser beam with the velocity of $0.2\,m\,min^{-1}$ relative to the substrate (HS6-5-2), while argon was used to prevent oxidation and to feed the powder of CrB or VB_2. These parameters yielded laser-melted tracks with the width and depth of about 2.0 and 1.0 mm, respectively. Microstructure investigations indicated, that most of the added CrB or VB_2 particles were dissolved into the superheated melt at the zone of interaction between the laser beam and the substrate. Small amounts of undissolved powder particles were present predominantly in the higher-alloyed tracks. The concentration of chromium varied within the range between 5.4wt.%Cr and 15wt.%Cr in CrB laser-alloyed tracks, while the concentration of vanadium varied within the range between 3.8 and 9.5wt.%V in VB_2 laser-alloyed tracks. Solidification starts from a nearly homogeneous melt. This alloyed melt decomposed during rapid solidification into various types of metallic phases, carbides and borides, resulting in complex microstructure. Boride additions proved to have a remarkable effect on the solidification microstructure in laser-alloyed tracks (Figs. 7.17 and 7.18).

Low CrB and VB_2 additions produced boride phases that were predominantly precipitated in the interdendritic space together with various types of carbides. Boride precipitation occurred even when boride additions were very small, which was due to low solubility of boron in the metallic crystals. The precipitation of carbides was similar to that in laser-melted HS6-5-2 and seemed to be unaffected by low boride

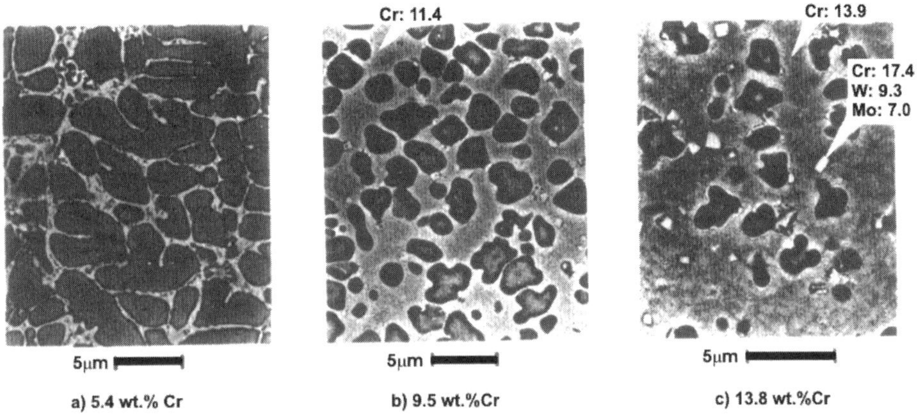

Figure 7.17: SEM microstructures of CrB laser-alloyed HS6-5-2.

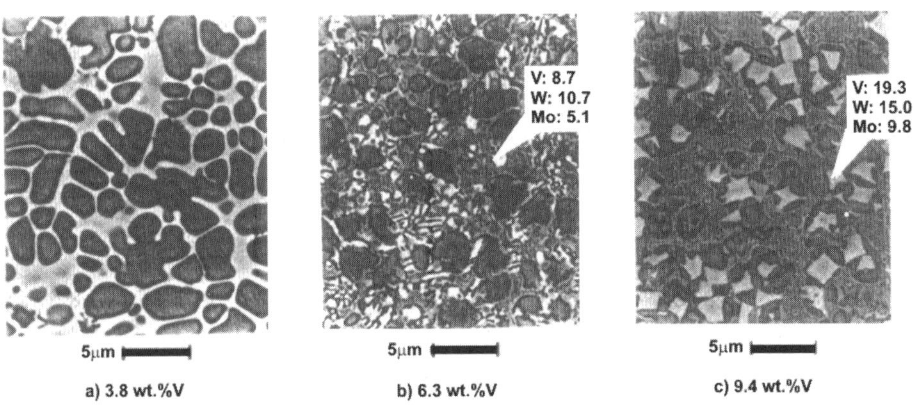

Figure 7.18: SEM microstructures of VB_2 laser-alloyed HS6-5-2.

additions. Medium-sized CrB and VB_2 additions led to the presence of δ-ferrite in the core of the metallic dendrites. This resulted in a severe loss of hardness at the point of exceeding 3wt.%V or 6wt.%Cr in the melt track. Further increase of the amount of boride addition led to an improvement of hardness as a result of the increasing volume fraction of hard phases. High CrB and VB_2 additions resulted in the precipitation of $M_{23}B_6$ and MB borides, whereas the amount of carbides was reduced due to the dilution of the base material. Primary borides with a blocky shape were present in the melt tracks exceeding 15wt.%Cr or 9wt.%V.

7.3.4. Titanium Alloys

Titanium and its alloys are very attractive materials due to their low density, high mechanical strength, good corrosion resistance and biocompatibility, and they could be widely used in many branches of modern industry. Moreover, due to the abundance of titanium ores in the earth crust and permanent progress in extraction technology of this metal, the price of pure titanium tends to decrease and is comparable to that of some alloyed steels. For this reason, the demand of titanium and its alloys is continuously increasing. Nevertheless, the spectrum of applications of titanium alloys is actually rather narrow. They are used mostly for constructional purposes. The main reason is that these alloys have poor tribologic characteristics during dry sliding, that is high wear and high coefficient of friction. Many studies have focused on surface modification of titanium alloys in order to overcome that obstacle and to improve their tribologic properties by means of numerous chemical vapor deposition (CVD) and physical vapor deposition (PVD) coating techniques (e.g. see Refs. [39–41]). Microstructure modification through the formation of a new surface layer can also be performed through laser processing. A chemical interaction of the liquid pool with the reactive shielding gas of the laser beam could be performed due to high chemical affinity of titanium to nitrogen or oxygen. Titanium alloys belong to modern constructional materials. As it was mentioned above, their satisfactory corrosion resistance combined with biocompatibility gives good prognosis as to their application in the function of medical materials. Therefore, much scientific effort is aimed to improve their surface properties due to, in general, a respectively low hardness and poor tribologic properties of titanium alloys. The simplest method of improving these properties seems to be the fabrication of a gradient material with the titanium nitride (TiN) in the surface layer. Such process could yield materials with hard and wear-resistant phases in the surface layers, while the ductile core of the material is maintained. Experiments were performed on titanium alloys of the following grades: Ti1Al1Mn (wt.%, alloy of α type) and Ti6Al4V (wt.%, alloy of $\alpha + \beta$ type) [42]. Two types of lasers were used, namely cw Nd:YAG laser (1064 nm) and cw diode laser (810 nm). Nitrogen environment was used in laser melting. Overlapping 1.5 mm wide laser tracks and individual laser track 7 mm wide were obtained for Nd:YAG or diode laser, respectively. The microstructure of the cross-section of the Ti1Al1Mn alloy, melted with Nd:YAG laser in nitrogen environment is presented in Fig. 7.19.

Three zones, about 80 μm thick, could be distinguished in the surface layer, where the observed microstructure was different from the parent material. The first zone was located close to the substrate and it was about 20 μm thick. It was the HAZ, where remelting did not take place, and it was characterized with a fine microstructure. This zone transformed continuously into the intermediate zone with

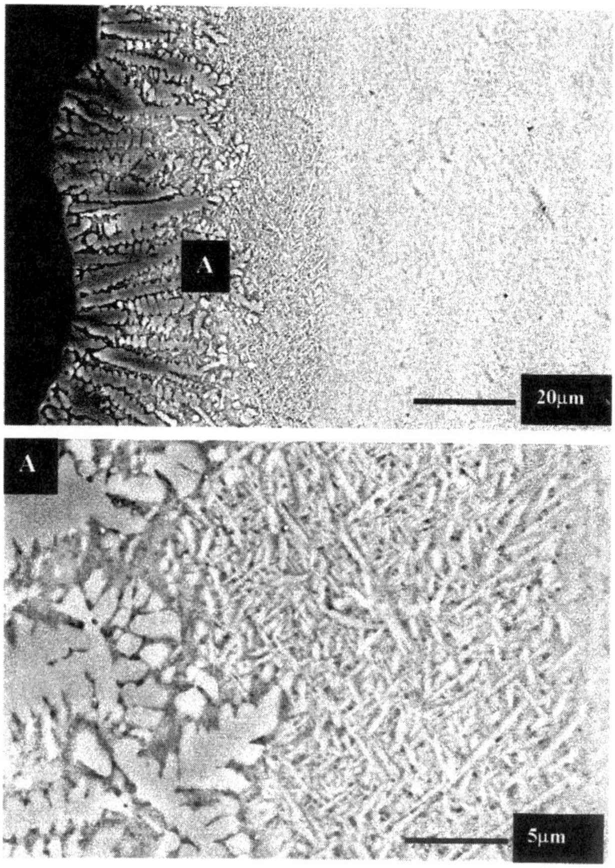

Figure 7.19: SEM micrographs of a cross-section of the Ti1Al1Mn alloy subjected to remelting with Nd:YAG laser in nitrogen environment by overlapping laser tracks.

a fine dendritic microstructure. This zone was about 20 µm thick. The third zone, close to the surface, contained well-developed dendrites with the tips directed from the surface into the parent material. This observation seems to be very important, because in the case of laser remelting, the dendrites formed close to the surface are usually directed at the surface, that is opposite to these observed in the examinations [7,8]. That could be related to the manner of changing the solidification rate in the laser remelting process, which varies from zero at the S/L interface to the maximum at the surface. It should be stressed, that the metallic titanium alloy was being remelted by a laser beam, while the TiN phase was solidified. Its melting point is much higher than the temperature of the liquid metallic pool; liquidus of TiN is in the range of 3000–3500°C, which is related to the nitrogen content in the TiN (Fig. 7.20). The chemical reaction between titanium and nitrogen could lead to the formation of the TiN phase, which was directly undercooled just after its formation. In this case, the solidification started from the surface and developed into dendrites starting at the surface and directed at the substrate. The previously electro-deposited nickel layer was also melted with the laser [42]. A general character of the morphology was similar to the one shown above. However, differences could be observed at the bottom of the remelted zone, where one could observe elongated forms of the cells enriched with nickel. Most of the dendrites in the surface layer assumed the surface-to-substrate direction, as it was observed previously (Fig. 7.19), while some tips of the dendrites aimed at the surface. The observed mechanism of solidification could be related to the appearance of two solidification fronts. The first one moved

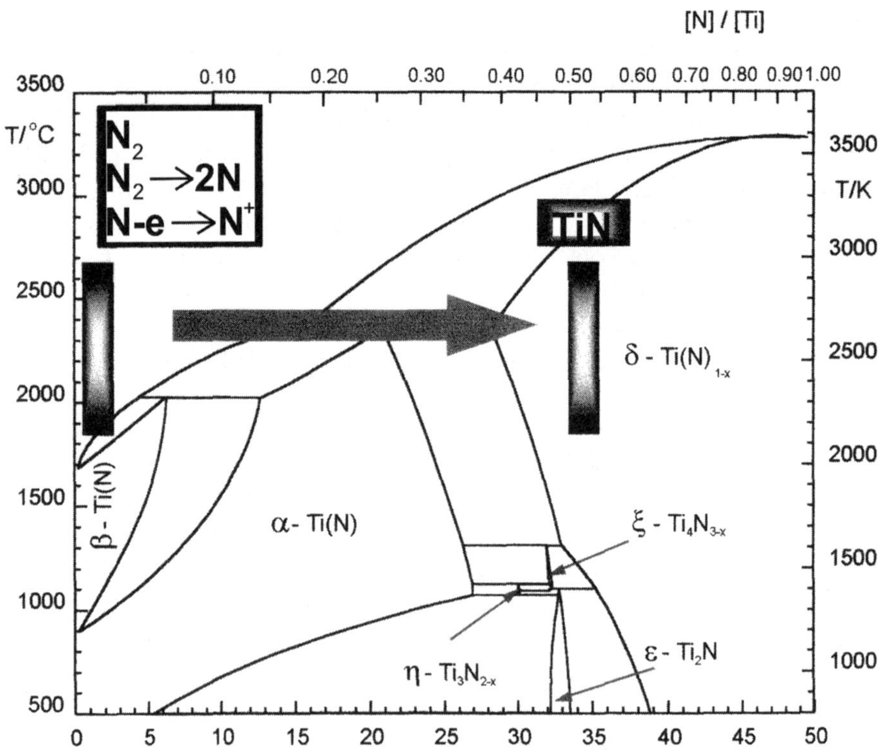

Rys.27. Układ równowagi fazowej Ti – N

Figure 7.20: TiN phase diagram with the scheme of nitriding in the melt pool of titanium alloys.

from the remelting boarder to the surface and the second one from the surface to the substrate. The latter, moving from the substrate, would be impoverished with nitrogen, due to difficult penetration of nitrogen down to the bottom of the molten pool. The fabrication of a nickel layer on the surface of titanium alloy and subsequent remelting led to the NiTi phase (verified by XRD), which changed the solidification conditions at the deep S/L interface. Elongated cells were formed and they subsequently transformed into dendrites directed at the surface, which was occasionally observed. Most of the metal volume which transferred into the TiN due to the chemical reaction with nitrogen solidified into dendrite form with the tips assuming surface-to-substrate direction, which was similar to the results obtained for the alloy without nickel. Ti6Al4V alloy was subjected to laser remelting with the diode laser. The microstructure in the surface layers is presented in Fig. 7.21. There can be observed similarity with the microstructure presented in Fig. 7.19, although titanium alloys differed in content ($\alpha + \beta$ alloy) and different types of the laser were used. It could be concluded that the solidification mechanism was similar to the previously described, based on two solidification fronts. The performed point microanalysis (EDS) in the region of a dendrite, between the dendrite arms and at the bottom of the remelted zone revealed differences in the content of Al, V and N. Detailed examinations (TEM) were performed on thin foils obtained from the cross-section (Fig. 7.21). Four different types of the microstructure of the TiN phase were observed (the TiN phase was verified by means of electron diffraction), namely the single-crystal type was observed inside the dendrite arms and it was characterized by a high dislocation density (a); in the interdendritic regions one could observe elongated cells (b) as well as equiaxial forms (c); moreover, observations identified a fine-grained structure (d) that yielded ring-shaped electron diffraction patterns.

An attempt was undertaken to explain the observed microstructure variation on the basis of the fact that the interaction of the laser beam with nitrogen environment could lead to the dissociation of N_2 into the atomic (N) or ionic forms (N^+), which led to three nitrogen forms in the environment (i.e. N_2, N and N^+). They could enter into reaction with the liquid metal pool. The ionic form was expected to be very active, and thus capable of leading to the formation of a fine-grained microstructure (even a nano-structure). This process was accompanied by a phenomenon characteristic for the unsteady solidification, where nano-particles were at first formed in the conditions of high solidification rate, to be pushed subsequently by the dendrites formed at a lower thermal gradient. Thus, the phenomenon of "pushing" instead of "joining" of nano-particles was more probable [10].

Examinations of the crystallographic texture (Fig. 7.22) were performed for the laser-remelted surface of titanium alloys. On the basis of the measured pole figures for TiN phase, it was stated that the orientation was close to be random despite of

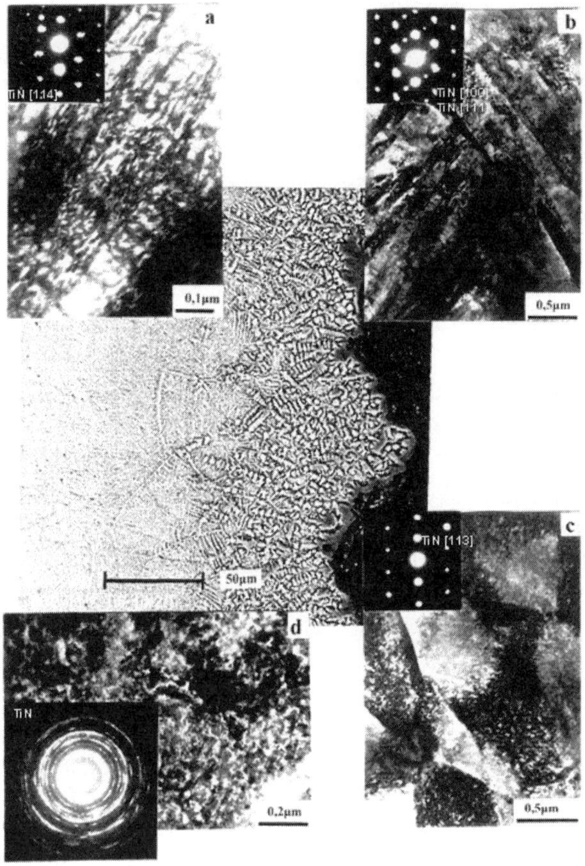

Figure 7.21: SEM and TEM micrographs of a cross-section of Ti6Al4V alloy subjected to remelting with the diode laser in nitrogen environment, showing the cross-section of the remelted layer (center) and details in the (a) dendrite arm and (b–d) between arms.

the fact that the dendritic morphology could suggest development of the preferred crystallographic orientation (Figs. 7.19 and 7.21).

7.4. Residual Stress in Laser Processing

Each laser process develops residual stress in laser-treated material. The measurements of stress in different types of laser processing performed by the author are summarized below. The determination and general mechanism as well as the method

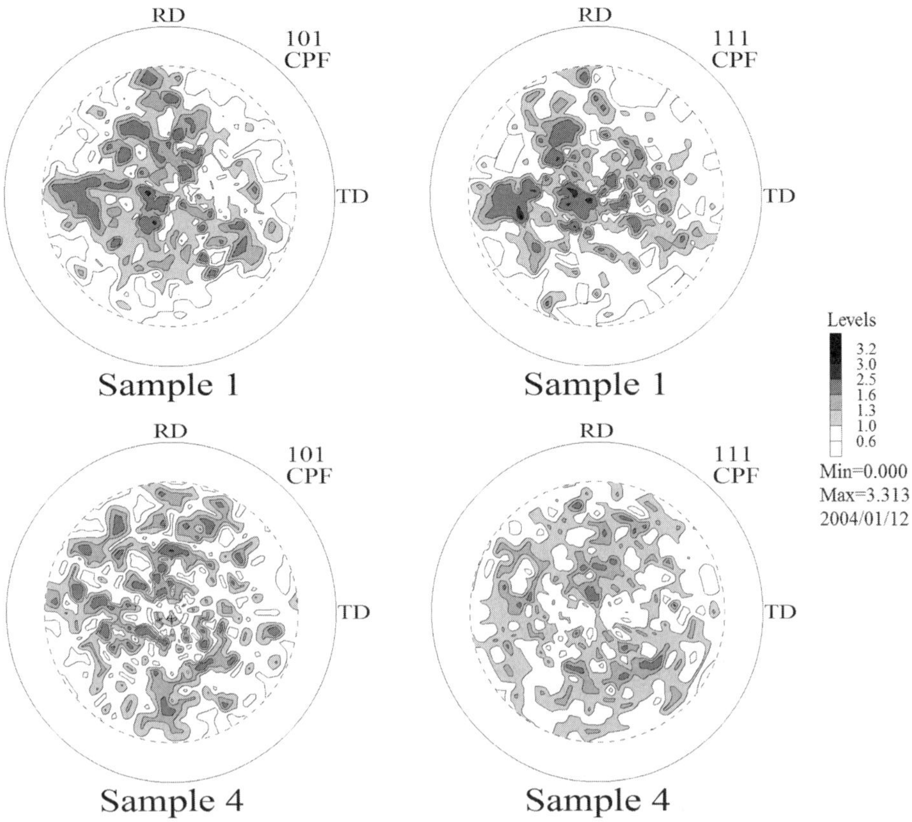

Figure 7.22: Pole figure of crystallographic texture measured in the surface layer of titanium alloys subjected to laser remelting in nitrogen atmosphere (nitride phases); (101) pole figure corresponds to Ti_2N tetragonal phase while (111) pole figure corresponds to TiN cubic phase: (a) sample 1 – Ti1Al1Mn alloy subjected to laser remelting by Nd:YAG laser using overlapping laser tracks and (b) sample 4 – Ti6Al4V alloy subjected to laser remelting by diode laser using individual laser track.

of measurement of residual stress is presented in Chapter 15. Therefore, the presentation below is limited to the measured macro-stress values obtained through the commonly known X-ray $\sin^2\psi$ method, based on the diffraction line shifting. The measurements were performed with a Philips PW1710 diffractometer by means of CoKα radiation. Splitting and fitting procedures was applied in Philips APD packet.

Laser remelting (overlapping tracks) in Ar atmosphere	
• Valve steel (ferrite)	−700 MPa (cw CO_2 laser)
• Manganese steel (ferrite)	−400 Mpa (cw CO_2 laser)
• Al12Si (α-Al)	−300 MPa (cw CO_2 laser)
Laser remelting (individual track) in N_2 atmosphere	
• Titanium alloy (Ti1Al1Mn) (TiN phase)	−400 MPa (cw diode laser)
Laser remelting (individual track) in Ar atmosphere	
• HSS (HS6-5-2) (conventional metallurgy)	−1500 Mpa (remelting; cw diode laser) −900 MPa (after tempering 530°C/3 × 2 h)
• HSS (HS10-2-5-8) (powder metallurgy)	−950 MPa (remelting; cw diode laser) −700 MPa (after tempering 530°C/3 × 2 h)
Laser alloying (individual track) in Ar atmosphere	
• HSS HS6-5-2 + VB_2 (V-8.7 wt.%, B-2 wt.%)	−330 Mpa (cw CO_2 laser)
• C45 + CrB (Cr-9.6 wt%, B-2 wt.%)	−570 MPa (cw CO_2 laser)
• C45 + VB_2 (V-5.7 wt.%, B-2.3 wt.%)	−730 MPa (cw CO_2 laser)

7.5. Concluding Remarks

Optical energy is a convenient tool for surface engineering and it can be used as a localized method, which avoids damaging the component yet is capable of modifying selected areas. Understanding the peculiarities of laser–surface interaction in laser melting and alloying, as well as understanding of the details of RSP both rely on the knowledge of intrinsic properties of lasers and surfaces. The use of laser technologies to study fundamental rapid solidification and diffusion control processes seems to be very practical. Materials treatment examines the possibility of practical application of fundamental changes of plain surface induced by laser radiation, such as heating, melting and plasma generation.

Moreover, structuring of the surface in terms of controlled crystallization processes, ablation, generation of periodic structure and generation of thin films opens great perspectives for technical applications. The advantages offered by laser processing are the following: highly localized and clean nature of the process, low distortion and good quality of finish, economy of the process. The development of laser devices and reliable more compact lasers is now both economically and technologically feasible to use lasers in industry for surface-treatment processes. The presented results in this chapter were mostly related to the own experiments performed on aluminum alloys, steels and titanium alloys.

Acknowledgments

A great part of the results presented in this chapter is the outcome of many years of co-operation between the Materials Center Leoben (MCL) in Austria and the Institute of Metallurgy and Materials Science at the Polish Academy of Sciences in Cracow, Poland.

The author is especially grateful to Prof. Reinhold Ebner (MCL) and his research group for active co-operation in research projects.

References

[1] D.S. Rickerby and A. Matthews (Eds.), *Advanced Surface Coating: A Handbook of Surface Engineering*, Chapman and Hall, New York, 1991.
[2] W.M. Steen, *Laser Material Processing*, Springer-Verlag, Berlin, 1992.
[3] H.-G. Rubahn, *Laser Application in Surface Science and Technology*, John Wiley, Chichester, 1999.
[4] T. Burakowski and T. Wierzchon, *Surface Engineering of Metals – Principles, Equipment, Technology*, CRC Press, Boca Raton/London/New York/Washington, DC, 1999.
[5] J.F. Ready, *Industrial Applications of Lasers*, Academic Press, San Diego, 1997.
[6] Y.C. Shin, *et al.*, *Mach. Technol.*, 11 (2000) 1.
[7] W. Kurz and D.J. Fischer, *Fundamentals of Solidification*, Trans Tech Publications, Switzerland/Germany/UK/USA, 1992.
[8] W. Kurz and R. Trivedi, *Mater. Sci. Eng.*, A179/A180, 46 (1994) 409.
[9] P. Gilgien and W. Kurz, in *Laser Processing; Surface Treatment and Film Deposition*, Eds. J. Mazunder, O. Conde, R. Villar and W. Steen, *NATO ASI Series E, Applied Sciences*, Vol. 307, Kluwer Academic Publisher, Dordrecht, The Netherlands, 1996, p. 77.
[10] C. Suryanarayana (Ed.), *Non-equilibrium Processing of Materials*, Elsevier Science Publication Co., Amsterdam, 1999.
[11] R. Vilar, Laser alloying and cladding, *Mater. Sci. Forum*, 301 (1999) 229.
[12] W.J. Boettinger, S.R. Corriel and R.F. Sekerka, *Mater. Sci. Eng.*, 65 (1984) 27.
[13] V. Synecek, B. Major, P. Bartuska, J. Lasek and M. Simerska, *Z. Metallk*, 83 (1992) 246.
[14] W.A. Tiller, K.A. Jackson, J.W. Rutter and B. Chalmers, *Acta Metall.*, 1 (1953) 428.
[15] C. Draper and J.M. Patoe, Laser surface alloying, *Int. Met. Rev.*, 30 (1985) 85.

[16] U. Luft, H.W. Bergmann, B.L. Mordike, in *Laser Treatment of Materials*, Ed. B.L. Mordike, Deutsche Gesellschaft f.Metallkunde, Oberurel, 1987, p. 147.
[17] H.J. Hegge and J.Th. De Hosson, *Scripta Metall.*, 24 (1990) 593.
[18] B. Major, R. Ciach, R. Ebner, F. Jeglitsch, B. Kriszt and K. Rabitsch, *Phys. Status Solid (a)*, 141 (1994).
[19] B. Major, R. Ciach, R. Ebner and F. Jeglitsch, *Acta Phys. Pol. A*, 89 (1996) 171.
[20] M. Gremaud, M. Carrard and W. Kurz, *Acta Metall.*, 38 (1990) 2587.
[21] B. Major, R. Ciach, Z. Handzel-Powierza and J. Radziejewska, *Proceedings of the Fifth International Conference on Welding and Melting by Electron and Laser Beams*, CISFFEL, La Baule, France, 1993, p. 409.
[22] R. Riedl, S. Karagöz, H.F. Fischmaister and F. Jeglitsch, *Steel Res.*, 58 (1987) 339.
[23] J. Kusinski, *Metall. Trans. A*, 19A (1988) 377.
[24] H.F. Fischmaister, R. Riedl and S. Karagöz, *Metall. Trans. A*, 20A (1989) 2133.
[25] R. Ebner, E. Brandstätter, G. Hackl and F. Jeglitsch, *Prakt. Met. Sonderband.*, 22 (1991) 569.
[26] R. Ebner, K. Rabitsch, B. Kriszt and B. Major, in *Laser Processing; Surface Treatment and Film Deposition*, Eds. J. Mazunder, O. Conde, R. Villar and W. Steen, *NATO ASI Series E, Applied Sciences*, Vol. 307, Kluwer Academic Publisher, Dordrecht, The Netherlands, 1996, p. 255.
[27] K. Rabitsch, R. Ebner and B. Major, *Prakt. Met. Sonderband.*, 25 (1994) 509.
[28] K. Rabitsch, R. Ebner and B. Major, *Scripta Metall.Mater.*, 30 (1994) 253.
[29] M.H. McCay, N.B. Dahortre, J.A. Hopkins and T.D. McCay, *J. Mater. Sci.*, 34 (1999) 5789.
[30] A. Woldan, PhD Thesis, AGH – University of Science and Technology, Cracow, Poland, 2003.
[31] J. Kusinski, A. Woldan and S. Kac, *Proc. SPIE*, 5229 (2003) 155.
[32] G. Hackl, PhD Thesis, Montanuniversität Leoben, Austria, 1991.
[33] F. Behr, F. Haberling and I. Schruff, *Thyssen Edelst. Techn. Ber.*, 5 (1990).
[34] R. Ebner, G. Hackl, E. Brandstätter and F. Jeglitsch, *Proceedings of the First International High Speed Steel Conference*, Leoben, Austria, 1990, p. 81.
[35] W. Bochnowski, PhD Thesis, Institute of Metallurgy and Materials Science, Polish Academy of Sciences, Cracow, Poland, 2002.
[36] B. Major, W. Bochnowski, A. Klimpel and R. Ebner, *Proc. SPIE*, 5229 (2003) 173.
[37] S. Kac, PhD Thesis, AGH – University of Science and Technology, Cracow, Poland, 2002.
[38] B. Kriszt, R. Ebner, W. Tragl and B. Major, *Prakt. Met. Sonderband.*, 26 (1995) 455.
[39] A.D. Wilson, A. Leyland and A. Matthews, *Surf. Coat. Technol.*, 62 (1999) 618.
[40] H. Dong and T. Bell, *Wear*, 238 (2000) 131.
[41] Y.X. Leng, *et al.*, *Surf. Coat. Technol.*, 138 (2001) 296.
[42] B. Major, R. Ebner, A. Klimpel and G. Kruzel, *Proc. SPIE*, 5229 (2003) 178.

Chapter 8

Growth of Coatings by Pulsed Laser Deposition

Francesco Fuso

8.1. Introduction

Presently, many physical methods are available for coating fabrication, featuring different basic methods for material vaporization, that is, for the formation of the elemental vapor that, once deposited onto the substrate, will produce the coating. Among them, pulsed laser deposition (PLD) uniquely exploits the ability of laser radiation, typically in the ultraviolet (UV) range, to be efficiently coupled with a solid-state (or liquid) target. Through a variety of microscopic processes, depending on both the material and the laser beam properties, such an intense coupling leads to local heating and, subsequently, to material vaporization. Depending on the nature of the target, photo-induced chemical processes leading to bond breaking can be involved as well. We will denote hereafter as *pulsed laser ablation* (or, shortly, laser ablation) the vaporization of *macroscopic amounts* of material induced by a laser shot impinging onto a target.

Laser ablation is a phenomenon known since the first introduction of pulsed sources in the 1960s. Initial experimental observations were probably associated with the annoying circumstance of material deterioration in coated optical components used to manipulate the pulsed beams. Rapidly, great efforts were devoted to understand the fundamental mechanisms of vaporization [1], but it took some time to fully realize the applicative potential of the intense interaction between a pulsed laser and a solid surface. Indeed, for a couple of decades the interest was restricted mostly to treatments aimed at surface modifications [2], as annealing or cutting processes. At the end of the 1980s, right after the discovery of a new, intriguing class of materials, the high-temperature superconductors (HTS), PLD turned out an extremely appealing technique for the deposition of superconductive thin films [3,4]. The main reason for the broad and rapid growth of interest in PLD was essentially related to the refractory character of HTS ceramics, in addition to their complex structure and to the sensitivity of the superconductive properties on the

Materials Surface Processing by Directed Energy Techniques
Y. Pauleau (Editor)

film composition. Deposition methods widely used in the area of microelectronics, for example, molecular beam epitaxy (MBE), sputtering, chemical vapor deposition (CVD), liquid-phase epitaxy (LPE), showed strong limitations in the achievement of good quality superconductive films, whereas PLD appeared as a relatively simple, straightforward and efficient tool for HTS deposition.

The many successful applications of PLD to HTS revealed soon two main features of the deposition technique, that is, its *efficiency* (meaning that virtually any kind of material can be efficiently laser ablated) and its *congruency* (meaning that the film composition can easily reflect, or reproduce, that of the starting target). This opened the way for an increasing list of applications (for a review, see, for instance, Ref. [5]), still growing at a remarkable rate. Furthermore, as long as suitable laser sources became available, and much progress in this sense was made during the 1980s by major laser companies, the experimental setup is typically not too complex and the process holds a flexible character, a key point for its exploitation with different materials also in view of multi-layered coatings.

Beyond the successful response to the demanding requirements of HTS and other innovative ceramics (e.g. ferroelectrics [6] and conductive perovskites [7]) that drove the initial diffusion of PLD in the community of material scientists, the above mentioned interesting features demonstrated their potential in the wide area of coating fabrication, especially when alloys or, more in general, complex and hard materials are to be deposited. At present, PLD can be regarded as a well-established technique for fabrication of thin films of virtually any kind of materials, with its pros and cons compared to other techniques, although its actual diffusion in the industrial context is still under debate.

In this chapter we will focus on the most appealing and distinctive features of PLD relevant for material coating, the basics of the technique being treated elsewhere in this book. In particular, we will discuss the interplay between the dynamical properties of the jet of ablated material, called *plume*, and its composition, which, in turn, can affect stoichiometry and structure of the deposited films. In our search for the origins of the PLD peculiarities, we will report on several diagnostic methods which have been widely employed in the past to investigate plume expansion either in a vacuum or in a background gas. Finally, we will briefly describe the processes involved in film growth and in the attainment of the required coating structure in the specific conditions of PLD. A huge variety of materials has been successfully deposited by PLD (including organics which will not be mentioned here): we will consider both simple (metals and metal alloys) and complicated (ceramics, hard coatings) materials, which can often be regarded as prototypal systems to understand the main phenomena involved in the deposition process.

8.2. Experimental Components

From the experimental point of view, PLD is one of the simplest techniques: basically, it requires only two main components, that is, a pulsed laser source and a deposition chamber. Remarkably, the vaporization system (the laser) must not be in contact, or in proximity, with the target, with advantages in terms of technical flexibility compared to other techniques. The laser beam is focused onto the surface of a solid target (liquid targets can be used as well, but coating applications usually require bulk solids); the beam direction is not critical (usually it is set at 45° with respect to the surface), providing additional flexibility. In our overview we will be mostly concerned with conventional PLD experiments, where laser pulses have duration in the 10 ns range and energy density (*fluence*) on the order of few $\mathrm{J\,cm^{-2}}$ (we will briefly report on the use of ultra short pulse lasers at the end of the chapter). Similar experimental conditions are customarily attained with excimer or Nd-YAG lasers, the latter being typically coupled to higher harmonic generators in order to get UV radiation. Such sources are commercially available with a variety of performances; in particular, the pulse repetition rate, a parameter relevant for determining the growth rate of the coating, can range through a few Hz to hundreds of Hz. Most PLD experiments are carried out at repetition rates around 10 Hz, a safe choice in terms of target deterioration, but in the industrial environment higher frequencies can be used. In any case, the target is kept in rotation, or moved, continuously in order to expose fresh portions of the surface at any consequent laser shot and prevent formation of craters.

A substrate is placed in front of the target (typically substrate and target surfaces are parallel to each other), at some distance customarily chosen in the centimeters range. Frequently, electrically operated heaters (sometimes replaced by laser radiation) are used to heat up the substrate favoring material crystallization. The deposition chamber hosts the target and the substrate. In most laboratories the chamber can be evacuated at ultra high vacuum (UHV) by means of mechanical or diffusion pumps in order to prevent impurities in the coating layer. If required, standard systems are used to allow controlled admission of background gases into the deposition chamber. Furthermore, the chamber can host *in situ* diagnostics of the process (e.g. plume analysis) and/or the film.

Similar experimental setups can be readily assembled with commercial or home made vacuum components; in the last decade, several companies have developed specific systems for PLD, which are available in a variety of configurations aimed at industrial exploitations. They include for instance carrousel multi-target holders for multi-layered coatings, and substrate manipulators to allow for a simple and efficient operation and large-scale coatings.

8.3. Formation and Properties of the Plume

If PLD can be regarded as a very simple and straightforward technique from the experimental point of view, the mechanisms underlying its operation are rather complicated. Indeed, a variety of physical, or physico-chemical, phenomena are involved in PLD; they rule essentially vaporization of the target material, its transfer to the substrate and the growth of the coating layer. According to a schematic picture of the process, dating back to models proposed in the early 1990s [8,9], PLD can be divided into four different stages. They deal with the vaporization of the solid target, the interaction of the plume with the laser pulse, the expansion of the plume after the end of the laser shot, and the transfer of the plume to the substrate, which may take place either in a vacuum or in a background gaseous environment (hereafter denoted as *ambient gas*). A fifth stage must be added in order to discuss the coating properties, which involves the arrival of the vaporized material onto the substrate and the growth of the film.

Considering the process as a whole is an obvious requirement to fully understand the potential of PLD. Moreover, one of the most striking feature of the technique is in its *abrupt* character, meaning that vaporization is initiated in a short time interval (typical timescales of tens to hundreds of picoseconds, depending on the nature of the bulk material, are considered for the initial transfer of energy from radiation to material electrons, as shown in the case of sub-nanoseconds laser pulses [10]) and that the plume expansion occurs in the microseconds timescale. As a consequence, each stage turns out to strongly affect the subsequent ones; nonetheless, treatment of the basic processes involved in PLD is strongly simplified by examining each stage separately. Within this approach, quality of the coatings and general features of the deposition process turn out to be mainly ruled by the final stages, that is, the expansion of the plume in a vacuum or in an ambient gas and its arrival onto the substrate. It must be noted that both processes are relevant in any physical deposition method where a vapor phase is concerned, but the peculiarities of PLD make the behavior of the vaporized material during its transfer toward and its arrival onto the substrate unique under several respects. The key words for understanding such a peculiar behavior are *energy* and *composition*.

Roughly speaking, in PLD the energy of the ablating laser pulse is coupled to the target material with an extremely large efficiency. This circumstance leads to an efficient vaporization even for refractory systems, with large vaporization temperature and specific heat. Coupling is typically accomplished in a small portion of the target, with a lateral size given by the dimension of the laser spot focused onto the target (in the millimeter range) and a thickness ruled by optical and thermophysical bulk properties (and by the occurrence of plasma absorption phenomena by the nascent plume). In typical PLD conditions, the final result is that the thickness of the ablated volume is much smaller, by four to five orders of magnitude, than its lateral size. As a consequence, the vaporized material gets high kinetic energy associated with the direction orthogonal to the surface, hence the dynamics of the plume acquires a peculiar behavior, with a strongly anisotropic expansion peaked along this direction (sometimes referred as the *plume expansion axis*). The space distribution of the ablated material as a function of the angle θ measured starting from the plume axis is known to follow a $\cos^n\theta$ law, where the exponent n can be greater than 10 [11]. Furthermore, as already mentioned, part of the energy transported by the laser pulse is available for further excitation and ionization of the plume, leading to a strong increase of the internal energy. The whole interaction process is then non-thermal, or super-thermal, meaning that both the *kinetic* and the *internal* energy distributions are far from the thermal equilibrium, as typically achieved with other deposition methods.

On the other hand, plume composition, and in particular the possible modifications of the composition during the process, play a dominant role in determining the quality and properties of the deposited coating. Though the formation of the thin film takes place onto the substrate, typically heated by electrical methods to temperatures up to the thousand Kelvin, which is a suitable energy reservoir for the chemical processes underlying the reconstruction of the film structure, the phenomena experienced by the plume can be relevant as well. In fact, the plume is the medium which allows the transfer of material from the target to the substrate, and any modification occurring during its expansion is expected to be reflected in the film.

The "surplus energy" acquired by the ablated material can actually affect its composition, a key point in a deposition method which is intended to be used with targets having a complex stoichiometry and, more in general, for the fabrication of complex films. The technical flexibility of PLD allows the user to exploit either a UHV or a background gas as the environment where the process takes place. In both cases plume composition can be modified according to the following lines:

(i) The adiabatic expansion experienced by the ablated material and the associated rapid cooling of the plume are relevant in promoting condensation of clusters in the vapor phase.

(ii) Whenever a reactive ambient gas is used (and only if an ambient gas is used), collisional processes can selectively occur resulting in the formation of molecular complexes, or clusters, with specific composition.

We point out that the *reactive* character of PLD can be efficiently controlled by the user through the admission of a specific gas into the deposition chamber at the desired partial concentration. Such a possibility is not always offered by other deposition methods. For instance, in sputtering-based techniques a foreign gas is in general needed to ensure the correct establishment of the process, but the choice of the gas (typically, inert) has to meet the requirements of the process itself more than those of the material to be deposited [12]. Furthermore, the reactive-collisional processes involving the plume and the background gas can be frequently enhanced by exploiting a variety of relatively simple, but efficient, approaches, as in the crossed-beam configuration [13].

8.4. Plume Expansion in a Vacuum

When a laser pulse with duration in the 10 ns range and fluence on the order of a few $J\,cm^{-2}$, which is typical for most PLD experiments, hits the target, a thin material layer is "abruptly" converted into the vapor phase. Lateral dimensions are typically in the millimeter, or fraction of millimeter, range, whereas the thickness is ruled by the optical absorption depth and the thermal diffusion within the bulk, which roughly depend on the class of material considered. For instance, metals typically show a strong absorption in the UV and a relatively large thermal diffusion coefficient, whereas in semiconductors and dielectrics the situation is reversed [14–16]. Absorption of radiation by the nascent plume is a relevant process whenever the pulse duration exceeds that of the formation of the vaporized layer. It can occur through many mechanisms, in particular multi-photon ionization and inverse Bremsstrahlung [17]; the result is that the vaporized layer gets a huge amount of energy. Such energy is "incubated" into a small volume, with a strongly anisotropic geometry. Just to give an idea of the quantities involved, more than 10^{15} particles are ablated per laser shot, being the ablation depth in the tens of nanometers range [18], the ionization degree can easily get close to unity [19], and temperature may rise above 20×10^3 K [20].

Assuming such an efficient exchange of energy between radiation and matter, which is by far larger than any binding energy of the bulk, the nascent plume is composed mainly of *elemental* particles, essentially neutral and ionized atoms and electrons. Such an idealized picture is based on a space homogeneous delivery of energy to the target, which is not always the case in real experiments. Aside from the role possibly played by "hot spots" in the space distribution of the radiation energy, thought to be responsible for feeding up the plume with electrons coming from intense multi-photon ionization [13], the signature of such non-homogeneity can be seen in one of the major drawbacks of PLD, the occurrence of *droplets*. For the moment, however, we will treat the nascent plume as composed mostly of elemental particles. A detailed analysis of the mechanisms leading to the initial plume formation is out of the scope of the present chapter. It is quite obvious, however, that the conditions occurring during the laser pulse, or right after its end, are far from equilibrium. The amount of energy incubated in the ablated layer, in the form of thermal and mechanical energy, affects the subsequent processes, that is, plume expansion and arrival onto the substrate. In other words, some kind of "memory" of the efficient energy exchange between laser and matter is maintained during the whole process.

The relevance of the plume dynamics for a general understanding of PLD was clear since the first experimental realizations, and efforts were soon made in order to have an experimental and quantitative confirmation of the theoretic expectations. As a matter of fact, one of the most striking feature of PLD is the appearance of a luminous jet (the name plume comes from that) of material leaving the target surface at the right angle. In fact, direct multi-photon excitation or, more likely, collisional processes can easily produce a massive excitation of the plume volume, hence a strong afterglow emission, which can be imaged in order to diagnose the expansion dynamics [21]. The transient character of vapor expansion can be efficiently investigated by time-of-flight (TOF) techniques triggered by the arrival of the laser shot onto the target. In particular, optical-TOF proved to be an efficient and simple experimental method. As schematically depicted in Fig. 8.1(a), emission can be easily collected at different distances from the target and analyzed in order to detect fluorescence lines of the different species composing the plume. This usually refers to neutral or ionized atoms and simple molecules [22], although broadband emission from nanoparticles can be collected and analyzed as well [23].

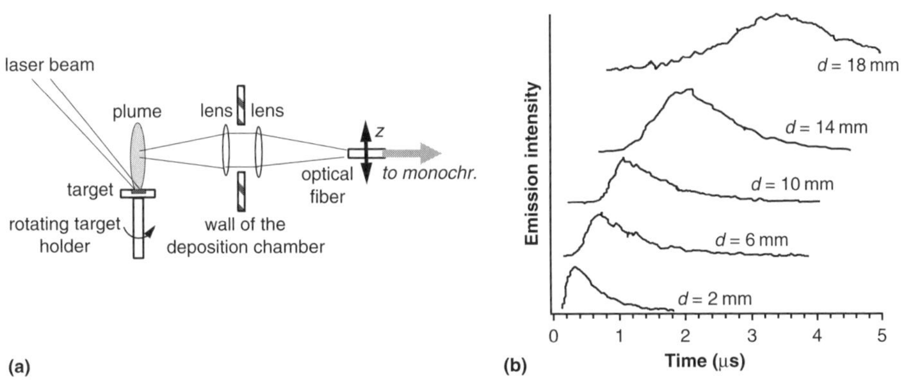

Figure 8.1: (a) Sketch of the experimental arrangement for optical-TOF measurements and (b) example of data acquired during ablation of a PZT target. In panel (a) a telescope made of two identical lenses is used to produce an image of the plume outside the deposition chamber; an optical fiber moveable on the image plane by means of a micrometric translator allows collecting radiation at an adjustable distance from target; the collected radiation is then sent to a monochromator for further analysis. Traces in panel (b) corresponding to different distances have been rescaled and shifted vertically for the sake of clarity.

In addition, especially at the beginning of the expansion process, a background ascribed to continuous (blackbody) plasma emission is detected [24]. Due to plume expansion the emission at some distance *d* from the target gets a transient behavior and measurements as a function of time *t* can provide information on the plume dynamics. Fig. 8.1(b) shows an example of such kind of measurements: emission from Pb atoms at 498.2 nm is acquired at different distances from target during laser ablation of a $Pb(Zr,Ti)O_3$ (PZT) target (a XeCl laser is used emitting at 308 nm, with pulse duration around 17 ns and fluence of 2.5 J cm^{-2}).

Specific data treatments can be used to account for the finite size of the region from where the radiation is collected and for the presence of a velocity distribution along the transverse direction of the plume [25]; in any case, by using the simple fluid conversion $v_z = d/t$, data can be used to derive the distribution function of the longitudinal velocity, $f(v_z)$. Many optical-TOF experiments have been carried out, and basically they all confirmed what can be inferred from the theory. Fluid dynamics models have been in fact applied to describe the plume expansion by many groups [8,9,26–28]. Basis of the application is the obvious fact that particles cannot escape freely, without collisions, from the target; a highly collisional layer (known as Knudsen layer) is formed right above the surface, which is able to strongly affect the velocity distribution function of the plume. This acquires the form of the translated Maxwellian distribution typical for supersonic beams [29]:

$$f(v_z) = Av_z^3 \exp(-m(v_z - v_0)^2/(2kT)) \qquad (8.1)$$

where *A* is a normalization factor accounting for the total number of particles, *k* is the Boltzmann constant, *T* is the translational temperature of the plume associated with the longitudinal expansion, and v_0 is the drift velocity, also known as center-of-mass velocity, of the fluid along this direction. Best fit to experimental data allows one to derive the v_0 and *T* parameters. At short distances from target, $d < 10$ mm, v_0 is typically found to be in the range of 10^5–10^6 cm s^{-1}, whereas the temperature *T* lies in the 10^4 K range, values much larger than those achieved with other techniques.

8.4.1. Condensation of the Plume in a Vacuum

From the point of view of elementary thermodynamics, the expansion of the plume in a vacuum resembles an *adiabatic* expansion [8,9]. Indeed, considering the level of vacuum typically achieved in PLD experiments (residual pressure below 10^{-8} mbar), the mean free path of ablated particles exceeds by far the typical target-to-substrate distance, customarily set around a few centimeters. In other words, when leaving the Knudsen layer and after the end of the laser pulse, the ablated material cannot receive nor exchange any additional energy with the external world, either through collisions or absorption of radiation. Thermodynamics then dictates a very well known relationship between the plume size, namely its volume V, and the temperature T, through the adiabatic coefficient γ:

$$TV^{\gamma-1} = \text{Constant} \tag{8.2}$$

Such a simple approach obviously fails in providing a detailed picture of the process. For instance, issues dealing with the attainment of equilibrium, as required by the thermodynamics, should be carefully considered. Furthermore, the definition of the adiabatic coefficient, associated with the effective degrees of freedom of the expanding gas, is a demanding task for the simultaneous presence of neutrals, ions, electrons in the plume [17]. Finally, the role of the internal energy (species in the plume are in excited states) and the possible transfer to kinetic energy available for the expansion (see, for instance, Ref. [30]) makes hard to describe the whole process in terms of elementary thermodynamics.

Nonetheless, this oversimplified picture suggests an important fact of PLD: during expansion, the plume undergoes a strong cooling. In order to roughly estimate the quantities entering the process, we can consider an initial volume for the nascent plume $V_0 = 10^{-11}$–$10^{-10}\,\text{m}^3$, which is a reasonable estimate for a ceramic target ablated with a laser shot 10 ns long. The volume occupied by plume can be roughly described by a cylinder, whose height grows up by more than two orders of magnitude during the expansion from the target to the substrate. Thus, assuming an initial temperature $T_0 \sim 5 \times 10^4\,\text{K}$ (reasonable for the early beginning of the process), and using the guess $\gamma \sim 1.7$, the temperature of the fully expanded plume will drop below 2000 K. The temperature decrease is clear for instance in data of Fig. 8.1(b), where the broadening of the time distribution of the signal at increasing distances from the target reflects the narrowing of the velocity distribution that is a temperature decrease. Using an alternative point of view, we can state that part of the initially incubated energy is spent for promoting the dynamics of the plume. Since the plume stems from the target, this implies that a non-negligible mechanical momentum is imparted to the target itself, exerting a transient pressure onto the surface: values as large as 10^8–$10^9\,\text{Pa}$ have been reported [31,32].

Among other effects, plume cooling leads to *condensation*. In its basic behavior, such a phenomenon is quite similar to what happens with macroscopic particles, which tend to stick together due to surface forces when their temperature, or kinetic energy, decreases. Since we are supposing the native plume as composed of elemental particles, condensation means as first formation of aggregates (*clusterization*): elements in the vapor-phase coalesce each other, and either stable or metastable clusters are produced involving a variable number of elemental components, ranging through a few units to the hundred. The connection between laser ablation and production of clusters in the vapor phase is evident when considering the efforts, started mostly in the eighty, devoted to the realization of the so-called cluster machines. Here, laser ablation serves as a tool to induce pulsed vaporization of the bulk, and a pulsed and synchronous injection of a carrier (inert) gas in the direction parallel to the target is exploited to produce a supersonic beam containing a mixture of carrier gas and vaporized species. The expansion of the supersonic beam, typically occurring in a specifically designed pipe, leads to a very efficient production of clusters. Among the very wide range of applications of the technique, one of the most striking results is the laboratory production of C_{60} (fullerene) molecules (see, for instance, Ref. [33]). Those experiments pointed out also the role played by the chemistry in favoring the formation of clusters with a specific number of elemental constituents, the so-called magic numbers, as, for instance in the case of carbon, 60, 72, and so on. Production of carbon nanotubes by PLD, started in the 1990s, exploits also similar features of the process, although in this case the ambient gas (typically injected continuously) and the metal nanoparticles used as the catalyst play a fundamental role [34].

It is important to stress that in any case, that is, even in the standard configuration of PLD experiments, formation of clusters may efficiently occur in the vapor phase. Several diagnostics have been developed in order to investigate inter-atomic aggregation during plume expansion. Very frequently, they rely on space and time resolved measurements of the ionized fraction of the plume. This can be easily carried out by using a TOF mass spectrometer (TOF-MS) collecting primary ions from a small volume, ~ few cubic millimeters, at a known distance from the target, as can be accomplished by using a geometrical configuration with the TOF-MS axis orthogonal to the plume axis and by displacing accordingly the target position. Fig. 8.2 shows the basic components of the diagnostics; in order to improve the instrumental resolution and decrease collection of background ions, pulsed extraction fields can be used, with adjustable delay and duration with respect to the laser pulse.

Formation of clusters can inherently affect plume composition during the expansion, leading to formation of *non-elemental* species. We will discuss in the following sections the processes occurring in the presence of a foreign (reactive) gas; in order to shed light on the plume modifications in UHV PLD, an instructive example deals with ablation of a simple binary system, the NiTi metal alloy. This alloy exhibits remarkable thermomechanical features; in particular, it belongs to the class of shape-memory alloys (SMAs), showing the unique ability to recover their shapes even after severe deformations [35]. Once deformed at low temperature, these materials will stay deformed until heated, when they will return spontaneously to the original shape. Such features are very appealing in the growing area of microactuation, which makes relevant the search for suitable deposition methods in view of

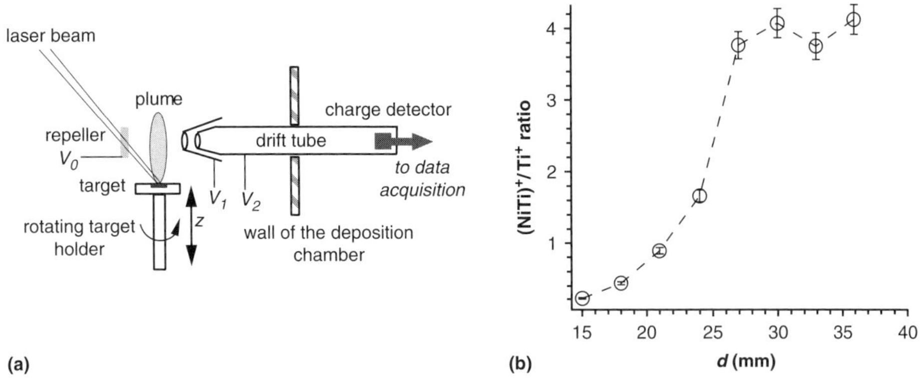

Figure 8.2: (a) Arrangement for ion mass spectroscopy of the plume by TOF-MS and (b) ratio between the signals corresponding to $(NiTi)^+$ and Ti^+ masses acquired during ablation of a NiTi target (a XeCl excimer is used with a laser fluence around $4\,J\,cm^{-2}$). Voltage of the repeller (V_0), the first (V_1) and the second collector (V_2) are set in order to extract ions from the plume and accelerate them along the drift tube; the distance between the extraction region and the target can be adjusted by displacing the target holder.

extensive exploitation of shape-memory driven miniaturized actuators [36]. The shape-memory effect implies a solid-to-solid-phase transition involving different lattice structures, hence a volume change and different geometrical shapes of the sample [35]. In order for the NiTi alloy to undergo complete and proper shape-memory transitions, the material must be *pure*, that is, it must be free of any non-stoichiometric precipitates, which would interfere in terms of energy dissipation with the shape-memory transition. Furthermore, the temperatures relevant for the transformation, in particular the so-called austenite and martensite start and finish points, are observed in the bulk to be extremely sensitive on the mass composition [37]. Thus, even if a simple binary alloy is concerned, NiTi films represent a very challenging prototypal system to check for the congruency of the deposition technique. In addition, incorporation of contaminants in the growing film can be prevented by operation in UHV conditions, which, as already stated, can be easily accomplished with PLD. Indeed, PLD proved to be a viable method for depositing shape-memory films [38] onto a variety of substrates (e.g. Si based, alumina), and also relatively thick (a few microns) free-standing films could be produced through chemical removal of the substrate after deposition. Both the measurement of the relevant temperatures and the occurrence of the shape-memory effect suggest the achievement of the proper structural properties in those films, as confirmed by X-ray diffraction (XRD) analyses, with minor contributions from the interfacial layer.

Space resolved TOF-MS measurements during laser ablation of NiTi showed a relative increase of the cluster content as a function of the distance from the target, beginning at a distance around 20–25 mm [39]. This is confirmed by data shown in Fig. 8.2(b), displaying the ratio between $(NiTi)^+$ and Ti^+ signals as a function of the distance from target (experimental parameters, reported in the caption, are typical for PLD of metal targets). The interpretation of this finding is straightforward: during the expansion, while the plume gets cooler, compound formation rate tends to overcome the associated reverse process, that is, fragmentation. However, special care must be devoted to the analysis of mass spectra acquired during PLD. Besides the possible metastable character of the ionized species, which might affect the quantitative evaluation of the plume components, the velocity selection introduced by the use of a spectrometer with its axis orthogonal to the plume direction and the nonlinear response of the apparatus may lead to a complicated behavior for the actual instrumental sensitivity. Furthermore, investigation of the total ion amount in the plume cannot provide information on the specific collisional pathways ruling the aggregation/dissociation processes.

8.4.2. Statistical Methods for the Investigation of Cluster Formation Processes

Thanks to the ability of the TOF-MS to collect full range mass spectra corresponding to a single laser shot, a statistical technique can be adopted which can partially overcome the mentioned limitations. The technique, first introduced by Frasinsky and co-workers for the analysis of gas-phase reactions [40], is based on the analysis of the covariance matrix built over a sequence of single shot measurements treated as a set of statistical data. Fluctuations of these data reflect the appearance, or disappearance, of certain species, and by deriving the covariance of the data set, statistical correlations between pair of masses can be detected, which are lost in pulse averaged spectra.

To this aim, we consider the normalized covariance matrix N_{ij}, defined as:

$$N_{ij} = C_{ij} / (C_{ii} C_{ij})^{1/2} \tag{8.3}$$

$$C_{ij} = \sum_{k=1}^{n} S_k(t_i) S_k(t_j)/n - \sum_{k=1}^{n} S_k(t_i) \sum_{k=1}^{n} S_k(t_j)/n^2 \tag{8.4}$$

where $S_k(t_i)$ and $S_k(t_j)$ are the signals at the TOF-MS detector at the time t_i and t_j, respectively, in the kth laser shot of the set containing a total number of n shots. Bidimensional plots of N_{ij} allow the user to observe the correlation (full correlation for $N_{ij} = 1$) or anticorrelation (full anticorrelation for $N_{ij} = -1$) between events occurring at different times. Since the events represent different masses detected

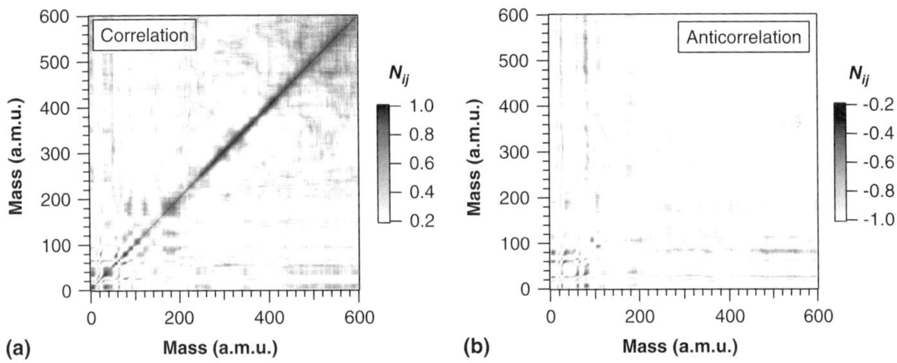

Figure 8.3: Covariance maps built with TOF-MS data acquired during laser ablation of NiTi (experimental parameters as shown in Fig. 8.2); for the sake of clarity, correlation and anticorrelation maps are reported in panel (a) and (b), respectively (adapted from [39]).

by the instrument, inspection of the maps provides information on elementary processes of fragmentation/aggregation occurring in the vapor phase [41].

Fig. 8.3 shows an example of covariance maps for UHV PLD of NiTi; a set of 100 single shot spectra acquired at a distance $d = 27$ mm is considered. For the sake of clarity, positive and negative values of the covariance are plotted in two different panels, (a) and (b), respectively; the highly correlated region close to the diagonal of panel (a) is due to the fact that each mass is obviously totally correlated with itself. The analysis strongly suggests [39] an efficient occurrence of processes based on the sequential capture of Ni and Ti ionized species leading to the formation of $(NiTi)_n^+$ compounds, with n up to 7, this limit being given by the mass range of the spectrometer in use. The process acquires efficiency at increasing distance from the target, where the plume temperature is expected to drop below 1000 K. Competitive processes involving dissociation of NiTi clusters into elemental components are also present, but their role becomes negligible at distances similar to those customarily employed for the substrate during film deposition. Thus, the measurements confirm that the relative abundance of clusters tends to increase during expansion; in addition, they show that not all compound species are equally probable, but *stoichiometric clusters*, with composition of the kind $(NiTi)_n^+$, are more likely to be formed. This behavior can be ascribed to energetic considerations, in agreement with the stability reported for some of the observed clusters [39].

As already mentioned, PLD is often referred to as being a technique able of an excellent degree of compositional congruency between target and film. A nice demonstration has been reported for PLD of a complex garnet containing six different elements and having a unit cell consisting of 160 atoms [16]. A fundamental point to explain congruency [13] is the unique ability to vaporize any kind of target material, meaning also that the nascent plume is composed by *all* elemental constituents. However, the transfer of the ablated material onto the substrate implies expansion, and all modifications experienced by the plume might affect the congruency. Data acquired in the simple binary alloy NiTi demonstrate that such modifications are mainly based on the appearance of simple clusters, having a composition "compatible" with that of the growing film. The arrival of such clusters onto the substrate can thus have obvious beneficial effects in the growth, especially when the substrate material does not allow heating or post-deposition annealing treatments. For instance, this is a key point for shape-memory films, where deposition at low temperature onto flexible polymeric substrates could open the way for appealing microactuator applications [36,39].

8.5. Plume Expansion in an Ambient Gas

Since the first pioneering applications of PLD to HTS, it has been clear that the possibility to operate in a controlled atmosphere consisting of molecular oxygen at partial pressure around a fraction of mbar could strongly enhance the features of the deposited samples [3,4]. We should note that, in the case of some superconductive cuprates, namely YBCO, oxygen deficiency ($x > 0.5$ in the stoichiometric formula $YBa_2Cu_3O_{7-x}$) can promote formation of a non-superconductive phase; hence an enrichment of the oxygen composition is of paramount importance in the fabrication of films with this prototypal system. The main mechanisms ruling the oxygen enrichment were discussed in the literature: basically, oxygen was thought to be incorporated in the film through diffusion and thermal annealing (at above 900 K). In addition, it was speculated that the presence of an ambient gas, even at a relatively low partial pressure, may contribute to prevent re-evaporation of the volatile species, including oxygen, from the film.

Aside from the reactive-collisional effects we will discuss in the following, the ambient gas turns out to strongly modify the plume expansion. We will not enter into the details of the dynamics, presented elsewhere in this book, but we will recall the basic elements essential for understanding the role played by the ambient gas in the coating process. The concept of *plume range*, corresponding to the distance from the target where plume and ambient gas have the same pressure [42], was initially introduced in order to account for the decrease in size of the visible plume afterglow as a function of the background pressure, suggesting some kind of "compression" of the cloud of ablated material due to collisional interactions with the background. On the other hand, it was soon clear that coating fabrication did not require placing the substrate within the plume range, but the best results in terms of film morphology could be attained by displacing the substrate out of the plume range. We also note that, PLD can be accomplished in air [43]; thus, the presence of a rather dense atmosphere is also "compatible" with the requirements of PLD, although the plume range decreases below a few millimeters.

The adiabatic expansion, which is still a reasonable approximation for in vacuum PLD, fails in this case due to highly collisional conditions. The energy initially incubated in the plume is partially lost in collisions with species of the foreign gas. The description, however, gets much more complicated if we take into account the transient nature of the expanding plume. In fact, a purely diffusional model can be used to describe only the final part of the expansion process, where the internal plume pressure approaches that of the surrounding medium and plume particles can move *through* the ambient. Like a piston which is moving fast in a gas, in the initial stage an abrupt displacement of ambient species is involved, rather than diffusion. Furthermore, a transition regime must be accounted for, in order to treat the dynamics in the intermediate regime between the early, highly collisional, stage and the final, diffusional, stage. Issues related to the development of shock waves [17] have been carefully considered, along with the peculiar phenomenon of plume splitting, which was observed thanks to sophisticated optical TOF and charge diagnostics (see, for instance, Refs. [44–49]). At the same time, theoretic descriptions have been improved (see, for instance, Refs. [46,50]) by using also numerical methods, demonstrating that a simple and straightforward interpretation based on the tools suitable for in vacuum case cannot provide satisfactory predictions.

Since the coating process involves arrival of material onto the substrate and the subsequent layer growth, it is quite obvious that the most striking effects of the modified plume dynamics are in the velocity distribution of particles when they impinge onto the substrate. In PLD of compound targets, some kinds of thermalization, or velocity moderation, effect [48,50,51] can be associated with the presence of an ambient gas. This is demonstrated, for instance, by optical-TOF measurements acquired for selected emission lines of different species composing the plume. Fig. 8.4 reports an example referring to laser ablation of PZT in 0.3 mbar O_2: the drift velocity of different components, derived as mentioned in Section 8.4, is plotted against the distance from the target. At some distance, velocity tends to get a similar value for all species, as a signature of the establishment of the diffusional regime. This may also have obvious consequences in terms of deposition congruency: all elemental constituents tend to have similar dynamical features at their arrival onto the substrate; hence they are all available to the film growth in a small time interval. Furthermore, the kinetic energy of such species may fall in a range which is favorable for species diffusion and rearrangement on the substrate, as we will mention in the following.

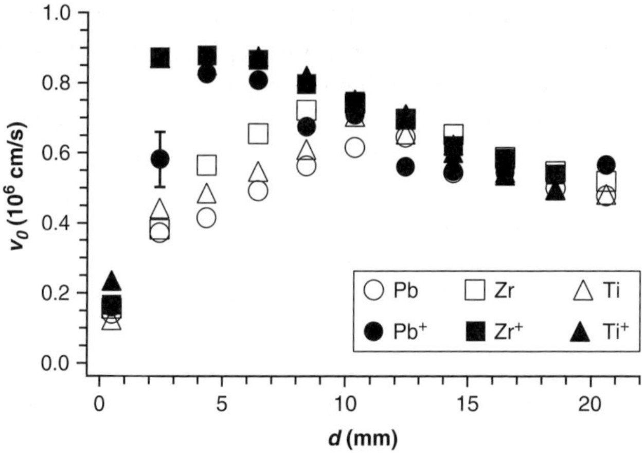

Figure 8.4: Drift velocity as a function of the distance from target as derived from optical-TOF data; laser ablation of PZT at a fluence of 2.0 J cm^{-2} (XeCl excimer laser) is considered in the presence of O_2 at a partial pressure of 0.3 mbar. Data for different species have been acquired by detecting emission at different wavelengths.

8.5.1. Compositional Modifications in an Oxygen Background

Effects are even more pronounced when the plume composition is considered. At the naked eye, injection of oxygen within the deposition chamber during PLD of ceramics, leads to a remarkable change in the color of the plume afterglow. Since plume excitation is due to collisional processes, the color change can be partly attributed to strong modifications of excited state populations, but compositional modifications are strongly suggested as well. Indeed, old experiments based on a careful spectral analysis of the optical emission from an YBCO plume faraway from target [52] showed features which were attributed to molecular oxides. Interpretation is straightforward: collisions involving the background oxygen and the ablated metal particles can promote molecule formation. If we consider that even in the diffusional stage the velocity of the plume species can be in the 10^5 cm s^{-1} range, collisions involves a remarkable amount of energy. Thus, even endothermic reactions can be activated. However, due to the typically weak oscillator strength and to the non-homogeneous excitation mechanisms, optical emission measurements did not allow deriving quantitatively and reliably the oxide abundance. Ion mass measurements were also extensively carried out. Contrary to the simple experimental situation mentioned in Section 8.4.1, where a binary alloy was considered, interpretation of the findings was partially hampered by the rich manifold of reactive pathways available to the ablated species. However, statistical approaches based on the covariance map technique

[53,54] demonstrated an efficient mechanism of oxygen incorporation and oxidized cluster formation during the plume expansion. In particular, strong correlations were found among ions having a mass difference equal to that of a single oxygen atom, suggesting a sequential addition of oxygen during the plume flight. Such oxygen-rich species, once arrived onto the substrate, may act as an oxygen reservoir for the growth of the ceramic coating, so that they can be seen as a viable medium for the ambient oxygen to be transported into the film.

From the point of view of the chemical reactions, oxygen incorporation requires, as the first step, dissociation of O_2 into O. Energy exchange is also involved in this part of the process in order to break the chemical bond of the molecule (binding energy ~5.1 eV). During the laser shot, energy is available in form of radiation, and multi-photon absorption of UV light can indeed result in dissociation. However, in the typical conditions of PLD the laser pulse duration is by far too short to ensure a suitable rate of atomic oxygen formation. Thus, a collisional mechanism must be envisioned to be investigated from the experimental point of view. A rather straight-forward diagnostic tool can be set up based on optical absorption spectroscopy of oxygen, both for the atomic and the molecular species. As a matter of fact, selectivity and sensitivity of this tool is large enough to meet the requirements for accurate analysis [55]. Furthermore, relatively strong absorption lines can be found lying in the near-infrared range for both species, which can be conveniently covered with diode lasers mounted in the external cavity configuration [56]: investigated lines are around 760 and 777 nm for molecular and atomic transitions, respectively. Optical transmission of the plume is measured, and by using a simple optical fiber system [55], able to generate a collimated probe beam with 1 mm diameter, the radiation can be sent at any desired position, so that acquired data get two-dimensional spatial resolution; in the direction of the probe beam, the absorption is obviously space integrated along the whole transverse size of the plume, which can be accurately measured, for instance by fluorescence imaging techniques. If the probe beam intensity is kept at a small value compared with the saturation intensity of the considered transition, the transmitted intensity I depends on the input intensity I_0 according to the Beer's law:

$$I = I_0 \exp(-kl) \tag{8.5}$$

where l is the transverse size of the investigated region and k represents the extinction coefficient. The latter is proportional to the actual density of the absorbing species, atomic or molecular oxygen; hence the method can provide quantitative information on the density of dissociated species. Values as large as 10^{14}–10^{15} cm^{-3} have been found at partial pressures of 0.5 mbar O_2, in the typical ablation conditions of PLD and at distances of a few millimeter above the target surface.

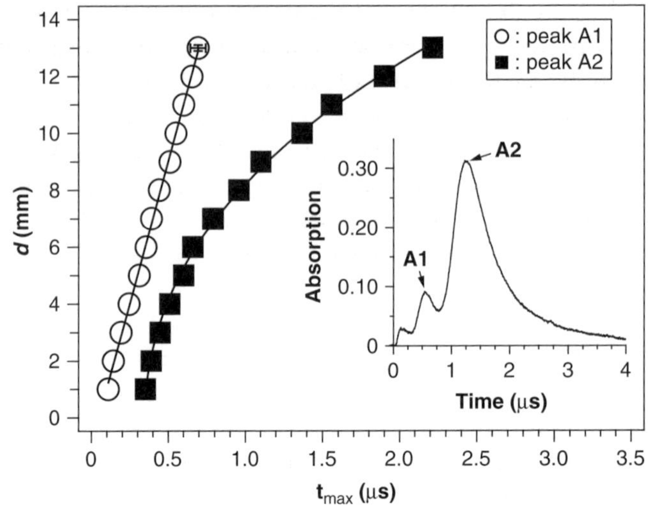

Figure 8.5: Comparison between the dynamics of the first and the second atomic oxygen absorption peaks (A1 and A2, respectively). In the inset: time resolved atomic oxygen absorption by the plume at a distance $d = 10$ mm from target during laser ablation of a NiTi in the presence of O_2 at a partial pressure of 0.35 mbar (a XeCl excimer laser is used at a fluence around 4 J cm^{-2}).

However, more than in the quantitative evaluation of the atomic oxygen abundance, the method holds a great potential in deriving time resolved pictures of the dissociation efficiency. Fig. 8.5 shows in the inset an example of the time resolved atomic oxygen absorption at the fixed distance of 10 mm. A strong absorption (marked as A2) is detected in a time window roughly corresponding to the arrival of the plume front to the observation region. This suggests a collision-driven dissociation mechanism involving the partial development of a shock wave [57]: an interface layer containing a high-density mixture of plume and ambient gas species occurs at the shock-wave boundary. Analysis of the absorption profile for molecular oxygen, carried out thanks to the excellent tunability and linewidth characteristics of diode laser sources, indicate that, right before dissociation, O_2 molecules are compressed in a layer a few millimeters wide, and their temperature can transiently exceed several thousands Kelvin. In such conditions, formation of atomic oxygen is very likely to occur even without assuming a direct multi-body collisional event with expanding species of the plume. As a consequence, atomic oxygen at relatively large local density is available for further reactive processes, including oxide and oxidized cluster formation. The dynamics of the absorption process can be studied by investigating the peak delay as a function of the distance from target, as in Fig. 8.5: the behavior of peak A2 turns out compatible with the shock-wave mechanism [57]. A careful analysis of the time resolved data shows the presence of an additional, *fast*, dissociation process, which takes place in a short time interval (peak marked as A1 in the inset of the figure). Several possibilities exist to explain the presence of this fast absorption peak showing a different dynamical behavior with respect to A2, including the above mentioned plume splitting effect: another interpretation relies on molecular dissociation through energetic electron bombardment, possibly initiated by the prompt release [58] of electrons from the metal target or hot spots in the laser beam distribution.

8.5.2. Reactive PLD

In the fabrication of coatings made of oxide ceramics, the need to produce oxygen-rich deposits combined with the high volatility of the oxygen component make the choice of the ambient gas rather mandatory. However, the possibility to use the energy incubated in the plume to promote chemical reactions leading to compound species in the vapor phase can be successfully exploited for a wide class of coatings. Such a possibility is at the basis of *reactive PLD* (this term refers sometime to conventional PLD in the presence of a background). The simplest configuration of reactive PLD foresees laser ablation of a single-component target; once vaporized, the target material reacts with the ambient gas, leading to the formation of the compound which is then deposited onto a proper substrate. A straightforward example is fabrication of copper oxide films by ablation of a copper target in the presence of N_2O [59]. Here the coating stoichiometry reflects that of the target and that of the ambient gas; in other words, one of the most appealing features of PLD, its congruency, is practically lost. On the other hand, advantages are found in the flexibility and ability to fabricate coatings of materials which are highly demanding in terms of composition, in particular hard coatings. The technique is in principle quite similar to one of the many reactive deposition methods, for example, plasma-enhanced CVD [12], which have been developed in order to produce solid films starting from a mixture of vaporized species and foreign gases. Also within this context the peculiarities of the laser ablated plume turn out to play an important role, especially if the key word energy is considered.

Oxides are clearly a good and successful example of reactive PLD applications, and actually a wide variety of oxide coatings have been produced by this technique; a complete list of materials is out of the scope of this chapter, and we just mention that oxide coatings for a wide range of applications, ranging, for instance, through microelectronics (e.g. SiO_2, GeO) to optics and mechanics (e.g. Al_2O_3 and titanium and vanadium oxides [60–63]) have been successfully deposited with excellent results in terms of homogeneity and smoothness of the obtained coatings. However, probably the most striking achievements of reactive PLD are to be searched in the area of carbides, silicides and nitrides, materials having important applications for their mechanical, tribologic and wear-protecting properties. Quite often, these appealing features are a direct consequence of proper chemical and structural configurations. Furthermore, practical exploitations usually require deposition on specific substrates, as workpieces designed for the use in mechanical applications, which sometimes cannot be heated. Thus, the ability to fabricate coatings at room temperature is often desirable, that makes technologic developments in hard coating fabrication highly challenging. In order to clarify these concepts, we will briefly mention here PLD of carbon nitride. A solid β-C_3N_4 phase has been predicted years ago [64]. Due to the short covalent length of this phase, hardness and bulk modulus are expected to be even larger than for diamond. Different deposition methods have been explored, including sputtering of a graphite target in N_2 and plasma-enhanced CVD, which represent two fairly common methods for refractory materials. Those techniques led generally to amorphous films, failing in providing the coating with the suitable chemistry and structure. In particular, a dramatic dependence of the grain dimension, shape and structure has been reported on the substrate properties. On the contrary, reactive PLD turned out a viable method for deposition of CN_x films onto Si(111) at room temperature, consisting of relatively large-sized grains coherently grown [65]. In the experiment, a graphite target was laser ablated in the presence of N_2 ambient gas at a pressure of a fraction of mbar. Similar to carbon nitride, many other hard coatings have been successfully deposited, including TiN, WN, SiC, TiC [66–69]; the list of successful exploitations of PLD and reactive PLD in the field of hard coatings is still continuously growing, at a rate which is probably the best demonstration for the capability of the technique in a highly challenging area.

As in conventional PLD, several diagnostic tools have been adopted for *in situ* monitoring the process. In particular, optical diagnostics, with spectral, time, and space resolution, has been widely exploited to follow the evolution of molecular compounds in the vapor phase. As a matter of fact, the presence of molecular lines was not observed in all experimental conditions. Several experiments (see, for instance, PLD of germanium in the presence of oxygen [70], but also ablation of $BaTiO_3$ [71]) demonstrated that molecular emission was below the sensitivity threshold, whereas in many other cases, involving always oxidization but for other material targets (e.g. silicon, titanium, magnesium [72–76]), relatively strong molecular emission was detected. A similar variety of results was also found with other ambient gases, as nitrogen or nitrogen oxide: no molecules were detected with silicon and titanium [77,78], but emission from AlN [79–81] and CN [82] compounds was observed. Even if, as already mentioned, the weakness of the oscillator strengths for rotational–vibrational emissions should be carefully considered before drawing any conclusion on the presence or absence of molecular species, the results show a controversial situation.

A possible interpretation is in the role played by the collisional energy in both the occurrence of selected reactions and in the population of atomic and molecular excited states responsible for fluorescence emission. Once again, optical diagnostics can be helpful in unraveling the process. One possibility to derive the temperature of the emitting particles from emission spectra is based on the application of the so-called local thermodynamic equilibrium (LTE) approximation [83]. Briefly, assuming that the system is in thermodynamic equilibrium and that atomic and molecular parameters are known, the temperature can be derived from the relative emission intensity for different transitions of the same species. The topic whether the so-obtained temperature is representative of the actual energetic of plume expansion is questionable (see, for instance, Ref. [84]), but the method holds the appealing ability to follow the temperature variations in the transient expansion process. Based on LTE, a model has been presented [85] able to describe the main features of the spectra observed in reactive PLD in terms of energy available for the chemical processes to occur, which can be regarded as a key point of the technique. The possibility of reactions upon arrival of the species onto the substrate must be also taken into account, especially in the case of heated substrates. However, the capabilities of reactive PLD are of paramount importance in the search for low-temperature deposition conditions.

8.5.3. Methods to Enhance Reactivity in the Vapor Phase

Having in mind the possibility to operate at low substrate temperatures, efforts have been devoted to enhance (and control) the collisional reactions between plume and background. Since the dynamical features of the plume cannot be easily modified, unless by drastically changing the properties of the laser beam (in particular the pulse duration), efforts have been focused on manipulating the ambient gas. Many methods have been proposed or implemented to improve reactivity. Basically, they aim at providing the reactions with a supplemental energy source; one of them, the *crossed-beam PLD*, based on pulsed synchronous injection of ambient gas into the plume, is gaining increasing momentum.

Historically, combination of PLD and pulsed gas beams is associated with the already mentioned cluster machines: plume cooling following expansion and multiple collisions promoted by the supersonic carrier gas enable cluster formation rates not accessible with conventional PLD [86]. The ability to promote also reactive collisions with the pulsed gas was already envisioned at the beginning of HTS laser deposition [87]. Pulsed gas injection can be accomplished by using piezoelectric valves able to produce short pulses with an adjustable delay, either positive or negative, with respect to the laser firing. Since pumping of the deposition chamber is kept on during the gas injection, the process gets a transient behavior. At commonly used laser repetition rates (tens of Hz) and by using stagnation pressures in the bar range, the gas pulse is evacuated from the chamber prior to the arrival of the subsequent laser shot. Thus, it is clear that the time averaged amount of foreign gas is reduced with respect to the conventional (continuous flow) configuration, which is still true even when the gas pulse is set to a much longer duration than the laser shot. As a matter of fact, typical durations are on the order of hundreds of microseconds [13], and the delay between gas and laser pulses must be carefully adjusted also to account for gas propagation in the pipes. Furthermore, the high vacuum typical of PLD apparatus (residual pressures below 10^{-8} mbar) affects the dynamics of the gas pulse [88], which can acquire a nearly supersonic behavior with a narrow velocity distribution (temperature values in the tens of Kelvin range have been reported [13]). The so-produced gas beam is directed toward the plume, crossing it at some angle in a region typically located a few millimeters above the target.

Aside from economic considerations that may be of interest in industrial applications, reduction of the background pressure can have other important implications. It must be noted that, in the usual PLD conditions, only a very small fraction of the reactive gas can be actually involved in reactions with the plume. Let us assume an ambient gas pressure of 0.1 mbar, which roughly corresponds, at the room temperature, to a particle density of 10^{15} cm^{-3}; by filling up at such a pressure a volume of tens of liters, typical for deposition chambers, a particle number around 10^{19}–10^{20} is obtained. The plume created by a single laser shot contains no more than 10^{15}–10^{16} particles, thus most of the ambient gas cannot be used for reactions. On the contrary, this excess of ambient gas can be detrimental in experimental terms. Combination of high temperature, due to the presence of substrate heaters, and oxidizing or aggressive environment can easily lead to degradation of the vacuum components. In addition, by keeping the average chamber pressure at a low level, film diagnostics based on electron or ion scattering can be successfully exploited [89] to monitor the layer growth in the real time.

Processes involved in crossed-beam PLD have been thoroughly investigated by means of ion and electron spectroscopy [90]. Roughly speaking, thanks to the large relative velocity of the colliding partners, and to the large particle density in the interaction region (10^{15}–10^{16} ablated particles, and up to 10^{17} gas particles interact in a volume below 1 cm^3, in usual experimental conditions), a strong fragmentation of the gas is first produced. Then, collisional reactions can take place, involving roughly 10% of the gas amount, which are accompanied by a remarkable increase in plume ionization. The latter phenomenon can be regarded as a key point in the technique. In fact, the so-produced ions, before impinging onto the substrate, have still a kinetic energy large enough to promote subsequent reactive processes. In other words, the effects of the interaction can persist even after the plume has crossed the gas beam, thus offering the opportunity for stoichiometric compounds to be formed in the vapor phase before reaching the substrate. Crossed-beam PLD has been so far applied to many materials, including silicides, carbides, nitrides (a list can be found for instance in [13]). One striking achievement in the fabrication of thin films for microelectronics applications was deposition of GaN using Ga ablation and N_2 pulses [91]; as a matter of fact, GaN films were not obtained in ordinary PLD conditions (and at the same substrate temperature), that is, by exploiting an N_2 continuous flow [92].

In addition to microelectronics applications, there are also several specific issues in other fields suggesting the use of the minimum allowed amount of ambient gas. In order to clarify this point, we will briefly discuss PLD fabrication of the so-called *coated conductors* [93]. In their simplest configuration, they consist essentially of a relatively thick HTS layer deposited onto a flexible metal substrate, with one or more dielectric films interposed as electron and inter-diffusion barrier. At the time of their appearance, around mid-1990s, coated conductors showed up as one of the more viable possibility to practically exploit HTS, envisioning a straightforward upgrade for the traditional technology based on superconducting metal alloys in the area of magnetic field generators and power transport. Thanks to the critical current density in excess of $10^6\,A\,cm^{-2}$ (for YBCO at zero field and liquid nitrogen temperature) coated conductors outperform the traditional technology with major benefits in terms of operating temperature. The above-mentioned performances, however, are achievable only with properly grown superconductive layers. This is relatively easy to attain with single crystals substrates, for example, MgO, $SrTiO_3$, and others having a lattice mismatch below a few percent with respect to the unit cell of the HTS material. Roughly speaking, the lattice matching ensures an epitaxial, or, more precisely, heteroepitaxial growth and the large-scale homogeneity of the substrate favors *texturing* of the ceramic layer [94]. We note that this is an additional requirement with respect to epitaxy: in a polycrystalline film with a high degree of texturing, a high degree of in-plane grain ordering is realized. As a consequence, grain borders are typically well aligned each other, and the inter-grain tunneling distance is kept to a minimum. Thus, transport mechanisms in the long range, the ones which are involved in power transmission and imply inter-grain tunneling, are favored.

Many routes have been proposed to achieve texturing in HTS films. Among the others, the so-called Rolled-Assisted Biaxially Textured Substrates (RABiTS, see for instance, Refs. [95,96]) emerged as one of the most promising also in view of possible industrial exploitations. In the technique, the polycrystalline metal substrate undergoes mechanical rolling and UHV annealing prior to deposition, leading to a high degree of texturing. In principle, substrate texturing can be carried out on line with the deposition, that is, a continuous ribbon with length in the meters range can be produced starting from a single metal ingot directly in the deposition chamber. The so prepared substrate is then coated by a dielectric layer, for example, CeO_2 or YSZ (yttrium stabilized zirconia), whose chemical and physical properties are compatible with the HTS film, through PLD or other techniques (e.g. sputtering). Quite often, in order to accommodate the lattice structure, the superconductive film grows up with its *c*-axis inclined with respect to the dielectric plane.

The choice of the metal to be employed as the substrate is critical: the material must be "soft" enough to allow for an efficient rolling, while maintaining a lattice structure compatible with the subsequent film deposition. In addition, the starting ingot must be available with a very high degree of purity; Ni and Ni-based alloys (see, for instance, Ref. [97]) show features suitable for this demanding task and excellent results in terms of critical current density have been achieved on relatively long samples. However, metal substrates show a very high sensitivity to the oxidizing atmosphere needed for PLD of the superconductive coating, which, for instance, prevented a widespread exploitation of iron-based material, exhibiting excellent properties in terms of economy and mechanics. Oxidization occurs basically through one "inherent" channel, oxygen diffusion across the dielectric barrier, and through contamination of the substrate surface by the ambient. We point out that effects of surface oxidization are highly detrimental for the system performance, since they can completely prevent the correct growth of the coatings.

Within this context, crossed-beam PLD can offer substantial advantages. Coupled to further radio frequency (RF) ionization of the gas pulse, composed of O_2, the technique offered encouraging results [98] also with NiFe substrates, a material extremely sensitive to oxidization. The effects of the pulsed injection can be understood, for instance, in Fig. 8.6, where atomic oxygen absorption by the plume at a

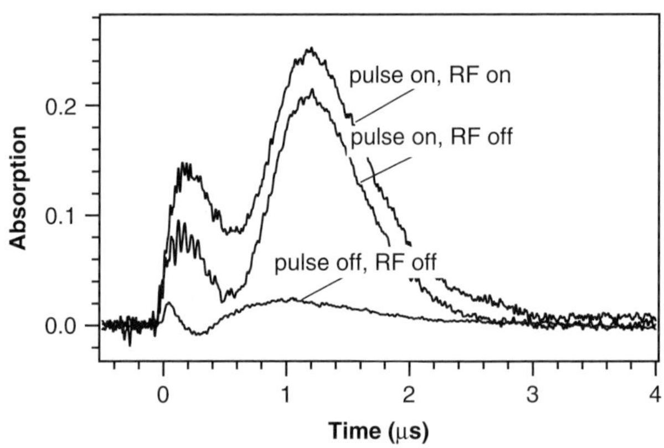

Figure 8.6: Atomic oxygen absorption during PLD of YBCO with or without pulsed gas injection as discussed in the text (pulse parameters: duration 300 μs, delay 200 μs).

distance 6 mm from the target is displayed as a function of time during ablation of YBCO with a XeCl excimer laser at 2 J cm^{-2}. The three curves refer to data acquired with a continuous flow of O_2 at the ordinary pressure of PLD and with pulsed injection of O_2 (in the absence or presence of a RF discharge at the end of the gas pipe). Data clearly show the enhancement in oxygen dissociation achieved by the use of pulsed injection (with a slight further increase associated with the RF discharge).

Finally, another crossed particle beam method is worth to be briefly mentioned, the so-called *ion-beam-assisted PLD* (IBAD). In this case, an energetic ion beam made of noble gases (e.g. Ar^+) is sent through the jet of ablated material right before its arrival onto the substrate. The goal is not to promote chemical reactions (the ion beam is inert), but to affect the growth of the coatings through modifications of the ablated particle dynamics as they impinge onto the surface. In fact, as we will discuss in the following section, particle dynamics is a key point in the achievement of the proper coating structure, being directly associated with the growth mechanisms. Moreover, by carefully choosing the incidence angle and the accelerating potential (typically hundreds of volts), beneficial effects can be found in the film texturing, which, among others, have been exploited in coated conductor fabrication [99]. The method allowed long samples on non-pretextured Ni substrates to be produced in a prototypal industrial apparatus [100].

8.6. Film Growth in PLD

The previous sections have addressed topics relevant to understand the dynamical and chemical behavior of the ablated material. When the result of the whole deposition process is considered, that is, fabrication of coating layers, issues associated with substrate/material interaction and film growth are of paramount importance. In the following discussion, we will restrict ourselves to experiments where a continuous coating is fabricated, neglecting all PLD applications where nanoparticles or small-sized islands are produced, although they are gaining increasing momentum, for instance as a viable method for the fabrication of nanotubes [34] or other mesoscopic structures (see, for instance, Ref. [101]). It can be noted that, all deposition methods involving a vapor-phase share similar problems to explain the growth, and that a general picture of the relevant phenomena, especially those related to surface processes, is far from being achieved. In fact, the vapor absorption by the substrate and the consequent growth are ruled by many effects, often specific for the actual choice of layer and substrate materials. A complete analysis of all processes is beyond the scope of the present chapter; interested readers are referred to surface physics textbook (see, for instance, Ref. [102]). We will here point out only a few features of the growth mechanisms in the specific conditions of PLD.

As a matter of fact, a huge literature demonstrates that PLD is capable of film fabrication with an enormous and still growing variety of materials (only a very restricted list has been considered in this chapter). Typically, layer thickness can range through tens of nanometers to an upper limit, mostly dictated by stress/strain structural problems, of a few microns. Depending on target and substrate materials and on experimental parameters, in particular the substrate temperature, either amorphous or epitaxial (heteroepitaxial) growth can be achieved. In the latter case, grains show typically a columnar shape, a dense space distribution, and an average lateral size in the hundreds of nanometers range. In general, the resulting films have a relatively small roughness, which make them suitable for coating applications, but, as discussed in the next section, the presence of surface defects, in particular material droplets, can pose a serious limit to the surface quality. An important parameter to assess the performance of the technique is the growth rate, which is also a function of the operating parameters, especially the laser fluence: values on the order of 0.5–1.0 Å per laser shot are customarily attained. The relatively weak surface coverage per pulse is often recognized as an advantage of PLD, allowing the user to easily control the thickness of the coating; nonetheless, laser sources designed for industrial applications may operate at repetition frequencies above 100 Hz, so that acceptable growth rates per unit time can be achieved, up to 0.5–1 $\mu m\,min^{-1}$.

Summarizing what we have presented in the previous sections, in PLD the substrate sees the arrival of elemental particles, small molecules, and aggregates. The physico-chemical processes ruling the plume expansion points at the following facts:

(i) Energy of the impinging species is typically super-thermal; at their arrival onto the substrate, after a flight of several centimeters, being this distance covered in a few microseconds, recombination of ionized species and de-excitation have been already occurred, so that material is almost completely made of neutrals and the energy is mainly associated with the dynamical degree of freedom along the plume axis.

(ii) Space distribution of the plume is highly peaked along the plume axis direction; since the substrate and target surfaces are usually parallel to each other, plume particles impinge onto the substrate with a large momentum along the orthogonal direction.

(iii) When ablation is carried out in a vacuum, cluster formation involves ablated species leading to aggregates which, in the case of multi-elemental targets and depending on energetic considerations, can have a composition "compatible" with that of the starting target.

(iv) When ablation is carried out in an ambient gas, dissociation of the background species occurs, providing the plume with reactive atoms.

(v) Collisional reactions take place during plume expansion in the ambient gas, leading to incorporation of foreign species in aggregates whose composition can be "compatible" with the target stoichiometry.

Once the material impinges onto a fresh substrate, depending on the sticking coefficient it can either be reflected back or start diffusing on the surface [103]. The sticking probability exhibits a very large dependence both on energy and on physical and chemical properties of the ablated and substrate materials: a ratio between the amount of ablated and deposited material indicates sticking probabilities below the unity, hence back-reflection of the plume particles cannot be ruled out. Particle diffusion on the surface is an issue inherently associated with any vapor deposition method. Elementary mechanics suggests that diffusion requires plume/substrate collisions able to efficiently change the direction of the velocity vector. Those collisions involve momentum and energy exchange with the phonons of the substrate, which depend on the kinetic energy of the plume and on the substrate temperature. Diffusion is a temperature-activated phenomenon, and, even if surface phenomena initiating the process are rather difficult to be modeled [102], we can expect that the high kinetic energy of the plume leads to a mobility of adatoms on the substrate greater for PLD than for conventional (thermal) deposition methods. In the early growth stage, adatoms stop diffusing when they reach suitable surface sites where they can loose kinetic energy, partly transferred to phonon excitation and partly used for activating chemical bonds (we refer here to chemisorption processes, involving strong adhesion, and neglect other effects where particles are weakly bound to the surface). Elected sites are typically defective, for instance surface steps, dislocations, and vacations. Nucleation then begins playing a role: the continuous arrival of new particles makes it possible to build up isolated islands which can be regarded as the initial growth stage.

Compared to other deposition methods, for example, MBE, which has been widely studied in the past, PLD shows the peculiar presence of aggregates, or clusters, in the plume. The amount of energy transported by a cluster is typically too small to foresee its fragmentation upon collision with the surface, since the energy available for any single cluster component is below the threshold for breaking chemical bonds; hence we can expect that clusters diffuse over the substrate along with lighter particles. However, due to their mass, diffusion timescale for clusters is longer than for elemental species, that is, their mean diffusion length is shorter [104]. Growth of nanostructures by cluster deposition has been thoroughly modeled in [105], where the role played by large mass aggregates in starting nucleation has been pointed out. In a simplified sketch, the simultaneous presence of heavy and light components may be regarded as a key point to achieve nucleation of clusters and incorporation of elemental species, leading to a smooth growth of islands.

Besides material re-evaporation from the substrate, which is obviously possible all during the deposition process depending on the substrate temperature (and on the ambient pressure), the next stage of layer growth is island coalescence. In the thin film technology, several relevant processes have been identified (see, for instance, Refs. [12,102], depending on the physical and chemical properties of the involved species, usually referred to as Volmer–Weber, Frank–van der Merwe, Stranski–Krastinov. They are different from each other for the mechanisms involved in the coalescence: islands can in fact either grow along three dimensions before coalescence, or they can get from the beginning a two-dimensional form, that is, formation of a single monolayer is preferred with respect to isolated islands. In addition, some kind of mixed regime can be observed, where the initial monolayer sustains growth of three-dimensional islands. Such mechanisms, which we will not treat here in details, are quite general for any physical deposition method, but, once more, we can identify the role played by some specific features of PLD. During coalescence, that is, after the initial stage of the growth, particle diffusion over the (partially) coated substrate is governed by a complicated set of rules [102]. In the simple model of terraced islands (see, for instance, Ref. [106]), which applies rather well to PLD, different energy barriers must be considered in order to account for intra-terrace or inter-terrace diffusion, the latter process being different when the particle climbs or drops down from one to another terrace: the mechanisms behind the concept of the so-called Ehrlich–Schwoebel barrier (see, for instance, Ref. [102]) clarify the physical origin of such energy-dependent processes. Within this context, it can be thought that the kinetic energy available to laser ablated particles may have some role in favoring all energy-dependent effects.

This point must be considered more carefully; indeed, if the kinetic energy of the impinging particles exceeds some threshold, a bombardment of the surface is accomplished with consequent re-sputtering. This is the basic mechanism of ion sputtering, and backscattering is also one of the main drawbacks for all sputtering techniques. However, the energy range typical for ablated particles is by far smaller than for those techniques where charge acceleration is exploited. More specifically, a molecular dynamics model [107] has pointed out that mobility of the uppermost layers of atoms during the island growth can be enhanced when they undergo collisions with impinging homonuclear species at moderate energy. Calculations made for silicon showed that ion-induced displacement was efficient for incident energies in the approximate range 10–40 eV, and that at 100 eV one bulk displacement with consequent defect formation was achieved for every two surface displacements. PLD turns out to provide particles with energies compatible with the bombardment-induced diffusion enhancement, which can (partly) explain the performance of the technique [13].

Finally, layer composition must be considered, which is governed by a variety of complex chemical and physical phenomena. In practice, chemical bonds must be produced to ensure a layer with proper stoichiometry and lattice structure. As in other deposition methods, also in PLD the substrate is usually kept at a temperature large enough to have a (partial or complete) annealing. In this way, energy is delivered to the growing lattice in order to favor its rearrangement. As already stated, part of the needed energy can be provided by the plume expansion, that makes generally possible a considerable decrease in the substrate temperature with PLD compared to other techniques. A pioneering example was the heteroepitaxial growth of Ge through PLD below 600 K [108], a temperature lower than that required in MBE. Besides technologic motivations connected to specific substrate materials, decreasing the temperature leads also to avoid excessive re-evaporation of deposited volatile species. On the other hand, as for oxygen in the case of ceramics, also diffusion of ambient species into the film is greatly suppressed, but, as already stated, oxygen deficiency can be overcome by exploiting chemical reactions in the vapor phase, able to provide the plume with species rich in the volatile component.

8.7. Some Drawbacks and Limitations of PLD

As any other deposition technique, PLD exhibits drawbacks and limitations; they are in part due to problems which might be solved by technologic progresses, but in part are of a more fundamental nature. We will mention here some of the main and most recognizable problems of PLD associated with the quality of the produced coatings, in particular in terms of homogeneity and surface coverage.

When presenting plume dynamics, we underlined the peculiar space distribution of the jet of ablated material, which is highly peaked along the direction normal to the target surface. This is a consequence of the efficient energy coupling between laser and target, in other words it is an essential feature of PLD, but at the same time it means that the surface covered by the plume is inherently rather small. As a matter of fact, typically the surface area is on the order of 1 cm^2, much smaller than with other techniques, for example, evaporation, MBE, sputtering, and others. Furthermore, the thickness homogeneity within the coated surface is not complete (percentage variations ~20–30% are typical on a 10 mm size [109]), which represents an additional problem both in microelectronics and in hard coating applications. However, solutions can be readily found by developing motorized systems to move the substrate position and/or the impact point of the laser shot onto the target, or by using line focal spots [110]. In this way, depositions over an area well above 100 cm^2 have been reported; furthermore, as already stated in the case of coated conductors, continuous PLD over moving substrates can be accomplished, although the experimental configuration must be carefully considered (see, for instance, Ref. [111]).

A more serious problem still dealing with coating homogeneity is the presence of surface defects in the film. Beside the occurrence of cracks and pits due to the localized impact of energetic particles, the most important issue is formation of material *droplets*, usually spherical or hemispherical in shape and with a transverse size up to the microns range. Droplets are not likely produced during plume expansion: in fact, their large dimensions are not compatible with the aggregation processes undergone by the plume, and their composition is normally different with respect to the deposited layer. A generalized interpretation of droplet formation able to describe processes occurring with any class of materials is still lacking; nonetheless, a few mechanisms can be identified. First of all, as already mentioned, the laser/target interaction produces mechanical forces on the order of 10^3–10^4 N over an area around 1 mm^2; these transient forces are able to promote mechanical removal of the uppermost layers of the target, leading to macroscopic fragments in some kind of exfoliation process [13]. Furthermore, internal explosion of the bulk can take place [112], in particular in materials where the time needed for evaporation is shorter than the time needed to convert laser energy into heat and to transfer it to the bulk. A rough explanation is the following: radiation penetrates deeply into the target and is coupled with non-homogeneous regions (the phenomenon is usually more pronounced with porous and polycrystalline materials). Extreme local heating leads to vaporization of the inner parts of the target, which in turn results in removal of macroscopic fragments from the surface. So-produced heavy fragments are usually re-deposited onto the target itself (as shown also by optical micrographs of its surface), but in some cases they can receive momentum from the plume particles and be transported onto the substrate. Another possibility is associated with partial vaporization of the target occurring in the outer regions of the laser focal spot. In fact, the space distribution of the beam is far from being homogeneous; as a consequence, the amount of energy coupled to the target is not uniform over all the irradiated area, leading to melt rather than vaporize the material. Liquid or semiliquid droplets are then formed right above the target, which can be also transported by the plume onto the substrate, where they are incorporated in the coating layer.

One obvious recipe to prevent droplet formation is an accurate control of the laser spot, in terms of fluence and space distribution, which can be greatly enhanced by using specifically designed laser sources. In addition, preparation of the target surface, aimed in particular at removing or avoiding inhomogeneous regions during the process [118], can be helpful. Also, other techniques have been proposed, which rely on the different dynamics of heavy and light components of the plume and aim at depleting the droplet content impinging onto the substrate while maintaining an acceptable density of plume particles. Off-axis deposition configurations have been proposed (see, for instance, Refs. [5,113,114]), where target and substrate surfaces are not parallel to each other; by carefully setting the substrate position and the operating parameters, it is possible to avoid, or limit, arrival of heavy fragments onto the substrate. Another possibility, which was investigated at the time of early PLD applications was based on the use of a velocity selector for the expanding material, consisting essentially on a mechanical rotating chopper stopping heavy fragments, which expand at a velocity one to two orders of magnitude smaller than the light components [5,115,116]. This method, which requires the introduction in the deposition chamber of motorized wheels rotating at high speed, demonstrated its validity in limiting the flow of particulates, at the expenses, however, of a reduced growth rate and of a rather complex setup.

8.8. PLD and Ultra Short Laser Pulses

It is quite obvious that the enormous progress of PLD in the last decades has been possible thanks to a widespread diffusion of suitable laser sources, in particular excimer and Nd-YAG systems able to deliver nanoseconds pulses. As we have described all through this chapter, such lasers still constitute the main component of any PLD set-up intended for coating applications. Nonetheless, the recent availability of lasers with ultra short pulses, in the sub-nanoseconds and even sub-picoseconds range (laser pulses as short as several femtoseconds can be presently produced), offers new perspectives to the technique, which are worth to be briefly mentioned. The most striking advantage of ultra short pulses is probably the increased efficiency of the ablation process, meaning that the threshold fluence which must be exceeded in order to have macroscopic removal of bulk material may be decreased by more than one order of magnitude [13]. This is a consequence of a different regime of radiation/matter coupling. When the pulse duration is shorter than the typical timescale for electron/lattice energy exchange [2,10], heat diffusion becomes practically negligible [117,118]. As a consequence, the target is concerned by the interaction only in the laser focus region; in such a small region, concentrated close to the surface, the extremely large laser intensity promote a fast and almost complete multi-photon ionization of the material. Ionization can also proceed through resonant channels due to the large bandwidth of ultra short pulses. The resulting electrons, further heated by the laser pulse, can start an avalanche ionization process leading to a rapid vaporization of a macroscopic amount of material in a process involving charge repulsion [119]. In principle, this occurs when the electron energy density (summed up over all electrons) approaches the binding energy of the target, that is, for electron densities around or above $10^{20}\,cm^{-3}$, depending on the material [117].

Due to the short pulse duration, absorption of radiation by the plume cannot follow the same mechanisms we have described for conventional PLD. In fact, as soon as the material is vaporized, the laser pulse vanishes, and no further energy is available for excitation, heating and ionization of the plume. We might expect as a consequence a reduced kinetic energy for the expanding species, but this is not the case: expansion diagnostics has been accomplished also in ultra short pulse ablation experiments, demonstrating that kinetic energies up to hundreds eV can be achieved with metal ions [120]. However, contrary to conventional PLD, in this case the nature of the ablated material can play a prominent role. In fact, much smaller energy are observed in ablation of semiconductors [121], which can be ascribed to a reduced effect of multiphoton ionization due to the presence of a band gap. Furthermore, effects which are typically negligible in conventional PLD, as, for instance, electron heat diffusion [122], may become relevant in ruling plume dynamics when ultra short pulses are used.

The relative infancy of the technique does not allow drawing a detailed picture of all relevant features; however, advantages can be foreseen in fabrication of coatings by using ultra short pulses. For instance, the reduced heat diffusion in the bulk will further reduce target damage, leading to suppress, or strongly prevent, the problem of droplet formation. On the other hand, reduction of the ablation threshold and the consequent increased efficiency of the process will definitely allow ablation of even ultra hard targets with acceptable rates.

8.9. Concluding Remarks

In this chapter we have demonstrated that nowadays PLD can be regarded as a mature technique for the deposition of thin films and coatings of a wide variety of materials. Energy and stoichiometry are the key words for understanding the distinctive features of PLD, which are reflected into a large degree of versatility, an extremely broad range of possible applications, and the very appealing ability to achieve congruency between target and film also in case of complex compounds. Even if PLD involves many different aspects of physics, chemistry, materials science, a large part of the basic mechanisms is now understood and interpreted, also thanks to many diagnostics experiments carried out in the past. The foreseeable introduction of ultra short laser sources will possibly further improve the performance of the technique, while limiting the relevance of the main problems which at present hamper the diffusion of PLD.

The interest in laser ablation by the research community is demonstrated by the huge number of publications appeared in the topic in the last two decades (a very restricted part of them is quoted here). It must be noted that, though the yearly rate of publication is decreased with respect to 10 years ago, there are still many hundreds of papers reporting on new applications or new implementations of the technique. Nonetheless, possibilities for an effective large-scale exploitation of PLD in a purely industrial environment are still under debate. Main motivations are probably to be found in the relatively slow diffusion of laser systems in the industry, where they are employed mostly for micro-processing (see, for instance, Ref. [123]) rather than for micro-fabrication. The well-established presence of other techniques, as well as the mentioned disadvantages of PLD, posed so far serious limitations in those sectors of manufacturing which are more sensitive to purity and homogeneity of the coating, as the microelectronics industry. However, also thanks to the progresses in laser sources, including the development of reliable ultra short pulse sources, it can be expected that during this decade the interest in PLD will definitely emerge and its application will be no longer restricted to niches as presently occurs.

References

[1] See, for instance, J.F. Ready, *J. Appl. Phys.*, 36 (1965) 462.
[2] D. Bäuerle, *Laser Processing and Chemistry*, Springer, Berlin, 2000; and references therein.
[3] D. Dijkkamp, T. Venkatesan, X.D. Wu, *et al.*, *Appl. Phys. Lett.*, 51 (1987) 619.
[4] J. Narayan, N. Biunno, R. Singh, *et al.*, *Appl. Phys. Lett.*, 51 (1987) 1845.
[5] D.B. Chrisey and G.K. Hubler, *Pulsed Laser Deposition of Thin Films*, John Wiley and Sons, New York, 1994.
[6] D. Roy, S.B. Krupanidhi and J.P. Dougherty, *J. Appl. Phys.*, 31 (1992) 2965.
[7] A. Iembo, F. Fuso, E. Arimondo, *et al.*, *J. Mater. Res.*, 12 (1997) 1433.
[8] R.K. Singh, O.W. Holland and J. Narayan, *J. Appl. Phys.*, 68 (1990) 233.
[9] R.K. Singh and J. Narayan, *Phys. Rev. B*, 41 (1990) 8843.
[10] D.S. Ivanov and L.V. Zhigilei, *Phys. Rev. B*, 68 (2003) 064114.
[11] I. Weaver and C.L.S. Lewis, *J. Appl. Phys.*, 79 (1996) 7216.
[12] M. Ohring, *The Materials Science of Thin Films*, Academic Press, San Diego, 1992.
[13] P.R. Willmott and J.R. Huber, *Rev. Mod. Phys.*, 72 (2000) 315.
[14] R. Singh and J. Narayan, *Mater. Sci. Eng. B*, 3 (1989) 217.
[15] D. Bhattacharya, R.K. Singh and P.H. Holloway, *J. Appl. Phys.*, 70 (1991) 5433.
[16] P.R. Willmott, P. Manoravi and K. Holliday, *Appl. Phys. A*, 70 (2000) 425.
[17] Ya.B. Zel'dovich and Yu.P. Raizer, *Physics of Shock Waves and High Temperature Hydrodynamic Phenomena*, Academic Press, New York, 1966.
[18] R. Timm, P.R. Willmott and J.R. Huber, *J. Appl. Phys.*, 80 (1996) 1794.
[19] R.W. Dreyfus, *J. Appl. Phys.*, 69 (1991) 1721.
[20] S. Amoruso, R. Bruzzese, N. Spinelli and R. Velotta, *J. Phys. B*, 32 (1999) R131.
[21] D.B. Geohegan, *Appl. Phys. Lett.*, 60 (1992) 2732.
[22] See, for instance, D. Fried, G.P. Reck, T. Kushida and E.W. Rothe, *J. Appl. Phys.*, 70 (1991) 2337.
[23] D.B. Geohegan, A.A. Puretzky, G. Duscher and S.J. Pennycook, *Appl. Phys. Lett.*, 72 (1998) 2987.
[24] F. Fuso, L.N. Vyacheslavov, G. Masciarelli and E. Arimondo, *J. Appl. Phys.*, 76 (1994) 8088.
[25] J.C.S. Kools, T.S. Balier, S.T. De Zwart and J. Dieleman, *J. Appl. Phys.*, 71 (1992) 4547.

[26] R. Kelly, A. Miotello, B. Braren, A. Gupta and K. Casey, *Nucl. Instrum. Method. Phys. Res. B*, 65 (1992) 187.
[27] D. Sibold and H.M. Urbassek, *Phys. Fluids A*, 4 (1992) 165.
[28] R. Kelly and A. Miotello, *Appl. Phys. B*, 57 (1993) 145.
[29] J.B. Anderson and J.B. Fenn, *Phys. Fluids*, 8 (1965) 780.
[30] H. Wu and E. Arimondo, *J. Phys. D*, 31 (1998) 2218.
[31] See, for instance, J.F. Ready, *Industrial Applications of Lasers*, Academic Press, New York, 1978.
[32] R.E. Russo, *Appl. Spectrosc.*, 49 (1995) 14.
[33] H.W. Kroto, J.R. Heath, S.C. O'Brien, R.F. Curl and R.E. Smalley, *Nature*, 318 (1985) 162.
[34] A.A. Puretzky, D.B. Geohegan, X. Fan and S.J. Pennycook, *Appl. Phys. Lett.*, 76 (2000) 182; and references therein.
[35] See, for instance, C.M. Wayman and H.K.D.H. Bhadeshia, in *Physical Metallurgy*, Eds. R.W. Cahn and P. Hassen, Elsevier, Amsterdam, 1996.
[36] See, for instance, *Proceedings of ICOMAT98, Special Issue of Mater. Sci. Eng. A*, 273–275 (1999).
[37] J.L. Murray, *Phase Diagrams of Binary Titanium Alloys*, ASM International, Metal Parks, 1987.
[38] F. Ciabattari, F. Fuso and E. Arimondo, *Appl. Phys. A*, 64 (1997) 623.
[39] A. Camposeo, F. Fuso, E. Arimondo and A. Tuissi, *Appl. Phys. A*, 76 (2003) 927.
[40] L.J. Frasinsky, K. Codling and P.A. Hatherly, *Science*, 246 (1989) 1029.
[41] V. Berardi, N. Spinelli, R. Velotta, M. Armenante, *et al.*, *Phys. Lett. A*, 179 (1993) 116.
[42] P.E. Dyer, A. Issa and P.H. Key, *Appl. Surf. Sci.*, 46 (1990) 89.
[43] J. Wild, J. Macl, H. Sichova, P. Bohacek, *et al.*, *Supercond. Sci. Technol.*, 11 (1998) 1341.
[44] D.B. Geohegan, *Thin Solid Films*, 220 (1992) 138.
[45] D.B. Geohegan and A.A. Puretzky, *Appl. Phys. Lett.*, 67 (1995) 197.
[46] D.B. Geohegan and A.A. Puretzki, *Appl. Surf. Sci.*, 96–98 (1996) 131.
[47] B. Thestrup, B. Toftmann, J. Schou, B. Doggett and J.G. Lunney, *Appl. Surf. Sci.*, 208–209 (2003) 175.
[48] S. Amoruso, B. Toftmann and J. Schou, *Appl. Phys. A*, 79 (2004) 1311.
[49] S. Amoruso, B. Toftmann, J. Schou, R. Velotta and X. Wang, *Thin Solid Films*, 453–454 (2004) 562.
[50] N. Arnold, J. Gruber and J. Heitz, *Appl. Phys. A*, 69 (1999) S87.
[51] T.E. Itina, W. Marine and M. Autric, *J. Appl. Phys.*, 82 (1997) 3636.
[52] See, for instance, G. Masciarelli, F. Fuso, M. Allegrini and E. Arimondo, *J. Mol. Spectrosc.*, 153 (1992) 96.
[53] V. Berardi, S. Amoruso, N. Spinelli, M. Armenante, *et al.*, *J. Appl. Phys.*, 76 (1994) 8077.
[54] V. Berardi, S. Amoruso, N. Spinelli, M. Armenante, *et al.*, *Int. J. Mass Spectrom. Ion Proc.*, 144 (1995) 1.
[55] A. Camposeo, F. Cervelli, F. Fuso, M. Allegrini and E. Arimondo, *Appl. Phys. A*, 69 (1999) S509.

[56] C. Wieman and L. Hollberg, *Rev. Sci. Instrum.*, 62 (1991) 1.
[57] A. Camposeo, F. Cervelli, F. Fuso, M. Allegrini and E. Arimondo, *Appl. Phys. Lett.*, 78 (2001) 2402.
[58] S. Amoruso, M. Armenante, R. Bruzzese, N. Spinelli, R. Velotta and X. Wang, *Appl. Phys. A*, 69 (1999) S483.
[59] P.R. Willmott, R. Timm, P. Felder and J.R. Huber, *J. Appl. Phys.*, 76 (1994) 2657.
[60] E. Fogarassy, A. Slaoui, C. Fuchs and J.P. Stoquert, *Appl. Surf. Sci.*, 54 (1992) 180.
[61] C.N. Afonso, F. Vega, J. Solis, F. Catalina, *et al.*, *Appl. Surf. Sci.*, 54 (1992) 175.
[62] A. Misra, H.D. Bist, M.S. Navati, R.K. Thareja and J. Narayan, *Mater. Sci. Eng. B*, 79 (2001) 49.
[63] M. Foltin, G.J. Stueber and E.R. Bernstein, *J. Chem. Phys.*, 111 (1999) 9577.
[64] A.Y. Liu and M.L. Cohen, *Science*, 245 (1989) 841.
[65] G. Barucca, G. Majni, P. Mengucci, *et al.*, *J. Appl. Phys.*, 86 (1999) 2014.
[66] O. Auciello, T. Barnes, *et al.*, *Thin Solid Films*, 181 (1989) 65.
[67] G. Soto, W. de la Cruz, F.F. Castillon, J.A. Diaz, R. Machorro and M.H. Farias, *Appl. Surf. Sci.*, 214 (2003) 58.
[68] G. Leggieri, A. Luches, M. Martino, A. Perrone, *et al.*, *Appl. Surf. Sci.*, 96–98 (1996) 866.
[69] G. Leggieri, A. Luches, M. Martino, *et al.*, *Thin Solid Films*, 258 (1995) 40.
[70] F. Vega, C.N. Afonso and J. Solis, *J. Appl. Phys.*, 73 (1993) 2472.
[71] J. Gonzalo, C.N. Afonso and I. Madariaga, *J. Appl. Phys.*, 81 (1997) 951.
[72] J. Hermann, F. Coursimault, O. Motret, S. Acquaviva and A. Perrone, *J. Phys. B*, 34 (2001) 1917.
[73] W. Marine, M. Gerri, J.M. Scotto d'Aniello, *et al.*, *Appl. Surf. Sci.*, 54 (1992) 264.
[74] J. Hermann, A. Perrone and C. Dutouquet, *J. Phys. B*, 34 (2001) 153.
[75] Y. Matsuo, T. Nakajima, T. Kobayashi and M. Takami, *Appl. Phys. Lett.*, 71 (1997) 996.
[76] E. Ishitani, S. Yoshimoto, H. Higashide, *et al.*, *Chem. Lett.*, 7 (1993) 1203.
[77] M.L. De Giorgi, G. Leggieri, A. Luches, *et al.*, *Appl. Phys. A*, 60 (1995) 275.
[78] J.C.S. Kools, C.J.C.M. Nillesen, S.H. Brongersma, *et al.*, *J. Vac. Sci. Technol. A*, 10 (1992) 1809.
[79] A.G. Guidoni, A. Mele, T.M. Di Palma, *et al.*, *Thin Solid Films*, 295 (1997) 77.
[80] A.K. Sharma and R.K. Thareja, *J. Appl. Phys.*, 88 (2000) 7334.
[81] C. Dutouquet and J. Hermann, *J. Phys. D*, 34 (2001) 3356.
[82] C. Vivien, J. Hermann, A. Perrone and C. Boulmer-Leborgne, *J. Phys. D*, 32 (1999) 518.
[83] A.P. Thorne, *Spectrophysics*, Chapman and Hall, London, 1974.
[84] H.R. Griem, *Plasma Spectroscopy*, McGraw-Hill, New York, 1964.
[85] J. Hermann and C. Dutouquet, *J. Appl. Phys.*, 91 (2002) 10188.
[86] D.E. Powers, S.G. Hansen, M.E. Geusic, *et al.*, *J. Phys. Chem.*, 86 (1982) 2556.
[87] A. Gupta and B.W. Hussey, *Appl. Phys. Lett.*, 58 (1991) 1211.
[88] G. Scoles (Ed.), *Atomic and Molecular Beam Methods*, Vol. I, Oxford University Press, New York, 1988.
[89] See, for instance, P.R. Willmott, R. Timm and J.R. Huber, *Appl. Surf. Sci.*, 127–129 (1998) 105.
[90] P.R. Willmott, R. Timm and J.R. Huber, *J. Appl. Phys.*, 82 (1997) 2082.

[91] P.R. Willmott and F. Antoni, *Appl. Phys. Lett.*, 73 (1998) 1394.
[92] R.F. Xiao, H.B. Liao, N. Cue, *et al.*, *J. Appl. Phys.*, 80 (1996) 4226.
[93] See, for instance, D.P. Norton, A. Goyal, J.D. Budai, *et al.*, *Science*, 274 (1996) 755.
[94] N.A. Rutter and B.A. Glowacki, *Supercond. Sci. Technol.*, 14 (2001) 680.
[95] A. Goyal, D.P. Norton, J.D. Budai, *et al.*, *Appl. Phys. Lett.*, 69 (1996) 1795.
[96] M. Paranthaman, D.F. Lee, R. Feenstra, *et al.*, *IEEE Trans. Appl. Supercond.*, 9 (1999) 2268.
[97] E. Varesi, V. Boffa, G. Celentano, *et al.*, *Physica C*, 372–376 (2002) 763; and references therein.
[98] M. Cantoro, N. Coppedé, A. Camposeo, *et al.*, *Surf. Coat. Technol.*, 180–181 (2004) 591.
[99] V. Betz, B. Holzapfel and L. Schultz, *Thin Solid Films*, 301 (1997) 28.
[100] Y. Iijima, K. Kakimoto, Y. Sutoh, *et al.*, *Supercond. Sci. Technol.*, 17 (2004) S264.
[101] J. Gonzalo, D. Babonneau, C.N. Afonso and J.-P. Barnes, *J. Appl. Phys.*, 96 (2004) 5163; and references therein.
[102] J.A. Venables, *Introduction to Surface and Thin Film Processes*, Cambridge University Press, Cambridge, 2000.
[103] J.A. Venables, G.D.T. Miller and M. Hanbücken, *Rep. Prog. Phys.*, 47 (1984) 399.
[104] P. Jensen, H. Larralde and A. Pimpinelli, *Phys. Rev. B*, 55 (1997) 2556.
[105] P. Jensen, *Rev. Mod. Phys.*, 71 (1999) 1695.
[106] H. Metiu, *Nature*, 366 (1993) 111.
[107] D.K. Brice, J.Y. Tsao and S.T. Picraux, *Nucl. Instrum. Method. Phys. Res. B*, 44 (1989) 68.
[108] H. Sankur, W.J. Gunning, J. De Natale and J.F. Flintoff, *J. Appl. Phys.*, 65 (1989) 2475.
[109] V.A. Yakovlev, G. Mattei, A. Iembo, *et al.*, *J. Appl. Phys.*, 78 (1995) 6321.
[110] B. Schey, T. Bollmeier, M. Kuhn, *et al.*, *Rev. Sci. Instrum.*, 69 (1998) 474.
[111] A.S. Kuzanyan, *IEEE Trans. Appl. Supercond.*, 13 (2003) 2868.
[112] D.L. Lin, X. Li, Z.D. Liu and T.F. George, *J. Appl. Phys.*, 72 (1992) 4227.
[113] R.J. Kennedy, *Thin Solid Films*, 214 (1992) 223.
[114] N.J. Appleyard, T.J. Jackson, R.G. Welch, *et al.*, *Meas. Sci. Technol.*, 6 (1995) 337.
[115] H.J. Dupendant, J.P. Gavigan, D. Givord, *et al.*, *Appl. Surf. Sci.*, 43 (1989) 369.
[116] T. Yoshitake, G. Shiraishi and K. Nagayama, *Appl. Surf. Sci.*, 197–198 (2002) 379.
[117] B.C. Stuart, M.D. Feit, A.M. Rubenchick, *et al.*, *Phys. Rev. Lett.*, 74 (1995) 2248.
[118] M.J. Lenzner, J. Krüger, S. Sartania, *et al.*, *Phys. Rev. Lett.*, 80 (1998) 4076.
[119] R. Stoian, D. Ashkenasi, A. Rosenfeld and E.E.B. Campbell, *Phys. Rev. B*, 62 (2000) 13167.
[120] H. Varel, M. Wähmer, A. Rosenfeld, D. Ashkenasi and E.E.B. Campbell, *Appl. Surf. Sci.*, 127–129 (1998) 457.
[121] A. Cavalleri, K. Sokolwski-Tinten, J. Bialkowski and D. Von der Linde, *Appl. Phys. Lett.*, 72 (1998) 2385.
[122] S. Amoruso, R. Bruzzese, M. Vitiello, N.N. Nedialkov and P.A. Atanasov, *J. Appl. Phys.*, 98 (2005) 449071.
[123] J.H. Brannon, *SPIE Proc.*, 4637 (2002) 474.

Chapter 9

Thermal Plasmas Surface Treatment

Pierre Fauchais and Armelle Vardelle

9.1. Principles of Thermal Plasma Surface Treatment

9.1.1. Introduction

Thermal plasmas produced by direct current (d.c.) arcs or radio frequency (RF) discharges at atmospheric pressure and reduced pressure (between 10 and 50 kPa) have been used since the 1960s for surface treatment. They are one of the thermal-spraying processes that also include flame- or detonation-based techniques. The worldwide thermal spray industry represented about €3B of sales in 2003.

However, the ranking of plasma processes among all the thermal-spraying processes is difficult because the sales of coatings are not well known as the sales of integrated shops are not necessarily individualized as coatings. The ranking presented in the following is based on the study of Magetex [1] related to the number of equipments used in Europe. The equipment prices, to the exclusion of spray booth and handling robot, are indicated in brackets:

15% for twin-wire arcs (€10–25 K).
8% for plasma spraying (€75–185 K at atmospheric pressure and €600–1500 K under controlled atmosphere). More than 97% correspond to d.c. plasmas and less than 3% to RF plasmas.
7% for plasma transferred arc (PTA) reclamation (€50–75 K).

In addition to these processes, two other ones can be considered to be related to surface treatment:

Powder spheroidization by RF plasma. This market is dominated by a Canadian company (Tekna).
Surface modification by using a d.c. plasma torch.

Such a technology is used in the turning process to lower the metal hardness by preheating the rod surface just prior to the tool to transform, for example, the bainite, pearlite, etc. in austenite for low-carbon steels. It may also be used to harden

Materials Surface Processing by Directed Energy Techniques
Y. Pauleau (Editor)

steels. The surface is heated in a few milliseconds by the plasma torch and, then, quenched by the bulk material below. For instance, this makes it possible to form martensite. However, in the following the surface modification processes, rather confidential, will not be tackled.

9.1.2. Plasma Deposition

In plasma spraying, finely divided metallic and non-metallic materials are deposited *in a molten or semi-molten state* on *a prepared substrate* [2,3]. Particles have to be in a molten state or semi-molten state at impact because the highest velocities obtained with plasma techniques are below 650 $m\,s^{-1}$ while to achieve coatings with low-temperature particles (as in the cold spray technique) velocities over 800–900 $m\,s^{-1}$ are mandatory [4]. Only materials which melting temperature is at least 300 K below its vaporization or decomposition temperature can be sprayed.

The coatings deposited by conventional plasma processes exhibit thickness ranging from 50 μm and a few millimeters. They are either sprayed or welded. In the first case, the substrate is not connected to the power supply and can be made of metal, ceramic, cermet, plastic, wood, etc. provided it is adequately cooled. In the second case, the substrate plays the role of anode and must be in metal or alloy.

Plasma torches enable the spraying of any material (ceramics, cermets, metals and alloys) while wire arc guns or PTA torches make it possible to spray mostly metals and alloys that are deposited in a molten or semi-molten state even if un-melted ceramic particles can be included within coatings.

Recently [5,6], new plasma-based techniques have been developed where solutions of precursors or suspensions of micron or submicron particles are injected in the plasma flow. They result in finely structured coatings with thickness varying between 1 and 100 μm. Such coatings fill the gap existing between vapor deposition techniques (chemical vapor deposition (CVD) assisted or not by plasma, etc.) and conventional plasma spraying.

Compared to other thermal spray techniques, the main advantages of plasma are the high temperatures (over 8000 K) allowing melting the most refractory materials and the possibility to control the plasma atmosphere: neutral, reducing, nitriding for d.c. plasmas and almost any one for RF plasmas. However, high temperatures come along with rather low plasma specific mass (one-thirtieth to one-fortieth of that of cold plasma-forming gases) resulting in a poor momentum transfer to powder particles. Moreover, d.c. plasmas are also characterized by very steep temperature and velocity gradients [3,6] and the treatment of particles will be governed by the trajectories followed by the particles in the plasma flow.

When the plasma jet issues in air, the latter is entrained within the flow where it reacts with the non-oxide molten particles that are often overheated. For instance, the oxidation of low-carbon steel can reach 15% [3]. The only way to limit or avoid particle oxidation is to use gaseous or mechanical shrouding systems to reduce the mixing of the plasma jet with the surrounding gas or spray in a controlled atmosphere chamber. On the other hand, the second solution involves an increase in the equipment price by a factor of 5–10.

In plasma spraying, the coating is made up of piled-up lamellae or splats that result from the flattening and solidification of single particles. For example, a powder feed rate of $6\,\mathrm{kg\,h^{-1}}$ injected in a d.c. plasma torch corresponds to about 10^8 particle $\mathrm{s^{-1}}$ for 30-μm alumina particles and the area covered by the particle jet on the substrate is about 2–4 cm in diameter. Therefore, the particles impact on already solidified lamellae. When the torch and substrate are fixed, coatings exhibit roughly a Gaussian shape. When the torch is moved with regard to substrate with a velocity, v_t, the formed bead has a Gaussian cross section. Its thickness depends on powder mass flow rate, deposition efficiency and v_t. Thus, the deposition of a coating with a uniform thickness requires a torch movement pattern that depends on the geometry of the part to be covered. This moving pattern must ensure an overlapping of the beads resulting in a pass and then coating with a constant thickness. Moreover, as explained later, the torch axis must be kept as orthogonal as possible to the substrate. Thus, as soon as the shape of the coated part is not flat or cylindrical, the use of a robot becomes mandatory. If necessary the sprayed part can also be moved to complete or simplify the robot movements. The problem is more complex when using an RF plasma torch as the latter is not movable and all the movements have to be achieved by moving the part to be coated.

Conventional coatings obtained by thermal plasma routes exhibit the morphology showed in Fig. 9.1. It comprises piled lamellae, voids and pores, oxides, cracks

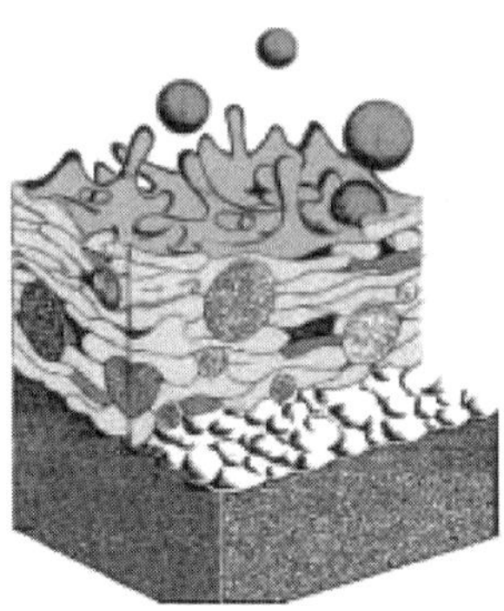

Figure 9.1: Sketch of the coating structure with splats, un-melted particles, voids, pores.

and un-melted particles and shows a hierarchy of microstructures with various length scales: nanoelements, micro-sized grains or columns contained within meso-scale splat structures and a variety of nano…, micro- and meso-scale defects including voids, micro-cracks and oriented boundaries. Coating properties are, to a great extent, linked to the real area of contact between the lamellae. In addition, according, to the way coatings are manufactured, they are strongly anisotropic.

9.2. Plasma Deposition Systems

9.2.1. General Remarks About Plasmas

The working conditions of thermal plasma devices are highly linked to the plasma thermodynamic and transport properties that depend on the plasma-forming gas composition. The mandatory condition to achieve sustainable plasma is that its electrical conductivity is higher than $10^3\,\text{mho}\,\text{m}^{-1}$ at atmospheric pressure. It is achieved for plasma gases (Ar, Ar–He, N_2–H_2, Ar–H_2, Ar–O_2, air) as soon as the plasma temperature is higher than 8000 K corresponding to a minimum enthalpy dissipated in the gas [7]. The plasma gas composition has, then, to be tailored to the momentum and heat to be imparted to the particles injected in the flow. The momentum depends on the one hand on the gas-jet velocity linked to the nozzle internal diameter (i.d.), plasma gas composition and flow rate, and, on the other hand, on the gas molar mass. Therefore, "heavy" gases as Ar, N_2 and air are often used as primary plasma-forming gas. The heat transfer to particles is essentially controlled by the gas plasma thermal conductivity, particle size and residence time in the flow. The increase in the gas thermal conductivity is ensured by the secondary plasma-forming gas [8]: hydrogen and/or helium, hydrogen being the most efficient at $4000 < T < 10{,}000\,\text{K}$. However, all these parameters are strongly intertwined and, thus, the range in which the working conditions of a plasma spray device can be varied is rather narrow and each process will have its specific applications.

9.2.2. Plasma Torches

9.2.2.1. d.c. Plasma Torches

Fig. 9.2 gives a sketch of a d.c. plasma-spraying process with a low-power (<60 kW) plasma torch. The particles are injected orthogonally to the plasma jet downstream of the arc root, thanks to a carrier gas. The molten particles impact on the substrate where they flatten and solidify and resulting splats pile up to form the coating.

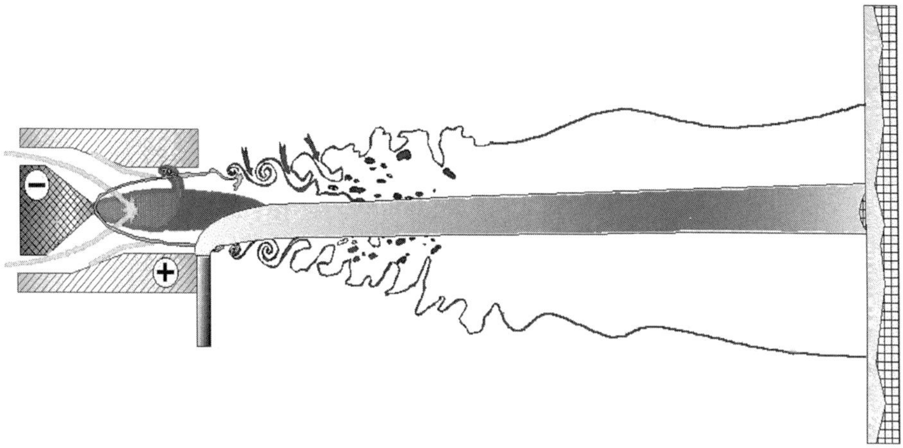

Figure 9.2: Sketch of the d.c. plasma-spraying process.

The arc strikes between a thermionic cathode and the anode. The latter has the passive function to collect the electrons and, so assure the current continuity. The stick-type cathode (8–12 mm in diameter) has a conical tip and is made of thoriated (2 wt.%) tungsten while the anode is made of high-purity oxygen free copper and can have an insert of tungsten to limit the anode wear. The arc attaches to the anode by a high-temperature, low-density gas column through the cold gas boundary layer that develops on the water-cooled anode wall under a pressure of 1.5–2 MPa. The cold gas flow in the boundary layer exerts a pulling down drag force on the hot connecting column while the Lorentz forces may act in the same or opposite direction depending on the curvature of the arc-connecting column. Under the combined actions of these forces and thermal effects, the connecting column lengthens and the voltage drop in the column increases up to a value where breakdown occurs creating a new arc root. This movement induces periodic variations in arc voltage. Such variations result in fluctuations in the enthalpy input to the gas, the plasma jet velocity, length and width and the way it mixes with the ambient gas when issuing from the torch. Three different arc fluctuation modes have been identified [6,9]: the steady mode, the takeover mode and the restrike mode. They differ in the movement of the arc root on the anode wall and, therefore, in the time-evolution of the arc voltage. In the restrike mode that corresponds to a large range of plasma parameters used for plasma spraying, the arc is stretched by the cold gas flow until an electric breakdown occurs through the colder and electrically insulating layer surrounding the arc. Each breakdown initiates a short circuit and a new arc attachment at the nozzle wall. Thus, the arc voltage exhibits large fluctuations and a high mean value. The frequency of arc root fluctuation ranges between 2000 and 8000 Hz [9] and the voltage fluctuation amplitudes can reach 50%.

The ranges of torch working conditions are summarized in Table 9.1. Gas velocities can reach $2300\,m\,s^{-1}$ with Ar–H_2 mixtures (subsonic) [10] or more than $3000–3500\,m\,s^{-1}$ with adapted anode-nozzle (de Laval profiles for supersonic velocities) [11]. The temperatures in the extinguishing plasma jet (particles are injected downstream of the arc root) are between 13,000 K at the nozzle exit on the torch axis and 8000 K in the jet fringes or at the jet core extremity but with radial gradients up to $5000\,K\,min^{-1}$. The erosion of the electrodes, controlling their mean lifetime (about 30–60 h depending on the working conditions and the number of restarts) is not negligible and has to be compensated, especially during the spraying of large parts. The high-velocity jet issuing in the steady surrounding atmosphere generates large scale eddies that entrain air bubbles 40 times denser than the plasma gas (see Fig. 9.2). They mix once the bubbles have been heated enough and the plasma jet sufficiently cooled down. In air atmosphere, oxygen dissociation cools rapidly the plasma jet. Two centimeters downstream of the nozzle exit, the jet core contains about 3–10 vol% of oxygen [12]. The modeling of the torch operation is still a challenge because of the interactions between the fluid and the electromagnetic fields and the three-dimensional transient phenomena at the anode [6]. The first model has been presented in 2003 [13].

With the interest of the automotive industry for the spraying of cylinder bores of automotive engine Al-13 Si blocks, small torches with a diameter less than 30 mm,

Table 9.1: d.c. arc spray torch main characteristics

Plasma torch type [Ref]	Conventional [3]	High power [17]	Triplex [15]
Cathode (W + 2 wt.%ThO_2)	Stick cathode	Button cathode	Stick cathode
Type and number	One	One	Three
Plasma gas injection swirl number S(−)	Axial or vortex (S < 4)	Vortex (S > 8)	Vortex (S < 4)
Anode material	OFHP copper or sintered W	OFHP copper	Segmented OFHP copper
Internal diameter (mm)	6–8	8	6–8
Plasma gas	Ar, Ar–He, Ar–H_2 Ar–He, N_2–H_2	N_2–H_2	Ar–He, Ar
Flow rate (slm)	40–100	40–200	30–60
Maximum arc current (A) Depending on plasma gas	1000, 900, 700 700, 500	500	300
Maximum voltage (V) Depending on plasma gas	30, 50, 80 90, 80	500	80–90
Maximum power (kW)	30, 45, 55 60, 40	250	20–55

S: swirl number; OFHP: oxygen free high purity.

a stick-type cathode and power level below 15 kW (300 A arc current) have been developed. They are used in production shops in Germany and Japan [14].

To limit arc root fluctuations a new torch called Triplex has recently been manufactured by Sultzer Metco [15]. It has three counter insulated cathodes supplied by independent sources distributing the energy to three parallel arcs striking at a unique anode. The voltage is increased by using an anode made up of a stack of insulated rings which the last has the function of anode. The operating conditions of this torch are given in Table 9.1.

It can also be mentioned the development of a three anodes–three cathodes plasma torch with a converging nozzle. Particles are axially injected at the point where the plasma streams converge. At that point the plasma/powder flow is accelerated through the nozzle [16]. With N_2–H_2 plasmas, the electric power level can range between 50 and 150 kW and powder flow rates can reach 30 kg. The claimed advantage is a better heating of oxide powders resulting in very dense coatings with porosity less than 1%.

At last, to cover large surfaces high-power torches have been developed [17]. They have a button-type cathode and a rather long anode-nozzle (up to 100 mm against less than 30 mm for conventional d.c. plasma torch) that can have either a convergent conical shape or exhibit a step change. They also use high gas flow rates with N_2–H_2 mixtures (see Table 9.1).

9.2.2.2. RF Plasma Torches

Tekna® is now the only manufacturer of RF spray torches. In the latter, the plasma is confined by a ceramic tube as shown in Fig. 9.3. The coil is inserted in the torch body and allows a perfect alignment and a closer distance between the coil and the discharge, thus a better coupling [18]. Spray torches generally work at 3.6 MHz with power levels up to 400 kW. Their characteristics are given in Table 9.2. Over 125 kW, high-power levels are obtained by coupling two coils at different frequencies: one at 3.6 MHz to run the starting plasma and the other at 400 kHz to increase drastically the power in the already existing plasma. Due to the inductive coupling the current only flows in a ring in the coil area. Thus, the ceramic water-cooled wall has also to be protected by a sheath gas. The area close to the torch axis is only heated by conduction–convection and this allows locating a water-cooled probe along the axis, even in the coil area. Downstream of the coil, an extinguishing plasma jet is formed as in d.c. plasma spraying. The gas velocity is roughly inversely proportional to the square of the torch i.d. and, thus, is below 100 m s^{-1} in a 60 kW torch. Such gas velocities result in long residence times of the particles in the plasma flow and allow the spraying of large particles: up to 250 μm in diameter with 400 kW torch and 150 μm with 60 kW torch. They also make it possible to use plasma gas with low thermal conductivity as argon. The latter enables an easy coupling

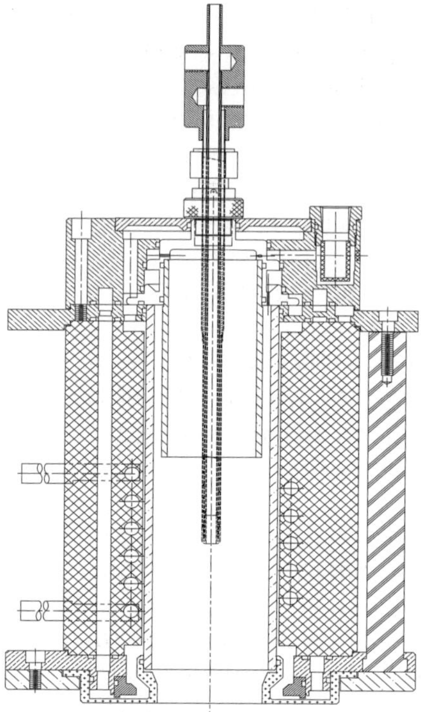

Figure 9.3: RF plasma torch PL50 of Tekna®.

Table 9.2: RF spray torch main characteristics

Torch type [Ref]	Conventional [18–20]	Supersonic [21,22]
Torch i.d. (mm)	35–50	35–50
Plasma gas nature	Ar	Ar
Flow rate (slm)	30–60	25
Sheath gas	Ar–H_2, Ar–O_2, Ar–Air, Air, O_2	Ar
Flow rate (slm)	90–150	80
Powder injection	Axial	Axial
Coil chamber pressure (kPa)	10–50	30–50
Spray chamber pressure (kPa)	Idem coil chamber pressure	5–10
Power levels (kW)	30–400	30–50
Maximum powder flow rate ($kg\,h^{-1}$)	Up to 50	2–3

at reasonable power levels [18,20]. The sheath gas close to the wall can be pure oxygen if necessary, allowing, for example, to spray materials very sensitive to oxygen losses. However, because of the tendency of the plasma to constrict at atmospheric pressure, such torches work at pressures in the 30–40 kPa range and require the use of controlled atmosphere. Plasma temperatures are between 8000 and 10,000 K with much lower gradients than in d.c. plasma jets.

Recently [21] a supersonic nozzle has been adapted to an RF torch, the pressure downstream of the nozzle being in the kPa range. Particles are heated and melted in the coil region and accelerated in the nozzle divergent where gas velocities can reach up to 1500–2500 $m\,s^{-1}$[22].

Compared to d.c. plasma torches, the modeling of RF torches is by far more developed. Most of the models assume two dimension and steady state flow. The predictions of the models are in rather good agreement with measurements [6,23] for both, plasma flow fields and particle in-flight properties.

9.2.3. Plasma Transferred Arc

In this technique, the work piece or substrate is part of the electrical circuit as the anode and it must be in metal or alloy. This process uses the high heat flux transferred to the anode and the energy utilization is drastically improved compared to non-transferred plasma torches [24]. The arc melts the substrate surface. The depth of the molten pool depends on the transferred arc current and residence time of the arc root at the same location. To limit the depth the arc has to be moved with respect to the substrate at velocities in the range of 10 $cm\,min^{-1}$. In addition, to distribute the heat more uniformly, the torch is oscillated along a distance between 5 and 15 mm at a frequency limited to 0.5–1 Hz in order to avoid the transition from arc dominated by cathode jet to arc dominated by anode jet. Fig. 9.4 gives a sketch of the process [25].

To have a better control of the transferred arc current, the pilot arc used to start the plasma is kept working during the process. It maintains ionized the gap between the torch nozzle and work piece and thus allows working with transferred arc currents below 100 A, and avoid the quenching of the arc by the powder. The pilot arc is run between the cathode and nozzle of the torch that acts a secondary anode at a lower potential than that of the work piece. As the plasma must not blow out the liquid of the molten pool, its velocity must be as low as possible. This is achieved by limiting the pilot arc current between 50 and 150 A with a nozzle i.d. of 3–4 mm and by using plasma gas flow rates between 1 and 10 slm (usually 2 slm). The transferred arc current is between 50 and 350 A and the total dissipated power is below 15 kW. The 32–48% of the total energy goes into the substrate; the heating by the pilot arc is below 20% [26]. The plasma temperature, due to the

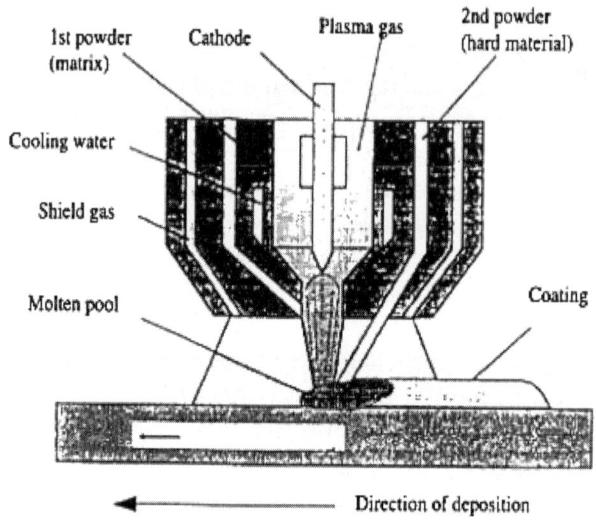

Figure 9.4: Sketch of the PTA process.

transferred arc, is higher than 11,000 K and, close to the cathode can reach 18,000 K on the torch axis. Particles are introduced forward to the jet displacement (see Fig. 9.4). They are mainly heated by the transferred arc [26] and thus land in the molten pool in a plastic state, as the arc length is less than 20 mm. They are melted in the pool by the transferred arc. In addition, oxide ceramic particles can be introduced in a solid state in the molten pool (see Fig. 9.4) to be imbedded in the welding material. A sheath gas flowing around the arc makes it possible to drastically decrease the oxidation of the high-temperature bead. This gas is generally composed of a mixture of argon with 5–15 vol% of hydrogen. Powder flow rates can go up to 300 g min^{-1} and the resulting coating is welded to the substrate.

9.2.4. *Twin-Wire Arc Spraying*

In this process (see Fig. 9.5(a)), two continuously fed wires made of ductile material serve as consumable electrodes. The d.c. arc strikes between both wires and melts the tips. A high-velocity gas flow removes the molten metal and atomizes it into large droplets that are propelled toward the substrate. The melting of the two electrodes is different. The arc attachment at the anode is rather diffuse while it is constricted at the cathode. It results in a larger heating zone at the anode inducing larger drops issued from the anode wire. Finally, a secondary atomization gas (see Fig. 9.5(b)) can be used to constrict the droplet spray jet. It is blown outside the

(a)

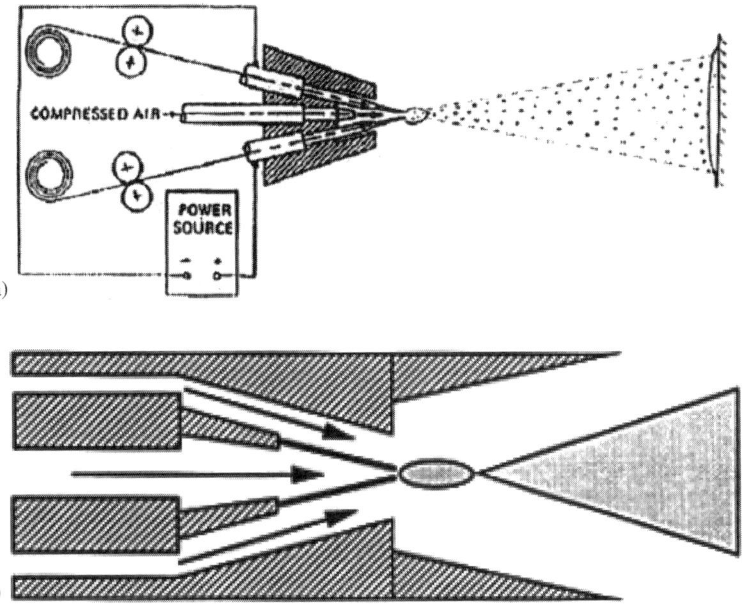

(b)

Figure 9.5: (a) Sketch of the wire-arc spraying process and (b) primary and secondary atomization gas in wire-arc spraying.

wires and generates the final droplet size distribution for a given arc current [27]. A wide variety of ductile metal wires can be utilized depending on the application used. Also cored wires containing non-ductile metal and/or ceramic particles in a ductile envelop. The metal particles are overheated in the arc and, thus, when compressed air is used as atomization gas, particle oxidation can be very high. It depends mainly on the nature and size of the droplets. For example, it can be as high as to 25 wt.% for aluminum wires.

The twin-wire arc spray technique allows high deposition rates (10–40 $kg\,h^{-1}$) with a good thermal efficiency: 60–70% of the electric power is used for material melting. The torch does not require any water cooling but uses very high atomization gas flow rate: up to 1.5 $m^3\,min^{-1}$ supplied at pressures between 0.25 and 0.6 MPa. This explains why most of the coatings are deposited with compressed air. The technique also allows a drastic limitation in the substrate and coating heating contrarily to plasma spray devices.

Typical arc currents range between 100 and 300 A, the voltage being limited to 25–35 V due to the short gap between electrodes. The difference between anode and cathode melting rates generate oscillations of 0.2–2 kHz with voltage peaks corresponding to current minima. The current setting controls the wire feed rate.

The starting of the arc does not require any of a starting unit as with plasma or PTA processes. It is initiated by bringing the two wires together until they touch. The power supply being a low voltage device (~60 V), the cost is rather low. The most recent studies [27–29] have shown the drastic influence of the fluid dynamics in wire-arc spraying.

9.3. Plasma–Particle Heat, Momentum and Mass Transfer

9.3.1. Case of a Single Particle

Many works have been devoted since the 1970s to a better understanding of the plasma–particle interactions. The problem is very complex due to the strongly non-linear properties of plasma [7,8], steep temperature gradients ($>5.10^6\,\mathrm{K\,m^{-1}}$) in the particle boundary layer, non-continuum effects (mean free paths in the plasma can be a few μm for condensed particles sometimes in the 10 μm range) [30], high evaporation rate (buffer effect of the vapor cloud surrounding a particle, particle radiation [7], particle turbulent dispersion, etc.). These effects modify the momentum and heat transfer to particles. However, it is still difficult to choose or establish the correction factors that can take into account these phenomena as it is hard to compare predictions and measurements. Indeed, the plasma flow temperature and velocity distributions can be measured as well as the diameter, surface temperature and velocity of single particle in flight (see for review [31,32]), but it is impossible to follow the trajectory of a single particle from its injection point to its impact on the substrate (injecting $10\,\mathrm{g\,s^{-1}}$ corresponds to about 10^6 particle $\mathrm{s^{-1}}$!).

The modeling of particle acceleration and heating has been first developed in two dimensions for d.c. and RF plasmas and, more recently, in three dimensions to account for the transverse injection of the particles, three-dimensional character of turbulence structures and turbulent dispersion of particles (see for review of [6]). Lately, the effect of transient phenomena (as arc root fluctuations) on the plasma field and particle behavior has been taken into account in models [33] and predictions have been successfully compared with measurements [34].

The models and especially the two dimensions and steady ones make it possible to predict the effect of a variation in operating parameters (nature and flow rate of the plasma-forming gas, arc current, particle size, etc.) on particle behavior.

Another puzzling problem is the mass transfer between the plasma and particles. It is mainly controlled by diffusion but under certain conditions, also by the convection inside the liquid particles [35,36]. When such a convection process takes place, it increases the mass transfer and possible chemical reactions (oxidation, nitridation, etc.) by almost an order of magnitude compared to the diffusion process.

9.3.2. Ensemble of Particles and Powder Injection Problem

Measurement techniques have been developed to determine the parameters (radial distribution of trajectories, velocity and temperature distribution or average values) of an ensemble of particles distributed, during a given time, along the radius of the jet at a given axial position [31,32]. The sophisticated laboratory set-ups have been transformed in sensors able to work in the harsh conditions of spray booths. At the present time, they allow to achieve on-line monitoring of the process and, in the future, on-line control [31,32,37] of the particle parameters at impact that control, to a great extent, the coating properties.

Both models and measurements have emphasized the key role of particle injection [38]. It controls particle trajectories and, therefore, acceleration and heating. The particle momentum ($m_p \cdot v_p$) must be adapted to that imparted by the plasma flow to the particle. Thus, both particle size, on which depends its mass m_p, and injection velocity v_p have to be carefully chosen. However, all powders have a size distribution. When the ratio of the particle maximum and minimum sizes is x, the mass ratio is x^3 (for $x = 2$, which is a minimum in commercially available powders, $x^3 = 8$). The injection velocity vector is determined by the carrier gas flow rate and the injector i.d., thus v_p can only be adapted to the mean particle size. Moreover, particles in the injector tube (usually with an i.d., between 1 and 2 mm) undergo collisions with the injector wall and also with other particles when the powder loading is high. The number of collisions increases when the particle diameter is below 20 μm and powder specific mass less than 6000–8000 kg m^{-3}, increasing the divergence of the particle jet at the injector exit. Presently the modeling of particle injection in the jet has not yet found a sound solution [6].

The deposition efficiency is around 50% for d.c. and 70% for RF (axial injection) plasma spraying. If the in-flight evaporation and the non-sticking of the un-molten particles onto the substrate take part in the decrease of deposition efficiency, it has been found that the problems at injection play the major role. Some particles do not penetrate in the plasma jet and land on the floor or the spray booth or they travel in the jet fringes and are still in a solid or partially solid state when impacting on the substrate where they rebound. It also explains the success of on-line monitoring of the particle mean trajectory and trajectory distribution to optimize the former and adapt the carrier gas or the plasma power level when a change in these parameters is observed. Such a variation is mainly due to electrode erosion for d.c. plasmas torches.

The problem is slightly different for wire-arc spraying where the high momentum of the atomizing gas induces the formation at the molten extremity of both wires of sheets of metal [27]. The eddy structure in the flow leads to sheet disintegration in a shower of drops with varying trajectories. The size and trajectory distributions of the drops strongly depend on the atomization and secondary flows and, internal nozzle geometry.

With the PTA process the injection problem is less critical. The plasma momentum density (ρv) is low as the flow velocity, v, is below 200 m s^{-1} and the gas specific mass, ρ, about 1/40, 1/50 that of the cold gas. In addition, the main treatment of the particles occurs in the molten pool via the transferred arc.

9.3.3. Finely Structured Coatings or Glasses

With the increasing interest for finely or nano-structured coatings different plasma spraying solutions have been searched:

(a) Spraying of agglomerates of nano-particles (made by high-energy ball milling) with conventional size distributions (e.g. 22–45 μm). The trick consists in either melting only the external shell of the particles [39,40] or spraying a particle made of materials with different melting points such as TiO_2 and Al_2O_3 [41]. Nano-particle agglomerates are trapped within the molten material and coatings have a bi-modal distribution of properties. However, the molten part of particles must be sufficient to bestow a certain amount of adhesion and cohesion on coatings.

(b) Spraying solutions of different types of liquid precursors: nitrates, isopropoxides, butoxides, etc. dissolved in various solvents [42,43]. The solutions are atomized with a gas, resulting in a wide range of particle diameters (20–50 μm) and injection velocities (up to 100 m s^{-1}). Precursors in various stages of heating reach the substrate surface and contribute to form different features to the microstructure. The coating exhibits powder structures, splats with an ultra-fine morphology, micro- and nano-meter sized interconnected porosity. It has been also observed that zirconia coatings show vertical cracks that go across the coating thickness.

(c) Spraying of suspensions. The process is based on the plasma-spraying of suspension of micron or submicron-size particles (between 0.1 and 3 μm) [44–47]. Suspension is the only way to inject such fine particles which would require carrier gas flow rates as high as 50–80 slm that would perturb drastically the plasma jet [38]. The gas atomization of suspensions has been used to spray with an RF plasma torch cobalt spinel [44] or zirconia [45]. In the latter case, a supersonic nozzle is used to achieve very dense coatings. In RF spraying with gas atomization of solution droplets, in the tens of micrometer range, the solvent is progressively vaporized and then the solid particle content is melted, producing a conventional coating when impacting on the substrate. With d.c. plasma jets rather large drops (100–300 μm) are mechanically produced by ultrasonic [48] or high-pressure injection [5,46,47] and fragmented by the plasma jet in micro-size droplets. Fragmentation requiring times in the order of μs occurs far before vaporization (one to two orders of magnitude faster for droplets bigger than 3 μs). That is why large drops with a sufficient momentum can penetrate within the plasma jet. After 10–15 mm trajectory, the droplets are completely vaporized and the solid particles contained in the droplets are accelerated and melted, as in conventional spraying but with particle below 3 μm in diameter.

(d) Metallic glasses. An alternative to produce fine-structured coatings is a two-step approach that combines the deposition of amorphous coatings and subsequent heating of these coatings above the crystallization temperature to develop nano-phase structures. Indeed, plasma and amorphous metal production and commercial alloys were developed in the 1980s to produce coatings with a high content of amorphous phases. These alloys exhibit a high glass forming ability that is favored by a number of elements higher than three with an atomic size ratio higher than 12% and a negative heat of mixing between elements. The resulting metallic glass has a low density of defects such as tiny cracks or holes [49].

9.4. Powder Spheroidization

Generally speaking, the pressing need for an industrial process to turn to spherical particles, is to seek at least one of the following benefits [50]:

- Improve the flow ability for a homogeneous flow without clogging of the feeding mechanism.
- Increase the powder packing density to produce dense coatings or parts.
- Eliminate particle internal cavities and fractures.
- Change the surface morphology of particles: smoother macroscopic surface causing less wearing and contamination of the feeding pipes.
- Enhance the powder purity through the selective reaction/vaporization of specific impurities.

The main advantage of using RF conventional plasmas (see Table 9.2) is the longer residence time of the particles that is about two orders of magnitude higher than in d.c. plasma flows. It allows to achieve a smooth transfer with argon and thus limit the heat propagation phenomenon inside particles. It, also, enables to control the sheath gas composition and limit or favor chemical reaction as oxidation.

The process developed by Tekna® is schemed in Fig. 9.6. Basically it consists in the in-flight heating and melting of thc material fed in the plasma flow in the form

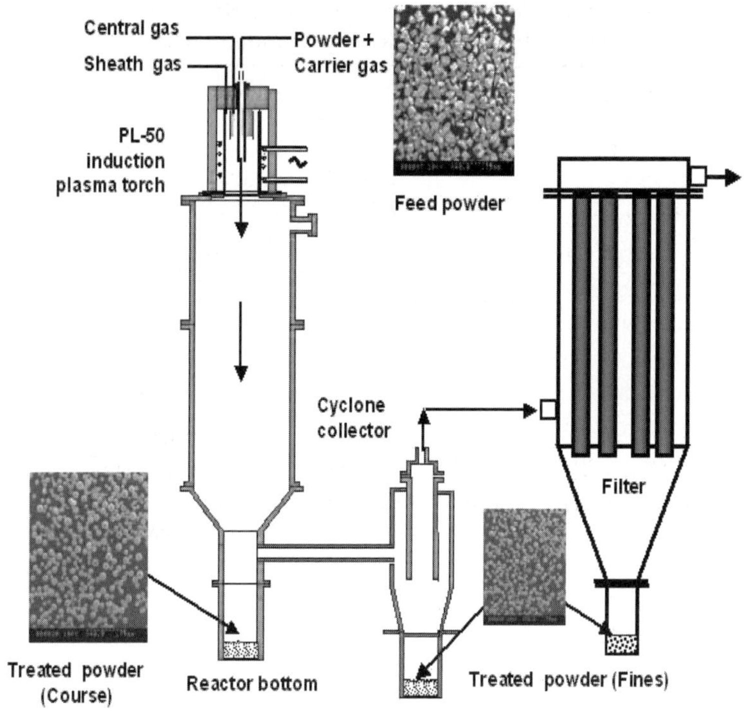

Figure 9.6: Schematic of the process of spheroidization.

of sintered, agglomerated and calcinated powders or fused and crushed powders. The formed molten spherical droplets are gradually cooled under free fall conditions. Depending on the size and apparent density of the treated powder, their time of flight is controlled so that the molten droplets have time for complete solidification before reaching the bottom of the primary reactor chamber. The finer particles, entrained by the plasma gases are recovered in a filter downstream. To decrease the costs, the exhaust gases and cooling water are recycled and the process is highly automatized.

The heating of the particles depends on the proper balance between conductive and convective heat transfer from the plasma to particles and radiative losses from the particle surface and vapor cloud surrounding the individual particles. The radiative losses increase drastically with particle temperature and size and, thus, the heating and melting of refractory metal or ceramic particles become more difficult for high melting temperature materials and large particle size. The typical powder materials spheroidized in industrial scale are presented in Table 9.3. The costs of the powder treatment vary in a ratio of 10 depending on the nature of the

Table 9.3: Typical powder materials spheroidized in industrial scale by Tekna® integrated system

Powder category	Powder name
Ceramics	
Oxides	SiO_2, ZrO_2, YSZ, Al_2TiO_5, glass
Non-oxides	WC, WC-Co, CaF_2, TiN
Pure metals	Re, Ta, Mo, W
Alloys	Cr/Fe/C, Re/Mo, Re/W

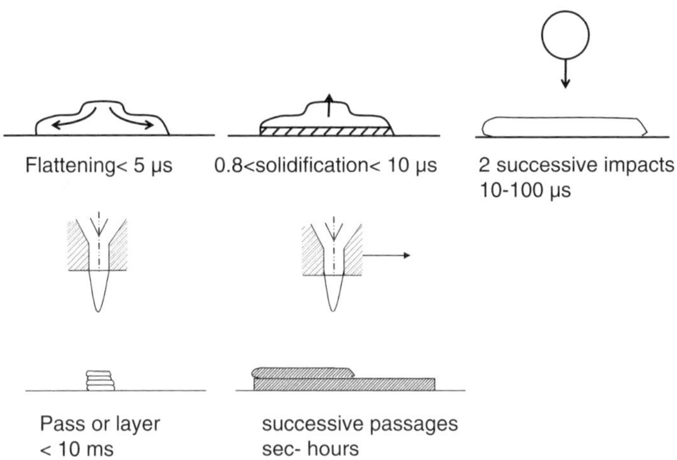

Figure 9.7: Time ranges in coating formation.

material, size of the particles (up to 250 μm), desired degree of spheroidization and batch size.

9.5. Coating Formation

In plasma and wire-arc spraying, three successive steps condition the coating formation: the individual particle flattening and solidification process to form a splat on the substrate or previously deposited layers, the splats layering forming a pass and the layering of the successive passes. All these steps are characterized by a very wide range of characteristics times for particles impacting in a molten or semi-molten state at velocities between 20 and 600 m s^{-1}. The characteristics times are presented in Fig. 9.7 for a flat substrate [6,51].

9.5.1. Splat Formation

9.5.1.1. Conventional Spraying

The diameters of the splats D_p range between 10 and a few hundreds of μm depending on the particle original diameter d_p and parameters at impact. The flattening degree, defined as (D_p/d_p) varies between 1.5 and 6. It can be, in a first approximation, written as:

$$\frac{D_p}{d_p} = K \times \left(\frac{d_p \rho_p v_p}{\mu_p}\right)^{0.2} \tag{9.1}$$

where ρ_p, μ_p and v_p are the specific mass, viscosity and impact velocity of the particle, respectively. The quantity in bracket is the Reynolds number of the particle at impact. The splat thickness, linked to D_p, varies between 0.8 and a few μm.

Fig. 9.8 summarizes the parameters that control the splat formation. They are two folds as follows:

- The parameters related to the particles: T_p, v_p, d_p, impact angle and oxidation stage which play a drastic role on material properties. It also affects the material viscosity that varies in accordance with temperature as $\mu = \mu_o \exp(E/kT)$.
- The parameters related to the substrate: temperature T_s, roughness, thickness and composition of the oxide layer at its surface (most substrates are metals and alloys) which in turn depends on T_s, the surrounding atmosphere as well as the rate and time of preheating.

Due to the shortness of the impact and solidification times (see Fig. 9.7), the studies are very complex [51] even if very sophisticated measuring and imaging devices have been recently developed [52,53]. The influence of the particle parameters at impact is now rather well understood and their effect can be described through dimensionless numbers: Reynolds, Weber, Peclet, etc. [4,51].

A systematic study (see the review [51]) of the effect of the impact angle θ (see Fig. 9.8) both on the splat and coating adhesion has shown that θ values higher than 30° induce, in most cases, a drastic decrease of the splat and coating adhesion. However, still many works are mandatory to understand what happens at the interface between splat and substrate. Anyhow the main tendency which has been emphasized [51] is that the substrate has to be preheated over a temperature called transition temperature, T_t (e.g. for zirconia on stainless steel $T_t \sim 500$ K). There is still a controversy about the physical meaning of T_t: temperature at which adsorbates and condensates are desorbed and/or above which the molten material wettability is modified, etc.

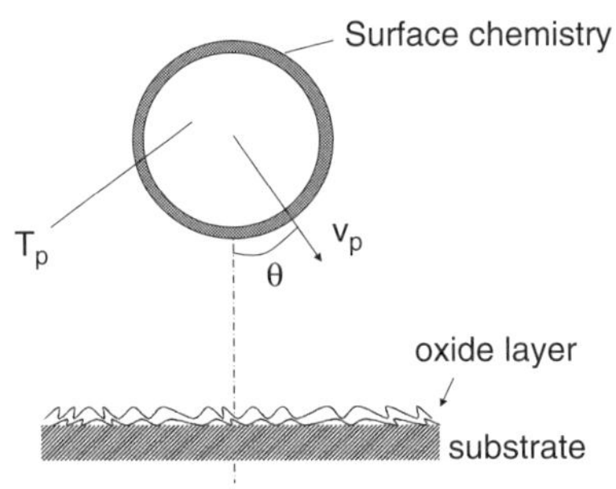

Figure 9.8: Parameters controlling splat formation.

However, for smooth substrates (Ra $< 0.05\ \mu m$) below T_t splats are extensively fingered and over it, they are disk shaped and for rough substrates preheated over T_t the coating adhesion is promoted by a factor 2.5. Substrates have to be roughened because in most cases the adhesion is of mechanical type with the shrinking of the splats upon cooling around the peaks of the roughened substrate. Of course, the size of the peaks has to be adapted to that of the splats: $D_p \sim 2 - 3 \times R_t$ (R_t is the distance between the highest peak and the deepest undercut). In addition, the surface roughening is generally achieved by grit blasting and the grit residues must be eliminated and the substrate cleaned of greases, adsorbates, condensates, etc. Other mechanisms may help the mechanical adhesion of the coating such as the melting of the substrate oxide layer when it is thin enough (in the 10 nm range) or the crystalline structure accommodation through the oxide layer, etc. [51]. The composition and thickness of the oxide layer at the substrate surface is a key parameter of the adhesion. As it depends on the preheating temperature over T_t, the way this preheating is achieved is very important.

The other adhesion mechanisms are the diffusion and chemical reaction. The former can occur only at high temperature ($>0.6T_m$, T_m being the substrate melting temperature) thus under soft vacuum and once the oxide layer has been removed. This is the case for superalloys that are plasma sprayed under vacuum. The substrate is previously cleaned by a reverse polarity transferred arc [3]. Chemical reactions can occur when Molybdenum or Niobium particles impact on steel or aluminum alloys. At last, upon cooling each splat contracts and, if its adhesion to the substrate is good, it creates a tensile quenching stress within the splat [54,55]. This quenching stress σ_q is relaxed by micro-cracking (mainly for ceramics) through splat thickness, interface sliding, creep, edge relaxation [54]. Depending on the sprayed material the final quenching stress is in the range 10 MPa (for ceramics) to 400 MPa (superalloys). As σ_q of each splat is additive when layering, the total quenching stress can reach very high values limiting the thickness of the coating to a few hundreds of μm with superalloys. The substrate and coating have to be treated at a temperature in the order of $0.4T_m$ when T_m is the substrate melting temperature, in a controlled atmosphere chamber, before other layers can be added.

9.5.1.2. Suspension and Solution Spraying

In this case, no measurements have been performed yet on individual particles during their flattening. The splat diameters are between 0.2 and 5 μm corresponding to flattening degrees below 2.5 and their thickness ranges between 40 and 200 nm. With such dimensions, no crack due to quenching stress relaxation can be observed in ceramic splats.

9.5.2. Pass Formation

When moving the torch relatively to the substrate in one direction at velocity, $v_{t,}$ a bead is formed (see Fig. 9.9(a)). Its thickness, h, depends on the powder flow rate, deposition efficiency and v_t. The highest is v_t the thinnest is the bead. The bead overlapping is defined (see Fig. 9.9(b)) by choosing a spray pattern to move the torch. After a few bead overlapping the thickness is constant and the pass is formed.

9.5.2.1. Conventional Coatings

The thermal history of the splat layering to form a pass is very complex [55] according to the time scales varying between 10^{-7} to about 10^{-2} s. Anyhow at the end of the splats layering to achieve the pass thickness h_p a mean temperature of the pass is reached. It depends on h_p linked to the bead thickness and overlapping, and heat flux imposed by the plasma flow (below 3 MW m^{-2} at spray distances of 100–120 mm). When the pass cools down, an expansion mismatch between the pass and substrate exists and the corresponding stress can be compressive or tensile

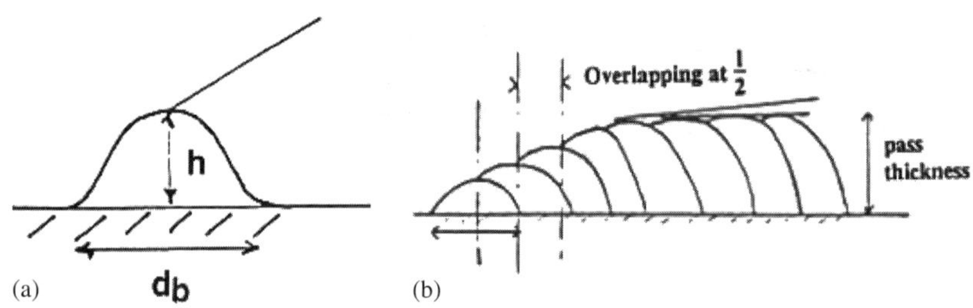

Figure 9.9: (a) Bead generated by moving the torch/substrate in one direction and (b) beads overlapping for a given spray pattern.

[54]. Moreover, if temperature gradients exist within the pass another stress is created. Thus the control of the pass temperature during spraying is a key issue for coating quality [54,55]. This is achieved by a careful choice of the spray pattern and torch velocity v_t and by the cooling by air jets of the coating surface. Moreover, the temperature of the surface has to be continuously monitored by Infra Red pyrometery. It is obvious, especially for materials with a low thermal conductivity that the pass temperature control is easier when the pass thickness is low. Beside, the pass temperature must not be too high to limit the sticking at the pass surface of submicron particles resulting from the recondensation of material that has been evaporated during particle treatment in the gas flow. The sticking of these very fine particles that create defects between two successive passes increases with the surface temperature.

9.5.2.2. Suspension or Solution Coatings

In this deposition processes, the spray distances can be as short as 40 mm and the heat flux from the plasma as high as 20 MW m^{-2} (about 10 times that in conventional plasma spraying). Depending on the substrate thermal properties, it can result in passes kept in a molten state under the plasma jet yielding to fully dense coatings with either columnar or granular structure [46,47].

9.5.3. Coating Formation

The coating formation results from pass layering. Once the coating is achieved, it cools down as it has been kept at a temperature higher than the transition temperature and as constant as possible. This cooling down induces the expansion mismatch stress. The residual stress distribution within the coating, on which its service conditions depend, is, thus, function of the expansion mismatch and quenching stresses as well as the compressive stress induced close to the substrate surface by the grit blasting [54] process that is used to roughen the substrate surface before deposition. The typical thickness of coatings ranges between 50 and 500 μm but thickness of a few mm or more can be achieved for self-standing or free forming bodies. Fig. 9.10 shows typical cross sections of plasma-sprayed metal and ceramic coatings. Fig. 9.10(a) represents a stainless steel coating plasma sprayed in air. The black areas correspond to oxides and pores. Also, un-melted particles can be seen within the lamellar structure. Fig. 9.10(b) corresponds to an alumina coating. The structure magnification in Fig. 9.10(c) shows the lamellar structure with the columnar structure within splats. At last, Fig. 9.10(d) represents wire-arc sprayed steel with 13 wt.% Cr. The denser structure is similar to that of Fig. 9.10(a).

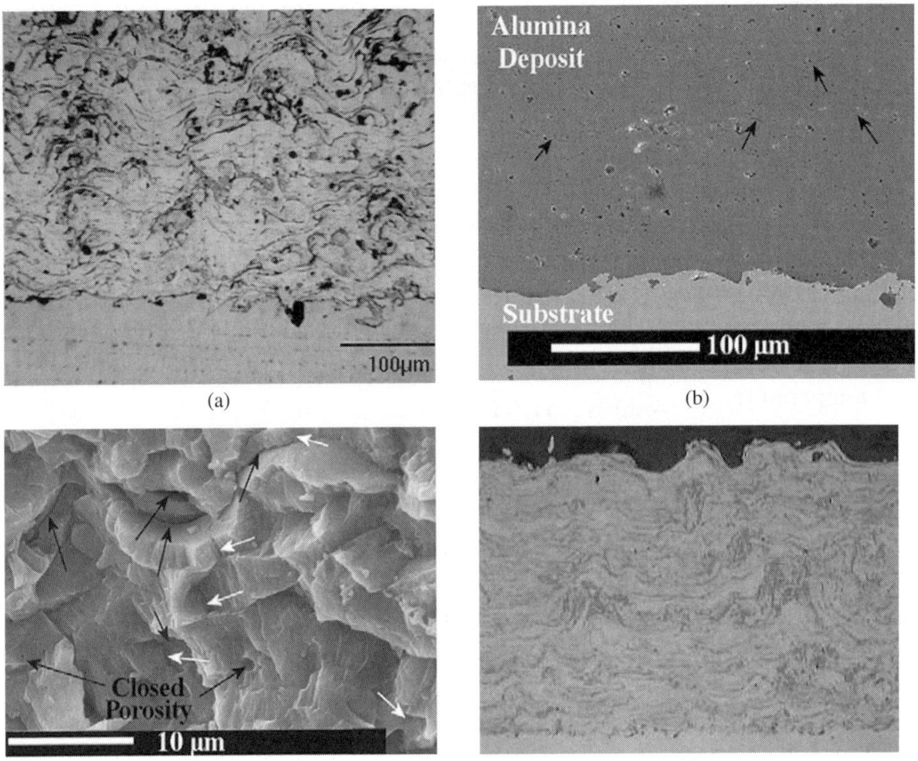

(a) (b) (c) (d)

Figure 9.10: Typical cross sections of coatings (a) stainless steel 316L d.c. plasma sprayed in air, (b) alumina d.c. plasma sprayed in air, (c) detail of (b) and (d) 13 wt.% Cr steel sprayed by wire arc.

9.5.4. Plasma Transferred Arc

As already underlined in Section 3.2, in PTA the particle melting is achieved at the surface of the coating or pass under construction by the transferred arc. Thus, the whole pass is in a molten state and cools down, after the torch passing over, by heat exchange with the substrate. A metallurgical compatibility is needed between coating and substrate as the pass is welded to it. A very important point is the dilution between the substrate material and that of the coating. Dilution results from the mixing of the deposited material and molten part of the substrate. It is defined as the ratio $B/(A + B)$ where B is the quantity of alloy formed between coating and substrate while A is the quantity of deposited material. Typical values are between 5% and 15%. To achieve such values, the molten pool created at the substrate surface must not be too deep, which can be obtained by reducing the transferred arc

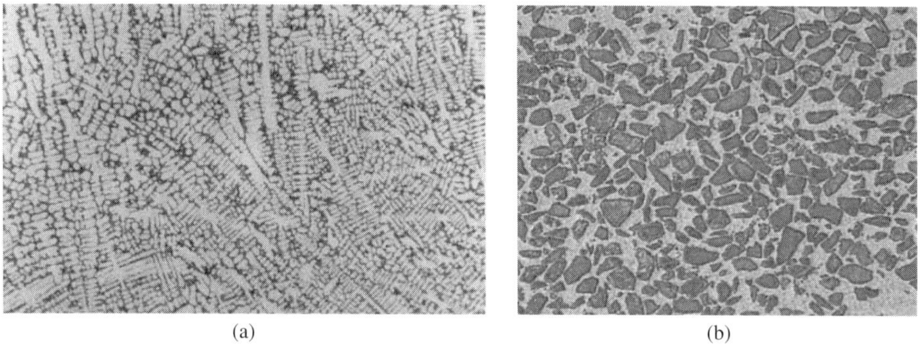

Figure 9.11: PTA deposit (a) Stellite® grade (cobalt-base alloy) and (b) Ni-based alloy with WC particles for excavation tooth.

current, moving the torch at higher velocity and increasing the powder flow rate. It means that coatings exhibit less dilution when the number of passes increases.

The total coating thickness can reach 6–10 mm and its microstructure is that of a metal or alloy with dendrites as illustrated in Fig. 9.11(a). The injection of carbide or oxide particles (see Fig. 9.4) improves significantly the wear resistance of coatings provided they are uniformly distributed as shown in Fig. 9.11(b). The metal evaporation during spraying results in the formation of submicron-size particles, usually oxides. They deposit on the torch and fringes of beads and coatings which is a real problem for large area applications. The vaporization can be minimized by injecting the particles in the fringes of the transferred arc, using large particles ($d_p > 60\,\mu m$), increasing the $Ar–H_2$ shield gas flow and the torch-substrate distance to avoid the deposition on the torch. Actually the weak point of PTA is the reclamation of aluminum alloys with high diffusivity which allows to evacuate rapidly through the substrate the heat transfer from the transferred arc up to a point where a rather thick area of the substrate melts and flows due to the low viscosity of the liquid alloy.

9.5.5. Auxiliary Equipment

Beside the spray torch, its power source, cooling water system and gas feeders, thermal spraying in general requires:

- A grit blasting system, if possible automated.
- A system to move the torch from a very simple one: rotation of the part and translation of the torch to a sophisticated six axes robot.

- A spray booth for the protection against noise, radiation, dust (particles below a few μm) and untreated particles. Aspiration systems up to 50,000 $m^3 h^{-1}$ with air velocities up to 150 $m\,min^{-1}$ and filters are required by the legislation.
- A powder feeder that can include two bowls to use two powders, a powder pre-heating system and a regulation system by continuous weighing.

9.6. Coating Properties and Uses

In general, coatings of high-performance materials – metals, alloys, ceramics, or cermets – are applied by plasma routes to relatively easy to work and more economical materials. Wire arc spray allows the spraying of rather low-cost materials such as aluminum or zinc for corrosion protection on large surface (bridge or battle ship). The base material/coating combination can be tailored to provide resistance to heat, wear, erosion and/or corrosion, as well as unique sets of surface characteristics. Coatings are also used to restore worn or poorly machined parts to original dimensions and specifications. Historically a surface technology, plasma spraying has moved from coating to materials processing technology [3]. In spite of competing technologies, thermal spray process sales are regularly increasing since 1990 especially those working in air atmosphere (the majority of them). It seems that, both, new "multi-materials" or "surface engineering" features are the driving forces. This development has also been boosted by the development of the on-line control of the process that has allowed diminishing by a factor five the number of rejected parts. This result has pushed the equipment manufacturers to "open" their control panel in order to connect it to the available industrial sensors, which is an important step for a real on-line control. At last it must be underlined that spraying techniques are relatively harmless to the environment.

9.6.1. Industrial Requests

The main industrial request for coatings is in the fields of:

(a) Wear:
   - Abrasive which is the most important (about 58%) and is due to the movement of hard particles along the surface. Coatings must be harder than the abrasive particles and cermets with hard particles are well adapted.
   - Erosion (about 10% with cavitation) due to the impact of particles transported by a gas or a liquid and strongly depending on the particle impact angle. Tough and hard coatings are requested.

- Cavitation due to the velocity difference between the solid and liquid requiring coatings with a high toughness and a good resistance to crack propagation, for example, cobalt-base alloys or amorphous materials.
- Adhesion/friction (about 17%). Dry friction can be avoided by using lubricants. The local heating can be limited especially in dry friction by using materials with a low friction coefficient (e.g. Sn, Pb, Cu alloys, chromium oxides, coatings with solid lubricants such as MoS_2, CaFe, C, etc.) sprayed on high thermal conductivity bulk material.
- Fatigue and fretting (10%) corresponding to very low relative movements and high pressures or cyclic movements with plastic deformation. Again tough and hard coatings are required.
- Corrosion–oxidation (about 5%) either in air, or water or in chemical products or at high temperature. The coatings have to be adapted to the environment, to be as dense as possible and achieve anodic protection with no discontinuity.

(b) Heat with thermal barriers (diesel engines, gas turbines) made of low-conductivity material: zirconia partially or totally stabilized depending on the working temperature. Materials such as W, Mo, Nb, Ta, Al_2O_3, Al_2O_3–TiO_2 can also be used but refractory metals have a poor resistance to oxidation.

(c) Insulation to high voltage is achieved with dense alumina coatings, the electrostatic protection with copper, magnetic coatings with ferrites, oxides, carbides, Sm/Co, FeNdB, oxygen sensors with zirconia, etc.

(d) Implants for human bodies. Of course bio-compatible materials are requested such as hydroxyapatite, alumina, TiAlV.

(e) Free standing spray formed components made of ceramics, metals (from Ni- or Fe-based alloys to refractory metals) or cermets. Often their engineering properties are similar or superior to conventionally produced materials.

(f) Decoration mainly with wire-arc spraying by using masks.

(g) Replacement of hard chromium.

The interested reader can find many details in [56–116].

It has, however, to be kept in mind that coatings have usually to resist to different effects. For example, thermal barriers must achieve a thermal protection but also a good resistance to oxidation and they usually "die" of the thermally grown oxide at the surface of the bond coat.

9.6.2. Techniques Adapted to These Requests

(a) *Plasma spraying (air plasma spraying (APS) or vacuum plasma spraying (VPS))*
Coatings exhibit a rather low porosity (from 20% to 3%) and a good adhesion value (up to 60 MPa in APS and a few hundreds of MPa in VPS of metals and

alloys). APS results, in general, in a non-negligible oxidation (up to 13 wt.% of oxide with low-carbon steel and 3 wt.% with stainless steel). The sprayed materials are numerous. However, metal oxides are probably the most important in APS while carbides, nitrides, borides, etc. can only be sprayed in controlled atmosphere. Almost any substrate can be used provided the cooling system is adapted. The deposition efficiency has a medium value of 50% and spraying has to be performed in spray booths (noise up to 130 dB, presence of dusts and gases). They are mainly used for:

- Resistance against abrasion but at low load: oxides (Al_2O_3, TiO_2, Al_2O_3–TiO_2, Cr_2O_3, etc.), zirconia, zircon ($ZrSiO_4$) up to 900°C, carbides, nitrides, borides (very specific applications and they can be sprayed only in VPS or controlled atmosphere).
- Resistance against adhesion under low loads (friction): metal carbides with a Co, Ni, Cr, etc. matrix sprayed in APS to decompose partially the carbide and form free carbon, Mo and its alloys, copper-base alloys, Cr_2O_3 with or without TiO_2, SiO_2, etc., $Al_2O_3 + TiO_2$, Ca–Ba fluorides (900°C), cobalt-base alloys (Stellite®).
- Their resistance to high-temperature corrosion, tantalum, niobium, tungsten, titanium (all sprayed in VPS) aluminum, nickel-base alloys (inconel, NiCr) in APS, superalloys mostly in VPS, stainless steels (316L, 304L, etc.), intermetallic (NiAl) cobalt-base alloys (Stellite®), metal oxides, sprayed in APS.
- Thermal barriers mainly made of zirconia partially or totally stabilized: for diesel engines stabilized with CaO or MgO, for high temperature (>900°C) stabilized with yttria, ceria, disporia, etc.
- Dielectric barriers, essentially alumina (highest dielectric properties), alumina–magnesia, partially stabilized zirconia for its ionic conduction (exhaust pipe sensors).
- Medical applications (implants), hydroxyapatite, titanium alloys (TA6V).
- To produce abradable coatings which wear preferentially upon contact with the mating part to establish automatically the clearance. Different matrices: aluminum, cobalt, copper, nickel base are used with different fillers: plastic, graphite, bentonite, boron nitride, etc. the choice being strongly linked to the service temperature.

(b) *Plasma Transferred Arc*

Coatings exhibit a very low porosity (<1%) and a metallurgical adhesion. The deposited material must be either a metal or an alloy and the metallic substrate must sustain high temperatures and thus, aluminum alloys are excluded. The deposition efficiency is excellent (up to 95%) and coatings have a rather low oxide content (below 3 wt.%). The deposition requires a protection against radiation (UV). They are mainly used:

- Against abrasion under low or high loads (erosion): stainless steel with 13–28 wt.% Cr, high-alloyed steels (Mg > 6 wt.%), self-fluxing alloys: Cr or Ni base with or without carbide particles, cobalt-base alloys (Stellite®), steels for tools.

- Against adhesion under low and high loads (friction): alloys with molybdenum, cobalt-base alloys (Stellite®).
- Against corrosion at high temperatures: nickel-base alloys (inconel, NiCr), stainless steels (316L, 304L, etc.), cobalt-base alloys (Stellite®), alloys with a high percentage of molybdenum.
- For electrical applications: copper (for electrical contact), aluminum, tin, NiAl (for resistors).
- For electro-magnetic shielding: copper, aluminum, zinc, zinc-tin.

(c) *Wire arc*
Coatings exhibit a low porosity (below 2–3%). Ductile metals or alloys can be used and with cored wires non-ductile alloys can be deposited as well as cermets. Spraying is mostly performed on metals or alloys but also on concrete. The deposition efficiency is good and the amortizing of the equipment is low. They are mainly used against:
- Abrasion (under low load): stainless steel (13–28 wt.% Cr) – high-carbon steel ($C > 0.8\%$) – high allied steels ($Mg > 6$ wt.%) – amorphous cored wires with iron bases, cast irons with Cr.
- Adhesion (under low load): molybdenum pure or alloyed, brass, tin or lead-base alloys, intermetallic NiAl.
- Atmospheric corrosion: tin, Al, Zn-Al, Zn-Mg, Zn; Ni-base alloys (inconel, NiCr), stainless steels (316L, 304L), intermetallic (NiAl).

9.6.3. Users and Economics

Fig. 9.12 from the survey of Magetex [1] summarizes the distribution of thermal coating activities in 2001 across the end-users in industry in Europe. Compared to the 1980s where the dominant activities were in aircraft and nuclear industries, thermal spray has spread widely in industries with much less added values and large production such as automotive industry. The development of coatings is, of course, strongly linked to their cost and the parameters that affect the coating cost are the following:

- The sprayed material (powder or wire). It is probably one of the most important elements that can represent more than 50% of the total cost. That is why any gain in process deposition efficiency is very important.
- The part and its surface preparations: about 5% of the cost.
- The gas and energy: about 6–8%.
- The capital cost which varies drastically depending on the technique used (see Section 9.1). For example, in APS, it represents 11–15%.

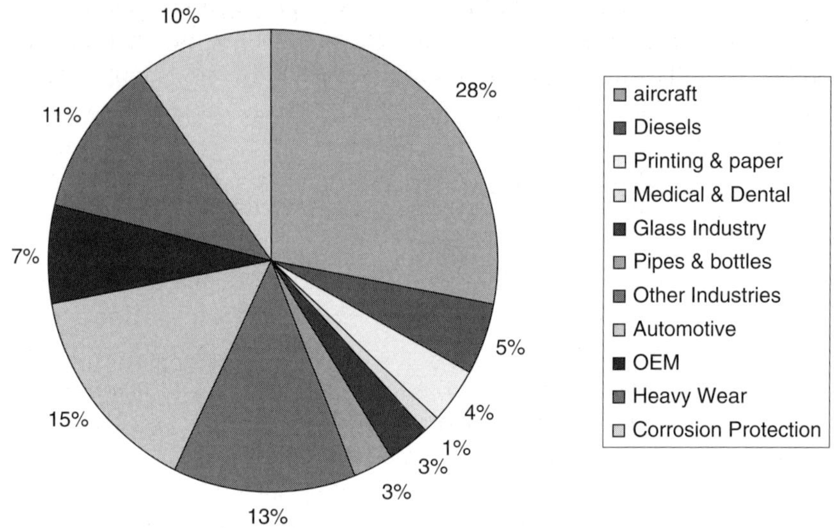

Figure 9.12: Repartition of the €61M coating activities in 2001 across end-user sectors in Europe (source Magetex study) [1].

- The labor cost which also depends strongly on the degree of automation and can range between 20% and 30%.
- The coating machining which depends strongly on the coating material and the means used: turning, drilling, grinding. It can represent 10–11% of the cost with ceramics or cermets.

For example, when spraying the same coating of NiAl by APS and wire arc with the same substrate preparation and no finishing, according to Magetex the price is 605€ for APS and 369€ for wire arc. In the total cost, the amortizing of APS is 191€ while that of wire arc is only 36.5€.

9.7. Conclusions

Compared to the 1980s, thermal spraying in general and plasma spraying in particular have dramatically evolved from an art to a science. It is now possible to achieve coatings tailored to specific service conditions and thermal spraying has become an integral part of the industrial commodity making process. The development of process on-line monitoring set-ups has greatly improved the coating reliability and reproducibility. Fundamental work linking coating properties, and especially service properties to the coating formation parameters has only started since the beginning of the new millennium.

New powders formed by mecano-fusion, self-propagating high-temperature synthesis (SHS), and fiber reinforcement are increasingly used and are opening new possibilities. The development of cored wires is also very promising in wire arc. The new techniques of suspension and solution plasma spraying open very interesting possibilities. Even if they are yet still in the research stage as they cover the gap between CVD and spraying and allow achieving finely structured coatings, their development will probably be rather fast because they use conventional thermal spray equipment.

Plasma spraying makes it possible to produce parts and manufacture forms and volumes that are difficult or even impossible to obtain using conventional processes. It also makes it possible to manufacture parts with complex shapes and material gradients, with associated protection against wear, heat and other advances challenging operating conditions. A promising possibility is the replacement of the hard chromium that actually represents a market of €400M in Europe. However, for example, on a jack the price of a hard chromium coating 25 μm thick is about 18€ m^{-2} while that of Al_2O_3–TiO_2 plasma-sprayed coating reaches 45€ m^{-2}. It means that plasma coating will become really competitive only when the environmental regulations will be reinforced, however such coatings have already been authorized in aeronautic industry.

New developments for producing metal glasses and then transform them in nanostructure are now used in industry. At last plasma spraying has demonstrated its ability to produce in a row the complete Solid Oxy Fuel Cell (connector, cathode, electrolyte and anode) with a very good energy conversion. The industrialization of the process is in progress. Moreover, interesting developments are foreseen with suspension plasma spraying with which dense and thin (about 20 μm) electrolytes can be achieved.

Thus, to summarize plasma spraying and deposition technologies are now part of industrial processes and new coatings or techniques regularly show up explaining the continuous development of these techniques.

References

[1] M. Ducos and J.P. Durand, in *Thermal Spray: New Surfaces for a New Millennium*, Eds. C.C. Berndt, K.A. Khor and E.F. Lugsheider, ASM International, Materials Park, OH, USA, 2001, pp. 1267–1271.

[2] L. Pawlowski, *The Science and Engineering of Thermal Spray Coatings*, John Wiley and Sons, New York, NY, 1995.

[3] P. Fauchais, A. Vardelle and B. Dussoubs, Invited review, *J. Therm. Spray Technol.*, 10 (1) (2001) 44–66.

[4] S.Q. Armster, J.P. Delplanque, M. Rein and E.J. Lavernia, *Int. Mater. Rev.*, 7 (2002) 265–301.

[5] P. Fauchais, V. Rat, C. Delbos, J.-F. Coudert, T. Chartier and L. Bianchi, Understanding of suspension d.c. plasma spraying of finely structured coatings for SOFCs, *IEEE Trans. Plasma Sci.*, April (2005).
[6] P. Fauchais, Topical review, *J. Phys. D Appl. Phys.*, 37 (2004) R1–R23.
[7] M. Boulos, P. Fauchais and E. Pfender, *Thermal Plasmas*, Vol. 1, Plenum Press, New York, 1994.
[8] P. Fauchais, M.-F. Elchinger and J. Aubreton, *J. High Temp. Mater. Process.*, 4 (2000) 21–42.
[9] J. Heberlein, *J. High Temp. Mater. Process.*, 6 (3) 321–328.
[10] P. Fauchais, J.-F. Coudert and M. Vardelle, *J. High Temp. Mater. Process.*, 6 (2) (2002) 247–267.
[11] Praxair Surface Technology. Brochure SG 100(11/01/1998) Praxair Surface Technologies, Inc. Thermal Spray Product, N671 Communication Drive, Appleton, WI 59916, USA.
[12] O. Lagnoux, A. Denoirjean, M.-F. Elchinger, L. Biancchi and P. Fauchais, in *Progress in Plasma Processing of Materials*, Eds. P. Fauchais and J. Amouroux, Begell House, NY, USA, 1999, pp. 339–345.
[13] C. Baudry, Contribution to the transient and 3D modelling of the dynamic behaviour of the arc within a plasma spray torch, PhD Thesis, University of Limoges, France, November 24th (in French).
[14] G. Barbeza, *J. Automot. Technol.*, 2 (2) (2001) 47–52.
[15] J. Zierhut, P. Halbeck, K.D. Landes, G. Barbezat, M. Müller, M. Schutzl, in *Thermal Spray: Meeting the Challenges of the 21st Century*, Ed. C. Coddet, Vol. 2, ASM International, Materials Park, OH, USA, 1988, pp. 1375–1379.
[16] Matthäus, The tri-electrode axial injection plasma torch Axial III, in *48th Internationales Wissenschaftiches Kolloquium, Illemenau*, Tech. University Illemenau, Germany, 2003, Session 7.7 – electronic version.
[17] T. Morischita, in *Plasma Technick 2nd Symposium*, Eds. S. Blein-Sandmeier, H. Eschnauer, P. Huber and A.R. Nicoll, Vol. 1, Plasma Technik, Wohlen, 1991, CH, 1, pp. 137–145.
[18] M. Boulos, *J. Therm. Spray Technol.*, 1 (1) (1992) 33–40.
[20] M. Boulos, *J. High Temp. Mater. Process.*, 1 (1) (1997) 17–39.
[21] K. Mailhot, F. Gitzhofer and M. Boulos, in *Thermal Spray: a United Forum for Scientific and Technological Advances*, Ed. C.C. Berndt, ASM International, Materials Park, OH, USA, 1997, pp. 21–24.
[22] V. Leveillé, M. Boulos and D. Gravelle, in *Thermal Spray 2003: Advancing the Science and Applying the Technology*, Eds. C. Moreau and B. Marple, Vol. 2, ASM International, Materials Park, OH, USA, 2003, pp. 1011–1016.
[23] S. Xue, P. Proulx and M. Boulos, in *Thermal Spray 2003: Advancing the Science and Applying the Technology*, Eds. C. Moreau and B. Marple, Vol. 1, ASM International, Materials Park, OH, USA, 2003, pp. 993–999.
[24] E. Pfender, Electric arcs and arc gas heaters, in *Gaseous Electronics*, Eds. M.N. Hirsch and J. Oskaur, Vol. 1, Academic Press, New York, 1978, pp. 291–398.

[25] M. Ducos, Rechargement par plasma d'arc transféré, in *Plasmas in Industry*, Eds. G. Laroche and M. Orfeuil, Dopér Collection, Avon, France, 1989 (in French).
[26] U. Dilthey, J. Ellermeir, P. Gladkij and A.V. Pavlenko, *Schweissen und Schneiden*, 45/5 (1993) 241, 244.
[27] N.A. Hussary and J.V.R. Heberlein, *J. Therm. Spray Technol.*, 10 (4) (2001) 604–610.
[28] M. Kelkar and J.V.R Heberlein, *J. Phys. D Appl. Phys.*, 33 (17) (2000) 2172–2182.
[29] M. Boulos, P. Fauchais, A. Vardelle and E. Pfender, *Plasma Spraying: Theory and Applications*, Ed. R. Suryanarayananan, World Scientific, Singapore, 1993, pp. 3–10.
[31] M. Vardelle and P. Fauchais, *Pure Appl. Chem.*, 1 (10) (1999) 1909–1916.
[32] P. Fauchais and M. Vardelle, in *Thermal Spray 2003: Advancing the Science and Applying the Technology*, Eds. C. Moreau and B. Marple, ASM International, Materials Park, OH, USA, 2003, pp. 1165–1173.
[33] E. Legros, Contribution to 3D modelling of the spray process and application to a two-torch set-up, PhD Thesis, University of Limoges, France, November 24 (in French).
[34] J.F. Bisson and C. Moreau, *J. Therm. Spray Technol.*, 12 (2) (2003) 256–265.
[35] H. Zhang, A. Vardelle and N.J. Themelis, *J. High Temp. Mater. Process.*, 3 (2003) 277–298.
[36] A.A. Syed, A. Denoirjean, P. Denoirjean, J.C. Labbe and P. Fauchais, In-flight oxidation of metallic particles in plasma spraying, *J. Therm. Spray Technol.*, 2004.
[37] C. Moreau, R. Lima and B. Marple, Diagnostics for advanced materials processing by plasma spraying, *Pure appl. Chem.*, 2004.
[38] M. Vardelle, A. Vardelle, P. Fauchais, K.-I. Li, B. Dussoubs and N.J. Themelis, *J. Therm. Spray Technol.*, 10 (2) (2001) 267–289.
[39] R. Lima, A. Kucuk and C.C. Berndt, *J. Surf. Coat. Technol.*, 135 (2001) 166–172.
[40] H. Chen and C.X. Ding, *Surf. Coat. Technol.*, 150 (2002) 31–36.
[41] M. Gell, E.H. Jordan, Y.H. Sohn, D. Goberman, L. Shaw and T.D. Xiao, *Surf. Coat. Technol.*, 144–147 (2001) 48–54.
[42] E. Bouyer, D.W. Branston, G. Lins, M. Müller, J. Verlegen and M. von Bradke, in *Progress in Plasma Processing of Materials*, Ed. P. Fauchais, Begell House, NY, USA, 2001, pp. 501–506.
[43] E.H Jordan, L. Xie, X. Ma, M. Gell, N.P. Padture, B. Cetegeon, A. Oztruck, J. Roth, T.D. Xiao and P.E.C. Bryant, *J. Therm. Spray Technol.*
[44] E. Bouyer, F. Gitzhofer and M. Boulos, in *Progress in Plasma Processing of Materials*, Ed. P. Fauchais, Begell House, NY, USA, 1997, pp. 735–750.
[45] F. Gitzhofer, M.-E. Bonneau and M. Boulos, in *Thermal Spray. New Surfaces for a New Millenium*, Eds. C.C. Berndt, K.A. Khor and E. Lugscheider, ASM International, Materials Park, OH, USA, 2001, pp. 61–68.
[46] V. Rat, C. Delbos, C. Bonhomme, J. Fazilleau, J.-F.Coudert and P. Fauchais, *J. High Temp. Mater. Process.*, 8 (1) (2004) 95–118.
[47] C. Delbos, Contribution to the understanding of the injection of suspension or solution in d.c. plasma jets to spray finely structured coatings, PhD, University of Limoges, France Nb 48-2004, November 2004 (in French).

[48] P. Blazdell and S. Kuroda, *Surf. Coat. Technol.,* 123 (2000) 239–246.
[49] Nanosteel Co, 485 North Keller Rd. Suite 100, Maitland, FL 32751, USA.
[50] M. Boulos, Powder densification and spheroidization using induction plasma technology, Teckna Plasma Systems Inc., Sherbrooke, Québec, Canada.
[51] P. Fauchais, M. Fukumoto, A. Vardelle and M. Vardelle, Invited review, *J. Therm. Spray Technol.*, 13 (3) (2004) 337–360.
[52] C. Escure, M. Vardelle and P. Fauchais, *Plasma Chem. Plasma P.*, 23 (2003) 185–222.
[53] J. Cedelle, M. Vardelle, B. Pateyron and P. Fauchais, *J. High Temp. Mater. Process.*, 8 (3) (2004) 397–407.
[54] T.W. Clyne and S.C. Gill, *J. Therm. Spray Technol.*, 5 (4) (1996) 401–418.
[55] S. Kuroda, T. Dendo and S. Kituhara, *J. Therm. Spray Technol.*, 4 (1) (1995) 75–84.
[56] R.W. Smith and R. Novak, *Powder Metall. Int.*, 3 (1991) 147–165.
[57] M.L. Thorpe, *Adv. Mater. Process.*, 3 (1993) 50–61.
[58] S. Dallaire and H. Levert, *J. Therm. Spray Technol.*, 6 (4) (1997) 456–462.
[59] K. Ghosh, T. Troczynski and A.C.D. Chaklader, *J. Therm. Spray Technol.*, 7 (1) (1998) 78–86.
[60] D.T. Gawne, Z. Qiu, Y. Bao, T. Zhang and K. Zhang, *J. Therm. Spray Technol.*, 10 (4) (2001) 599–603.
[61] Yi Zeng, Soo W. Lee, and Chuanxian Ding, *J. Therm. Spray Technol.*, 11 (1) (2002) 129–133.
[62] S. Dallaire and G. Cliché, *J. Therm. Spray Technol.*, 2 (1) (1993) 39–44.
[63] J. Li, Y. Zhang, J. Huang and C. Ding, *J. Therm. Spray Technol.*, 7 (2) (1998) 242–246.
[64] B.R. Marple and J. Voyer, *J. Therm. Spray Technol.*, 10 (4) (2001) 626–643.
[65] S. Dallaire and J.-G. Legoux, *J. Therm. Spray Technol.*, 5 (1) (1996) 43–48.
[66] S.C. Deevi, V.K. Sikka, C.J. Swindeman and R.D. Seals, *J. Therm. Spray Technol.*, 6 (3) (1997) 335–344.
[67] L. Pawlowski, *J. Therm. Spray Technol.*, 5 (3) (1996) 317–334.
[68] K. Kishitake, H. Era and F. Otsubo, *J. Therm. Spray Technol.*, 5 (4) (1996) 476–482.
[69] E.M. Leivo, M.S. Vippola, P.P.A. Sorsa, P.M.J. Vuoristo and T.A. Mäntylä, *J. Therm. Spray Technol.*, 6 (2) (1997) 205–210.
[70] K.N. Lee, C.A. Barrett and J. Smith, *J. Therm. Spray Technol.*, 9 (1) (2000) 121–124.
[71] T.E. Lister, R.N. Wright, P.J. Pinhero and W.D. Swank, *J. Therm. Spray Technol.*, 11 (4) (2002) 530–535.
[72] R. Brousseau, M. Arnott and B. Baldock, *J. Therm. Spray Technol.*, 5 (1) (1996) 49–52.
[73] R. Brousseau, B. Arsenault, S. Dallaire, D. Rogers, T. Mumby and D. Dong, *J. Therm. Spray Technol.*, 7 (2) (1998) 193–196.
[74] P. Chraska, J. Dubsky, K. Neufuss and J. Pisacka, *J. Therm. Spray Technol.*, 6 (3) (1997) 320–326.
[75] K.A. Gross and A. Kovalevskis, *J. Therm. Spray Technol.*, 5 (4) (1996) 469–475.
[76] Z. Zurecki, D. Garg and D. Bove, *J. Therm. Spray Technol.*, 6 (4) (1997) 417–421.
[77] P. Chraska, K. Neufuss and H. Herman, *J. Therm. Spray Technol.*, 6 (4) (1997) 445–448.

[78] N. Mesrati, H. Ajhrourh, N. Du and D. Treheux, *J. Therm. Spray Technol.*, 9 (1) (2000) 95–99.
[79] H. Nakahira, Y. Harada, N. Mifune, T. Yagoro and H. Yamane, *J. Therm. Spray Technol.*, 2 (1) (1993) 51–58.
[80] R.A. Miller, *J. Therm. Spray Technol.*, 6 (1) (1997) 35–42.
[81] J. Musil, M. Alaya and R. Oberacker, *J. Therm. Spray Technol.*, 6 (4) (1997) 449–455.
[82] F.O. Soechting, *J. Therm. Spray Technol.*, 8 (4) (1999) 505–511.
[83] J.A. Haynes, M.K. Ferber and W.D. Porter, *J. Therm. Spray Technol.*, 9 (1) (2000) 38–48.
[84] Y. He, K.N. Lee, S. Tewari and R.A. Miller, *J. Therm. Spray Technol.*, 9 (1) (2000) 59–67.
[85] I. Sevostianovand and M. Kachanov, *J. Therm. Spray Technol.*, 9 (4) (2000) 478–482.
[86] C. Friedrich, R. Gadov and T. Schirmer, *J. Therm. Spray Technol.*, 10 (4) (2001) 592–598.
[87] G.A. Kool, *J. Therm. Spray Technol.*, 5 (1) (1996) 31–34.
[88] S. Bose and J. DeMasi-Marcin, *J. Therm. Spray Technol.*, 6 (1) (1997) 99–104.
[89] Z. Mutasim and W. Brentnall, *J. Therm. Spray Technol.*, 6 (1) (1997) 105–108.
[90] T.M. Yonushonis, *J. Therm. Spray Technol.*, 6 (1) (1997) 50–56.
[91] M.B. Beardsley, *J. Therm. Spray Technol.*, 6 (2) (1997) 181–186.
[92] S. Ahmaniemi, J. Tuominen, P. Vuoristo and T. Mäntylä, *J. Therm. Spray Technol.*, 11 (3) (2002) 320–332.
[93] R.A. Neiser, G.R. Smolik, K.J. Hollis and R.D. Watson, *J. Therm. Spray Technol.*, 2 (4) (1993) 393–400.
[94] W. Cai, H. Liu, A. Sickinger, E. Muehlberger, D. Bailey and E.J. Lavernia, *J. Therm. Spray Technol.*, 3 (2) (1994) 135–141.
[95] L.E. Weiss, F.B. Prinz, D.A. Adams and D.P. Siewiorek, *J. Therm. Spray Technol.*, 1 (3) (1992) 231–238.
[96] A. Geibel, L. Froyen, L. Delaey and K.U. Leuven, *J. Therm. Spray Technol.*, 5 (4) (1996) 419–430.
[97] K. Neufuss, P. Chraska, B. Kolman, S. Sampath and Z. Travnicek, *J. Therm. Spray Technol.*, 6 (4) (1997) 434–438.
[98] T. Valente and C. Bartuli, *J. Therm. Spray Technol.*, 3 (4) (1994) 63–68.
[99] N.X. Tan, A.J. Bourdillon, Y. Horan and H. Herman, *J. Therm. Spray Technol.*, 1 (1) (1992) 71–74.
[100] S.J. Yankee and B.J. Pletka, *J. Therm. Spray Technol.*, 2 (3) (1993) 271–282.
[101] K.A. Khor and P. Cheang, *J. Therm. Spray Technol.*, 3 (1) (1994) 45–50.
[102] Y. Wang, K.A. Khor and P. Cheang, *J. Therm. Spray Technol.*, 7 (1) (1998) 50–57.
[103] X.B. Zheng and C.X. Ding, *J. Therm. Spray Technol.*, 9 (4) (2000) 520–525.
[104] K. Barthel, S. Rambert and St. Siegmann, *J. Therm. Spray Technol.*, 9 (3) (2000) 343–347.
[105] H.C. Chen, J. Heberlein and R. Henne, *J. Therm. Spray Technol.*, 9 (4) (2000) 348–353.
[106] S. Takenoiri, N. Kadokawa and K. Koseki, *J. Therm. Spray Technol.*, 9 (3) (2000) 360–363.

[107] H. Tsukuda, A. Notomi and N. Hisatome, *J. Therm. Spray Technol.*, 9 (3) (2000) 364–368.
[108] M. Lang, R. Henne, S. Schapper and G. Schiller, *J. Therm. Spray Technol.*, 10 (4) (2001) 618–625.
[109] S. Usmani and K.N. Tandon, *J. Therm. Spray Technol.*, 1 (3) (1992) 249–256.
[110] A.C. Savarimuthu, H.F. Taber, I. Megat, J.R. Shadley, E.F. Rybicki, W.C. Cornell, W.A. Emery, D.A. Somerville and J.D. Nuse, *J. Therm. Spray Technol.*, 10 (3) (2001) 502–510.
[111] M. Müller, R.B. Heimann, F. Gitzhofer, M.I. Boulos and K. Schwarz, *J. Therm. Spray Technol.*, 9 (4) (2000) 488–493.
[112] C.C. Berndt, *J. Therm. Spray Technol.*, 10 (1) (2001) 147–182.
[113] S. Thiele, R.B. Heimann, L.-M. Berger, M. Hermann, M. Nebelung, T. Schnick, B. Wielage and P. Vuoristo, *J. Therm. Spray Technol.*, 11 (2) (2002) 218–225.
[114] D. Sacriste, N. Goubot, J Dhers, M. Ducos and A. Vardelle, *J. Therm. Spray Technol.*, 10 (2) (2001) 352–358.
[115] S. Dallaire, *J. Therm. Spray Technol.*, 10 (3) (2001) 511–519.
[116] F.N. Longo, *J. Therm. Spray Technol.*, 1 (2) (1992) 143–146.

Chapter 10

Thin Film Growth by Ion-Beam-Assisted Deposition Techniques

Raúl Gago, Ignacio Jiménez and José M. Albella

10.1. Introduction

The change and control over the surface properties is a critical aspect for the performance and, hence, technological applications, of thin films and coatings. In particular, intrinsic properties of the material in thin solid form as well as issues related with the film-substrate interface, such as adhesion, have to be addressed.

As it is well known, the bombardment of solid targets with energetic particles (10^2–10^4 eV) can induce beneficial modifications in the surface and near-surface regions [1]. For this reason, many growth processes, referred commonly as *ion-beam-assisted deposition* or *IBAD* methods, have incorporated the assistance of energetic particle bombardment during the film growth [2]. In this way, properties such as chemical composition, density, hardness, roughness, structure (grain size, crystalline orientations, intrinsic stress, etc.), can be manipulated. Complementarily, the use of ion bombardment in the first stages of the growth can also be used to modify the film-substrate interaction and improve the film adhesion to the substrate.

Although the term *energetic particle* refers mainly to ions or, at least, particles whose kinetic energy is supplied when they are charged, energetic neutrals are expected to induce similar effects as those described here. The technologic relevance of ions is that, unlike neutrals, they can be easily accelerated to any kinetic energy between 1 and 10^7 eV. The effects induced upon ion bombardment depend strongly on the ion energy range, which, according to the kinetic energy, can be referred to as *thermal* (<1 eV), *hyperthermal* (1–500 eV), *low energy* (0.5–10 keV), *medium energy* (10–500 keV) and *high energy* (>0.5 MeV).

The hyperthermal range is typical of IBAD processes, where the relevant effects are ballistic, that is, the accelerated ions have enough energy to induce bond breaking or relocation processes in the atomic lattice. The key point in the control of the process is the spatial distribution of the initial energy that is shared to the atoms in

Materials Surface Processing by Directed Energy Techniques
Y. Pauleau (Editor)

the solid in the collision events. The ions may additionally react with the target atoms or activate certain chemical reactions to create new species or compounds. The effects of chemical nature in IBAD processes mainly include the use of reactive gases (N_2, O_2, etc.) and the formation of compounds (nitrides, oxides, etc.).

In this chapter, we will first introduce the fundamental aspects that govern the interaction of energetic particles with solid targets. Second, the main properties that can be modified by the previous processes will be presented. The most usual experimental IBAD configurations, including a description of the common ion sources used in the deposition of films, will be shown thereafter. It is important to note that the IBAD process is very versatile, being possible to incorporate it to many technologic processes. Finally, illustrative examples of the potential use of these techniques for the growth of several materials in thin solid form will be briefly described.

10.2. Ion Interaction with Surfaces and Growing Films

10.2.1 Energy Loss and Ion Range in the Solid

As energetic particles or ions travel through the solid, they interact with the atoms of the target. As a result of these interactions, many different effects can be induced as described in detail in Section 10.2.2. Some of these effects imply certain energy loss of the incoming particles and, therefore, they are slowed down until they come to rest in the solid. Alternatively, the incoming ions may also be reflected with a lower energy.

The rate at which the particle loses its energy, E, as it travels a distance, r, is referred as the *stopping power* (dE/dr) and can be produced by either nuclear or electronic loss mechanisms [3]. The energy losses due to nuclear stopping account for atom displacements and they can induce lattice vibrations (phonons) and/or lattice disorder. The nuclear processes also imply a significant angular deflection of the particle trajectory. The electronic stopping accounts for energy loss associated to electronic excitations, such as collective excitations like plasmons and ionization processes. As a general rule, the nuclear stopping dominates for low energies and high atomic number (Z), while the electronic stopping becomes more relevant for high energies and low Z.

The stopping power defines the *range* of an incident particle in the solid, which is the integrated distance along the path of the particle as it travels through the target. The particle trajectory is normally known as *ion track* and, due to the collisions with the target atoms, it is not straight. In addition, the stochastic (random) nature of the collision processes implies that particles with the same characteristics do not necessarily come to rest in the same place.

In addition to the ion range, the distribution of the ion energy deposited along the ion track is a relevant parameter for the evaluation of the ion bombardment effects. Particularly, this distribution can be directly correlated with the spatial distribution of atomic displacements. For typical ion-assisted energies (below a few keV), most of the deposited energy is located at the beginning of the ion track and, therefore, most of the induced effects take place in the near-surface region.

10.2.2. Different Ion Bombardment Effects on Solid Targets

During the ion trajectory along the ion track, the ions interact with the target atoms resulting in different chemical and physical effects. The identification and understanding of these effects is crucial for the final control of the target or growing film properties upon ion bombardment. A sketch of main effects induced by the ion bombardment is shown in Fig. 10.1.

At very low energies (<0.1 eV), the ions do not have enough energy to penetrate in the solid or to induce lattice displacements. In this case, the effects of the ion bombardment are mainly induced at the surface of the solid. Some of these effects are directly related to the growth of thin films since chemical activation of the species at the surface or the increase of the *reactivity of surface* sites can be induced. In addition, the *surface mobility* of adatoms can be increased as well and, therefore, more stable configurations can result from the ion bombardment.

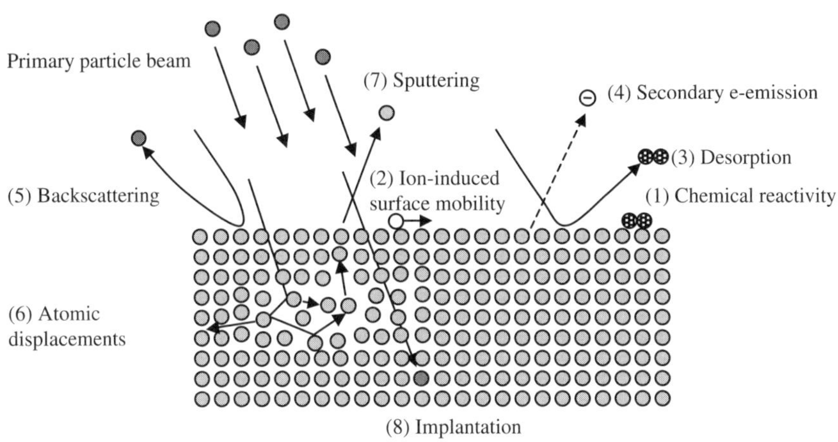

Figure 10.1: Scheme of ion-bombardment effects occurring at the surface and near-surface region of a solid (incoming ions = ●, chemical and adsorbed species = ⊕⊕, material atoms = ○).

In the range between 0.1 and 10 eV the processes are still at the surface but the energy might be high enough for *desorption* of weak-bonded species. This process is very interesting as treatment to clean surfaces in numerous technologic applications.

For energies above 1 eV, *secondary electrons* are emitted from the target upon ion impact. The energy of the emitted electrons are of the same order as the projectile. The emission of electrons are fundamental to sustain electrical discharges when the target acts as an electrode. However, for ion-assisted processes, this fact may be detrimental when bombarding insulating targets due to charging effects (added to the incoming positive charges). The charging effects can be mitigated by the incorporation of neutralization units in the ion sources that provide electrons to the incoming beam (see Section 10.4).

The collision or knock-on processes are relevant for energies high enough to induce *atom displacements* in the solid (10^2–10^5 eV). This is the typical energy range of interest in many ion-assisted processes. In order to induce the displacement of an atom from a lattice position, the energy transferred to the atom in the collision should be high enough to overcome the so-called displacement energy. These energies are typically of a few eV and are proportional to the binding energy of the atom in the solid. In the simplest case, the displaced atom leaves a vacancy and occupies an interstitial position. This vacancy-interstitial defect is referred as a *Frenkel pair*. For increasing energies above the displacement threshold, the primary knock-on atom can also displace other atoms of the network, creating a *collision-cascade*. The cascade results in a large disorder or damage of the atomic structure as a consequence of the production of a large number of vacancies, interstitials, and other type of defects. This damage can be quantified by the distribution and number of Frenkel pairs generated.

For surface atoms, the binding energy is lower than the bulk position, since, the coordination number is lower. Therefore, when the deposited energy close to the surface is high enough, the atoms in this region can be ejected as a result of the collision cascades. The process is known as *sputtering* and, typically, the yield increases with the energy range (in the 10^2–10^5). The sputtered atoms that result from the ion-assistance process can have different implications, since they can be used as the source of material for deposition in a remote substrate or to remove less-energetic bonding configurations (e.g., to favored metastable bonding configurations). However, the sputtering process can also be considered as detrimental as the origin of re-emission of material from a growing film, resulting in an effective lower or even negative growth rate.

In the case of particle energies above several keV, the probability of direct incorporation of the incoming ions into the material is approximately unity. This process is called *ion implantation*. The ions are normally implanted in a certain and well defined depth of the solid which is determined by the ion energy. Obviously, as discussed in the previous section, the implantation depth scales with the ion energy and ranges from several tens of nanometer to some microns in the typical energy range of 10^3–10^7 eV. Ion implantation is used in many technologic applications for doping a material or compound formation. In the latter case, high ion doses are required and the formation of a layer compound is achieved by the simultaneous contribution of sputtering (etching of the non-implanted surface region) and diffusion processes.

Finally, for very high energies (>10^5 eV), the electronic interaction between the projectile and the atoms in the solid dominates over nuclear interaction (knock-on processes) and *backscattering* of the ions is more likely to occur. Since the energy of the backscattered atom carries information about the collision event, it can be used to determine the nature of the atom in the solid and the depth (through the energy loss) where the interaction took place. This principle is the base of ion beam analysis techniques like Rutherford Backscattering (RBS).

In summary, we can identify different ion-induced effects according to the energy range of the incoming ions. To reinforce this point, the different effects described above are sorted according to their energy range in Fig. 10.2. This classification should be considered only qualitative since the combination of the ion and of the target material has to be taken into account. As discussed in Section 10.2.1, the classification of Fig. 10.2 also attends to the ion range and, therefore, the region of the solid affected by the ion bombardment.

As a final remark, in the discussion we have considered atomic ions for the description of the ion-induced effects. In the case of molecular or polyatomic ions, we should consider two regimes. First, for sufficiently low energy, the ions may be

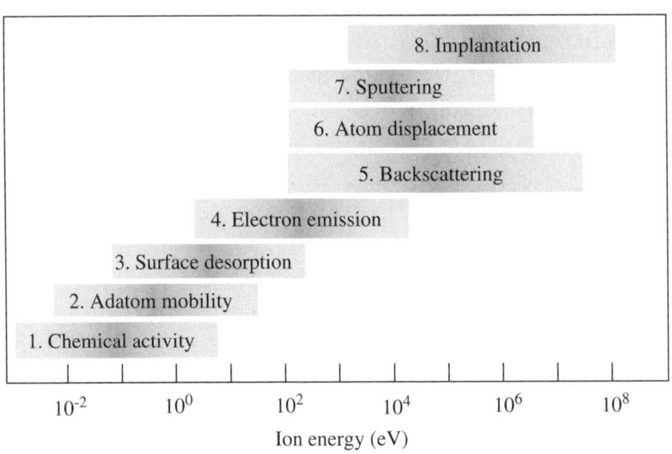

Figure 10.2: The different effects displayed in Fig. 10.1 sorted by their energy range.

attached intact to the surface and, in this way, transfer their chemical functionality to the interface. Second, for high enough energy, the ions are fragmented upon the impact with the surface. In the latter case, the total energy is shared between the individual atoms. The effective energy per incoming atom should be taken into account for the classification of the induced effects with molecular ions.

10.2.3. Ballistic Processes During Ion Bombardment

The relevant effects in IBAD are ballistic and, therefore, the understanding of the interatomic interaction plays a primary role. In this section, the basic equations governing the collision process between an incoming energetic particle and the solid are presented. This description is used to quantify and model the effects of ion bombardment in IBAD processes.

10.2.3.1. Dynamics of Binary Elastic Collisions

From the point of view of an incoming energetic particle, the main consequences of the interaction with the solid target are deflection of its trajectory and loss of energy. In order to determine the particle trajectory and final atomic configuration, the nature of the interatomic interaction of the incoming particle with the atoms in the solid should be known. In a simple approximation, we can consider a binary collision to describe the collision process, where a fast moving particle collides with an atom in rest at its lattice position. This situation is displayed schematically in Fig. 10.3. In the energy range for ion-assisted processes (below a few keV), the typical penetration depth is of a few nanometers [3]. Therefore, the time scale for a particle in this energy range to be stopped within the solid is of $\sim 10^{-13}$ s. This time scale is of the same order of that for lattice vibrations and, therefore, the description of the ion bombardment considering a binary collision between a fast moving particle and a rest atom is justified.

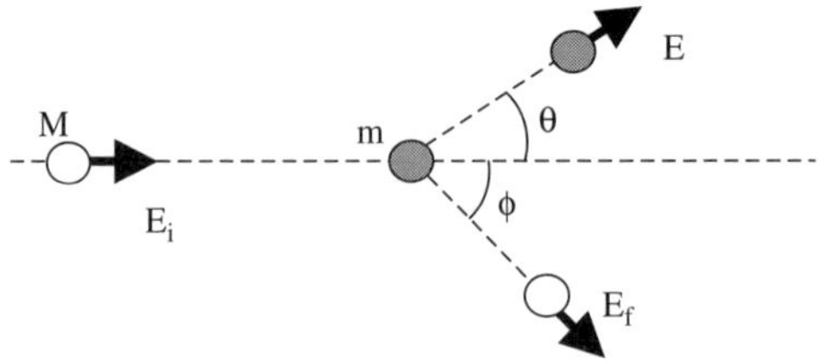

Figure 10.3: Binary collision between an incoming particle (M) with energy E_i and an atom at rest (m).

Due to the typical distance of the particle–atom interaction, the Coulomb repulsion force between the nuclei is the greatest. Therefore, the Coulomb potential is appropriate to describe the scattering process, but multiplied by a certain function that considers the electrostatic screening by the space charge of the innermost electron shells. Under this classic approach, the collision event can be quantified by introducing the reduced energy, ε, given by:

$$\varepsilon = 0.69 \frac{a}{Z_i Z} \left(\frac{m}{M + m} \right) E_i \tag{10.1}$$

where E_i is the energy in eV of the incoming ion, *of* atomic number Z_i, and mass M, and a accounts for the typical screening length (in nm) of the interatomic potential. Z and m are the atomic number and mass, respectively, of the target material. The reduced energy is a non-dimensional parameter that measures the energy involved in the collision and how close the ions get to the nucleus of the target atoms. As illustrative example, the typical values for some of the effects described in Section 10.2.2 are:

1. Ion assisted deposition: $\varepsilon < 0.2$
2. Ion implantation: $0.1 < \varepsilon < 2$
3. RBS: $\varepsilon > 10$

From the point of view of the atom at rest, the important factor in the collision process is the energy conferred from the incoming ion. The same classical approach as above leads to the following expression for the *transfer energy*, E:

$$E = k\,E_i = \frac{4\,M\,m}{(M + m)^2} \cos^2 \phi\, E_i \tag{10.2}$$

where ϕ represents the scattering angle as displayed in Fig. 10.3. The ratio between E and E_i, k, is known as the *kinematic factor* and, which has a maximum value ($k = 1$) for a head-on collision ($\phi = 0$) and when $M = m$. In other cases, the kinematic factor will be always below 1. As already discussed, the value of E has to be larger than the corresponding displacement energy in order to induce lattice displacements, which are relevant in IBAD to control the film properties (see Section 10.3).

Due to the nature of the collision process, not only the energy transfer should be considered to quantify the bombardment but also the *momentum transfer* [2]. The calculation of the collision momentum transfer, ΔP, yields:

$$\Delta P = \sqrt{2mkE_i} \tag{10.3}$$

The units for momentum which are normally used in ion-assisted processes are $(\mathrm{eV} \times \mathrm{amu})^{1/2}$. since they can be directly related with experimental parameters.

10.2.3.2. Parameters for Ion-Assisted Growth

As presented above, the dynamics of the assistance process can be described by the nature of the atoms involved, the incoming ion energy, E_i, and the geometrical arrangement. However, the assistance process is concurrent with the film growth and, therefore, other values apart from ion energy, flux, and species should be considered. In particular, an important parameter for the quantification of the ion bombardment during film growth is the *transport ratio*, *I/A*, which corresponds to the arrival ratio of incoming ions and neutral atoms:

$$\frac{I}{A} = \frac{\mathrm{ions/cm^2 s}}{\mathrm{atoms/cm^2 s}} \tag{10.4}$$

This term acts as a normalization factor of the bombardment parameters to the rate of material deposition. Typical values of the transport ratio in ion-assistance processes range between 0.1 and 5.

The effects of the ion energy on the film growth can be normalized introducing the *average energy deposited* per incoming ion, E_d, which can be easily calculated as:

$$<E_d>= \frac{I}{A} E_i \tag{10.5}$$

The E_d value has been recognized as a relevant parameter for describing the effects of low-energy (100 eV) ion irradiation, but it cannot be used as a universal parameter [4]. A more precise description should incorporate the kinetics of the collision process. In this way, the *average transfer energy*, $<E>$, and the *average momentum transfer* per incoming ion, $<\Delta P>$, must be introduced to account for the quantification of the ion bombardment:

$$<E>= \frac{I}{A} k\, E_i; \quad <\Delta P>= \frac{I}{A} \sqrt{2mkE_i} \tag{10.6}$$

In the case of the assistance with several ionic species, the above expressions can be expanded as the sum of the individual terms, corresponding to each species.

The above equations assume that the collisions are binary and elastic. This is only an approach for the quantification of the ion bombardment since many-body collisions have to be considered in a real case. The actual values of energy and momentum transfer are, therefore, expected to be less that those obtained from the present simplifications. However, in a practical case the above equations can be used to describe qualitatively the effects of the assisting process in a series of similar samples.

As a final remark, when looking at the effects of ion bombardment, the ion energy and the transport ratio effects should not be considered separately. The resulting modifications at a given transport ratio may vary depending on the energy of the ions (see Section 10.2.2) and, likewise, the effect of bombardment at a given ion energy will also depend on the number of bombarding ions (e.g. stoichiometry). As will be shown in Section 10.3, most of the experimental results can be described by the concurrence of both the parameters.

10.3. Effects Induced by Ion Bombardment on Film Properties

From the materials science standpoint, the interest in using ion-assisted deposition stems from the new properties that are conferred to the growing films, due to the structural changes induced by the ion irradiation. The different physical and chemical processes activated by the hyperthermal particles have been described separately in the preceding section, although in reality it is the combination of several processes which affect the film structure.

To predict the influence of ion bombardment on the film properties, three main theoretical approaches have been followed. The first one is the *thermal spike* continuum model based on assuming that the local energy deposited in the collision cascades relaxes in a very short time (typically 1 ns) giving rise to a high local temperature, well beyond the melting point of any solid, for a very short time and limited to the volume surrounding the ion tracks. The second one is the *ballistic or collision-cascade* model. This is a discrete model based on accounting for the atomic displacements induced by the energetic ions due to the transfer of energy and momentum in the collision. In many cases, a simple version of the ballistic model is considered by using ion implantation models, like the TRIM code, where the ions reach a pre-formed material and not a growing film. The third model is based on *molecular dynamic* calculations, an atomistic model that accounts for the impingement of both atoms and hyperthermal ions, and the interactions among them. The last one is the state-of-the-art model, requiring large computation power. The accuracy of their predictions largely remains on the description of the interaction potentials between particles.

The ion bombardment induces three main types of effects on the film growth that will be described throughout this section. The most general effect induced by ion assistance is the modification of the thin film microstructure (grain size and orientation, void fraction, density, texture) that always takes place when ion bombardment is used. A second possible effect is the modification of the bonding structure, enabling the growth of metastable phases not achievable by conventional physical vapour deposition (PVD) techniques. The third one is a chemical effect, particularly important in the case of reactive ions like nitrogen and oxygen. In practice, three ion species are mainly used in IBAD: Ar^+ as an inert ion that provides energy and momentum, and the reactive ions O_2^+, and N_2^+.

10.3.1. Modification of the Microstructure

The microstructure of a film includes several factors, like the grain size, grain orientation, void fraction, density, texture, or roughness. It is the result of the details of the events of nucleation, growth of nuclei, and coalescence, which are mediated by the adatom mobility and adatom nucleation rates.

10.3.1.1. Nucleation, Growth, and Grain Size

The atoms reaching a surface from the vapor can experience processes of desorption (return to the vapor phase), surface diffusion, formation of nuclei, or bonding to pre-existing nuclei. These four processes control the rate of nuclei formation, growth, and coalescence.

The use of energetic ions influences the balance between nucleation and nuclei growth in several ways. First, energetic atoms have larger *mobility* giving rise to a lower nuclei density, and larger island size. Second, the energetic ions create *defects* on the substrate that act as nucleation sites, hence increasing the nuclei density and decreasing the island size. Third, the energetic ions may *sputter* away the smallest islands, favoring the presence of larger islands. Fourth, the ions *disrupt the columnar growth* of tall and narrow grains, favoring the growth of smaller islands. The balance between these four processes finally controls the result, which depends on the ion energy and transport ratio and is specific of each system [5].

In general, there is a decrease in grain size with ion bombardment for low energies (50–100 eV). The grain size increases for energies exceeding 10 keV due to recrystallization of the film due to the thermal spike associated with the high-energy ions.

10.3.1.2. Surface and Interface Roughness

In a first approximation the surface roughness can be considered equivalent to that resulting from the coalescence of the growing particles that form the film; consequently, a smaller grain size would give a lower roughness. However, additional factors make the determination of surface roughness a complex problem. If the increased mobility induced by the ions favors a preferential two-dimensional island growth, the roughness is largely reduced. When sputtering processes take place ($E_i > 200\,\text{eV}$), the roughness increases by film erosion, a process that is angle dependent with the minimum at normal incidence. Finally, some textures or crystal orientations yield smoother surfaces than others. In general, when the more densely packed crystallographic planes are parallel to the surface, the roughness decreases [6].

The final roughness is due to the balance between thermodynamics, that tends to create smooth surfaces by arranging all atoms in the position of maximum coordination, and kinetics, that tends to stick the arriving particles in random positions. In general, there is a decrease in surface roughness for low-ion beam energies (50–100 eV), followed by a small increase with increasing energies beyond this value, as is shown in Fig. 10.4.

The ion bombardment can also affect the roughness of the interface between substrate and layer, or the various interfaces that are produced during the growth of multilayer coating. In the second case, the phenomena taking place are similar to those described for the surface roughness. In multilayer coatings, one normally desires smooth interfaces, which are obtained by maximizing the adatom mobility without triggering roughening mechanisms such as sputtering. Hence a low energy ion beam is required [7].

Regarding the film-substrate interface, the modification of the interfacial roughness requires the use of energetic particles that penetrate the growing film. Although interface mixing has to be minimized in the case of high-quality epitaxial layers,

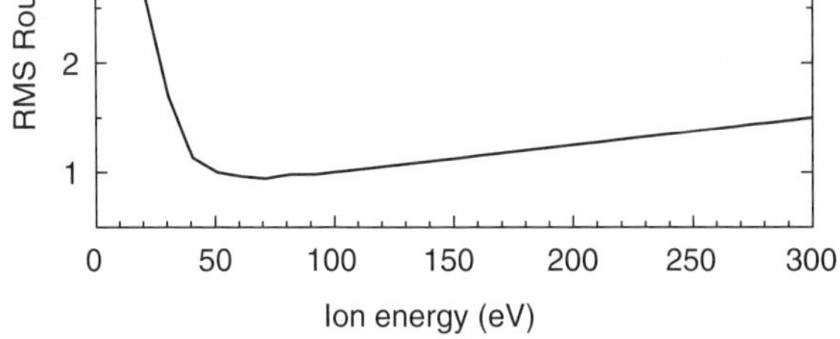

Figure 10.4: Typical dependence of the film roughness with the ion energy in IBAD.

a rough interface with atomic mixing is beneficial for film adhesion and is important in applications where mechanical stability is required, as in the use of hard and protective coatings.

10.3.1.3. Film Density

The films produced by IBAD are normally denser than those produced by conventional evaporation, a property of paramount importance in optical applications due to the correlation between refraction index and density [8]. This effect will be explained in more detail in Section 10.5.1. The main reason for the increased density is the capability of the hyperthermal particles to disrupt the columnar growth that normally takes place in thin films. Columnar growth appears by the impingement of arriving atoms on top of the formed nuclei, with shadowing of the lateral regions, producing the growth of columnar crystallites with empty space or voids in-between. To account for the void fraction, it is commonly defined the *volume packing density* of a film, p_d, as the ratio between the volume of solid part over the total film volume, including both solid and voids. For evaporated films, p_d typically ranges between 0.7 and 0.95.

Films grown by ion-assisted deposition are denser due to a more compact arrangement of the crystallites, compared to the columnar growth. With increasing energy and transport ratio of the ions, there is a decrease in void space until a maximum density is obtained for the optimal values of energy and *I/A* ratio. Beyond the optimal values, there is an increasing damage in the films, with crystallographic defect formation that decreases the density. Based on the ballistic model, calculated optimal energy ranges between 20 and 30 eV for oxides and nitrides, and 60–80 eV for metals.

A second process affecting the film density is the denser packing of atoms in the films, both due to the shortening of bonding distances and to the increasing number of atoms in interstitial positions, effects that also yield an internal compressive stress that is discussed below. A third mechanism of densification takes place when phase transition towards a denser metastable configuration takes place, as is the case of hexagonal to cubic transition in carbon or boron nitride (BN). This is discussed in Section 10.3.2.

10.3.1.4. Defect Generation

Ion irradiation of solids creates a large number of vacancies, interstitials, and dislocations by atomic displacements, although the thermal spike associated with the ion irradiation can also anneal out many of these defects. The case of irradiation during film growth is less studied, although the defect densities are some five orders of magnitude higher than in bulk materials can be found.

Defect generation has to be minimized in highly ordered films, as those used in microelectronics [9], by using low-energy ions and post-deposition thermal treatments to anneal out the point defects. Rapid thermal annealing treatments are specially important. An important issue is the fact that the energy required to produce a surface defect is typically between 20 and 30 eV, about half the value required for subsurface or bulk defects, which ranges between 50 and 100 eV [10]. Since surface defects can be compensated by adatoms with high surface mobility, the use of ions-assisted deposition with energy below 50 eV can be used to achieve epitaxial growth.

A type of defect specific of ion assistance is the incorporation of atoms or molecules that are present as ions in the bombarding beam. Inert atoms like Ar, or molecules like N_2 have been found embedded in the films grown by ion assistance, with typical concentrations of 1–5 at.% for films grown at room temperature. Increasing the substrate temperature normally reduces the concentration of these incorporated species. The incorporation of bombarding particles into the film, modifies properties like the electrical resistivity or the hardness, partly due to an increase of the compressive stress in the films.

10.3.1.5. Internal Stress

Thin films accumulate an internal stress that increases with the film thickness, being a limiting factor for possible applications. When the internal stress exceeds a critical value, the films delaminate. The stress is referred as *tensile* when it increases the length of a solid, and is considered as a positive value. In the case of thin films, tensile stress corresponds to an in-plane enlargement. The stress is *compressive* when it shortens the solid (taking a negative value), and in films it corresponds to an in-plane shrinkage. A large void fraction in the film induces a tensile stress, since such a low-density film is actually occupying more volume that would correspond to the number of atoms present in the film. On the contrary, a high-density film containing a large proportion of interstitial atoms presents compressive stress. More details on this subject can be found in Chapter 15 of this book.

For IBAD films grown with very low-energy ions, typically below 30 eV, the adatom mobility is enhanced without disrupting the columnar growth, and this results in a slight increase of the tensile stresses due to a larger size of the voids. More energetic ions disrupt the columnar growth, releasing the tensile stress associated to the voids. IBAD at ion energies about 100–500 eV normally results in compressive stress, since the atomic packing increases with the ion bombardment by occupation of interstitial positions. Finally, for more energetic ions, the thermal spike releases the compressive stress and the film densification. Based on these facts, the use of ion assistance permits the control of the film internal stress, from a positive stress in the case of non-ion-assisted films, to a compressive stress in IBAD films, as shown in Fig. 10.5 for a typical metal [11].

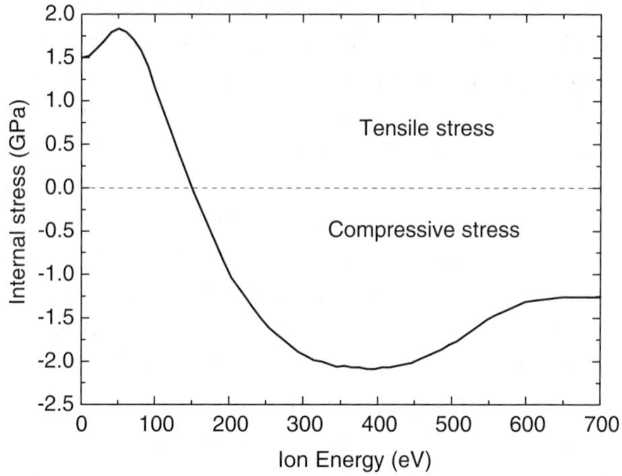

Figure 10.5: Typical dependence of the internal stress of metal films with the ion energy in ion-assisted growth.

10.3.1.6. Film Texture

Texture appears when the crystallites in the film are not randomly oriented, but showing a preferential orientation for certain crystal planes, normally expressed with regards to the film surface. The presence of textures is easily determined by X-ray diffraction (XRD) by comparing the intensity of the different Bragg peaks from a film with the tabulated intensities corresponding to a polycrystalline sample. A particular case occurs for the growth of two-dimensional islands, where all the crystallites have a common axis, and the possibility of obtaining epitaxial growth stems from the success in pinning an in-plane preferential orientation.

In the more general case of three-dimensional growth, the films obtained by evaporation or conventional PVD techniques normally present the most densely packed planes parallel to the surface, which are the (111) planes in the case of face centered cubic (fcc) metals, the (110) for body centered cubic (bcc) metals and the (0001) for hexagonal closed-packed (hcp), due to thermodynamics constraints related to the fact that the arriving particles tend to bond in a configuration of maximum coordination. However, under ion bombardment, the closest packed planes tend to move out of this geometry, with new textures in the films [12]. The following table indicates the preferential textures for films grown without ion assistance and with IBAD techniques.

Structure	Texture without ions	Texture for IBAD deposition
fcc	(111)	(110), (100)
bcc	(110)	(111), (100)
hcp	(0001)	(11–10)

Two main different mechanisms can account for the change in texture with the ion assistance: compressive stress in the films and preferential sputtering during growth. The sputtering model applies well to the growth of grains with well-defined crystallinity, whereas, the compressive stress model also applies to films of less-ordered material. In addition, a third mechanism is the thermal spike induced by irradiation, that also has implications on the orientation more connected with the thermodynamics, but is only dominant at high ion energies.

The compressive stress that is build up during IBAD tends to orient perpendicular to the surface the planes with the lowest in-plane compressibility (i.e. the closest packed ones). A clear example occurs in the IBAD growth of highly anisotropic layered materials, like graphite or hexagonal boron nitride, where the compressive stress tends to orient the basal planes perpendicular to the surface, even in films lacking long-range order. The degree of texturing increases with the ion energy up to energies of 1 keV due to the increased compressive stress induced in the films. For energies well above 1 keV the structural damage is large and the compressive stress is lost.

The *preferential sputtering model* assumes that the arriving ions find a well-defined crystalline structure, with more or less open channels in the direction of motion depending on the grain orientation. If the texture is not closed packed, the channels are wide and the ions can penetrate deeply in the bulk without collisions. However, for closed packed structures the channels are narrow and the ions collide near the surface producing sputtering and removal of the closed packed planes. This model has been supported by molecular dynamic simulations [13]. A very important point of the sputtering model is that the surviving texture is the most open in the arriving direction of the ions. Therefore, by tuning the ion arrival angle, one can select the dominant film texture.

10.3.2. Modification of the Bonding Structure and Stabilization of Metastable Phases

The film material growth by IBAD techniques can differ in the bonding structure from the stable bulk materials or from thin films obtained without the use energetic particles. These differences may appear in the coordination number and geometry, or in the bonding distances. When the change affects the number of neighbors or the geometrical arrangement, it can be considered as a chemical modification. In the case of slight variations in the bonding distances, it is normally considered as a bond distortion that creates intrinsic stress in the films.

10.3.2.1. Amorphization

To define a true amorphous material, it is necessary to express the extent of the local order. A material maintaining the coordination number and bonding distances of the crystalline reference, but exhibiting different bonding angles beyond the second nearest neighbor coordination sphere, would have a well-defined bonding structure similar to the crystalline reference, but without any long range order. This is, for instance, the case of amorphous SiO_2. A more extreme amorphous material occurs when even the coordination number is not similar to any crystalline reference, and hence the bonding distances and angles are not maintained. This situation occurs in amorphous or diamond-like carbon (DLC), where direct bonding between trigonal sp^2 hybrids and tetragonal sp^3 hybrids takes place, unlike the reference graphite and diamond crystalline materials.

Ion assistance affects the crystallinity of the material, and a significant broadening of the Bragg reflections or even the loss of the Bragg peaks can be found with increasing bombardment (both by ion energy and transport ratio). However, beyond a certain point, the thermal spike produced by ion assistance can induce the film re-crystallization.

10.3.2.2. Stabilization of Metastable Phases

Nowadays, one of the main applications of IBAD techniques is the stabilization of certain phases at growth temperatures well below the phase transition, or even the synthesis of films in a metastable phase, not achievable by conventional techniques. One example of phase stabilization is the case of TiO_2, that grows in the anatase phase for $T < 600°C$ and in the rutile phase for $T > 600°C$. However, by IBAD techniques the rutile phase can be grown at temperatures below 600°C. Regarding the synthesis of metastable phases, some examples are the growth of diamond-like films with tetrahedral coordination (ta-C), cubic BN (c-BN) or cubic GaN (c-GaN), instead of the more stable phases graphite, h-BN and wurzite GaN, respectively. The binding energy difference between w-GaN and c-GaN is only 0.2 eV; therefore the transfer of energy and momentum in IBAD to promote the synthesis of c-GaN is not difficult. However, the binding energy difference for carbon or BN in sp^3 and sp^2 coordination is very large, of the order or 1 eV, and the growth conditions to obtain the sp^3 phases are rather stringent.

In the case of ta-C, it is required the use of a monoenergetic beam of carbon atoms of about 100 eV while maintaining the substrate at room temperature. In this case, the conventional IBAD setup with Ar^+ assistance is not optimal, since it provides a broad range of energies to the carbon atoms. Instead, the filtered cathodic vacuum arc (FCVA) technique is used, where a vapor of carbon atoms is produced by arching and the mass and energy is filtered so that only carbon atoms with the optimal energy of about 100 eV bombard the substrate. In the case of c-BN, conventional IBAD with the assistance of argon and nitrogen ions is a well-suited technique, being the optimal conditions a transport ratio about unity, ion energy of about 500 eV and substrate temperature near 500°C (see Section 10.5.2).

The mechanism permitting the synthesis of metastable phases is based on a local increase of temperature and pressure due to the ion bombardment that mimics the macroscopic conditions obtained in the high-temperature high-pressure (HTHP) synthesis of such materials from starting powders. The most accepted theoretical model accounting for such phase transformation is the sub-plantation model proposed by Robertson [14].

10.3.3. Chemical Effects

Ion bombardment can affect the composition of compound materials, allowing the synthesis of different stable stoichiometries, as in the case of iron oxides, or the formation of non-stoichiometric compounds. Here, one must consider two different cases.

When non-reactive ions are used, mainly Ar^+ and Ne^+, the formation of the compound takes place by chemical reaction between the different atomic species evaporated or sputter deposited. The impinging ions play a double role:

1. They communicate energy and momentum to the neutral atoms or molecular moieties arriving at the substrate surface, favoring the chemical reactivity.
2. They modify the composition due to the preferential sputtering of certain chemical species, an effect that is energy dependent. Also, there is a small incorporation, typically of 1–5%, of the neutral ions in the film.

The use of reactive ions, normally oxygen or nitrogen, is a common method of forming thin films of oxides (SiO_2, PbO_x, CuO_x MgO, ITO, etc.) and nitrides (TiN, AlN, GaN Si_3N_4). In this case, the reactive ions can react with the neutral atoms yielding stable species that are incorporated to the film or they can form unstable moieties that are desorbed from the substrate, producing the so-called chemical sputtering. This latter process consumes part of the evaporated atoms and should be minimized to make the IBAD technique efficient. All of these effects are also dependent on the ion energy, and normally there is an optimal value for the formation of stable moieties that incorporate to the film. The amount of reactive gas that is incorporated to the film depends linearly on the actual supply of reactive gases that is, on the ion/atom ratio, until a saturation level is obtained. The saturation is obtained for I/A values below the compound stoichiometry, indicating that not all of the evaporated atoms incorporate to the film due to chemical sputtering. For I/A values above the saturation point, chemical sputtering is enhanced, reducing the growth rate.

A special case is the use of low-energy IBAD for the reactive deposition of epitaxial single crystal nitrides in microelectronics. The use of low-energy IBAD for the synthesis of group III-nitride semiconductors, like GaN, AlN is specially important. Obtaining highly crystalline thin films requires the use of very-low energy nitrogen ions ($<50\,eV$), so that the reactivity is large, but the probability for defect formation, mixing, and sub-plantation is low. The use of ion bombardment can be performed by IBAD or by plasma-assisted MOCVD and molecular beam epitaxy (MBE).

The stabilization of compounds with unusual composition or stoichiometry, for example, superoxides with a coordination number larger than the common oxide phase, can also take place in the IBAD film growth, a mechanism that can be considered as the stabilization of metastable phases from the chemical point of view, in contrast with the geometrical metastable phases considered in Section 10.3.2.

10.4. Implementation of IBAD Systems

The use of ion beams has been extended to many physical processes as it permits a flexible control of the type and number of ions (dose), and its angle of incidence with a well-defined energy on the target material. Additionally, most of the sources operate in a high vacuum level, which provides a clean atmosphere necessary for the purer films. Its main limitation lies on the maximum diameter available with a homogeneous beam profile, which makes difficult the application in large-scale systems. Ideally, one can assume a cylindrical symmetry, with a flat profile near the centre or, in a more realistic case, following a gaussian profile. Generally, the beam is collimated to eliminate the outer ring with a current density variation larger than $\pm 5\%$ with respect to the average. In any case, it is of the utmost importance to characterize the beam homogeneity by determining the radial profile of the current density.

10.4.1. Experimental Configurations

Different techniques have been devised to benefit from the ion bombardment during the deposition of films and coatings that differ in the special arrangement of the ion source (or gun) and in its interaction with the atoms of the substrate or with other gas molecules, as illustrated in Fig. 10.6. In all of these cases, the beam energy rather than the ionization state of the bombarding species is exploited, so the terms direct ion beam deposition (IBD), ion beam sputtering (IBS), or IBAD, which emphasize the use of charged particles, can be misleading [15]. Anyway, they will be used here to describe the deposition processes based on the bombardment of the growing film with ion beams for control or modification of its properties. A brief description of these techniques is given below.

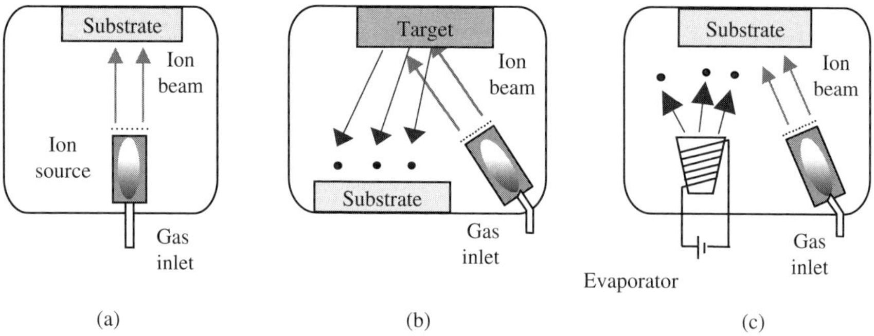

Figure 10.6: Different experimental configurations in thin film deposition with ion-beam assistance: (a) Direct bombardment on the substrate either for IBD or etching (IBE); (b) IBS; and (c) ion-beam-assisted deposition (IBAD).

10.4.1.1. Direct Ion Beam Deposition (IBD)

This technique is commonly used with reactive ions, giving rise to the formation of non-volatile compounds. The gases are introduced in the ion source where they are ionized and accelerated to a given energy. After impinging on the substrate surface they may remain directly adsorbed or react in the surface region to give a new compound layer. Generally, the reactivity of the ionized gases may degrade the hot filaments. Hence, the use of ion sources based on cathode filament emission should be avoided in this configuration.

One of the major problems with this technique is the availability of appropriate gas precursors for the deposition of specific compound layers. In the case of metals, organo-metallic compounds have been tried, though the deposited film can be contaminated with a large fraction of carbon and hydrogen. More often, this technique has been applied for the preparation of hydrogenated amorphous carbon, starting from methane or other hydrocarbon precursors. A special method, relevant for the deposition tetrahedral amorphous carbon (ta-C) or DLC, is represented in Fig. 10.7, where a FCVA is used instead of a conventional ion source. In this arrangement, the material cathode (carbon) is vaporized by an arc discharge, producing a high-density plasma consisting of neutral atoms, atomic clusters, and a high proportion of very energetic ions. By means of a magnetic filter, made of a curved coil, only the ions with a definite charge to mass ratio (e/m) are driven throughout the "magnetic tube" towards the biased substrate whereas the rest of particles (neutrals and clusters) are filtered out. High deposition energies are needed to obtain carbon films with appropriate atomic bonding (sp^3 hybrid orbitals), which are only attainable with this technique.

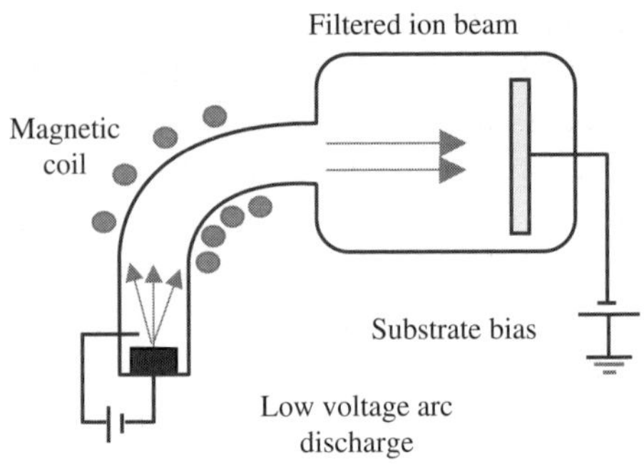

Figure 10.7: Scheme of the FCVA deposition technique.

10.4.1.2. Ion Beam Sputtering (IBS)

In this technique, a high-energy ion beam (accelerated to several hundred eV) is used to sputter a target material, so that the sputtered atoms can be condensed on a substrate. Though the fundamentals of this method are similar to the conventional cathodic sputtering technique, it offers unique advantages associated to the use of ion beams, that is, lower pressures, and therefore cleaner vacuum atmosphere, and more defined energy and bombardment angle (due to the lower scattering of the ions with the residual gases), which allows a better control on the sputtering process. Other advantages lie on the flexibility in the selection of the target geometry and morphology, and absence of thermal or electrical contacts to the cathode. Besides, there are no limitations regarding the magnetic properties of the material, since no magnetic fields are imposed in the target area. However, the deposition rate reached by IBS is relatively low ($\approx$100 nm/min) which inhibits the application for large area substrates. For the deposition of compound materials, a second ion source, fed with reactive gases, can be mounted for direct bombardment of the target surface.

10.4.1.3. Ion Beam Assisted Deposition (IBAD)

This is the more versatile process for ion beam deposition of films since it allows a large variety of material compounds to be deposited even in large-scale production. It combines the high-rate deposition by any PVD technique (thermal evaporation, sputtering, etc.) with simultaneous bombardment of the substrate surface by beams of controlled energy, intensity and incidence angle, supplied by the ion source. In this process, there is a strong interaction between the arrival atoms and bombarding ions at the surface, leading to energy and momentum transfer. This exchange of energy and momentum makes it possible to increase the arrival energy of the deposited atoms, which otherwise would be difficult to ionize and, consequently, to accelerate. After the collision, these hyperthermal atoms may give rise to different phenomena such as the formation of metastable atomic bonds, or interstitial atoms giving denser films, or even new compound molecules when reactive ions are used, as discussed in the preceding sections.

Non-reactive IBAD is a very useful technique for the formation of dense films with improved properties. In this case, noble gases, generally low cost Ar, are dispensed to the ion source. Apart from the energy and momentum transfer during the collision with the arrival atoms, some Ar ions (in a proportion lower than 5%) can be incorporated in the films as inert contaminant. An interesting example is that of carbon deposits, which may show a density between that of graphite, $2.0\,g\,cm^{-3}$, and diamond, $3.5\,g\,cm^{-3}$. By optimizing the IBAD parameters it is possible to grow coatings with intermediate density and other physical properties [16].

Reactive IBAD is generally used when compound films are to be deposited. In this case, apart from the ion energy and transport ratio, the reactivity between the ion and the evaporated atoms controls the deposition process, particularly the stoichiometry of the deposits. The method is very attractive for the formation of metal oxides and nitrides where the assistance of O_2^+ or N_2^+ ions is required, as discussed in Section 10.5.1.

10.4.2. Ion Sources

The ion sources are able to supply a beam of ions, generally positive, accelerated to certain energy within a definite solid angle. They are based on plasma discharges in gases, where positive ions and electrons co-exist in approximately the same proportion along with un-ionized neutrals. The ions from the discharge are extracted from the plasma and accelerated with a specific energy so that they can be directed to a target. The different source types differ in the way the discharge is provoked and sustained (either by DC or AC voltages, generally under the presence of a magnetic field), and also in the extraction and acceleration system for the ions. Obviously, the design is always aimed at optimizing the ion current extracted from the discharge, as well as the acceleration energy and beam quality (beam diameter, spatial dispersion, etc.). In fact, there is a direct relationship between the plasma density and the ion current that can be drawn from the plasma [17]. So, the efforts have been focussed on the search for efficient methods to generate dense plasmas.

In the following, we describe the fundamentals and operation mode of some ion sources commonly used in such surface processes, particularly those employed either for direct sputtering, IBS techniques or for IBAD techniques, discussed in this chapter. A characteristic of these sources is the ability to supply a broad beam of ions, necessary to bombard large surface areas. A more detailed description of the ion source technology can be found in specialized books and articles (quoted at the end of this chapter).

10.4.2.1. Kaufman Source

Initially developed by Kaufman [18] for space propulsion (ion thrusters), this type of sources has become very popular in the thin film technology due their ability to supply a broad and uniform ion beam (up to 40 cm diameter) in a wide range of energies (100–1200 eV). They are based on a low pressure (<1 mbar) DC glow discharge between two electrodes: the anode made of a cylindrical plate and the cathode consisting in an incandescent filament placed in the centre. The electron emission from the cathode and the presence of a magnetic field help to sustain the discharge at relatively low voltages (50 V) and pressures (10^{-4} mbar).

There are a wide variety of discharge chambers, depending on the final application. Fig. 10.8 shows the scheme of a typical source, along with the corresponding electrical supply, typically used for research purposes. The chamber case and the screen grid confine the plasma inside the anode enclosure and they are generally connected to the cathode potential. A magnetic coil or permanent magnet (not shown) provides a uniform magnetic field, between 50 and 120 G, parallel to the axis of the chamber. The magnetic field deviates the plasma electrons in helical trajectories around the field lines and prevents the electrons to travel directly to the anode, thereby increasing the ionization events in their path to the anode. Alternatively, a multipole configuration can also be used in larger sources (*multicusp system*) to create a strong magnetic field at the periphery of the source, near the anode, thus avoiding electron losses at the chamber walls. Primary electrons of the discharge are then

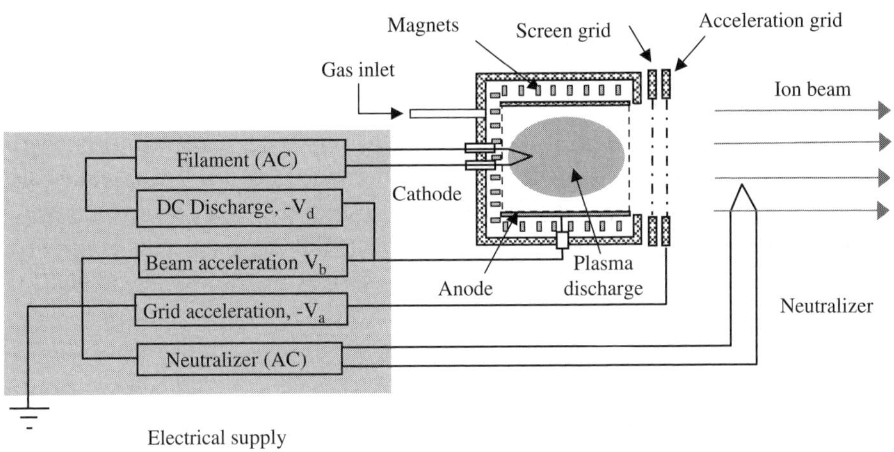

Figure 10.8: Scheme of the Kaufman broad beam ion source.

confined within this "magnetic bucket" where they oscillate, thus leading to a more homogenous and isotropic plasma. As we shall see later, the enhancement of the plasma density by the application of magnetic fields is a common technique utilized in most of the ion sources.

Since the plasma electrons are accelerated approximately at the plasma discharge voltage, V_d, this voltage is chosen above the value of the first ionization potential, V_{i1}. However, in order to avoid double charged ions, V_d should be kept below the sum of the first and second ionization potential (i.e. $V_{i1} < V_d < V_{i1} + V_{i2}$). In the case of argon, for instance, the first and second ionization potentials are 15.8 and 27.6 eV, respectively. Therefore, the discharge voltage should be <43.4 eV. Actually, the voltage is usually even lower (35–40 V) to offset other side effects. In the case of N_2, higher voltages are required (typically 50–60 V).

Whereas the proportion of ions in the plasma with respect to neutral gas atoms or molecules is generally very low, the ion beam is made mostly of charged species (50–90%). An important advantage of this type of sources is the independent control over the beam energy and beam current (given by the electron emission at the discharge). Since the acceleration grid eliminates the electrons coming from the plasma, it is necessary to add an external source of electrons at the exit, generally a hot filament, as a neutralizer of the positive charge of the ion beam, thus avoiding the beam divergence. When reactive gases are used in the discharge, a hollow cathode electron source can be utilized as beam neutralizer. A second electron emitter may also be required, instead of the filament cathode of the discharge.

The maximum ion beam current, I_b, that can be extracted from this and other types of sources is imposed by the spatial charge accumulated at the grids, and can be calculated approximately by *Child's law*:

$$I_b \cong (4\varepsilon_0/9)A_b(e/m)^{1/2}(V_b + V_a)^{3/2}/d_g^{\,2} \qquad (10.7)$$

where ε_0 is the permittivity of vacuum, A_b is beam area, e/m is charge-to-mass ratio of the accelerated ions and d_g is distance between the screen and accelerator grids. Actually, the ion beam current is usually 20–50% of the theoretical value due to several factors, among them the reduced effective extraction area, which is much less than the geometrical area. Typical values of the ion beam current density are 1–2 mA cm^{-2} for an Ar^+ ion source working at 10^{-3} mbar pressure, with 100 cm^2 beam area and accelerated at 500–2000 eV energy. The beam is practically monoenergetic within 1%.

10.4.2.2. End-Hall Sources

In contrast with the broad-beam ion sources, the so-called End-Hall sources do not have any grid for the ion extraction and acceleration, so that they do not present the current density limitations associated to it. The construction is simple, and is based also on a DC plasma discharge between an annular anode and a filament cathode [19], arranged as sketched in Fig. 10.9.

The plasma discharge operates in the near glow regime, at relatively low voltages (200 V) and high plasma currents (several mA cm^{-2}), due to the electron emission supplied by the hot filament as well as to the presence of the axial magnetic field, which drives the electrons in an orbital motion towards the anode, as in the Kaufman source. In their way to the anode they collide with neutral atoms producing ionization mostly in the anode vicinity. The ions formed in the discharge are then accelerated towards the cathode travelling both in the radial and axial directions, thus resulting in a divergent beam exiting the source. The cathode filament is operated in an excess electron emission in order to neutralize the positive charge of ion beam current.

The acceleration of the ions towards the cathode is also enhanced by the inhomogenous distribution of the magnetic field lines, which drives the electrons, and hence the ions, in the direction of lower magnetic field values, that is, towards the *end* of the magnetic field. The ion energy may range between 0–100 eV and is approximately 60% of the anode potential, with an rms deviation of the mean energy value around 30%. The scattering of energetic ions with the background gases in the vacuum chamber produces an additional radial spread of the beam and, consequently, the ions away from axis have lower energies. Low-energy ions produced downstream in the source tend also to be more deflected in the radial direction of the beam.

The End-Hall sources are able to supply significant beam currents, up to several amps, which cannot be attained with grided ion sources. Nevertheless, the angular spread of the ion emission leads to a decrease of the current density with the square

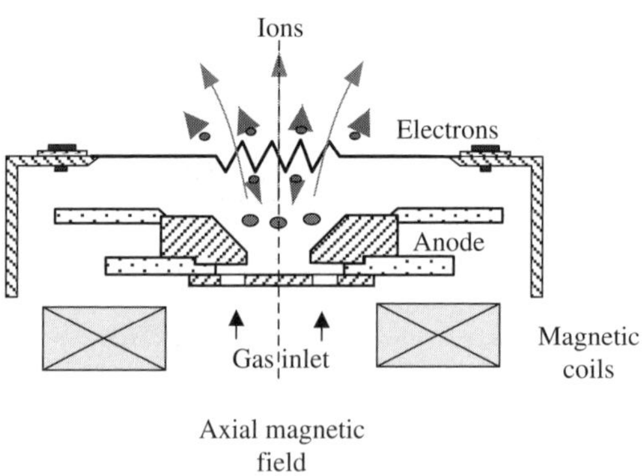

Figure 10.9: Scheme of the End-Hall ion source (adapted from Ref. [19]).

of the distance. The density profile also decays with the angle from the beam axis, being practically 0 at 60°. One limitation of the source, when compared with other types, is the interdependence between the ion current and the accelerating voltage.

Despite this spatial distribution in both the ion energy and current density it is worthwhile to remark that are many physical processes (surface cleaning, film densification, etc.) that depend more on the arrival rate of the ions than in their energy, and these processes can advantageously benefit from the large surplus of low-energy ions provided by the End-Hall sources. For these reasons, these sources have become very popular in the deposition of optical films and coatings with ion assistance.

10.4.2.3. Radio frequency Ion Sources

High frequency discharges are able to accelerate the electrons by the electromagnetic field to energies sufficient for ionization of gas atoms, and for this reason they have been also used since long to create dense plasmas. The radio frequency (RF) sources are based on this concept, using either a coil external to the source case or an electrically isolated inner antenna to induce the discharge. Capacitively coupled sources can be also used in a similar approach, in this case with the cathode electrode coated with an insulating layer. The absence of metal electrodes in direct contact with the plasma allows the operation with any type of gas feed, particularly oxygen, which can easily degrade tungsten filaments, thus rendering a unique advantage over DC sources. This characteristic has found important applications in plasma and reactive ion etching or when long-life operation in a clean atmosphere is needed.

Fig. 10.10 shows the scheme of an RF capacitively coupled source, made of metal chamber and an inner electrode connected to the RF supply. In addition, there is an external magnetic coil, which induces a constant axial magnetic field inside the chamber, avoiding electron bombardment and losses to the anode lateral walls. The bottom wall is protected with an insulating plate, made of Al_2O_3, which acquires a

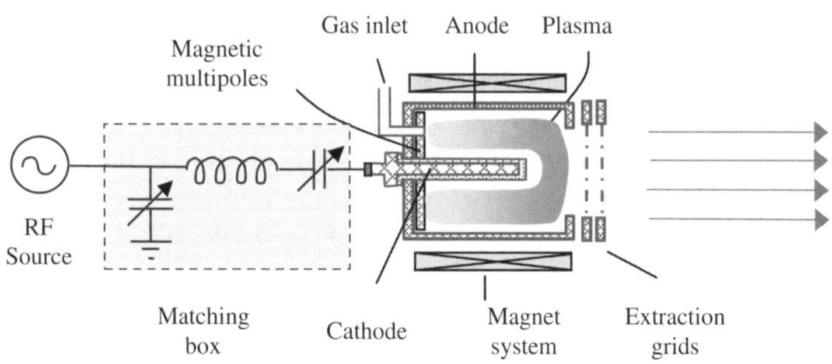

Figure 10.10: Scheme of the RF ion source.

negative potential thus repelling the electrons from the discharge (electrostatic plasma limiter).

In principle, the electrodes in a capacitively induced plasma act alternatively as anode and cathode in every cycle (symmetrical configuration). However, plasma discharges with electrodes of different sizes (asymmetrical configuration), and coupled through an external blocking capacitor, develop a negative bias in the active electrode connected to the source, so that this electrode becomes the cathode of the discharge. For this reason, this electrode is protected with a ceramic Al_2O_3 envelope to prevent ion bombardment and sputtering on its surface. As indicated in Fig. 10.10, the coupling of the RF power to the source requires a matching network to adapt the impedance of the load to that of the source, generally 50 ohm, otherwise it can be severely damaged by the reflected power.

Electrons present in the discharge are easily excited by the RF field oscillating at the same frequency and acquiring enough energy to ionize the gas particles forming the plasma. The source operates at a gas pressure of about 10^{-3}–10^{-5} mbar, with an RF power of a few hundred watts. The RF frequency may vary between one to several tens of megahertz, although most of the sources work at 13.56 MHz, that is, one of the allowed windows in the electromagnetic spectrum for technical applications. With an adequate extraction system, it is possible to obtain from the RF sources beam current densities up to several mA cm^{-2}, in an energy range of 100–2000 eV [20].

10.4.2.4. Electron Cyclotron Resonance Ion Sources

The need of denser plasmas for high emission currents has led to the development of plasma discharges at higher frequencies, with the advent of microwave and electron cyclotron resonance (ECR) sources, initially implemented for space craft propulsion and, latterly, for material processing. As the RF sources, the ECR ion emitters operate with not hot filaments, but at lower pressures, which makes them very attractive for both thin film deposition and etching where reactive gases are generally used.

As it is well known, an electron moving in the presence of a magnetic field B undergo an orbital motion transverse to the field, with a frequency given by: $\omega_c = eB/m$ (cyclotron frequency), m and e being the electron mass and charge, respectively. When an oscillating electric field is applied, the energy transfer from the electric field to the moving electron attains a maximum value when the frequency coincides with ω_c (resonance condition). Most of the ECR sources work at 2.45 GHz because this frequency is also in an allocated window of the electromagnetic spectrum, and the plasma densities achievable at these frequencies are high enough for material processing. The magnetic field intensity required for the resonance conditions is 875 G, which can be easily obtained with conventional electromagnets.

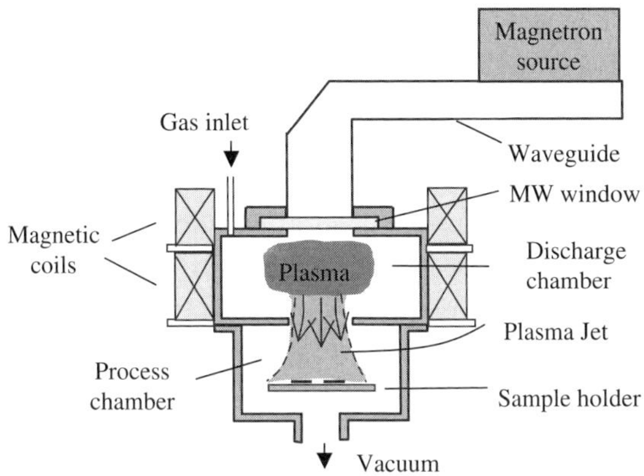

Figure 10.11: Scheme of the ECR ion source provided with a process chamber for thin film deposition or etching.

Though the fundamentals of the microwave sources are in principle similar to that of RF sources, the design and construction of the circuit elements are more complex due to the different propagation and absorption modes of microwaves. In addition, the load impedance of the plasma may vary over several orders of magnitude as the discharge is ignited or adjusted to the operating conditions. A sketch of a typical microwave ion source used for film deposition or etching is shown in Fig. 10.11. The microwave discharge on the chamber is induced through a quartz window, which is located under a coil close to the point of the highest magnetic field induced by the coil, since the microwaves are strongly absorbed and propagated in the plasma for fields slightly above the ECR resonance field of 875 G. So, in order to allow the microwaves to propagate before they are damped in the plasma, a higher magnetic field is required in the source region above the vacuum window [21]. Input powers up to several kW can be coupled with an 85–90% efficiency to create a high-density plasma of 10^{10}–10^{12} cm^{-3} in the chamber (i.e. some 10–100 times larger than in an RF source). The ionization degree may reach up to 10%, allowing the work at vacuum levels down to 10^{-4} mbar. As in other type sources, the use of a multipole system for magnetic confinement has led to the termed *distributed ECR* or DECR sources, with increased electron densities and enhanced plasma homogeneity [22].

ECR ion sources also need an extraction grid system to draw the positive ions from the discharge with a definite energy. However, the ECR discharge can advantageously be used as a plasma reactor for chemical vapor deposition (CVD) by extracting a plasma jet towards a second process chamber, as indicated in Fig. 10.11. For this purpose, a divergent magnetic field B is induced by an additional coil, which pushes the plasma electrons with a force, given by $-\mu\nabla B$, along the field lines in the direction of lower fields (μ is the electron magnetic moment induced by the field). Positive ions are then dragged to the deposition chamber as a consequence of the electric field setup by the charge separation, though they can also be lost to the chamber walls by the perpendicular component associated to their large cyclotron resonance. In this configuration, named *remote plasma*, the possible damage on the substrate surface arisen by the plasma radiation is avoided. The bombardment ion energy reached by this method is in the range 10–50 eV, which is very adequate for the excitation of molecular species in CVD reactions. When more intense ion bombardment on the surface is needed, the substrate can be DC or RF biased, thus increasing the ion energy to a definite value.

10.5. Examples

Owing to the denser films that are obtained by IBAD, this has been a preferred technique in the field of optical coatings. Extensive research work in this field was performed in the 1980s, that allowed the optimization of the refraction index of optical films, and made the IBAD techniques the preferred option in the optical coating industry. Considering the success of IBAD in this field, we have included a few examples of optical coating application.

However, the research interest of IBAD in the last decade has moved to a more complicated issue, the synthesis of metastable phases that are not achievable with other techniques. In the case of optical coatings the use of IBAD allowed the optimization of properties, although the material forming the films existed in nature or could be synthesize in thin film form by other routes. Now the problem is more challenging, since IBAD is allowing the synthesis of out-of-equilibrium phases, as is the case of cubic boron nitride. In this section we will consider first the case of optical coatings, followed by the most recent results on BN, carbon nitride (CN_x), and ternary boron carbonitrides.

10.5.1. Optical Coatings

Among the PVD techniques, IBAD is perhaps the one that has gained major acceptance in optical applications. This is because the possibility to modify the microstructure to obtain denser films at ambient or relatively low temperatures. Films deposited by conventional thermal evaporation generally exhibit voids in the columnar structure, which has a detrimental effect in the optical application of the coatings. Porous films, upon exposure to water or moisture, absorb the water vapor irreversibly by capillarity through the pores, thus modifying the optical properties and the long-term stability of the whole system. In the earlier optical applications (such as antireflective or selective coatings), where only a few layers were implied, this did not pose any serious problem. But with the advent of new optical devices, such as those used in high-power lasers, strict requirements have been imposed on the optical properties, that is, refractive index, absorbance, reflectivity, stability in humid atmospheres, etc.

The effect of water absorption is apparent in the shift toward greater wavelengths, λ, of the maxima and minima of the transmission curves of the optical films with respect to the dense films. In fact, this displacement has been used since long to calculate the packing density of the films, p_d, defined in Section 10.3.1, by assuming that the pores are completely filled with water. Once the refractive index of the film, n_f, is known, p_d can be derived from the empirical formula of Kinoshita–Nishibori [23] based on a simple rule of mixtures (valid for materials with low refractive index):

$$n_f = (1 - p_d)n_v + p_d n_s \tag{10.8}$$

More accurate expressions, based on the classical Maxwell–Garnett model or Bragg and Pippard mode can been also used. In the above formula, n_v and n_s are, respectively, the refractive indexes of the void filled material and of the solid columnar material of the film.

Owing to the importance of the ion bombardment, the densification mechanisms in optical materials have been the subject of numerous studies and theoretical modelling. Two different models have been followed, namely the continuum "*thermal spike*" and the atomistic "*collision-cascade*" approaches, described in Section 10.3. Based on the last model, Müller [24] has elucidated the ion densification effects in thermally grown films by assuming that the vacancies created near the surface, after the cascade collisions produced by the bombarding ions, are partially refilled by the new arrival atoms. When the transport ratio is high enough, this effect produces a densely packed structure under the surface due to inward atom recoil. The results obtained by Martin *et al.* [25] on the variation of the density for ZrO_2 films as a function of the transport ratio, show a good agreement with this model. These authors have also proved that ZrO_2 coatings are less prone to water absorption when deposited under O_2^+ bombardment and, consequently, more stable than thermally grown films after exposure to water vapor.

At present, there is a strong demand in the technology of deposits with high refractive index and band gap for applications either as antireflective or reflective coatings and selective coatings. However, materials meeting these two requirements are difficult to synthesize since the empirical data for non-metallic materials seem to follow an inverse relationship between the index and band gap [26]. Typical materials are transition metal oxides (TiO_2, ZrO_2, CeO_2, etc.), but these compounds need to be deposited with the appropriate stoichiometry because they are generally accompanied with less-stable sub-oxides due to their unfilled atomic d-shells. It has been observed that low-energy bombardment suffices to modify the stoichiometry whereas, higher energies are needed changes in the microstructure of the films.

TiO_2 films are widely used as optical material and represent a good example of the above assert. In its crystalline form, rutile has a very high refractive index, is transparent in wide region of the visible and near infrared (IR) spectra, and is chemically stable. However, there are many difficulties in the synthesis of titania films with the appropriate stoichiometry and crystalline structure. As a consequence, the optical properties are strongly dependent on the deposition conditions (temperature, oxygen partial pressure, deposition rate, etc.). In addition, thermally grown of TiO_2 films present very low packaging density that can be only improved by deposition at high temperatures. Fig. 10.12 shows comparative results for the dispersion curve of the refraction index obtained by thermal evaporation, IBAD and IBS, and evidences the increase in the index values when the deposition is carried out by ion-assistance techniques. The high values of n for the IBS films are attributed to the high packing density of the deposited films [25].

Giving a detailed account of the abundant literature on this subject is beyond the scope of this chapter. For further details (see Refs. [27,28]). From the above and similar results on other metal oxides (Ta_2O_5, CeO_2, SiO_2, etc.) and nitrides (TiN, CrN, etc.),

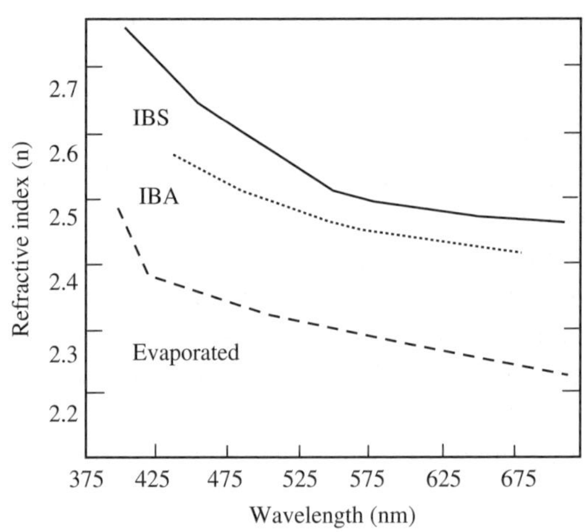

Figure 10.12: Comparison of the refractive index of evaporated, IBAD and IBS of titania films (adapted from Ref. [25]).

it follows that IBAD has gained wide acceptance for the production of denser coatings, with lower moisture pick-up, due to the disruption of the natural columnar structure of the evaporated films. Surface smoothness, internal stresses, and substrate adhesion are also benefited which is an important issue for the mechanical stability of the films. This allows also the deposition of thicker coatings at reduced temperatures. In fact, IBAD is based on pre-existing technologies, no adding high investment expenses. The optical applications of IBAD films run practically in the whole light spectra, from IR to X-rays, and include filters, optical wave-guides, high-power laser coatings, and multilayers for selective or protective coatings, etc.

10.5.2. Nucleation and Growth of Cubic Boron Nitride

BN, as in the case of carbon allotropes, can be found in either amorphous (a-BN), or layered hexagonal (h-BN), wurzite (w-BN), and cubic (c-BN) phases. c-BN is the most attractive allotrope due to its outstanding physical and chemical properties. In particular, c-BN is a promising candidate as an alternative to diamond in microelectronic and mechanical applications, since, its hardness is comparable to diamond, does not react with metals (relevant for coating applications on tools), presents higher thermal stability (up to 1000°C), and can be either *p*- or *n*-type doped [29].

In the different routes for the synthesis of c-BN in thin solid form, a common feature is the requirement of ion bombardment for the nucleation and subsequent growth. Therefore, IBAD techniques have been preferred for the study of the c-BN growth process due to the fine control offered by this method. Normally, a flux of B atoms is provided by electron evaporation under simultaneous assistance with a mixture of N_2 and Ar ions. The use of the heavier Ar ions reduces the ion energy and flux threshold required for the nucleation of the cubic phase.

The growth mode of c-BN normally takes place in a layered sequence, the c-BN phase growing on top of an h-BN seed layer [30]. The thickness of the h-BN layer depends on the growth conditions and, normally, it is of the order of some tens of nm. The top most layer can contain different percentages of the c-BN phase embedded in an h-BN matrix. In order to increase the c-BN content, substantial ion bombardment (ion energies between 100–1000 eV and ion current densities in the range of 0.1–1 mA cm^{-2}) and high substrates temperatures (200–600°C) are required. As an example, Fig. 10.13 shows the resulting phase, as derived from IR spectroscopy, for BN films grown by IBAD at different substrate temperature. The cubic content can be derived from the relative intensity of the corresponding h-BN ($\sim$1400 cm^{-1}) and c-BN ($\sim$1100 cm^{-1}) IR absorption bands. Apart from the substrate temperature, the ion energy and the transport ratio are decisive parameters for the phase

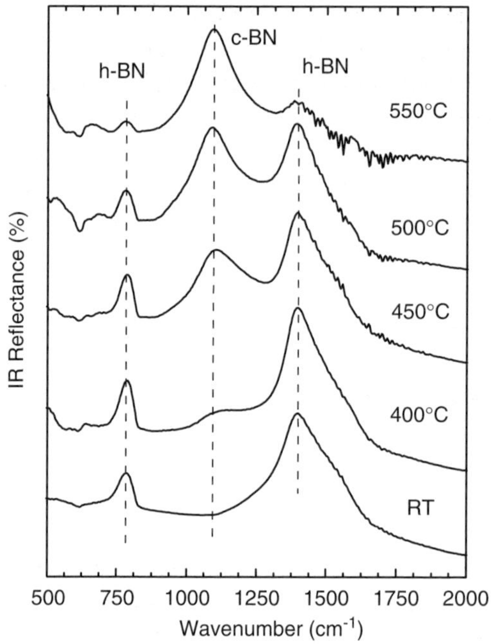

Figure 10.13: IR spectra of BN samples grown by IBAD under different substrate temperatures (500 eV).

control in c-BN. Moreover, the momentum transfer per incoming atom has been shown to be the controlling factor for the final phase configuration [31].

The substantial high ion bombardment necessary for the nucleation of c-BN has additional detrimental implications, since a high intrinsic compressive stress in the film is generated up to 10–20 GPa. This fact has precluded the actual application of c-BN despite its outstanding properties. Therefore, many current attempts are focused on the development of mechanisms for stress relaxation.

The generation of stress by ion bombardment in c-BN fits well with the predictions of Davis's model [32]. In this case, the state of compressive stress should be maximum for the ion energies involved in the growth of c-BN (100–500 eV). According to Davis's model, the stress value could be reduced by increasing the ion energy above this range ($\sim$1 keV), however, this results in a decrease in the growth rate or complete re-sputtering of the film. Another approach to reduce the stress considers the different window parameters for nucleation and growth, so the requirements for subsequent growth are gentler than for nucleation such as lower substrate temperatures ($>$100°C) and low-ion energy ($>$100 eV) [31]. However, the level of intrinsic stress released in this case is still not sufficient for practical applications.

Recently, the use of post-deposition treatments like ion implantation (10–100 keV) and thermal annealing (up to 900°C) has shown effective results [33]. Therefore, the incorporation of these methods in a sequential or simultaneous process during the growth are the most promising alternatives for the growth of thick c-BN films. However, there are still other issues to be addressed such as the reduction of stress at the interface and, in this way, improving the adhesion of a stress-free c-BN layers to the film structure.

10.5.3. Binary and Ternary Carbon-Based Materials

In this section we will consider some examples of reactive IBAD, to produce compound materials of the B-C-N system that do not exist in nature and cannot be synthesized with conventional PVD techniques.

10.5.3.1. Carbon Nitride

The synthesis of CN_x by IBAD using carbon atoms and nitrogen ions represent a clear example of reactive IBAD using chemical species with low reactivity between them. Therefore, chemical sputtering is a very competitive process with the film growth.

The growth of CN_x thin films has been an intense research topic due to the predictions of outstanding mechanical properties for a hypothetical cubic phase that would be harder than diamond. Although the synthesis of such a cubic phase has not been achieved, the research effort has produced new materials with interesting mechanical properties that are used as the protective coatings of magnetic hard disks in the present technology. Such CN_x are based on a graphitic carbon network with substitutional nitrogen atoms. Depending on the growth parameters, one can obtain a hard and elastic fullerene-like structure, due to the folding and cross-linking of the basal planes [34].

The synthesis of CN_x requires the use of energetic ions, either by direct assistance with an ion beam, or by biased magnetron sputtering. Here, we focus on the synthesis by IBAD using carbon evaporation assisted with mixtures of N_2^+ and Ar^+ ions. The optimal ion energy range to obtain a dense film, with the minimum ion-induced damage, is about 100 eV, with a transport ratio close to unity. The reactivity between carbon and nitrogen is low, with a high tendency to form volatile moieties that are desorbed from the surface, a process known as chemical sputtering. As a result, the film growth rate decreases significantly with N_2^+ assistance, as compared to bombardment with Ar^+ ions of the same energy and transport ratio, as it is shown in Fig. 10.14. However, the incorporation of nitrogen into the film is only found for conditions that also induce chemical sputtering, as is shown in

Fig. 10.15. Both processes, nitrogen incorporation and chemical sputtering are energy dependent [36].

Nitrogen is incorporated in the film mainly in pyridine-, cyanide-, and graphite-like environments. The ratio of nitrogen atoms in these environments determines the structure and properties of the film, being the desired one the three-fold graphitic that yields continuity to the graphene sheets. Substrate temperatures of ~500°C are

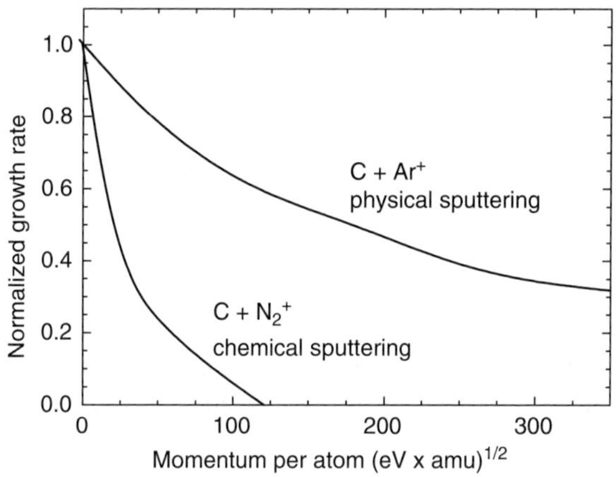

Figure 10.14: Growth rate in IBAD assisted with Ar and nitrogen ions, normalized to the carbon evaporation rate, as a function of the momentum per impinging carbon atom (adapted from Ref. [35]).

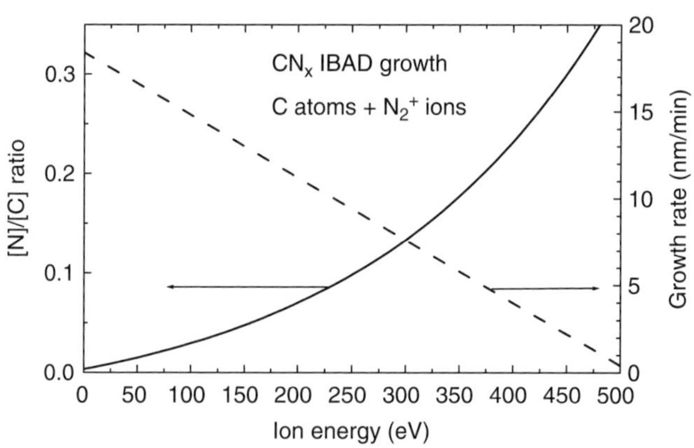

Figure 10.15: Growth rate and N incorporation in CN_x films grown by carbon evaporation with nitrogen assistance, as a function of the ion energy (adapted from Ref. [36]).

required for the optimal fullerene-like structure. Regarding the microstructure, the material ranges between amorphous and nanocrystalline, without XRD-long range order, although transmission electron microscopy (TEM) pictures show a certain orientation of basal planes or a fullerene-like structure [34]. Neither crystalline grains can be recognized, nor crystallographic defects, although vacancies in the graphene sheets must be common according to the structural models proposed. The films are dense, without any columnar growth or voids. However, textured films are common, despite the lack of long-range order, due to the compressive stress that tends to orient the basal planes perpendicular to the surface.

Regarding the application of CN_x as protective coating in hard disks, it is necessary to obtain the hardest structure with the lowest friction coefficient, using growth conditions that do not produce any damage on the supporting magnetic medium. Therefore, a compromise is found between the mechanical performance of the coating and the processing conditions.

10.5.3.2. Ternary Boron–Carbon–Nitrogen Compounds

The synthesis of ternary Boron–Carbon–Nitrogen (BCN) films is a representative example of the use of IBAD techniques with two independent evaporation sources, in this case boron and carbon, assisted by nitrogen ions. Fig. 10.16 shows a scheme of the author's setup for production of ternary BCN coatings based on two electron beam evaporators plus a Kaufman ion gun for assistance.

The structural similarity between the allotropic forms of carbon and BN, and the fact that BN pairs are isoelectronic to CC pairs, has stimulated intensive research in the last decade towards the synthesis of ternary BCN compounds that may reunite the advantages of C and BN. Cubic BCN compounds (c-BCN) are expected to combine the properties of diamond and cubic boron nitride (c-BN), namely, extremely

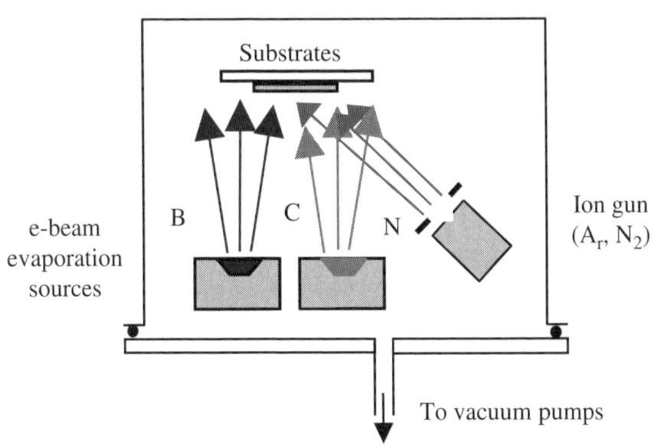

Figure 10.16: Scheme of a dual evaporator IBAD system.

high hardness, chemical stability at high temperatures and electronic gap of ~6 eV, whereas hexagonal structures (h-BCN) are expected to behave as semiconductors with various band gap energies, ranging from metallic character for carbon-rich to wide gap semiconductors for carbon-depleted compounds. Moreover, the hexagonal phase can exhibit the fullerene-like structure observed in CN_x.

According to the above isostructural and isoelectronic premises, a stable ternary BCN compound should have a BC_xN composition, that is, equal number of boron and nitrogen atoms, to fulfill charge neutrality conditions. Otherwise, the ternary compound is not stable and will decompose into segregated phases (C, BN, B_4C, etc). Apart from the chemical routes, that generally produce a hydrogenated BCN with the hexagonal structure, the synthesis of ternary BCN films has been studied by magnetron sputtering and by IBAD techniques. The first reports on IBAD synthesis of BCN were based on the evaporation of B_4C assisted with N_2^+ and Ar^+ ions, that yielded films with an intrinsic limitation in the carbon content $[C]/[B] < 0.25$ due to the B_4C source, far away from the more promising BC_2N and BC_4N compositions [35]. Depending on the ion energy, transport ratio and substrate temperature, it is possible to obtain an sp^3 phase of the ternary compound, although with a limited carbon content of 5 at.% [37].

More recent work has exploited the use of a dual evaporation system that allows the independent evaporation of boron and carbon, and the assistance with N_2^+ and Ar^+ ions. In this way, independent control on the atomic fluxes of B, C, and N is achieved, permitting a fine control of the BCN composition [38]. Although the synthesis of ternary BCN compounds is possible, only in a few reports the production of a real ternary compound is clear, since the presence of BC, BN, and CN bonding is not enough to discern between a true ternary phase and a mixture of segregated binary phases. Despite the ease to obtain hexagonal BCN with controlled composition, the synthesis of cubic BCN with large carbon contents has not been successful up-to-date, remaining an open research field.

References

[1] M. Nastasi, J.W. Mayer and J.K. Hirvonen, in *Ion–Solid Interactions: Fundamentals and Applications*, Cambridge University Press,1996.

[2] A.R. González-Elipe, F. Yubero and J.M. Sanz, in *Low-energy Ion Assisted Film Growth*, Imperial College Press, 2003.

[3] J.F. Ziegler, J.P. Biersack and U. Littmark, in *The Stopping and Range of Ions in Solids*, Pergamon, 1985.

[4] I. Petrov, F. Adibi, J.E. Greene, L. Hultman and J.-E. Sundgren, *Appl. Phys. Lett.*, 63 (1993) 36.

[5] H.A. Durand, K. Sekine, K. Etoh, K. Ito and I. Kataoka, *Thin Solid Films*, 336 (1998) 42.

[6] H. Ji, G.S. Was, *Nucl. Instrum. Methods.*, B 148 (1999) 880.

[7] N.D. Telling, M.D. Crapper, D.R. Lovett, S.J. Guifoyle, C.C. Tang and M. Petty, *Thin Solid Films*, 317 (1998) 278.

[8] R. Thielsh, A. Gatto, J. Heber and N. Kaiser, *Thin Solid Films*, 410 (2002) 86.

[9] V. Kirchner, H. Heinke, U. Birkle, S. Einfeld, D. Hommel, H. SelFe and P.L. Ryder, *Phys. Rev.*, B 58 (1998) 15749.

[10] Z.Q. Ma and Y. Kido, *Thin Solid Films*, 359 (2000) 288.

[11] F.M. d'Heurle and J.M.E. Harper, *Thin Solid Films*, 171 (1989) 81.

[12] H.J. Shin, Y.J. Cho, J.Y. Won, H.J. Kang, C.H. Baeg, J.W. Hong and M.Y. Wey, *Nucl. Instrum. Methods.*, B 190 (2002) 807.

[13] L. Dong and D.J. Srlolovitz, *J. Appl. Phys.*, 84 (1998) 5261.

[14] H.J. Robertson, *Diam. Relat. Mater.*, 3 (1994) 361.

[15] F.A. Smidt, *Int. Mater. Rev.*, 35 (1990) 62–128.

[16] G.K. Wolf, *J. Vac. Sci. Technol.*, A10 (1992) 1757–1764.

[17] N. Gravilov, in *The Physics and Technology of Ion Sources*, Chapter 7, Ed. I.G. Brown, Wiley-VCH (2004).

[18] H.R. Kaufman, J.J. Cuomo and J.M.E. Harper, *J. Vac. Sci. Technol.*, 21 (1982) 725–736.

[19] R.S. Robinson and H.R. Kaufman, in *Handbook of Ion Beam Processing Technology*, Chapter 4, Eds. J.J. Cuomo, S.M. Rossnagel and H.R. Kaufman, Noyes Publications, 1989.

[20] K. Leung, in *The Physics and Technology of Ion sources*, Chapter 9, Ed. I.G. Brown, Wiley-VCH, 2004.

[21] J.E. Stevens, in *High Density Plasma Sources*, Chapter 7, Ed. O.A. Popov, Noyes Publications, 1995.

[22] J.L. Peletier, in *High Density Plasma Sources*, Chapter 8, Ed. O.A. Popov, Noyes Publications, 1995.

[23] K. Kinoshita and M. Nishibori, *J. Vac. Sci. Technol.*, 6 (1969) 730–733.

[24] K.-H. Müller, *J. Appl. Phys.*, 59 (1986) 2803–2807.

[25] J.P. Martin and R.P. Netterfield, in *Handbook of Ion Beam Processing Technology*, Chapter 19, Eds. J.J. Cuomo, S.M. Rossnagel and H.R. Kaufman, Noyes Publications, 1989.

[26] U. Gibson, in *Physics of Thin Films*, Vol. 13, Academic Press, 1987.

[27] S. Mohan and M.G. Krishna, *Vacuum*, 46 (1995) 645–659.

[28] F. Flory and L. Escoubas, *Prog. Quant. Electron.*, 28 (2004) 89–112.

[29] P. Mirkarimi, K. McCarty and D. Medlin, *Mat. Sci. Eng. R*, 21 (1997) 47.

[30] D.J. Kester, K.S. Ailey, D.J. Lichtenwalner and R.F. Davis, *J. Vac. Sci. Technol.*, A 12 (1994) 3074.

[31] W. Kulisch and S. Ulrich, *Thin Solid Films*, 423 (2003) 183.

[32] C.A. Davis, *Thin Solid Films*, 226 (1993) 30.

[33] J. Ullmann, A.J. Kellock and J.E.E. Baglin, *Thin Solid Films*, 341 (1999) 238.

[34] L. Hultman, J. Neidhardt, N. Hellgren, H. Sjöstrom and J.E. Sundgren, *MRS Bull.*, 28 (2003) 194.
[35] R. Gago, I. Jiménez, F. Agulló-Rueda, J.M Albella, Zs. Czigány and L. Hultman, *J. Appl. Phys.*, 92 (2002) 5177.
[36] R. Gago, I. Jiménez, J.M. Albella, F. Agulló-Rueda, D. Cáceres, I. Vergara, A. Climent-Font, T. Sajavara, J.Räisänen and E. Raühala, *Chem. Mater.*, 13 (2001) 129.
[37] R. Gago, I. Jiménez, T. Sajavara, E. Raühala and J.M. Albella, *Diam. Relat. Mater.*, 10 (2001) 1161.
[38] I. Caretti, I. Jiménez and J.M. Albella, *Diam. Relat. Mater.*, 12 (2003) 1079.

Chapter 11

Cathodic Arc Evaporation and its Applications to Thin-Film Synthesis

Marie-Paule Delplancke-Ogletree

11.1. Introduction

The development of arc plasma physics is closely related to the development of electricity and thus can be traced back to the 18th century. Recent articles by Anders [1,2] and Boxman [3] are tracking down the origin of arc plasma in pulsed, oscillating and continuous modes, and provide extensive references to early work.

Cathodic arc plasma deposition is among the oldest energetic thin-film deposition method. T.A. Edison and A.W. Wright were the first to use the arc discharge to process materials [4–6]. Despite this early start, the first commercial applications of the vacuum arc were only developed in the early part of the 20th century. The use of arc to process materials has very quickly matured in recent decades. High-speed computers, wireless communications, high-current switching devices, optical and electronic applications would not have been developed without modern materials technologies.

An arc discharge can be defined as an electrical discharge of relatively high current at relatively low voltage. It is characterized by a collective mechanism of electron emission from the cathode. Vacuum arcs can be created either in a high-vacuum or in a low-pressure gaseous environments, and films composed of metals, ceramics, diamond-like carbon (DLC), some semiconductors and superconductors, and other materials can be prepared. The method is a versatile and powerful tool for synthesizing novel and technologically interesting surfaces. The interest in the cathodic arc technique arises from the high deposition energy of the deposited particles which is important to promote the adhesion of the film and to disrupt columnar growth that is so detrimental for corrosion resistance for example.

Excellent books, devoted exclusively to arc physics, deposition and technology were published in the last few years [7,8]. The present chapter aims to introduce the characteristic features of this deposition technique and the related implications for the properties of the deposited films.

Materials Surface Processing by Directed Energy Techniques
Y. Pauleau (Editor)

11.2. Cathodic Arc Nature and Characteristics

An energetic deposition can be defined as a deposition process in which a large fraction of the depositing particles arrive on the substrate with a kinetic energy greater than the bulk displacement energy. This energy is not well defined as it depends on the crystallography of the substrate and the relative impinging direction of the bombarding particles. The bombardment induces the formation of defects such as Frenkel pairs. Displacement energy vary between 10 and 40 eV depending on the material [9]. The flux of energetic particles can either be supplied from an independent source which assists the deposition process or the film-forming particles themselves can be energetic and the growth does not require assistance. Different techniques are available for energetic film deposition: ion-beam deposition, ion-beam-assisted deposition (IBAD), ionized and pulsed magnetron sputtering, pulsed laser plasma deposition, cathodic vacuum arc plasma deposition, and metal plasma immersion ion implantation and deposition (MePIIID). The vacuum arc and MePIIID will be developed in the present chapter while the other methods are presented in other chapters of the book. Reviews on deposition technology with vacuum arcs were published recently [10–13] and contain extensive references to earlier work.

11.2.1. Energy

A vacuum arc is a plasma discharge between two metallic electrodes in vacuum. Arc spots on the cathode are strange and fascinating phenomena. Tiny bright points of light run rapidly and irregularly across the surface. The consumable cathode is transferred to the plasma state at these microscopic (diameter of the order of 10 μm) non-stationary spots characterized by extremely high current density ($\sim 10^{11}\,\mathrm{A\,m^{-2}}$), power density (over $10^{12}\,\mathrm{W\,m^{-2}}$), and plasma density [14,15]. These characteristics are due to a regime of electron emission that is totally different from the abnormal glow discharges used for example in sputtering. In arcing conditions, the thermo-ionic and field emission of electrons grows rapidly and gives rise to collective processes caused by many plasma particles. To observe a strong thermo-field emission, the surface temperature of the spot root and the electric field strength at the surface must be high enough. Both conditions can be fulfilled by the discharge itself due to the ion flow from the plasma to the cathode surface:

1. The ions accelerated in the cathode sheath deliver their kinetic, thermal, and recombination energy during impact with the cathode surface [16]. The temperature of this surface increases strongly. In addition, resistive heating by the arc current contributes to the increase in surface temperature.

2. The space charge due to the ions flowing toward the cathode generates a high electric field.

A high ion current density is only possible if the plasma density in front of the cathode is very high. Intense evaporation from the surface is necessary to provide this condition. These conditions for sufficient emission are only fulfilled if the arc spots contract.

Arc spots move across the surface of the cathode very quickly, 0.1–100 $m\,s^{-1}$, in most cases, and in random directions [17]. There are two possible explanations for this movement. The spot itself modifies local cathode properties by its own action until it can no longer survive. It then reignites nearby where local conditions are better. Another possibility is that the conditions remain satisfactory on the spot but better conditions develop suddenly in the neighborhood. This could be due to the explosion of a pre-existing micro-protrusion, or formed by ejected liquid matter from the first spot. The arc spot leaves behind, a crater that demonstrates the extremely intense interaction between the cathode spot and the solid surface [18]. In a few nanoseconds, intense heating and strong mechanical forces induce melting and excavation.

The ions leave the cathode spots with a very high kinetic energy. They are accelerated by three forces:

1. The pressure gradient within the cathodic plasma.
2. The electron–ion friction.
3. The electric field that has the opposite direction in the plasma expansion zone.

Multiply charged ions are formed by thermal and pressure ionization in the cathodic plasma [19].

The plasma jets reach speeds of 10^3–$10^4\,m\,s^{-1}$ with ion energies of 20–200 eV, depending on the nature of the evaporated material.

11.2.2. Charge

Table 11.1 gives the ion charge distribution and energy as well as the cohesive energy and displacement energy for various materials and selected cathodic (vacuum) arc plasmas [9]. Higher average ion charge states are observed for the materials with higher cohesive energies. The ions are usually multiply charged, although this depends on the cathode material. The burning voltage, power density, average ion charge state, electron temperature, and ion kinetic energy are all related to the cohesive energy of the cathode material. Brown [20] measured these values for a large number of elements and Anders [19] presented them in the form of a periodic table. The flux of neutral atoms is usually negligible compared to the flux of ions. The ion charge states can be manipulated by magnetic fields and other methods [21].

Table 11.1: Cohesive energy, displacement energy, burning voltage at 300 A, kinetic energies, charge states (particle fractions), and ionization energies for selected cathodic arc plasmas

Cathode material	Cohesive energy (eVatom^{-1})	Displacement energy (eV)	Arc burning voltage (V)	Kinetic ion energy (eV)	E_0 (eV)	$f(1+)$ (%)	E_1 (eV)	$f(2+)$ (%)	E_2 (eV)	$f(3+)$ (%)	E_3 (eV)	$f(4+)$ (%)	E_4 (eV)	$f(5+)$ (%)
Li	1.63		23.5	19	5.39	100	75.64	0		0		0		0
C	7.37	25	29.6	19	11.3	100	24.38	0		0		0		0
Graphite														
Mg	1.51	10	18.8	49	7.65	46	15.04	54	80.14	0		0		0
Al	3.39	16	23.6	33	5.99	38	18.83	51	28.45	11	120.0	0		0
Si	4.63	13	27.5	34	8.15	63	16.35	35	33.49	2	45.14	0		0
Ti	4.85	19	21.3	59	6.83	11	13.76	75	27.49	14	43.27	0		0
V	5.31	26	22.5	70	6.75	8	14.66	71	29.31	20	46.71	1	65.28	0
Cr	4.1	28	22.9	71	6.77	10	16.49	68	30.96	21	49.16	1	69.46	0
Fe	4.28	17	22.7	46	7.90	25		68	30.65	7	54.80	0		0
Co	4.39	22	22.8	44	7.88	34	17.08	59	33.50	7	51.30	0		0
Ni	4.44	23	20.5	41	7.64	30	18.17	64	35.19	6	54.90	0		0
Cu	3.49	19	23.4	57	7.73	16	20.29	63	36.84	20	57.38	1	79.80	0
Zn	1.35	14	15.5	36	9.39	80	17.96	20	39.72	0		0		0
Ge	3.85	15	17.5	45	7.90	60	15.93	40	34.22	0		0		0
Y	4.37		18.1	80	6.22	5	12.24	62	20.52	33	60.60	0		0
Zr	6.25	21	23.4	112	6.63	1	13.13	47	22.99	45	34.34	7	80.35	0
Nb	7.57	28	27.0	128	6.76	1	14.32	24	25.04	51	38.30	22	50.55	2
Mo	6.82	33	29.3	149	7.09	2	16.16	21	27.13	49	46.40	25	54.49	3
Pd	3.89	26	21.3	131	8.34	23	19.43	67	32.93	9	60.87	1	78.25	0
Ag	2.95	23	23.0	69	7.58	13	21.49	61	34.83	25	60.52	1	80.01	0
Cd	1.16	19	16.0	27	8.99	68	16.91	32	37.48	0		0		0
In	2.52	15	17.5	21	5.79	66	18.87	34	28.03	0		0		0
Sn	3.14	22	17.5	30	7.34	47	14.63	53	30.50	0		0		0
Ta	8.1	32	28.7	136	7.89	2	14.47	33	23.49	38	36.32	24	49.14	3
W	8.9	38	31.9	117	7.98	2	15.08	23	25.43	43	39.29	26	53.15	6
Pt	5.84	33	22.5	67	9.00	12	19.24	69	35.25	18	51.27	1	67.28	0
Au	3.81	36	19.7	49	9.23	14	20.50	75	37.37	11	54.80	0		0

Data from Ref. [9].

11.2.3. Time Scale

As mentioned before, the cathode spots are moving very fast (residence time in the sub-microsecond range) and their characteristic size is smaller than 1 μm. The specifics depend on the cathode material and surface conditions [22]. Two main types of arc spots have been identified and are historically labeled types 1 and 2 (Fig. 11.1). Type-1 arc spots are comparatively fast, have a low brightness and arc current per spot. They are associated with surface contaminants and adsorbed gases on the cathode surface. Due to their characteristics, they do not contribute significantly to ablation of the cathode material. Type-2 arc spots are slower (around five times), have a high brightness and high current per spot. This type of spot is responsible for the ablation of the material. The transition between the two types of arc spot is especially important in the case of pulsed vacuum arc plasma sourced operating in reactive environments. It has been shown [23] that the composition of the plasma operated in an oxygen atmosphere depends on the arc pulse frequency. This is due to the formation of adsorbed gas monolayers on the cathode surface between pulses. The pulse frequency necessary to avoid the adsorption of significant amounts of contaminants and thus to operate in a stable regime will depend strongly on the background gas pressure.

The plasma generated at the cathode spots expands rapidly into the ambient atmosphere. Electron–ion coupling results in observed final ion velocities in the range of $1–2 \times 10^4\,\mathrm{m\,s^{-1}}$ which are more or less independent of the cathode material and ion charge state (i.e. Ref. [24]).

(A) (B) (C)

Figure 11.1: CCD images of an aluminum cathode showing transition form type-1 spot mode to type-2 mode. Image (a) is taken 8 pulses after the cathode was inserted into the vacuum system. After 49 pulses (b) the arc begins to operate initially in a type-2 mode before reverting to type-1 as contaminated regions are reached. After 62 pulses (c) the cathode surface has been cleaned of contaminants and the arc is operating purely in type-2 mode. Image size 6 cm × 6 cm (from Ref. [22]).

11.2.4. Macroparticles

Unfortunately, the cathode spots not only generate a flux of ionized plasma, they are also the source of droplets and debris particles in the micrometer and sub-micrometer range commonly referred to as *macroparticles* (Fig. 11.2). This is due to the explosive nature of plasma formation in a cathode spot. The formation and size distribution of macroparticles have been extensively studied (i.e. Refs. [25–28]). The term "macroparticles" covers a wide range of sizes and should be understood for all particles that are much larger than atom clusters. For applications such as decorative or tool coatings, the presence of macroparticles could be tolerated. For higher end applications, such as precision optics and electronics, macroparticles must be removed or suppressed without significantly reducing the ion flux reaching the substrate. This is the main challenge for the development of industrial applications of cathodic arc plasma deposition. The most successful means to reduce film contamination by macroparticles are based on the great difference in charge-to-mass ratio between ions and macroparticles. Magnetic filtering is the most classical and commonly used method and will be presented in Section 11.3.4.

11.3. Types and Modes of Cathodic Arc Evaporation

Choosing the continuous or pulsed mode of operation for a cathodic arc depends mainly on the aims to be achieved. DC systems are able to provide very high deposition rates on large areas. They are especially convenient to deposit thick films for applications that do not require reduction or elimination of macroparticles. Most of

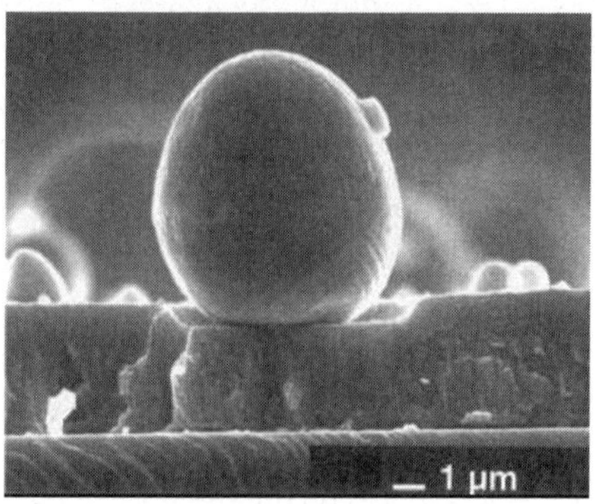

Figure 11.2: Cathodic arc erbium macroparticle deposited during growth of an erbia film (from Ref. [8]).

the commercially used arc systems operate in the DC mode. On the other hand, pulsed systems have relatively small deposition rates with significant advantages in terms of cooling, plasma density, stability, and controllability.

11.3.1. Continuous Arc (DC)

This mode of operation requires a suitable arc power supply and adequate cooling due to ohmic heating. The cathode and the anode of the source are usually water cooled. DC sources typically operate from 30 up to 300 A. The requirements on cables and electrical feedthroughs are very demanding. The cathode is often relatively large. Without magnetic fields, cathode spots move randomly on a homogeneous surface. However, they have a tendency to reside longer on inhomogeneities such as contamination sites, grain boundaries, and inclusions. The resulting erosion rate is not smooth. To prevent this phenomenon, most DC sources are equipped with magnetic fields that steer the cathode spots within well-defined erosion areas. A more detailed description of DC source design can be found in the literature [29].

Unfortunately, DC arcs can display density fluctuations on the millisecond time scale due to arc spot instabilities and on the second time scale due to variations in the degree of plasma coupling to the filtering system. The production of ultra-thin films with reproducible performance under these conditions is difficult [22].

11.3.2. Pulsed Arc

Pulsed cathodic arc plasma sources are characterized by currents up to 5000 A, pulse length of few milliseconds to few nanoseconds and relatively high repetition rates (from few Hz to few kHz). A major advantage of pulsed operation is the relatively low average power which reduces the cooling requirements. No electrode cooling is necessary if the source is operated at low duty factor. The duty factor δ is defined by the ratio of the arc on-time (t_{on}) to the duration of the pulse cycle:

$$\delta = t_{on}/(t_{on} + t_{off}) \tag{11.1}$$

In addition, pulsed sources are relatively easy to build and cathode spots can be easily retained at the front cathode surface. High-current arc systems (in the kA range) can work at high pulse repetition rates and provide a stable and reproducible high plasma density [30–32]. The Siemroth group has reported that high-current pulsed operation produces less macroparticles and in the case of carbon, a higher energy of ejected particles [33].

From the point of view of the depositing film, pulsed operation provides flexible deposition conditions. Amorphous carbon films deposited using the pulsed mode tend to have a higher sp^3 content than DC-deposited films. On the other hand, pulsed deposition can enhance the incorporation of contaminants.

11.3.3. Reactive Arc Deposition

Continuous and pulsed cathodic arc sources are extensively used to prepare compound coatings (see Section 11.6). In many cases, one of the elements is introduced as a gas in the deposition chamber. This gas interacts with the ion flux emitted by the cathode. Bilek and co-workers experimentally studied the influence of gas pressure and flow rate and cathode composition on ion energy distribution in filtered cathodic vacuum arcs [34,35]. In absence of gas, the ion energy distribution is described by a shifted Maxwellian curve. When a background gas is added, the distribution changes to one with higher Maxwellian temperature and smaller directed velocity. At still higher pressures, the distribution splits into two components: one of thermalized material and another of ions that had little interaction with the background gas (Fig. 11.3). The energy distribution of background gas ions shows a single contribution corresponding to a "cold" distribution. Simultaneously, the charge state distribution is modified: charge exchange collision processes and electron impact ionization both play an important role in determining the number of ions in each charge state. The total ion current is reduced.

Boxman *et al.* modeled the interaction between a vacuum arc plasma jet and a neutral background gas [36].

The modifications of the ion energy and charge state distributions have an impact on the growth mechanisms of the films.

11.3.4. Filtered Arc Deposition

Various types of macroparticle filters have been proposed in the literature and a few of them are patented. The basic principle of filtering is to prevent line-of-sight macroparticle trajectories between cathode and the substrate and to efficiently guide the plasma to the substrate [11].

The transport must be as efficient as possible. Transport efficiency is defined as the ratio between the number of ions leaving the filter and the number entering it. As these numbers are difficult to determine experimentally, the ratio is replaced by the "system coefficient", κ, that characterizes not the filter alone, but the coupled

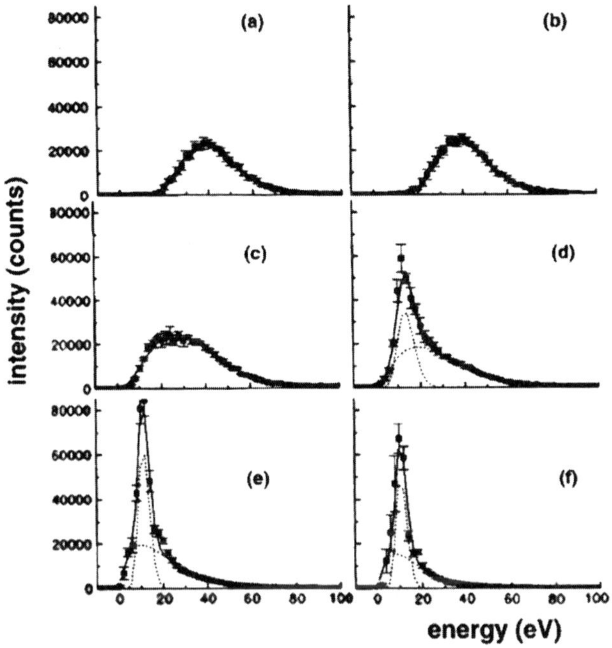

Figure 11.3: Ion energy distributions for Ti^+ as a function of background nitrogen gas pressure for pressures of (a) 3×10^{-7}, (b) 10^{-4}, (c) 5×10^{-4}, (d) 10^{-3}, (e) 1.5×10^{-3}, and (f) 2×10^{-3} mbar. The solid curve is the fit to the data points (from Ref. [34]).

arc source and filter system:

$$\kappa = I_{i,\text{filtered}}/I_{\text{arc}} \tag{11.2}$$

where $I_{i,\text{filtered}}$ is the filtered ion current and I_{arc} is the arc current. Hypothetically the filter plasma transport efficiency may reach 100% but the system coefficient can never reach 100%, since the electrons carry most of the arc current. κ is typically around 1%.

Plasma transport is obtained by a mechanism combining an applied magnetic field and a self-generated electric field. The magnetic field, usually axial, is used to "drive" the plasma to the substrate. The electrons, having a gyration radius smaller than the filter size and a collision frequency smaller that the gyration frequency, follow and spiral around the magnetic field lines. The ions, having a larger gyration radius, do not follow the magnetic field lines. Instead, they electrostatically follow the electrons to maintain plasma quasi-neutrality. Due to their larger mass-to-charge ratio macroparticles move along straight trajectories and arrive at the walls of the filter. They can either stick to the walls or leave the filter through opening if the filter has an open architecture. Unfortunately, a small proportion of the macroparticles can be reflected by the walls in the direction of the substrate, so filtering does not guarantee a particle-free film.

The filter geometry, mode of operation, and architecture allow us to categorize them. The architecture can be open or closed. The former is preferred for cathode materials such as carbon which forms macroparticles that bounce from surfaces [37]. The macroparticles can leave the plasma region through the spaces between the solenoid turns. The closed architecture is preferred for DC-operated filters.

Simple, straight, magnetic systems can reduce the macroparticle content of the deposited films [38]. Most macroparticles are ejected at low angles to the cathode surface and impact the walls of the tube. The plasma is streamed toward the substrate. Shielding of the substrate by an intermediate deflection plate (Fig. 11.4(a))

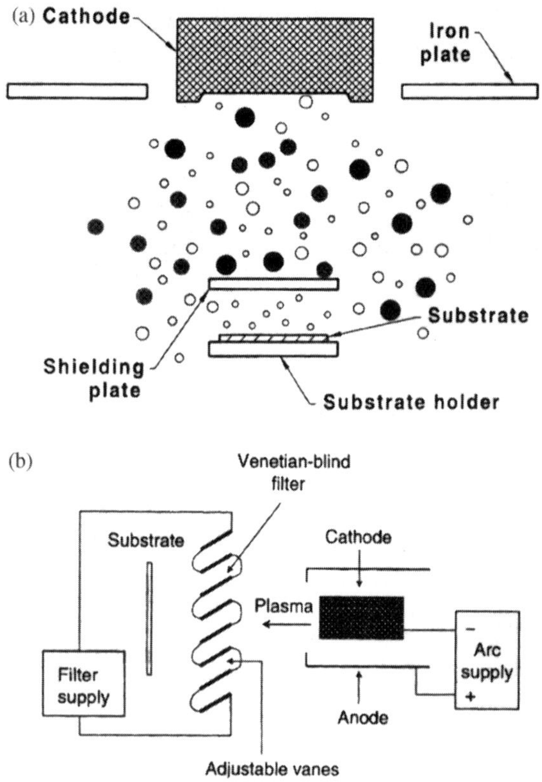

Figure 11.4: Line-of-sight macroparticle filters: (a) simple shielding method [39] and (b) Ryabchikov Venitian Blind filter [41,42] (from Ref. [11]).

reduces the number of large macroparticles but reduces drastically the deposition rate [39]. This effect can be reduced by adding a magnet or an electromagnetic coil behind the substrate [40]. Ryabchikov *et al.* [41,42] proposed a filter in the form of a Venitian blind (Fig. 11.4(b)), geometry that has also been studied by other groups [43]. A series of plates deflect the plasma through the application of a high current while the macroparticles are limited by the lack of line-of-sight trajectories between the cathode and substrate.

Real non-line-of-sight devices provide more effective filtering than straight geometries. Aksenov and co-workers introduced a toroidal 90° filter in their pioneering work in the 1970s [44,45]; 45° "knee" and 30° filters are also commercially available. Circular and rectangular section have been proposed [46]. Complex combinations such as in-plane S and out-of-plane double bend filters have also been proposed [47–50] (Fig. 11.5). T-shape filters have been recently introduced [51]. The key issue with all these geometries remains the maximization of ion transmission, however the filtering performance of such complex systems are remarkable.

Several modeling and experimental studies have been conducted to optimize the transport efficiency [52,53]. It can be maximized by focusing the plasma as a jet into the duct and by biasing the duct with respect to the plasma by about +10 to +20 V.

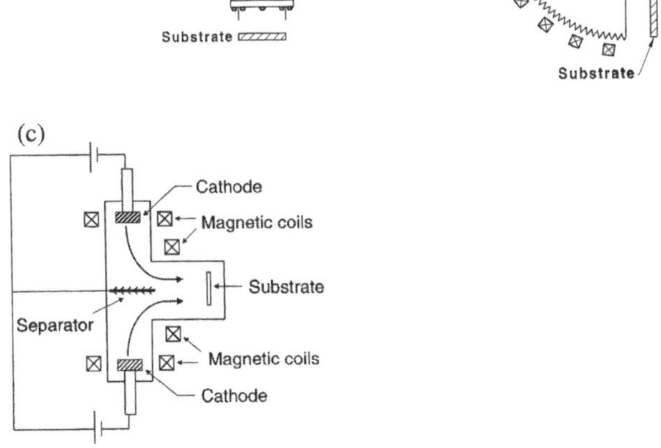

Figure 11.5: Non-line-of-sight filtering methods: (a) toroidal magnetic filter [45], (b) "S"-shape filter [47–50] and (c) industrial scale system [56,57] (from Ref. [11]).

If biasing is difficult or impossible, a similar effect can be obtained by inserting a curved strip electrode inside the duct along the outer major radius [54]. The influence of the magnetic field strength and bias voltage on the ion charge state fractions, mean charge state and flux of ions has also been modeled and measured [55]. Linear filtered arc source have been developed during the last 5 years. The Gorokhovsky team's source represented in Fig. 11.5(c) provides a large area deposition with a high transmission efficiency. Two arc sources share the same linear plasma filter [56,57]. The separator plate bias is one of the variables that can be adjusted. This plate physically separates the two plasma streams but contributes to the plasma steering mechanism in the same way as a biased strip in the toroidal duct.

Magnetic *bottles* or *buckets* also called homogenizers are not macroparticle filters but can be used to improve the deposition uniformity from filtered cathodic arc sources. It is possible to expand and flatten the plasma profile over a short distance beyond the exit of the filter by using one of these homogenizer. McKenzie and co-workers [58,59] demonstrated that it is possible to establish a multicusp magnetic wall using permanent magnets (Fig. 11.6(a)).

The region near the axis has only a weak fringing field. The ions are injected at one end and are moving in a more or less homogeneous, isotropic space, expanding and filling this space. The plasma assumes a flat density profile, is macroscopically stable and has a low fluctuation level. This approach has been developed and used extensively. Brown and co-workers used the current-driven bucket approach (Fig. 11.6(b)) and obtained similar results [60]. The current necessary to drive the multipole conductors is conveniently provided by the arc current. Better results were obtained with a more complex systems. Two coils and a plasma homogenizer has been combined to achieve a thickness homogeneity better than 10% over a 10-cm diameter area [61].

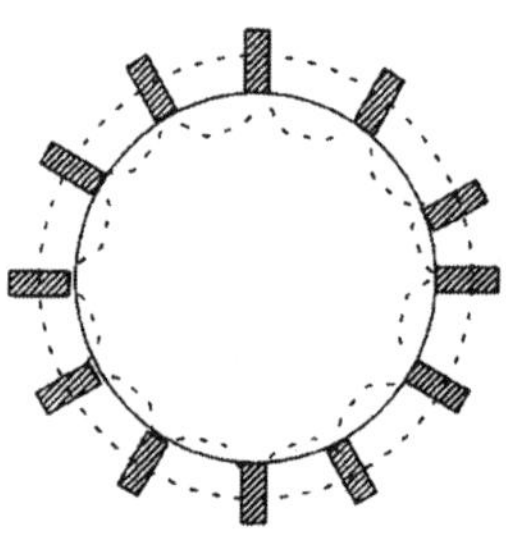

(a)

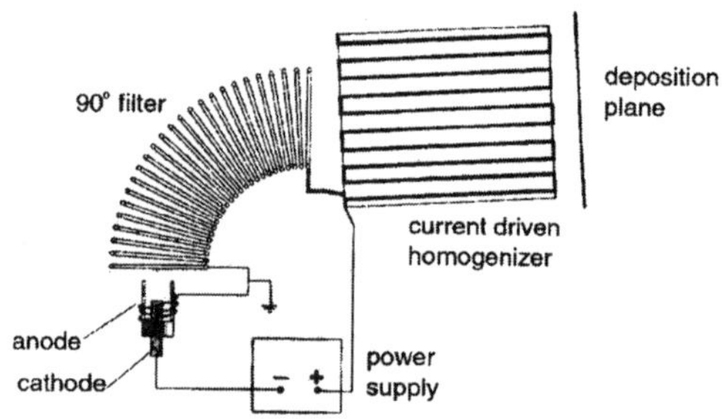

(b)

Figure 11.6: Schematic diagram of (a) a magnetic bucket and (b) the vacuum arc plasma generating system together with magnetic filter and current-driven homogenizer (from Ref. [61]).

11.4. Interaction between the Arc Plume and the Growing Film

As mentioned before, the cathodic arc plasma is mostly ionized. A few neutral species may be present due to evaporation from macroparticles or from previously active sites on the cathode, but they do not represent a large fraction of the stream. After filtering, the flux directed toward the substrate consists of ions and electrons. Any modification of the substrate potential relatively to the plasma potential will induce the development of an electric sheath around the substrate that will act on all charged particles. By controlling the substrate bias, it is thus possible to tailor the energy of plasma particles arriving at the substrate surface. This is a major difference from other physical vapor deposition (PVD) techniques such as evaporation and sputtering where most of the particle flux is neutral.

The biasing technique has been widely used and extended over the years: bias can be pulsed or continuous and the applied voltage ranges from a few volts to tens of kilovolts. By combining a cathodic arc plasma source with high-voltage biasing, it is possible to implant species in the substrate. This type of plasma immersion ion implantation (PIII) overcomes many disadvantages of classical line-of-sight ion implantation. This approach eliminates the need for ion-beam extraction, focusing, transport, scanning, and other manipulations. Controlling and modifying the substrate bias voltage during the treatment can also lead to deposition if the plasma flux is condensable. In all cases, the conditions in the sheath will control the resulting surface treatment.

11.4.1. Plasma Sheath

The basic physics of plasma sheath has been extensively reviewed in the literature (e.g. see Ref. [62]) but is mainly focused on steady-state conditions. When using pulsed high-voltage bias, it is necessary to consider transient sheaths.

When negative high-voltage pulses are applied to a sample, electrons are repelled from the sample surface leaving the ions to form an ion sheath. The ions are accelerated toward the sample surface. To sustain the ion flux, the ion sheath expands outwards until the end of the voltage pulse or until the vessel walls stop its progress. The development and stability of these transient sheaths have been modeled and characterized experimentally using Langmuir probes [63–69].

If we consider an electrically conductive substrate that is repetitively pulse biased, the ion density profile across the sheath depends on the repetition rate of the pulse. If the pulse frequency is high enough and the sheath sufficiently wide, the time between pulses is too short to allow the plasma to totally replenish the depleted region by diffusion. The current reaching the substrate during the second pulse is a lot smaller than the current observed during the first pulse (Fig. 11.7). The lower plasma density in the depleted region allows a quicker expansion of the sheath during the first few pulses. Quickly, a multiple-pulse steady state will be established: the charge flowing continuously into the depleted region is completely used up during the pulse.

The electrical conductivity of the substrate modifies the temporal sheath development. With an electrically insulating substrate, the sheath will expand and quickly collapse due to the accumulation of charges on the substrate. The collapse time depends on the plasma density and the magnitude of the high voltage applied to the substrate holder [67,68].

11.4.2. Plasma Immersion Ion Implantation and Deposition

If condensable plasma species such as ions of cathodic arc plasma are present, pulsing the substrate bias implies high-energy phases during which ion implantation takes place and low-energy phases during which deposition occurs. This novel surface modification technique combining PIII with cathodic arc deposition was introduced by Brown and co-workers at the beginning of the 1990s [70,71] and was called plasma immersion ion implantation and deposition (PIIID). If metal ions are involved, the method is also called MePIIID, metal plasma ion immersion implantation and deposition. The bias voltage can be adjusted during the surface treatment process. High voltages (a few kilovolts) can be applied during the early

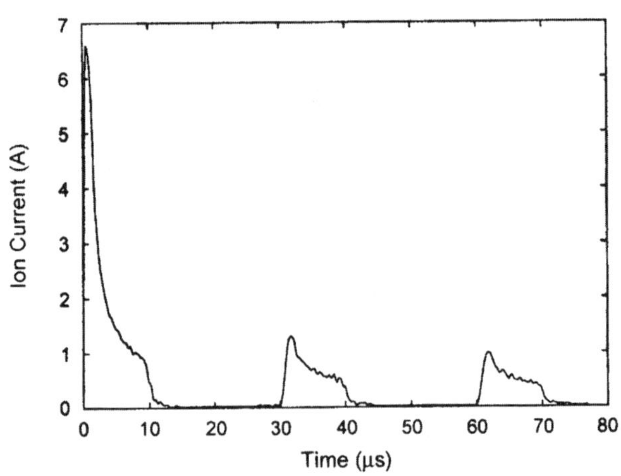

Figure 11.7: Ion implant current for three 10 μs voltage pulses separated by 20 μs (from Ref. [8]).

stages of the deposition process to sputter clean the substrate and to form a mixed interface by implantation. Low-voltage bias can be applied to provide ion assistance during the growth process. In this case, no foreign species are added, growth is self-assisted. The importance of the assistance low bias values or of sputtering and intermixing for high bias, is controlled by the biasing duty cycle. The duty cycle (δ_b) is defined as the ratio between the bias on-time (τ_p) and the total duration:

$$\delta_b = \tau_p / (\tau_p + \tau_{off}) \qquad (11.3)$$

Various experimental set-ups have been designed to provide a wide choice of growth parameters. Fig. 11.8 presents the set-up used at the Lawrence Berkeley National Laboratory. It is based on a pulsed vacuum arc source. Each long arc pulse is synchronized with bursts of bias pulses [72].

Depth profiles that are not attainable by pure ion implantation (due to the sputter limit) or other vapor deposition (due to the difficulty of forming an intermixed region) can be synthesized by PIIID.

If a DC substrate bias is applied or if the bias is constant during the length of the arc pulse, pure ion implantation can be achieved.

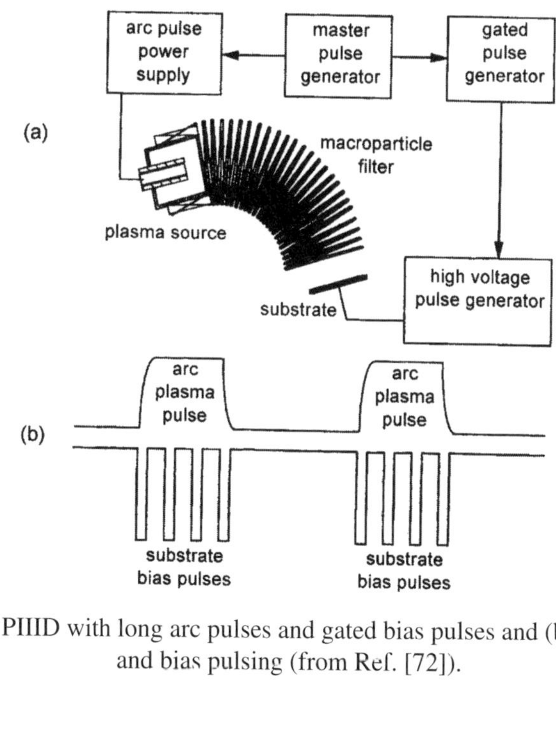

Figure 11.8: (a) PIIID with long arc pulses and gated bias pulses and (b) sequences of arc and bias pulsing (from Ref. [72]).

A few aspects of PIIID using cathodic arc plasmas should be emphasized [63]. These points are related to the characteristic features of PIIID arc plasmas: multiply charged ions are present, the plasma is fully ionized, and the plasma flow velocity is supersonic:

(i) Multiple ion charge states give rise to multiple ion energies. If E_0 is the ion energy corresponding to the flow velocity ($E_0 = Mv_0^2/2$) and V_b is the negative bias voltage, in a collisionless situation, the ions acquire discrete energies depending on their charge state:

$$E_Z = |ZV_b| + E_0 \tag{11.4}$$

The energy input for the growing film depends on the composition of the plasma and on the charge distribution of each species. The sputter rate depends on the energy, and the observed sputter rate is an average determined by the abundance of the different ions. The sheath thickness is also influenced by the mean ion charge.

(ii) The ion flux, ion current densities, and deposition rates are relatively high in comparison with other implantation and techniques. The sheath is thus very thin and the potential drop is concentrated across a short distance. This can lead to the sheath breakdown by arcing to the substrate.

(iii) PIII is considered to be a three-dimensional process. This is only true if the plasma velocity is small compared with the thermal ion velocities. Vacuum arc plasmas do not fulfill this condition. So this advantage of an isotropic three-dimensional treatment is partially lost [65].

11.4.3. Energy Release and Thermal Effects

PIIID is an energetic deposition technique. A large fraction of the incident particles have enough energy to penetrate under the surface. The impinging ions deliver a large quantity of energy (kinetic and potential) to the growing surface that will contribute to interface mixing, subplantation, texture and stress development or reduction, atomic scale heating, secondary electron emission and sputtering. The kinetic energy of condensing ions is taken into account in most of the published work, but potential energy is usually ignored despite its importance. Anders [73] has shown that potential energy of the impinging ions contributes significantly to atomic scale heating.

An ion emitted by a cathodic arc acquires kinetic energy in three steps: first at the cathode spot, second in the sheath, and third extremely close to the growing surface (acceleration by the image charge). The last contribution is the smallest of the three. The ion kinetic energy determines the displacements and short collision cascades in the substrate. The different mechanisms involved in the loss of kinetic energy in the solid are well documented [74]. The displaced solid atoms vibrate around their new site with large amplitudes and are considered as "hot". Thermal conduction quickly quenches them.

The ions potential energy includes the excitation energy (E_{exc}) of electrons but also rotational and vibrational energy for molecular ions, cohesive energy (E_c), and ionization energy ($E_{Q'}$). So the total energy of an impinging ion can be expressed as follows:

$$E(Q) = E_{\text{kin},0} + AQeV_{\text{sheath}} + E_{\text{ic}} + E_{\text{c}} + E_{\text{exc}} + \sum_{Q'=0}^{Q-1} E_{Q'} \qquad (11.5)$$

where Q is the ion charge state, $E_{\text{kin},0}$ is the kinetic energy corresponding to the initial velocity, and V_{sheath} is the potential drop across the sheath. The largest contribution to the potential energy is the ionization energy. For a multiply charged ion, each ionization energy must be taken into account. For example, in a tungsten cathodic arc plasma, the majority of the ions are triply charged. The neutralization of these ions brings to the surface 48.5 eV per ion. This value is not negligible when compared to the $E_{\text{kin},0}$ for tungsten, 117 eV. The mechanisms of potential energy release are not well studied but include photon radiation, Auger processes, local electronic excitation, and secondary electron emission.

Kinetic and potential energies delivered to the growing surface contribute to atomic scale heating, and usually exceeds the binding energy and activation energy for surface diffusion. They have significant effects on film formation and evolution.

11.5. Relations Film Properties: Deposition Conditions

A detailed description of the effect of energetic particles on film growth is presented in other chapters of the present book. Most of the considerations can be extended to PIIID coatings. Messier modified the classical structure-zone model of Thornton [75] to show the effects of both bombardment and thermal-induced mobility [76]. In general, the deposition of dense films at practical growth rates requires sufficient energy as the atoms need to diffuse to form a coating microstructure more similar to the bulk material. A high degree of ionization is also often desirable in order to promote the formation of compound films. Ion-assisted deposition processes including cathodic arc PIIID can fulfill these conditions. In the present paragraph, we will only focus on few aspects that are characteristic of this latter technique. In comparison with evaporation and sputtering technologies, arc technology often offers outstanding film properties and a wider process window for reactive deposition. The film properties can be tuned and tailored because cathodic arc plasmas are fully ionized. The plasma density and the ion energy distributions can be manipulated with magnetic and electric fields. Non-equilibrium bonding states can be obtained in a coating by manipulating the ion kinetic energy.

11.5.1. Adsorption, Intermixing, and Adhesion

Improved adhesion has been often observed for films deposited by the cathodic arc process. This is due to the higher energy of impinging ions which has multiple consequences. First the physisorbed impurities are desorbed either by direct impact or by local heating around the impact, leaving a cleaner surface for the condensing ions. In the case of reactive deposition, Hubler and Sprague [77] pointed out that ion bombardment influences gas adsorption by increasing the sticking coefficient of gases such as nitrogen and changing the nature of adsorption sites from lower-energy physisorption sites to higher-energy chemisorption sites. This could explain why the cathodic arc process is so effective in forming nitrides and carbides.

The advantages of diffuse interfaces for enhancing coating adhesion do not need to be demonstrated here. The fully ionized flux emitted by cathode spots is ideal for forming this type of interface. The implantation profile can be tuned by varying the bias conditions as described in a previous section. Implantation induces displacement of substrate atoms, breaking of bonds and formation of new bonds across the interface. A diffuse interface also spreads the stresses over a wider volume therefore decreasing the probability of delamination at the interface between the substrate and the coating. The bombardment by energetic ions can also, under certain conditions, roughen the interface and improve adhesion by mechanical interlocking.

However, care must be taken in choosing the initial deposition conditions to avoid significantly damaging the substrate, which would be detrimental for the performances of the coated piece.

11.5.2. Stress Development, Relaxation, and Texture

The magnitude of internal stresses in coatings has a significant impact on their physicochemical properties and consequently on their ability to meet performance requirements. However, the presence of stresses in coatings is not necessarily a negative characteristic, and some compressive stress is usually beneficial for protective coatings. On the other hand, similar levels of stresses in membranes used for MEMS would results in unacceptable bending. This illustrates the importance of being able to control the level of intrinsic stress during the deposition process.

The evolution of intrinsic residual stress in thin films has been extensively studied (e.g. see Refs. [78–83]). The deposition conditions, such as the working pressure, substrate bias, etc., control the degree of stress developed in the films.

The intrinsic stress generated by ion impact seems to follow a universal curve [84]. At low thermal energies (<1 eV), the stress is tensile. The stress reverses to compressive with increasing ion energy and reaches a maximum from around 20 to a few hundreds of electron volts before starting to decrease at higher energies. This trend has been observed for a wide variety of materials such as DLC on silicon [85], tungsten nitride on tungsten carbide [81], and boron carbide on silicon [86].

Thermal annealing [87,88], ion bombardment [89], and laser annealing post-treatments have been applied to reduce intrinsic stress in films. Recently, Bilek and Delplancke have demonstrated that a careful tuning of the biasing voltage, duration, and repetition rate can induce stress relaxation in films grown predominantly with lower energy in a high stress regime. This has been demonstrated for amorphous carbon [90], titanium nitride (TiN) [90], boron nitride (BN) and aluminum nitride (AlN) [92,93], and DLC [94,95].

Non-hydrogenated DLC has been plagued by the excessive level of intrinsic stresses in particular for films with a high content of sp^3 content. An extensive review article on the fabrication methods and properties of DLC has been recently published [96]. The development of hardness is usually accompanied by the development of very high compressive stress. The films prepared by cathodic arc evaporation follow this trend. Attempts have been made in the past to reduce intrinsic stresses in DLC films by producing multilayers consisting of alternating layers of sp^3-rich ("hard") and sp^3-not-so-rich ("soft") carbon layers [97]. Unfortunately, the final properties are an average of the properties of the individual layers. *In-situ* determination of stress during DLC deposition provides some insight on how to optimize stress relaxation by high-energy ion bombardment. The experimental set-up described in Fig. 11.8 was used to perform an experiment in which the bias voltage changed between -100 and -2000 V (duty cycle 25%). The deposition was carried out for a large number of pulses (500) with bias voltage of -100 V, followed by -2000 V for a short number of pulses (50 in one case and 100 in another). Fig. 11.9 shows the results of a Monte-Carlo simulation (T-DYN) [98,99] of the profile of C atom implantation–deposition in this experiment. The atomic fraction of the C atoms produced in each phase of the deposition is plotted and shows that the tail of the C concentration produced at -2000 V bias voltage penetrates through the entire -100 V layer and thus can alter the properties of the entire layer. On the contrary, the 100 V C^+ ions caused very little modification of the immediately preceding layer.

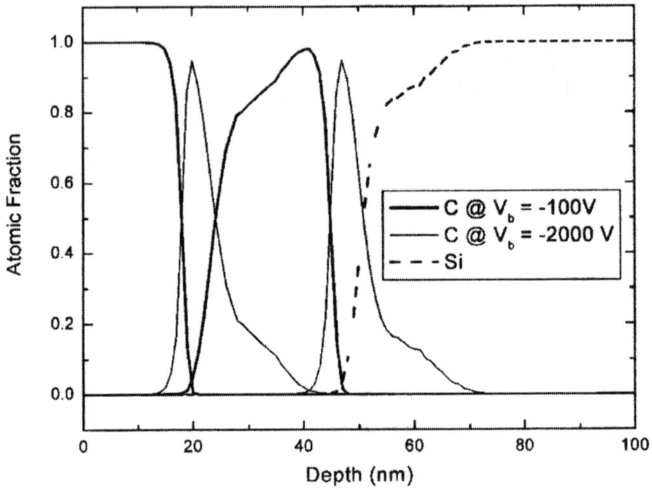

Figure 11.9. Profile of carbon atoms sequentially deposited under two deposition conditions (−100 and −2000 V) obtained by Monte-Carlo simulation. This profile indicates the depth of interaction between the incident ions and the film (from Ref. [94]).

The average stress evolution during deposition is given in Fig. 11.10. It is readily seen that after each high-energy step, the average stress in the film is brought to approximately zero. Film density was used as a parameter to indicate the degree of change in sp^3 content [100] in DLC as a consequence of the high-energy bombardment. The DLC density of the films produced at −100 and −2000 V were 2.81 and 2.47 $g\,cm^{-3}$, respectively. The density of the stress-relaxed film was 2.79 $g\,cm^{-3}$. This reduction in density is small and should not account for large changes in sp^3 content and therefore in film properties. Increasing the number of high-voltage pulses led to greater density reduction. It is possible to relax stresses in a layer-by-layer process so that the final total stress is determined solely by the stress in the last DLC layer. The results obtained by Delplancke and co-worker on DLC films agree with the model developed by Bilek and co-workers [90]. The key parameter in the stress-relaxation process by energetic bombardment is the total energy delivered to the growing film by the high-energy impacts. The total energy delivered is proportional to the product of applied voltage by the pulse length and the frequency. It should be possible to reach similar results with different combination of these three parameters. Nevertheless, care must be taken to avoid conditions that produce excessive substrate heating that could independently affect the microstructure.

Another path to change the total energy delivered during the impacts is to introduce in the plasma stream ions with higher atomic number and multiply charges. Fig. 11.11 shows how the introduction of tungsten ions in the plasma stream drastically

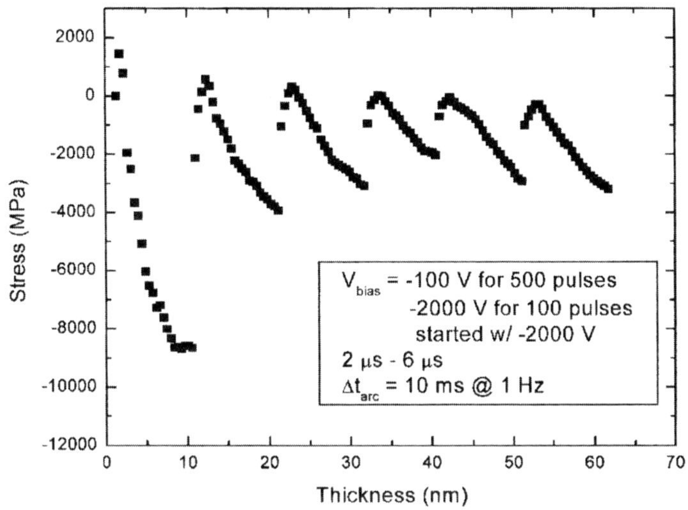

Figure 11.10. Evolution of average intrinsic stress in a DLC multilayer during deposition, in which we alternate long periods with low bias voltages, with short periods of deposition with high bias voltages (from Ref. [94]).

reduces the intrinsic stress in DLC layers without reducing the hardness in a similar way. The mean charge of tungsten ions in cathodic arc plasma is higher than 3. The energy input per tungsten ion is important especially as the bias is increased.

Bilek and co-workers [90–93] studied the impact of the stress relaxation by high-energy bombardment on the microstructure. They observed a strong correlation between the residual stress and the preferred orientation of AlN, BN, and TiN coatings. In their amorphous carbon films, they observed a correlation between the level of graphitization and the product of the bias with the repetition rate. This picture is consistent with a thermal-spike relaxation model if we assume that local volumes annealed by thermal spikes tend to graphitize. For crystalline films, the phases most stable under low-stress conditions replace phases stable under high-stress conditions during the local thermal-spike annealing process.

11.6. Trends in Applications and Research

A wide variety of materials can be deposited by cathodic arc evaporation (metals, carbon, semiconductors, ceramics, etc.) for various applications (mechanical, optical, electronic applications, corrosion protection). It is impossible to provide here an exhaustive review of the latest developments. Only few examples will be given.

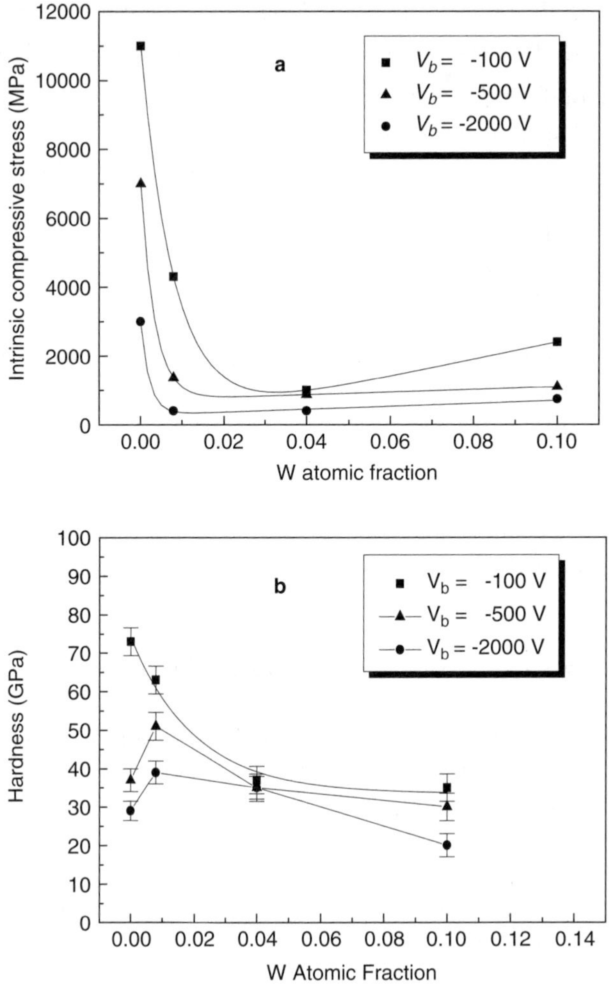

Figure 11.11. (a) Variation of the intrinsic stress of DLC films with the concentration of incorporated tungsten for three difference substrate biases and (b) variation of the film hardness in the same conditions.

11.6.1. Ceramic Coatings for the Tool Industry

While cathodic arc technology is still extensively used in the industry to coat tools, plumbing fixtures, door knobs, and lamp reflectors with nitrides (mostly TiN), the tool industry is getting interested in more complex coatings for dedicated applications. The (Ti,Al)N coatings that became common in the middle of 1990s still show significant development potential. During the last decade arc technology has grown to become the most important deposition technology for Al-containing nitrides. According to L. Karlsson [101], the trends are the following:

- Incorporation of different alloying elements such as Si, Cr, and Y to increase the oxidation resistance of the coating.
- Increasing the thermal stability by fine-tuning the composition and deposition conditions.
- Optimization of the micro- and nanostructures.
- Designing new combinations of layers that will give optimized coatings based on the demands of specific applications.

One of the factor slowing down the industrial development of these new coatings is the availability of high-quality targets at reasonable prices. For industry, aluminum oxide is still the preferred coating material for many application areas. Unfortunately, PVD-Al_2O_3 is not yet available on the market. At the laboratory scale, the formation of alumina films by cathodic arc deposition has been extensively studied because it is one of the few viable routes for deposition on heat sensitive substrate materials [102].

11.6.2. Nanostructured and Nanocomposite Coatings

Pure PIII and PIIID technology or hybrid technologies combining for example an arc source and magnetron sputtering are used to prepare nanocomposite or nanostructured coatings. As the preparation of atomically smooth coatings by arc technology is now possible due to improvements in filter technology, new coating designs can be envisioned. The deposition of functionally graded, multilayer, super-lattice, and nanocomposite architectures of multi-elemental coatings over large surface areas has been recently demonstrated [103].

In addition, the highly ionized source allows us to create new materials, often in an amorphous or glassy form and to impart new properties to existing materials and surfaces.

DLC and modified DLC films deposited by arc technology are used to protect read–write heads and magnetic hard disks. They are ultra-thin, continuous, dense, hard, and tough. But the improvement of the properties of DLC coatings for other applications remains a major research subject. The range of carbon-based materials grows continuously. Doped monolithic films [104], multilayered carbon films [105], carbon-based chameleon coatings (material able to rearrange their structure and chemistry on demand to adapt to variable surface conditions) [106] are being developed in many laboratories.

We should see these new coatings on market in a relatively near future.

11.7. Perspectives and Conclusions

Despite an early start, cathodic arc deposition has long been limited to applications where macroparticles could be tolerated such as coatings for machine tools or for decoration purposes. During the last 15 years, the combination of progress in different fields has opened the window of applications for this coating technology. Arc deposition is now applied in more demanding technologic areas in optics and electronics. This is partially due to the development of efficient macroparticle filters with high throughput and to a deeper understanding of the cathode processes and of the plasma interaction with a biased substrate. The interest in cathodic arc evaporation arises primarily from the fully ionized nature of the plasma. The film properties can be tailored by manipulating the plasma density and ion energy distributions with magnetic and electric fields. The superior properties of arc-deposited coatings have been recognized and advanced tribologic, electronic, and optical materials are being constantly developed.

Source design has also progressed during recent years. New filtered sources for uniformly coating larger substrates are available. The control of the cathode spot has been improved, which allows a better use of the cathode material. Hybrid methods combining cathodic arc sources with other deposition techniques such as DC magnetron sputtering are being developed.

Cathodic arc plasma deposition science and technology are very active research fields at the present time. It brings together physicists, chemists, material scientists, and engineers to develop new materials and coating devices. We can speculate that the next 15 years will see a strong expansion of the field of application for this method.

Acknowledgments

The author would like to express their gratitude to Othon R. Monteiro, André Anders, Marcela M.M. Bilek, and I.G. Brown for helpful discussions and to R.A. MacGill for his assistance.

References

[1] A. Anders, *IEEE Trans. Plasma Sci.*, 31 (2003) 1052.
[2] A. Anders, *IEEE Trans. Plasma Sci.*, 31 (2003) 1060.
[3] R.L. Boxman, *IEEE Trans. Plasma Sci.*, 29 (2001) 759.
[4] A.W. Wright, *Am. J. Sci. Arts*, 14 (1878) 169.

[5] A.W. Wright, *Am. J. Sci. Arts*, 13 (1877) 49.

[6] T.A. Edison, Art of plating one material with another, US Patent 526,147, September 18, 1894.

[7] *Handbook of Vacuum Arc Science and Technology*, Eds. R.L. Boxman, P.J. Martin and D.M. Sanders, Noyes Publications, Park Ridge, 1995.

[8] *Handbook of Plasma Immersion Ion Implantation and Deposition*, Ed. A. Anders, John Wiley & Sons, New York, 2000.

[9] A. Anders, *Vacuum*, 67 (2002) 673.

[10] D.M. Sanders and A. Anders, *Surf. Coat. Technol.*, 133–134 (2000) 78.

[11] P.J. Martin and A. Bendavid, *Surf. Coat. Technol.*, 142–144 (2001) 7.

[12] P.J. Martin and A. Bendavid, *Thin Solid Films*, 394 (2001) 1.

[13] R.L. Boxman, *IEEE Trans. Plasma Sci.*, 29 (5) (2001) 762.

[14] B. Jüttner, V.F. Puchkarev, E. Hantzsche and I. Bellis, in *Handbook of Vacuum Arc Science and Technology*, Eds. R.L. Boxman, P.J. Martin and D.M. Sanders, Noyes Publications, Park Ridge, 1995, pp. 73–281.

[15] E. Hantzsche, *IEEE Trans. Plasma Sci.*, 31 (5) (2003) 799.

[16] A. Anders, *Appl. Phys. Lett.*, 80 (2002) 1100.

[17] B.E. Djakow and B. Jüttner, *J. Phys. D Appl. Phys.*, 35 (2002) 2570.

[18] S.A. Barengolts, G.A. Mesyats and D.L. Shmelev, *IEEE Trans. Plasma Sci.*, 31 (2003) 809.

[19] A. Anders, *Phys. Rev. E*, 55 (1997) 969.

[20] I.G. Brown, *Rev. Sci. Instrum.*, 65 (1994) 3061.

[21] E.M. Oks, A. Anders, I.G. Brown, M.R. Dickinson and R.A. MacGill, *IEEE Trans. Plasma Sci.*, 24 (1996) 1174.

[22] T.W.H. Oates, J. Pigott, D.R. Mckenzie and M.M.M. Bilek, *Rev. Sci. Instrum.*, 74 (2003) 4750.

[23] J.M. Schneider, A. Anders, I.G. Brown, N. Hjorvarsson and L. Hultman, *Appl. Phys. Lett.*, 75 (1999) 612.

[24] E. Hantzsche, *IEEE Trans. Plasma Sci.*, 23 (1995) 893.

[25] J.E. Daalder, *J. Appl. D Appl. Phys.*, 9 (1976) 2379.

[26] R.L. Boxman and S. Goldsmith, *Surf. Coat. Technol.*, 52 (1992) 39.

[27] D.A. Marpov, *Surf. Coat. Technol.*, 96 (1997) 22.

[28] T. Schülke, T. Witke, H.-J. Scheibe, P. Siemroth, B. Schultrich, O. Zimmer and J. Vetter, *Surf. Coat. Technol.*, 120 (1999) 226.

[29] S. Falabella and D.A. Karpov, in *Handbook of Vacuum Arc Science and Technology*, Eds. R.L. Boxman, P.J. Martin and D.M. Sanders, Noyes Publications, Park Ridge, 1995, pp. 397–412.

[30] P. Siemroth, T. Schülke and T. Witke, *Surf. Coat. Technol.*, 68 (1994) 314.

[31] T. Wilke, T. Schülke, B. Schultrich, P. Siemroth and J. Vetter, *Surf. Coat. Technol.*, 126 (2000) 81.

[32] T.W.H. Oates and M.M.M. Bilek, *Plasma Sources Sci. Technol.*, 12 (2003) 508.

[33] T. Schülke, T. Wilke, H.-J. Scheibe, P. Siemroth, B. Schulrich, O. Zimmer and J. Vetter, *Surf. Coat. Technol.*, 120 (1999) 226.

[34] M.M.M. Bilek, P.J. Martin and D.R. McKenzie, *J. Appl. Phys.*, 83 (1998) 2965.
[35] R.N. Tarrant, M.M.M. Bilek, T.W.H. Oates, J. Pigott and D.R. McKenzie, *Surf. Coat. Technol.*, 156 (2002) 110.
[36] E. Gidalevich, S. Goldsmith and R.L. Boxman, *J. Appl. Phys.*, 90 (2001) 4355.
[37] T. Schülke and A. Anders, *Plasma Sources Sci. Technol.*, 8 (1999) 567.
[38] S.Y. Chun, A. Chayahara, A. Kinomura, N. Tsubouchi, C. Heck, Y. Horino and H. Fukui, *Jpn. J. Appl. Phys.*, 38 (1999) L467.
[39] H. Brandolf, US Patent 4,511,593 (1985).
[40] H. Takikawa, M. Nagayama, R. Miyano and T. Sakakibara, *Surf. Coat. Technol.*, 169–170 (2003) 49.
[41] A.I. Ryabchikov and I.B. Stepanov, *Rev. Sci. Instrum.*, 69 (1998) 810.
[42] A.I. Ryabchikov, I.B. Stepanov, S.V. Dektjarev and O.V. Sergeev, *Rev. Sci. Instrum.*, 69 (1998) 893.
[43] M.M.M. Bilek, A. Anders and I.G. Brown, *IEEE Trans. Plasma Sci.*, 27 (1999) 1197.
[44] I.I. Aksenov, V.A. Belous, V.G. Padalka and V.M. Khoroshikh, *Sov. J. Plasma Phys.*, 4 (1978) 425.
[45] I.I. Aksenov, V.A. Belous, V.G. Padalka and V.M. Khoroshikh, *Instrum. Exp. Tech.*, 21 (1978) 1416.
[46] R.P. Welty, US Patent 5,480527 (1996).
[47] S. Anders, A. Anders, M.R. Dickinson, R. MacGill and I.G. Brown, *IEEE Trans. Plasma Sci.*, 25 (1997) 670.
[48] A. Anders and R.A. MacGill, *Surf. Coat. Technol.*, 133–134 (2000) 96.
[49] A. Anders, *IEEE Trans. Plasma Sci.*, 30 (2002) 108.
[50] G.F. You, B.K. Tay, S.P. Lau, D.H.C. Chua and W.I. Milne, *Surf. Coat. Technol.*, 150 (2002) 50.
[51] H. Takikawa, K. Izumi, R. Miyano and T. Sakakibara, *Surf. Coat. Technol.*, 163–164 (2003) 368.
[52] D.B. Broecker, D.M. Sanders, J. Sorer and S. Falabella, *J. Appl. Phys.*, 69 (1991) 115.
[53] C.A. Davis and I.J. Donelly, *J. Appl. Phys.*, 72 (1992) 1740.
[54] T. Zhang, B.Y. Tay, Q.C. Chen, Z.M. Zeng, P.K. Chu, M.M.M. Bilek and I.G. Brown, *Rev. Sci. Instrum.*, 70 (1999) 3329.
[55] D.T.-K. Kwok, P.K. Chu, M.M.M. Bilek and I.G. Brown, *IEEE Trans. Plasma Sci.*, 28 (2000) 2194.
[56] V.I. Gorokhovsky, D.G. Bhat, R. Shivpuri, K. Kulkarni, R. Bhattacharya and A.K. Rai, *Surf. Coat. Technol.*, 140 (2001) 215.
[57] V.I. Gorokhovsky, R. Bhattacharya and D.G. Bhat, *Surf. Coat. Technol.*, 140 (2001) 82.
[58] R. Limpaecher and K.R. McKenzie, *Rev. Sci. Instrum.*, 44 (1973) 726.
[59] E.R. Ault and K.R. McKenzie, *Rev. Sci. Instrum.*, 44 (1973) 1697.
[60] P. Machina, M.M.M. Bilek, O.R. Monteiro and I.G. Brown, *Rev. Sci. Instrum.*, 71 (2000) 3373.
[61] M.M.M. Bilek, A. Anders and I.G. Brown, *Plasma Sources Sci. Technol.*, 10 (2001) 606.
[62] M.A. Lieberman, in *Handbook of Plasma Immersion Ion Implantation and Deposition*, Ed. A. Anders, John Wiley & Sons, New York, 2000, pp. 29–123.

[63] B.P. Woods, D.J. Rej, A. Anders, I.G. Brown, R.J. Faehl, S.M. Malik and C.P. Munson, in *Handbook of Plasma Immersion Ion Implantation and Deposition*, Ed. A. Anders, John Wiley & Sons, New York, 2000, pp. 243–301.
[64] I.G. Brown, O.R. Monteiro and M.M.M. Bilek, *Appl. Phys. Lett.*, 74 (1999) 2426.
[65] A. Anders, *Surf. Coat. Technol.*, 136 (2001) 85.
[66] D.T.-K. Kwok, T.W.H. Oates, D.R. McKenzie and M.M.M. Bilek, *IEEE Trans. Plasma Sci.*, 31 (2003) 1044.
[67] T.W.H. Oates, J. Pigott, D.R. McKenzie and M.M.M. Bilek, *IEEE Trans. Plasma Sci.*, 31 (2003) 438.
[68] T.W.H. Oates and M.M.M. Bilek, *J. Appl. Phys.*, 92 (2002) 2980.
[69] X. Tian, R.K.Y. Fu, J. Chen, P.K. Chu and I.G. Brown, *Nucl. Instrum. Method. Phys. Res. B*, 187 (2002) 485.
[70] I.G. Brown, W. Godechot and K.M. Yu, *Appl. Phys. Lett.*, 58 (1991) 1392.
[71] I.G. Brown, A. Anders, S. Anders, M.R. Dickinson, I.C. Ivanov, R.A. MacGill, X.Y. Yao and K.-M. Yu, *Nucl. Instrum. Method. Phys. Res. B*, 80–81 (1993) 1281.
[72] A. Anders, S. Anders, I.G. Brown, M.R. Dickinson and R.A. MacGill, *J. Vac. Sci. Technol. B*, 12 (1994) 815.
[73] A. Anders, *Appl. Phys. Lett.*, 80 (2002) 1100.
[74] M. Nastasi, J.W. Mayer and J.K Hirvonen, *Ion–Solid Interactions*, Cambridge University Press, Cambridge, UK, 1996.
[75] J.A. Thornton, *J. Vac. Sci. Technol.*, 11 (1974) 666.
[76] R. Messier, A.P. Giri and R.A. Roy, *J. Vac. Sci. Technol. A*, 2 (1984) 500.
[77] G.K. Hubler and J.A. Sprague, *Surf. Coat. Technol.*, 81 (1996) 29.
[78] H. Windischmann, *Crit. Rev. Solid State Mater. Sci.*, 17 (1992) 547.
[79] M.F. Doerner and W.D. Nix, *Crit. Rev. Solid State Mater. Sci.*, 14 (1988) 25.
[80] A. Misra, S. Fayeulle, H. Kung, T.E. Mitchell and M. Nastasi, *Nucl. Instrum. Method. Phys. Res. B*, 148 (1999) 211.
[81] A. Misra and M. Nastasi, *Nucl. Instrum. Method. Phys. Res. B*, 175–177 (2001) 688.
[82] H. Ljungcrantz, L. Hultman, J.-E. Sundgren and L. Karlsson, *J. Appl. Phys.*, 78 (1995) 832.
[83] M.M.M. Bilek, D.R. McKenzie, R.N. Tarrant, S.H.M. Lim and D.G. McCulloch, *Surf. Coat. Technol.*, 156 (2002) 136.
[84] D.R. McKenzie, *J. Vac. Sci. Technol. B*, 11 (1993) 1928.
[85] J.W. Ager III, S. Anders, A. Anders and I.G. Brown, *Appl. Phys. Lett.*, 66 (1995) 3444.
[86] O.R. Monteiro, M.-P. Delplancke-Ogletree and C.C. Klepper, *J. Mater. Sci.*, 38 (2003) 3117.
[87] T.A. Friedman, J.P. Sullivan, J.A. Knapp, *et al.*, *Appl. Phys. Lett.*, 71 (1997) 3820.
[88] O.R. Nmonteiro, J.W. Ager III, D.H. Lee, R. Yu Lo, K.S. Walter and M.J. Nastasi, *J. Appl. Phys.*, 88 (2000) 2395.
[89] R.N. Tarrant, C.S. Montross and D.R. McKenzie, *Surf. Coat. Technol.*, 136 (2001) 188.
[90] M.M.M. Bilek, R.N. Tarrant, D.R. McKenzie, S.H.N. Lim and D.G. McCulloch, *IEEE Trans. Plasma Sci.*, 31 (2003) 939.

[91] S.H.N. Lim, D.G. McCulloch, M.M.M. Bilek and D.R. McKenzie, *J. Appl. Phys.*, 93 (2003) 4283.
[92] D.R. McKenzie and M.M.M. Bilek, *Thin Solid Films*, 382 (2001) 280.
[93] B.K. Gan, M.M.M. Bilek, D.R. McKenzie, Yang Shi, D.A. Tompsett, M.B. Taylor and D.G. McCulloch, *J. Phys. Condens. Mat.*, 16 (2004) 1751.
[94] M.-P. Delplancke-Ogletree and O.R. Monteiro, *Diamond Relat. Mater.*, 12 (2003) 2119.
[95] M.-P. Delplancke-Ogletree and O.R. Monteiro, in *Nanostructured Thin Films and Nanodispersion Strengthened Coatings 3*, Eds. A.A. Voevodin, D.V. Shtansky, E.A. Levashov and J.J. Moore, Kluwer Academic Publishers, Dordrecht, 2004, pp. 167–174.
[96] J. Robertson, *J. Mater. Sci. Eng. R*, 37 (2002) 129.
[97] S. Anders, D.L. Callahan, G.M. Pharr, T.Y. Tsui and C.S. Bhatia, *Surf. Coat. Technol.*, 94–95 (1997) 189.
[98] J.P. Biersack, *Nucl. Instrum. Method. Phys. Res. B*, 19–20 (1987) 32.
[99] J.P. Biersack, *Nucl. Instrum. Method. Phys. Res. B*, 27 (1987) 21.
[100] A. LiBassi, A.C. Ferrari, V. Stolojan, B.K. Tanner, J. Robertson and L.M. Brown, *Diamond Relat. Mater.*, 9 (2000) 771.
[101] L. Karlsson, *International Conference on Metallurgical Coatings and Thin Films*, San Diego, California, 2005.
[102] J. Rosen, S. Mraz, U. Kreissig, D. Music and J.M. Schneider, *International Conference on Metallurgical Coatings and Thin Films*, San Diego, California, 2005.
[103] V.I. Gorokhovsky, C. Bowman, P.E. Gannon, D. VanVorous, A. Rutkowski, A.A. Voevodin, R.J. Smith, A. Kayani and V. Shutthanandan, *International Conference on Metallurgical Coatings and Thin Films*, San Diego, California, 2005.
[104] X.-Z. Ding, B.K. Tay, S.P. Lau, P. Zhang and X.T. Zeng, *Thin Solid Films*, 408 (2002) 183.
[105] D.R. McKenzie, R.N. Tarrant, M.M.M. Bilek, T. Ha, J. Zou, W.E. McBride, D.J.H. Cockayne, N. Fujisawa, M.V. Swain, N.L. James, J.C. Woodard and D.G. McCulloch, *Diamond Relat. Mater.*, 12 (2003) 178.
[106] A.A. Voevodin and J.S. Zabinski, *International Conference on Metallurgical Coatings and Thin Films*, San Diego, California, 2005.

Chapter 12

Spectroscopic Analyses of Surfaces and Thin Films

Jean-Paul Deville and Costel Sorin Cojocaru

12.1. Introduction

The first methods devoted to the characterization of surfaces and to the investigation of their properties were developed as early as in the 19th century. Although mostly phenomenologic they used physical properties which are among the most important to understand the evolution of surfaces and crystal growth. One can think, for example, of the Brunauer–Emmet–Teller (BET) method, based on the physisorption of rare gases at low temperature correctly described by Langmuir, and of surface-angle measurements that depend on the surface tension of materials and rely on equations derived by Young, using the Gibbs approach of the thermodynamics of surfaces. These two properties (sorption and surface tension) are well known to be critical in thin film deposition and their surface modification. This explains perhaps why these methods are still widely used in surface science, especially in industry-related laboratories. However, nowadays surface scientists and thin film elaborators prefer using methods based on spectroscopies for the chemical characterization and on diffraction for the structural properties. In Chapter 13, the methods used for structural investigation will be dealt thoroughly and we shall focus here on the spectroscopic methods which indeed go much farther than the chemical characterization. One has indeed to think that structural methods and spectroscopic methods cannot be separated so strictly. There are deep relationships between one and the other.

These spectroscopic methods were mainly developed in the second half of the 20th century, much later than the structural methods such as transmission electron microscopy (TEM) or X-ray diffraction. The reasons were essentially the need for technologic developments such as ultra-high vacuum (UHV) which was not readily obtainable before the second half of the past century or better electronics to increase the signal to noise ratio. The signal characteristic of the surface is indeed always much lower than the one issued by the bulk of the material and its detection

Materials Surface Processing by Directed Energy Techniques
Y. Pauleau (Editor)

will be difficult. For structural methods, it has inhibited methods based on surface scattering of X-rays used as a probe (grazing incidence X-ray diffraction, GIXD, grazing incidence small angle X-ray scattering, GISAXS) until the development of synchrotron radiation utilities.

12.2. Ideal Surfaces and Real Surfaces

During the early years of modern surface science, the necessity to validate the new structural methods based on electron diffraction, namely low-energy electron diffraction (LEED) and reflection high-energy electron diffraction (RHEED) implied working on clean (or supposedly clean) surfaces with a very good structural quality. These ideal surfaces were cut out of single crystals and only the simplest planes such as the low Miller indices planes viz. (100), (111) or (110) were studied. It was already very useful for investigating semiconductors which, for electronic applications, needed in those days to be made out of single crystals. By the end of the sixties of the previous century the technologic and theoretic advances of LEED allowed to study what was called vicinal or stepped surfaces, that is, surfaces cut at a small angle from the low Miller indices planes. These surfaces can be described as a succession of terrasses made out of low Miller index planes separated by steps of atomic (or bi-atomic) height. These surfaces with controlled defects were used in this period of time to investigate some catalytic properties of metals [1]. About at the same time, appeared methods such as Auger electron spectroscopy (AES) which allowed to investigate the surface composition of materials. In this case, it was not any more necessary to study single crystals and even the work on insulating materials was made possible however difficult because of surface charging of the samples. If coupled with ion-beam ablation of the sample, AES gave the possibility to study real surfaces as obtained industrially, for example, in metallurgy or thin film applications. However, working on polymer surfaces was not possible since the electron beam heavily damaged the surfaces. One had to wait for the appearance of photoelectron spectroscopy in the early seventies of the past century to be able to analyze about every type of material.

In such industrial materials, there exist what is called interfaces, that is, boundaries between solids, either between the same solid (grain boundaries) or between different solids (interphase boundaries). They are inside the material and, of course, their spectroscopic analysis may be difficult. We shall assume that as a general concept, interfaces can be described as a surface and, generally, we shall not make any difference between them, speaking the most often of *surfaces* in general but always keeping in mind that *interfaces* are included in this concept. Of course when there will be important differences between surfaces and interfaces this will be stated in the text.

12.2.1. A Brief Description of Phenomena and Properties at the Surfaces

A sample obtained after an industrial process such as laser deposition is rather complex. The morphology, the crystalline structure and the chemical composition and bonds are all affected by the process. If hopefully the bulk of the sample has generally the expected chemical composition, it may often contain impurities trapped during the process or outcoming from the material source used for the deposition if it is not pure enough. Its crystalline structure is characterized by a high disorder of the grains and its morphology is far from being a simple one. It is not flat on an atomic point of view, not even at a larger scale.

In case of metallic or semiconducting materials, the extreme surface can be capped with an oxidized area (native oxide), the thickness of which depends on the material itself and on the oxidation kinetics. For example, in the case of silicon, the native oxide thickness is around 2 nm and increase very slowly as a function of time. It is also the case with passive metals. But for aluminum, titanium or zinc, it is possible to get surface oxide or hydroxide layers thicker than several hundreds of nanometers. Moreover, the very top of the surface is often covered with layers of organic molecules due to the chemical or physical adsorption of the surrounding atmosphere. These molecules can be inert or even live. In this case the layer is named a *biofilm*. All these chemical defects may have their importance in the properties of the films.

Surface analysis has to take into account the entire area which is affected by the various kinds of degradation processes: surface stress and strain, physical and mechanical modifications, chemical reactions with the surrounding atmosphere, controlled or not. So it should be clear that when we are speaking of a *surface* analysis we mean investigating a region which is not clearly defined in depth and which can extend up to several hundreds of nanometers. This is probably the largest difficulty in surface science: one never knows at which extent the sample is homogeneous in depth either crystallographically speaking, or in terms of chemical composition and bonds. These defects can affect the materials properties. In Table 12.1 are listed some physically and chemically driven properties which can be affected by the different phenomena occurring in the surface region as it has been just defined here. This table is of course non-exhaustive and could be extended at will.

12.2.2. Stability and Evolution of Surfaces: Basic Thermodynamics

A naive view of daily life could induce the idea that surfaces are very stable. However, surface scientists know that it is not true. For example, obtaining a clean and ordered surface as a template for deposition of nanostructures can need hours

Table 12.1: Some properties which are depending on surface phenomena

Electrical properties	Mechanical properties	Kinetically driven properties	Magnetic properties
Heterojunctions	Metal fatigue	Adsorption	Magneto-resistance
Recombination processes	Adhesion	Segregation	Perpendicular anisotropy
Diffusion and doping	– Metal/ceramics	Catalysis	
Electromigration	– Metal/polymers	– Activation	
MIS, MOS, SOI	Hard layers	– Poisoning	
	– Nitrides	Corrosion	
	– Carbides	Deposition	
		Growth modes	

of work and in a few seconds all this work can be ruined if the sample is annealed at a too high temperature. The author has observed some very complex industrial multilayered architectures which simply did not work because of a wrong choice on the protective coating materials, not because this coating was not protective against oxidation but because it induced such a high surface strain that the active nanometer scale parts of the device were destroyed by the relaxation of strain. As a matter of fact surfaces reach more or less quickly a state of stability which can be somewhat predicted. Thermodynamics allows to show that a specific surface parameter, the surface tension γ, is related to a thermodynamical potential. Knowing that a system in evolution tries to minimize such a potential one can predict how will be the evolution of the surface when the thermodynamical variables will be modified through the deposition process.

It is not necessary to describe here surface thermodynamics which can be found in textbooks or in Ref. [2]. This last paper is rather old but is still totally relevant. Two important concepts have however to be pointed out: (i) surfaces are open systems where work and chemical species can be exchanged, (ii) the exchange of work cannot be reduced as it is too often thought to the mechanical work of pressure against volume ($\mathrm{d}W = -P\mathrm{d}V$). One cannot neglect electric, magnetic and elastic forces nor surface stress and surface tension. One cannot neglect either the chemical potentials of all impurities present in the system. An easy derivation of the classical equations of thermodynamics shows that the surface tension of a material tends to decrease when the surface concentration of species having higher chemical potentials is increased and that the change in surface tension depends on the chemical potentials of every chemical entity that is present in the system (even in the bulk). The surface will reach the highest degree of oxidation (this is good for the stability of oxides which is well known). Nitrides and carbides will be also very stable (but will generally add large strains). Adsorption of organic molecules will be easy and a surface will be always covered with organic contamination coming from our environment. On the contrary, a polymer surface consisting of organic molecules will be thermodynamically so stable that it will be difficult to proceed to metal deposition on it. Surface modifications of these materials will be necessary to allow metal wetting. Many plasma processes are known to make such processes (Corona discharge, …).

In the case of interfaces, it will not be possible to adsorb on them molecules coming from the surrounding atmosphere. But segregation will occur, that is, diffusion of impurities from the bulk toward the boundary. Segregation can be thermally activated or induced by strain leading to stress relaxation. It is known, for example, that impurities such as sulfur, phosphorus and antimony tend to migrate towards the grain boundaries in metal films. This is good to diminish the energy of the system but it can be a disaster for some mechanical properties. It is worth noting that surface segregation does also occur and in the case of metal alloys this effect depends on three factors: the surface tension of the alloy-forming elements, the respective size of the atoms and the possibility to build ordered phases. Thermodynamics allows to predict that:

- The chemical species having the smallest surface tension will segregate toward the surface or the interface.
- The chemical species having the largest atomic diameter will try to reach the surface through segregation and atomic exchange since this will release the strain. One has to note here a difference in the case of interfaces where this is not true. In this case, the strain would be enhanced by inserting large atoms.
- Ordered phases will form at the surface if the mixing enthalpy has the appropriate value.

In most cases, there will be a competition between these three segregation factors. This will make difficult any prediction and one will need more sophisticated calculations using, for example, molecular dynamics [3]. And finally, one has to remember that thermodynamics predicts evolutions but does not tell anything about kinetics effects. In this case numerical simulations will be also necessary to predict surface evolutions.

12.3. The "Dream Machine" for the Surface Scientist vs. the Reality of Cost (Time and Money)

Many surface scientists have tried to design special apparatus to study surfaces during the growth of thin films and to perform most of thc characterization *in situ* to be sure of well understanding the properties. They would think of a universal machine able to perform several analyses and test many properties during the same experiment. Although some of these machines do exist they are expensive and generally difficult to run safely. Surface analysis is still a costly science, as far as materials science is concerned. Many experiments need the use of UHV with especially designed setups. Surface analysis is also time-consuming since the overall signal is always quite weak: there are much fewer atoms in a surface than in the bulk! So it does not seem reasonable to design an analytic apparatus having in the same time morphologic, structural and surface chemistry possibilities of investigation. Imaging and mapping techniques are often very popular, there are time-consuming both during measurement and data processing. On the other hand, it would be unreasonable to use a single technique. So, depending on the quality of the samples which have to be analyzed a surface analysis setup should at least have LEED (or RHEED), scanning tunneling microscopy (STM) (or atomic force microscopy (AFM)), AES (or X-ray photoelectron spectroscopy (XPS)) facilities. An access to electron microscopes (scanning or transmission) and to X-ray fluorescence spectroscopy (XRF) will be always very useful.

Then it will be advisable to be able to perform other characterizations in dedicated laboratories, especially for techniques which need some expertise to be used with a reasonable efficiency such as secondary ion mass spectrometry (SIMS), keeping in mind that these techniques should be used only to investigate specific and crucial problems. The access to techniques available in synchrotron facilities is nowadays nearly mandatory. The development of methods needing high photon fluxes such as GIXD, grazing incidence X-ray small angle scattering not mentioning more classical experiments which are very useful in the particular case of thin films such as strain and stress measurements, anomalous X-ray scattering to investigate materials having atoms which are neighbors in the periodic table of elements and so on.

12.4. Classifying Surface Science Methods and their Specificities

12.4.1. Global Methods vs. Local Methods

All analytic methods need using a probe (an electromagnetic radiation, a beam with charged particles, a mechanical actuator, ...) interacting with a sample. The different phenomena taking place during the interaction with the surface are absorption, transmission, reflection, diffusion, emission, mechanical response, ... The result of this interaction can be another radiation, various particles, the change of a field or it can be a comportmental law. These are converted and/or observed in a measuring system. Various means of measurement can be used (counts, energy or wavelength discrimination, ...). To obtain the best surface analysis method, one has to study the thinner thickness that is possible, the diameter of the probed surface area should be the smallest as possible, one would like to be able to scan the entire surface under investigation and possibly get an image of it which could be topologic, structural, chemical or related to a property (e.g. magnetic). One would also like to investigate the part of the sample which is underneath the surface. And last but not least, one would like to follow the sample evolution against time and thus one would be very happy to have time-resolved spectroscopies, ranging from femtoseconds to hours, depending on the property which is under investigation! Indeed, this dream machine does not seem to be a reasonable one.

Of course, no method can afford simultaneously all these performances and one will have to adopt compromises or use several techniques at different times. One will favor first techniques in which the interaction process between the probing "radiation" (true radiation or particles) and the sample has the largest possible cross section. This is the reason why methods using ion or electron beams as an incident probe are so frequently used. The penetration of ions in a solid is negligible if their energy is not too high (less than a few keV). Due to their electric charge ions are easy to focus and ion beams can be easily scanned on the surface. Electrons have a penetration which varies with their kinetic energy and they can be also easily focused. One can use photons as an incident beam. Due to their small cross section of interaction they will enter deeply in the sample (several micrometers) but due to photon absorption they will interact with matter in such a way that ions or electrons will be created. These particles will not cross the sample if they are emitted too deeply and so only electrons or ions coming from the vicinity of the surface will be collected and measured. In the case of fields of interaction, things will be somewhat different since their interaction potentials are $1/r^n$ functions. They will be always extremely surface sensitive. A first classification of surface analytic methods can be given, choosing as a function of the probe and "radiations" (Table 12.2).

For methods using electrons either as a probe or as the particle analyzed in the detection system it is easy to get a simple evaluation of the depth under analysis. One needs to know their energy and using a diagram giving the penetration depth of electrons or their mean free path taking care of various effects [4]. The depth of analysis is roughly three times the mean free path of the electrons having a given energy. In this case, 99% of the electrons will be either absorbed (if it is the probe) or transferred out of the sample (if they are detected). Photons (X-rays or UV radiation) have mean free paths orders of magnitude larger than that of electrons and

Table 12.2: Surface analytic method as a function of probe and detection (see appendix for the meanings of abbreviations or acronyms)

Probe ⇒	Photons	Electrons	Particles	Fields
Detection ⇓	IR-FTIR	EPMA		
Photon	SERS	BIS	GDS (ions)	
	XRD		LAMMA (ions)	
	XRF			
Electron	XPS-ESCA	AES		
	SEXAFS	EELS		
	UPS	LEED		
	XANES	SEM		
		RHEED		
Particles			ISS (ions)	
– Ions			NAA (neutrons)	AP
			RBS (ions)	
– Neutrons			SIMS (ions)	
– Atoms			Helium scatt.	
Fields				
– Force				AFM
– Electrical				STM
– Magnetic				MFM

their depth of interaction will never be limited by adsorption in a thin film. One will have to use grazing incidence methods at very small angles to enhance the surface signal for X-rays (Table 12.3).

A second classification takes into account the kind of information which is looked for. One may investigate as a priority either morphology, physical and chemical properties, analytic composition, or structure (electronic or crystallographic). For practical reasons, we give the kind of vacuum needed since some samples need to be studied in controlled atmosphere.

One could also have a classification depending on the area which is probed by the method, either in terms of depth of information or lateral resolution, separating global and local methods. The most surface sensitive methods having a large spatial resolution are STM and AFM. They probe less than 0.1 nm in depth and from 0.1 to 100 nm in lateral length. Then comes AES and SIMS probing 1 or 2 nm in depth and 100 nm to 1 μm in lateral length. XPS probes 1–10 nm in depth and several micrometers in lateral length.

Table 12.3: Surface methods as a function of the information sought in priority

Information	Method	Vacuum
Analytic	AES	UHV
	XRF	Normal vacuum up to atmosphere
	NAA	None
	RBS	Normal vacuum
	SERS	Atmosphere + liquid
	LAMMA	Normal vacuum
	GDS	None
	SIMS	UHV + gas
	XPS – ESCA	UHV
Morphology	SEM	Normal vacuum
	AFM	None
	STM	Normal vacuum up to atmosphere
Crystal	XRD	None
Structure	LEED	UHV
– Long distance	RHEED	UHV
– Short distance	STM	UHV
	EXAFS	UHV
Electronic	UPS	UHV
Structure	EELS	UHV

Finally one could like to know which method is the most useful, for which refer to Table 12.4.

12.4.2. Morphology and Surface Topography

The methods giving informations on topography are beyond the scope of this chapter which is more closely related to spectroscopic methods. But investigating a surface needs generally in a first approach a study of the morphology and the topography. There are global methods, such as scanning electron microscopy (SEM) and local methods such as AFM or STM. One can note that some of these apparatus can be used also to develop spectroscopic studies. In SEM, there is generally the possibility to proceed to a chemical analysis of the sample through electron probe

Table 12.4: Various analytic methods

• An elemental surface analysis	AES
With a knowledge of the chemical bounds	ESCA
• Quantitative analysis	NAA
	XRF
• Very small concentrations	SIMS
	AP
• Fast in-depth analysis (destructive)	SIMS
With many elements in the sample	GDS
• Non-destructive in-depth analysis	RBS
• Knowledge of the surface morphology	SEM
Atomically resolved	STM
At molecular scale	AFM
• Knowledge of the crystal structure	ISS
	LEED
	RHEED
	STM
	GIXS
• Short-distance order	EXAFS
	SEXAFS
• Cartography of elements	SAM
. . . spatially resolved	XRF
	SIMS
• Knowledge of the electronic structure	ARUPS
	EELS
• Work without UHV constraints	GDL
	XRD
	AFM
• *. . . and even with liquids!*	STM
	AFM
• Investigating organic materials	FTIR
	XPS

ARUPS: Angle resolved ultra violet photo electron spectroscopy; SAM: Scanning Auger microscopy.

microanalysis (EPMA) (a technique closely related to XRF which will be described in a following section). With STM some spectroscopic studies are also possible to get informations on the chemical state of the tested atoms. This will be shown in Section 12.7. There are many textbooks on these methods.

12.4.3. Chemical Analysis (Elements and Bonds)

It is very important to know the composition of a surface and that of its vicinity. Spectroscopic methods are very useful for that. We shall see in a further section that several methods are able to give the composition of the surface with a reasonable accuracy. In the case of thin films, due to the small thicknesses of the samples XRF or EPMA will offer chemical composition informations quickly and safely. But this will be possible only if their architecture is not too complex. At any rate, XRF will not give informations about the chemical bonds and there are very few methods which are useful to get a good idea on the chemical bonds. Some structural methods such as extended X-ray absorption fine structure (EXAFS) or surface EXAFS (SEXAFS) give indirect evidence of the nature of the chemical bonds but one will essentially have to rely on XPS to get a good and reliable surface information.

12.4.4. Can Properties be also Studied by Surface Science Analytic Methods?

Spectroscopic methods are also useful to understand properties. One can think of simple contact angle measurements which can give the adhesion properties of a surface. These measurements which are not very popular among surface scientists are however quite sensitive, fast and easy to obtain and a great number of samples can be analyzed which is helpful when dealing with samples grown on a industrial scale. One can think also to ellipsometry and its related technique magneto-optical Kerr effect (MOKE). It may give excellent informations on optical and/or magnetic properties. More recently were developed magnetic measurements based on magnetic dichroism using photoelectrons and thus surface sensitive.

12.5. Some Experimental Methods

This section does not pretend to be exhaustive. We have selected only a few methods which seem to be the most efficient to study the surface of thin films.

12.5.1. Methods Based on X-ray Emission: XPS vs. XRF

XPS often called also electron spectroscopy for chemical analysis (ESCA) by chemists and industrials. One could think that this acronym is not clear enough since many electron spectroscopies such as AES give informations about the chemistry of the samples. However, the constant use of this acronym can be explained by the fact that the main advantage of this method is its ability to investigate surface composition in a semiquantitative approach and to have a very good information about the chemical bonds at the surface. Its only disadvantage is the need for UHV. For thin films analysis XRF is often preferred to XPS, even if it is not strictly speaking a surface analysis method since it gives an analysis of the whole film.

As seen in Table 12.1, XPS-ESCA and XRF are based on a probe using X-rays (energy less than 3 keV for XPS, less than 10 keV for XRF) and the detection of electrons in the case of XPS and photons for XRF. The interaction of the incident X-ray beam with matter induces an absorption of a photon leading to the ionization of atomic valence or core levels characterized by the emission of a photoelectron, the energy of which is characteristic of the atoms of the sample. This X-ray-induced photoelectron reaches an excited final state having a single electron hole which is rather simple to describe in terms of energy. XRF is based on one of the possible processes of relaxation of this excited state through a radiative process. Another de-excitation process will be a non-radiative process, that is, the emission of an Auger electron. This will lead to a surface analysis method which will be described in Section 12.5.3.

For XPS, the X-ray beam, generally a Kα radiation, needs not to be monochromatized but in given applications (high resolution) the use of a monochromator will be helpful. The exciting radiations generally used are the following:

Al Kα	hν = 1486.6 eV	ΔE = 1.0 eV	(0.5 eV with a monochromator)
Mg Kα	hν = 1253.7 eV	ΔE = 0.7 eV	(0.4 eV with a monochromator)

For XRF, the X-ray beam is also a Kα radiation (i.e. Cu, Co, ...) obtained with a conventional X-ray tube. The emitted X-rays are analyzed with an energy wavelength-dispersive XRF spectrometer (WDX) or an energy-dispersive detector (EDX). One measures the intensity of the X-ray fluorescent lines which are emitted during the radiative relaxation process. This is a quantitative spectroscopy which is not suitable for light elements ($Z < 9$), moderately sensitive for Na, Mg, Al, Si and P but very sensitive for heavier elements (around 10 ppm for elements with $Z > 19$). The lateral resolution is poor, the best being nowadays $10 \times 10\,\mu m^2$. The depth of analysis is in the order of 1 μm which makes this method unsuitable for surface analysis. For thin films it will give the overall composition of the sample which is useful if the thin film is made out of a single chemical product but unsuitable for layered samples. So we shall focus now on XPS (ESCA).

In the case of XPS, the kinetic energy E_k of the emitted photoelectrons is equal to the difference between the incident photon energy (hν) and their binding energy, E_B, in the atom where they are coming from. To index these peaks one uses the quantic state of the ionized level ($1s_{1/2}$, $2s_{1/2}$, $2p_{1/2}$, $2p_{3/2}$, $3d_{3/2}$, $3d_{5/2}$, $4f_{5/2}$, $4f_{7/2}$, ...). This kinetic energy ranges between 200 and 1500 eV and corresponds to thickness of analysis ranging from 2 to 10 nm.

Binding energies are potential energies depending on the interactions between the electrons and the nucleus. If there is an electron charge transfer between two atoms which are engaged in a chemical bond a modification of the binding energy will occur since electrons will not sense the same potential. A chemical shift will result from that and will be characteristic in its value of the intensity of the chemical bond. As a matter of fact chemical shifts are due to electron relaxation phenomena which are much complex than simple models of potentials and electrostatic interaction and they will be sometimes difficult to predict. If several different chemical bonds are present in the sample this will lead to several photoelectron lines for the same element. In the case of organic materials and especially polymers the C1s, F1s, O1s and S2p photoelectron lines are well separated in the energy spectrum but each photoelectron lines (respectively C1s, F1s, O1s and S2p) will consist of the addition of the lines characteristic of all bonds in which the particular element is included. Of course these lines have a finite width which is measured at half maximum (full width at half maximum (FWHM)). This width is a convolution of two effects: (i) the natural width of the photoelectron line which is related to the lifetime of the core hole, it is described by a Lorentzian function; (ii) the instrumental widening of the line described by a Gaussian function. A particular software is often included with the apparatus to perform the curve fitting. Some groups prefer to develop their own software. Good examples of curve fittings will be found in papers reviewed in Section 12.6.1. At any rate, one will need to use standard samples or special tables to obtain a good interpretation. These tables [5,6] are easy to hand, very precise and well documented if one has some basic knowledge in chemistry and some practical sense.

All elements, but hydrogen, lead to characteristic lines. The surface of oxides may be slightly destroyed by the outcoming photoelectrons and secondary electrons and one would observe suboxides which are in fact artifacts. One can work on insulators but a surface charge of the sample appears because of the secondary emission of electrons. One can use a flood gun sending on the surface electrons with a kinetic energy ranging from 10 to 100 eV. They afford a compensation of the surface space charge but, in this case, the evaluation of the chemical shifts may be somewhat imprecise.

In-depth resolution is in the 2–5 nm range, depending on the photoemission lines which are excited. When using an ion gun it is possible to obtain a chemical profiling of the subsurface region but one can hardly investigate depths larger than 0.5 μm since the ion bombardment process is time-consuming and finally modifies the chemistry of the surface. Space resolution is not easily affordable with this method. In conventional apparatus, the area of the probe is at best $3 \times 3\,\text{mm}^2$. In expensive systems one can reach $15 \times 15\,\mu\text{m}^2$.

As a quantitative method one can detect one atom out of one hundred or out of one thousand depending on the cross section of the process which is better for heavy atoms than for light ones. The sensitivity is thus around 10^{12} to 10^{13} at.% cm^{-2}. If the element under investigation is evenly dispersed in the whole sample this is around 0.1 at.%. Of course this is not a very high sensitivity and we recall that XRF can afford much better sensitivities. If the element is segregated or deposited on the surface with small thicknesses (around several nanometers or less) the sensitivity is much better and, in this case, XRF has no efficiency. At any rate, XPS with AES is one of the rare methods allowing to investigate light elements such as carbon, nitrogen and oxygen. It will thus very useful for investigating organic films, especially on the chemical bond point of view since chemical shifts are very well documented for these elements.

The accuracy in the quantitative treatment of XPS is generally around $\pm 10\%$. This accuracy can be enhanced when using standard samples or if there are not too many different kinds of atoms in the vicinity of the surface. The commercial apparatus have softwares allowing to derive the various concentrations: they have to be handled with care. To save time only the intensities of the main lines are generally measured and this induces large distortions in the calculation.

In conclusion XPS is best suited if:

- the surface under investigation is homogeneous to avoid problems due to the poor lateral resolution;
- the surface atoms lead to chemical shifts easily understandable;
- the thickness of the interesting layer under investigation is in the 10–100 nm range.

12.5.2. Methods Based on X-ray Absorption: SEXAFS

EXAFS can be used as a method of surface analysis (SEXAFS) and is based on the measure of the variation of the X-ray absorption coefficient of a material sample. Although clearly spectroscopic in its principles this method gives both structural and chemical informations.

It uses the properties of photon absorption by matter which has been described earlier under the term of core-hole ionization. When scanning the adsorption coefficient as a function of energy, one can observe oscillations which follows the adsorption edge. These oscillations also called Kronig oscillations are characteristic of the local structure of the material. One can find an extensive description of this spectroscopy in the book by Koningsberger [7]. An intense X-ray beam (generally synchrotron radiation) impinges the material and the incident intensities are measured simultaneously with the transmitted (EXAFS) or reflected (SEXAFS) ones. Their ratio gives the absorption coefficient μ. The oscillating part of this coefficient μ is an energy or wavevector dependent function.

$$\chi(k) = \frac{\mu(k) - \mu_0(k)}{\mu_0(k)} \tag{12.1}$$

If one uses a simple diffusion model this function contains the interference between the wave function of the photoelectron of the absorbing atom and that of electrons backscattered by the neighboring atoms i, at a distance R_i from the adsorbing atom:

$$\chi(k) = -\sum_i \frac{N_i^*}{kR_i^2} A_i(k) \mathrm{e}^{-2R_i/\lambda(k)}\, \mathrm{e}^{-2k^2\sigma_i^2} \sin[2kR_i + \varphi_i(k)] \tag{12.2}$$

with

$$N_i^* = 3\sum_{i=1}^{N_i} \cos^2 \alpha_j^i \quad \text{for } i \neq j$$

The exponential factors are damping terms (λ, lifetime of the photoelectron related to the mean free path and σ, the Debye–Waller factor describing the disorder due to thermal scattering). A is the amplitude of backscattering, φ a change of phase induced by the absorbing and scattering atoms, N_i is related to the number of atom neighbors through the angle α between the chemical bond and the polarization vector of the incident beam. EXAFS oscillations then mix crystallographic parameters (R, σ, N) and electronic ones (A, φ, λ).

For surface applications, instead of measuring directly the absorption coefficient, one uses a parameter proportional to μ, sensitive to the first nanometers. Since after a photoemission process (photon absorption) there is a de-excitation process, either radiative (X-ray fluorescence) or non-radiative (Auger electron) the products of this de-excitation will be related to the photon absorption coefficient.

SEXAFS is a very specific method and it is not very often used. In the case of thin films, the bulk method EXAFS can be very useful for the same reason as that given earlier for XRF analysis. The fact that the film is thin allows the X-ray beam to cross the sample, provided the template substrate is not too thick. But one needs to investigate an homogeneous sample.

12.5.3. Methods Based on Electrons: AES

AES is one of the most useful technique for surface analysis. It gives a chemical analysis of the surface on a thickness ranging from 1 to 2.5 nm.

AES is an energy analysis of electrons emitted from a sample when struck by an electron probe of low to medium energy (100 eV to 10 keV). The energy distribution of electrons emitted and reflected from a surface is somewhat complicated but it is very rich in information. One obtains elastically scattered electrons which are used for electron diffraction (LEED or RHEED). There are also primary electrons which have suffered quantified losses of energy and which are useful to obtain informations on the electronic structure (electron energy loss spectroscopy, EELS), secondary electrons and Auger electrons. These Auger electrons are outcoming from a non-radiative de-excitation process after an ionization in a core atomic level as described in the previous paragraph. This de-excitation process needs three energy levels of an atom. The incident beam induces a core-level ionization (hole) in level *W*. The extra energy is given to an electron of a less-bounded level *X* to occupy this hole. Now there are possible processes: either a fluorescence process (as described for XRF which is also used in EPMA) which is a radiative process and leads to photons, either the ejection of another electron from a less-bounded level *Y* called an Auger electron. The sum of the yields of the two processes ϖ_F and ϖ_A is equal to 1 but for the same initial ionized core level *W* the Auger yield is always higher for the lightest chemical elements. This explains why Auger spectroscopy is better than XRF for light elements such as carbon, nitrogen and oxygen.

In a first approximation the energy of an Auger electron is $E_A = E_W - E_X - E_Y$. This energy does not depend on the primary beam energy and is characteristic of a given atom. By tradition, an Auger transition is called after the three spectroscopic levels which are successively used in the process (*K*, *L*, *M*, valence and conduction bands, respectively, noted *V* and *C*). If there is a chemical shift of the levels equal to $+\Delta E$ (respectively, $-\Delta E$) the Auger energy shift will be $-\Delta E$ (respectively, $+\Delta E$). If the Auger transition takes place with valence electrons, the changes in chemical bonds will induce a modification of the peak shape. This confers to AES a extremely high sensitivity to chemical effects.

Finally, since electrons can be easily focused it will possible to scan the surface to obtain point analysis with a spatial lateral resolution around 20 nm. A mapping of the various chemical elements present on a surface will be possible in a special apparatus, the scanning Auger microscope. Since the electronic yield of the process is not equal to 1, studying insulators can be difficult. In the energy range where the secondary electron emission yield is not much higher than 1, the surface space charge stabilizes and analyses are possible.

As it we already said, surface analysis with electron probes depends on the *inelastic mean free path of electrons (IMFP)* in the sample. It is the distance λ followed without any energy loss by *N/e* electrons if *N* is the number of incident-electrons incidents and *e* the Neper number equal to 2.718. To favor the Auger yield the incident energies are less than a few keV and thus the collected Auger electrons will have an energy smaller than 3 keV, the smallest energy being a few tens eV. Since the absorption of electrons follows a Beer–Lambert law, it is easy to show that 99% of the signal will outcome from less than 1 nm. This means that it will be necessary to proceed with an ion bombardment to clean the surface when investigating a sample coming from the external atmosphere. This ion bombardment will also be used as with XPS for depth profiling the sample. This process is very slow and if this kind of information is needed one would rather choose other techniques as SIMS, glow discharge spectroscopy (GDS) or Rutherford backscattering (RBS).

AES is a semiquantitative method and as for XPS it will be useful to use tables not only to correctly index the Auger peaks but to derive the surface composition [8]. One has finally to recall that AES is very sensitive to chemical modifications at the surface through the shape of the peaks. In that case, one has to rely on standards: predicting the shape of an Auger through theoretic calculations is indeed very difficult and beyond the daily possibilities of users.

12.5.4. Methods Based on Ions: SIMS

SIMS is a direct analysis of atoms or molecules outcoming from the surface of a sample as ions. There is a parent technique which analyzes neutral species secondary neutral ion mass spectroscopy (SNMS) which will not be described in this chapter since the principles are not very different. SIMS has been set up in the fifties of the past century but it is only during the seventies and eighties that it reached its full development. SIMS is not very easy to handle and needs generally to be done in dedicated laboratories. It is very useful to detect impurities at a very low level, to obtain in-depth concentration profiles, to be able to separate isotopes, and last but not least to detect hydrogen. For polymers, one should note that the method is sensitive to conformation [9].

When a surface is bombarded by incident ions (primary beam) such as rare gas ions or alkaline ions having an energy ranging from a few tens eV to a few hundreds eV several processes occur at the surface. First occur crystalline disorder, atomic displacements and implantation of rare gases. Then there are emissions of particles such as positively or negatively charged ions, neutral atoms, secondary electrons, photons, … With SIMS only ions (negative or positive) are collected. Of course it is clear that SIMS is a destructive method but it should be noted that the primary beam has a very small diameter and the surface will not be entirely destroyed.

The measured intensity of a SIMS signal obeys the following equation:

$$I_{\mathrm{i}}^{\pm} = I_{\mathrm{p}} \times Y_{\mathrm{i}} \times \alpha_{\mathrm{i}}^{\pm} \times \eta_{\mathrm{i}} \qquad (12.3)$$

where I_p is the primary ion current, Y_i the sputtering yield for species i, $\alpha_i^{\mp}$ the probability to create a negative or a positive ion from species i and η_i a geometric transmission factor depending on the apparatus.

SIMS is very surface sensitive since the emitted ions outcome from the very first atomic layers (less than 1 nm). The in-depth resolution is very high, all elements can be detected even hydrogen and its isotopes. Sensitivity can be as high as parts per million or even in given specific cases parts per billion. One can obtain informations on the chemical composition of the surface but it is a very tedious job.

SIMS needs argon or oxygen ions primary beams (sometimes caesium beams) and the beam diameter can be less than 10 μm. The primary beam can be scanned over the surface. Detection is performed with different kinds of mass spectrometers (quadupoles, electrostatic, or time of flight, …).

12.5.5. Ellipsometry and Optical Kerr Effect

Ellipsometry is a non-destructive optical method which allows to obtain the thickness and the optical index of a thin film. It is based on the modification of the polarization of a light beam when interacting with the sample surface. Generally it is performed in the reflective mode. When working in transmission one can analyze the optical properties of the medium which is crossed by the light but in this case one does not speak of ellipsometry. One knows a extension of ellipsometry based on the MOKE. It is indeed well known that the polarization of light can be modified when the incident beam is reflected by a magnetic material. There are three different geometries for measuring the Kerr effect: (i) polar if the magnetization vector M is perpendicular to the sample and is in the incidence plane, (ii) longitudinal if the M is in the plane of the sample and in the plane of incidence, (iii) transverse if M is in the plane of the sample and perpendicular to the incidence plane. In the first two configurations an electromagnetic radiation, linearly polarized, impinges on the surface and is transformed in an elliptical wave after reflection on the magnetic sample surface.

Let us recall some basic principles about the polarization of light. An electromagnetic wave (visible light, infrared (IR) light, X-rays, …) is characterized by its amplitude, its frequency (related to its energy), its direction of propagation and its Poynting vector (orientation of the electric field). Generally, light is elliptically polarized and the ellipse described by the Poynting vector is fully determined by: (i) its plane in space (polarization plane) which is generally normal to the direction of the propagation of light in the incidence plane, (ii) its orientation in this plane given by the angle θ between the large axis of the ellipse with a given direction in the plane which is often chosen perpendicular to the incidence plane, (iii) its amplitude which is the length of the large axis of the ellipse (a), (iv) the ellipticity which is the ratio between the lengths of the two axes of the ellipse (a,b). Polarization is said *linear* if this ratio is zero and *circular* if it is equal to ± 1.

In most ellipsometers one gets ellipsometric angles ψ and Δ which are related to the ratio of the Fresnel coefficients R_p and R_s for the light parallely polarized (p) or perpendicularly polarized (s). This ratio is:

$$r = R_p/R_s = tg\ \Psi\ \exp(i\Delta) \tag{12.4}$$

where Δ is the phase difference between s and p polarizations, tg Ψ being the attenuation between the s- and p-waves. The couple of values cos Δ, tg Ψ is characteristic of the dielectric function of the layer deposited on a sample and of its thickness. One has to note that even if Δ and Ψ are often indicated on ellipsometers, modern ellipsometers give directly the real and imaginary part of $r = \rho e^{i\Delta}$.

A drawback of classical ellipsometers was that the signal/noise ratio of the light detectors is bad when the signal is very weak. So, modern ellipsometers use a photo-elastic phase modulator and the signal is detected through a lock-in amplifier. The sensitivity is thus 10^{-3} degree of angle for a time constant in the order of 1 s. This corresponds to about 1% of an oxide monolayer on a silicon substrate. If one accepts a larger limit of detection, one will gain in time resolution and study kinetics in the millisecond range.

It is possible to use ellipsometry in another mode: the spectroscopic one. Instead of illuminating the sample with monochromatic light, one uses a white source (e.g. a xenon lamp) which allows to investigate the light spectrum from UV (0.192 μm) to IR (2.05 μm). The sample is fixed on a goniometer to be able to modify the angle of incidence and a monochromator allows one to select a given wavelength. The measure gives $\Delta.f(\theta, \lambda)$ and $\psi.f(\theta, \lambda)$. Data processing allows then to analyze complex structures such as multilayers, rough interfaces, inhomogeneous or non-isotropic layers. Finally it is possible to focus the incident beam and to scan the surface. A charge coupled device (CCD) camera allows then to obtain an image of the optical properties of the surface. The lateral resolution is in the order of 1 μm. Ellipsometry is a non-destructive method and is used *in situ*. It needs a surface with a very good reflection factor and the deposit has of course to be transparent to the particular light used in the apparatus. It has a very good time resolution and it can continuously test surfaces as a function of the deposition time, going from one nanometer to several hundreds nanometers. As a drawback, one has to note that the interpretation of what is called the optical trajectory is not straightforward.

For an extensive description of the method and of the calculations needed in ellipsometry, one can refer to the rather old book by Azzam and Bashara [10] taking into account that the technical developments are of course quite obsolete. The author has been fascinated by Internet links on these techniques which are very useful and easy to apply.

12.5.6. Contact Angle Measurements

Contact angle measurements are used to obtain a parameter which is characteristic of the surface, *viz.* the surface energy or the adhesion energy.

Contact angle measurements are based on daily observation about wetting, that is, the possibility of liquids to flow on a surface or to form droplets. In laboratory one can test also wetting of solids on other solids, that is crystal growth and deposition. Full wetting, that is, creating a continuous layer on a surface without any drop is rare. Drops, more or less large, more or less spread on the surface, have a shape that minimizes the total energy of the system. The angle θ of a liquid/solid droplet (L) surrounded either by a gas or its vapor (G) with a solid/liquid substrate (S) is called the *contact angle* and depends on the interfacial tensions of these various phase. If one writes the equilibrium of forces one gets the Young equation:

$$\gamma_{SG} - \gamma_{LG} \cos\theta = \gamma_{SL} \quad \text{or} \quad \gamma_{SG} = \gamma_{SL} + \gamma_{LG} \cos\theta \qquad (12.5)$$

where the various γ are the interfacial tensions.

This relationship can be simplified in the case of partial wetting and the Young–Dupré equation is obtained: $W_{SL} = \gamma_{LG}(1 + \cos\theta)$, where W_{SL} is the adhesion energy. One can also use γ_L instead of γ_{LG} and γ_S instead of γ_{SG}.

This energy can be separated in two parts: one called *dispersive* since it is due to the London dispersion forces between two non-polar molecules, another one is called *non-dispersive* and is due to dipolar interactions, hydrogen bonds, …

Then we obtain:

$$W_{SL}^{d} = 2(\gamma_S^{d}\,\gamma_L^{d})^{1/2} \quad \text{and} \quad W_{SL}^{nd} = 2(\gamma_S^{nd}\,\gamma_L^{nd})^{1/2} \qquad (12.6)$$

The two γ_S values can be derived through measurements of contact angles obtained with at least two different liquids, the various components of the surface tension of which are known.

However this approach of the calculation of the surface energy is too simple since it mixes in the non-dispersive component various interactions which have nothing to do one with another. One has to split the Lifshitz–van der Walls interactions (LW) from acid–base interactions (AB). One obtains the adhesion work (or surface energy) through the following formula:

$$\begin{aligned} W_{SL} &= \gamma_L(1 + \cos\theta) \\ &= 2\left[(\gamma_S^{LW}\,\gamma_L^{LW})^{1/2} + (\gamma_S^{+}\,\gamma_L^{-})^{1/2} + (\gamma_S^{-}\,\gamma_L^{+})^{1/2}\right] \end{aligned} \qquad (12.7)$$

where γ^+ and γ^- are respectively, the contributions to the surface tensions having either a «Lewis acid» character (electron acceptor) or a «Lewis base» character (electron donor) [11].

Using at least three different liquids for which the various contributions to surface energy are well known one can finally obtain the values of the surface tension of the solid substrate.

Now, we shall describe the experimental measurement. As it has been said earlier one has to measure the dispersive and non-dispersive contributions using various liquids as a probe. One needs a first measurement with a liquid probe which is strictly non-polar and other measurements with known polar liquids. This is of course a rather empirical measurement since the polar part of the measurements mixes several contributions which are in their physical chemistry principles quite different. One can follow Van Oss *et al.* who developed the chemical description of surface energy in terms of Lewis acids and bases [12]. Good reviews of contact angle measurements can be found in Ref. [13].

12.5.7. Raman Spectroscopy

Raman spectroscopy based on a special scattering process of photons is an optical method and thus the depth of penetration of the incident beam will be too large for investigating surfaces in the proper sense. But, as for XRF, in thin film analysis, Raman spectroscopy can be useful if the sample is homogeneous enough. This technique, sensitive to the structure, chemistry and strains of the material can give some important results and should not be neglected. It is another spectroscopic method which covers various fields of materials properties.

The interaction occurring between a laser source and the sample leads to photons. When the process is elastic (no change in energy) one speaks of Rayleigh scattering, when it is inelastic (losses in energy) it is Raman scattering. Raman scattering leads to a very weak signal and relies on vibration modes in the solid. It can be of first or second order depending on the number of photons engaged in the process. In the first-order approximation, the energy difference between the incident (initial state *i*) and emitted photons (final state *f*) is equal to the energy of a phonon E_ν: $E_\mathrm{f} = E_\mathrm{i} \pm E_\nu$. Depending on the sign there will be an anti-Stokes process (phonon annihilation) or a Stokes process (phonon creation). The various vibration modes that are active in the sense of Raman scattering in a single crystal are limited but when a crystal is distorted these vibration modes are modified. The Raman line will be displaced in energy and widened. If the sample is amorphous all the vibration modes possible in the Brillouin zone can also be excited. The Raman spectra is thus characteristic of the density of the vibrational states in the solid.

Raman spectroscopy has been found to be very useful in carbon-based samples since there are large modifications between diamond, graphite, carbon nanoparticles, nanotubes, … Another advantage of Raman spectroscopy is that micro-beam apparatus exist and the laser beam can be focused on small particles having a diameter in the nanometer range.

12.6. Some Examples on Surfaces Grown or Modified by Plasma Interaction

The present section intends to give some examples of surface analysis in the case of thin films grown with plasma-enhanced techniques or surface modified by some kinds of plasmas such as plasma etching. Of course, all fields cannot be described in this chapter and even in the fields which have been selected only a few examples will be given. The author has selected the references on the basis of the various techniques which are used and developed, trying to show that many other spectroscopic techniques can be used in conjunction with those which have been more thoroughly described in Section 12.5. In such a way, one will be aware of many surface science techniques and will learn how surface scientists have to be clever to use as many complementary techniques as possible within the possibility of time and … money. For fields which will not been dealt with one can find interesting descriptions and results for silicon-based integrated circuits in Ref. [14] and for magnetic storage technology in Ref. [15]. Surface treatment of metals has been also extensively studied and one can refer to Ref. [26] and the references quoted therein.

12.6.1. Surface Modification of Polymers

Surface treatment and modification of polymers are nowadays very important in numerous technologic applications. The plasma technique is known to be powerful since it applies to a large domain of materials and is a relatively cold process. It can be used in microelectronics as seen in the introduction of this section [14], for the improvement of surface adhesion [16], in membrane permeation selectivity [17] and biocompatibility [18]. One of the peculiar quality of plasma treatment is to uniformly modify the surface without affecting the bulk properties. Most generally the surface analysis of these plasma modified polymers uses a multi-technique approach. One can cite several recent papers. In Ref. [19], the topography of several polymers such as polypropylene (PP), polymethylmethacrylate (PMMA), polytetrafluoroethylene (PTFE) and polyethyleneterephthalate (PET) has been studied. It is shown by AFM measurements that in PP and PMMA plasma etching induces nanostructures. A correlation with XPS allows to conclude that these nanostructures are due to a melting process, involving chain scission effects, dangling bonds recombination and that the increase in temperature plays only a little role in this melting process.

Synthesis of teflon-like thin layers known for their chemical inertness, their thermal stability, their low coefficient of friction and low-dielectric constant due mainly to their low-surface energy is difficult with conventional technologies. It is possible to use plasma-enhanced chemical vapor deposition (PECVD). In Ref. [20], are described the results of deposition of such thin films through a modification of this technique which is pulsed to favor a better flourine/carbon stoichiometry. ESCA is used to check this stoichiometry together with IR spectroscopy. Differences in topography depending on the fluorine content are shown by AFM. This shows that topography and surface chemistry can be deeply related as it has been said earlier and that one should not focus its investigations on spectroscopic studies alone but has to look for the structure and morphology of the thin films. The same group investigated the synthesis of highly fluorinated thin surface layers with a continuous SF_6 plasma on paper and some polymer surfaces as substrates [21]. Beyond a classical characterization of the plasma by actinometry, the authors proceed to an extended XPS analysis or the various XPS lines (C1s, O1s and F1s). This paper is very useful to understand how can be used the chemical shifts given by the various bonds with an apparatus having a high-resolution power: 11 sublines are taken into account for the C1s line, 8 in the case of the F1s line, 3 in the case of the oxygen line which is much less sensitive to chemical shifts that fluorine and carbon. Frustrated total internal reflection *and* Fourier transform IR (FTIR) and AFM are also used. The result is that SF_6 plasmas are possibly the best fluorination media for macromolecular structure containing carbon- and oxygen-based bonds through a mechanism of preoxidation of the surface.

Polymers are used also as barrier materials, for example, in packaging technology and protective coating applications. The barrier effect can be obtained with plasma-enhanced chemical vapor processes and silicon dioxide or silicon nitride layers have been obtained with organic precursors such as hexamethyldisiloxane (HMDSO). Polyamides such as PA12 are studied in Ref. [22] as barriers against water. These polymers have low water permeabilities and good mechanical properties. However, their barrier efficiency against organic vapors is not sufficient and it can be improved with a CF_4 plasma treatment. Surface analysis is largely used in this study with AFM, contact angle measurements and XPS. The contact angle measurements are well described and can be taken as a good illustration of Section 12.5.6 with the use of different testing materials for surface chemical titration.

The same authors [23] investigate the surface modification of polyethylene in CO_2, H_2O and CO_2/H_2O plasmas used to increase surface carboxylic functions on the surface. They use XPS measurements and in a related paper they describe SIMS analyses used as a correlation technique [24].

Polymer decomposition under oxygen plasma treatments has been recently questioned by Kumagai *et al.* [25] to see which bonds were destroyed by the process on PET. Emission spectroscopy and FTIR which have not been described in the present paper are used in conjunction with XPS. They show that the oxygen species attacked the PET surface, then the abstracted hydrogen atoms. After the initial step in which ester bonds are broken, polymer decomposition proceeds to form a stable antiplasma layer.

12.6.2. Anchoring of Liquid Crystals

Liquid crystals are important not only for fundamental reasons but also for their display applications. Their interaction with solid substrates is known as *anchoring*. This property is difficult to predict since it depends on several factors which are described in [27] by Fonseca *et al.* This group has investigated the modification of self-assembled monolayers made out of octadecyl-trichlorosilane and deposited on indium tin oxide (ITO) covered glass in the postdischarge region of a cold plasma. Surface modification is observed by XPS and the angle of anchoring by optical methods. It is found that depending on the concentration of oxygen species contained in the postdischarge region one can tune the transition between homeotropic and planar anchorings.

12.6.3. Applications in the Biomedical Field

Protein adsorption to implanted biomaterial surfaces is believed to mediate cellular adhesion and activation that influence the long-term performance of medical devices. This sentence is a quotation from Ref. [28]. Surface modification of substrates used in the human body (biomaterials) will thus be necessary. It can be made with plasmas such as radio frequency glow discharge. In this paper the authors describe the modification of a FEP sample (a tetrafluorethylene – hexafluorethylene copolymer) followed by the deposition of tetraglyme before protein adsorption. Surface analysis has been performed coupling Tof-SIMS and ESCA (XPS). Correlations with specific biologic tests have been made to demonstrate that plasma deposition of tetraglyme at different plasma powers enables to vary the surface chemistry and to tune protein and cell resistance or uptake. The same gas in the reactor can give surfaces ranging from non-fouling to strongly cell adhesion. This offers possibilities to pattern templates for various biomedical applications.

12.6.4. Carbon-based Nanostructures and Films

Carbon-based nanostructures are of increasing interest to obtain materials with novel electronic, structural or mechanical properties. Among these one can cite carbon nanotubes and carbon nitrides. In the first case, the potentialities are enormous since carbon nanotubes are structurally very versatile and since they can be doped with metal particles to obtain nano-objects with specific properties such as field emission and magnetic properties. In the second case theoretic predictions, either with quite empirical methods or with first-principle calculations, have lead to an extensive experimental work to grow β-C_3N_4. It was predicted that this species could have a bulk modulus of 422 GPa and a hardness of 73 GPa, values which are comparable to that of diamond. In both cases, nanotubes and carbon nitrides, the experimental results, also already important, are somewhat beyond the hopes. The reason is the extreme difficulty to grow the exact entities which are wished. This may explain the large number of experiments using extensive surface characterization in order to try and explain why the properties are not so easily reached.

The first way to solve these problems can be to understand the initial growth of films through cluster deposition [29]. In this paper, the authors describe the use of a carbon plasma generated by an electric discharge between two graphite rods mixed with an inert buffer gas. The layers, deposited on silicon or on copper, are analyzed with a set of techniques (Tof-SIMS for the mass distribution of clusters, Raman scattering and XPS). It is shown that the films are essentially graphite-like with the presence of a high number of distorted sp^3 bonds. What is interesting in this paper, as far as surface science is concerned, is the possibility to split the C1s photoelectron lines in two components related to sp^3 and sp^2, respectively. This shows the sensitivity of XPS to hybridation. Another paper dealing with carbon-assembled carbon films with different nanostructures uses several spectroscopic methods [30]. The apparatus is based on a pulsed microplasma cluster source eroding a graphite target. Raman scattering, EELS and XPS either in the valence band region or for core holes are currently used in this work. A delicate interplay exists between the structure and the electronic properties of cluster-assembled carbon.

It is then possible to investigate carbon nanotubes. These entities discovered in the beginning of the nineties of the previous century [31] have about all the possible qualities either electronic, catalytic or mechanical. They can be used as molecular quantum wires in nanocircuits if their electronic properties are well controlled. They can be doped or structurally modified with elements such as metals or nitrogen. A comprehensive spectroscopic study of nitrogenated carbon nanotubes has been described by Droppa *et al.* [32] since nitrogen is believed to modify the nanotube conductivity and to enhance the mechanical properties. The carbon nanotubes are obtained by the arc-discharge technique but in this case the obtained material has a large quantity of residual materials such as amorphous carbon or clusters. Beyond structural techniques TEM and high-resolution TEM (HRTEM) or the investigation of morphology (SEM), photoelectron spectroscopy, EELS, Raman spectroscopy have been used to characterize the samples. One of the problems here addressed is the purification of the materials of clusters and amorphous carbon either by heating the sample with an intense laser beam or by a chemical purification process, another one is the presence of bamboo-like structures which are generated at relatively high nitrogen partial pressures. If the nitrogen content is low or equal to zero single wall nanotubes are readily obtained and they are found to be ordered.

The growth of aligned nanotubes has been described in Ref. [33]. Many CVD processes, involving classical thermal and activated CVD processes, to grow films of aligned carbon nanotubes and other carbon nanostructures (nanofibers, nanocones, ...) with a high selectivity have been compared. Activation of the gas phase was carried out by hot filaments CVD (HF CVD), by a gas discharge (DC CVD) or by the use of both activation pathways (DC HF CCVD). Prior to the CVD growth, a dispersion of either Co or Fe catalytic particles on the silicon substrate was performed. A narrow size distribution down to 2 nm of the catalytic particles is thus obtained. Samples before and after CVD were characterized by SEM, TEM, HRTEM, X-ray absorption (XAS) and Raman spectroscopy and *in situ* XPS, AES, EELS. It is found that the DC HF CVD process is by far the most efficient process to get aligned nanotubes. This allows to propose a mechanism of nanotubes nucleation and growth: nucleation occurs through a nest of carbon that cannot further develop due to etching and sputtering. With highly dispersed Fe nanoparticles, it is possible to grow individual and aligned single or double wall nanotubes with density higher than $10^{12}\,cm^{-2}$. The alignment of the films of carbon nanotubes has been quantitatively studied by angular X-ray absorption spectroscopy (XAS) on the C K edge. It is found that the $p \Rightarrow p^*$ transition at 285.2 eV is highly sensitive to the mutual orientation of the carbon nanotubes.

12.7. New Trends: Methods Based on Synchrotron Radiation

Synchrotron radiation has been one of the major advances in materials science during the last decades. Due to the high flux of photons, to its polarization, to its well-defined beam, to the possibility to tune the wavelength in a large range and to its time structure allowing time resolution in the picosecond range, many investigations have been made possible. It is not in the scope of this chapter to enter in all details of the new techniques which mix spectroscopies, structural studies and morphologic investigations. Some are already used by scientists developing thin films with plasma-induced deposition techniques, some others will be certainly of interest if the near future. Most methods using synchrotron radiation are not fundamentally different from that used with conventional X-rays and they will lead to techniques very near to those described in this chapter and in the following one. But with synchrotron radiation facilities grazing incidence is easily reached, giving rise to methods enhancing the surface sensitivity. Well-documented review articles can be found in the literature and on the web. One can cite for magnetic materials the paper by Kortright *et al.* [34]; for microelectronics the paper by Jordan-Sweet [35] is a very good review.

Many methods can be useful for surface (interface) scientists since they mix various kinds of information obtained on the same sample. One can think of diffraction anomalous fine structure (DAFS) which provides informations on atoms which have nearly the same atomic number and are thus difficult to investigate with conventional X-rays. One can refer to Ref. [36] to have an idea of the principles and results given by this method. One can think also to circular (linear) magnetic dichroism (XMCD) [37] which is a spectroscopic technique which can be used for surface analysis since it can be based on the detection of photoelectrons. It gives mainly indications on magnetic properties such as the orientation of the magnetization vector.

Another technique has been developed recently and has been used by scientists growing metal particles through plasma deposition. This technique is GISAXS and is very similar in its principle to a technique which has been widely used for liquids, liquid crystals and polymers: SAXS. The bases for SAXS can be found in [38], a very old book but containing all the theoretical tools needed for this method. Broadly speaking, there are no differences in the general principles between conventional X-ray scattering and SAXS. What is special in terms of the distinction between SAXS and regular wide-angle X-ray scattering is the location of the scattering of interest. This is typically at small angles in the vicinity of the primary beam and extending to less than two degrees for standard wavelengths. The scattering features at these angles correspond to structures ranging from nanometers to a few hundreds nanometers depending on the incident wavelength of the photons and on the distance between the sample and the collector (generally a CCD camera). SAXS and GISAXS give the correlation between the particles, their distribution in size and a rough idea of their shape. SAXS and GISAXS are sensitive to the difference in electron density and are thus the most sensitive when the particles under investigation have an atomic number very different from that of the substrate. Thus it will be very interesting for metal deposits on polymers, carbon or silicon.

GISAXS uses grazing incidence and works in reflection close to the critical angle, inducing small X-ray penetration depths. The incident beam is nearly totally reflected from the substrate and the refracted beam is an evanescent wave, confined in the vicinity of the surface layer. A set-up of this method can be found in Ref. [39] which describes several fields of application in the growth of nanostructures and especially the problem of real-time monitoring of growing nanoparticles. This is one of the rare technique able to probe with time resolution (in the order of the second and longer times) the growth of nanoparticles. By varying the angle of incidence of X-rays it is also possible to investigate particles which are not too deeply embedded in the substrate. If one is interested in the theoretical approach and in the way images can be treated, Ref. [40] gives detailed informations.

GISAXS has been used recently to investigate silver nanoparticles encapsulated in carbon cages obtained by co-sputtering of silver and graphite with a Kaufmann source [41] and the growth mechanism of clusters obtained by plasma deposition [42,43]. The nucleation and initial growth of platinum islands by plasma sputter deposition on amorphous surface has been recently studied [44] coupling GISAXS with GIXD and RBS, a non-destructive technique which has not been detailed in Section 12.5 and that can be used to obtain concentration profiles. Self-organization of cobalt dots and pillars on a gold substrate has been studied by GISAXS coupled STM and scanning tunneling spectroscopy (STS) [45]. The advantage of coupling structural information given here by GISAXS and a spectroscopic technique such as STS is that it was possible to give a direct chemical proof of the vertical growth of cobalt pillars in a substrate from cobalt dots self-organized on the surface which in the case of Au(111) is a template for growing ordered nanostructures.

References

[1] G.A. Somorjai, *Introduction to Surface Chemistry and Catalysis*, John Wiley and Sons, New York, 1994.

[2] M.J. Spaarnay, *Surf. Sci. Rep.*, 4 (1984) 101–270.

[3] C. Goyhenex, H. Bulou, J.P. Deville and G. Tréglia, *Appl. Surf. Sci.*, 134 (2002) 188; C. Goyhenex and G. Tréglia, *Surf. Sci.*, 446 (2000) 272.

[4] A. Jablonski and H. Ebel, *Surf. Interf. Anal.*, 11 (1988) 627. The following link can be useful: www.lasurface.com.

[5] L.E. Davis, N.C. MacDonald, *et al.*, *Handbook of X-ray Electron Spectroscopy*, Perkin Elmer Corporation, Eden Prairie, Minnesota, USA.

[6] G. Beamson and D. Briggs, *High Resolution of Organic Polymers, the Scienta ESCA 300 Database*, Wiley, Chichester, 1992. It seems that this book is out of print and has been replaced by a CD version <<The XPS Database of Polymers>> (same editors).

[7] D.C. Koningsberger and R. Prins (Eds.), *X-ray Absorption. Principles, Applications, Techniques of EXAFS, SEXAFS, and XANES*, John Wiley and Sons, New York, 1988.

[8] C.D. Wagner, W.M. Riggs, *et al.*, *Handbook of Auger Electron Spectroscopy*, Perkin Elmer Corp., Eden Prairie, Minnesota, USA.

[9] D. Léonard, *et al.*, *J. Adh. Sci. Technol.*, 10 (1996) 1165–1197.

[10] R.M.A. Azzam and N.M. Bashara, *Ellipsometry and Polarized Light*, North-Holland, Amsterdam, 1977.
[11] E. Darque-Ceretti, *Rev. Métall. Cah. Inf. Tech.*, 94 (1997) 617–633.
[12] C.J. Van Oss, R.J. Good and M.K Chaudhury, *J. Coll. Interf. Sci.*, 111 (1986) 378–390.
[13] R.J. Good, *J. Adh. Sci. Technol.*, 6 (1992) 1269–1302; J.F. Padday, *J. Adhes. Sci. Technol.*, 6 (1992) 1347.
[14] G.S. Oehrlein, M.F. Doemling, *et al.*, *IBM J. Res. Dev.*, 43 (1999) 181–199. www.research.ibm.com/journal/rd/431/oehrlein.html.
[15] J. Windeln, C. Bram, *et al.*, *Appl. Surf. Sci.*, 179 (2001) 167–180.
[16] S. Vallon, A. Hofrichter, *et al.*, *Thin Solid Films*, 68 (1996) 290.
[17] F. Poncin-Epaillard, N. Médard and J.C. Soutif, *Macromol. Chem. Phys.*, 201 (2000) 212.
[18] H. Biedermann and Y. Osada, *Adv. Polym. Sci. Polym. Phys.*, 95 (1990) 57.
[19] M. Collaud Coen, *et al.*, *Appl. Surf. Sci.*, 207 (2003) 276.
[20] C. Naum, M. Manolache and F. Denes, *Langmuir*, 16 (2000) 749.
[21] L. Emilio Cruz-Barba, S. Manolache and F. Denes, *Langmuir*, 18 (2002) 9393–9400.
[22] F. Dreux, *et al.*, *Langmuir*, 18 (2002) 10411.
[23] N. Médard, *et al.*, *Langmuir*, 18 (2002) 2246.
[24] N. Médard, *et al.*, *Surf. Interf. Anal.*, 31 (2001) 1042.
[25] H. Kumagai, *et al.*, *J. Vac. Sci. Technol. A*, 22 (2004) 1.
[26] M.C. Kim, *et al.*, *Surf. Coat. Technol.*, 174–175 (2003) 839.
[27] J.G. Fonseca, J. Hommet and Y. Galerne, *Appl. Phys. Lett.*, 82 (2003) 58–60.
[28] M. Shen, *et al.*, *Langmuir*, 19 (2003) 1692–1699.
[29] A. Libassi, *et al.*, *Surf. Sci.*, 402–404 (1998) 441–444.
[30] E. Riedo, *et al.*, *Solid State Commun.*, 116 (2000) 287–292.
[31] S. Iijima, *Nature*, 354 (1991) 56.
[32] R. Droppa Jr., *et al.*, *Phys. Rev. B*, 69 (2004) 045405.
[33] C.S. Cojocaru, *et al.*, *International Conference on the Science and Application of Nanotubes*, San Luis Potosi, Mexico, 2004.
[34] J.B. Kortright, *et al.*, *J. Magn. Magn. Mater.*, 207 (1999) 7.
[35] J.L. Jordan-Sweet, *IBM J. Res. Dev.*, 44 (2000) 457–476 www.research.ibm.com/journal/rd/444/jordansweet.html.
[36] O. Ersen, *et al.*, *Phys. Rev. B*, 67 (2003) 94116.
[37] C. Boeglin, *et al.*, *Phys. Rev. B*, 66 (2002) 014439/1-6.
[38] A. Guinier and G. Fournet, *Small-angle Scattering of X-rays*, John Wiley and Sons, New York, 1955.
[39] G. Renaud, *et al.*, *Science*, 300 (2003) 1416–1419 and for supporting on online material. www.sciencemag.org/cgi/content/full/300/5624/1416/DC1.
[40] R. Lazzari, *Appl. Cryst.*, 35 (2002) 406–421.
[41] D. Babonneau, *et al.*, *Surf. Sci.*, 409 (1998) 358–371.
[42] D. Babonneau, *et al.*, *Appl. Phys. Lett.*, 76 (2001) 2892.
[43] D. Babonneau, *et al.*, *Phys. Rev. B*, 63 (2001) 195401.
[44] P. Andreazza, *et al.*, *Surf. Coat. Technol.*, 151–152 (2002) 122–127.
[45] O. Fruchard, *et al.*, *J. Cryst. Growth*, 237–239 (2002) 2035–2040.

Appendix

If one would like to understand the acronyms used in Tables 12.2–12.4 and in titles of some chapters in the literature such as the following (real) ones …

* Studies of polymorphic systems using the methods of FTIR, DRIFT, ^{13}C-, CP/MAR/NMR, DTA and XRD.
* An investigation of the structures of the chemisorbed species from hydrocarbons on metal surfaces using FTIR and EELS in conjunction with LEED, AES and TPD.

… a little dictionary can be helpful. The author has collected more than one hundred acronyms but here only the most useful ones are given.

AAS:	Atomic absorption spectroscopy
AES:	Auger electron spectroscopy *but also* Atomic emission spectroscopy
AFM:	Atomic force microscopy
AFS:	Atomic fluorescence spectroscopy
AP:	Atomic probe
ATR:	Attenuated total reflection
BET:	Brunauer, Emmet, Teller
BIS:	Bremsstrahlung Isochromat Spectroscopy (also bioelectrical impedance spectroscopy)
CP/MAR/NMR:	Cross polarization magic angle rotation NMR
CVD:	Chemical vapor deposition
EBIC:	Electron beam-induced current
EDS:	Energy-dispersive spectrometer
EDX:	Electron dispersion X-ray (spectroscopy)
EELS:	Electron energy loss spectroscopy
EPMA:	Electron probe microanalysis
ESCA:	Electron spectroscopy for chemical analysis
(S)EXAFS:	(Surface) Extended X-ray absorption fine structure
FTIR:	Frustrated total internal reflection *and* Fourier transform infrared
GDL:	Glow discharge lamp
GISAXS:	Grazing incidence small angle X-ray scattering
GIXS:	Grazing incidence X-ray scattering
IBAD:	Ion-beam-assisted deposition
IBS:	Ion-beam sputtering
ISS:	Ion scattering spectroscopy
LAMMA:	Laser-assisted microprobe analysis
LEED:	Low-energy electron diffraction
MBE:	Molecular beam epitaxy
MFM:	Magnetic force microscopy

MOKE:	Magneto-optical Kerr effect
NAA:	Neutron activation analysis
NMR:	Nuclear magnetic resonance
PIXE:	Proton-induced X-ray emission
PLD:	Pulsed laser deposition
RBS:	Rutherford backscattering
RHEED:	Reflection high-energy electron diffraction
SEM:	Scanning electron microscope
SERS:	Surface-enhanced Raman spectroscopy
SIMS:	Secondary ion mass spectroscopy
(ToF) – SIMS:	Time-of-flight SIMS
SNMS:	Secondary neutral ion mass spectroscopy
SNOM:	Scanning near field optical microscopy
STEM:	Scanning transmission electron microscopy
STM:	Scanning tunneling microscopy
TEM:	Transmission electron microscopy
(HR)TEM:	High-resolution TEM
UPS:	Ultraviolet photoelectron spectroscopy
WDS:	Wavelength-dispersive spectrometer
XANES:	X-ray absorption near edge structure
XAS:	X-ray absorption spectroscopy
XPS:	X-ray photoelectron spectroscopy
XRD:	X-ray diffraction
XRF:	X-ray fluorescence

Chapter 13

Formation and Characterization of the Structure of Thin Films and Coatings

György Radnóczi and Péter B. Barna

13.1. Introduction

Application-oriented preparation of thin films needs reproducible technologies. As long as setting up a reproducible production is possible by a trial and error technique, the preparation of films of a designed structure in the sense of morphologic, textural and phase characteristics and, consequently, in the sense of the desired physical properties, needs the knowledge of the driving forces and available kinetics of the processes of structure formation for each material system. When available, this type of knowledge is usually summarized in Structure Zone Diagrams (SZD), first proposed by Movchan and Demshichin for micro-structural morphology of films as the function of the growth temperature [1]. However, a complete SZD has to show how the morphology, phase composition, crystallographic texture and other structural properties depend on the preparation conditions and has to specify the main processes leading to the described structures in the form of the pathway of structure formation [2].

To describe and characterize the structure and morphology of coatings we need microscopic techniques. The reason to this is manifold. Two of the most important reasons are that the object of observation and characterization is the atomic processes taking place during the formation of thin films, and that most of the film structures owes its functional properties to the nanostructure or nano-sized morphologic elements.

Usually, films or coatings are intentionally grown far from equilibrium conditions. This is because the crystallographic phases, providing morphologies, grain size, texture and properties essential for the selected application can be realized only at these conditions. Long-term stability of these metastable systems is also a key problem of their applicability.

Film growers use intuition, experience and knowledge to synthesize certain structures and properties, needed for application. In this chapter we want to make a contribution to the knowledge part of this sophisticated know-how by discussing the

Materials Surface Processing by Directed Energy Techniques
Y. Pauleau (Editor)

fundamental phenomena and processes of structure formation. First the results obtained for the pure, one-component systems are summarized then an attempt is made to extend this model to the multi-component (mainly binary) systems. Only growth from vapor phase will be discussed, solid phase processes like crystallization of amorphous materials, solid phase reactions will not be considered.

13.2. Overview of the Processes Taking Place During Film Growth

Definitely, all processes of thin film formation are governed by the thermodynamic driving forces and realized by the atomic processes, the kinetics, available for the system at the selected preparation conditions. In some sense this paragraph will serve as a dictionary, defining the terminology, used throughout this work to help the qualitative understanding of the processes taking place during thin film growth. First we shall shortly describe the properties of the substrate, affecting film growth, then the processes, as they are taking place during deposition and growth in a time sequence (growth stages).

13.2.1. Substrates, Surface Energies, Interface Energies and Misfit

When describing the process of the formation of thin films and coatings it has to be considered that they grow on a substrate. Some aspects of the development of the film will be determined by the properties of the substrate.

The thin films and coatings, we want to produce, have to be laterally uniform and smooth, usually. One precondition of this is, that the substrate, on which they are growing should also be uniform and smooth. The best way to fulfil these conditions is to use polished surfaces of single crystals like a Si, sapphire (Al_2O_3) or SiC wafer and cleaved ionic crystals (NaCl, MgO), or amorphous substrates like SiO_2 covered Si, glass, amorphous carbon. For example polycrystalline substrates, like tools and metal parts are generally non-uniform in the sense, that crystallites of different phases and different orientations are displayed on the surface. The effect of non-uniformity can be minimized by surface treatments or by selecting such growth parameters that it does not make crucial influence on the structure of a growing coating. The smoothness of the surface, however, can be assured.

The **substrate surface inhomogenity** (or non-uniformity) means different adsorption (E_a), surface (σ_s) and interface (σ_i) energies on different kinds of grains or domains. Local contaminations are also considered as inhomogenities. All these physical quantities are figuring in the equations, describing the process of condensation, nucleation and growth, accordingly. For selective deposition purposes, on the contrary, inhomogeneous substrates are required, where on one kind of domains condensation is possible and on the other kind it does not occur.

Substrate surface roughness will result in shadowing effects and nodular growth leading to non-uniform thickness and morphology. In the cases of low adatom mobility the substrate surface roughness leads also to porosity [3].

The **surface energy** (σ_s) equals to the half of the work necessary for cutting the bonds across a surface, when separating the two halves of the crystal. From this it follows, that higher atomic density planes have higher surface energies. The surface energy is also influenced by the relaxation of the cut bonds. The anisotropy of the surface energy can be demonstrated in NaCl-type lattices. The {100} planes are the cleavage planes (i.e. the lowest energy planes), because they are electrically neutral, and relaxation of the ionic bonds can take place in the cleavage plane. Cleavage along the {111} planes is very difficult, because the cleaved surfaces are charged. One half contains only positive while the other only negative ions and relaxation of the cut bonds is hardly possible.

The **interface energy** (σ_i) is mainly determined by the strength of the bonds across the interface between two phases or two grains. In this chapter the interface energy will be used to describe the energy of the boundary between the two solids (in general, the surface energy is also interface energy). If the interface energy of a coating to its substrate is much smaller than the sum of the surface energies of the coating and of the substrate, then the coating is strongly bonded to the substrate, and its adhesion is strong.

In the interface of two crystalline phases, as a result of improper matching or **misfit** of crystallographic planes, high stress can build up in the growing film. The strain energy of the film increases with increasing film thickness. The formation of misfit dislocations increases the energy of the interface at the expense of the decrease of the volume energy of misfit strain in the coating itself.

13.2.2. Adsorption, Surface Mobility, Desorption and Clustering

An atom condensing on the substrate surface possesses a kinetic energy of typically 10 eV in the case of physical vapor deposition. This kinetic energy dissipates during adsorption in the interaction with the atoms of the surface and the adsorbed atoms (thermal accommodation). The accommodation process is very fast, around 10^{-14} s, and is possible, when the kinetic energy of the arriving atom is not very much higher than ten times of the adsorption energy (E_a). When adsorbed on the surface the atom can migrate or be desorbed as long as it has not built in into the surface. As soon as it occupied a fixed position on the growth surface, that is it has become the part of the growing crystal lattice or amorphous network we can consider that the state of the atom changed from an adatom to a surface atom.

Adsorption energy (E_a) of an adsorbed atom is the energy, which has to be delivered to the adsorbed atom to be desorbed from the surface. This energy can be related to self-atoms or foreign atoms, and the energy for different atoms is different (typically a few eV) [4,5]. The adsorption energy plays a key role in the condensation process and residence time (t_r) of the atoms on a given surface at a given temperature (T). In a non-energetic deposition process most atoms are adsorbed on the substrate, so the arrival rate and deposition rate are practically equal, that is the condensation coefficient (sticking coefficient) is close to unity. When accelerated species arrive at the surface, and their energy exceeds $10E_a$ (50–100 eV), more and more of them will be reflected from the surface.

The average residence time (t_r) of the adatoms can be written in the form:

$$t_r = t_o \exp(E_a/kT) \tag{13.1}$$

where $t_o = 1/f$ is the time between successive vibrations of the adatom on the surface, f is the vibration frequency ($\sim 10^{14}\,s^{-1}$) and k is the Boltzmann's constant.

The deposition rate (R, atoms/cm^{2}s) and desorption rate $1/t_r$ together determine the steady state adatom concentration (N) on the substrate [6]:

$$N = Rt_r = Rt_o \exp(E_a/kT) \tag{13.2}$$

An important consequence is that the surface adatom concentration (N) decreases as T increases or R and t_o decreases.

The adsorption energy (E_a) differs from the activation energy of the **surface diffusion** (E_S).

The average length l_a, a migrating adatom covers before re-evaporation is:

$$l_a \sim (D_S t_r)^{1/2} \tag{13.3}$$

and D_S is the surface diffusion coefficient,

$$D_S = D_{S0} \exp(-E_S/kT), \tag{13.4}$$

where D_{S0} is proportional to the vibration (jump attempt) frequency f.

However, l_a can be limited instead of t_r by the deposition rate (R). If the arrival time necessary for burying an adatom is shorter than the residence time t_r, then the diffusion length in Eq. (13.3) will be determined by the deposition rate (R).

E_S is usually not uniform along the surface. It can locally differ from the value, valid for large and smooth terraces of the substrate or growth surface. For example, at defects and surface steps the value of E_S can increase or decrease creating by this way fast diffusion "pipes" or diffusion barriers. Step edges or channels of a reconstructed surface can be considered as diffusion pipes. An important diffusion barrier of this kind is related to the edge of the surface steps, the so-called Schwoebel–Ehrlich barrier [7,8], which separates terraces from each other in an asymmetric way. Incorporation of adatoms into the step across this barrier is difficult downwards, but it is easy from the lower terrace. This influences the movement of growth steps, the step bunching and through this the surface morphology of the growing crystal. The effect of growth inhibitors and surfactant promoters on the surface processes is mainly related to the modification of the barrier height or effectiveness of the diffusion pipes by addition of a small quantity of a foreign material [8]. The fast adatom movement along step edges or reconstruction channels can be retarded by inhibitors. Another mechanism of the growth inhibitors is the initiation of the nucleation of a second phase, resulting by this in the pinning of the growth steps or grain boundaries and leading to corresponding surface morphologic changes. A surface smoothening effect can be achieved, when the height of the Schwoebel–Ehrlich barrier is decreased, and step bunching can be avoided [8].

When surface diffusion (l_a) is very limited either due to the factors figuring in Eq. (13.3) or due to high R, then ballistic effects in the condensation of depositing species become important. The consequent most pronounced phenomenon is shadowing, resulting in tilted morphologies in the obliquely deposited films and in the formation of pores in the shadowed regions (decrease of the density of the film) [9,10].

13.2.3. Nucleation, Nucleation Density, Average Distance of Nuclei, Nucleating Phase, Orientation of Nuclei, Nucleation Texture and Growth of Stable Nuclei

The precondition of nucleation is the formation of clusters, which are temporary formations of interacting adatoms forming and falling apart statistically. Nucleation sets in when the adatom concentration reaches a critical value N^*. At this concentration the clusters can grow up to a critical size at which their free energy decreases, when binding a new adatom. At low condensation rates a steady state adatom concentration (Eq. 13.2) can never reach N^* (i.e. $N^* > N$).

Expressing the free energy ΔE of the formation of a cluster as a function of size we obtain that this energy initially increases, then having reached a maximum, decreases.

$$\Delta E(r) = \Sigma F_j(r)\sigma_j - V(r)\Delta G \qquad (13.5)$$

where F_j and σ_j are the surface (facet, interface) area and surface (facet, interface) energy of the cluster, respectively, V is the volume of the cluster and ΔG is the difference between the Gibbs free energy of the unit volume of the two phases (the adatom phase and the condensed (liquid or solid) phase).

The size corresponding to the maximum value of ΔE is the critical nucleus size r_c (Fig. 13.1). Capturing one more atom by the critical nucleus a stable nucleus forms. All the smaller formations are unstable clusters. The size of the critical nucleus is given by:

$$r_c = F(r)\sigma/\Delta G \qquad (13.6)$$

Here, $F(r)$ is a shape factor, depending on the shape of the nucleus; σ is the effective surface energy of the nucleus.

σ is composed from the energies of different facets, edges and substrate-nucleus interface. It contains a negative part as well which is the surface energy of the substrate area occupied by the nucleus. All these are taken into account by the shape factor and the effective value of σ, used in Eq. (13.6).

The typical values of r_c are of 1–2 nm. However, there can be situations, when the critical nucleus is one atom. In these cases the thermodynamic approach fails and an atomistic model is appropriate [11,12].

The nucleation is a rather abrupt event occurring all over the surface of the substrate simultaneously, as long as the adatom distribution is uniform. Once a stable

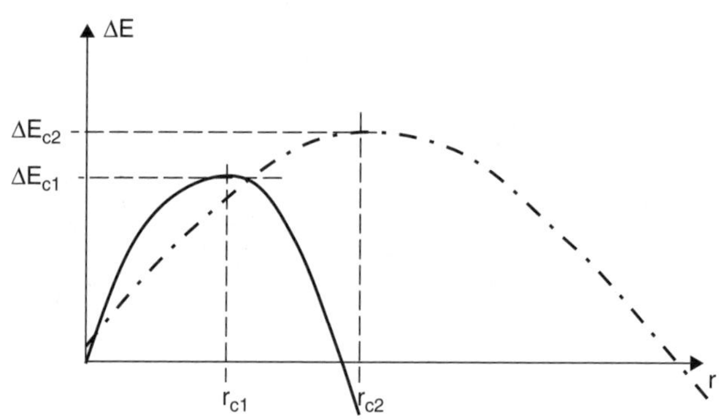

Figure 13.1: Schematic curves of ΔE as a function of the cluster size r for two different phases competing in nucleation.

nucleus is formed, it captures the adatoms from a circular area of radius l_a, surrounding the nucleus. The adatom concentration will drop and no adatom concentration can build up to initiate new nucleation in the capture area.

Theoretically, the stable nucleus can grow unlimited. On a homogeneous substrate the stable nuclei will form at an average distance l_a and will be practically of the same size. However, if the substrate is inhomogeneous, r_c will be different on sites of different σ_s. Consequently preferred nucleation takes place on surface sites of high σ_s (Eq. 13.6). Provided that l_a is sufficiently large, these nuclei will collect all adatoms from large areas, preventing nucleation in domains of radius l_a (selective nucleation). This is the mechanism of the formation of the so-called decoration patterns, when the selective nucleation is used to reveal inhomogenities of σ_s and r_c of a surface, correspondingly (Fig. 13.2). Such a typical inhomogenity is the

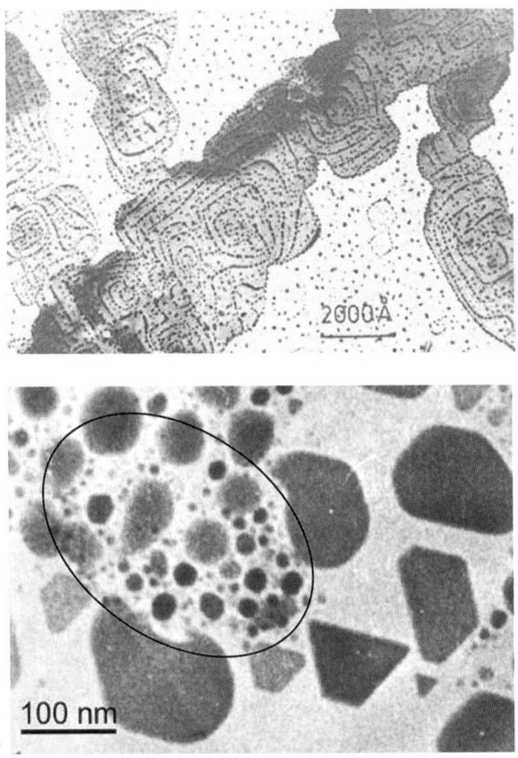

a) b)

Figure 13.2: Decoration patterns achieved by selective nucleation of Au on the growth steps of SnSe [13] (a); and by condensation of In in crystalline and liquid phase on pure MoS_2 and on a contaminated domain (encircled), respectively. On the pure area the nucleation is epitaxial (b).

edge of a surface step, which is usually a preferred nucleation site (Fig. 13.2(a)). Inhomogenities in σ_s can lead to other effects as well. It can alter, for example, the phase state of the forming nuclei, that is when generally crystalline nuclei are expected, amorphous or liquid nuclei can form on these domains of the substrate (Fig. 13.2(b)). This can be generally understood from Eqs. (13.6) and (13.7) describing the height of the maximum ΔE_c in Fig. 13.1.

$$\Delta E_c = F'(r)\sigma^3/\Delta G^2 \tag{13.7}$$

Here $F'(r)$ depends on the shape of the nucleus.

That nucleus will form first, which has the smallest r_c at also smallest ΔE_c.

The quantity n, where

$$1/n = l_a^2 \tag{13.8}$$

is called nucleation density. This is the number of the nuclei formed on a unit area.

When nucleation occurs on a substrate it is called heterogeneous nucleation, contrary to the homogeneous nucleation when the nucleation takes place in the volume of a medium. In the case of heterogeneous nucleation not only a new surface for the new nucleus has to be created, but also a fragment of the open surface of the substrate is eliminated and the nucleus-substrate interface is created instead. In the case of crystalline materials this interface energy depends on the mutual orientation of the substrate and nucleus. The interface energy determines the orientation of the nucleus relative to the substrate by assuring the most advantageous conditions for this selected orientation in terms of r_c and ΔE_c. Epitaxial nucleation takes place, if the conditions are such that two crystallographic directions of the nucleus align relative to the substrate (Fig. 13.2(b)). If these two directions are not perfectly aligned, then a two axis texture forms instead of epitaxy. If only one direction is aligned, then a one axis texture forms. Textures forming as a result of oriented nucleation are called **nucleation textures**.

If material is continuously supplied, the nuclei keep growing collecting all adatoms from the capture area (of radius l_a) and directly from the vapor beam. There is no interaction between the growing particles. If a crystalline phase nucleates, the islands (particles) are mostly single crystalline, and free from defects, though in materials of low planar defect energy twinning and stacking fault formation can occur. The mechanism of this defect formation is not yet clear, for example, it may be connected to contamination or built in internal stresses.

Three **growth modes** are distinguished in the literature, which in fact would be more adequate to call **nucleation and growth modes**, because the growth mode is selected already at the nucleation:

Volmer–Weber (or three-dimensional (3D) nucleation and growth),

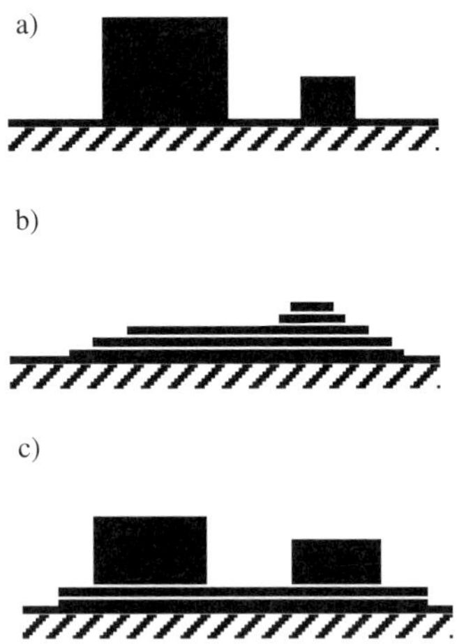

Figure 13.3: Nucleation and growth modes: Volmer–Weber or 3D (a), Frank–Van der Merwe or 2D (b), Stranski–Krastanov or mixed (c).

Frank–Van der Merwe (or two-dimensional (2D) nucleation and growth), Stranski–Krastanov (or mixed, starting in 2D and continued in 3D growth (Fig. 13.3)).

Which of them will operate will depend on the ratio of the surface energy of the film (σ_f), of the substrate (σ_s) and substrate-film interface energy (σ_i) if the appropriate kinetics are available.

The condition for Volmer–Weber type growth is $\sigma_s < (\sigma_f + \sigma_i)$. If $\sigma_s > (\sigma_f + \sigma_i)$ the Frank–Van der Merwe mechanism operates. The Stranski–Krastanov growth mode is observed when $\sigma_s \sim (\sigma_f + \sigma_i)$. All crystals grow mainly by the movement of steps of atomic height. In the case of 2D growth the lateral speed of steps is much larger than the normal growth rate. In the case of 3D growth the normal growth rate is the higher. As we implicitly considered in $F(r)$ and in using an effective value for σ, the nuclei can have different shapes. These shapes can be hemispherical or faceted, making different contact angle to the substrate. The shape and even phase state of the nucleus can change during growth, since surface energy anisotropy and melting or crystallization temperatures are size dependent [14,15].

Far from equilibrium conditions the island shape will be determined by the mobility of the self-adatoms on the surface of the growing particle having a migration length l_f [16]. As long as this migration length is not much smaller than the diameter of the growing particle (r_f), the growing island will be bordered by facets or have the shape of a hemispherical cap, depending on whether the phase is crystalline or liquid/amorphous, respectively. However, when $r_f > l_f$ a branching morphology develops with convex and concave segments, resembling dendrites or fractals (Fig. 13.4).

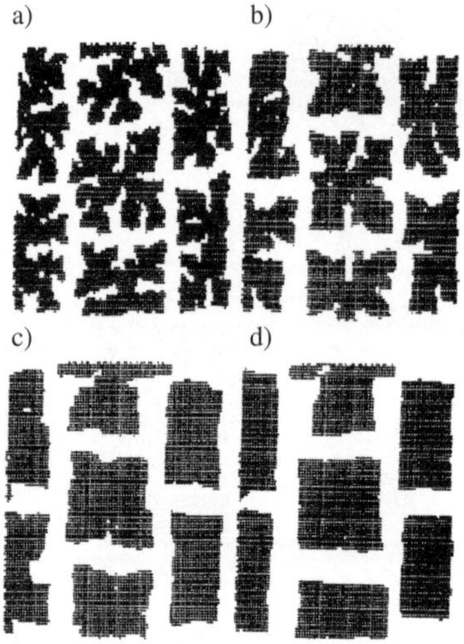

Figure 13.4: Shape of 2D islands as a function of the migration length of self-adatoms on the island edge [16]. The corresponding diffusion lengths (l_f) are in the units of atomic steps: (a) 0, (b) 2, (c) 4 and (d) 8 steps.

13.2.4. Coalescence

Detectable interactions between the growing particles start, when the overlap of their capture zones or stress fields becomes perceptible. One form of interaction is the Oswald ripening when the atoms dissolved from smaller islands can condense on the larger ones due to the differences in the equilibrium pressure around these particles [5]. However, this can be effective only in the very first stage of growth. The event, when two individually growing islands join into one island is called coalescence.

The process of coalescence can be **complete**, when no interface between the merging particles remains, independently of the way, how the forming interface disappears. The coalescing particles can melt completely or partly (utilizing the released surface energy) and merge by liquid flow. **This is called liquid-like coalescence**. The process of complete coalescence can take place also in solid state. Then the forming grain boundary (GB) will move out by GB migration. **This is called solid-state coalescence**.

In the case of liquid-like coalescence substrate areas occupied by a growing island become open again, where condensation, nucleation, growth and coalescence can repeat several times. This second nucleation, which takes place on the substrate after coalescence started, is called **secondary nucleation**. One of the most straightforward consequences of that is the bimodal size distribution of the growing islands.

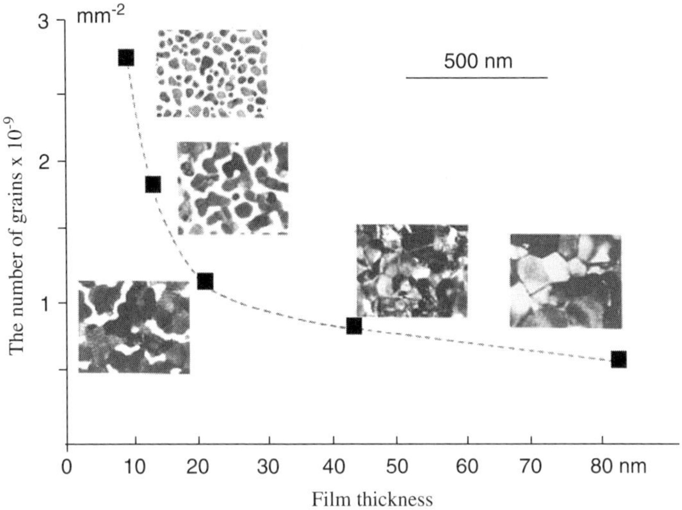

Figure 13.5: The dependence of the number density of grains in Al films grown at 300°C as a function of film thickness. Approaching 100 nm of thickness, about 25% of grains remains.

As one can imagine, many crystallites formed in the nucleation process will disappear during complete coalescence. Fig. 13.5 shows, that about 75% of the crystallites disappears in the thickness range of 10 to 100 nm. However, the percentage of the disappearing crystallites can be much higher, since the effect of the very early stages of liquid-like coalescence cannot be seen in Fig. 13.5. So, there is a chance to select a certain orientation for further growth, while diminishing others. The driving force for selection is again the minimisation of the film surface energy and/or substrate-film interface energy. This will lead to the formation of crystallographic textures, which are called the **textures of coalescence or restructuring** [2].

The process of **coalescence is incomplete**, when the interface between the touching particles does not disappear. When the two coalesced islands are separated by an interface they are called grains of a polycrystalline island, while the interface between them a GB in crystalline materials. In the case of amorphous materials the interface forming at incomplete coalescence is called column wall/boundary. In case of incomplete coalescence no open substrate areas develop, so no secondary nucleation occurs.

Naturally, there are different driving forces, to move out the GBs from the polycrystalline islands. However, there are also different obstacles which block the GB movement. A number of geometrical obstacles, where the GB is trapped, are shown

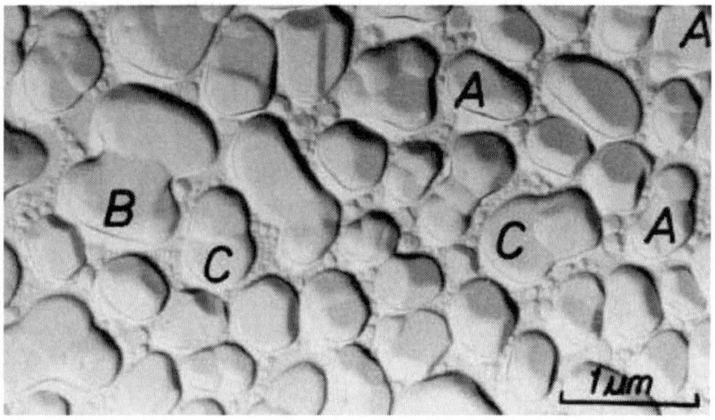

Figure 13.6: Trapping of GB in coalescing crystals by GB grooving (A), deep neck (B) and height/width difference (C).

in Fig. 13.6. Alternatively, the driving forces are the decrease of the GB-, interface- or surface energies. As more and more grains compose an island, the possibility of moving out for a GB decreases. As the film thickness increases, the islands reach the percolation threshold and form a continuous network, interrupted in a maze-like manner by channels. The film growth reaches the so-called channel stage.

It is important to emphasize the difference between grain growth (recrystallization) and crystal growth in polycrystalline thin films. When the growth of individual grains takes place through GB movement or by liquid-like coalescence the process is called grain growth. During grain growth usually new crystallographic orientation occurs and/or existing orientation disappears. When the growth of crystallites occurs from vapor by direct impingement or by adatom migration on the substrate it is called crystal growth.

13.2.5. Continuous Film, Growth of the Thickness

By continuing the film growth the channels will be filled up and a continuous film forms. From this moment GBs start to play a more important role in the growth processes. The GBs are basically perpendicular to the film plane, and strongly influence the surface morphology of the coatings. They can be active in the nucleation of monolayer growth steps. In this case hills can develop on the GBs relative to the grain interior especially in large grain size materials (e.g. soft metal coatings). When the nucleation of monolayer growth steps occurs in the grain interior and the growth steps proceed towards GBs, grooving of the GBs takes place. However, these grooves are very shallow in a one-component system and the grooves do not belong to the main features of the surface roughness. It must be underlined that the deep GB groves generally observed are related to contamination effects, discussed in Section 13.4.1. Namely a covering layer of contaminants forms at the GB and it retards the crystal growth in the GB region [17–19]. These grooves now play an important role in the formation of the surface roughness of the coatings.

As the film thickness increases, the total GB energy stored in the film also increases linearly with the film thickness and provides a driving force of the GB movement. As kinetic preconditions are fulfilled, this will result in a continuous grain growth during film thickening [20]. As the movement of high-energy GBs is more probable, first those will disappear from the film, and special, low-energy GB configurations remain [21]. This contributes to the formation of a smooth surface through shallow GB grooves, belonging to these types of GBs.

Besides of the GB energy the decrease of the substrate-film interface energy as well as the decrease of the surface energy of the film can be the driving force for the GB migration. This again will lead to a selection and survival of grains with low surface or interface energy, and consequently to a texture called **competing grain growth texture** [17] in polycrystalline thin films. The degree of the texture will increase as the film thickness increases [22]. The number density of grains as a function of the thickness is shown in Fig. 13.5. This figure illustrates that numerous grains and consequently orientations disappear during the film growth, leaving only a few percent of the born crystallites. This is a strong selection process of grains and orientations, which leads to the formation of texture through the successive selection mechanisms.

On the basis of the governing processes four main growth stages (and the corresponding processes) of the development of a film can be distinguished: nucleation (condensation, surface diffusion on the substrate and cluster formation), island growth (surface diffusion), coalescence (island interaction, surface diffusion, GB movement) and thickness growth of a continuous film (GB movement, surface and volume diffusion). The grain growth can be continued after finishing the deposition when the kinetic conditions of GB mobility are still fulfilled. This process is a grain coarsening process often called recrystallization in the literature, especially when occurring during a post deposition annealing.

13.3. Structure Evolution of Elemental Films: Morphology and Texture

It is an early experience of film growers that the thickness, at which continuity of a growing film occurs, depends on the growth temperature. This fact can be well understood by considering the processes taking place during the early stages of film formation. These stages, the nucleation, island growth, coalescence and the thickness growth in a continuous film [17,23], follow each other at different equivalent thickness, characteristic for the film and substrate material and the deposition temperature. At higher temperatures nucleation density is lower, so coalescence sets in at a larger quantity of condensed material and the liquid-like coalescence keeps proceeding to larger particle sizes. The incorporation of GBs occurs at relatively high effective thickness. Channels and their filling, leading to the continuous layers takes place at high deposited thickness. To summarize all these processes and resultant morphologies a 3D graph, so-called SZD [1], can be reasonably used, where the axes are the homologous temperature (T_s/T_M), film thickness and most often the morphology of the film (Fig. 13.7).

The first successful attempts to compile a SZD for thin films were made for materials which, despite the inevitably existing impurities, were sufficiently pure to grow in a way, characteristic for one-component materials. Movchan and Demshichin in their pioneering work [1] used both metals and compounds. Their observation that in real thin films four morphologic zones (Zones I, T, II and III) exist was correct. However, the so-called Zone III (equiaxed grains), noted in their work for high temperature growth shows the evidence of the uncontrolled impurity effect. This Zone III never occurred in experiments carried out at high-purity conditions. The phenomenon leading to the formation of Zone III structure and taking place due to contamination, that is blocking of the growth of already existing crystals by a contamination layer, and forming new nuclei on their surface leading to equiaxial and overlapping grain structure [24] was not realized in [1].

In successive reviews [2,18,25–28], to explain and further develop the achievements made in [1] the following **fundamental phenomena** were considered: nucleation,

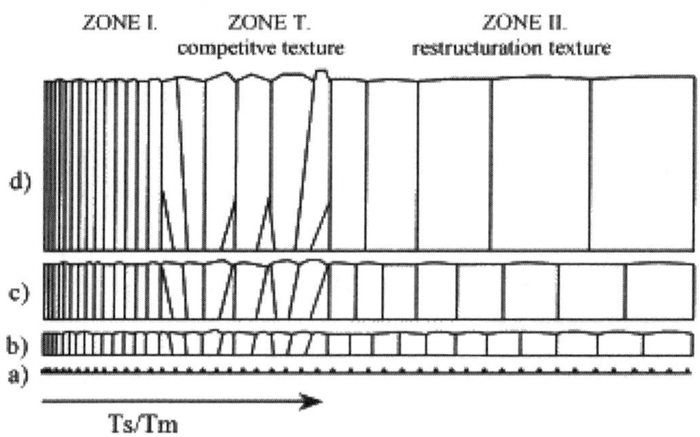

Figure 13.7: The schematic zone diagram of a single component (pure) system as a function of film thickness (a–d) [18].

crystal growth, coalescence and grain growth. The corresponding **atomic processes** are condensation, adsorption, surface diffusion and clustering for nucleation; surface diffusion, formation and movement of surface steps for crystal growth; surface diffusion, plastic/liquid flow and GB migration for coalescence; GB migration and bulk self-diffusion for grain growth, often called recrystallization. The corresponding main **driving force** is the decrease of bulk free energy (ΔG in Eq. 13.5) for nucleation and crystal growth, minimization of interface/surface and GB energies for coalescence and grain growth/recrystallization. Naturally, these are only the main phenomena and processes, more phenomena are playing role in bringing about the final bulk and surface morphology, phase composition and texture.

A typical SZD of a one-component one-phase system is shown in Fig. 13.7 [2,18] and can be read as follows. The SZD of a one-phase system consists of three zones: Zone I, Zone T and Zone II. No Zone III is observed, contrary to [1]. At starting thickness (case (a) in Fig. 13.7) nucleation occurs at a density, decreasing with increasing temperature. For all temperatures, nucleation occurs only on the substrate in one-component films! No nucleus of a new crystal can form on the surface of the already growing crystals, since there is no driving force to stop the crystal growth and to create a GB and differently oriented nucleus on a pure growth surface. As a result, all one-component (and one phase) films must be columnar in their nature (Figs. 13.8–13.10). However, as illustrated in Fig. 13.7, the development of characteristic features of the zones is a function of the overall thickness of the film. As a continuous film develops (Fig. 13.7(b)) the grain size and the orientation of the crystals will depend on the effectiveness of the coalescence processes.

At low temperatures (Zone I) each nucleus survives as no kinetics for merging and competitive growth is available and a fibre or columnar structure develops. The crystallographic orientations formed during nucleation are preserved. The

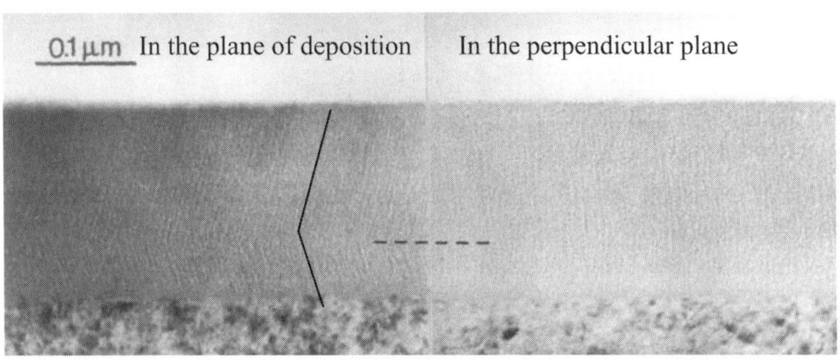

Figure 13.8: TEM cross section of a Zone I obliquely deposited amorphous Si film, showing the kinking of columns when evaporation direction was changed (marked by the dotted line in the images).

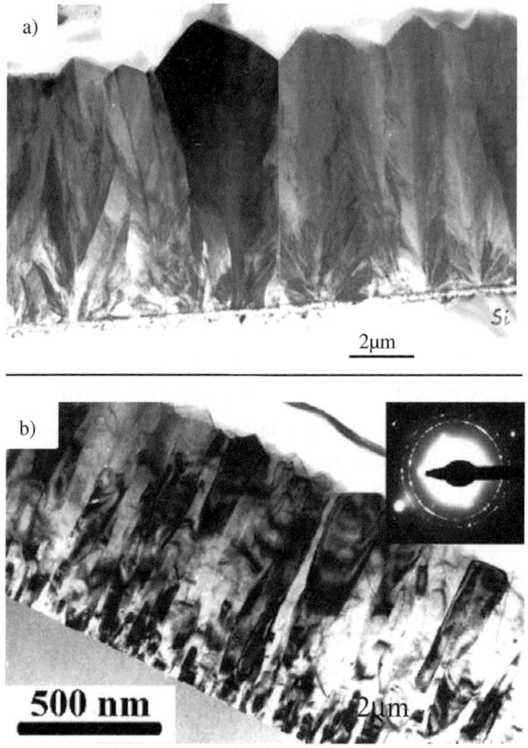

Figure 13.9: TEM cross sections of the structures, corresponding to Zone T in Fig. 13.7. (a) chemical vapour deposition (CVD) grown diamond and (b) evaporated Al film.

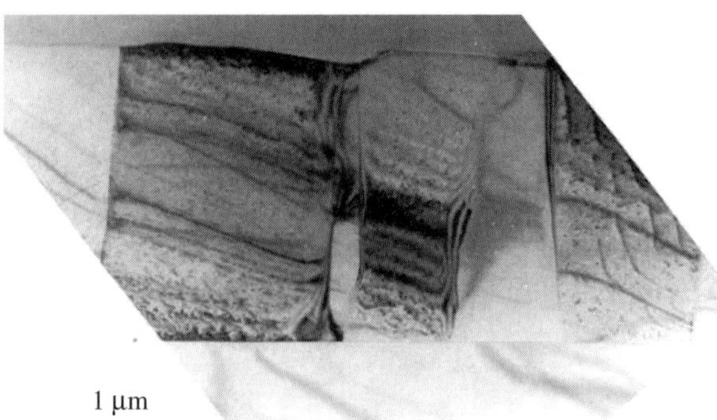

Figure 13.10: TEM cross section of a thick, Zone II Al film.

column diameter is practically determined by the diameter of the capture zone l_a. The lowest temperature excludes any surface mobility also on the growth surface, and only ballistic effects play role in the growth leading to column tilt at oblique deposition [10]. If the deposition direction is changed during growth, the growth direction also changes, resulting in kinked columns, no matter the columns are amorphous or crystalline. The situation is illustrated on the growth of amorphous Si film at room temperature [30], where mobility is below the crystallization threshold (Fig. 13.8). The column walls are the analogs of the grain boundaries in crystalline material, being composed from a lower density, defective material. The tilt of the columns changes with the direction of deposition as marked in the figure.

In the transition region (Zone T ($0.1T_M < T_s < 0.3T_M$)) the adatom mobility becomes significant. This results in the competing growth of neighboring crystals of different orientation. Namely crystals of lower surface energy broaden and grow faster at the expense of others by capturing more adatoms from the GB region. As a result V shape grains develop (Fig. 13.9). This competitive growth leads to diminishing of crystals of high surface energy, and a texture of low surface energy orientations develops. These textures are called **competing crystal growth textures**. A considerable surface roughness develops which scales with the average in-plane grain size. The consequence of competitive growth is a continuous change in morphology, texture and surface topography as a function of film thickness [22].

At high temperature (Zone II, $T_s > 0.3T_m$) more crystallites, each grown from an individual nucleus, merge to form one grain resulting in a large, comparable to the thickness of the film grain size and in texture [2,17]. As shown schematically (Fig. 13.7), the grain size increases with increasing thickness. This process is driven by the decrease of the GB energy, free surface and the substrate-film interface energy and kinetically realized by GB movement. The stored GB energy grows linearly with film thickness. In continuous films the microstructure is composed of columnar grains with diameter in the range of film thickness (Fig. 13.10). The surface of continuous films is smooth.

A SZD contains direct information besides of the film morphology also on the crystallographic texture and surface roughness. Namely all structural features develop simultaneously through the same processes and in strong interrelation. Once the processes are known, the SZD of a one-phase system can be deduced [18].

When a coating of a given crystallographic texture is to be grown, then only those crystallites must have the possibility to grow, which correspond to this texture. As we have seen above, all crystallites are nucleating on the substrate (Figs. 13.8–13.10). So, either we take care, that only crystallites of the wanted orientation nucleate (nucleation texture) for example, by carefully selecting a single-crystalline substrate, or, we lead the growth process in such a way, that even in the case of random nucleation, the competing processes eliminate or diminish the crystallites of unwanted orientations.

Consequently, we promote either liquid-like coalescence, where the surface and interface energies will determine the texture or grain growth driven by the decrease of the GB/surface/interface energy in Zone II (Fig. 13.10). Another possibility is to allow only competing crystal growth in Zone T, where the decrease of the surface energy will be the driving force (Fig. 13.9). Correspondingly, we obtain coalescence or restructuring/recrystallization textures in Zone II and competing crystal growth textures in Zone T [17].

Consequences, concerning the surface roughness, can be deduced from the processes taking place during growth as well. In Zone I all the columns form and grow with the same rate as the adatoms have no mobility. Therefore no roughening mechanism exists, and the films will be smooth. In Zone T faceting of the growth surface of the crystals and the competitive growth of crystals of various orientations lead to significant surface roughness (Fig. 13.9). Zone II films are normally smooth, as the growth surface of the grains is of the same surface energy due to their texture, and no driving force is present to create roughness at higher thicknesses. However, the films are rough at smaller thicknesses where liquid-like coalescence exists.

According to the discussions in Fig. 13.3 in a one-phase system the Zone III structure of equiaxial grains cannot be the result of any growth process. In such a system Zone III structure can be developed only by the recrystallization of a Zone I structure driven mainly by the decrease of the GB energy. The occurrence of Zone III in as deposited films indicates that repeated nucleation took place and the film cannot be considered a one-component one-phase system any longer.

13.4. Structure Evolution in Multiphase (Composite) Films and Coatings

13.4.1. Composite Thin Films

Nanocomposites are successfully applied to increase hardness, oxidation resistance, to improve wear behavior, and other (e.g. magnetic) properties of thin films. The simultaneous growth of the amorphous and nanocrystalline phases by co-deposition of different compositions results in different morphologies, which in turn affect the coating's properties. The aim is to intentionally design and prepare nanostructure controlled functional coatings. For its realization the key issues are the understanding of the evolution of nanostructure and morphology, the clarification of the SZD and the processes leading to certain morphologies as well as the selection of technologic parameters determining the possible processes [2,31]. Examples of controlling hardness through nanostructure and morphology are presented in Refs.

[32–34]. To obtain enhanced hardness, the grain size and shape of the nanocrystalline phase must be below 10 nm and columnar, respectively, while the thickness of the separating GB layer must not exceed 1 nm. The main reason for the enhanced hardness is the absence of a working deformation mechanism.

If a system is composed of different chemical elements forming only one phase, its structure evolution will be identical to that of the elemental films. However, when more compound or elemental phases appear, the structure evolution will be significantly different. Nevertheless, the processes of structure formation will be the same, as described for the one-phase systems, and the growth of the individual phases proceeds through the stages described in Fig. 13.3. However, the phases, growing in the same coating may have different pathways of their own development which can differ from that predicted by their own SZD. This makes the morphologic development of such systems rather complicated.

If atoms of two components impinge on the substrate surface a 2D alloy layer of adatoms forms. The composition of this layer will depend on the corresponding residence times of the species of each component and on the corresponding deposition rates (see Eqs. (13.1) and (13.2)). The cluster formation takes place in this 2D layer, but the forming clusters do not conserve generally the composition of the adsorption layer [35,36]. The selection of the phase forming at first depends on the critical size (r_c) and ΔE_c (Eqs. (13.6) and (13.7)) of the possible phases (Phase 1 in Fig. 13.1). Only the stable clusters, that is the nuclei of the first phase will grow, extracting those species from the 2D adsorption layer which are consumed by the growth of the first phase. The concentration of these species will abruptly decrease in the 2D adatom layer. The remaining adatom layer offers new compositional conditions for the formation of nuclei of the next phase(s). Metastable phases can be formed in this way.

This can be illustrated by the behavior of the co-deposited Al–Pt system. The formation of large Al crystallites observed in Pt–Al films of Al_xPt ($x = 2$) composition can be understood in this view [37]. The actual composition of the adsorption layer facilitates the formation of Al_2Pt as the first phase, the growth of which extracts all Pt species from the adatom layer. While the excess Al species due to their large diffusion length (l_a) and to their probably small critical radius (r_c) form Al crystal nuclei faster than the expected Al_5Pt.

In general, after nucleation of the first phase an excess of one component occurs in the adsorption layer, relative to the original composition. The excess species can behave in four different ways:

1. re-evaporate,
2. remain dissolved in the adsorption layer,
3. remain on the surface of the nucleus of the first phase as adatoms or
4. form a new nucleus of a second phase.

Table 13.1: The different growth modes of the growing phases

	Growth mode		
Growth stage	$3D_1$–$2D_2$	$3D_1$–$3D_2$	$2D_1$–$3D_2$
Nucleation	Ni-C [38]	Co-SiO_x [39]	———
Coalescence	In-C [24]	SnSe/Au [13]	Al_2Au/Al_2O_3 [42]
	Al-O [18]	Al/Sn [41]	
Continuous film	Bi-O [29]	Al/Sn [41]	Al-C [37]
	Al-Pt [43]	Al-C [37]	C-Cr [40]

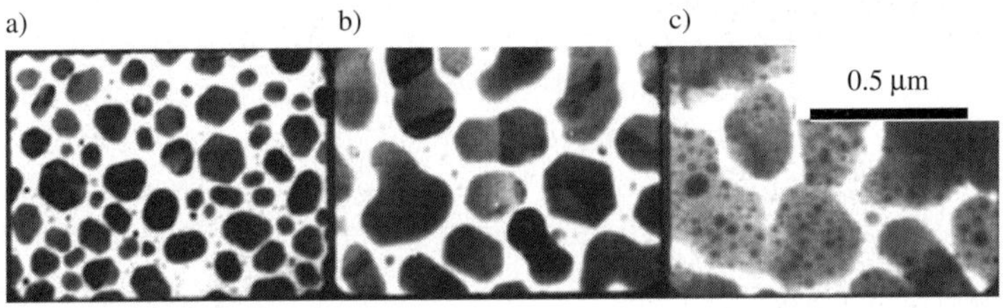

Figure 13.11: 2D growth of co-depositing carbon on 3D in islands resulting in the formation of a surface covering layer rounding of crystal shapes (a), inhibiting coalescence (b); leading to repeated nucleation of In on the carbon layer covering completely the surface of In crystals (c) (*in situ* TEM experiment, $T_s = 75°C$) [24].

In the first case they are lost for further processes. The further three cases can also be considered as the successive stages of the development of the adsorption layer leading to the incorporation of excess species into the structure in the form of a second phase. So, the system will be a two (or more) phase system. On one hand, the resulting morphology will depend on the relative quantities of the two phases and on the growth stage of the first phase, at which the second phase nucleates. On the other hand, the morphology depends on the growth modes of the growing phases. Some examples for the growth stages of the first phase (subscript 1) at which the second phase (subscript 2) can nucleate, and the possible growth modes (2D and 3D) of the phases are summarized in Table 13.1.

An example of the development of the 2D phase is shown in Fig. 13.11(a–c) for the case of the In-C system ($3D_1$–$2D_2$). The 2D growing amorphous C layer is the secondary phase forming on the surface of 3D growing In islands [24]. This situation is illustrated schematically in Fig. 13.12(a). The 2D phase is an inhibitor-type component and starts forming immediately after or together with the first 3D majority phase.

Consider a low concentration of an inhibitor-type component, leading to 2D second phase formation in the stage of the continuous film growth. The growth of

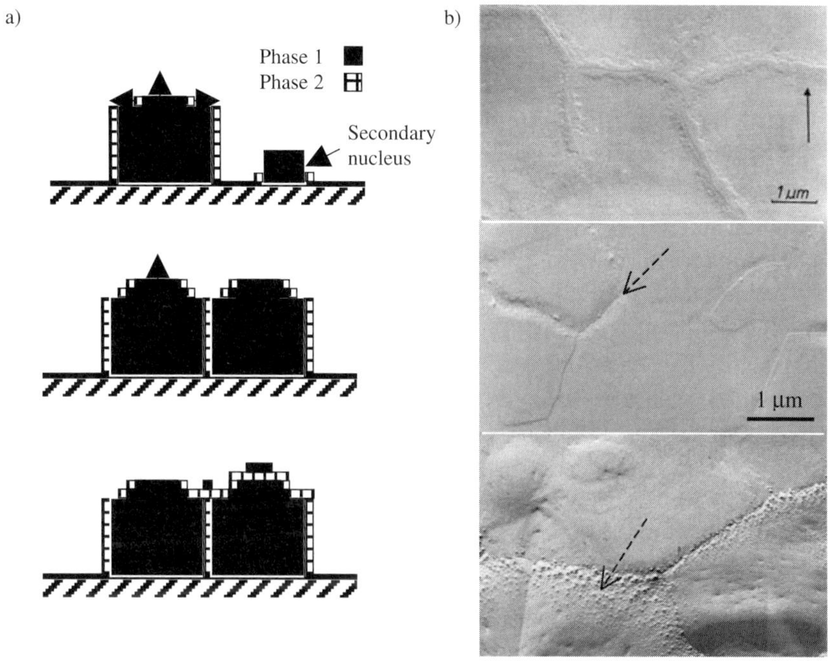

Figure 13.12: Schematic growth morphologies (a), the corresponding surface morphologies in Al-oxygen films (b) for the case of 3D growth of phase 1 and 2D growth of the secondary phase ($3D_1$–$2D_2$).

individual crystals of the continuous film in which the grain boundaries are pure occurs by nucleation of the growth step at the GBs. In this case the species of the inhibitor-type component are rolled by the growth steps to the grain interior. The accumulating species hinder the movement of the growth steps and block them at a certain distance from the GB by nucleating the 2D growing phase. Step bunching occurs at this place. The grain boundaries are smooth, or even hills can form along them. The GB groves are hardly detectable (top image in Fig. 13.12(b)). As an example, the Al-O system can be considered [17,18,44].

At higher concentration of the inhibitor type component the formation of the 2D growing secondary phase occurs in the island growth stage of the primary phase. The surface of the islands will be partly covered by the forming 2D secondary phase at first. These processes lead simultaneously to a retarded coalescence resulting in smaller grain size and consequently less-developed texture since the orientation selection during coalescence is also limited [2,18]. Incomplete coalescence takes place in that case and the GBs will be covered by the 2D phase. These GBs will be no more active in the nucleation of growth steps. The growth of the crystals in the continuous film proceeds by nucleation of growth steps in the interior of the crystals. The 2D phase situated at the grain boundaries will collect the species of the second component, and as a first consequence, the growth of the 2D phase extends to the surface of the adjacent crystals. Pronounced GB grooves form at these GBs (Fig. 13.12(b) middle).

A further increase of the quantity of the inhibitor-type component will lead to the spreading of the 2D secondary phase from the GBs to the top surface of the growing 3D crystals (called covering layer). Here the movement of the growth steps will be stopped. As a result deep and broad GB grooves develop and repeated nucleation of the majority 3D phase starts on the covering layer (Fig. 13.12(b), bottom). In the case of a complete covering layer, there is no contact between the lower and upper layers. The grain size will be determined by the growth time until the covering layer closes. A change in the morphology from columnar to globular (Zone III) develops. The formation of Zone III of equiaxial grains starts developing first at high temperatures and extends to lower temperatures with increasing concentration of the inhibitor-type component.

The corresponding schematic SZDs are shown in Fig. 13.13. The diagrams show how the increasing quantity of the 2D second phase is shifting the structure from

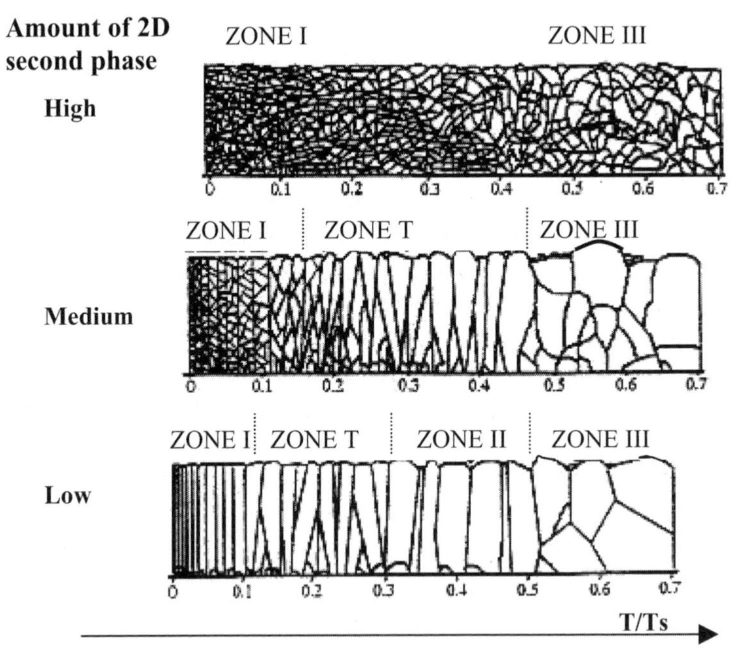

Figure 13.13: Schematic SZD of two-component systems, for the case of 3D growth of phase 1 and 2D growth of minority phase 2 ($3D_1$–$2D_2$).

columnar to globular (Zone III) grains by decreasing at first the width of Zone II, then Zone T and causing a substantial decrease of the grain size. When the concentration of the inhibitor-type component is high, the growth of the 3D grains is stopped already at small grain size and a nanocomposite forms in which traces of columnar growth can only hardly be recognizable. A typical example of this behavior is the Al-O system for which the above discussed SZD has been worked out [18].

From the above consideration we can conclude that nanocrystalline thin films must be two-phase materials, especially when they are grown from vapor phase and not by a transformation in solid state. This can be understood by the mechanism of the formation of the surface-covering layer of the 2D phase leading to repeated nucleation of the primary 3D phase. Consequently, a nanocomposite is energetically possible only if the growth surface is covered by a continuous layer of a second phase, blocking the growth of the already existing crystals and serving as substrate for the repeated nucleation.

If the 2D growing phase is the majority phase, the development of the structure can be schematically illustrated as shown in Fig. 13.14. The primary phase is the 2D phase and the crystallites of the 3D secondary phase are embedded [33,34].

The generality of the building principles of the SZD can be illustrated by the morphologic behavior of the Ni-C system where also $3D_1$–$2D_2$ situation exists. No stable compound exists in the equilibrium phase diagram of this system. The structure obtained by co-sputtering of C and Ni below 400°C is a nanocomposite [38], consisting of the metastable Ni_3C crystalline phase of columnar and globular grains and a fullerene-like, about 1 nm thick carbon 2D tissue phase (Fig. 13.15). In this case the amount of the two components is such that the forming Ni_3C crystallites are surrounded by a nanotube of 3–6 graphene layers at very small grain size (5–1 0 nm, Fig. 13.15(b)). The carbon 2D tissue phase is situated in the GBs. It could form by segregation of C species from the growing Ni_3C phase, nucleated at first, and by collecting C species also from the intergrain substrate area. The structure, shown in Fig. 13.15(a), corresponds already to Zone III in the SZD (Fig. 13.13), since not all the grains are nucleated at the substrate, showing clear evidence of repeated nucleation.

In the case of CN_x-Ni nanocomposite, grown at 300°C the system morphology may be considered to correspond to Zone T since all grains practically start on the substrate. The competing growth in this case is strongly influenced by the GBs.

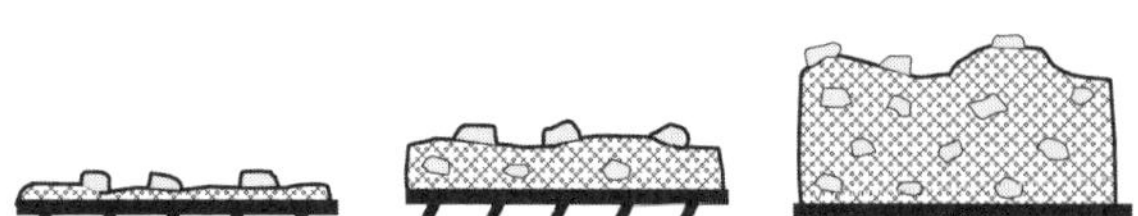

Figure 13.14: The schematic growth morphology of $2D_1$–$3D_2$ system.

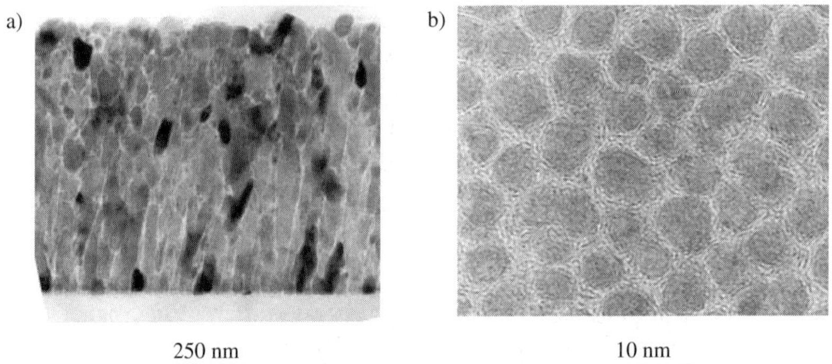

Figure 13.15: Cross section (a) and high-resolution lateral view (b) of C-Ni_3C composite films, prepared by co-deposition of C and Ni in Ar at 200°C. The thickness of the film shown in (b) is 20 nm.

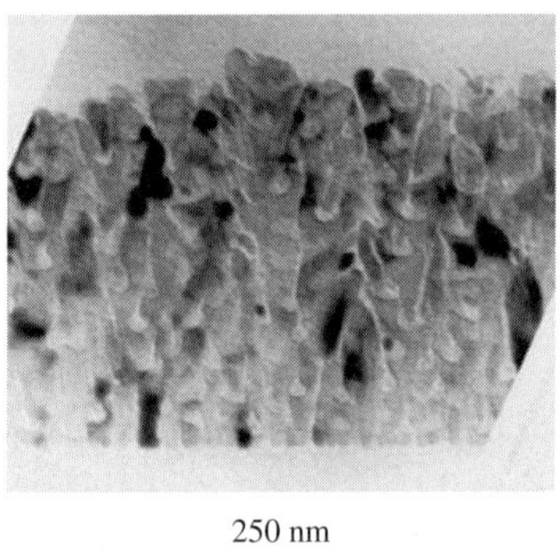

Figure 13.16: CN_x-Ni_3C nanocomposite prepared by co-deposition of C and Ni in N_2 at 300°C temperature. Columnar (Zone II) growth of Ni_3C crystallites is observed.

The GBs are unusually wavy due to the formation of the CN_x pockets in them, besides of the carbon 2D tissue phase. The growth of the amorphous CN_x phase resembles the Stranski–Krastanov growth mode (Fig. 13.16).

Above 400°C the crystalline phase is face centered cubic (f.c.c.) Ni in the Ni-C system. If the growth temperature in the C-Ni system is increased into the region of volume diffusion (800°C) the coalescence of the growing Ni particles will be possible and thicker stacks of graphene layers will be necessary to stop this process. As a result, a nanocomposite of larger f.c.c. Ni grains and thicker graphene sheets forms resulting in a globular morphology (Fig. 13.17) [38] corresponding to the Zone III structure of high second-phase quantity in Fig. 13.13. Systems showing similar behavior are: TiN-aSi_3N_4 [32,33], Al-C [37], Co-Cr-SiO_2 [39] and Cr-C [40].

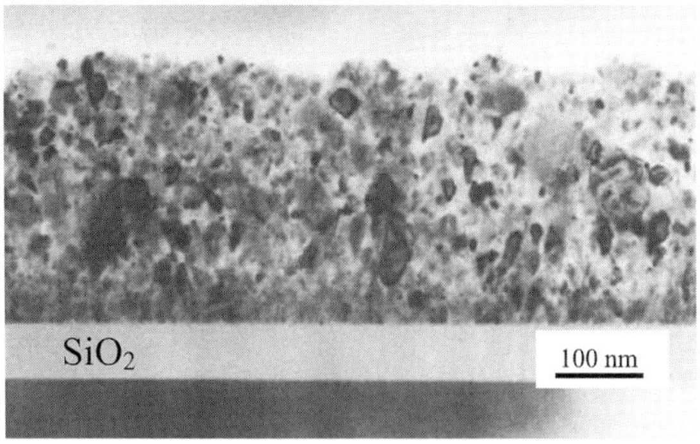

Figure 13.17: Morphology in cross section of a C-Ni nanocomposite film co-sputtered at 800°C. The temperature of deposition is large enough to provide kinetics for coalescence of Ni crystals, resulting in the formation of some large Ni grains. The structure corresponds to Zone III of the SZD in Fig. 13.13.

a) 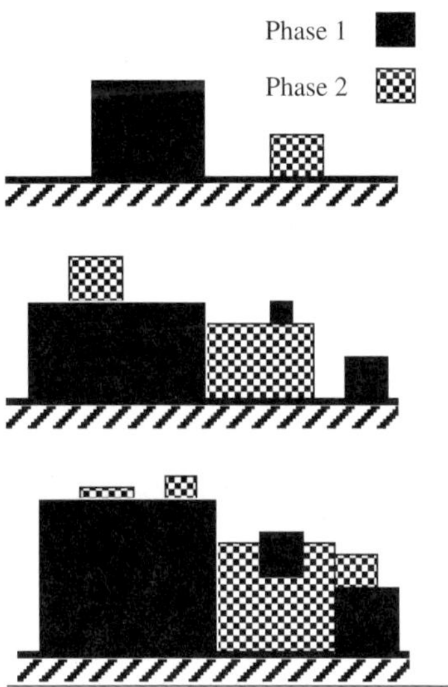

b) 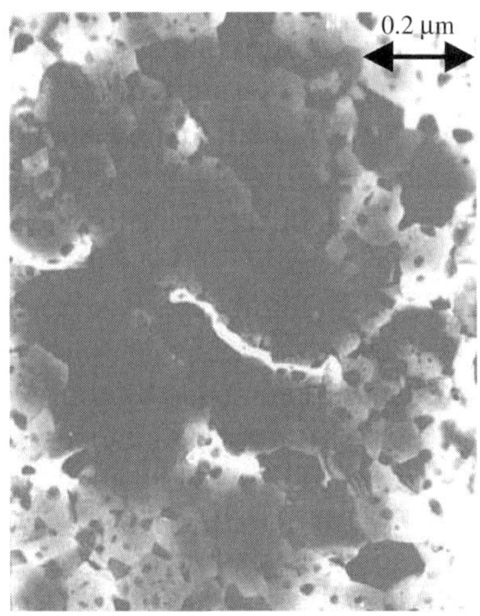

Figure 13.18: Schematic growth morphologies (a) and TEM image of co-deposited Al and Ni (5 vol%) (b) of a two-component system, for the case of 3D growth of both phases 1 and 2.

Similarly to the $3D_1$–$2D_2$ case in Fig. 13.13, the schematic SZD can be built for the case, when both phases are growing in the 3D form (Figs. 13.18 and 13.19).

The morphologic development of a phase can be strongly influenced not only by limiting the crystal or grain growth by the development of a 2D second phase but also by adding a small quantity of a component which is present as an adatom layer

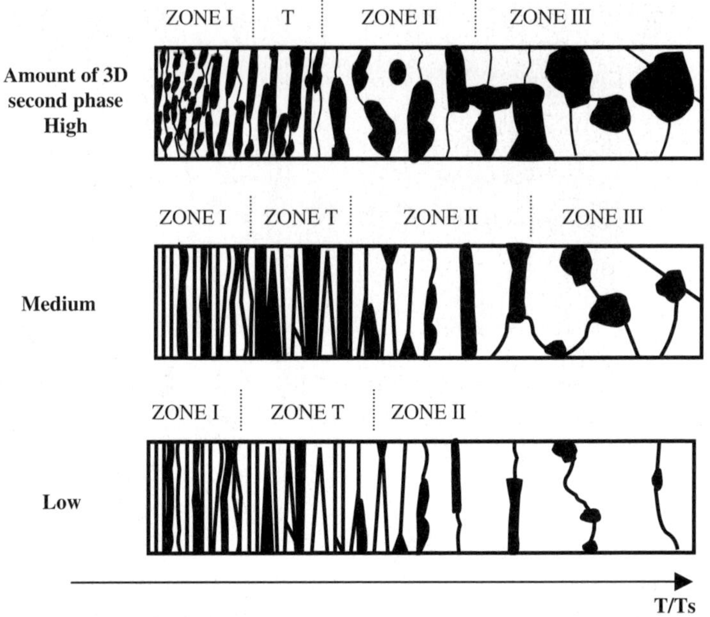

Figure 13.19: Schematic SZD of a system in which both phases grow in 3D mode ($3D_1$–$3D_2$ in Table 13.1) as a function of the quantity of the second phase.

on the growth surface. This adatom layer can influence the growth mode (Fig. 13.3) by changing it from 3D mode to 2D growth mode or vice versa. These kinds of additives are called surfactants. One way of their action is changing the height of the Schwoebel barrier at the growth steps, making easy or difficult the atomic diffusion across the steps influencing the movement of steps [8,45].

Oxygen is an inhibitor for Al film growth processes, as it hinders all growth processes even before forming stable second phase [18,46]. Crystal growth of Si is inhibited by co-depositing SiO, leading to the change of 2D growth mode of Si to the 3D case through formation of {311} facets on the (100) growth surface [47]. There are additives, which act in a reverse way, by increasing the adatom mobility on the growth surface [8]. These additives are called promoters. Such a promoter is Pb or In on Cu [8] or Sn on Al [48]. Pb on Cu is enhancing 2D growth of Cu, while Sn on Al is enhancing coalescence and, by this way, increasing the grain size of the film compared to the pure Al, that is shifting the border between Zone T and Zone II to lower temperatures. Promoters can act also in the/nucleation stage, like for example the presence of Ag on NaCl surfaces aids the condensation of Cd or Zn.

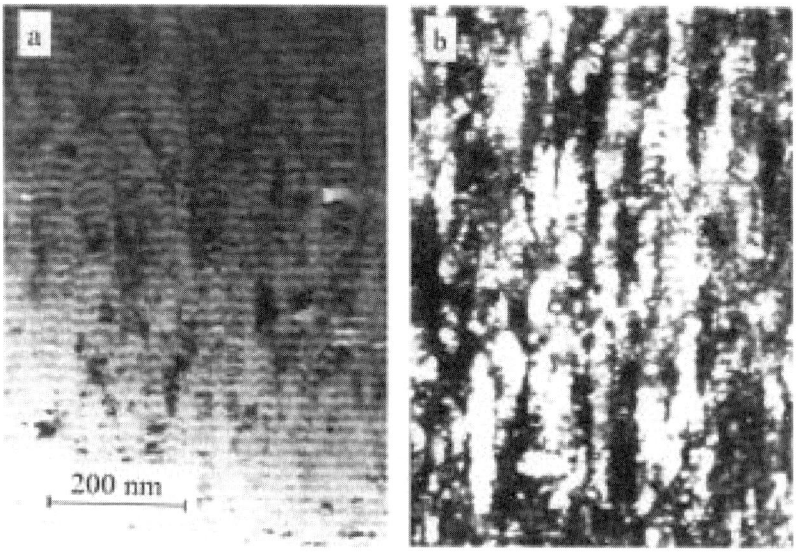

Figure 13.20: Structure of 12 nm period symmetric polycrystalline Ag/Cu multilayer with a columnar structure of Zone I. (a) Bright field image and (b) dark field image [3].

13.4.2. Multilayers

If we look at the structure of the film shown schematically in Fig. 13.7, we can see that it changes with thickness. Using very small film thicknesses the competitive processes cannot fully develop. This is the case in the growth of multilayers composed from very thin amorphous or polycrystalline sublayers. Each sublayer develops the morphology corresponding to its own SZD and thickness. The interaction of the sublayers is an additional effect. This interaction is manifested in providing nucleation sites, smooth or curved surfaces for growth and epitaxy. In multilayers one of the layers usually plays dominant role in the development of the morphology while the other layer replicates this morphology [3,50]. According to [3] this dominant role is played by the thicker layer, and the grain size or column width is determined by its thickness. Simultaneously an increasing surface roughness can be observed with increasing film thickness [3], in accordance with the development of the surface morphology of a one-component film.

This behavior of morphologic development of multilayers hardly depends on the nature of the film material (amorphous or crystalline) and does not depend on the possible epitaxial relation between the successive layers. Since the thin layers can be continuous only at high nucleation densities, most of the multilayers of polycrystals are grown at limited kinetic conditions, that is at Zone I or Zone T conditions. The

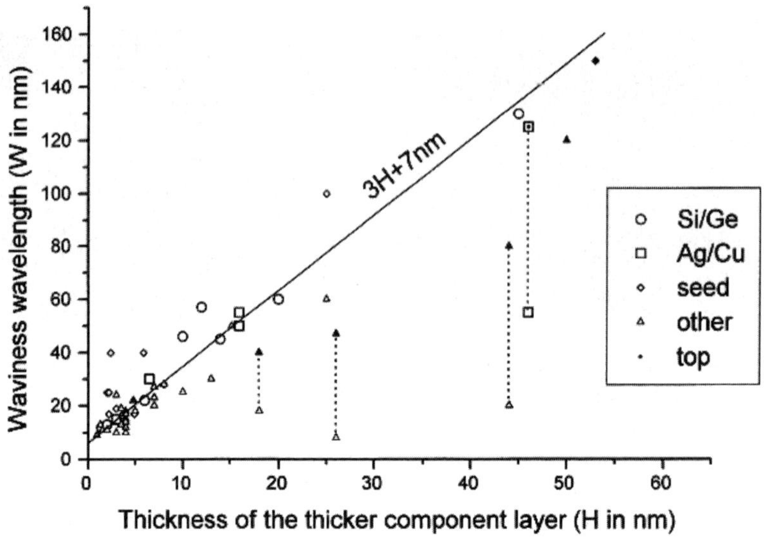

Figure 13.21: Wavelength of interface waviness as a function of the thickness of the thicker sublayer in different multilayers. The data, denoted with open diamonds and triangles were measured for films grown with or without seed layer, respectively. Dashed lines connect the points belonging to the same multilayer, measured in the vicinity of the substrate and far from it [3]. The straight line is a fitting to the measured points.

limited kinetics and the small layer thicknesses lead to columnar structures in multilayers as shown in Fig. 13.20(a) and (b). It is a cross section TEM image of a Cu/Ag multilayer, where columnar morphology (Zone I in SZD) developed, and the Cu and Ag layers are grown epitaxial on each other. Each column is composed of alternating 6 nm thick Cu and Ag layers building a single crystal of 12 nm multilayer period. The bright domains in Fig. 13.20(b) correspond to these single crystals.

In multilayers of different layer thickness the period of the interface waviness, what practically equals the lateral grain size increases linearly with increasing thickness of the thicker sublayer. This rule (Fig. 13.21) was found in different multilayer systems, composed from crystalline and/or amorphous materials [3].

The development of the cup-shaped surface of the individual columns can be explained by the low temperature deposition. Incomplete coalescence sets in rather early, and the GBs will be trapped at very small grain sizes. These GBs will have grooves due to the cup shape of the coalescing islands. As the surface mobility of adatoms on the growth surface is low, shadowing effects will enhance the GB grooves even in the case of deposition at normal incidence (Fig. 13.20a). Oblique deposition will further enhance this effect. The amplitude of the waviness increases with the thickness of the film, as shown in Fig. 13.7.

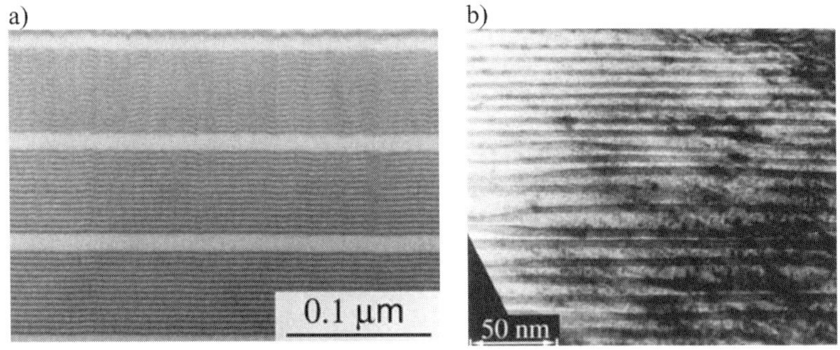

Figure 13.22: Variation of the waviness as a function of bias voltage in aSi/aGe multilayer prepared by ion-assisted deposition. The bias is 0 V for the upper, −20 V for the middle and −40 V for the bottom layer system. The layer systems are separated by a thicker amorphous Si layer as shown in (a). Inhomogeneous structure developed in Mo/V single crystal multilayer due to the roughening transition of V (brighter layers) at 700°C. The multilayer period increases from 5 nm to 10 nm downward in the image as shown in (b).

Waviness occurs also in amorphous Si/Ge multilayers, when grown by DC magnetron sputtering (Fig. 13.22(a)). Providing additional surface kinetics, for example by ion-assisted deposition, the shadowing effects and consequently waviness of layers can be avoided as long as the main driving force is the decrease of the surface energy. Deposition of aSi/aGe multilayers by DC sputtering at a bias beyond −40 V resulted in smooth interfaces (Fig. 13.22(a), bottom) [49] due to the removal of shadowing effects as a result of enhanced adatom mobility.

In Mo/V single crystal multilayers an inhomogeneous structure develops at certain thickness ratio of the sublayers [50]. In the layer system shown in (Fig. 13.22(b)) the thickness of the Mo layers is constant, while the V layers have strong thickness variation. The thickness fluctuation within the V layers is due to the driving force of the strain fields of misfit dislocations and to the kinetics provided by the surface melting or roughening of the V layers at the growth temperature (700°C). Since no surface melting of Mo occurs at the growth temperature the Mo layers grow with constant thickness.

13.4.3. Self-organized Multilayers

Multilayers can form in a self-organized manner by lamellar growth, as illustrated in the co-deposited Al–Sn thin film doped with oxygen. The thickness of the lamellae is seen in Fig. 13.23 while the slabs of a new lamella on the surface.

The proposed mechanism is that, discussed in Section 13.4.1. Sn and O are segregating to the surface of the growing Al crystal where SnO_2 nucleates and

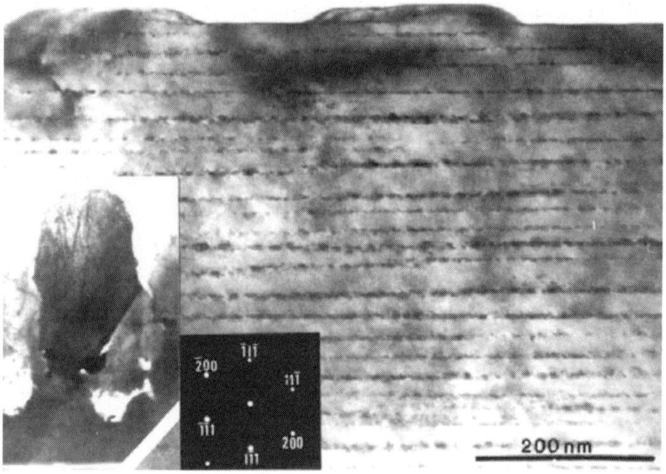

Figure 13.23: Self-organized multilayer of Al and SnO_2 (dark dots) in a grain (insert) of co-deposited Al–Sn thin film doped with oxygen.

develops a covering layer. Repeated nucleation of Al crystals, formation of new slabs and their lateral growth take place on this layer. At steady growth conditions this process regularly repeats at a distance of the lamella height. The result is a self-organized multilayer of Al lamellae separated by SnO_2 covering layers [41].

13.5. Summary

The experience, that the morphology of elementary polycrystalline thin films is determined by the available kinetics during growth is basically valid for multi-component systems as well.

In multi-component films the formation of second-phase particles (3D) or layers (2D) of an inhibitor-type component will hinder the lateral growth and coalescence of crystals. A morphology, which would correspond to lower temperature in a pure system of the first phase, is stabilized in this way. It results in shifting of the Zone T–Zone II boundary to higher reduced temperature (T_S/T_M). Another consequence is the formation of the covering second-phase layers on the growth surface. This leads to further phenomena, not observed in pure systems. The two most important phenomena are the repeated nucleation (of both phases), which leads to the break down of the columnar structure (multigrain or globular structure in the thickness of the film), and the stabilisation of grain boundaries by the second phase. As a result metastable nanocomposites can be stabilized by inhibitor-type second components.

The morphology, grain size, texture and surface roughness of the films develop simultaneously in strong interaction with each other. Orientation selection and, consequently, texture development can be controlled in multi-component systems by selecting appropriate components.

Acknowledgment

This work was partly supported by the Hungarian National Science Foundation (OTKA T-048699) and also by the EU under the IHP contract New Fullerene-like Materials: HPRN-CT-2002-00209

References

[1] B.A. Movchan and A.V. Demshichin, *Fiz. Met. Metalloved.*, 28 (1969) 83.
[2] I. Petrov, P.B. Barna, L. Hultman and J.R. Greene, *J. Vac. Sci. Technol.*, A 21(2003) S117–S128.
[3] Zs. Czigány and G. Radnóczi, *Thin Solid Films*, 347 (1999) 133–145.
[4] M. Harsdorff and E.A. Knabbe, *Surf. Sci.*, 86 (1979) 36–41.
[5] M. Zinke-Allmang, L.C. Feldman and M.H. Grabow, Clustering on surfaces, *Surf. Sci. Rep.*, 16 (1992) 377–463.
[6] B. Luis and J.C. Anderson, *Nucleation and Growth of Thin Films*, Academic press, Inc., London, 1978, p. 6.
[7] R.L. Schwoebel and E.J. Shipsei, *J. Appl. Phys.*, 37 (1966) 3682.
[8] J. Camarero, V. Cros, M.J. Capitán, J. Álvarez, S. Ferrer, M.A. Niño, J.E. Prieto, L. Gómez, J. Ferrón, A.L. Vázquez de Parga, J.M. Gallego, J.J. de Miguel and R. Miranda, *Appl. Phys.*, A69 (1999) 553–557.
[9] R.N. Tait, T. Smy and M.J. Brett, *Thin Solid Films*, 226 (1993) 196–201.
[10] Paritosh, D.J. Srolovitz, *J. Appl. Phys.*, 91 (2002) 1963.
[11] D. Walton, *Phil. Mag.*, 7 (1962) 1671; *J. Chem. Phys.*, 37 (1962) 2182.
[12] G. Zinsmeister, *Thin Solid Films*, 2 (1968) 497; 7 (1971) 51.
[13] W.M. Kosewich, A.A. Sokol and G. Radnóczi, *Acta Physica Hungarica*, 33 (1973) 353–361.
[14] Á. Barna, P.B. Barna and J.F. Pócza, *J. Vac. Sci. Technol.*, 6 (1969) 472.
[15] Ph. Buffat and J.-P. Borel, *Phys. Rev. A*, 13(6) (1976) 2287.
[16] Á. Barna, P.B. Barna, G. Radnóczi, H. Sugawara and P. Thomas, *Thin Solid Films*, 48, (2) (1978) 163–174.
[17] P. Barna and M. Adamik, in *Science and Technology of Thin Films*, ed. By F.C. Matacotta and G. Ottaviani, World Scientific Publishers, Singapore 1995, p. 1.
[18] P.B. Barna and M. Adamik, *Thin Solid Films*, 317 (1998) 27.
[19] M.J. Rost, D.A. Quist and J.W.M. Frenken, *Phys. Rev. Lett.*, 91 (2003) 26101.
[20] C.V. Thomson, *Annu. Rev. Mater. Sci.*, 20 (1990) 245.

[21] G. Radnóczi, *Proceedings of the 8th European Congress on Electron Microscopy*, Budapest 13–18 August, 1984, pp. 1229–1238.
[22] R. Lindsay, J.N. Chapman, A.J. Craven and D. McBain, *Ultramicroscopy*, 80 (1999) 41.
[23] D.W. Pashley, M. Stowell, M. Jacobs and T. Law, *Phil. Mag*., 10 (1964) 127.
[24] J.F. Pócza, Á. Barna, P.B. Barna, I. Pozsgai and G. Radnóczi, *Jap. J. Appl. Phys*., (Suppl 2), Part 1 (1974) 525–532.
[25] J.A. Thornton, *Ann. Rev. Mater. Sci*., 7 (1977) 239.
[26] R. Messier, A.P. Giri and A.R. Roy, *J. Vac. Sci. Technol*., A2(1984) 500.
[27] C.R.M. Grovenor, H.T.G. Hentzell and D.A. Smith, *Acta Metall*., 32 (1984) 773.
[28] G.H. Gilmer, H. Huang and C. Roland, *Comput. Mater. Sci*., 12 (1998) 354.
[29] Á. Barna, P.B. Barna, R. Fedorowich, G. Radnóczi and H. Sugawara, *Thin Solid Films*, 36 (1) (1976) 75–79.
[30] G. Radnóczi, M.-A. Hasan and J.-E. Sundgren, *Thin Solid Films*, 240 (1–2) (1994) 39.
[31] D.W. Hoffman and J.A. Thornton, *J. Vac. Sci. Technol*., 20 (3) (1982), 355.
[32] R. Hauert and J. Patscheider, *Adv. Eng. Mater*., 2/5 (2000) 247–259.
[33] J. Patscheider, MRS Bulletin, March, 2003; p. 180, J. Patscheider, T. Zehnder and M. Diserens, *Surf. Coat. Technol*., 146–147 (2001) 201–208.
[34] S. Veprek, *Thin Solid Films*, 297, (1997) 145.
[35] A. Schmidt, V. Schünemann and R. Anton, *Phys. Rev*., Part B 41 (17) (1990) 11875.
[36] H. Mori and H. Yasuda, *Mat. Sci. Eng*., A312 (2001) 99–103.
[37] A. Kovács, P.B. Barna and J.L. Lábár, *Thin Solid Films*, 433 (2003) 78.
[38] Gy.J. Kovács, G. Sáfrán, O. Geszti, T. Ujvári, I. Bertóti, G. Radnóczi, *Surf. Coat. Technol*., 180–181 (2004) 331.
[39] G. Sáfrán, J.T. Suzuki, J.K. Ouchi, H. Muraoka and Y. Muraoka, *Tech. Rep*., IEICE, MR98-1, 2002.
[40] S. Yang, X. Li, N.M. Renevier and D.G. Teer, *Surf. Coat. Technol*., 85 (2001) 142–144.
[41] P.B. Barna, M. Adamik, G. Sáfrán, B. Pécz, A. Bergauer and H. Bangert, *Phys. Stat. Sol*., a 146(1994)317.; P.B. Barna, M. Adamik, G. Sáfrán, B. Pécz, H. Bangert and A. Bergauer, *Proc. ICEM-13*, Eds. B. Jouffraey, C. Colliex, *et al*., Paris, 1994, Vol. 2, pp. 207–208.
[42] G. Radnóczi and P.B. Barna, *Thin Solid Films*, 116 (1984) 143–149.
[43] Zs. Radi and P.B. Barna, *Surf. Coat. Technol*., 100–101 (1998) 90–93.
[44] P.B. Barna, G. Radnóczi and F.M. Reicha, *Vacuum TAIP* 38 (1988) 527.
[45] J. Ferrón, L. Gomez, J.M. Gallego, J. Camarero, J.E. Prieto, V. Cros, A.L. Vázquez de Parga, J.J. Miguel and R. Miranda, *Surf. Sci*., 459 (2000) 135.
[46] Á. Csanády, Y. Pitton, H.J. Matthieu, K. Kessler, R. Fuchs and M. Textor, *Surf. Interface Anal*., 21 (1994) 546.
[47] Wei-Xin Ni, G. Radnóczi and G.V. Hansson, *Thin Solid Films*, 184 (1990) 403–414.
[48] C. Eisenmenger-Sittner, H. Bangert, H. Störi, J. Brenner and P.B. Barna, *Surf. Sci*., 489 (2001) 161.
[49] K. Jarrendahl, G. Radnóczi, Zs. Czigány, I. Ivanov, J.-E. Sundgren and J. Greene, *J. Mater. Res*., 12(7) (1997) 1806–1815.
[50] J. Birch, L. Hultman, G. Radnóczi and J.-E. Sundgren, *Phys. Rev. B*, 53, (1996) 8114–8123.

Chapter 14

Mechanical Characterizations of Surfaces and Coatings

Anthony Fischer-Cripps

14.1. Introduction

Two of the most important physical properties of surfaces and coatings are elastic modulus and hardness since it is these properties that dictate the response of a material when subjected to a mechanical loading. The mechanical loading may be external (as in in-service loads) or internal (such as residual stresses arising during manufacture). In practice, the avoidance of failure of the surface or coating is desired. Failures may take the form of pitting, delamination, scratches, plastic deformation (e.g. blunting) not to mention discoloration, chemical, and thermal property changes. The elastic modulus determines the extent a material extends or contracts under an applied load. The hardness of a material is a measure of its resistance to permanent deformation, or yield. Both these properties can be easily measured for relatively large-scale specimens, but special techniques are required for the testing of thin films and surfaces.

Traditional hardness testing, such as the Brinell test or the Vickers test, involves indenting the surface of the specimen with a diamond or hard metal ball or pyramid. The establishment of the Brinell, Knoop, Vickers, and Rockwell tests all follow from a refinement of the original Mohs' hardness test where one material is scratched or indented with another. The purpose of the test is to cause sufficient plastic deformation in the specimen to leave a residual impression in its surface upon the removal of load. The applied load, divided by the area of the impression, is usually taken to be a measure of the hardness of the material. When the projected contact area is used (as distinct from the actual curved sides of the impression) the hardness is equivalent to the mean contact pressure at full load. The confined nature of the stresses under indentation loading means that the hardness of nominally brittle materials can be measured without catastrophically fracturing the specimen.

For hardness measurements on thin films, traditional hardness testing is unsuitable for two reasons. Firstly, it is not generally possible for large-scale hardness test

Materials Surface Processing by Directed Energy Techniques
Y. Pauleau (Editor)

instruments (including micro-hardness testers) to apply a sufficiently small enough load to the indenter so that the penetration depth or indenter displacement is kept well within the film thickness. Secondly, even if such indentations could be made, the usual method of measuring the size of the residual impression using optical methods introduces large levels of uncertainty in the results. For thin films of say less than a couple of microns thickness, the diagonals to be measured are sometimes only a few microns in extent.

To overcome these difficulties, hardness testing of thin films is usually done using depth-sensing indentation in which the depth of penetration of the indenter into the specimen surface is recorded as a function of the indenter load and then, using the known geometry of the indenter, the contact area is calculated. The added benefit of the technique is that the load-displacement curve that results from the test can be used to calculate the elastic modulus of the specimen. In practice, the depths of penetration are usually in the nanometer range and the technique is commonly referred to as "nanoindentation" [1].

14.2. Contact Mechanics

In order to gain an appreciation of the issues involved in nanoindentation, it is necessary to become familiar with the mechanics of contact between solids. Formal studies of contact mechanics begins with Hertz who, in 1881, was interested in determining the elastic deformation of glass lenses placed in contact in optical instruments. In the 20th century, several workers studied the mechanics of the ball and pyramid indentation test which involved plastic deformation beneath the indenter and several theories exist which account for the observed deformations in this type of contact. In general, contact between an indenter and a specimen involves both elastic and plastic deformations. Fortunately, these deformations can often be treated separately in the development of analytical models that yield values for material properties of interest.

14.2.1. Elastic Contact

For an elastic contact between two curved surfaces, the radius of the circle of contact a is related to the indenter load P by [2,3]:

$$a^3 = \frac{3}{4}\frac{PR}{E^*} \qquad (14.1)$$

where R is the combined radius of the contacting surfaces. For the case of a sphere on a flat surface, R is the radius of the indenter. The quantity E^* is the combined elastic modulus of the indenter and the specimen is given by:

$$\frac{1}{E^*} = \frac{(1-\nu^2)}{E} + \frac{(1-\nu'^2)}{E'} \tag{14.2}$$

where ν is Poisson's ratio and the primed terms apply to the indenter and the unprimed terms to the specimen. Usually, the modulus of the indenter (diamond $\approx$ 1000 GPa) is very much higher than that of the specimen and so the value of E^* is dominated by E. However, for some thin films (e.g. diamond-like carbon (DLC), titanium nitride (TiN), etc.) the elasticity of the specimen can be significant compared to that of the indenter and the elastic deformation of the indenter must be taken into consideration through Eq. (14.2).

It is interesting to express Eq. (14.1) in terms of the mean contact pressure:

$$p_m = \frac{P}{\pi a^2} = \frac{4}{3}\frac{E^*}{\pi}\frac{a}{R} \tag{14.3}$$

whereupon it can be seen that, for an elastic contact, the mean contact pressure, or the indentation stress, is linearly proportional to the quantity a/R which is termed the indentation strain.

Within the circle of contact, the displacement h of the specimen surface as a function of r from the axis of symmetry is given by:

$$h = \frac{1}{E^*}\frac{3}{2}\frac{P}{4a}\left(2 - \frac{r^2}{a^2}\right) \quad r \le a \tag{14.4}$$

where for $r = 0$, we obtain the distance of mutual approach of the indenter and the specimen h_t and for $r = a$, we obtain the depth of the circle of contact beneath the specimen-free surface h_a.

The most popular indenter geometry for nanoindentation tests is the three-sided Berkovich indenter. The face angle θ of the Berkovich indenter is 65.3° and gives the same projected area-to-depth ratio as the more commonly known Vickers indenter used in micro-hardness testing. The quality of the indenter is often measured by the sharpness of the tip, the tip radius for a typical Berkovich indenter being in the order of 100–200 nm. In the analysis of data obtained with a Berkovich indenter, it is customary to treat the indenter as an equivalent cone, the angle of which provides the same area-to-depth ratio as the actual indenter.

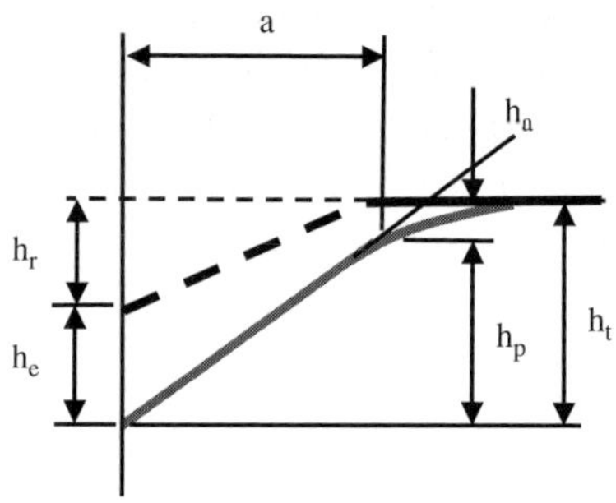

Figure 14.1: Geometry of loading with a conical indenter. The full line shows a section of the deformation at full load, and the dotted line is full unload. When plastic deformation occurs, a residual impression of depth h_r is left in the surface. Reloading the residual impression results in an elastic deflection h_e.

An equivalent conical indenter has the advantage of possessing axial symmetry and the equations of contact are well known. For a conical indenter, such as that shown in Fig. 14.1, the radius of circle of contact is related to the indenter load by [4]:

$$P = \frac{\pi a}{2} E^* a \cot \alpha \tag{14.5}$$

where α is the cone semi-angle. The quantity $a \cot \alpha$ is the depth of penetration h_p measured at the circle of contact. The depth beneath the specimen-free surface within the circle of contact is given by:

$$h = \left(\frac{\pi}{2} - \frac{r}{a} \right) a \cot \alpha \qquad r \leq a \tag{14.6}$$

With $r = 0$, and substituting into Eq. (14.5), we obtain:

$$P = \frac{2E^* \tan \alpha}{\pi} h_t^2 \tag{14.7}$$

where h_t is the depth of penetration of the tip of the indenter beneath the original specimen-free surface ($r = 0$). When plastic deformation occurs, a residual impression of depth h_r is left in the surface. Reloading the residual impression results in an elastic deflection h_e.

The equations given above apply to an elastic contact between the indenter and the specimen, even for the case of the conical indenter that has a stress singularity at the tip. In practice, the singularity is avoided because the tip of the indenter is usually rounded, no matter how carefully the indenter is made. It is this rounding of the tip that limits the indentation technique in the measurement of hardness of very thin films since, as we shall see, the hardness measurement relies on the formation of a fully developed plastic zone within the specimen material.

14.2.2. Elastic–Plastic Contact

In general, contact between an indenter and a specimen may result in both elastic and plastic deformations. Consider first the case of a spherical indenter. The physical quantity of interest is the mean contact pressure, which as shown previously in Eq. (14.3), is linearly dependent upon the ratio a/R for an elastic contact. As the load, and hence the mean contact pressure, is increased from zero, the contact is initially elastic but at higher loads, the level of shear in the indentation stress field may increase to the point such that plastic flow occurs within the specimen material. As the load is increased further, the relationship between the mean contact pressure and the indentation strain, a/R, becomes non-linear. Eventually, and in the absence of strain hardening, the mean contact pressure approaches a constant value as the area of contact begins to increase at the same rate as the increase in applied load. At this condition, the plastic zone has grown in size sufficiently to be considered fully developed.

The value of the mean contact pressure p_m at which there is no increase with increasing indenter load is usually taken to be the indentation hardness H of the material and is computed from the load P divided by the projected area of the contact A:

$$H = \frac{P}{A} \tag{14.8}$$

At a condition of full plasticity, experiments show that the mean contact pressure, or hardness H, is directly proportional to the material's yield stress Y and can be expressed as:

$$H \approx CY \tag{14.9}$$

The factor C is called the "constraint factor" and is $\approx$3 for materials with a large value of the ratio E/Y (e.g. metals) and $\approx$1.5 for low values of E/Y (e.g. glasses [5,6]).

For an ideal (infinitely sharp) conical indenter, a condition of full plasticity is achieved immediately upon contact and the geometry of the deformation has the property of geometrical similarity. Under these conditions, the mean contact pressure is independent of load. For real indenters, the tip rounding results in an initial elastic contact and the hardness cannot be measured until the indenter load (and hence penetration depth) is increased. The sharper the tip radius, the lower the load required to produce a fully developed plastic zone. For this reason, extremely sharp tips ($<$100 nm) are desirable for the measurement of properties of thin film materials. The penetration depth usually has to be approximately $<$10% of the specimen thickness to minimize the effect of the substrate or mounting although this figure depends upon the material being tested. In practice, nanoindentation instruments can provide valid data to penetration depths of about 10 nm as a minimum and so the minimum film or coating thickness that can be measured reliably is in the order of 100 nm.

14.3. Analysis of Load-Displacement Curves

Optical measurements of the contact area in an indentation test are not usually convenient when the depth of penetration is less than a few microns. For this reason, nanoindentation tests are usually performed using a depth-sensing technique whereby the load and depth of penetration are recorded as load is applied and then withdrawn from the indenter. The resulting load-displacement curve enables the area of contact to be determined from the known geometry of the indenter.

Usually, a pyramidal indenter is represented by an equivalent axis-symmetric cone, as shown in Fig. 14.1, and the associated load-displacement curve is shown in Fig. 14.2. The radius of the circle of contact is usually located below the plane of the specimen-free surface and the depth of the circle of contact measured from the tip of the indenter, h_p, is the important quantity to be measured. This quantity

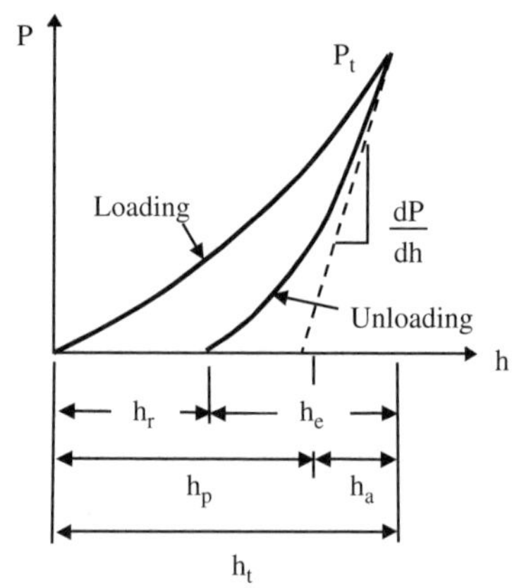

Figure 14.2: Load-displacement curve for elastic–plastic contact with a conical indenter.

cannot be estimated from the total depth of contact using the elastic equations of contact because of the plastic deformation in the specimen material as the load is applied. However, it is important to know that as load is withdrawn, the response of the material is elastic; that is, the unloading response represents elastic recovery of the specimen material. At full unload, there is a residual impression in the surface resulting from the plastic deformation that occurred during loading. This means that the elastic equations of contact can be applied to the unloading data and, as will be shown, enables the quantity h_p, and hence, area of contact to be measured. The most popular method for doing this is that developed by Doerner and Nix [7] and further refined by Oliver and Pharr [8].

A three-sided pyramidal Berkovich indenter has a face angle of 65.3° whereby the equivalent axis-symmetric semi-cone angle is $\alpha = 70.3°$. For the elastic unloading, the load and displacement are related by Eq. (14.7). The derivative of the load P with respect to the displacement h is called the contact stiffness and is given by:

$$\frac{dP}{dh} = 2\frac{2E^* \tan\alpha}{\pi}h \qquad (14.10)$$

Substituting back into Eq. (14.7), we have:

$$P = \frac{1}{2}\frac{dP}{dh}h \qquad (14.11)$$

If the total depth of penetration is h_t, at load P_t, then as the load is removed, the indenter moves through a distance h_e. At $P = P_t$ the displacements $h_{r=0} = h_e$ and $h_{r=a} = h_a$. Making use of Eq. (14.6) at $r = a$, the plastic (or contact) depth h_p is found from:

$$h_p = h_t - \left[\frac{2(\pi - 2)}{\pi}\right]\frac{P_t}{dP/dh} \qquad (14.12)$$

where P_t and dP/dh are measured during an experiment. The square-bracketed term in Eq. (14.12) is often given the symbol ε and evaluates to 0.72 but it is common practice to use a value of 0.75 since this has been shown to account for non-uniformities in the material response as the load is withdrawn. Once a value for h_p has been determined, the area of contact is found from:

$$A = \pi h_p^2 \tan^2\alpha \qquad (14.13)$$

where, for a Berkovich indenter, Eq. (14.13) evaluates to:

$$A = 24.5h_p^2 \tag{14.14}$$

The elastic modulus is found from [9]:

$$E^* = \frac{dP}{dh}\frac{1}{2}\frac{\sqrt{\pi}}{\sqrt{A}} \tag{14.15}$$

Unlike conventional micro-hardness tests, the depth-sensing method of measurement and analysis described above does not require any direct measurement of the contact area. The measurement instead relies upon an accurate measurement of the load-displacement response together with a precise knowledge of the indenter angle.

14.4. Corrections to Load-Displacement Data

The scale of testing in nanoindentation testing is such that matters ordinarily considered unimportant in conventional micro-hardness testing become very important sources of error if not corrected for. The three most common corrections required are the estimation of the initial penetration depth, the compliance of the instrument, and a correction to account for the non-ideal shape of the indenter.

14.4.1. Initial Penetration

In theory, the load-displacement curve obtained in depth-sensing indentation test begins at zero load and zero depth. In practice, this does not occur due to the need to make an initial contact with the specimen surface in order to zero the displacement sensor (Fig. 14.3).

At the initial contact load (usually made to be as small as possible), there is a corresponding physical penetration of the specimen surface by the indenter that is not measured directly by the displacement sensor. This depth offset, h_i, must be estimated from the load-displacement data and then added to all the displacement readings as a correction to the raw data. This has the effect of shifting the experimental data to the right on the load-displacement curve. The initial displacement, h_i, can be obtained by simply fitting a smooth curve to the initial data and extrapolating the curve back to zero load. The displacement intercept reading at zero load

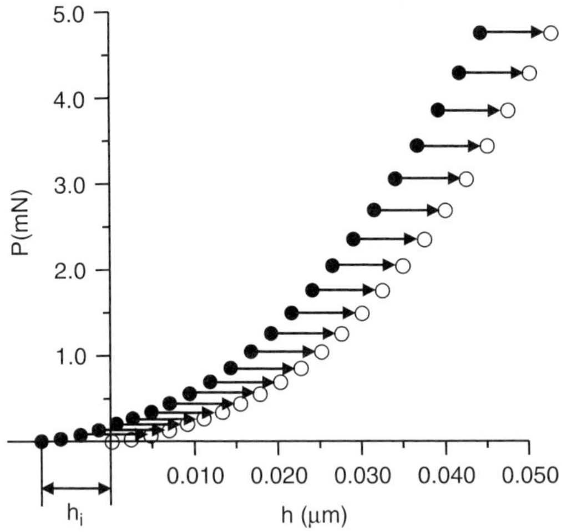

Figure 14.3: Initial penetration correction shifts the load-displacement curve to the right an amount given by the extrapolation of the original data (●) to zero load.

becomes h_i, and all the data is then shifted to the right by this amount. Typically, h_i is in the range 1–10 nm.

14.4.2. Instrument Compliance

Application of load to the indenter may, depending on the design of the instrument, result in an appreciable deflection of the instrument frame in the manner of a reaction force. This deflection is registered by the depth sensor and if not corrected for, may be incorrectly treated as depth of penetration into the specimen by the indenter. The amount of deflection is dependent upon the load applied and the compliance of the loading frame. The compliance, C_f, is usually taken to include the compliance of the frame, the indenter shaft, the specimen mounting and everything else in the load path with the exception of the localized deflections of the actual indentation contact. Since the deflections are linear with applied load, the contact stiffness $S = dP/dh$ and the compliance C_f can be combined to give the total compliance which is obtained from the load-displacement curve as measured by the instrument:

$$\frac{dh}{dP} = \frac{1}{S} + C_f \qquad (14.16)$$

For the case of a Berkovich indenter, where $A = 24.5h_p^2$, we obtain:

$$\frac{dh}{dP} = \sqrt{\frac{\pi}{24.5}}\frac{1}{2E^*}\frac{1}{h_p} + C_f \tag{14.17}$$

A value for C_f for contact with a Berkovich indenter may be obtained by plotting experimental values of dh/dP vs. $1/h_p$ over a range of indenter loads. The intercept gives the compliance of the instrument C_f directly as implied by Eq. (14.16).

Correction for instrument compliance is very important, especially when testing materials with a relatively high value of elastic modulus. Very often, nanoindentation instruments are calibrated using a fused silica standard, which has a relatively low elastic modulus, and although the instrument may appear to be functioning properly under those conditions, the results may be significantly in error when a high modulus material is tested. The issue of compliance is also important when it is desired to undertake nanoindentation testing using an atomic force microscopy (AFM). The compliance of the cantilever in an AFM is purposely made relatively small so that deflections due to surface forces are relatively large and can be readily measured – it is primarily an imaging device. When one penetrates the surface with an AFM tip, most of the displacement signal corresponds to the bending of the cantilever and only a small proportion represents the signal of interest – penetration into the specimen surface. In contrast, a nanoindenter is an instrument with a very stiff loading mechanism whereby the displacement sensor is positioned and optimized to record penetration into the specimen surface.

14.4.3. *Area Function*

The analysis method above relies on the geometry of the indenter being ideal. In practice, this is never the case due to the inevitable rounding of the indenter tip and variations in angle of the faces as a result of the manufacturing process. It is therefore necessary to apply a correction factor to determine the real area of contact at depth h_p. The correction factor takes the form of a ratio A/A_i at a given value of h_p where A is the actual area of contact at h_p and A_i is the ideal area of contact (that which would be obtained with an ideal indenter) at that value of h_p. The ratio A/A_i can be determined by direct measurement using an AFM or scanning electron microscopy (SEM), but is more usually obtained by measurements using a range of indenter loads on a standard material of known modulus. With E^* as an input, the actual area of contact is given by:

$$A = \pi\left[\frac{dP}{dh}\frac{1}{2E^*}\right]^2 \tag{14.18}$$

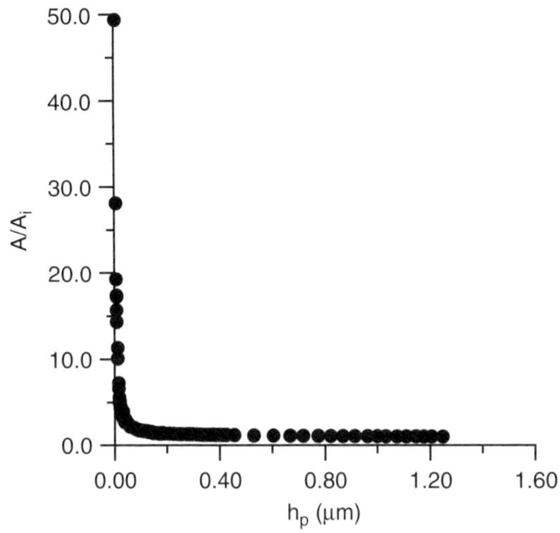

Figure 14.4: Area function for a Berkovich indenter. A is the actual contact area, A_i is the ideal contact area. The sharp rise at low values of h_p indicates tip rounding.

The ideal area for a given depth h_p is found from geometry (Eq. (14.13)). The ratio is sometimes called the area function for the indenter. As shown in Fig. 14.4, for a Berkovich indenter, A/A_i is usually quite large at very small penetration depths and is an indication of the tip radius. When A/A_i is determined from the area function, the corrected hardness is obtained from:

$$H = \frac{P}{A}\left[\frac{A_i}{A}\right] \tag{14.19}$$

and the corrected modulus is given by:

$$E^* = \frac{dP}{dh}\frac{\sqrt{\pi}}{2\sqrt{A}}\sqrt{\frac{A_i}{A}} \tag{14.20}$$

14.5. Factors Affecting Nanoindentation Test Data

The corrections described in Section 14.4 are a product of imperfections and the nature of the testing instrument. In addition to these, nanoindentation test data is very much influenced by the physical nature of the specimen. Surface roughness, micro-structural variations, and strain hardening are just some of the issues that may require consideration in an indentation test at this scale.

14.5.1. Surface Roughness

Indentation depths for most thin film systems and surface-modified layers can routinely be in the order of <100 nm. It is important therefore that some consideration be given to surface roughness of the specimen since the analysis methods given above assume an atomically smooth surface. In practice, the surface roughness is often a product of the specimen preparation procedure. For vacuum-deposited thin films, the as-deposited surface is usually tested directly. For other materials, conventional metallographic polishing procedures can be used if the indentation load is increased so that the total penetration depth is large compared to the root mean square (rms) value of the asperity height.

14.5.2. Piling-Up and Sinking-In

Theoretical analysis and finite element modeling shows that, for an elastic contact, the surface of the specimen beneath the indenter is drawn inwards towards the axis of symmetry as the load is applied. The actual circle of contact lies beneath the original specimen-free surface. This is called "sinking-in" to distinguish this type of deformation from "piling-up" which can occur for non-strain hardening materials with a large value of the ratio E/Y such as that shown in Fig. 14.5.

When piling-up occurs, the area of contact computed from the analysis methods given above is underestimated. This has the effect of making the specimen both stiffer and harder than it really is. Errors of up to 60% can be observed. There have been

Figure 14.5: AFM image of a residual impression of an indentation with a three-sided Berkovich indenter in steel. The dimension of one side of the impression is about 3 μm.

some attempts in the literature [10] to account for piling-up but no one method is satisfactory for all types of materials.

14.5.3. Micro-structural Events

The small scale of deformation in nanoindentation testing is such that events on a micro-structural scale become significant. In some materials, the hardness is observed to increase with the decreasing size of the plastic zone due to the density of dislocations in the material beneath the indenter. This is called the indentation size effect or ISE. The nature of the effect depends in many cases on the strain gradient in the material.

In some crystalline materials, the extremely high hydrostatic pressure in the indentation stress field is sufficient to cause phase changes that result in variations in the experimentally observed load-displacement response. In Silicon, a volume dilation occurs during unloading at a contact pressure of approximately 8 GPa [11]. At this condition, a body centered cubic (b.c.c.) structure is thought to form (Silicon III). Raman spectroscopy of the residual impression indicates the presence of Silicon III and Silicon XII phases [12]. This volume change results in a characteristic step in the unloading curve (see Fig. 14.9 in Section 14.8.3).

The scale of contact and resolution available from nanoindentation instruments allows a detailed study of events at the micro-structural scale that lead to plasticity, particularly in crystals. Schuh, Nieh, and Kawamura [13] show that load-displacement curves on a bulk metallic glass exhibit a series of pop-in events corresponding to the motion of individual shear bands within the material. Chinh, Horváth, Kovács, and Lendvai [14] show that plastic instabilities in metal alloys and bulk metallic glasses are a result of a negative strain rate sensitivity originating mainly from the interaction between mobile dislocations and diffusing solute atoms during plastic deformation. These types of deformations show as periodic serrations or steps in the load-displacement curve.

14.5.4. Residual Stress

The deposition of thin films and modifications of surface layers often results in considerable amounts of residual stress in the modified layer. These stresses, which may be either tensile or compressive, are often able to cause the bending of the structure. It is often desired to measure the level of residual stress or at least have some appreciation of its effect on calculations of modulus and hardness in nanoindentation testing. There are no universally accepted procedures for the measurement of residual stresses using nanoindentation techniques although there have been some attempts to do so which depend upon the nature of the material being tested.

In one method, examination of the shape of the pile-up occurring at the edge of the contact circle provides information about the level and sign of residual stress within the specimen [15]. Another method uses measurements of the critical load to initiate cone cracks to determine the magnitude and direction of surface residual stresses in brittle materials [16]. Another method uses measurements of the size of median cracks generated with sharp indenters in brittle solids as a means of determining the level of residual stresses arising from tempering [17,18]. In another method, deviations in shape of the load-displacement response from the ideal shape can be used as an indication of residual stresses. Some experiments [19] show that the effect is too small to be measured accurately although other studies [20] generally report a smaller value of depth h_t for a maximum load P_t for a compressive stress and a larger value of h_t for a tensile stress compared to the stress-free state. The residual stress for a thin film can be found from a quantitative difference between the load-displacement curves for stressed- and stress-free specimens [21]. Another method uses a comparison of the features of experimentally derived load-displacement responses obtained with a spherical indenter to those obtained by finite element calculation [22]. In this method, it is recognized that for an elastic contact, any residual stress in the specimen serves to alter the stress distribution from the Hertzian case and thus causes a change in the indentation strain (a/R) at which yield might first occur. The most common method is not a nanoindentation test at all, but uses flexure of wafer specimens. In this method, the radius of curvature of the structure is measured before and after deposition of the film. Residual stresses are usually determined using the Stoney equation [23]:

$$\sigma_f = \frac{E_s}{1-\nu_s}\frac{t_s^2}{6t_f}\left(\frac{1}{R}-\frac{1}{R_o}\right) \tag{14.21}$$

where σ_f is the stress in the film, E_s and ν_s refer to the properties of the substrate, t_s is the substrate thickness, t_f is the film thickness, R_o is the initial radius of curvature, and R is the final radius of curvature of the specimen.

The influence of residual stresses on nanoindentation test results was investigated by Tsui, Oliver, and Pharr [19]. A bulk specimen of aluminum alloy was bent so as to introduce a controlled state of stress and conventional nanoindentation tests were carried out. It was found that the hardness increased slightly with applied compressive stress and decreased slightly in the case of tensile stress over that measured in the stress-free state. This variation in hardness arises from the conditions for initial yielding within the material. The elastic modulus, determined from the slope of the elastic unloading, showed a similar trend but the variations in modulus were far greater, up to 10% change in going from tensile to compressive stress. This variation arises from errors in estimating the true area of contact from the unloading data. When the area of contact was measured optically, the elastic modulus was found not to vary in the presence of either compressive or tensile stress.

14.5.5. Thermal Drift

Thermal stability in a nanoindentation test is extremely important since changes in dimension of just a few nanometers per second can be recorded by the depth sensor and interpreted as penetration into the specimen. Errors in the displacement signal due to thermal expansion at the contact is called thermal drift. Thermal drift can be minimized with the use of a highly insulating environmental chamber and ensuring that the thermal equilibrium has been reached before the test begins. In some cases, displacement data can be corrected for thermal drift by measuring the thermal drift rate and then adjusting the displacement readings accordingly. In many cases, thermal drift in the displacement data cannot be distinguished from genuine creep within the specimen material when it is placed under load. For this reason, some practitioners prefer to obtain thermal drift correction data (hold at constant load) at the final unloading point rather than the maximum load, the intention being that the level of stress which might induce creep would be less at the smaller value of load. This might be reasonable for the case of a spherical indenter, and to some extent a fairly blunt pyramidal indenter where the rounding of the tip results in elastic, rather than elastic–plastic contact.

14.6. Nanoindentation Instruments

The specifications of nanoindentation instruments may be quoted as theoretical resolutions at specified load and depth ranges, or as noise floor figures – those which would be obtained in a real indentation test. The difficulty is that theoretical specifications can be useful in comparing different instruments, but the noise floor figures will describe what can actually be achieved in practice. Often, the noise floor is very heavily dependent on the laboratory environment – temperature stability, mechanical isolation, etc. and so any noise floor figures quoted by a manufacturer are likely to be under the best possible conditions. Some typical specifications for a nanoindenter [24] are shown below:

Maximum load	50 mN or 500 mN
Minimum contact force	2 μN
Force resolution/noise floor	500 nN/750 nN
Maximum depth	2 μm or 20 μm
Depth resolution/noise floor	0.03 nm/0.05 nm
Sample positioning	±0.5 μm (±0.1 μm optional)
Field of testing	100 mm × 100 mm
Load frame compliance	0.1 nm mN^{-1}

It should be noted that the noise floor of the specifications is the most important factor for the end user. The minimum contact force is an important specification since this, amongst other things, determines the minimum thickness of thin film sample that can be usefully tested. The specifications for specimen positioning are important if it is desired to accurately position the indentation, for example, on an individual grain in a ceramic or composite. A step size of less than a micron is generally required for this type of testing.

The forces and displacements measured in this type of testing are at the extreme end of the limits of measurement instruments and so calibration is very important. It is essential that an instrument be calibrated with traceable reference standards. In practice, the performance of an instrument can be verified by testing a standard specimen of known properties. To be effective, several specimens of different properties (modulus and hardness) should be tested so that the full range of the instrument is tested. Force calibrations are usually done using an electronic pan balance of the required resolution. When using a balance, it is important to convert the mass readings to forces using the local value of the gravitational constant. Displacement calibrations are performed using a laser interferometer or other displacement device traceable to standards of length. It is important for all measurement instruments, nanoindentation instruments included, to be calibrated with traceable reference materials and methods. It is easy to neglect the importance of calibration and rely on variations in the indenter area function to remedy any deficiencies in the force and depth readings. Such a procedure is certain to lead to problems when testing materials different from those used in the indenter area function calibration.

14.7. Experimental Technique

A typical nanoindentation test cycle consists of an application of load followed by an unloading sequence. Load is usually applied as a series of small increments. At each increment, a partial unloading may be programmed that provides information about the stiffness of the contact (dP/dh), which is useful for measuring changes in modulus or hardness with penetration depth. Contact stiffness may also be found by superimposing a small oscillatory motion onto the loading signal. Most analysis methods, such as those described previously, rely on the initial unloading data although there is some value in collecting data throughout the full loading and unloading portions. For example, the initial contact depth is found by fitting a curve to the initial loading data. Thermal drift is measured using data obtained at maximum load (or sometimes at the final unloading point); events such as cracking and phase changes can also be identified as features anywhere on the load-displacement curve. It is customary to present load-displacement data with the load on the vertical axis even though most nanoindentation instruments are load-controlled devices. Specimens are usually prepared as small, flat, polished, or mirror finish pieces bonded on to a steel mounting which is firmly clamped to the frame of the instrument. It is very important to minimize any compliance in the specimen mounting. Testing usually proceeds when thermal equilibrium is established. Mechanical vibration is usually minimized by mounting the instrument on an anti-vibration table.

The International Standards Organisation has formulated a draft standard for depth-sensing indentation: ISO 14577. The draft standard describes the method by which indentation hardness of a material is measured using depth-sensing indentation where both the force and displacement during plastic and elastic deformation are measured. ISO 14577 consists of four parts. Part 1 of the standard contains a description of the method and principles of the indentation test. Part 2 of the standard specifies the method of verification and calibration of the test instruments. Part 3 of ISO 14577 specifies the method of calibration of reference blocks that are to be used for verification of indentation testing instruments. Part 4 of the standard provides recommended procedures for the indentation testing of thin films.

14.8. Applications

Nanoindentation, despite being available since 1989, is becoming more popular as interest in surface properties of materials increases with the downsizing of structures in the nanotechnology. In this section, some examples of the technique are given below.

14.8.1. Precise Specimen Positioning

One of the advantages of the nanoindentation technique is the ability to measure the mechanical properties of very small volumes of materials. Although the most

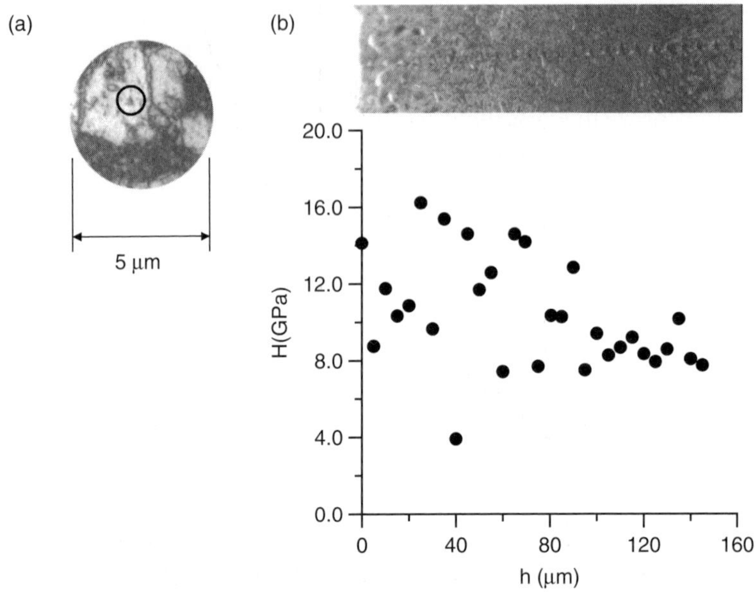

Figure 14.6: (a) Precise positioning of an indentation on two-phase metal alloy (b) indentations made in cross-section of a heat-treated steel and calculated hardness at each indentation.

popular application of the technique is the measurement of thin films, the scale of the contact in a typical indentation test, together with the availability of precise positioning systems, enables the indenter to be positioned onto very small structures (e.g. in Fig. 14.6(a)).

The precise specimen positioning capability of a nanoindentation test instrument also permits the properties of sectioned materials to be accurately measured. Fig. 14.6(b) shows a series of indentations spaced at 5 μm spacing through a cross-section of a heat-treated steel. There is a detectable increase in hardness in the surface layer compared to the material deeper within the bulk of the specimen.

14.8.2. Diamond-Like Carbon (DLC)

DLC is a popular choice for a low friction, hard-wearing coating for steel and silicon structures. Fig. 14.7 shows a typical load-displacement curve for a 2 μm DLC coating on silicon. The mechanical properties measured using the techniques described previously are also shown.

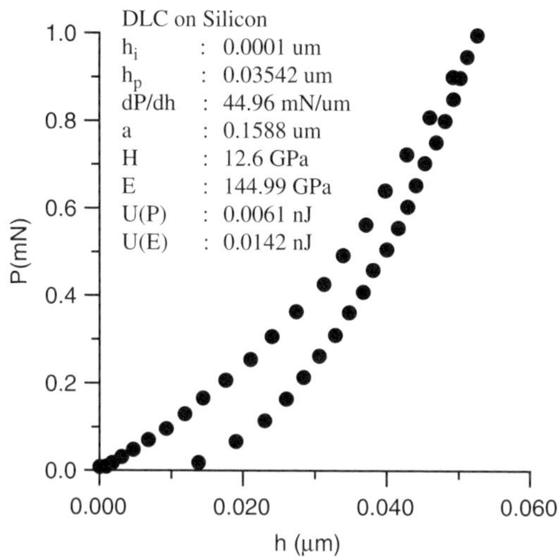

Figure 14.7: Load-displacement curve on a 2 μm DLC coating on silicon. Note that the maximum penetration depth in this case is <60 nm.

14.8.3. Silicon

One of the more interesting physical phenomena observed in nanoindentation testing is a change of phase of the test material in the indentation stress field. Such phase changes are readily observed in silicon. Fig. 14.8 shows the load-displacement curve for a nanoindentation test on a silicon specimen. The characteristic step in the unloading response indicates a phase change to a b.c.c. structure.

14.8.4. Surface Profiling

There are several methods of measuring mechanical properties as a function of indentation depth. The most straightforward method is to simply make a series of indentations at an ever-increasing load. This method can be sped up considerably by the simple expedient of partially unloading the indenter at each load step, thus enabling the contact stiffness to the measured throughout the range of total penetration. When undertaking tests of this kind, it is important to note that an initial increase in hardness is expected as the contact, even with a nominally sharp indenter, is initially elastic. For a hard coating on a soft substrate, such as that shown in Fig. 14.9, there is a plateau of hardness just beneath the surface (corresponding to a fully developed plastic zone) followed by a decrease in hardness as the indenter

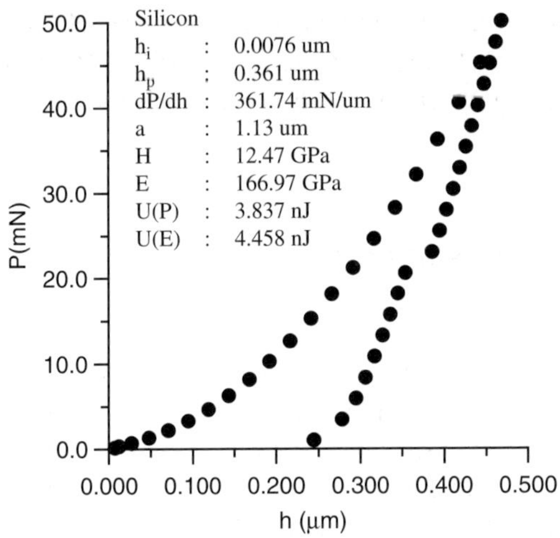

Figure 14.8: Load-displacement curve on silicon. Characteristic step in the unloading response indicates a phase change corresponding to a volume increase in the material.

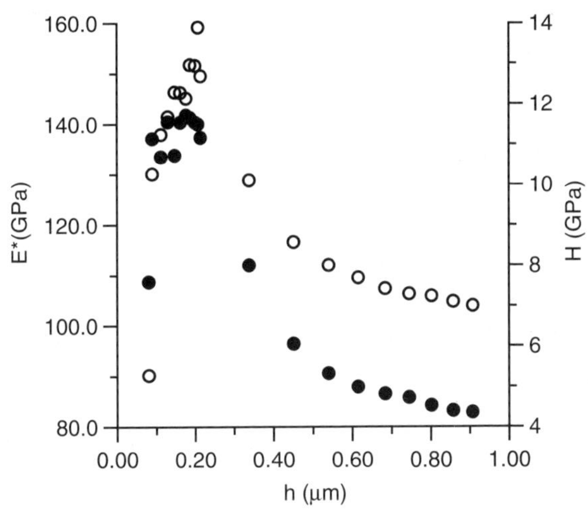

Figure 14.9: Modulus (●) and hardness (○) as a function of indenter penetration depth for a 2 μm tantalum oxide coating on a glass substrate.

begins to probe the substrate. Modulus measurements are not so affected by the sharpness of the indenter and reliable estimates of modulus are expected as long as the area function for the indenter being used has been accurately prepared. However, unlike the hardness measurement, the value measured for the modulus of the coating will always have some contribution of that of the substrate. It is common practice to ensure that only penetrations $<10\%$ of the film thickness are used to determine the coating modulus, but this practical rule of thumb has little formal justification.

14.9. The Future of Nanoindentation

Hardness testing using an indentation technique begins with the scratch test of Mohs in 1822. In the 20th century, hardness testing was more commonly applied in an engineering context through the use of hardness testers which used a Vickers pyramid or a Brinell sphere – both methods relying on an optical measurement of the size of the residual impression. As interest in smaller-scale phenomenon grew, micro-hardness measurements became very popular in the 1970s. Modern techniques are capable of measuring the mechanical properties of materials to within a few nanometers depth into the surface. As the scale of mechanisms becomes smaller, interest in mechanical properties on a nanometer scale, and the nature of surface forces and adhesion, will continue to increase. There remains considerable scope for basic research into contacts and deformations on the nanometer scale, particularly with respect to friction and tribologic applications.

14.10. Surface Acoustic Waves

Another technique finding increasing popularity is the determination of mechanical properties of surfaces and coatings by ultrasonic surface, or Rayleigh waves, often referred to as simply the surface acoustic wave (SAW) technique. The literature on this method extends back to the mid-1980s [25]. In this technique, a wave pulse is induced in the sample surface by a laser. The pulses are then detected at some distance away from the source and the dispersion of the wave is used to determine the elastic modulus of the medium (i.e. the surface – which might be a coating) and also its thickness.

14.10.1. Theoretical Background

The SAW method works particularly well for coatings because the amplitude of the wave decreases exponentially with distance into the material. The penetration depth of the wave decreases with increasing frequency. To make the measurement on a coating, the propagation velocity of the wave is measured as a function of frequency, and the elastic modulus of the surface is found from the inverse solution of the dispersion relation for the material.

In a homogenous material, the propagation velocity c is given by:

$$c = \frac{0.87 + 1.12\nu}{1 + \nu}\sqrt{\frac{E}{2\rho(1 + \nu)}} \quad (14.22)$$

where ρ is the density of the material. Note that the velocity does not depend on the frequency of the wave in this case. Since the penetration depth depends upon the frequency of the wave, in a coating material, the substrate has an effect on the wave that depends upon the frequency of the incident wave. The effect of the substrate means that the phase velocity of the wave depends upon the frequency instead of being a function of material properties of the medium (as in Eq. (14.22)). This dependence on frequency is called dispersion and, for a coated material, can be expressed in functional terms as:

$$c = \frac{\omega}{k} = c\left(\frac{d}{\lambda}, E', \nu', \rho', E, \nu, \rho\right) \quad (14.23)$$

where the primed terms refer to the properties of the substrate and d is the coating thickness. To extract material properties from the dispersion information, the phase velocity of the wave is measured as a function of frequency and from readings of two different pulses that have traversed two different path lengths x_1 and x_2.

$$c(f) = \frac{(x_2 - x_1)2\pi f}{\phi_2(f) - \phi_1(f)} \quad (14.24)$$

The phase spectra $\phi(x_1,f)$ and $\phi(x_2,f)$ are calculated from Fourier transforms of the amplitude of the voltage impulses measured at the receiver. The coating thickness, and the elastic modulus of both the film and the substrate are calculated using a non-linear regression of the dispersion equation. Reasonably fast convergence can be obtained if the dispersion curve is measured over a sufficiently large frequency range.

14.10.2. Apparatus

In a typical SAW apparatus (see Fig. 14.10), a pulsed laser induces short-time local heating of the specimen surface which produces an ultrasonic impulse which travels

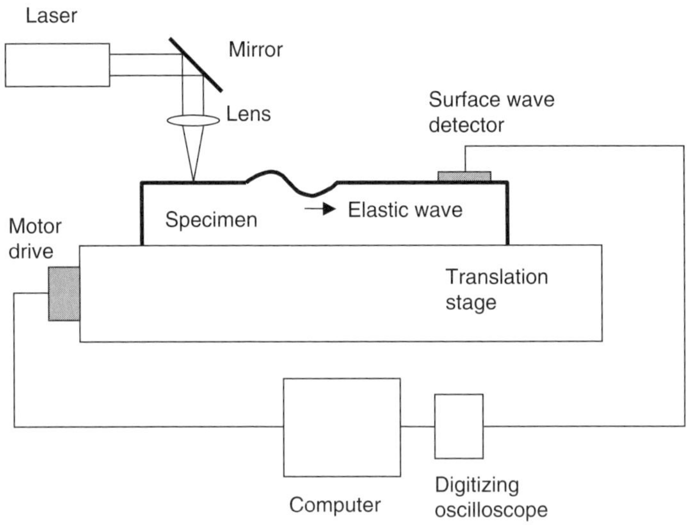

Figure 14.10: Typical surface wave measurement apparatus. Surface waves are generated by a pulsed laser beam. A piezo-electric transducer bonded to the specimen surface detects the surface waves and the amplitude is recorded as a function of time.

long the surface to the receiver. The duration of the pulse is typically 0.5 ns. The laser beam is focused by a cylindrical lens. Approximately 70% of the acoustic pulse energy is generated as surface waves. The piezo-electric transducer is bonded to the specimen surface a few millimeters from the laser focus line. During the measurement, the distance between the laser focus line and the receiver is precisely varied by a translation stage. Surface wave impulses as a function of time and hence distance are captured and analyzed to provide the readings of interest.

14.10.3. Applications

The SAW technique has been used to determine the elastic properties of thin films which include cemented carbide coated with nickel (Ni) or titanium carbide (TiC) [26], plasma sprayed ZrO_2 on Al [27] amorphous carbon films [28], TiN-coated steels [29], laser-hardened steels [30], and DLC films [31]. A particular advantage of the technique is that it is non-destructive and is suitable for films of thickness in the order of 100 nm. The results on a wide range of materials indicate that the elastic modulus measured using the SAW method is comparable to that measured by other means. The results, however, are sensitive to film quality (presence of microflaws, pores, lattice imperfections, etc.).

The sensitivity of the technique can be put to good use in the assessment of film quality. Ollendorf and Schneider [32] undertook a comprehensive comparison of film adhesion measurement techniques for TiN films on steel. They compared adhesion assessments from friction work, acoustic emission, scratch testing, four-point bending, cavitation test, impact, and elastic modulus measurements using the SAW method. These authors showed that the measured elastic modulus correlated with the quality of the film in terms of adhesion when measured using other conventional methods.

Despite an impressive decade of achievement, the SAW technique is often still described as "new" in the literature. The technique is one of only a very small number that offers considerable advantages for measurements of coatings and near-surface properties of materials.

References

[1] A.C. Fischer-Cripps, *Nanoindentation*, 2nd edition, Springer-Verlag, NY, 2004.
[2] H. Hertz, *J. Reine Angew. Math.*, 92 (1881) 156. Translated and reprinted in English in Hertz's Miscellaneous Papers, Macmillan & Co., London, 1896, Chapter 5.
[3] H. Hertz, *Verh. Ver. Beförderung Gewerbe Fleisses*, 61 (1882) 410. Translated and reprinted in English in Hertz's Miscellaneous Papers, Macmillan & Co, London, 1896, Chapter 6.
[4] I.N. Sneddon, *Proc. Cambridge Philos. Soc.*, 44 (1948) 492.
[5] M.C. Shaw, in *The Science of Hardness Testing and its Research Applications*, Eds. J.H. Westbrook and H. Conrad, American Society for Metals, Cleveland, OH, 1973, pp. 1–15.
[6] M.V. Swain and J.T. Hagan, *J. Phys. D. Appl. Phys.*, 9 (1976) 2201.
[7] M.F. Doerner and W.D. Nix, *J. Mater. Res.*, 1 (4) (1986) 601.
[8] W.C. Oliver and G.M. Pharr, *J. Mater. Res.*, 7 (4) (1992) 1564.
[9] G.M. Pharr, W.C. Oliver and F.R. Brotzen, *J. Mater. Res.*, 7 (3) (1992) 613.
[10] A. Bolshakov and G.M. Pharr, *J. Mater. Res.*, 13 (4) (1998) 1049.
[11] E.R. Weppelmann, J.S. Field and M.V. Swain, *J. Mater. Res.*, 8 (4) (1993) 830.
[12] V. Domnich, Y. Gogotsi and M. Trenary, *Mater. Res. Soc. Symp.*, 649 (2001) Q8.9.1.
[13] C.A. Schuh, T.G. Nieh and Y. Kawamura, *J. Mater. Res.*, 17 (7) (2002) 1651.
[14] N.Q. Chinh, Gy Horváth, Zs Kovács and J. Lendvai, *Mater. Sci. Eng. A*, 324 (2002) 219.
[15] J.H. Underwood, *Exp. Mech.*, 30 (1973) 373.
[16] S.G. Roberts, C.W. Lawrence, Y. Bisrat and P.D. Warren, *J. Am. Ceram. Soc.*, 82 (7) (1999) 1809.
[17] M.M. Chaudhri and M.A. Phillips, *Philos. Mag. A*, 62 (1990) 1.
[18] S. Chandrasekar and M.M. Chaudhri, *Philos. Mag. A*, 67 (6) (1993) 1187.
[19] T.Y. Tsui, W.C. Oliver and G.M. Pharr, *J. Mater. Res.*, 11 (3) (1996) 752.
[20] A. Bolshakov, W.C. Oliver and G.M. Pharr, *J. Mater. Res.*, 11 (3) (1996) 760.

[21] Y.-H.Lee and D. Kwong, *J. Mater. Res.*, 17 (4) (2002) 901.
[22] A. Taljat and G.M. Pharr, *Mater. Res. Soc. Symp. Proc.*, 594 (2000) 519.
[23] G.G. Stoney, *Proc. Roy. Soc. A*, 9 (1909) 172.
[24] UMIS® CSIRO Division of Industrial Physics, Lindfield, Australia.
[25] D. Schneider and B. Schultrich, *Exp. Technik der Physik*, 33 (1985) 251.
[26] D. Schneider, T. Schwarz and B. Schultrich, *Thin Solid Films*, 219 (1992) 91.
[27] D. Schneider and T. Schwarz, *Thin Solid Films*, 224 (1993) 177.
[28] B. Schultrich, H.-J. Scheibe, G. Grandremy and D. Schneider, *Phys. Stat. Sol. A*, 145 (1994) 385.
[29] D. Schneider, H. Ollendorf and T. Schwarz, *Appl. Phys. A*, 61 (1995) 277.
[30] D. Schneider, B. Brenner and T. Schwarz, *J. Nondestruct. Eval.*, 14 (1) (1995) 21.
[31] D. Schneider, C.F. Meyer, H. Mai, B. Schoneich, H. Ziegele, H.J. Scheibe and Y. Lifshitz, *Diam. Relat. Mater.*, 7 (1998) 97.
[32] H. Ollendorf and D. Schneider, *Surf. Coat. Tech.*, 113 (1999) 86.

Chapter 15

Determination and Generation Mechanisms of Residual Stresses in Thin Films Produced by Physical Vapor Deposition Techniques

Yves Pauleau

15.1. Introduction

A wide variety of conductive, semiconductor and insulating thin films can be prepared by physical vapor deposition (PVD) techniques including direct vacuum evaporation, sputtering and ion plating. These films are used as active or passive components in various devices for advanced technologies such as microelectronics, optics, micromechanics, etc. Physically vapor-deposited (PVD) thin films must be tightly adherent to substrate surfaces to ensure performance and reliability of devices. Usually, PVD thin films prepared under appropriate conditions are firmly bonded to massive substrates (i.e., much thicker than the films). As a result, any change in length along the film plane, which is not matched, exactly by an equal change in length of the substrate will generate a stress in the film. Excessive stresses in thin films may cause defect formation and delamination at the film-substrate interface, mechanical damage in the film (film fracture), adhesion failures and defect formation in the substrate [1]. On a less catastrophic scale, large stresses can affect the optical, electrical and magnetic properties of PVD films. The elimination and release of stresses are also the cause of undesirable formation of hillocks, whiskers, holes and defects, which have a detrimental effect on the properties, and reliability of films. The successful implementation of PVD processes and performance of thin films for various applications are partially governed by the formation and development of residual stresses. Therefore, there is a considerable need to understand the origin of stresses in PVD films and control their magnitude, which depends on deposition process parameters.

Materials Surface Processing by Directed Energy Techniques
Y. Pauleau (Editor)

15.2. Overview of Residual Stresses

A stress can exist in the bulk of a polycrystalline material which is not subjected to external forces. All residual stress systems developed in a material have zero resultant force, the material being in equilibrium. This balance is achieved over various distances, which can be used to distinguish three kinds of residual stress. The first kind of residual stress, referred to as *macroscopic stress* or stress of *order I*, is long range in nature and constant over a large number of grains. This stress resulting from thermal treatments, mechanical operations, chemical processes and combinations of each [2] can be measured by mechanical means. The stress of *order II*, often termed *structural microstress*, develops over sizes of one grain and can arise from a difference in mechanical properties of different phases in the material. The effects of stresses of orders I and II are additive. As a result, the separation of the role and measurement of the first two kinds of residual stress is often difficult or even impossible. The third kind of residual stress or stress of *order III*, often called *internal stress*, is caused by the interaction of dislocations inside one grain. This type of stress may develop around second phase particles such as in dispersion-hardened materials due to plastic incompatibilities between hard particles and elastic–plastic matrix.

Usually, thin films prepared by PVD techniques at room or moderate temperatures exhibit amorphous or nanocrystalline structures. Therefore, the major stress, which can be investigated in these disordered materials, is the macroscopic residual stress. The magnitude of residual stresses, σ, in PVD thin films results from three different contributions and can be expressed as follows:

$$\sigma = \sigma_{\mathrm{th}} + \sigma_{\mathrm{i}} + \sigma_{\mathrm{e}} \tag{15.1}$$

The thermal stress, σ_{th}, arises from the different thermal expansion coefficients of the film and substrate combined with the difference between the deposition or substrate temperature, T_{s}, and the temperature of samples, T_{r}, during determination of the magnitude of stresses, which is generally performed at room temperature. The intrinsic stress, σ_{i}, is introduced in PVD films during deposition and is not directly related to thermal mismatch between the deposited material and substrate. This stress contribution may originate from incomplete structural-ordering process occurring while the film grows. The magnitude of the intrinsic stress can be related to the microstructure or morphology of films, which, in turn, depends on various process parameters. The structural characteristics of PVD films are directly dependent on two major process parameters, that is, substrate temperature and kinetic energy of particles striking the film surface during deposition; these incident particles including atoms, molecules, ions, radicals can be either condensed on the film surface and incorporated in the film lattice or simply reflected at the film surface and backscattered to the gas phase. As a result, considerations on intrinsic stresses in thin films produced via non-energetic particles with an average kinetic energy below 0.5 eV (energy of species produced by thermal evaporation) can be separated conveniently from those developed for films deposited by condensation of energetic species and/or under energetic particle bombardment with kinetic energies ranging from a few eV to a few hundreds of eV [3]; these PVD processes in which are involved the simultaneous deposition and energetic particle bombardment of films are currently named ion-assisted deposition (IAD) or ion beam-assisted deposition (IBAD) processes [4]. The third component in residual stresses, often termed extrinsic stress, σ_e, is induced by external factors and results from interactions between the deposited material and environment which are subsequent to deposition of films. For instance, extrinsic stress may develop in porous films, which are prone to adsorb water vapor molecules during exposure to room air. In general, the variation of residual stress resulting from the development of the extrinsic stress induced by adsorption phenomena occurs instantaneously when the sample is exposed to room air. This stress variation is also reversible (i.e., the magnitude of residual stress returns readily to its initial value as the film is moved from air to vacuum).

This chapter deals with the effects of PVD process parameters on the microstructure of films and magnitude of residual stresses developed in continuous films, that is, much thicker than films resulting from the coalescence of islands on flat substrate surfaces [5]. The average value of residual stresses in continuous films is observed to be essentially independent of the film thickness. The continuous films under consideration are deposited onto non-epitaxial substrates, which are likely to induce an island mode of growth with no crystallographic relationship between the orientation of the islands, and the substrate. In addition, these PVD films are grown on substrates at reduced temperatures (or ambient temperature) at which thermally activated annealing processes such as grain growth, defect annihilation and mass transport are excluded. The goal of this chapter is to present briefly the major technique used for determination of residual stresses in PVD thin films deposited on thick substrates and to review the major models proposed in the literature to explain the origin of residual stresses in PVD films, in particular the origin of compressive intrinsic stresses which develop in thin films produced from energetic particles. Stress data for various PVD films such as amorphous-carbon films (a-C) deposited by conventional and unbalanced magnetron sputtering, silicon dioxide films prepared by direct thermal evaporation, arc presented and discussed to illustrate the applicability, validity and limits of the models described previously.

15.3. Determination of the Magnitude of Residual Stresses in Thin Films

Various methods developed for measuring the magnitude of residual stresses in thin films are described in detail in the literature [6–9]. With films tightly adherent to thin substrates such as Si wafers, a biaxial stress in the films of a sufficiently high intensity causes the substrates to bend elastically. A tensile stress (with a positive conventional sign) bends the film-substrate structure so that the film surface is concave whereas a compressive stress (with a negative conventional sign) bends the sample so that the film surface is convex. The most common methods for determining residual stresses in thin films are based on measurements of the deformation of film-substrate structures. These methods are well designed for the determination of residual stresses in amorphous and nanocrystalline thin films for which the adherence on substrate surfaces depends directly on the magnitude of the macroscopic residual stresses. The substrates can be in form of either thin-cantilevered beams or disks. Cantilevered beam-type substrates are generally employed for *in situ* measurements of stresses in the films during deposition. The magnitude of residual stresses is deduced from measurements of the radius of curvature of the beam, the deflection of the free end of the beam or the displacement of the center of a circular disk-type substrate (Fig. 15.1). The average value of macroscopic residual stresses can be

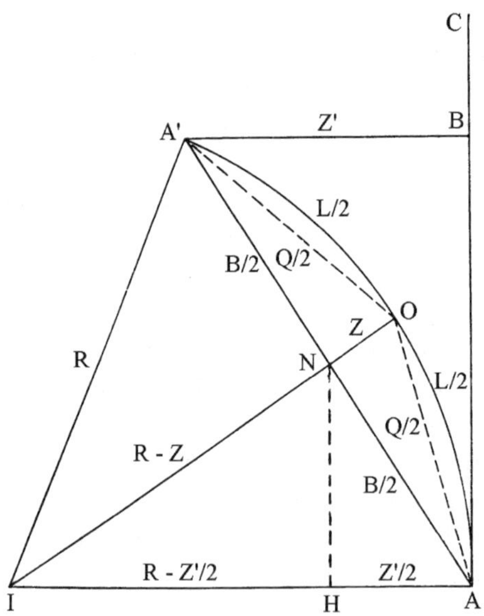

Figure 15.1: Schematic representing the position of a curved strip with respect to the position of the flat strip.

calculated from the characteristic length value (radius of curvature, deflection or displacement) using appropriate equations.

The equation for stress calculation can be established on the basis of the fundamentals of the theory of elasticity and elementary considerations of beam theory [10]. The residual stresses in thin films deposited on an initially flat substrate will bend elastically the film-substrate structure. The magnitude of residual stresses, σ, in thin films can be calculated from the equation derived by Stoney [11]:

$$\sigma = \frac{1}{6}\left(\frac{E_s}{1-\nu_s}\right)\left(\frac{t_s^2}{t_f R}\right) \tag{15.2}$$

where E_s and ν_s are Young's modulus and Poisson's ratio of the substrate, respectively; t_s is the substrate thickness, t_f the film thickness and R the radius of curvature of the film-substrate structure. When the substrate is not absolutely flat before deposition of the film; that is, the initial radius of curvature of the substrate is R_1, the stress is given by:

$$\sigma = \frac{1}{6}\left(\frac{E_s}{1-\nu_s}\right)\left(\frac{t_s^2}{t_f}\right)\left(\frac{1}{R_2}-\frac{1}{R_1}\right) \tag{15.3}$$

where R_2 is the radius of curvature of the film-substrate structure.

The derivation of Eq. (15.2) or Stoney equation is based on various assumptions, in particular, the thickness of the film is considered to be infinitesimal compared to the thickness of the substrate [12]. However, Eq. (15.2) or Eq. (15.3) can be used without much error for film thickness, which does not amount to more than a few percentage of the thickness of the substrate. Usually, the uncertainty on stress values calculated from Eq. (15.2) is less than the experimental error, which is about 5–10%. The error resulting from the approximate Eq. (15.2) or Eq. (15.3) begins to exceed the experimental error for films with thicknesses >5% of the substrate thickness. In addition, the stress value calculated from previous equations is valid for samples in which the moduli of the substrate and film are the same. A negligible effect of the difference in moduli is observed when the thickness of the film amounts to only a few percentage of the thickness of the substrate. Thus, major conditions must be fulfilled before accurate stress calculations from Eq. (15.2) or Eq. (15.3). Accurate stress determinations require knowledge of Young's modulus, E_s, and Poisson's ratio, ν_s, of substrates, in particular, for specific orientations within the crystallographic plane defining the surface of substrates. For stress determinations performed within these various limitations, the accuracy on the stress value would be essentially related to the experimental error on the measurements of the radii of curvature, R, R_1 and R_2.

The radius of curvature of the film-substrate structure can be determined from measurement of either the camber of the curved strip or the deflection of the end of the strip. Initially, the flat strip or substrate is located at a position represented by the straight line ABC in Fig. 15.1 whereas the position of the strip covered with the stressed film corresponds to the arc AOA′. The radius of curvature, R, is determined by measuring the sagitta Z of the arc AOA′. The curved strip is set on a flat surface with the convex side up; the camber of the arc can be measured with a microscope or other means. The radius of curvature of the film-substrate structure is related to the sagitta Z:

$$\frac{1}{R} = \frac{8\,Z}{Q^2} \approx \frac{8\,Z}{L^2} \tag{15.4}$$

The length, Q, of the broken line AOA′ is approximately equal to the length, L, of the curved strip AOA′. The replacement of Q by the arc length, L, in Eq. (15.4) leads to an error of less than about 2% provided that the sagitta, Z, of the strip-type substrate is not more than 10% of the length of the substrate. Therefore, Eq. (15.2) for stress calculations can be written as:

$$\sigma = \frac{4}{3}\left(\frac{E_\mathrm{s}}{1 - \nu_\mathrm{s}}\right)\left(\frac{t_\mathrm{s}^2}{t_\mathrm{f} L^2}\right) Z \tag{15.5}$$

The deflection, Z', of the end A' of the strip can be measured when the end A is firmly maintained in a fixed position (Fig. 15.1). The radius of curvature of the film-substrate structure, R, can be expressed as a function of the deflection, Z':

$$\frac{1}{R} = \frac{2\,Z'}{B^2} \approx \frac{2\,Z'}{L^2} \tag{15.6}$$

The length, L, of the arc AOA′ can be substituted for the chord, B, without much error and the expression of residual stresses, σ, can be derived from Eqs. (15.2) and (15.6):

$$\sigma = \frac{1}{3}\left(\frac{E_\mathrm{s}}{1 - \nu_\mathrm{s}}\right)\left(\frac{t_\mathrm{s}^2}{t_\mathrm{f} L^2}\right) Z' \tag{15.7}$$

From Eqs. (15.4) and (15.6), it can be noted that the deflection Z' is four times higher than the sagitta Z, that is, $Z' = 4\,Z$. The error on the value of R calculated from Eq. (15.6) is $<2\%$ of the length of the strip type substrate. Finally, for equal values of the radius of curvature of the film-substrate structures, the error involved in stress calculation from Eq. (15.7) is about twice as large as that coming from Eq. (15.5).

15.4. Origin of Residual Stresses in Thin Films

According to Eq. (15.1), three major contributions in residual stresses in PVD thin films bonded to rigid substrates can be schematically distinguished. There is no single model accounting for the origin of all the residual stresses observed for a wide variety of films. Major models invoked to explain the origin of tensile and compressive intrinsic stresses have been already reviewed extensively [3,6,7,12–17]. Therefore, the intend of the present treatment is to concentrate on models often used to explain the origin of macroscopic residual stresses in amorphous and polycrystalline PVD thin films that have found many proponents. Then, recent investigations and results that have substantiated the prediction of these models are described in more detail.

15.4.1. Thermal Stresses

The biaxial strain, ε, in films bonded to substrates at a temperature, T_r, higher or lower than the substrate or deposition temperature, T_s, is given by:

$$\varepsilon = (\alpha_s - \alpha_f)(T_r - T_s) \tag{15.8}$$

where α_s and α_f are the thermal expansion coefficients of the substrate and the film, respectively. Without any plastic deformation in the film-substrate structure during temperature change, the elastic strain can be directly related to the thermal stress in the film through Hooke's law:

$$\sigma_{th} = \left(\frac{E_f}{1 - \nu_f}\right)(\alpha_s - \alpha_f)(T_r - T_s) \tag{15.9}$$

where E_f and ν_f are Young's modulus and Poisson's ratio of the film, respectively. In general, the deposition temperature is higher than room temperature used for residual stress determination, that is $(T_r - T_s) < 0$; as a result, the thermal stress developed during temperature change is tensile ($\sigma_{th} > 0$) or compressive ($\sigma_{th} < 0$) depending on the relative values of α_s and α_f. For a film with $\alpha_f > \alpha_s$, during cooling down the film is trying to contract much more than the substrate, and the thermal stress is positive according to Eq. (15.9) (i.e., the thermal stress is tensile).

The contribution of the extrinsic stress, σ_e, is negligible when the interactions of the film-substrate structure and environment can be avoided. In this case very frequently observed, the residual stress value is the result of the intrinsic stress built up during deposition of the film and the thermal stress which may develop during temperature change occurring between deposition of the film and residual stress determination. The thermal stress is nil and the residual stress, σ, is equal to the intrinsic stress, σ_i, in films deposited at room temperature ($T_s = T_r$). In general, the magnitude of the intrinsic stress in films deposited at a given temperature is independent of the nature of the substrate. For films deposited at moderate temperatures ($T_s \neq T_r$) on substrates with different thermal expansion coefficients; for example, Ge and Si substrates, the difference in residual stresses, $\Delta\sigma$, is equal to the difference in thermal stresses, $\Delta\sigma_{th}$, which reflects the disparity of thermal expansion coefficients:

$$\Delta\sigma = (\sigma_{Ge} - \sigma_{Si}) = \Delta\sigma_{th} = \left(\frac{E_f}{1 - \nu_f}\right)(\alpha_{Ge} - \alpha_{Si})(T_r - T_s) \quad (15.10)$$

As a result, the biaxial modulus of films can be determined from the thermal stress difference, $\Delta\sigma_{th}$.

The determination of thermal and intrinsic stress contributions in the residual stress value is often required for elucidation of growth mechanisms of films and origin of stresses; however, it is not easy to achieve a clear separation of these respective contributions. The main difficulty encountered in the calculation of the thermal stress from Eq. (15.9) arises from the determination of the thermal expansion coefficient of films; the value of α_f can be very different from the value for the bulk material given in the literature. The more straightforward method for the determination of the thermal expansion coefficient (and biaxial modulus) of films is to investigate the annealing temperature effect on the residual stresses in films deposited at a given temperature, T_s, on two types of substrates, for example, Ge and Si substrates, with different thermal expansion coefficients. In general, the annealing cycles are performed at temperatures ranging from room temperature to a maximum temperature, T_a, higher than the deposition temperature. Usually, during the first annealing cycle, the variation of residual stresses is found to be irreversible because of the release of the intrinsic stress caused by the annealing treatment on the microstructure of the deposited material. For subsequent annealing cycles, the residual stresses vary linearly and reversibly with annealing temperature, T_a:

$$\sigma = (\sigma_i - \sigma_{th}) = \sigma_i + \left(\frac{E_f}{1 - \nu_f}\right)(\alpha_s - \alpha_f)(T_a - T_s) \qquad (15.11)$$

where the intrinsic stress, α_i, is independent of the annealing temperature and nature of the substrate. The slope of the straight line corresponding to the variation of the residual stresses for annealing cycles of stabilized samples can be derived from Eq. (15.11):

$$\frac{d\sigma}{dT} = \left(\frac{E_f}{1 - \nu_f}\right)(\alpha_s - \alpha_f) \qquad (15.12)$$

The thermal expansion coefficient, α_f, of the film can be deduced from Eq. (15.12) using the biaxial modulus value determined previously from Eq. (15.10). However, for films deposited on two different substrates, for example, Ge and Si substrates, Eq. (15.12) can be used to determine both the biaxial modulus and the thermal expansion coefficient of films.

Finally, it is worthy to note that the experimental approach adopted to investigate the thermal stress in thin films allows to determine mechanical constants of stabilized films in a given temperature range. In other words, the values of the biaxial modulus and thermal expansion coefficient of films deduced from the variation of residual stresses as a function of the annealing temperature are valid for stabilized films in which the intrinsic stress was totally or partially released and was independent of the annealing temperature. The biaxial modulus and thermal expansion coefficient of as-deposited films (just after quenching of samples) cannot be deduced from these experimental data and may be different from those valid for annealed films.

15.4.2. Intrinsic Stresses

The magnitude of intrinsic stresses in PVD films is related to the microstructure of films, that is, morphology, texture, grain size depicted by the structure-zone diagrams [18–20]. The microstructure and thereby the intrinsic stress value depend on two major process parameters, namely the substrate temperature and pressure in the deposition chamber which affect directly or indirectly the mobility of adatoms or species adsorbed on the surface of growing films. An increase in substrate temperature leads to enhanced surface diffusion of adatoms which can migrate to energetically favorable sites before being covered with additional condensing atoms and films with densely packed microstructures can be obtained. Moreover, a decrease in gas pressure in the deposition chamber results in larger mean free paths of evaporated or sputtered atoms and higher kinetic energy of atoms condensed on the film surface. With sufficiently energetic interactions between the film surface and species coming from the gas phase, adatoms can escape from the potential energy well corresponding to adsorption sites and diffuse on the surface from one site to another until a more favorable energetic site is reached. The resulting atomic rearrangements lead to the growth of films with mass densities close to the bulk density, in particular for films produced by IBAD. Particle–film interactions induce various phenomena such as collision cascades in the film, creation of vacancy-interstitial pairs and thermal spikes [21]. The densification of films produced by IBAD may be attributed to the distortion of the growing surface layer by energetic particles; these phenomena can be analyzed on the basis of the ion-peening model [22]. Energetic particles striking and penetrating the film surface randomly displace atoms from their equilibrium positions through a series of primary and recoil collisions, which lead to a volumetric distortion. At low normalized growth temperatures, that is, for absolute substrate temperature to absolute melting temperature of the deposited material ratios, T_s/T_m, below 0.25, mass transport phenomena and defect mobility are low enough to freeze the volumetric distortion in place. The model based on the knock-on linear cascade theory of transmission sputtering developed by Sigmund [23] predicts the dependence of the volumetric film distortion on the flux and kinetic energy of incident particles [24]. The volumetric distortion is assumed to be proportional to the fractional number of atoms displaced from equilibrium positions in the growing film. The expression of the number of displaced atoms per unit volume, n, calculated by Windischmann [24] is given by:

$$n = 4.79\ \delta\ \phi_p \sqrt{E_p} \tag{15.13}$$

with

$$\delta = \left[\frac{M_t^{1/2} (Z_p\ Z_t)^{1/2}}{U_0 (M_p + M_t)^{1/2} (Z_p^{2/3} + Z_t^{2/3})^{3/4}} \right] \tag{15.14}$$

The factor δ depends on the cohesive energy, U_0, of atoms in the film, masses (M_p and M_t) and atomic numbers (Z_p and Z_t) of incident particles (projectile) and film atoms (target), respectively; ϕ_p and E_p are the flux and kinetic energy of incident particles, respectively. According to Eq. (15.13), the energetic particle bombardment effect on the microstructure of films is rather related to the momentum of incident particles (dependence of n on $(E_p)^{1/2}$).

Various theoretical attempts were conducted to elucidate the densification mechanism of films produced by IBAD. Monte Carlo calculations were developed to investigate the effect of thermal spikes on film density [25]; for these ion-induced localized annealing events, the relevant parameter is the kinetic energy of incident particles. The major conclusion of this work was that the thermal spikes do not significantly improve the mass density of films under typical IBAD conditions. In fact, the energetic particle bombardment of growing films results in surface depletion due to sputtering and inwardly recoil atoms, and also leads to an improved mass density deeper in the bulk because of particle and recoil implantation. A collision cascade model showed that the film density improvement could be caused by the recoil implantation of near-surface atoms which is a momentum-dependent process [26].

The momentum transfer was experimentally proved to be the dominant factor involved in the densification mechanism by Targove and Macleod [27]. To separate the kinetic energy effect from the momentum effect of incident energetic particles, lanthanum fluoride films deposited by thermal evaporation were bombarded with three different inert gas ions (Ne, Ar and Kr) of 500 eV and the refractive index of the films was used as a measure of their packing density. According to the simple thermal spike model, the number of atoms rearranged in the films should be constant for all three experiments and the rate of atomic rearrangement should be proportional to the ion current density (J), on the film surface. In fact, the data show that the refractive indices and, therefore, the packing density of films are not proportional to the ion current density (J), and depend on the mass of ions striking the film surface.

The theoretical model proposed by Müller [26] shows that ion implantation and recoil implantation of surface atoms lead to an improvement of the film density slightly below the surface of a growing film. The densification of films depends on the ability of condensing atoms to refill surface vacancies, which are created by sputtering and driven-in atoms. Therefore, the mechanism responsible for the film densification may involve a momentum transfer from incident energetic particles to atoms incorporated in the film. The implantation depth is about proportional to the momentum transferred to the film atoms. With a negligible ion energy loss before an implantation event, the maximum momentum transfer per projectile, P_{max}, is given by [27]:

$$P_{max} = \sqrt{2\, M_p\, \gamma\, E_p} \tag{15.15}$$

where M_p and E_p are the mass and kinetic energy of ions (projectiles), respectively; γ is the energy transfer coefficient which determines the maximum energy transferred from the incident ion to the film atom of mass M_t. The expression of γ established from equations resulting from the kinetic energy and momentum conservation laws for an elastic head-on collision between a projectile atom and target atom can be written as follows:

$$\gamma = \frac{4\, M_p\, M_t}{(M_p + M_t)^2} \tag{15.16}$$

A mean value of γ was adopted by Targove and Macleod since two elements, La and F, are found in LaF_3 films [27]. The total momentum transfer rate is approximated to:

$$P_{total} = P_{max}\, J = J\, \sqrt{2\, M_p\, \gamma\, E_p} \tag{15.17}$$

or for ions of charge q used as projectiles:

$$P_{total} = q\, \phi_p\, \sqrt{2\, M_p\, \gamma\, E_p} \tag{15.18}$$

where P_{tot} is expressed in (μA cm^{-2}) (uma)$^{1/2}$ (eV)$^{1/2}$. An excellent linear correlation was found between the refractive indices of films measured at a wavelength of 350 nm and the total momentum transfer [27]. These data demonstrate that the packing density of films produced by IBAD is related to the momentum transfer rather than to energy transfer.

However, with a given flux, ϕ_p, and kinetic energy, E_p, of incident ions, the efficiency of the IBAD process for film densification may be dependent on the deposition rate of films. For a given flux of incident energetic particles, the bombardment effect on the densification of films is expected to decrease with the increasing deposition rate. As a result, the normalized momentum, P_n, appears to be a more relevant factor governing the densification of films produced by IBAD; this normalized momentum, P_n, can be approximated to:

$$P_n = \frac{\phi_p}{\phi_t}\sqrt{2\, M_p\, \gamma\, E_p} \tag{15.19}$$

where ϕ_p and ϕ_c are the fluxes of incident energetic particles and condensing atoms on the film surface, respectively.

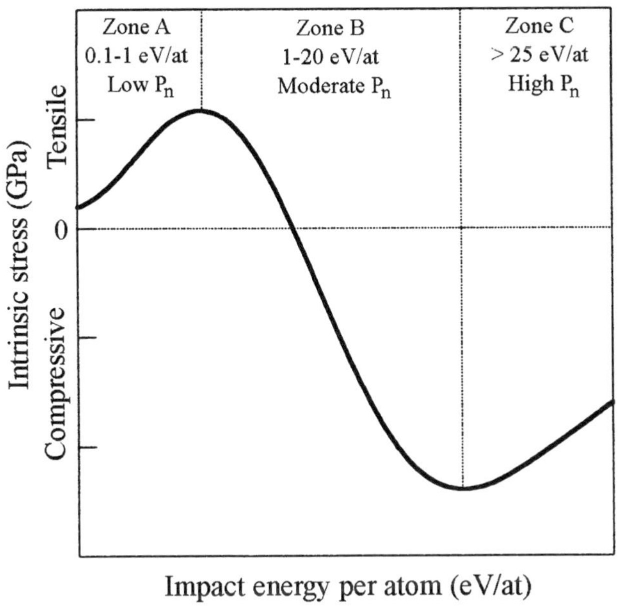

Figure 15.2: Idealized intrinsic stress vs. normalized momentum, P_n. According to Eq. (15.19), P_n is related to the kinetic energy, E_p, of energetic particles impinging on the film surface per condensing atom.

The variation of the intrinsic stress with the normalized momentum can be depicted by an idealized curve (Fig. 15.2) corresponding to the state of continuous, amorphous or polycrystalline films deposited at low temperatures ($T_s/T_m < 0.2$) for which thermally driven diffusion-induced strain relief is negligible. Three distinct zones or regimes appear in this diagram depending on the normalized momentum given in a qualitative manner or impact energy which is the total kinetic energy of species colliding with the film surface per atom condensed and incorporated in the film. The calculation of the normalized momentum is not always straightforward or possible since the determination of the kinetic energy and flux of incident particles can be problematic. Typical impact energy values can be more easily indicated in the intrinsic stress diagram. At a low normalized momentum, tensile intrinsic stresses are negligible since the microstructure of films (zone 1 of structure-zone diagrams proposed by Grovenor *et al.* [19] and Thornton [20]) is not sufficiently dense to support stress. As the normalized momentum increases in zone A (Fig. 15.2), the microstructure of films changes progressively from zone 1–zone T of the structure-zone diagrams, the film density increases and the tensile intrinsic stress reaches a maximum value. For moderate normalized momenta (zone B in Fig. 15.2), the tensile intrinsic stress decreases with increasing momentum as compensating compressive stress caused by the atomic peening mechanism becomes operative [28]. When the normalized momentum increases in zone B, a transition from tensile to compressive stress is observed as a result of the densification of films and a maximum value of compressive stresses can be reached if the films remain tightly adherent to substrates. At high-normalized momentum values (zone C in Fig. 15.2), intrinsic stresses decrease as the yield strength of the deposited material is attained. There is no model able to explain all the observations. Various models can be invoked for the origin of intrinsic stress; each of them is valid in a given zone of the stress-momentum diagram. Since tensile stress develops in films produced from non-energetic particles and compressive stress appears in films obtained from energetic particles, it is convenient to separate the presentation of models on the basis of the kinetic energy of incident species.

15.4.2.1. Intrinsic Stresses in Films Produced from Non-Energetic Particles

The model most often used to explain the origin of tensile intrinsic stresses in films produced from non-energetic particles (zone A of the idealized diagram in Fig. 15.2) is the grain boundary relaxation (GBR) model proposed by Hoffman *et al.* [7,13,29–33]. This model was developed for polycrystalline films; however, it is adaptable to amorphous films exhibiting a columnar or fibrous microstructure very often observed for PVD films. Voids and intercolumnar spaces are assumed to play a role similar to grain boundaries in polycrystalline films.

This GBR model is based on various physical arguments and hypotheses. As the film grows from isolated atomic clusters, interatomic attractive forces acting across grain boundaries, gaps between grains or intercolumnar spaces cause an elastic deformation (or relaxation) of the grain walls. These attractive forces are counterbalanced by the intragrain or intracolumn tensile forces imposed by the constraint caused by the adhesion of the film to the substrate surface. The adhesion forces exceed the tensile forces developed between adjacent grains or columns. The intragrain tensile forces are prone to reduce the volume of the film as the film is detached from the substrate. The induced stress is tensile since the film is trying to contract. The intragrain strain energy can be related to the difference in the surface energy of adjacent crystallites and the energy of the resultant grain boundaries [33]. A fraction of the energy produces a constrained relaxation of the lattice in the grain. The elastic strain is given by:

$$\varepsilon = \frac{x - a}{a} = \frac{\Delta}{L_g} \tag{15.20}$$

where a is the unstrained lattice constant and $(x - a)$ is the variation of the lattice constant; Δ and L_g are the GBR distance and the final grain size, respectively. The average value of Δ can be calculated using the method of the interaction potential between the two atoms [34]:

$$\Delta \approx \frac{\bar{a}\, r_i - \frac{r_i^2}{2}}{2\,\bar{a} - r_i} \qquad (15.21)$$

where $\bar{a}$ is the average bulk lattice constant and r_i the ionic radius; the distance of closest approach is the sum of the ionic radii. The elastic deformation is responsible for the macroscopically observed tensile intrinsic stress, σ_i, according to Hooke's law:

$$\sigma_i \approx \left(\frac{E_f}{1 - \nu_f}\right)\frac{\Delta}{L_g} \qquad (15.22)$$

The GBR distance, Δ, can be computed from the grain separation potential for each grain diameter, L_g. The intrinsic stress in nickel films was obtained by inserting the values of Δ and L_g in Eq. (15.22); the calculated stress values were found to reproduce the form of the measured stress vs. substrate temperature curve satisfactorily and numerical agreement was within roughly 30% [31]. This approach can be simplified if the GBR distance is approximated to the ionic radius, r_i, that is, the tensile intrinsic stress is given by:

$$\sigma_i \approx \left(\frac{E_f}{1 - \nu_f}\right)\frac{r_i}{L_g} \qquad (15.23)$$

The inverse grain size dependence of the intrinsic stress predicted by the GBR model has been observed for many film-substrate structures. This simplified approach leads to a reasonable good agreement between calculated and experimental values of tensile intrinsic stresses in chromium [33,35], nickel [31], diamond [36] and magnesium fluoride [37] films.

The GBR model was adapted by Itoh *et al.* [38] to the calculation of intrinsic stresses developed in tungsten films sputter-deposited on Si substrates at room temperature under an argon pressure in the range 0.25–2.5 Pa. A Morse potential was chosen to calculate the force acting across the grain boundaries. The average distance between the grain boundary faces was calculated from the mass density of the film consisting of rectangular grains. This average distance is usually higher than the interatomic distance. Therefore, interatomic attractive forces exist between two adjacent grain faces. The theoretical stress values are in good agreement with stress values determined experimentally for films deposited under argon pressures higher than 5 m Torr or 0.66 Pa. In this pressure range, a thermalization of *W* atoms ejected from the target may occur between the target and the substrate by collisions with argon atoms. These films were probably grown from non-energetic *W* atoms. Below 5 m Torr, compressive intrinsic stresses developed in *W* films since under low argon pressures energetic *W* atoms may condense on the film surface.

The origin of tensile intrinsic stresses in metal films and the existence of a long-range attractive component in the interatomic potential were re-examined recently by Machlin [3]. Various considerations and physical arguments based on the relation provided by Rose *et al.* [39] for the energy between metal surfaces are presented to determine the maximum separation of metal surfaces across which an adequate tensile force can be generated to account for the observed intrinsic stresses in thin films. A tensile intrinsic stress can be induced either by a crack narrower than 1.7 times the spacing between close-packed planes or by a region that has less density than the equilibrium density, but in which the atomic distance is less than 1.4 times the equilibrium spacing. In other words, it has been demonstrated that the atomic interactions across voids (grain boundaries or intercolumnar spaces) thicker than about 1.7 Å are nil or not sufficient to account for the magnitude of the observed tensile intrinsic stresses. For grains or columns separated by <1.7 Å, the attractive interaction leading to a decrease in volume can result in a tensile strain. The maximum value of the tensile intrinsic stress is the yield strength of the deposited material (transition between zone A and zone B of the idealized diagram in Fig. 15.2).

15.4.2.2. *Intrinsic Stresses in Films Produced from Energetic Particles*

For film surfaces which sense energetic particle bombardment during deposition, the adequate surface diffusion of adatoms leads to the removal of voids in the films and the microstructure is modified progressively from zone 1 to zone T of the structure-zone diagrams. The volume reduction of voids, grain boundaries and intercolumnar spaces removes the origin of the tensile intrinsic stress; the magnitude of the tensile intrinsic stress decreases and tends to be equal to zero even though the yield strength of the deposited material is greater than zero. Beyond the sudden drop in tensile intrinsic stress, a transition between tensile and compressive intrinsic stresses is observed and the magnitude of the compressive intrinsic stress increases progressively up to a maximum value with increasing normalized momentum (zone B in Fig. 15.2). Two major models have been proposed to account for the generation of compressive intrinsic stresses in PVD films produced from energetic particles [17,24,40].

15.4.2.2.1. Forward Sputtering Model Proposed by Windischmann

The model proposed by Windischmann is based on three major assumptions: (i) energetic bombardment of the film surface causes displacements of atoms in the film from their equilibrium positions through a series of primary and recoil collisions, producing a volume distortion, (ii) for films deposited at low-substrate temperatures ($T_s/T_m < 0.25$), mass transport and defect mobility are sufficiently low to freeze the volumetric distortion in place and (iii) the relative volumetric distortion, d, which corresponds to a strain is proportional to the fractional number of atoms, n/N, displaced from equilibrium positions (i.e., $d = K(n/N)$ where K is the proportionality factor). The calculation developed by Windischmann [17,24] leads to the expression of the volumetric distortion or elastic strain:

$$d = 4.79\left(\frac{K\,\phi_p}{N}\right)\delta\sqrt{E_p} \tag{15.24}$$

The compressive intrinsic stress, σ_i, in the film is given by Hooke's law:

$$\sigma_i = \left(\frac{E_f}{1-\nu_f}\right)d = \left(\frac{E_f}{1-\nu_f}\right)4.79\left(\frac{K\,\phi_p}{N}\right)\delta\sqrt{E_p} \tag{15.25}$$

The atomic number density, N, is given by:

$$N = \frac{N_{Av}\,D_f}{M_t} \tag{15.26}$$

where N_{Av} is Avogadro's number, D_f the mass density of the film and M_t the mass of film atoms (target atoms). By substituting Eq. (15.26) in Eq. (15.25), the expression of the compressive intrinsic stress becomes:

$$\sigma_i = \left[\frac{E_f\,M_t}{(1-\nu_f)\,D_f}\right]4.79\left(\frac{K\,\phi_p}{N_{Av}}\right)\delta\sqrt{E_p} = k\,\phi_p\sqrt{E_p}Q \tag{15.27}$$

where ϕ_p is the flux of energetic particles (projectiles) and E_p is the kinetic energy of projectiles. The constant k is equal to $(4.79\,K\,\delta/N_{Av})$ and the term Q which represents the stored elastic energy per mole is given by:

$$Q = \left(\frac{E_f}{1-\nu_f}\right)\left(\frac{M_t}{D_f}\right) \tag{15.28}$$

The units for Q are erg mol^{-1}, for E_f given in dyn cm^{-2}, M_t in amu and D_f in g cm^{-3}. To convert erg mol^{-1} to eV atom, the value of Q is divided by 10^{12}.

The compressive intrinsic stress depends on the atomic volume of the film material, M_t/D_f. This dependence means that the interaction between the energetic particles (projectiles) and the deposited material (target) gives rise to a variable strain depending upon the atomic arrangement in the film. The linear relationship between the compressive intrinsic stress and the factor Q has been observed for films deposited by various sputtering techniques [17,24].

The forward sputtering model is not applicable for films produced with very light energetic particle bombardment, that is, for $Z_p/Z_t \ll 1$, since the assumptions involved in the sputtering theory of Sigmund are not valid. In addition, the model cannot be invoked to explain the intrinsic stress when the film surface senses particles having very low or very high kinetic energies. The lower energy limit of applicability corresponds approximately to the energy for atomic displacement in the films which is about 18–30 eV for most materials [41,42]. With low energy projectiles, the assumption of isotropic cascades in the target (film) involved in the sputtering theory is no longer valid. Bombardment of growing films by ions with a normalized energy, E_n, of less than a few eV per atom condensed on the film surface leads to films containing a significant fraction of voids resulting in tensile intrinsic stress developed in the films rather than compressive intrinsic stress [43]. The upper energy limit of applicability is related to resputtering of the deposited material and possible mechanical damage of the film such as void formation and plastic flow [44]. For instance, under krypton ion bombardment, the fraction of platinum films resputtered increases linearly with increasing normalized energy, E_n, and can reach 50% at $E_n = 50$ eV/atom. Moreover, the magnitude of the compressive intrinsic stress predicted by this model increases continuously with increasing projectile energy, E_p. However, the maximum compressive stress observed in PVD films is usually less than the yield strength. In addition, beyond this maximum, the decrease in compressive intrinsic stress with increasing normalized momentum increases (zone C in Fig. 15.2) is not predicted by the forward sputtering model.

15.4.2.2.2. Model Proposed by Davis

This model proposed to explain the formation of compressive intrinsic stresses in relatively dense films is based on two major assumptions [40]. The compressive intrinsic stress is assumed to be caused by film atoms implanted below the surface of the film by knock-on processes in accordance with the model proposed by Windischmann. In addition, thermal spikes are assumed to reduce the stress by causing displacement of the implanted atoms. The implanted atoms in metastable position which acquire more than some excitation energy, E_d, will escape from their metastable position to the surface of the film. Intense local heating resulting from the energy of bombarding particles transferred to film atoms in the very small area of the impact, that is, thermal spike is supposed to provide the energy required for releasing implanted atoms from their metastable position within the film [45]. Assuming that there is a balance between implantation and relaxation processes, the density, n, of the implanted atoms (or atoms displaced by energetic particle bombardment) is constant with time. On the basis of these assumptions, the expression of the ratio, n/N, where n and N are the number of atoms displaced by energetic particle bombardment per unit volume and the atomic number density in the film, respectively, was established by Davis [40]:

$$\frac{n}{N} = \left(\frac{13.36}{U_0}\right)\left[\frac{(E_p)^{1/2}}{\frac{\phi_c}{\phi_p} + k_a(E_p)^{5/3}}\right] \tag{15.29}$$

where U_0 is the sublimation energy of the deposited material and ϕ_c the flux of atoms condensed on the film surface and incorporated in the growing film; ϕ_p and E_p are the flux and kinetic energy of incident energetic species, respectively. The factor k_a in Eq. (15.29) is equal to $[0.016\,\rho\,(E_d)^{-5/3}]$ where ρ is a material-dependent parameter which is of the order of unity [40]. According to Hooke's law, the compressive intrinsic stress in the film can be expressed as follows:

$$\sigma_i = \left(\frac{E_f}{1-\nu_f}\right)\left(\frac{K\,n}{N}\right) = \left[\left(\frac{13.36\,K}{U_0}\right)\left(\frac{E_f}{1-\nu_f}\right)\right]\left[\frac{(E_p)^{1/2}}{\frac{\phi_c}{\phi_p} + k_a(E_p)^{5/3}}\right] \tag{15.30}$$

or

$$\sigma_i = \kappa\left[\frac{(E_p)^{1/2}}{\frac{\phi_c}{\phi_p} + k_a(E_p)^{5/3}}\right]$$

where K is the proportionality factor between the relative volumetric distortion, d, and the fractional number of atoms, n/N, displaced from equilibrium positions in the films, that is, $d = K\,(n/N)$. The parameters κ and k_a in Eq. (15.30) are generally unknown. One method for determining these parameters is to use them as fitting parameters for a least-squares fit to data for which the kinetic energy, E_p, ion flux, ϕ_p and deposition rate of films, ϕ_c, are known. As a result, a strong dependence of the magnitude of compressive intrinsic stresses on the kinetic energy, E_p, of ions used for IBAD of films produced by evaporation or sputtering is predicted by Eq. (15.30).

When the normalized flux, ϕ_p/ϕ_c, is low, the ratio ϕ_c/ϕ_p in Eq. (15.30) is large with respect to the term $[k_a\ (E_p)^{5/3}]$ and the compressive intrinsic stress can be approximated to:

$$\sigma_i \approx \left(\frac{E_f}{1-\nu_f}\right)\left(\frac{13.36\,K}{U_0}\right)\left[\frac{\phi_p\ (E_p)^{1/2}}{\phi_c}\right] \tag{15.31}$$

This expression is comparable with Eq. (15.27) which gives the compressive intrinsic stress predicted from the forward sputtering model. In the theory developed by Davis, the compressive intrinsic stress is found to be proportional to ϕ_p/ϕ_c whereas the forward sputtering model leads to a compressive intrinsic stress intensity linearly depends on ϕ_p only. For a very large normalized flux, ϕ_p/ϕ_c, Eq. (15.30) can be approximated to:

$$\sigma_i \approx \left(\frac{E_f}{1-\nu_f}\right)\left(\frac{13.36\,K}{U_0}\right)\left[\frac{1}{k_a(E_p)^{7/6}}\right] \tag{15.32}$$

Two different strategies can be adopted to reduce the magnitude of compressive intrinsic stresses in films produced by IAD techniques. The first possibility is to produce PVD films with low normalized fluxes, ϕ_p/ϕ_c, that is, at high deposition rates and low ion fluxes. The compressive intrinsic stress predicted by Eq. (15.31) is also reduced with decreasing ion energy. However, in this case, the benefits of IAD may be considerably diminished. The alternative approach is to prepare PVD films with high-normalized fluxes, ϕ_p/ϕ_c, or at low deposition rates with high ion fluxes. The magnitude of compressive intrinsic stress expressed by Eq. (15.32) is reduced with increasing ion energy. However, too high ion energies may lead to resputtering phenomena and significant damage with void formation in the films. Eventually, films produced under these conditions exhibit reduced mass densities and tensile intrinsic stress.

15.4.3. Extrinsic Stresses

In addition to the thermal stress and intrinsic stress, a third term, named extrinsic stress, σ_e, can be distinguished in the overall residual stresses in PVD films (Eq. (15.1)). Impurities such as oxygen, hydrogen and inert gas atoms can be incorporated in PVD films during or after deposition by evaporation or sputtering. A lattice distortion may result from incorporation of atoms of a size different from the host, reaction such as oxidation or hydrogenation, which produces a new phase with a different molar volume, and grain surface energy reduction. These phase transformations lead generally to volume expansion and compressive stress can be observed [46]. Consequently, for not fully dense films exhibiting tensile intrinsic stress, an impurity-induced compensating compressive stress may develop during deposition. This compressive stress related to impurity incorporation may reduce the tensile stress in a non-overtly manner or produce a net compressive stress if the tensile intrinsic stress is totally compensated and overwhelmed even though the atomic peening mechanism is ineffective. Therefore, the interpretation of stress data must be performed cautiously for PVD films with compositions or impurity contents which vary with process parameters as well as for PVD films produced by reactive evaporation and reactive sputtering processes which may involve the formation of a new phase with a different molar volume by surface reactions such as oxidation, nitridation or hydrogenation.

Furthermore, various molecules can penetrate open voids or pores present in not fully dense films and adsorb on pore walls thereby interaction forces between adsorbed species, in particular, between polar species such as water molecules can act to modify residual stresses. These interaction forces may be responsible for extrinsic stress generation in porous films, which were exposed to room air or chemical agents contained in various environments. Depending on the relative orientation of adsorbed polar molecules, nature of adsorbed species and composition of pore walls, attractive or repellent interaction forces may develop and the resulting extrinsic stress in the films can be tensile or compressive. A model to explain the origin of extrinsic stress based on the adsorption of polar species on pore walls was proposed by Hirsch [47]. This model is not of great numerical precision; however, dipole interactions between adsorbed molecules are demonstrated to be responsible for the observed forces and stresses produced by realistic amounts of adsorbate.

Hirsch assumes that the adsorbed molecular dipoles are arranged on the cylindrical pore wall with their axes normal to the surface and with charges of the same sign pointing inward and outward, respectively. In addition, the circumference of the adsorbing surface is assumed to remain at the constant value of $2\pi a$ (where a is the radius of cylindrical pores), irrespective of the amount of material adsorbed on pore walls. The interaction force between two dipoles was calculated from elementary electrostatic theory. The total interaction force acting on the cross-section of the film per unit width was calculated to be [47]:

$$F_{\text{total}} = \frac{3\, n^2\, \alpha^2\, I(a, d_0)\, t_f}{5\, \pi\, \varepsilon\, a^2} \tag{15.33}$$

where n is the number of dipoles per unit area, α the dipole moment, t_f the film thickness and ε the dielectric constant of the free space. The interaction integral, $I(a,d_0)$, can be numerically computed. Hirsch has observed that for pore radius, a, of 10 Å, changes of 10–20% in diameter, d_0, of polar molecules do not affect significantly the interaction integral value; in addition, for molecular diameter, d_0, in the range 3–6 Å, the interaction integral can be expressed as follows:

$$I(a, d_0) - 5.8 \times 10^8\; a \tag{15.34}$$

where the pore radius, a, is expressed in meter. Substituting $I(a, d_0)$ by Eq. (15.34) in Eq. (15.33), the total interaction force becomes:

$$F_{\text{total}} = \frac{1.74 \times 10^9\, n^2\, \alpha^2\, t_f}{5\, \pi\, \varepsilon\, a} \tag{15.35}$$

According to Eq. (15.35), for very porous films with relatively large pore radii, a, the extrinsic stress resulting from adsorption of molar molecules would be rather low. In these films, the distance between adsorbed polar molecules on opposite pore walls is too large for efficient attractive or repellent interactions. For dense films, the pore radii tend to be nil and the interaction force would be infinite. This situation is unrealistic since for too low pore sizes, the coverage of the pore walls by adsorbed polar molecules is expected to be non-uniform or eventually molecules cannot be adsorbed if the pore diameter is less than the polar molecule diameter. As the pore diameter decreases, the number of dipoles per unit area, n, adsorbed on the pore walls decreases concomitantly and the interaction force tends to be negligible. The magnitude of extrinsic stress originating from adsorption of polar molecules in not fully dense films would be maximum for pore sizes comparable to the diameter of polar molecules.

15.5. Residual Stresses in a-C Films Deposited by Conventional Magnetron Sputtering on Grounded Substrates

Residual stresses were investigated in sputter-deposited a-C films to be used for lubrication in hostile environments [48]. Experiments were designed to estimate the flux and kinetic energy of particles impinging on the surface of a-C films deposited on <100> – oriented Si substrates of (6×1) cm^2 at ambient temperature by direct current (dc) magnetron sputtering from a graphite target in pure argon discharges [49]. The magnitude of residual stresses in the films was calculated from the change of the radius of curvature of Si substrates using Eq. (15.3).

The compressive residual stresses in these a-C films decreased with increasing argon pressure whereas a reverse trend was observed for the mass density of films [49]. The residual stresses in a-C films sputter-deposited on Si substrates may result essentially from thermal stress and intrinsic stress developed during the growth of films. Since the maximum substrate temperature was $<60°$C, the contribution of the thermal stress to the residual stresses was neglected and the magnitude of intrinsic stress was approximately equal to that of residual stresses.

Intrinsic stresses are generated by energetic particle bombardment of the surface of growing films. The surface of films deposited by magnetron sputtering can be exposed to four types of energetic particles, namely (i) neutral target atoms sputtered from the target surface and condensed on the substrate, (ii) neutral sputtering gas atoms previously implanted in the target as sputtering ions which are ejected from the target surface together with neutral target atoms, (iii) ions of the sputtering gas coming from the discharge and (iv) neutralized sputtering ions backscattered from the target surface by elastic collisions. The flux of the second type of energetic particles is negligible with respect to the flux of carbon atoms ejected from the target. The reflection coefficients, R_0 and γ_0, corresponding to the flux and energy of sputtering ions neutralized and backscattered from the target surface (fourth type of energetic particles) are dependent on the atomic mass ratio, M_C/M_{Ar} [50,51]. For the carbon–argon system, the atomic mass ratio is equal to 0.3; hence, the reflection coefficients R_0 and γ_0 are $<10^{-3}$ and 0.1, respectively [12]. Therefore, for this system, the contribution of backscattered neutral argon atoms to the energy deposited on the surface of growing films is negligible with respect to the contribution of neutral carbon atoms and argon ions originating from the plasma.

The argon pressure effect on the compressive intrinsic stress developed in sputter-deposited a-C films was analyzed using the forward sputtering model [17,24]. The compressive intrinsic stress in a-C films is given by [49]:

$$\sigma = 4.79 \left(\frac{E_f}{1 - \nu_f} \right) \left(\frac{K}{N} \right) \left[2.79\, (E_{Ar})^{1/2}\, \phi_{Ar} + 2.64\, (E_C)^{1/2}\, \phi_C \right] \quad (15.36)$$

where E_{Ar} and E_C are the argon ion and carbon atom energies on the surface of a-C films, respectively; ϕ_{Ar} and ϕ_C are the argon ion and carbon atom fluxes on the surface of growing films, respectively. K is a proportionality factor and N the atomic number density in the deposited material. The dependence of the compressive intrinsic stress on energetics of the deposition process is in good agreement with the prediction of the model proposed by Windischmann (Fig. 15.3).

The contribution of argon ions to the energy deposited on the film surface is relatively small compared to that of neutral carbon atoms, in particular, at low-argon pressures when the thermalization phenomena of sputtered carbon atoms are not very efficient. In other words, in Eq. (15.36), at low-argon pressures, the term 2.79 $(E_{Ar})^{1/2}$ ϕ_{Ar} is negligible with respect to the term 2.64 $(E_C)^{1/2}$ ϕ_C. These thermalization phenomena involving collisions between sputtered carbon atoms and sputtering gas atoms play a major role in the deposition process. The number of collisions of carbon atoms in the gas phase and the thermalization of sputtered carbon atoms depend on the product between the target-substrate distance and the argon pressure. This product is a relevant factor affecting the development of the intrinsic stress and governing their magnitude in sputter-deposited films (Table 15.1). The compressive intrinsic stress

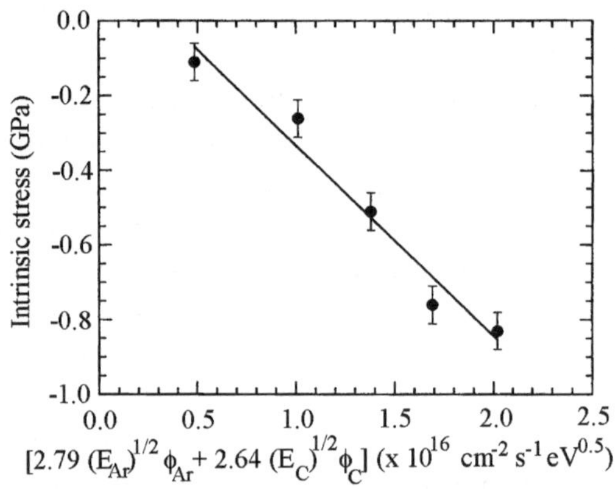

Figure 15.3: Dependence of the intrinsic stress in a-C films on the energetics of the deposition process; the solid line was deduced from a least-squares fit to the experimental data [49].

Table 15.1: Effect of the product (target-substrate distance) × (argon pressure) on the residual stresses in a-C films produced under various deposition conditions [49]

Argon pressure, P (Pa)	Target-substrate distance, d (cm)	$d \times P$ (cm Pa)	Growth rate of films (nm/min)	Residual stresses (GPa)
0.125	14	1.75	4.36	−0.75
0.250	7	1.75	12.2	−0.75
0.500	3.5	1.75	27.6	−0.75

was found to remain at a constant value of −0.75 GPa in a-C films produced under various experimental conditions with the same value of the product ($d \times P$).

15.6. Residual Stresses in a-C Films Deposited by Conventional and Unbalanced Magnetron Sputtering on Biased Substrates

The magnetron target was placed in the center of a cylindrical magnetic coil of 23 cm in inner diameter and the unbalanced magnetron sputtering mode operated by varying the current in the coil. The water-cooled substrate holder was either grounded or biased to negative voltages up to 300 V. The characteristics of the deposition process such as substrate temperature, ion flux, ϕ_{Ar} and carbon atom flux, ϕ_C, on the film surface were determined as functions of the substrate bias voltage for sputtering of a-C films using conventional and unbalanced magnetron modes [48,49,52].

The substrates mounted on the water-cooled substrate holder were maintained at room temperature. As a result, the thermal stress in these films can be neglected. The residual stresses correspond to the intrinsic stress built up during sputter-deposition of a-C films. The compressive intrinsic stress in films sputter-deposited by conventional magnetron mode at an argon pressure of 0.25 Pa and a sputtering power of 0.5 kW increased progressively up to a maximum value of −2.8 GPa with increasing bias voltage [53]. This maximum value is about three times higher than that obtained from films sputter-deposited on grounded substrates. By contrast, a-C films with a reduced mass density sputter-deposited at an argon pressure of 2 Pa exhibited negligible compressive intrinsic stress whatever the bias voltage value although the argon concentration in these films was dependent on the substrate bias voltage.

The compressive intrinsic stress in a-C films produced by unbalanced magnetron sputtering also increased with increasing substrate bias voltage [53]. The films produced with a negative bias voltage higher than 50 V were not adherent to Si substrates since the compressive intrinsic stress in these films was probably higher than 2.5 GPa. The value of the compressive intrinsic stress in a-C films produced on grounded substrates using the unbalanced magnetron mode was of the same order of magnitude as that for a-C films deposited by conventional magnetron sputtering, that is, in the range −0.8 to −1 GPa. The compressive intrinsic stress in a-C films produced by unbalanced magnetron mode increased more rapidly with increasing bias voltage than the compressive intrinsic stress in films deposited by conventional magnetron sputtering.

The impact energy available on the film surface originates essentially from the kinetic energy of positive ions created in the argon discharge and accelerated by the negative substrate bias voltage. The maximum value of the average kinetic energy of neutral carbon atoms condensed on the substrate is about 10 eV whereas the kinetic energy of positive ions is in the range 10–300 eV. The energy deposited on the growing films and resulting from the positive ion bombardment of the surface appears to be the relevant factor affecting the magnitude of the compressive intrinsic stress in a-C films deposited by conventional and unbalanced magnetron sputtering. Since the flux ratio, ϕ_{Ar}/ϕ_c, is also dependent on the negative substrate bias voltage or kinetic energy of positive ions impinging on the film surface, this flux ratio can be considered as an additional factor affecting the value of the intrinsic stress in a-C films produced under the experimental conditions investigated.

15.6.1. Applicability of the Forward Sputtering Model Proposed by Windischmann

According to Windischmann's model, the compressive intrinsic stress is predicted to be proportional to the product of the particle flux, ϕ_p, and the square root of the particle energy, E_p (Eq. (15.27)), that is, to the product of the flux, ϕ_{Ar}, and square root of the kinetic energy, E_{Ar}, of argon ions striking the surface of growing a-C films. For a-C films sputter-deposited on Si substrates biased to a negative voltage in the range 20–300 V under an argon pressure of 0.25 Pa with a sputtering power of 0.5 kW, the intrinsic stress vs. product $\phi_{Ar}\ (E_{Ar})^{1/2}$ is plotted in Fig. 15.4. The experimental values of the compressive intrinsic stress are in good agreement with

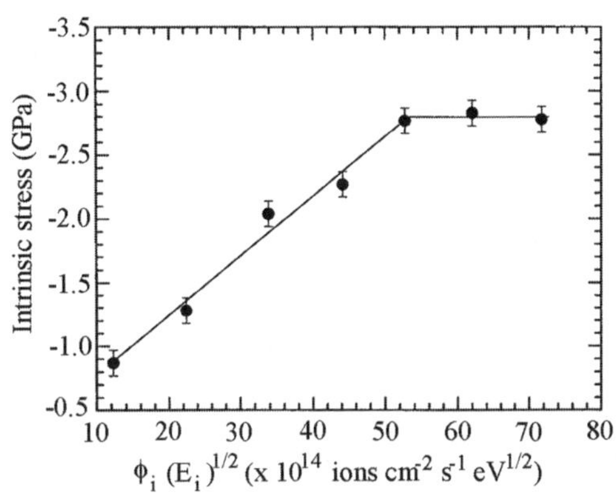

Figure 15.4: Effect of the flux and energy of positive ions on the intrinsic stress in a-C films deposited with the conventional magnetron mode at a sputtering power of 0.5 kW under an argon pressure of 0.25 Pa; the substrate bias voltage for sputter-deposition of these films were in the range −20 to −300 V [53].

the prediction of Windischmann's model as the product $\phi_{Ar}\,(E_{Ar})^{1/2}$ is less than 5×10^{15} ions cm^{-2}s^{-1}eV$^{1/2}$. Beyond this value, a large deviation between experimental and predicted results can be observed. As a result, additional phenomena resulting in stress relaxation must be considered for interpreting and modeling the effect of the ion flux and energy on the intrinsic stress in a-C films deposited by sputtering under intense energetic particle bombardment.

15.6.2. Applicability of the Model Proposed by Davis

On the basis of the model proposed by Davis [40], the compressive intrinsic stress in films for which the film surface senses energetic particle bombardment during deposition can be calculated from Eq. (15.30). The flux ratio ϕ_C/ϕ_{Ar} was determined as a function of the kinetic energy, E_{Ar}, of Ar^+ ions striking the surface of negatively biased substrates for sputter-deposition of a-C films by conventional and unbalanced magnetron modes [53]. The magnitude of the compressive intrinsic stress in a-C films predicted by Davis's model is related to the kinetic energy of argon ions by the following equations:

$$\sigma = \kappa \left[\frac{(E_{Ar})^{1/2}}{\left(\dfrac{22.75}{2.4 + 0.1\,(E_{Ar})^{1/2}}\right) + k_a (E_{Ar})^{5/3}} \right] \tag{15.37}$$

$$\sigma = \kappa \left[\frac{(E_{Ar})^{1/2}}{\left[\dfrac{22.75}{26.6 + 5.72\,(E_{Ar})^{1/2}}\right] + k_a (E_{Ar})^{5/3}} \right] \tag{15.38}$$

which are valid for a-C films sputter-deposited by conventional and unbalanced magnetron modes, respectively.

These equations contain two undetermined parameters, κ and k_a, which were used as fitting parameters for a least-squares fit to experimental results. The theoretical curves and the experimental values of the compressive intrinsic stress in the a-C films sputter-deposited by conventional and unbalanced magnetron modes are represented

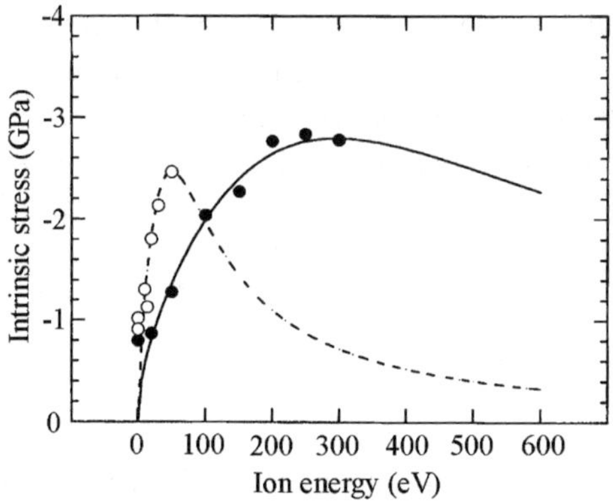

Figure 15.5: Effect of the positive ion energy on the intrinsic stress in a-C films deposited at a sputtering power of 0.5 kW under an argon pressure of 0.25 Pa using: (•) the conventional magnetron sputtering mode, and (○) the unbalanced magnetron sputtering mode with a current intensity in the coil of 5 Å; the curves in solid and dashed lines represent the intrinsic stress values predicted by the model proposed by Davis [53].

in Fig. 15.5. The experimental results and predicted values of the compressive intrinsic stress are in very good agreement for a-C films sputter-deposited by conventional magnetron mode. For a-C films produced by unbalanced magnetron mode, the theoretical curve also fits the experimental values; however, these values correspond to relatively low ion energies since the films produced under more energetic ion fluxes were found to be not adherent to Si substrates.

15.7. Residual Stresses in Silicon Dioxide Films Prepared by Direct Thermal Evaporation

Stoichiometric silicon dioxide films have been deposited on various substrates by electron-gun evaporation of a SiO_2 source under an oxygen pressure varying from 2×10^{-5} mbar (base pressure during evaporation) to 4×10^{-4} mbar [54,55]. The deposition temperature was varied between 20°C and 285°C. The magnitude of residual stresses was calculated from the radius of curvature of Si substrates determined by interferometric measurements while the samples were placed either in room air or under vacuum (5×10^{-5} mbar). The residual stresses in films deposited on Si substrates at 200°C measured under vacuum were less compressive than in room air. In addition, the variation of residual stresses, $\Delta\sigma_w$, observed when the samples were moved from air to vacuum was independent of the oxygen pressure in the deposition chamber [55].

The residual stresses, σ, and the stress component, $\Delta\sigma_w$, were dependent on the mass density of films [55]. Films with a reduced mass density exhibited tensile residual stresses. The magnitude of the tensile residual stresses decreased with increasing mass density; the tensile-to-compressive stress transition was observed for a mass density of about 1.7 g cm^{-3}. Compressive residual stresses were found in relatively dense SiO_2 films. The stress component, $\Delta\sigma_w$, was minimum for films with a low mass density and increased slowly as the film density varied from 1.5 to 1.9 g cm^{-3}. This stress component increased again as the film density approached the bulk density of SiO_2.

Water molecules incorporated in porous SiO_2 films were detected by infrared (IR) absorption spectroscopy. The intensity of the IR absorption band at 2.9 μm which characterizes H–O–H bonds was measured in room air and under vacuum for SiO_2 films deposited at 250°C under an oxygen pressure of 1×10^{-4} mbar; the mass density of these not fully dense films was equal to 1.98 g cm^{-3}. The intensity of the absorption band at 2.9 μm was considerably reduced as the samples were analyzed under vacuum. This decrease in intensity confirms that water molecules desorb from the porous films when the samples are placed in vacuum. In addition, when the films were exposed anew to room air, the intensity of the absorption band increased rapidly to reach the initial value, that is, the phenomenon responsible for the intensity variation is rapid and reversible. The model proposed by Hirsch [47] involving the interaction forces between the permanent electric dipole moments associated with H_2O molecules adsorbed on the pore walls is well designed to explain the origin of the stress difference $\Delta\sigma_w$. In addition, on the basis of this model, the extrinsic stress is expected to vary with the nature of adsorbed molecules, in particular with the square of the dipole moment, α, of molecules (Eq. (15.35)).

Silicon dioxide films of 1.98 g cm^{-3} in mass density were placed under vacuum (for H_2O desorption) and, then exposed to various gas phase environments such as propanol 2 (CH_3–CHOH–CH_3), acetone, water vapor and nitrogen [12]. The sagitta, Z, of the Si substrate, (28×1) cm^2 and 0.6 mm-thick, covered with an evaporated SiO_2 film of 560 nm in thickness was determined by interferometry and investigated as a function of the sample environment (Table 15.2). The film surface was always convex, that is, the stress was compressive. The stress variation is related to the sagitta variation expressed as a fraction of the wavelength, λ, used for interferometric measurements. Two air-vacuum cycles were performed before exposure of the sample to acetone vapor. The first two cycles corresponding to a sagitta variation of about 0.5 λ reveal a perfect reversibility of phenomena responsible for stress variation. When the sample was exposed to acetone vapor, the sagitta variation was smaller (0.3 λ); however, the phenomenon was always perfectly reversible. The propanol exposure of the sample led to a sagitta variation of about 0.2 λ. Since the sagitta remained constant when air was introduced in the chamber, the adsorption

Table 15.2: Sagitta measured at the center of a Si substrate covered with a 560 nm-thick SiO_2 film deposited by direct thermal evaporation at 250°C under an oxygen pressure of 1×10^{-4} mbar as the sample was exposed to various environments [12]

Experiment #	1	2	3	4	5	6	7
Environment	Air	Vacuum	Air	Vacuum	Acetone	Vacuum	Propanol
Sagitta ($\times \lambda$)	0.555	0.075	0.565	0.105	0.405	0.110	0.340
Experiment #	8	9	10	11	12	13	14
Environment	Air	Vacuum	Nitrogen	Vacuum	Water vapor	Vacuum	Propanol
Sagitta ($\times \lambda$)	0.340	0.088	0.080	0.075	0.655	0.090	0.570

site in SiO_2 films were probably totally saturated or occupied by adsorbed propanol molecules. The initial sagitta value was obtained anew when the sample was in vacuum. Therefore, the adsorption–desorption phenomena of propanol molecules were reversible. The variation of sagitta was nil when the sample in vacuum was then exposed to nitrogen. As a result, N_2 molecules either were not adsorbed in the silicon dioxide film or the nitrogen adsorption has no effect on the stress. The maximum sagitta variation of about 0.65 λ obtained from the sample exposed to water vapor suggests that the relative humidity of the environment may affect significantly the magnitude of the extrinsic stress. After the series of experiments, the curvature of the Si substrate in air or in vacuum returned to its initial value. Major findings have been obtained from these experiments. Firstly, the exposure of evaporated SiO_2 films to non-polar species such as nitrogen molecules does not modify the residual stresses. The adsorption and desorption phenomena of polar molecules investigated are totally reversible at room temperature. The magnitude of the extrinsic stress depends on the nature of the polar molecules in agreement with the prediction of the model proposed by Hirsch. The extrinsic stress resulting from film exposure to polar molecules was always compressive; as a result, repellent forces develop between polar molecules adsorbed on pore walls.

The dipole moment, α, of molecules and the number of dipoles, n, per unit area are needed to verify the applicability of the model developed by Hirsch [47]. The dipole moment values of vapor molecules investigated are given in the literature [56]. The number of dipoles adsorbed per unit area may be approximated to be inversely proportional to the area of the polar molecules. The effective surface area, S_{eff}, of a molecule is given by [57]:

$$S_{eff} = 2\sqrt{3}\left(\frac{M}{4\sqrt{2}\,N_{Av}\,\rho}\right)^{2/3} \tag{15.39}$$

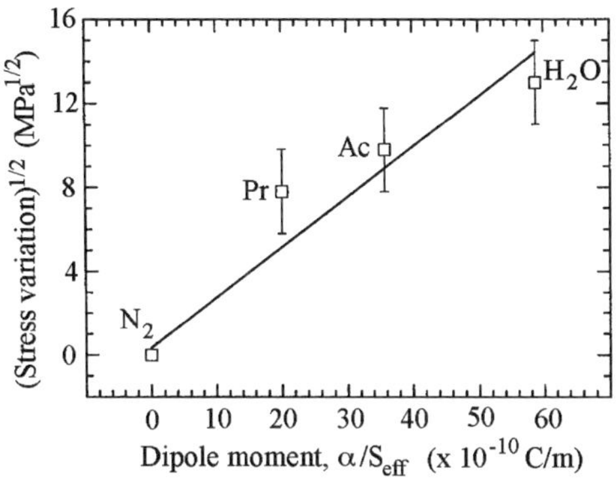

Figure 15.6: Residual stress variation induced by adsorption of various polar molecules (H_2O, acetone, propanol) in porous SiO_2 films as a function of the dipole moment per unit area [12].

where M and ρ are the molar mass and the mass density of the molecules, respectively; N_{Av} is Avogadro's number. The values of S_{eff} and those of the dipole moment per unit area, α/S_{eff}, can be calculated from Eq. (15.39). According to Eq. (15.35), the square root of the extrinsic stress is expected to be proportional to the dipole moment per unit area, α/S_{eff}. The diagram in Fig. 15.6 displays a linear dependence of $(\Delta\,\sigma_w)^{1/2}$ on the dipole moment per unit area expected from Eq. (15.35). Although the number of data points is low, the general trend illustrated by these data is in accordance with the prediction of Hirsch's model.

The validation of the model would be completed from the investigation of the extrinsic stress as a function of the pore radius in SiO_2 films. According to Eq. (15.35), the extrinsic stress is expected to be inversely proportional to the pore radius, a. The radii of pores or the size distribution in the films were not measured directly; however, the radius of pores may be assumed to be dependent on the mass density of films. The extrinsic stress, $\Delta\sigma_w$, measured as the sample was moved from air to vacuum vs. mass density of films is plotted in Fig. 15.7. The compressive extrinsic stress value exhibits a maximum value in SiO_2 films having an intermediate mass density of about 1.9 g cm^{-3}. The extrinsic stress decreases either with decreasing mass density from 1.9 to 1.5 g cm^{-3} or with increasing mass density up to the bulk density value of 2.2 g cm^{-3}. The magnitude of the extrinsic stress reaches a maximum value in films with a packing density of about 0.87. Assuming that the pore size decreases progressively as the mass density of films increases, the dependence of the extrinsic stress induced by adsorption of water molecules on the film density is quite compatible with the prediction of Hirsch's model.

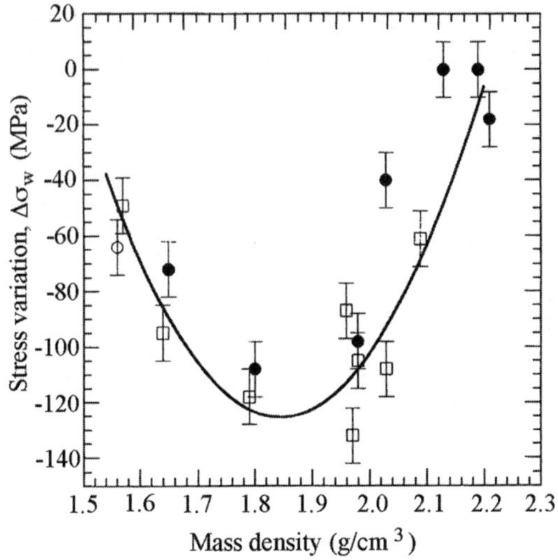

Figure 15.7: Residual stress variation as the sample is moved from vacuum to air as a function of the mass density of SiO_2 films deposited on Si substrates using two different equipments: (□) films prepared by direct thermal evaporation in equipment I and (●) films obtained under oxygen ion beam bombardment or without any ion beam assistance in equipment II [12].

15.8. Conclusion

Residual stresses in PVD films produced at moderate or low temperatures result from the contribution of thermal, intrinsic and extrinsic stresses. The residual stresses in amorphous, nanocrystalline or polycrystalline films can be determined by the measurements of the deformation of film-substrate structures. Thermal stresses in films deposited at temperatures higher or lower than room temperature arise from the difference of thermal expansion coefficients of the film and substrate. Various models can be invoked to explain the development of intrinsic stress in PVD films. Tensile intrinsic stress is normally observed in films deposited on substrates at low temperatures by thermal evaporation in the absence of any energetic particle bombardment. These not fully dense films exhibit a zone 1-type microstructure of the structure-zone diagram and the GBR model can explain the origin of the stress. Compressive intrinsic stress develops in films deposited at low temperatures under energetic particle bombardment. These films exhibit dense microstructures corresponding to zone T in the structure-zone diagram and their compressive intrinsic stress results from the atomic peening mechanism. In addition to thermal and intrinsic stresses, extrinsic stress may develop in thin films subsequently to the deposition process. Water vapor molecules may penetrate in porous films exposed to room air. Adsorption phenomena and adsorbed dipole interactions can be invoked to explain the origin of extrinsic stress.

Stress data illustrating the applicability of models proposed for the origin of stresses have been obtained from sputter-deposited a-C films and SiO_2 films prepared by thermal evaporation. The magnitude of compressive intrinsic stress predicted from the models proposed by Windischmann and by Davis are clearly substantiated by experimental results obtained from a-C films produced by energetic deposition techniques. The variation of the extrinsic stress developed in porous SiO_2 films prone to adsorb polar molecules was in good agreement with the prediction based on the model proposed by Hirsch.

References

[1] M.D. Thouless, *Thin Solid Films*, 181 (1989) 397.

[2] M.R. James and J.B. Cohen, *The Measurement of Residual Stresses by X-Ray Diffraction Techniques, Treatise on Materials Science and Technology*, Ed. H. Herman, Vol. 19A, Academic Press Inc., London, 1980, p. 1.

[3] E.S. Machlin, *Materials Science in Microelectronics, The Relationships between Thin Film Processing and Structure*, Chapter VI, Giro Press, Croton-on-Hudson, NY, 1995, p. 157.

[4] J.K. Hirvonen, *Materials and Processes for Surface and Interface Engineering*, Ed. Y. Pauleau, *NATO-ASI Series, Series E: Applied Sciences*, Vol. 290, Kluwer Academic Publishers, Dordrecht, The Netherlands, 1995, p. 307.

[5] J.E. Greene, *Multicomponent and Multilayered Thin Films for Advanced Microtechnologies: Techniques, Fundamentals and Devices*, Eds. O. Auciello and J. Engemann, *NATO-ASI Series, Series E: Applied Science*, Vol. 234, Kluwer Academic Publishers, Dordrecht, The Netherlands, 1993, p. 39.

[6] D.S. Campbell, *Handbook of Thin Film Technology*, Eds. L.I. Maissel and R. Glang, McGraw-Hill, New York, 1970, pp. 12–21.

[7] R.W. Hoffman, *Physics of Non-Metallic Thin Films,* Eds. C.H. Dupuy and A. Cachard, *NATO-ASI Series, Series B: Physics*, Vol. 14, Plenum, New York, 1976, p. 273.

[8] M.R. James and O. Buck, *Crit. Rev. Solid State Mater. Sci.*, 9 (1980) 61.

[9] S. Tamulevicius, L. Augulis, G. Laukaitis and L. Puodziukynas, *Medziagotyra*, 2 (1998) 40.

[10] A. Brenner and S. Senderoff, *J. Res. Natl. Bur. Stand.*, 42 (1949) 105.

[11] G.G. Stoney, *Proc. Roy. Soc. London Ser. A*, 82 (1909) 172.

[12] Y. Pauleau, *Deposition and Processing of Thin Films, Handbook of Thin Film Materials*, Ed. H.S. Nalwa, Vol. 1, Chapter 9, Academic Press, San Diego, CA, 2002, p. 455.

[13] R.W. Hoffman, *Physics of Thin Films – Advances in Research and Development*, Eds. G. Hass and R.E. Thun, Vol. 3, Academic Press, New York, 1966, p. 211.

[14] D.S. Campbell, *Basic Problems in Thin Film Physics*, Eds. R. Niedermayer and H. Mayer, Vandenhoeck and Ruprecht, Gottingen, 1966, p. 223.
[15] K. Kinosita, *Thin Solid Films*, 12 (1972) 17.
[16] M.F. Doerner and W.D. Nix, *Crit. Rev. Solid State Mater. Sci.*, 14 (1988) 225.
[17] H. Windischmann, *Crit. Rev. Solid State Mater. Sci.*, 17 (1992) 547.
[18] B.A. Movchan and A.V. Demchishin, *Phys. Met. Metallogr.*, 28 (1969) 83.
[19] C.R.M. Grovenor, H.T.G. Hentzell and D.A. Smith, *Acta Metall.*, 32 (1984) 773.
[20] J.A. Thornton, *J. Vac. Sci. Technol.*, 11 (1974) 666.
[21] P. Sigmund, *Phys. Rev.*, 184 (1969) 383.
[22] F.M. D'Heurle, *Metall. Trans.*, 1 (1970) 725.
[23] P. Sigmund, *Topics in Applied Physics: Sputtering by Particle Bombardment I*, Ed. R. Behrisch, Vol. 47, Chapter 2, Springer-Verlag, Berlin, 1981.
[24] H. Windischmann, *J. Appl. Phys.*, 62 (1987) 1800.
[25] K.-H. Müller, *J. Vac. Sci. Technol. A*, 4 (1986) 184.
[26] K.-H. Müller, *J. Appl. Phys.*, 59 (1986) 2803.
[27] J.D. Targove and H.A. Mcleod, *Appl. Opt.*, 27 (1988) 3779.
[28] F.M. D'Heurle and J.M.E. Harper, *Thin Solid Films*, 171 (1989) 81.
[29] J.D. Finegan and R.W. Hoffman, *J. Appl. Phys.*, 30 (1959) 597.
[30] R.E. Rottmayer and R.W. Hoffman, *J. Vac. Sci. Technol.*, 8 (1971) 151.
[31] F.A. Doljack and R.W. Hoffman, *Thin Solid Films*, 12 (1972) 71.
[32] R.W. Hoffman, *Thin Solid Films*, 34 (1976) 185.
[33] H.K. Pulker, *Thin Solid Films*, 89 (1982) 191.
[34] R.W. Hoffman, J.D. Finegan, F.A. Doljack and R.W. Springer, *AEC Tech. Rep.*, Atomic Energy Commission, Case Western Reserve University, Cleveland, OH, 18; 1961, 64; 1970, 76; 1971, 79; 1972; 82; 1975; 83; 1975.
[35] R. Berger and H.K. Pulker, *Proc. SPIE*, Vol. 401, SPIE, Bellingham, WA, 1983, p. 69.
[36] H. Windischmann, G.F. Epps, Y. Cong and R.W. Collins, *J. Appl. Phys.*, 69 (1991) 2231.
[37] H.K. Pulker and J. Mäser, *Thin Solid Films*, 59 (1979) 65.
[38] M. Itoh, M. Hori and S. Nadahara, *J. Vac. Sci. Technol. B*, 9 (1991) 149.
[39] J.H. Rose, J. Ferrante and J.R. Smith, *Phys. Rev. Lett.*, 47 (1981) 675.
[40] C.A. Davis, *Thin Solid Films*, 226 (1993) 30.
[41] E.A. Kenick and T.E. Mitchell, *Phil. Mag.*, 32 (1975) 815.
[42] M.J. Makin, S.N. Buckley and G.P. Walters, *J. Nucl. Mat.*, 68 (1977) 161.
[43] J.E. Yehoda, B. Yang, K. Vedam and R. Messier, *J. Vac. Sci. Technol. A*, 6 (1988) 1631.
[44] B. Window and K.-H. Müller, *Thin Solid Films*, 171 (1989) 183.
[45] K.-H. Müller, *J. Vac. Sci. Technol. A*, 40 (1986) 209.
[46] E.S. Machlin, *Materials Science in Microelectronics, The Relationships between Thin Film Processing and Structure*, Chapter VI, Giro Press, Croton-on-Hudson, NY, 1995, p. 179.
[47] E.H. Hirsch, *J. Phys. D: Appl. Phys.*, 13 (1980) 2081.
[48] Y. Pauleau, E. Mounier and P. Juliet, *Protective Coatings and Thin Films: Synthesis, Characterization and Applications*, Eds. Y. Pauleau and P.B. Barna, *NATO-ASI*

Series, Partnership Sub-Series 3: High Technology, Vol. 21, Kluwer Academic Publishers, Dordrecht, The Netherlands, 1997, p. 197.
[49] E. Mounier and Y. Pauleau, *J. Vac. Sci. Technol. A*, 14 (1996) 2535.
[50] J. Bottiger, J.A. Davies, P. Sigmund and K.B. Winterbon, *Radiat. Effect.*, 11 (1971) 69.
[51] W.R. Gesang, H. Oechsner and H. Schoof, *Nucl. Instrum. Methods*, 132 (1976) 687.
[52] E. Mounier, E. Quesnel and Y. Pauleau, New Diamond and Diamond-Like Films, *Proceedings of the Topical Symposium II on New Diamond and Diamond-Like Films of the 8th CIMTEC-World Ceramic Congress and Forum on New Materials, Advances in Science and Technology 6*, Vol. 6, Ed. P. Vincenzini, Techna srl, Faenza, Italy, 1995, p. 183.
[53] E. Mounier and Y. Pauleau, *Diamond Relat. Mater.*, 6 (1997) 1182.
[54] H. Leplan, B. Geenen, J.Y. Robic and Y. Pauleau, *Optical Interference Coatings, Proceedings SPIE*, Ed. F. Abelès, Vol. 2253, SPIE, Bellingham, WA, 1994, p. 1263.
[55] H. Leplan, B. Geenen, J.Y. Robic and Y. Pauleau, *J. Appl. Phys.*, 78 (1995) 962.
[56] *Handbook of Chemistry and Physics*, 60th edition, Eds. R.C. Weast and M.J. Astle, CRC Press, Inc., Boca Raton, FL, 1980.
[57] S. Dushman, *Scientific Foundations of Vacuum Technique*, 2nd edition, John Wiley & Sons, New York, 1962, p. 460.

Chapter 16

Hard Coatings Based on Metal Nitrides, Metal Carbides and Nanocomposite Materials: PVD Process and Properties

Teresa Vieira, José Castanho and Cristina Louro

16.1. Introduction

In the last 30 years the hard coatings deposition technology was strongly improved. Besides the traditional surface treatments, new deposition technologies emerged and were implemented industrially. Such surface technologies answered the increasing demands imposed by the automation of the processes and the requirements of more performing materials to be used in aggressive environments. Deposition temperature, thickness of coatings, working temperature and atmosphere, mechanical solicitations, cutting speeds, workpiece material and lubrication are the parameters that should be considered in the selection of the surface modification technology. Among the available surface modification technologies, the physical vapor deposition (PVD) has been elected as the most reliable, environmental friendly and the one with the best performance for depositions of new materials, which cannot be produced using conventional surface-deposition technologies. The properties of thin films, such as the hardness of structural and protective coatings, depend on the grain size, morphology, density and internal stress and this dependence makes it important to understand *processing–microstructure–property* relationships. Considering the hard coatings, the main requirements for advanced ones are the following:

- The optimum surface quality, in order to improve the tribologic behavior, decreasing the need of coolants and lubricants.
- The highest surface hardness, which ensures the best wear strength.
- Maximum endurance in load shocks, which is attained by associating hardness and fatigue behavior of the coating.
- The best oxidation/corrosion resistance under conditions of dry/wet environment.

Materials Surface Processing by Directed Energy Techniques
Y. Pauleau (Editor)

In practical applications binomial relationships such as hardness/toughness, hardness/thermal resistance and hardness/oxidation must be satisfied. During the last decennia, an important variety of new compositions of hard coatings emerged from binary to higher-order transition metal carbide/nitride. In order to elucidate *composition–structure–properties* relationship, carbide/nitride-based on titanium and tungsten are used as *leitmotiv* in this chapter. Therefore, it will be focus the physical and mechanical properties of the hard coatings based on nitrides and carbides of titanium (group IV) and tungsten (group VI). The titanium carbides/nitrides due to their enormous quantity of studies, and tungsten carbides/nitrides because of being a consolidated research performed by the authors of this chapter. Even though nitrides and carbides of transition metals are very similar in main characteristics, there are, however, important differences resulting from the tri-valence of nitrogen and tetra-valence of carbon and from the lower dimension of nitrogen, which explain different physical and chemical behaviors of these interstitial compounds. After a previous outlook of the main characteristics and properties of hard coatings, their role will be highlighted first of all on hard binary nitrides and carbides. This will be the subject of the second section and the starting point to the next one, which concerns the discussion of the characteristics vs. properties of ternary nitrides (hard and ultra-hard coatings used in mechanical applications of extreme severity such as high-speed cutting). Before the conclusions, it will be discussed the properties of multilayer nanocomposite coatings of short and long period. The main characteristics and properties, and coating types studied are summarized as follows:

Main characteristics and properties	Coatings
Thickness	Binary metal nitrides and carbides
Chemical composition	Titanium nitrides and carbides
Structure	Tungsten nitrides and carbides
Morphology and grain size	Ternary nitrides
Roughness and surface morphology	Titanium aluminum nitrides
Residual stress	Titanium tungsten nitrides
Adhesion	Nanocomposite multilayer nitride-based coatings
Hardness	
Young's modulus	Multilayer coatings of short and long bilayer period
Fracture strength	
Thermal–chemical stability	Nanolaminate $Ti_{1-x}Al_xN$ coatings with metallic layers
	Nanolaminate $Ti_{1-x}W_xN$ coatings with metallic layers

16.2. Characteristics and Properties of Hard Coatings

A hard coating can be defined as a region with a reduced dimension, limited above by a surrounding atmosphere and below by a substrate [1], or a portion of condensed matter restricted in one dimension [2], presenting hardness values higher than 15 GPa. The hard coating's deposition and characterization fit in a technology named as surface engineering, which Subramanian *et al.* [3] had defined as the field that embraces the techniques and the processes used to modify and improve the performance of the surfaces relatively to wear, fatigue and corrosion, based on high hardness. The optimization of the physical, chemical and mechanical properties of the composite coating/substrate contributes to the overall improvement of performance and lifetime of components/parts. Along this section the importance of an accurate control of the coating's characteristics such as thickness, chemical composition, structure, morphology, residual stress and adhesion will be emphasized [1–7], and their role in the main properties of a hard coating, such as hardness, Young's modulus and oxidation resistance, will be highlighted.

16.2.1. Thickness

Optimizing the modified surface layer thickness can attain the best performance of the coatings on structural applications. If the coating is a very thin layer, the effect induced by its presence is easily vanished. However, if the coating is too thick, it behaves as a fragile bulk material [3]. The thickness of the coatings to be used in structural applications can be estimated according to Halling's equation [8]:

$$t = R\frac{H}{E'} \tag{16.1}$$

$$\frac{1}{E'} = \frac{1}{E_1} + \frac{1}{E_2} \tag{16.2}$$

where t is the coating's thickness, R is roughness parameter, H is the hardness of the coating, E', E_1 and E_2 are the equivalent, coating and counterbody and Young's modulus, respectively. Considering, for example, the Young's modulus and hardness values of the hard coatings and metallic alloys (counterbody), in high-speed cutting applications, the optimal thickness of the coatings was found to be in the order of 5 μm [9]. Singer [10] also established an optimal coating thickness (1–3 μm) on tribologic applications, to which the friction coefficient was minimal. Bearing in mind the use of hard coatings in different structural applications, the thickness of the coatings must be limited to a maximum of 5 μm and they are, sometimes, designated as thin films.

16.2.2. Chemical Composition

The chemical composition strongly influences the mechanical properties of the coatings, particularly those depending on interatomic forces, such as the Young's modulus. Thus, materials with high Young's modulus possess high atomic bonding energies and low bonding lengths and Young's modulus decreases with the increasing of the ionic- or metallic-bonding type present in the coatings [11–13]. Transition metal nitrides and carbides have been extensively used in the production of hard coatings. These type of interstitial compounds, specially metal nitrides, have a high number of metal–non-metal bonds due to the hybridization of the sp and p orbitals, and each full-filled atomic orbital will create two bonds to a total of six [11]. Therefore, these interstitial compounds must be highly covalent and the metal/non-metal bonding directions can explain their Young's moduli as well as their fragile behavior [14].

The high-temperature behavior of the coatings is intimately related to their chemical stability; mainly the oxidation and corrosion resistance, which means the resistance to chemical degradation, depends on the stability of the oxide layer formed. Moreover, if the coating is chemically inert, the possibility to exist an adhesive interaction with the counterbody is strongly reduced [3,9,15–17].

16.2.3. Structure

The atomic radius ratio of the non-metal vs. transition metal induces simple structures, mainly of cubic system. However, this tendency decreases with the increase metal group numbers of Periodic Table involved in the bonding. Some deposition techniques, such as sputtering, can lead to the formation of metastable structures (*configurationally frozen metastable structures*) [18]. These structures hold a high density of defects that induces high compressive residual stress in the coatings, which can contribute to a good performance in working conditions. Nevertheless, these metastable structures must require a high potential energy to change to equilibrium structures, that is, the metastable-structured coatings must not suffer any structural change during their mechanical application or they will collapse [15–17]. The metastable structures in certain conditions are amorphous phases, with implications on the mechanical properties. Crystalline coatings are, generally, harder than amorphous ones [3,4,19–22].

16.2.4. Morphology and Grain Size

The mobility of the adatoms influences the morphology of thc coatings deposited by sputtering [23]. The low adatoms mobility favors the existence of coatings with columnar morphologies [1], molding peaks and valleys that can promote the shadow effect and increase of roughness. The morphology can be tailored by controlling the deposition rate, temperature, pressure and substrate bias. Different authors, among them Messier *et al.* [24], Thornton [25] and Cavaleiro *et al.* [26] proposed several models to identify the morphologic type of coatings deposited by sputtering, function of the pressure, the temperature of substrate (Fig. 16.1) and the substrate bias.

It is very well known [9,27] the dependence between hardness and grain size, and it is expressed, according to Hall [28] and Petch [29], as:

$$H = H_0 + \frac{k}{\sqrt{d}} \tag{16.3}$$

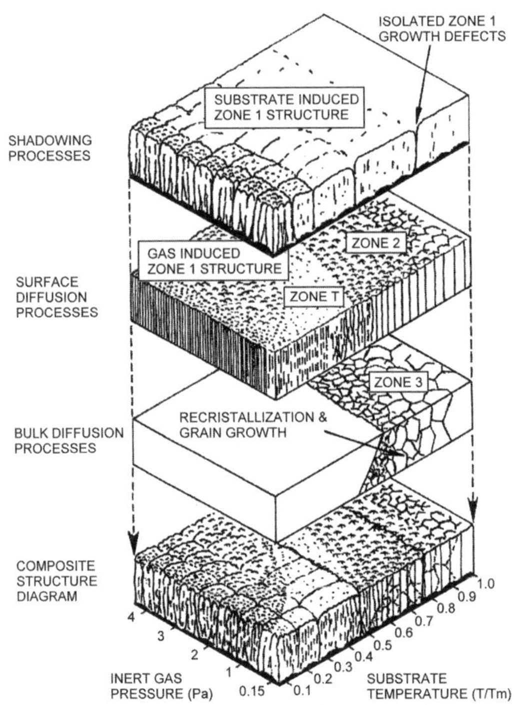

Figure 16.1: Thornton coating's morphologic model [23,25].

where H is the hardness of the coating, H_0 is the intrinsic hardness, k is a material constant and d is the grain size. However, the lower limit of application of this law is not yet completely known. The increase of the hardness with the decrease of the grain size is not exclusive of the bulk materials and this equation can also be used in coatings. However, according to several authors [3,30,31], the Hall–Petch equation is limited to monolithic coatings with grain size or to multilayer coatings with bilayer period higher than 10 nm. In order to attain ultra-hard coatings several researchers have been working on nanocrystalline materials, presenting grain sizes in range of 10–100 nm [32–34]. The evaluation of grain size in thin films causes some problems, resulting from artifacts of the morphology using scanning electron microscopy (SEM), the size evolution during the preparation for observation in transmission electron microscopy (TEM) or the overlapping effects in its calculation using X-ray diffraction (XRD) lines.

16.2.5. Roughness and Surface Morphology

The coating's roughness influences significantly the tribologic behavior of the tribosystems. Sputtered thin films follow the surface roughness of the substrate [1,3,4]. Moreover, either high substrate roughness or an extremely polished surface may contribute to a decrease of adhesion between the coating and the substrate [35,36], hence there is an optimal value of substrate roughness.

16.2.6. Residual Stress

Hard coatings deposited by sputtering on softer substrates can have high levels of compressive residual stress, attaining values as high as 10 GPa [37]. However, the compressive stress is welcome in almost all applications if it is kept at a not too high level. Hard coatings with compressive residual stress of 3 GPa had shown in some applications the best performance [38,39]. Residual stress in coatings is constituted by three main types of stress named according to its origin. The first contribution is *epitaxial stress*, due to the structural mismatch between the coating and the substrate. The second is attributed to the growing process of the coating, also known as *intrinsic stress*. And last but not the least, is the *thermal stress* due to the mismatch between coating and substrate thermal expansion coefficients, and is generated during the cooling process from the deposition temperature (which can attain temperatures in order of 400°C) to the room temperature. The contribution of the epitaxial stress can be neglected, when compared with others residual stress sources [40] and it is assumed that it is relieve by dislocations, still by acting directly in the interface, it affects significantly the mechanical behavior of coatings [41]. Therefore, the residual stress value expresses mainly the intrinsic and thermal stresses. Moreover, structural transformations occurring during or after deposition process also contribute to modify residual stress value of the coatings [42].

16.2.7. Adhesion

The ASTM standards (D907–70) define adhesion as the condition in which the surfaces are kept together by interfacial forces. This can be due to chemical bonding, electrostatic forces or both. It is necessary to establish the difference between the basic adhesion that corresponds to the maximum value of adhesion, and the practical adhesion, which can be measured by the several experimental methods [36,43–51]. Adhesion can be evaluated following a force or an energy approach. Considering the forces and defining the adhesion force as the force necessary to detach coating from substrate, or in terms of energy, that is, defining the work of adhesion as the energy spent to separate the materials maintaining a well-defined interface, and is expressed as:

$$W_{CS} = \gamma_C + \gamma_S - \gamma_{CS} \tag{16.4}$$

where W_{CS} is the work of adhesion, γ_C and γ_S are the specific surface energy of the coatings and the substrate, respectively, and γ_{CS} is the specific interfacial energy. The work of adhesion can be related with the force of adhesion (F_{CS}) by the equation:

$$W_{CS} = \int F_{CS}(x)\,dx \tag{16.5}$$

Plus, it is necessary, on adhesion evaluation, to distinguish between adhesive and cohesive failures. Cohesive failures occur inside the coating (crack perpendicular to the coating surface) or in the substrate, and the load applied to the indenter during the scratch test (used as adhesion test in hard coatings) is named Lc_1. But, if the failure occurs in the interface coating/substrate, the failures are adhesive and the load applied to the indenter named Lc_2. The existence of defects and surface irregularities can induce uncharacteristic failures at lower loads, usually 1 or 2 N lower than the normal values [44]. Several factors influencing the critical load values evaluated by scratch test are summarized in Table 16.1.

Table 16.1: Factors influencing the critical load

Intrinsic factors	Extrinsic factors
Load speed [46]	Substrate properties [44,47,49]
Indentation speed [46]	Coatings properties [44–51]
Indenter radius [44,46]	Friction coefficient [47,48]
Wear of the indenter [46]	

16.2.8. Hardness

The hardness of the materials reflects their ability to resist to plastic deformation imposed by indentation. This property depends on the characteristics described before, particularly structure, chemical bonding, grain size, defect density and residual stress of the coatings [4,6,7]. An ideal hard coating must be hard at low and high temperatures. The hardness of the coatings can be increased by the formation of solid solutions, by precipitation of a second phase, by nanocrystallinity or by the deposition of multilayers of nanometric dimensions. The hardness is also strongly dependent of the morphology of the coatings. The columnar morphology of coating generally induces lower hardness values than a compact one and can also contribute to anisotropy of this property [5], which is normally attributed to grain growth preferential orientations. Nevertheless, the effect of the residual stress in the hardness of the coatings should not be neglected, since the coatings with compressive residual stresses present the highest hardness values.

16.2.9. Young's Modulus

The Young's modulus is the property less dependent of the characteristics of the coatings (microstructure and morphology), but is strongly dependent of the chemical composition. Thus, the nitrides and carbides with high bond energy values, short bonding length and a high degree of covalence, are associated to high Young's moduli [52,53]. Almost all the coatings for high performance tribologic applications are oriented in searching high hardness but the Young's modulus must not be neglected as assured Matthews and Leyland [54,55]. In fact, the optimization of the hardness to Young's modulus ratio [54–57], well known by tribologists and defined as plasticity index (δ_H), is necessary to attain a good mechanical performance. Other formulation of plasticity index of coatings was proposed before by Milman *et al.* [58] by establishing a relation between the hardness, the

Young's modulus and the Poisson coefficient (ν) of coatings, and expressed as follows:

$$\partial_{\mathrm{H}} = 1 - 14.3(1 - \nu - \nu^2)\frac{H}{E} \tag{16.6}$$

16.2.10. Fracture Strength

The fracture toughness and, specially, the resilience of coatings are two parameters that assume an important role; especially in service conditions where the successive interruption and the restarting during the application and/or manufacturing processes like cutting, demand coatings with high fracture strength. Several mechanisms were identified to increase the fracture strength of the coatings, without decreasing its hardness, mainly those acting on the nanocrystallinity [27–29,54,55] and on the presence of interlayers in the coatings [3,54,55,59–62].

16.2.11. Thermal–Chemical Stability

In the study of the hard coatings, besides the mechanical properties evaluation, their thermal–chemical stability should not be underestimated. In fact, in real service applications of hard coatings, such as cutting and drilling operations, the temperature is a factor always present. Thus, besides the structural evolution with the increase of the temperature, the oxidation process should be also taken in consideration. Most of the time, the failure of hard coatings is not due to unsuitable mechanical properties, but to the degradation caused by oxidation/corrosion problems. Consequently, it has been the normal procedure in the development of hard coatings to study their behavior at increasing temperatures in oxidizing environments, after the envisaged value for the mechanical properties has been attained. These studies have been somewhat postponed for a time, namely the most studied system TiN, which was once the subject of much research work, only began to be seriously investigated in the last decade. From recent studies, several proposals to overcome the mediocre behavior at high temperatures have been reported. Among them, and without intending to be exhaustive, it is important to mention the following approaches:

- *The addition of "reactive elements" (RE)*: They can act either as preferential nucleation sites, in the oxidation process leading to the formation of a protective scale, or form an intermediate oxide layer acting as a diffusion barrier. Moreover, their presence in solid solution segregates or precipitates, as can modify the oxide morphology and microstructure, influencing the diffusion rates and the mechanical properties of the oxide scales. The most studied RE are Al, Cr and Si [63–76].

- *The deposition of multilayer coatings*: The scope is the association of two materials one of them with a higher oxidation resistance than the other, in order to gives the overall coating good oxidation behavior [77–80].

16.3. Binary Metal Nitrides and Carbides

In this section will be discussed the properties of the well-known titanium and tungsten nitride and carbide hard monolithic coatings.

16.3.1. Titanium Nitrides and Carbides

Titanium nitride and carbide coatings deposited by sputtering are stable over a wide chemical composition range but their structure and properties are strongly dependent of their divergence from stoichiometry [81,82]. Cubic TiN_x is stable over a broad composition range ($0.6 \leqslant x \leqslant 1.2$) and appreciable vacancy concentrations (up to 50 at.%) [83]. The TiN phase, indexed to a NaCl structure, has at the stoichiometric composition a lattice parameter of 0.4240 nm. Besides the chemical composition, also the thermal expansion coefficients mismatch between the coating and the substrate, which causes stresses in the coatings, can promote a shift in the lattice parameter value. Additionally, the lattice parameter deviations can be due to the presence of impurities in the structure; while oxygen induces a decrease, carbon and nitrogen originate an increase of the lattice parameter; and argon, generally, induces high values of residual compressive stress in the sputtered coatings causing a significant change in the lattice parameter (up to 1 at.% of argon can be trapped in TiN coatings) [84,85].

Coatings deposited by sputtering have, usually, preferential growth orientations. For the TiN coatings the most commonly observed growth orientation is (111), although the (200) and (220) are also reported [84,86]. Increasing ion current density, the growth orientation change from (111) to (200) and increasing the substrate bias a change from (111) to (220) are detected. However, for high deposition rates, (111) is the preferential orientation. The (111) has the lowest surface energy, and, since close-packed planes are slow growth planes, they survive at expenses of fast growth planes [84,87]. The increase of the substrate temperature promotes the increase mobility of the adatoms during growth and, also, the improvement of the migration of grain boundaries. These factors compel the grain size growth. Moreover, if coatings are exposed to ion bombardment during growth, the grain size is attempted to decrease as the energy of ions is increased. In fact, ions with high energy give rise to a high surface defect that will increase the number of preferential nucleation sites. Furthermore, an increased ion bombardment induces a preferential re-sputtering of adatoms, thus causing a low adatom mobility, whose effect is a reduction in the grain size. However, not only the energy of the ions, but also the number of ions or ion current density is important. In the cases of low voltages, but still with a relatively high ion current, the momentum transference from the ions to the adatoms can occur, giving rise to higher mobility and also to slightly larger grain sizes and/or few defects in the coatings. It is worth noting that deviations from stoichiometry can change the grain size. In the stoichiometric coatings no voids are generally observed, whereas in the non-stoichiometric ones these defects, located in the grain boundaries, are clearly detected. The existence of the voids can be evaluated by the density measurements and it is associated to a low density of the non-stoichiometric coatings. The growth conditions, such as substrate temperature or ion bombardment are not the only responsible for the structure and crystallite grain sizes of the coatings, other factors, like the substrate material can play an important role. The presence of different phases in the substrate can induce a localized epitaxial growth if one phase of the substrate has a similar structure to the coating (e.g. in high-speed steel the presence of V_xC_y, with a lattice mismatch to TiN of 1.6% [84], induces an epitaxial growing, and a microstructure with large grains surrounded by small ones). TiN and TiC coatings have a columnar morphology typical of a zone 1 or T according to Thornton [25] classification. However, if substrate bias is applied, the columnar growth morphology is, generally, destroyed, and for high bias voltages and current densities featureless morphologies have been reported [12,84].

The compressive residual stress of the titanium carbide and titanium nitride coatings are extremely high, achieving values of 8 GPa [88,89] being almost independent of the substrate bias [89] and increasing as the non-metal to titanium ratio approaches to the stoichiometric composition [90]. These values may result from the high density of grain boundaries and defects in TiC and TiN thin films, decreasing the mobility of the dislocations. The stress state could be more favorable if the morphologies were columnar. The highest hardness of the TiN are found on coatings with nitrogen to titanium ratio of approximately 0.8, attaining values of 30 GPa. For over-stoichiometric coatings, hardness drops abruptly to values of 3.5 GPa for nitrogen to titanium ratio of 1.2. Typically, the hardness of the titanium nitride is lower than the titanium carbide coatings [3,91,92]. Concerning the Young's modulus, the titanium carbide attains values around 500 GPa and the titanium nitride only 260 GPa [3,91,93]. In fact these results are in agreement with the relative strength of the TiN and TiC chemical bonding, if these coatings have the same morphology. However, some authors pointed out the opposite [92] with titanium nitride coatings attaining values higher than 500 GPa [88], which could be attributed to different morphologies of the coatings.

In spite of the wide range of applications of TiN coatings, they present some disadvantages, mainly the oxidation at relatively low temperatures ~500°C. Therefore, the role of oxidation is very important for practical applications and has been subject of several studies [63,94–101]. The great driving force for the transformation of TiN to titanium oxide:

$$TiN(s) + O_2(g) \leftrightarrow TiO_2(s) + \frac{1}{2}N_2(g) \tag{16.7}$$

is the largest negative free energy produced by this reaction. It is expected that the nitrogen evolved during oxidation to be released into the atmosphere. However, different researchers [96,98–101] observed that, after oxidation at low temperatures, a certain amount of nitrogen remains in the uppermost part of the coating forming an oxinitride scale, TiN_xO_y. Only with increasing temperature the nitrogen is completely released and the oxygen substitutes, progressively, the nitrogen in the crystalline TiN lattice. All investigators agree that the progress of oxidation of TiN films is parabolic in the temperature range from 400°C to 650°C, which means that the prolonged oxidation leads to the formation of a TiO_2 (rutile) layer, whose thickness growth is proportional to the square of time. The titanium oxides are non-stoichiometric compounds and it is generally concluded that the oxidation of TiN coatings is mainly controlled by the inward diffusion of oxygen through the TiO_2 layer, determining the parabolic kinetic of oxidation [63,94,95]. TiN coatings exhibit a superior resistance to oxidation in comparison to TiC thin films, whose oxidation begins at temperatures as low as 400°C [102]. Although, both Ti-based systems present the same rate-determining step of oxidation: diffusion of oxygen through the rutile scale, the TiC films show a substantially higher weight gains then TiN coatings. This behavior was attributed to the higher defect structure of TiC in comparison to TiN structure [102], being more permeable to the diffusion of the reactive species. As nitrogen, also carbon diffuses to the scale surface where it reacts with oxygen to form carbon oxides. Both nitrogen and carbon diffusion must be considered as a possible explanation for the morphologic variations observed in TiN and TiC oxidized surfaces, giving rise to spalling of oxide layer at high-temperature oxidation [95].

16.3.2. Tungsten Nitrides and Carbides

Tungsten nitride coatings have been used in the microelectronics industry as diffusion barriers between silicon and its oxides [103–105] and after applied as cutting tools coatings due to the low dependence of hardness with temperature.

The nitrogen content in the tungsten nitride coatings can be as high as 65 at.% [103], even though these values are lesser than the carbon content in the tungsten carbide, which can be as higher than 90 at.% [106]. The structure of the W–C or W–N coatings function of the chemical composition can be assembled in four major groups [19]:

- *Group I* – Comprises the coatings with a nitrogen and carbon content to 7.5 at.% and 17.6 at.%, respectively and presents a b.c.c. structure type α-W.
- *Group II* – In the coatings with nitrogen and carbon content of 16.9 at.% and 21.1 at.%, respectively, a new f.c.c. structural phase type W_2N and β-WC_{1-x} respectively, appears and the final structure becomes a mixture of these compounds with α-W.
- *Group III* – In this group there is no sign of a α-W structure and the coatings show only an f.c.c. structure type nitride or carbide for coatings with 54.9 at.% of nitrogen and with 36.9 at.% of carbon.
- *Group IV* – This last group includes coatings with high carbon content beyond 70 at.% and only exhibit a distorted f.c.c. structure type β-WC_{1-x}.

The deposition pressure influences the texture of the coatings, for example, the single-phase β-WC_{1-x} deposited with low deposition pressures showed a preferred (200) orientation, instead of (111) obtained for high pressure [107].

The morphologies of W–N coatings with increasing nitrogen contents [19,105] are generally more compact then the ones of W–C coatings for similar carbon contents, in spite of the several types of morphologies observed [19,107–112]. The deposition pressure also controls the residual stress of the single-phase coatings. Hence, Keller *et al.* [107] confirmed that increasing the deposition pressure decrease the compressive residual stress in β-WC_{1-x} coatings. Moreover, multiphase coatings possess lower compressive residual stress if one of the phases is amorphous.

The adhesion of the coatings decreases as the non-metallic element content increases. A catastrophic failure at low indentation loads was detected in the coatings with high carbon content, which is due to the amorphous phase [19]. The presence of the non-metallic element induces an increment of hardness of the coatings, from 15 to 35 GPa, if the structure remains b.c.c. α-W. However, further increases of non-metallic element content in the coating induce new structures but no significant change in the hardness can be noticed. In fact, coatings with high carbon content (>70 at.%) have low hardness values, which can be related to the presence of amorphous carbon in the film.

Considering the oxidation behavior of the binary W–N/C sputtered films, the addition of interstitial elements C or N to the W sputtered coatings results in a decrease of the weight gain during the oxidation tests, in comparison to a single W film [113,114]. First of all, it should be pointed out that these elements cannot be detected

a) b)

Figure 16.2: SEM micrograph showing the oxide layer destruction resulting from the gas evolution; (a) $W_{93}C_7$ film oxidized at 800°C/30 min [75] and (b) $W_{45}N_{55}$ film oxidized at 700°C/30 min [74].

in the oxide layers; that is, they are lost for the atmospheric environment. Thus, their removal from the coating counterbalances the incorporation of oxygen and limits the increases in the weight gain of the sample [74,75,113,114]. During the oxidation process, the oxygen ions substitute C/N in interstitial positions and these elements can then diffuse outwards either as C or N or forming CO/CO_2 or N_2. If the oxidation process proceeds slowly, the molecular gas diffuse outwards as it is formed. Otherwise, bubbles of these gases appear in the interface oxide/coating. When the C/N content in the film is high enough, the bubbles of molecular gas formed in the interface oxide/coating reach a significant size, creating stress in the oxide layers that can lead to their destruction. This follows several different morphologic aspects as shown in the Fig. 16.2. This phenomenon is in some cases, very severe and it is expressed in the oxidation curves by a sudden abrupt decrease in the weight gain values arising from the loss of flaked oxides from the sample holder in the thermobalance [74]. The oxidation behavior of W–C/N sputtered coatings is described, in the temperature range of 600–800°C, by a parabolic growth law [113,114], which indicates that the process is controlled by the diffusion of reactive species (oxygen ions). Concerning the type of oxides formed, it was possible to distinguish a compact internal layer, indexed as a WO_x and a porous outer scale corresponding to WO_3.

16.4. Ternary Nitrides

Ternary nitrides have been used in mechanical industry applications with increasing success, when compared with carbides. Among ternary nitrides, titanium aluminum nitride is now commercially available, namely as cutting tools coating, replacing successfully the titanium nitride coatings in high-speed cutting operations. In these applications, the high aluminum content in thc TiN lattice is advantageous due to the formation of a high adhesion thin aluminum oxide layer instead of titanium oxide, which is easily detached during the cutting operations. Additionally, a recent developed ternary nitride system based on Ti and W transition metals exhibits promising performance and is envisaged for applications in severe mechanical work conditions*.

16.4.1. Titanium–Aluminum Nitrides

In equilibrium, settled by the Ti–Al–N ternary diagram, the titanium, the aluminum and the nitrogen show insolubility in AlN, TiN and TiAl phases, respectively [114,115]. However, different authors [3,33,59,67,71,117–157] stated that TiN and AlN form solid solutions with aluminum and titanium, respectively, when the production process is far from the equilibrium conditions. Hence, these metastable compounds are named $Ti_{1-x}Al_xN$ and the covalent character of their chemical bonds increase with increasing x-value. Besides the nature of the chemical bonding, also, the structure of the $Ti_{1-x}Al_xN$ coatings changes as a function of x, varying from an f.c.c. NaCl type to an h.c.p. structure. Wurzite type structure are obtained, when x-value increases from $x = 0$ (TiN) up to $x = 1$ (AlN) (Table 16.2). This dependence was theoretically calculated by Makino and Miyake [152], who estimated that the maximum solubility of AlN in TiN, in optimal conditions, was 65.3%. However, this solubility decreases with increasing sputtering deposition temperature attaining the value of 50% at 450°C. Despite the widely recognized dependence of the titanium aluminum nitrides structure on the chemical composition, there still is no agreement about the limits of the phase stability ranges (Table 16.2). $Ti_{1-x}Al_xN$ sputtered coatings with small values of x present a B1-NaCl structure similar to TiN coatings, where the preferred growth orientation is (111). By increasing the Al content in the coating, some authors also detected the (200) orientation [122]. However, it should be kept in mind that not only the coating chemical composition, but also the deposition parameters influence the coating structure. In fact, some authors have reported the highest decrease of the intensity of the XRD peaks corresponding to the high Miller indices, with the increase of the substrate bias [131]. Further increase of x-value reveals a new second phase, h.p.c. B4-AlN, concomitantly

---

* "ECOSTAMP – Eco-efficient stamping process of sheet metal parts by development of innovative coatings for self-lubricating dies" – CE, Projecto GROWTH – Contract n GIRD-CT-2002-00712 (2002–2005).

Table 16.2: Structure of $Ti_{1-x}Al_xN$ coatings function of chemical composition

x ($Ti_{1-x}Al_xN$)	Structure	Preferential growth planes	Lattice parameters (nm)	Grain size (nm)	References
0.34	B1-TiN	(200)	0.425	–	Penttinen *et al.* [132]
0.5		(200)	0.417	–	
~0.5		(111)	0.416–0.417	–	Jehn *et al.* [118]
~0.5		(111)	0.417 → 0.422 V_S →	105 → 35 V_S →	Håkansson *et al.* [119]
~0.5		(111)	0.4157	–	Adibi *et al.* [122]
~0.5		(111)	–	–	Lii *et al.* [59]
$x \leqslant 0.5$		(200)	0.424		Zhou *et al.* [143]
0.5		(200) (111)	–	10	Andersen *et al.* [147]
$x \leqslant 0.52$		(002)	–		Wahlström *et al.* [123]
$0 \leqslant x \leqslant 0.60$		(111)	0.424 → 0.418		Suzuki *et al.* [133]
0.62		(111)	0.417	–	Zoestbergen *et al.* [150]
0.62		(111)	–	–	Witthaut *et al.* [71]
$x = 0.54$	δ-TiN	(111) → (200)	–	–	Wu *et al.* [146]
~0.55		(111)	–	–	Jiménez *et al.* [142]
$0.52 \leqslant x \leqslant 0.59$	B1-TiN + B4-AlN	(002) + (0002) $(10\bar{1}1)$	–	70 → 10 100 → 400°C	Musil and Hrubý [33]
$x > 0.60$			–	–	Suzuki *et al.* [133]
$0.6 \leqslant x \leqslant 0.7$			0.415	–	Zhou *et al.* [143]
0.73			–		Witthaut *et al.* [71]
$0.59 \leqslant x \leqslant 0.84$	B4-AlN	(0002) + $(10\bar{1}1)$	–	–	Wahlström *et al.* [123]
$0.59 \leqslant x \leqslant 0.86$		(0002) $(10\bar{1}1)$	–		Musil and Hrubý [33]
$x > 0.86$		(0002)	–		Wahlström *et al.* [123]
$x = 1$		(0002)			Musil and Hrubý [33]

with the TiN structure, where (0002) and ($10\bar{1}1$) are the preferred orientations. For the highest aluminum contents, the B4-AlN is the phase present and (0002) is the growth orientation. However, some authors also found, in $Ti_{0.4}Al_{0.6}N$ coatings, the presence of an f.c.c. AlN phase, which grows coherently with the f.c.c. TiAlN phase [155–157]. The lattice parameter of the $Ti_{1-x}Al_xN$ sputtered coatings in TiN monophasic domain is a function of the chemical composition and decreases as the aluminum content increases. In example, the lattice parameter for $x = 0.32$ is 0.424 nm (similar to TiN) and decreases up to 0.417 nm for $Ti_{0.38}Al_{0.62}N$ coatings [150]. Some authors [138] stated that the chemical composition of $Ti_{1-x}Al_xN$ sputtered coatings influences their nanocrystallinity and the decreasing of aluminum content showed to lead to a simultaneous decrease of the grain size. Nevertheless, for other authors it seems to observe the opposite [123,156].

Other deposition parameters can also reduce the grain size of $Ti_{1-x}Al_xN$ sputtered coatings, namely the increasing of the substrate bias [119] and/or the substrate temperature [33] with the subsequent improvement of the coating's mechanical properties.

Concerning the morphology, $Ti_{1-x}Al_xN$ sputtered coatings present, generally, a dense columnar morphology [118,119,133,136–139] (Fig. 16.3), which corresponds to type I or T of the Thornton's model [25]. Coating's density is strongly influenced by the morphology and it approaches the theoretic value of $Ti_{0.5}Al_{0.5}N$ density (4.6) [135,158] as the coatings columns get closer to each other. The flux of nitrogen in the deposition chamber changes the chemical composition and, consequently, the density of the coatings [158]. Generally, the increase of nitrogen flow promotes an improvement of density. However, there is an ultimate nitrogen flux from which the opposite effect is produced, leading to a decrease of compactness and hence to the decrease of the density of the coatings.

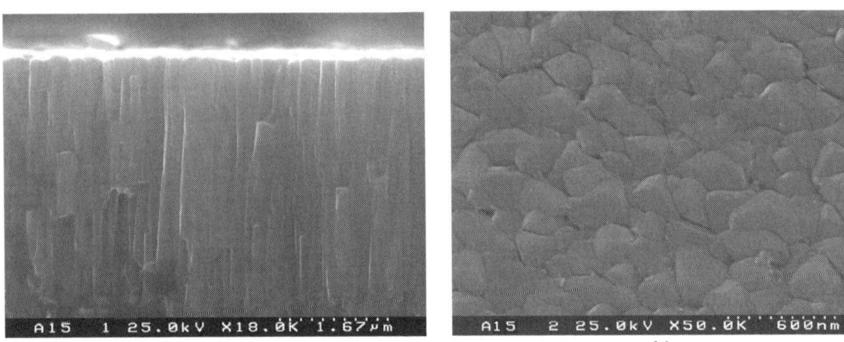

a) b)

Figure 16.3: Morphology of (a) cross section and (b) surface of $Ti_{1-x}Al_xN$ sputtered coatings [154,158].

Table 16.3: Mechanical properties of $Ti_{1-x}Al_xN$ coatings

x ($Ti_{1-x}Al_xN$)	Hardness (GPa)	Young's modulus (GPa)	Residual stress (GPa)	Adhesion (N)	References
0.4	33.8		−6.2		Santana *et al.* [156]
0.6	38.5		−11.0		
0.7	29.3		−4.9		
0.5	22–23	–	–	45–60	Münz [97]
~0.5	10 → 42 V_S →	–			Håkansson *et al.* [119]
0.5	22–24	–	–	~60	Knotek *et al.* [120]
~0.5	–	–	–	>70	Lii *et al.* [59]
0.5	~36	–			Shew *et al.* [131]
$x \leqslant 0.46$	20–23	–	–	30–35	Jarms *et al.* [137]
$x \geqslant 0.69$	12	–	–	23–18	
$0 \leqslant x \leqslant 0.60$	19 → 32	–			Suzuki *et al.* [133]
$x \geqslant 0.69$	28 → 12	–			
$x \leqslant 0.52$	35	–	–	≤85	Panjan *et al.* [138]
0.64	17	–	–	104	
$x \leqslant 0.5$	26 → 40	360 → 650	+0.1 → −1.1	–	Zhou *et al.* [143]
$0.6 \leqslant x \leqslant 0.7$	33 → 22	500 → 420	−0.2	–	
0.59	27	~400	–	–	Musil and Hrubý [33]
↑N	33				
	47 T_d = 200°C				
$x = 0.54$	16–34	–			Wu *et al.* [146]
0.64	–	–	−0.26	29	Tavares *et al.* [151]
0.5	36 → 40 ↑V_S 80 → 130	~450	−1.5	–	K.N. Andersen *et al.* [147]

$Ti_{1-x}Al_xN$ coatings with titanium to aluminum ratio closer to 1 exhibit the best performance in cutting applications. For several researchers, the selection of the most promising coating is performed from their hardness evaluation. However, the Young's modulus, as already stated, must not be neglected. Table 16.3 shows the values of the main mechanical properties of the $Ti_{1-x}Al_xN$ sputtered coatings, namely hardness and Young's modulus. Due to their importance in the cutting performance of the coatings, as referred in a previous section, the adhesion and residual stress values are also summarized in Table 16.3.

The hardness of the $Ti_{1-x}Al_xN$ coatings is strongly dependent on the aluminum content. Coatings with high aluminum content and an f.c.c. structure possess the highest hardness values [133,138,142,143]. In fact, the substitution of titanium by aluminum atoms decrease the interatomic distance of the supersaturated $Ti_{1-x}Al_xN$ coatings, and increase the covalent-bonding energy (E_B), which can be expressed by the equation [103]:

$$E_B = kd^{-2.5} \tag{16.8}$$

where k is a constant-dependent bonding and d is the interatomic distance. Consequently, the covalent-bonding degree (α_B) of the $Ti_{1-x}Al_xN$ also increases with the aluminum content, which associated to the decrease of the lattice parameter results in an increase of hardness of the coatings. This increase in the hardness (H) of the coating, according with Hones *et al.* [13], can be expressed as:

$$H = C\frac{E_B\alpha_B}{a^2} \tag{16.9}$$

where C is a morphology-dependent constant, E_B is the bonding energy and a is the lattice parameter.

Therefore, in $Ti_{1-x}Al_xN$ all the variables favor an increase of the hardness values. However, further increase of Al content leads to the formation of h.c.p. or a c.f.c. AlN structure, which induces a significant decrease [121,133,137,138,143] or increase [157] in the hardness, respectively.

The Young's modulus of the coatings follows the same tendency as their hardness. Reporting to stress values, $Ti_{1-x}Al_xN$ sputtered coatings have compressive residual stress which increases the increase of aluminum content incorporated in the TiN lattice in substitutional positions. Compressive stresses generally fall in the range from 0 to -3 GPa, but the coatings deposited on stainless steel substrates can achieve values as high as 4.2 GPa. Having in mind the adhesion values shown in Table 16.3, it seems clear that this characteristic increase with x-value up to $x = 0.64$. However, Santana *et al.* [156] observed residual stresses attaining -11.0 GPa, in the $Ti_{0.4}Al_{0.6}N$ coatings. The increase of these mechanical characteristics is due to the decrease of the grain size and the presence of second f.c.c. AlN phase.

Considering the oxidation behavior of these ternary systems, the introduction of Al in TiN as the purpose of establishing an outermost protective layer of Al_2O_3 should protect the nitride film from further oxidation and improve the wear resistance

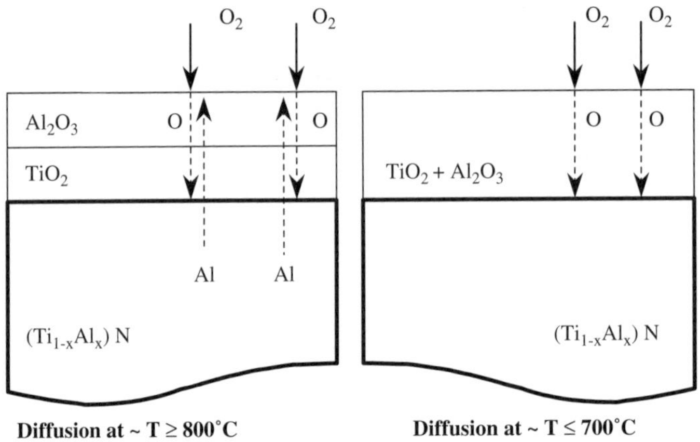

Figure 16.4: Schematic illustration of diffusion of O and Al in high and low temperatures ranges for the $Ti_{1-x}Al_xN$ films [66].

under operating conditions [63,64,66,98,159–161]. In fact, while TiN films start to be oxidized from 500°C, Ti–Al–N coatings showed much improved oxidation resistance up to 800°C [64,160]. Two different oxidation mechanisms are reported for $Ti_{1-x}Al_xN$ films, corresponding to the formation of a two-layer structure of Al_2O_3/TiO_2 at the surface or the formation of mixed of TiO_2 and Al_2O_3 oxides (Fig. 16.4).

At high temperatures above 800°C, the formation of two-layer structure was interpreted [63,64,66,160–162] as being mainly due to the preponderant out-diffusion of Al, which governs the establishment of the outer Al_2O_3 layer. Using inert marker experiments McIntyre *et al.* [64] have shown that Al is the mobile specie during the oxidation of $Ti_{0.5}Al_{0.5}N$ on the basis of scanning transmission electron microscopy (STEM) with energy dispersion X-ray (EDX) analysis. Assuming that the Al-diffusion rate in Al_2O_3 in the temperature range 700–1000°C is higher than that of oxygen, it is likely that Al diffuses through the Al_2O_3 layer as well and oxidizes near the free surface. On the other hand, the formation of titanium oxide below the Al_2O_3 layer can only be explained by oxygen transport through the outer Al oxide layer and the fact that practically no Ti oxide was observed on top of the Al_2O_3. This suggests an efficient trapping of Ti below the Al_2O_3 layer by the diffusing oxygen and a reduced diffusivity of Ti through this layer [63,118,160]. At low temperatures, such as in the 600–700°C range, the formation of mixed oxide layer [64,66,130,159] could only be explained by the low diffusivity of Al in the $Ti_{0.5}Al_{0.5}N$ coatings. Oxygen, on the other hand, diffuses rapidly inward through the growing oxide film and forms oxides of both Al and Ti at the oxide–nitride interface. The driving force for the migration of Al to the surface is its oxidation, which is associated with the substantial decreases in free energy when compared with formation of titanium

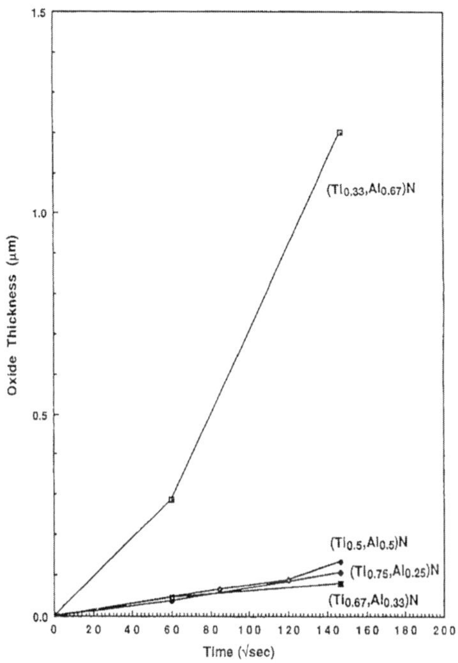

Figure 16.5: Oxidation kinetics of various $Ti_{1-x}Al_xN$ films at 700°C in air [66].

oxides. In spite of the oxidation behavior of $Ti_{0.5}Al_{0.5}N$ coatings be the most studied, the oxidation of other compositions was also studied [66,130,162] and those compositions sometimes showed different oxidation kinetics (Fig. 16.5).

As a final remark, stoichiometric $Ti_{0.5}Al_{0.5}N$ sputtered coatings have the best characteristics to be used under severe mechanical applications such as high-speed cutting.

16.4.2. Titanium–Tungsten Nitrides

The investigation of $Ti_{1-x}W_xN$ sputtered coatings has been a natural evolution from the research work developed concerning the W–N thin films [19,22,103,104,163–167]. The well-known effect of nitrogen content, in the improvement of the hardness of W–N coatings (35 GPa) [19,22,165,166], associated to the improvement of chemical stability resulting from the addition of titanium to the tungsten nitride, resulted in $Ti_{1-x}W_xN$ coatings suitable for application in aggressive environments. Several authors [34,167–172] are unanimous considering that the structure of the $Ti_{1-x}W_xN$ is isomorphous of TiN and W_2N structures. Therefore, it is possible to observe the occurrence of this last phase in $Ti_{1-x}W_xN$ coatings under-stoichiometric toward nitrogen. The titanium to tungsten ratio in $Ti_{1-x}W_xN$ coatings sputter deposited from

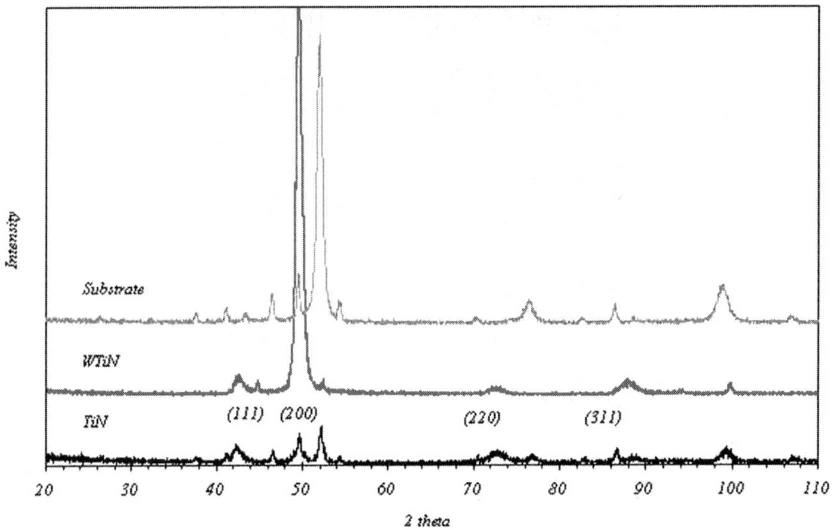

Figure 16.6: $Ti_{1-x}W_xN$ XRD pattern [158].

targets with Ti:W ratio equal to 1 significantly changes with the nitrogen content. In fact, the increase of nitrogen content produces a decrease of metals ratio to values of 0.85 [168–171], which is attributed to the preferential re-sputtering of titanium during the deposition process. The structure of these $Ti_{1-x}W_xN$ coatings is a B1-NaCl type, similar to what was mentioned before for TiN and W_2N phases, with a lattice parameter of 0.425 nm and they grow with a preferential (200) orientation (Fig. 16.6), as stated by some authors for $Ti_{1-x}Al_xN$ ($x \cong 0.5$).

The morphology of the $Ti_{1-x}W_xN$ sputtered coatings is denser than the one found in $Ti_{1-x}Al_xN$ coatings (Fig. 16.7). Some authors [170,171,173] claim that it is columnar, but highly compacted with low intercolumnar porosity. However, this statement is not unanimous and other authors [12,34,158,169,170,172] state that the morphology is featureless. When the Ti–W–N system is sputtered in a reactive atmosphere, the coatings are constituted by a huge quantity of small nanograins of titanium and tungsten nitrides, whose dimensions are in the range from 20 to 50 nm, surrounded by nanograins of 5 nm constituted by the excess of tungsten [168]. These ternary transition metal nitride coatings have high hardness (>35 GPa) and Young's modulus (≅600 GPa) [34,174]. The maximum hardness values (50–60 GPa) [34,171] are attained in the coatings with nitrogen content in the range from 20 to 30 at.% [34,171], and further increase of nitrogen content leads to the decrease of the hardness and the Young's modulus due to the increasing number of anti-bonding chemical bonds [12].

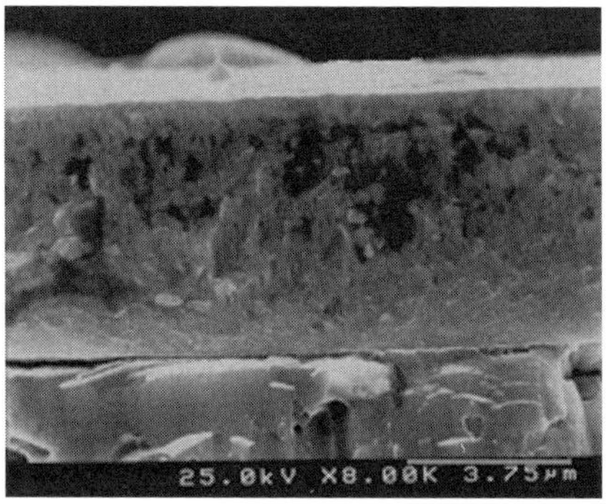

Figure 16.7: Cross-section morphology of $Ti_{1-x}W_xN$ sputtered coating [158].

Concerning compressive residual stress of $Ti_{1-x}W_xN$ coatings, they increase with nitrogen content, from 1.5 GPa in coatings for 0 at.% N to 7 GPa in the coatings with 40 at.% N [171,175].

The adhesion of the coatings, evaluated by scratch test, is also function of the titanium to tungsten ratio and of the nitrogen content [36,169,170,175]. Adhesion values of 40 N can be attained in the $Ti_{0.3}W_{0.3}N_{0.4}$ sputtered coatings [175].

The modification of W–N-sputtered coatings by the addition of selected elements (i.e. M = Ti, Ni) is known for leading to the improvement of oxidation behavior of the W–N–M thin films [75]. Fig. 16.8 presents the parabolic rate constants of oxidation as a function of the oxidation temperature for some different types of W–N–M films.

The influence of Ti on the oxidation behavior should be different on W–Ti ($E_a = 234\,\text{kcal mol}^{-1}$) and W–Ti–N ($E_a = 197\,\text{kcal mol}^{-1}$) [113] sputtered films. Different slopes in Arrhenius law curves (Fig. 16.8) suggest different oxidation mechanisms, due to the simultaneous presence of Ti and N or not. Moreover, the analysis of the XRD patterns obtained at the oxidation temperature range from 600°C to 800°C allowed the detection of TiO_2/Ti_3O_5 in W–Ti films which is not revealed in $Ti_{1-x}W_xN$ films [75,113]. The analysis by electron dispersion X-ray spectroscopy (EDXS) of the cross section of the oxide layers lead to the conclusion that there is no agglomeration of titanium oxide in the form of a protective layer, at least with the lateral resolution permitted by the EDXS technique. If the reactivity of Ti for oxygen is compared to that of W ($\Delta H_f^{TiO_2} = -219\,\text{kcal mol}^{-1}$, $\Delta H_f^{WO_3} = -200\,\text{kcal mol}^{-1}$ [14]), it is possible to conclude that titanium can be preferentially oxidized than tungsten, but at a very low degree; that is, tungsten is almost simultaneously oxidized as titanium. Thus, TiO_2/Ti_3O_5 oxide must appear in the form of very fine particles (for Ti content much less than W content), not forming a continuous layer inside/outside the tungsten oxide layer. For $Ti_{1-x}W_xN$ coatings thermodynamic shows

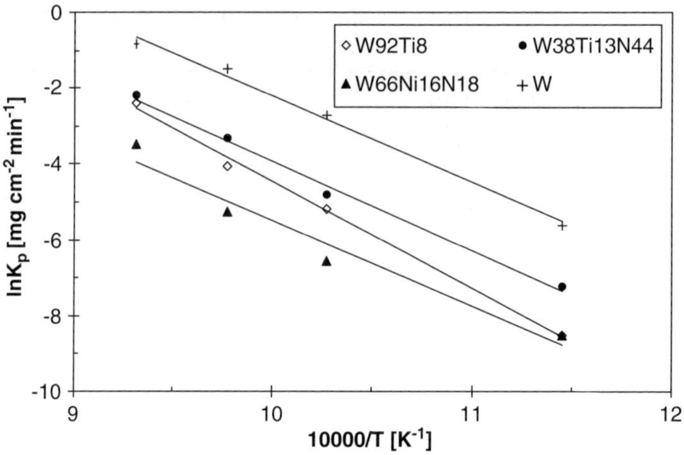

Figure 16.8: Arrehenius law for parabolic oxidation of W–N–M, (M = Ti, Ni) coatings annealed in air at increasing temperature [75].

that TiN oxidation is less favorable than W_2N oxidation ($\Delta H_f^{TiN} = -80\,\text{kcal mol}^{-1}$, $\Delta H_f^{W2N} = -17\,\text{kcal mol}^{-1}$ [14]) and no TiO_2/Ti_3O_5 has been detected by XRD.

16.5. Nanocomposite Multilayer Nitride-Based Coatings

In spite of the improvement of the characteristics and properties of conventional monolithic coatings of nitrides and carbides of transition metals by the addition of other elements to form either ternary or quaternary systems, that improvement has not been high enough to answer the increasingly demanding technological requirements. Nowadays, the nanostructured coatings constitute a base for the new approaches to overcome this challenge and research concerns to main types of nanostructured materials: coatings with nanocrystallites dispersed in an amorphous matrix, and nanocomposite multilayer coatings. Despite the high hardness attained by the first kind of coatings [32,33,176,177] its applicability on mechanical applications is very limited due to the insufficient adhesion and chemical instability at high temperatures.

16.5.1. Multilayer Coatings

Nanocomposite multilayer coatings belong to the new generation of the coatings with high hardness/toughness, and suitable residual stresses and adhesion properties. These multilayer coatings can present either short or long bilayer periods, depending

Table 16.4: Mechanical properties of the short-period multilayer coatings, function of the substrate bias [40]

Coating	Hardness (GPa)	Residual stress (GPa)	Adhesion (N)
TiAlN/CrN	55 → 60	−5.1 → −9.2	50 → 52
TiAlN/VN	39	−3.3	70
TiAlYN/VN	42 → 78	−4.0 → −8.5	30 → 39
TiAlN/ZrN	55	− 10	55
CrN/NbN	42 → 56	−1.8 → −7.0	50 → 62
CrN	11	–	–
VN	15	–	–
ZrN	15	–	–
NbN	14	–	–

on the property to be improved, hardness and/or toughness are enhanced by short-period coatings and adhesion by long-period ones.

16.5.1.1. Short Period

Multilayer coatings with short bilayer period (Λ), in the range of tenth of nanometer exhibit high hardness. The high hardness is induced by the presence of these nanometric layers and can be expressed by a modified Hall–Petch expression:

$$H = H_0 + k\Lambda^{-\beta} \tag{16.10}$$

where the β is a constant with values in the range 0.5–1 [31,178]. However, the hardness attains its highest value (critical hardness) for a critical value (Λ_c) of about 10 nm, which is almost constant for all multilayer coatings. In some situations, the increase of hardness detected in the multilayer coatings with ceramic/ceramic [30,37,129,132,178–182], ceramic/metal [60–62,123,183] and metal or intermetallic/metal [31,184–188] is accompanied by a significant increase of the Young's modulus [123], designated as supermodulus effect. However, several researchers [184,189] attribute this increase to incorrect experimental methodologies.

In order to attain the maximum hardness, short-range period multilayer coatings must have similar critical bilayer period and layer's thickness [132]. According to some researchers, the hardness increase is due to the formation of a superstructure (superlattice), in which the interference of the structure of each layer contributes to the generation of a new lattice [31,37,184–187]. Table 16.4 summarizes the main characteristics of multilayer coatings with short period. Besides hardening effect due to the multilayer configuration, the coatings exhibit good adhesion and high compressive residual stresses, even higher than monolithic coatings [30,140], which increase with the substrate bias [37,140].

Short bilayer period ceramic/ceramic multilayer coatings do not present the adequate properties for cutting operations. In fact, besides the low chemical stability, the existence of a high level of compressive stresses can enhance the possibility of catastrophic failures. The replacement of one of the ceramic layer materials for a metallic one enhances the fracture toughness of the multilayer coating and, for certain period values can even lead to high hardness values, as stated to TiAlN/Mo multilayer coatings [30,123,180,190].

The adhesion to substrate of the ceramic/metal multilayer coatings is similar to that of the ceramic monolithic coating, as shown by Ma *et al.* [60] when comparing monolithic TiN coatings with Ti/TiN multilayer coatings. Also, in the multilayer coatings, if the titanium layers are thin, the adhesive failures occur in the interface coating/substrate instead of in the Ti/TiN interfaces. However, based on chemical stability rather than in the adhesion of the coatings, Knotek *et al.* [191] do not advise the application of this kind of coatings on tribologic severe conditions, such high-speed cutting.

16.5.1.2. Long Period

Taking into account the improvement of the mechanical properties attained by the macrolaminate [56] and microlaminate [192] composites, laminated coatings with characteristics suitable for each application were also produced. In fact, in very thick coatings [56] and bulk microlaminate composites [192], the presence of a ductile layer increases the toughness. If after the propagation of a crack in the ceramic layer, the failure bumps into a ductile metallic barrier it activates the slipping mechanisms (generally oriented 45°) where part of the fracture energy is spent. And, normally, the subsequent failure propagation in the hard and fragile layer involves the nucleation of a new crack in a parallel plane and at a distance equal to the multiplication of the thickness of the ductile layer by the tangent of the angle of the slipping plane. Moreover, only for high loads, where the fragile layers fracture is catastrophic, there is the possibility of the fracture of the ductile layers, which are, generally, only plastically deformed [193–195]. In homothetic coatings, if the metal layers are thick, although nanometric to prevent the decrease of the hardness of the coating, the failures can occur in the metal/ceramic interfaces due to the metal plastic deformation. Thus, concerning toughness, the multilayer coatings with longer bilayer period can be advantageous relatively to short-period ones. The long bilayer period multilayer coatings are constituted by a small number of layers, whose thickness may range from tens of nanometers to micron [4,27,60,62,140,141,179, 181,191,192,196–199], that is, from nanolaminate to microlaminate coatings, in order to assure that hardness values suitable for cutting applications are always obtained. In order to explore the potential of this kind of nanostructured coatings, the concept of long-period multilayered coatings was applied to the different systems, namely Ti–Al–N and W–Ti–N, whose properties have already been discussed in the section concerning the study of hard coatings (*cf.* 16.4).

16.5.2. Nanolaminate $Ti_{1-x}Al_xN$ Coatings with Metallic Layers (Interlayers)

The morphology of nanolaminate coatings with thin-metal interlayers (tenths of nanometers) is similar to the one of the monolithic hard coating, whatever the metal layer composition is. Namely, the columns of the hard layers, which grown over metallic interlayers, maintain the same columnar geometry (Fig. 16.9).

The existence of the metallic interlayers is evidenced by the presence of the sharp fracture surfaces parallel to the substrate. A peculiar three-dimensional lattice is formed with the boundaries of the columns and the interlayers. This effect is

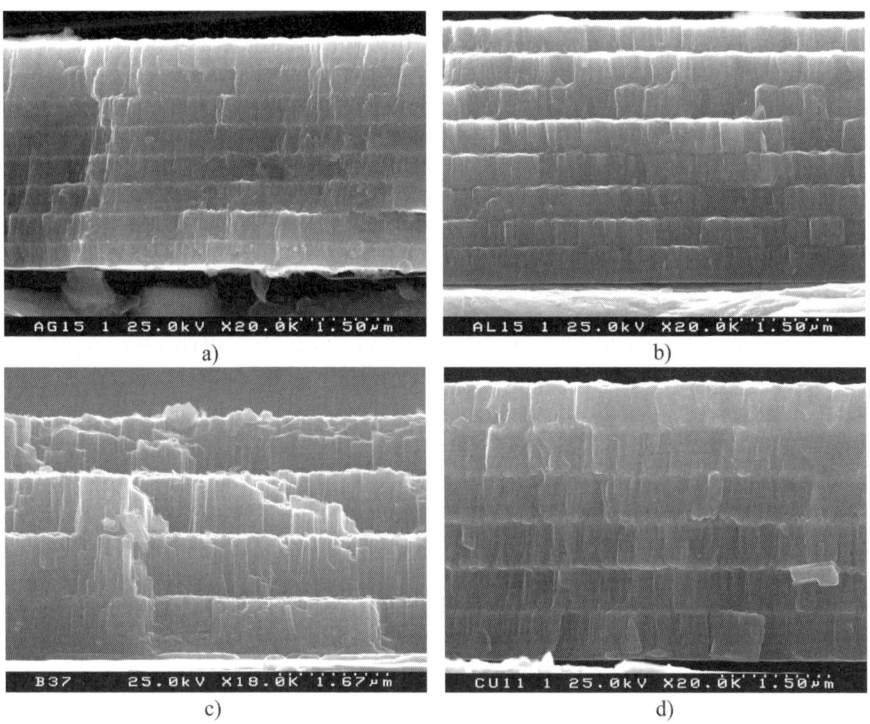

Figure 16.9: Cross-section morphology of nanolaminate coatings with (a) silver, (b) titanium, (c) aluminum and (d) copper interlayers [158].

more pronounced as distance between the interlayers approach the columns dimensions (Fig. 16.9(b)). In contrast to cross-section morphology, the surface morphology of the nanolaminate coatings differs significantly from the monolithic coatings (Fig. 16.10) and it is independent of the thickness and of the type of metal interlayer.

The hardness and the Young's modulus of the nanolaminate coatings are strongly influenced by the metal type and the number of the interlayers. Exception made for the case of aluminum and copper interlayers which produce a really pronounced decrease of the coating properties, the deposition of interlayers softer than the $Ti_{1-x}Al_xN$ monolithic coatings induces only a slightly decrease of the nanolaminate coating hardness and Young's modulus.

The nanolaminate coatings with three, five or seven titanium or silver interlayers have constant hardness and Young's modulus values of 28 GPa and 450 GPa, respectively, which are very similar to the values of the $Ti_{1-x}Al_xN$ monolithic coatings, 30 GPa and 456 GPa. Hence, the interfaces play an important role in properties of the nanolaminate coatings and the higher or the lower lattice mismatch between the metal interlayers and the ceramic layers will induce different mechanical behavior. In fact, besides the low thickness of the interlayers, the interfaces between the titanium and the $Ti_{1-x}Al_xN$ layers are almost certainly coherent, once there is the possibility of the reaction of titanium with nitrogen to form nitrides with similar ceramic structure. In the case of silver interlayers, although, silver has no chemical affinity to nitrogen, it crystallizes in an f.c.c. structure with a lattice parameter slightly lower than the $Ti_{1-x}Al_xN$.

Residual stress in nanolaminate coatings depends on the metal type, thickness and number of interlayers. The introduction of a single metal interlayer reduces the compressive residual stress of the hard monolithic coatings, however, increasing number of interlayers induces their increase up to values in the range of those measured in $Ti_{1-x}Al_xN$ monolithic coatings.

Figure 16.10: Surface morphology of nanolaminate multilayer coating [158].

The main effect resulting from the presence of metal (titanium, silver and aluminum) interlayers in hard coatings is related to the appreciable increase of the adhesion of the nanolaminate coatings compared to the monolithic coatings, attaining values of 70 N, which is more than 60% higher than the value for monolithic $Ti_{1-x}Al_xN$ coating (42 N). This behavior can be explained by the coherency of the interfaces and their ability to absorb the load energy and allow the sliding of the hard layers [38,154,158].

16.5.3. *Nanolaminate $Ti_{1-x}W_xN$ Coatings with Metallic Layers (Interlayers)*

As in $Ti_{1-x}Al_xN$ coatings, the incorporation of metallic interlayers in Ti–W–N system do not cause any morphologic disruption toward the featureless morphology observed in monolithic coatings. While the hardness is kept almost unchanged, relatively to monolithic coatings, even after the addition of metallic interlayers, the most important effect of this introduction in this kind of hard coatings is the reduction of the high compressive residual stress of the $Ti_{1-x}W_xN$ coatings to about one half [158,175].

16.6. Conclusions

The selection of the coatings reported in this chapter followed the progressive improvement of mechanical properties rather than their historic evolution. Thus, the first generation of coatings used in mechanical applications: monolithic titanium and tungsten carbides and nitrides coatings showed high hardness but low-oxidation resistance reducing their ability to be used in high-speed unlubricated cutting. The next generation of coatings, ternary nitrides coatings with a TiN structure, $Ti_{1-x}Al_xN$ and $Ti_{1-x}W_xN$, represented a significant improvement of the mechanical properties toward the binary nitride monolithic coatings. Finally, the last coatings generation, multilayer nanolaminate coatings with short and long bilayer period is yet in a development stage but their promising characteristics and properties pointed out to an increase of efficiency of the mechanical component/part operating in aggressive mechanical and environmental conditions. It is evident that the achievements attained in the research in thin film domain are closely related to the advances concerning the nanocrystalline materials performance. It seems clear that the establishment of bridges between 3D (bulk) and 2D (thin films) materials will contribute to a better understanding of the mechanisms of hardening and fracture of materials. Nowadays, much research effort continues to be performed, aiming to improve/optimize the mechanical, physical and chemical properties of the coatings, and also to develop new methodologies of prediction of coatings behavior, both for coatings selection as well as for prediction of their performance in real structural applications.

References

[1] R.F. Bunshah and D.M. Mattox, *Phys. Today*, 33 (5) (1980) 50.
[2] L. Ter Minassian-Saraga, *Pure Appl. Chem.*, 66 (8) (1994) 1667.
[3] C. Subramanian, K.N. Strafford, T.P. Wilks and L.P.Ward, *J. Mater. Process. Technol.*, 56 (1996) 385.
[4] C. Subramanian and K.N. Strafford, *Wear*, 165 (1993) 85.
[5] D.C. Rickerby and P.J. Burnett, *Thin Solid Films*, 157 (1988) 195.
[6] J.-E. Sundgren and H.T.G. Hentzell, *J. Vac. Sci. Technol. A*, 4 (5) (1986) 2259.
[7] R.F. Bunshah and C.V. Despandey, *Vacuum*, 39 (10) (1989) 955.
[8] J. Halling, *16th Leeds-Lyon Symp. Mechanics of Coatings*, Elsevier, 1990, p. 477 cited by C. Subramanian, K.N. Strafford, T.P. Wilks and L.P. Ward, *J. Mater. Process. Technol.*, 56 (1996) 385.
[9] H.G. Prengel, W.R. Pfouts and A.T. Santhanam, *Surf. Coat. Technol.*, 102 (1998) 183.
[10] I.L. Singer and H.M. Pollock (Eds.), *Fundamentals of Friction: Macroscopic and Microscopic Processes*, Kluwer Academic Publishers, 1992, p. 237.
[11] J.-E. Sundgreen and H.T.G. Hentzel, *J. Vac. Sci. Technol. A*, 4 (5) (1986) 2259.
[12] F. Lévy, P. Hones, P.E. Schmid, R. Sanjinés, M. Diserens and C. Wiemer, *Surf. Coat. Technol.*, 120–121 (1999) 284.
[13] P. Hones, R. Sanjinés, F. Lévy and O. Shojaei, *J. Vac. Sci. Technol. A*, 17 (3) (1999) 1024.
[14] H.J. Goldschmidt, *Interstitial Alloys*, Butterworth & Co., 1967, pp. 44–45, 235.
[15] D.G. Bhat and P.F. Woerner, *J. Met.*, 38 (2) (1986) 68.
[16] B. Mills, *J. Mater. Process. Technol.*, 56 (1996) 16.
[17] J. Kopac, *J. Mater. Process. Technol.*, 78 (1998) 95.
[18] D. Turnbull, *Metall. Trans. A*, 12 (1981) 695.
[19] J.M. Castanho, A influência do elemento intersticial no comportamento tribológico dos filmes do sistema W-N/C, MSc Thesis, Coimbra, Portugal, 1995.
[20] A. Inspektor, C.E. Bauer and E.J. Oles, *Surf. Coat. Technol.*, 68/69 (1994) 359.
[21] R.C. Dewes and D.K. Aspinwall, *J. Mater. Process. Technol.*, 69 (1997) 1.
[22] J.M. Castanho and M.T. Vieira, *NordTrib '98: Proceedings of the 8th International Conference on Tribology*, 1998, p. 889.
[23] Milton Orhing, *The Materials Science of Thin Films*, Academic Press, 1992, p. 223.
[24] R. Messier, A.P. Gri and R.A. Roy, *J. Vac. Sci. Technol. A*, 2 (1984) 500.
[25] J.A. Thornton, *J. Vac. Sci. Technol. A*, 4 (6) (1986) 3029.
[26] A. Cavaleiro, M.T. Vieira and G. Lemperiere, *Thin Solid Films*, 213 (1992) 6.
[27] Z.N. Farhat, Y. Ding, A.T. Alpas and D.O. Northwood, *J. Mater. Process. Technol.*, 63 (1997) 859.

[28] E.O. Hall, *Proc. Phys. Soc. B*, 64 (1951) 747.
[29] N.J. Petch, *J. Iron Steel Inst.*, 174 (1953) 25.
[30] C.J. Tavares, L. Rebouta, M. Andritschky and S. Ramos, *J. Mater. Process. Technol.*, 92–93 (1999) 177.
[31] S. Tixier, P. Böni and H. Van Swygnhoven, *Thin Solid Films*, 342 (1999) 188.
[32] S. Veprek, *Thin Solid Films*, 317 (1998) 449.
[33] J. Musil and H. Hrubý, *Thin Solid Films*, 365 (2000) 104.
[34] A. Cavaleiro, B. Trindade and M.T. Vieira, *Surf. Coat. Technol.*, 174–175 (2003) 68.
[35] H.E. Hintermann, *Wear*, 100 (1984) 381.
[36] H. Ollendorf and D. Schneider, *Surf. Coat. Technol.*, 113 (1999) 86.
[37] W.-D. Münz, D.B. Lewis, P.Eh. Hovsepian, C. Schönjahn, A. Ehiasarian and I.J. Smith, *Surf. Eng.*, 17 (1) (2001) 15.
[38] J.M. Castanho and M.T. Vieira, *Key Eng. Mater.*, 230–232 (2002) 635.
[39] J.M. Castanho, D. Pinheiro and M.T. Vieira, *Adv. Mater. Forum*, 455–456 (2004) 619.
[40] H. Oettel and R. Wiedemann, *Surf. Coat. Technol.*, 76–77 (1995) 265.
[41] Y. Miyamoto, K. Watanabe, S. Nakagawa and M. Naoe, *Vacuum*, 51 (4) (1998) 711.
[42] J.W. Gerlach, T. Kraus, S. Sienz, M. Moske, M. Zeitler and B. Rauschenbach, *Surf. Coat. Technol.*, 103–104 (1998) 281.
[43] B.N. Chapman, *J. Vac. Sci. Technol.*, 11 (1) (1974) 106.
[44] A.J. Perry, *Thin Solid Films*, 107 (1983) 167.
[45] M.T. Laugier, *Thin Solid Films*, 117 (1984) 243.
[46] P.A. Steinmann, Y. Tardy and H.E. Hintermann, *Thin Solid Films*, 154 (1987) 333.
[47] P.J. Burnett and D.S. Rickerby, *Thin Solid Films*, 154 (1987) 403.
[48] P.J. Burnett and D.S. Rickerby, *Thin Solid Films*, 157 (1988) 233.
[49] S.J. Bull and D.S. Rickerby, in *Plasma Surface Engineering*, Eds. E. Broszeit, *et al.*, Vol. 2, DGM Informationgesellschaft-Verlag, Oberursel, 1989, p. 1227.
[50] K.L. Mittal, *Adhesion Measurement of Films and Coatings*, Ed. K.L. Mittal, VSP, 1995, p. 1.
[51] S. Venkataraman, D.L. Kohlstedt and W.W. Gerberich, *J. Mater. Res.*, 11 (12) (1996) 3133.
[52] A. Kelly and N.H. MacMillan, The ideal strength, Chapter 1, *Strong Solids*, 3rd edition, Oxford Science Publications, 1986.
[53] Milton Orhing, *The Materials Science of Thin Films*, Academic Press, New York, 1992, p. 552.
[54] A. Leyland and A. Matthews, *Wear*, 246 (2000) 1.
[55] A. Matthews and A. Leyland, *Key Eng. Mater.*, 206–213 (2002) 459.
[56] A. Matthews, R. Jones and S. Dowey, *Tribol. Lett.*, 11 (2) (2001) 103.
[57] J. Musil, F. Kunc, H. Zeman and H. Poláková, *Surf. Coat. Technol.*, 154 (2002) 304.
[58] Y. Milman, B. Galano and S.I. Chugunova, *Acta Metall. Mater.*, 41 (9) (1993) 2523.
[59] D.-F. Lii, J.-L. Huang and M.-H. Lin, *Surf. Coat. Technol.*, 99 (1998) 197.
[60] K.J. Ma, A. Bloyce and T. Bell, *Surf. Coat. Technol.*, 76–77 (1995) 297.
[61] J.L. He, W.Z. Li, H.D. Li and C.H. Liu, *Surf. Coat. Technol.*, 103–104 (1998) 276.
[62] M. Ignat, T. Marieb, H. Fujimoto and P.A. Flinn, *Thin Solid Films*, 353 (1999) 201.

[63] S. Hofmann, *Thin Solid Films*, 193/194 (1990) 648.
[64] D. McIntyre, J.E. Greene, G. Hakansson, J.-E. Sundgren and W.-D. Münz, *J. Appl. Phys.*, 67 (3) (1990) 1542.
[65] Bruce G. McMordie, *Surf. Coat. Technol.*, 49 (1991) 18.
[66] A. Joshi and H.S. Hu, *Surf. Coat. Technol.*, 76–77 (1995) 499.
[67] C.-H.T. Huang and J.-G. Duh, *Surf. Coat. Technol.*, 81 (1996) 164.
[68] Y. Otani and S. Hofmann, *Thin Solid Films*, 287 (1996) 188.
[69] F. Vaz, L. Rebouta, M. Andritschky, M.F. Silva and J.C. Soares, *Surf. Coat. Technol.*, 98 (1998) 912.
[70] I.J. Smith, W.D. Münz, L.A. Donohue, I. Petrov and J.E. Greene, *Surf. Eng.*, 14 (1) (1998) 37.
[71] M. Witthaut, R. Cremer, A. Richthofen and D. Neuschütz, *J. Anal. Chem.*, 361 (1998) 639.
[72] M. Éekada, P. Panjan, B. Navinšek and F. Cvelbar, *Vacuum*, 52 (1999) 461.
[73] M. Diserens, J. Patscheider and F. Levy, *Surf. Coat. Technol.*, 120–121 (1999) 158.
[74] C. Louro and A. Cavaleiro, *Thin Solid Films*, 343–344 (1999) 50.
[75] A. Cavaleiro, C. Louro and F. Montemor, *Surf. Coat. Technol.*, 131 (2000) 441.
[76] C. Louro, *Surf. Coat. Technol.*, 180–181 (2004) 544.
[77] P. Panjan, B. Navinšek, A. Cvelbar, A. Zalar and I. Milosev, *Thin Solid Films*, 281–282 (1996) 298.
[78] I. Wadsworth, I.J. Smith, L.A., Donohue and W.-D. Munz, *Surf. Coat. Technol.*, 94–95 (1997) 315.
[79] J. Vetter, E. Lugscheider and S.S. Guerreiro, *Surf. Coat. Technol.*, 98 (1998) 1233.
[80] P. Panjan, B. Navinšek, A. Cvelbar, A. Zalar and J. Vlcek, *Surf. Coat. Technol.*, 98 (1998) 1497.
[81] L.E. Toth, *Transition Metal Nitrides and Carbides*, Academic Press, New York, (1971).
[82] W.D. Sproul, M.E. Graham, M.-S. Wong, P.J. Rudnik and K.O. Legg, *Handbook of Thin Film Process Technology*, IOP Publishing, 1995, X1.1:1–X1.1:4.
[83] L. Hultman, *Vacuum*, 57 (2000) 1.
[84] J.-E. Sundgreen, *Thin Solid Films*, 128 (1985)21.
[85] I. Petrov, L. Hultman, J.-E. Sundgreen, L.C. Market and J.E. Greene, *J. Vac. Sci. Technol. A*, 10 (1992) 265.
[86] J. Deng and M. Braun, *Surf. Coat. Technol.*, 70 (1994) 49.
[87] A. Bendavid, P.J. Martin, R.P. Netterfield and T.J. Kinder, *Surf. Coat. Technol.*, 70 (1994) 97.
[88] M.D. Tran, J. Poublan and J.H. Dautzenberg, *Thin Solid Films*, 308–309 (1997) 310.
[89] X.-Z. Ding, B.K. Tay, H.S. Tan, S.P. Lau, W.Y. Cheung and S.P. Wong, *Surf. Coat. Technol.*, 138 (2001) 301.
[90] E. Kusano, A. Satoh, M. Kitagawa, H. Nanto and A. Kinbara, *Thin Solid Films*, 343–344 (1999) 254.
[91] Z. Yu, K. Inagawa and Z. Jin, *Thin Solid Films*, 264 (1995) 52.
[92] H. Holleck, *J. Vac. Sci. Technol. A*, 4 (6) (1986) 2661.

[93] A. Karimi, O.R. Shojaei, T. Kruml and J.L. Martin, *Thin Solid Films*, 308–309 (1997) 334.
[94] M. Wittmer, J. Noser and H. Melchior, *J. Appl. Phys.*, 52 (1981) 6659.
[95] I. Suni, D. Sigurd, K.T. Ho and M.-A. Nicolet, *J. Electrochem. Soc.*, 130 (5) (1983) 1210.
[96] C. Ernsberg, J. Nickerson, T. Smith, A.E. Millewr and D. Banks, *J. Vac. Sci. Technol. A*, 4 (6) (1986) 2784.
[97] W.-D. Münz, *J. Vac. Sci. Technol. A*, 4 (6) (1986) 2717.
[98] H.A. Jehn, S. Hofmann and W.-D. Münz, *Thin Solid Films*, 253 (1987) 2173.
[99] I. Milosev, H.-H. Strehblow and B. Navinsek, *Thin Solid Films*, 303 (1997) 246.
[100] S. Logothetidis, E.I. Meletis, G. Stergioudis and A.A. Adjaottor, *Thin Solid Films*, 338 (1999) 304.
[101] I.G. Polyakova and T. Hubert, *Surf. Coat. Technol.*, 141 (2001) 55.
[102] O. Knotek, W.-D. Münz and T. Leyendecker, CEI Couse, *Vapour Deposited Wear Protective Coatings*, Vol. 82, Chapter 11, Miiltenberg am Main, Frankfurt, Germany, 1986, p. 64.
[103] K. Affolter, H. Kattelus and M.-A. Nicolet, *Thin Solid Films*, 158 (1988) 45.
[104] B. Matthes, E. Broszeit, J. Aromaa, H. Ronkainen, S.-P. Hannula, A. Leyland and A. Matthews, *Surf. Coat. Technol.*, 49 (1991) 489.
[105] G. Meier zu Köcker and K.-H. Habig, *Surf. Coat. Technol.*, 82 (1996) 294.
[106] Ph. Gouy-Pailler and Y. Pauleau, *J. Vac. Sci. Technol. A*, 11 (1) (1993) 96.
[107] G. Keller, I. Barzen, R. Erz, W. Dötter, S. Ulrich, K. Jung and M. Ehrhardt Fresenius, *J. Anal. Chem.*, 341 (1991) 349.
[108] C.E. Wickersham, E.L. Foster and G.H. Stickford, *J. Vac. Sci. Technol.*, 18 (2) (1981) 223.
[109] P.K. Srivastava, T.V. Rao, V.D. Vankar and K.L. Chopra, *J. Vac. Sci. Technol. A*, 2 (3) (1984) 1261.
[110] P.K. Srivastava, V.D. Vankar and K.L. Chopra, *J. Vac. Sci. Technol. A*, 3 (6) (1985) 2129.
[111] P.K. Srivastava, V.D. Vankar and K.L. Chopra, *Bull. Matter. Sci.*, 8 (1986) 379.
[112] K. Fuchs, P. Rödhammer, E. Bertel, F. Netzer and E. GorniK, *Thin Solid Films*, 151 (1987) 383.
[113] C. Louro and A. Cavaleiro, *J. Electrochem. Soc.*, 144 (1) (1997) 260.
[114] C. Louro and A. Cavaleiro, *J. Mater. Process. Technol.*, 92–93 (1999) 162.
[115] J.C. Schuster and J. Bauer, *J. Solid State Chem.*, 53 (1984) 260.
[116] F. Perdrix, M.-F. Trichet, J.-L. Bonnentien, M. Cornet and J. Bigot, *Intermetallics*, 9 (2001) 147.
[117] W.-D. Münz, *J. Vac. Sci. Technol. A*, 4 (6) (1986) 2717.
[118] H.A. Jehn, S. Hofmann and W.-D. Münz, *Thin Solid Films*, 153 (1987) 45.
[119] G. Håkansson, J.-E. Sundgreen, D. McIntyre, J.E. Greene and W.-D. Münz, *Thin Solid Films*, 153 (1987) 55.
[120] O. Knotek, W.-D. Münz and T. Leyendecker, *J. Vac. Sci. Technol. A*, 5 (4) (1987) 2173.
[121] A. Bosseboeuf and A. Fourrier, *Appl. Surf. Sci.*, 53 (1991) 353.

[122] F. Adibi, I. Petrov, J.E. Greene, L. Hultman and J.-E. Sundgren, *J. Appl. Phys.*, 73 (12) (1993) 8580.
[123] U. Wahlström, L. Hultman, J.-E. Sundgren, F. Adibi, I. Petrov and J.E. Greene, *Thin Solid Films*, 235 (1993) 62.
[124] C.-H.T. Huang and J.-G. Duh, *Surf. Coat. Technol.*, 71 (1995) 259.
[125] J.G. Han, J.S. Yoon, H.J. Kim and K. Song, *Surf. Coat. Technol.*, 86–87 (1996) 82.
[126] K.-L. Lin, M.-Y. Hwang and C.-D. Wu, *Mater. Chem. Phys.*, 46 (1996) 77.
[127] L.P. Ward, K.N. Strafford, C. Subramanian and T.P. Wilks, *J. Mater. Process. Technol.*, 56 (1996) 375.
[128] I.J. Smith, D. Gillibrand, J.S. Brooks, W.-D. Münz, S. Harvey and R. Goodwin, *Surf. Coat. Technol.*, 90 (1997) 164.
[129] H.G. Prengel, A.T. Santhanam, R.M. Penich, P.C. Jindal and K.H. Wendt, *Surf. Coat. Technol.*, 94–95 (1997) 597.
[130] F. Vaz, L. Rebouta, M. Andritschky, M.F. da Silva and J.C. Soares, *J. Eur. Ceram. Soc.*, 17 (1997) 1971.
[131] B.-Y. Shew, J.-L. Huang and D.-F. Lii, *Thin Solid Films*, 293 (1997) 212.
[132] I. Penttinen, J.M. Molarius, A.S. Korhonen and R. Lappalainen, *J. Vac. Sci. Technol. A*, 6 (3) (1998) 2158.
[133] T. Suzuki, D. Huang and Y. Ikuhara, *Surf. Coat. Technol.*, 107 (1998) 41.
[134] A. von Richthofen, R. Cremer, M. Witthaut, R. Domnick and D. Neuschütz, *Thin Solid Films*, 312 (1998) 190.
[135] S. Ikeda, S. Gilles and B. Chenevier, *Thin Solid Films*, 315 (1998) 257.
[136] L. Cunha, M. Andritschky, L. Rebouta and R. Silva, *Thin Solid Films*, 317 (1998) 351.
[137] C. Jarms, H.-R. Stock and P. Mayr, *Surf. Coat. Technol.*, 108–109 (1998) 206.
[138] P. Panjan, B. Navinšek, M. Éekada and A. Zalar, *Vacuum*, 53 (1999) 127.
[139] K.-D. Bouzakis, N. Vidakis, N. Michailidis, T. Leyendecker, G. Erkens and G. Fuss, *Surf. Coat. Technol.*, 120–121 (1999) 34.
[140] M.Z. Huq and J.P. Celis, *Wear*, 225–229 (1999) 53.
[141] A. Kimura, H. Hasegawa, K. Yamada and T. Suzuki, *Surf. Coat. Technol.*, 121 (1999) 438.
[142] C. Jiménez, C. Sánchez-Fernandez, C. Morant, J.M. Martínez-Duart, M. Fernández and J. Sánchez-Olías, *J. Mater. Res.*, 14 (7) (1999) 2830.
[143] M. Zhou, Y. Makino, M. Nose and K. Nogi, *Thin Solid Films*, 339 (1999) 203.
[144] M. Pinkas, J. Pelleg and M.P. Dariel, *Thin Solid Films*, 355–356 (1999) 380.
[145] D.-Y. Wang, C.-L. Chang, K.-W. Wong and W.-Y. Ho, *Surf. Coat. Technol.*, 120–121 (1999) 388.
[146] S.K. Wu, H.C. Lin and P.L. Liu, *Surf. Coat. Technol.*, 124 (2000) 97.
[147] K.N. Andersen, E.J. Bienk, K.O. Schweitz, H. Reitz, J. Chevallier, P. Kringhøj and J. Bøttiger, *Surf. Coat. Technol.*, 123 (2000) 219.
[148] E. Schäffer and G. Kleer, *Surf. Coat. Technol.*, 133–134 (2000) 215.
[149] M. Brogren, G.L. Harding, R. Karmhag, C. Ribbing, G.A. Niklasson and L. Stenmark, *Thin Solid Films*, 370 (2000) 268.
[150] E. Zoestbergen, N.J. Carvalho and J.Th. De Hosson, *Surf. Eng.*, 17 (1) (2001) 29.

[151] C.J. Tavares, L. Rebouta, M. Andritschky, F. Guimarães and A. Cavaleiro, *Vacuum*, 60 (2001) 339.
[152] Y. Makino and S. Miyake, *Trans. JWRI*, 30 (2) (2001) 39.
[153] C.-M. Shu, B.-W. Hwang and R.-I. Murakami, *Mater. Sci. Eng. A*, 1–7 (2003) 343.
[154] J.M. Castanho and M.T. Vieira, *J. Mater. Process. Technol.*, 143 (2003) 352.
[155] T. Vasco Boutos, R. Sanjinés and A. Karimi, *Surf. Coat. Technol.*, 188–189 (2004) 409.
[156] A. Escudeiro Santana, A. Karimi, V.H. Derflinger and A. Schütze, in *Thin Solid Films*, (2005) 469–470.
[157] A. Karimi, *Oral Presentation, 18th International Conference on Surface Modification Technologies*, 15–17 November 2004, Dijon, France.
[158] J.M. Castanho, A influência de intercamadas metálicas no comportamento mecânico de revestimentos duros, PhD Thesis, Universidade de Coimbra, 2004.
[159] L. Rebouta, F.Vaz, M. Andritshky and M.F. da Silva, *Surf. Coat. Technol.*, 76–77 (1995) 70.
[160] S. Hofmann and H.A. Jehn, *Surf. Interface Anal.*, 12 (1988) 329.
[161] Y. Tanaka, T.M. Gur, M. Kelly, S.B. Hagstrom, T. Ikeda, K. Wakikawa and H. Satoh, *J. Vac. Sci. Technol. A*, 10 (1992) 1749.
[162] S. Inoue, H. Uchida, Y. Yoshinaga and K. Koterazawa, *Thin Solid Films*, 300 (1997) 171.
[163] K.Y. Ahn, S.B. Brosky, C.Y. Ting and J. Kim, *J. Vac. Sci. Technol. A*, 4 (6) (1986) 3111.
[164] K. Affolter, H. Kattelus and M.-A. Nicolet, *Proc. Mater. Res. Soc.*, 47 (1985) 167.
[165] J. Castanho, A. Cavaleiro and M.T. Vieira, *Vacuum*, 45 (1994) 1051.
[166] J.M. Castanho and M.T. Vieira, *Surf. Coat. Technol.*, 102 (1998) 50.
[167] J.C. Oliveira, A. Cavaleiro and C.M.A. Brett, *Corros. Sci.*, 42 (2000) 1881.
[168] A.G. Dirks, R.A.M. Wolters and A.E.M. De Veirman, *Thin Solid Films*, 208 (1992) 181.
[169] J.H. Moser, F. Tian, O. Haller, D.B. Berstrom, I. Petrov, J.E. Greene and C. Wiemer, *Thin Solid Films*, 253 (1994) 445.
[170] M.T. Vieira, C.M. Pereira and J.M. Castanho, *Surf. Coat. Technol.*, 131 (1–3) (2000) 417.
[171] L.R. Shaginyan, M. Mišina, J. Zemek, J. Musil, F. Regent and V.F. Britun, *Thin Solid Films*, 408 (2002) 136.
[172] A. Cavaleiro, C. Louro, J.V. Fernandes and C.M.A. Brett, *Vacuum*, 52 (1999) 157.
[173] H. Ramarotafika and G. Lemperiere, *Thin Solid Films*, 266 (1995) 267.
[174] M.T. Vieira, A. Cavaleiro and B. Trindade, *Surf. Coat. Technol.*, 151–152 (2002) 495.
[175] J.M. Castanho and M.T. Vieira, Internal stresses in wear resistant ceramic/metal multilayer coatings, *Proceedings of 7th ICCE*, Ed. D. Hui, Denver, USA, 2000, p. 105.
[176] J. Musil, *Surf. Coat. Technol.*, 125 (2000) 322.
[177] C. Suryanarayana, *J. Mater. Res.*, September (2002), 24.
[178] S.L. Lehoczky, *Phys. Rev. Lett.*, 41 (26) (1978) 1814.
[179] M. Nordin, M. Larsson and S. Hogmark, *Surf. Coat. Technol.*, 106 (1998) 234.
[180] C.J. Tavares, L. Rebouta, B. Almeida, J. Bessa e Sousa, M.F. da Silva and J.C. Soares, *Thin Solid Films*, 317 (1998) 124.

[181] A. Stanishevsky and R. Lappalainen, *Surf. Coat. Technol.*, 123 (2000) 101.
[182] J.G. Han, H.S. Myung, H.M. Lee and L.R. Shaginyan, *Surf. Coat. Technol.*, 174–175 (2003) 738.
[183] S.P. Rawal, G.M. Swanson and W.C. Moshier, *J. Mater. Res.*, 10 (7) (1995) 1721.
[184] M.K. Small, B.J. Daniels, B.M. Clemen and W.D. Nix, *J. Mater. Res.*, 9 (1) (1994) 25.
[185] S.-G. Chang, J.-Y. Lee, G.-H. Kim and C.-H. Chum, *J. Mater. Res.*, 10 (6) (1995) 1555.
[186] R. Banerjee, R. Ahuja and H.L. Fraser, *J. Electron. Mater.*, 23 (1994) 1027, cited by J. Bonevich, D. van Heeredn and D. Josell, *J. Mater. Res.*, 14 (5) (1999) 1977.
[187] A.F. Jankowski and M.A. Wall, *NanoStruct. Mater.*, 7 (1996) 35, cited by J. Bonevich, D. van Heeredn and D. Josell, *J. Mater. Res.*, 14 (5) (1999) 1977.
[188] J. Bonevich, D. van Heeredn and D. Josell, *J. Mater. Res.*, 14 (5) (1999) 1977.
[189] S.P. Baker and W.D. Nix, *J. Mater. Res.*, 9 (12) (1994) 3145.
[190] C.J. Tavares, L. Rebouta and E.J. Alves, *Thin Solid Films*, 373 (2000) 287.
[191] O. Knotek, F. Löffler and G. Krämer, *Int. J. Refract. Met. Hard Mater.*, 14 (1996) 195.
[192] M. Li and W.O. Soboyejo, *Metall. Mater. Trans. A*, 31A, (2000) 1385.
[193] R.H. Dauskardt, M. Lane, Q. Ma and N. Krishna, *Eng. Fract. Mech.*, 61 (1998) 141.
[194] R.P. Singh and V. Parameswaran, *Opt. Laser. Eng.*, in press.
[195] E. Harry, A. Rouzaud, P. Juliet, Y. Pauleau and M. Ignat, *Surf. Coat. Eng.*, 116–119 (1999) 172.
[196] E. Kusano, M. Kitagawa, Y. Kuroda, H. Nanto and A. Kinbara, *Thin Solid Films*, 334 (1998) 151.
[197] J.C. Hu, T.C. Chang, L.J. Chen, Y.L. Yang and C.Y. Chang, *Thin Solid Films*, 332 (1998) 423.
[198] E. Harry, A. Rouzaud, P. Juliet and Y. Pauleau, *Thin Solid Films*, 342 (1999) 207.
[199] A. Kakanakova-Georgieva, Ts. Marinova, O. Noblanc, C. Arnodo, S. Cassette and C. Brylinski, *Appl. Surf. Sci.*, 121–122 (1997) 208.

Chapter 17

Friction Mechanisms and Fundamental Aspects in Solid Lubricant Coatings

Christophe Donnet and Ali Erdemir

17.1. Introduction

Solid lubricant coatings possess inherent lubricating properties and hence are very attractive for use under severe application conditions (such as high vacuum, aerospace, high speeds, high loads, and very low or high temperatures), where conventional materials and lubricants cannot provide the desired levels of performance or durability. A general design appraisal of the tribologic requirements on a coated-surface contact can be formulated as follows:

- The initial coefficient of friction (CF), the steady-state CF, and the friction instability must not exceed certain design values.
- The wear of the contacting surfaces, including the coated one, must not exceed certain design values.
- The lifetime of the system must, with a specified probability, be longer than the required lifetime. The lifetime limit of the system may be defined as the time when even one of the earlier requirements is not maintained.

The concomitant decrease of both friction and wear is not systematically required. In some applications, such as coated tools, the big challenge is to increase the wear resistance in order to extend the lifetime of the components. In other applications, a strong reduction of the friction in the 10^{-2} range is paramount to save energy. At present, no single coating can provide both low friction and high wear resistance over very broad use conditions, temperatures and environments.

During the past two decades or so, remarkable progress has been made in the design, development, and uses of solid lubricant films. The current trend in modern tribology is to limit or reduce the use of liquid lubricants as much as possible (mainly because of environmental concerns), but increase the use of solid materials and coatings with self-lubricating properties. However, in the near term, the best compromise may be to consider a combination of solid and liquid lubricants to

Materials Surface Processing by Directed Energy Techniques
Y. Pauleau (Editor)

Shortcomings	Requirements
Adherence	Interface optimization
Fluctuation in friction coefficient	Run-in control & wear debris removal
Finite wear life	Difficulty of replenishment
Tribo reactivity	Effect of temperature & environments
Ageing	Longer shelf-life and durability
Poor thermal conductivity	How to carry away heat?
Complex deposition procedures	Lowering costs

Figure 17.1: Major shortcomings and requirements of solid lubricant coatings.

meet emission or environmental requirements of future tribologic systems while providing the levels of desired friction and wear performance.

Solid lubricant coatings have come a long way in recent years, and they are now capable of providing extremely low friction and wear coefficients under certain or highly controlled test conditions. However, as depicted in Fig. 17.1, solid lubricant films still have shortcomings, in spite of considerable progress made during the past two decades.

Solid lubricant coatings still have problems with limited lifetime, difficulty in replenishment, and oxidation and aging-related degradation (in the case of certain lamellar solid lubricants, such as MoS_2). A comprehensive review of solid lubricant coatings can be found in Ref. [1]. In this chapter, the emphasis is on current practices and future trends for solid lubricant coatings. Our review will not be so exhaustive or inclusive as to cover all aspects of such coatings, but the main objective is to give a general sense of what has so far been accomplished and where the field is going.

17.2. Classification of Solid Lubricants

Naturally, certain solid materials possess low shear strength and when applied on a sliding surface, they can lower friction and wear. These materials are referred to as "solid lubricants". Unique lubricating properties of some of the solid lubricants (i.e. molybdenum disulfide, graphite, hexagonal boron nitride, boric acid) are mainly

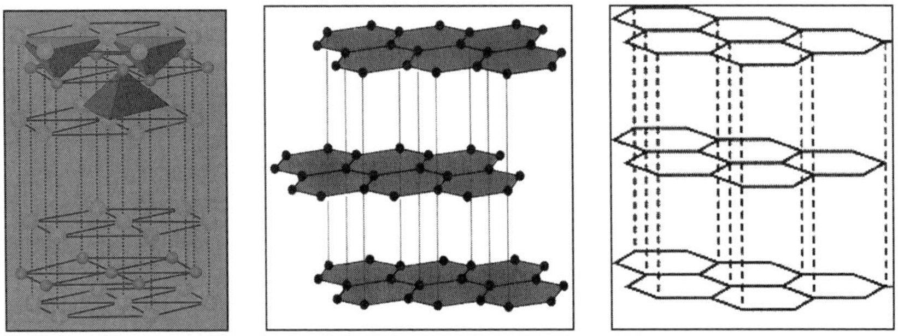

Figure 17.2: Layered lattice structures of (a) molybdenum disulfide, (b) graphite, and (c) hexagonal boron nitride.

associated with their layered or lamellar crystal structures as depicted in Fig. 17.2; while others (such as diamond and diamond-like carbon (DLC)) provide lubrication mainly because of their extreme chemical inertness. Based on their physical, chemical, structural, and mechanical properties, solid lubricant coatings can be classified into several sub-categories. For the sake of simplicity, we will divide them in two broad categories: soft (hardness less than 10 GPa) and hard (hardness more than 10 GPa) solid lubricants [1–3]. Fig. 17.3 lists some of the solid lubricants for each category.

The hard solid lubricants exhibit higher wear resistance in addition to lower friction when compared with soft lubricants, which can provide low friction but not always high wear resistance. Hard solid lubricant coatings include some of the carbon-based coatings (such as diamond and DLC) and certain oxides. Some of the nitrides (i.e. CBN), borides (TiB_2, $AlMgB_{14}$), and carbides (i.e. TiC, WC, Ti_3SiC_2, etc.) can also provide friction coefficients of 0.2–0.4; hence they can also be considered as solid lubricants. Some of these ceramics are rather hard and when their high mechanical hardness is combined with such levels of low friction coefficients, they can be very attractive for a wide range of tribologic applications. Among all the known borides, $AlMgB_{14}$ has been reported to provide very high hardness (i.e. over 40 GPa) as well as fairly low friction coefficients (less than 0.1) when used on sliding surfaces [4]. By way of a simple chlorination method, carbide-based ceramics and their coatings can partially be converted into a nanostructured carbon film that is called "carbide-derived carbon" and hence their solid lubrication properties can be further improved [5]. In recent years, self-lubricating properties of certain quasicrystals have also been reported and some of them are now used as non-stick coatings on certain cooking wares [6,7].

Soft solid lubricant coatings include polymers, soft metals, halides, and sulfates of alkaline earth metals, and the well-known lamellar solids, including transition-metal

Hard solid lubricant coatings **Hardness higher than ≈ 10 GPa**	**Soft solid lubricant coatings** **Hardness lower than ≈ 10 GPa**
Nitrides CBN, CN_x	**Soft metals** Ag, Pb, Au, In, Sn, Cr, Ni, Cu
Carbides TiC, WC, CrC, SiC, CDC, Ti_3SiC_2	**Lamellar solids** MoS_2, WS_2, Graphite H_3BO_3, HBN, GaS, GaSe
Oxides TiO_2, ZnO, CdO, Cs_2O, PbO, Re_2O_7	**Halides, sulfates, sulfides** CaF_2, BaF_2, PbS, $CaSO_4$, $BaSO_4$
Borides TiB_2, $AlMgB_{14}$	**Polymers** PTFE, PE, Polyimide Polymerlike DLC (high hydrogen content)
DLC & Diamond a-C, ta-C, a-C:H, ta-C:H, a-C:X(:H), (nc-)diamond	**Quasicrystals** Al-Pd-Mn, Al-Cu-Fe, Al-Ni-Co-Si

DLC = diamondlike carbon a= amorphous ta = tetrahedral amorphous
X = a metal nc = nanocrystalline CDC = Carbide derived carbon
PTFE = polytetrafluorethylene PE = polyethylene

Figure 17.3: Classification of tribologic coatings, depending on the nature of the constituting material.

dichalcogenides, graphite, and boric acid. Most of these self-lubricating materials have been extensively studied in recent years, and some of them have been further optimized to provide better lubricity when used under specific conditions or in environments such as high temperatures (see excellent reviews in [1,8–11]).

Oxide-based materials are generally hard to shear at room temperature, but some of them become highly shearable and hence can provide fairly low friction coefficients at elevated temperatures. These oxides are often referred to as "lubricious oxides". Recently, Erdemir proposed a crystal-chemical model to classify these oxides on the basis of their lubrication performance and operational limits [7]. Efforts to optimize these and other solid lubricants are still in progress. However, on the basis of the fundamental understanding as well as modeling approaches, great strides are being made in the formulation of novel solid lubricants that can meet the increasingly stringent operating conditions of future tribosystems. For example, Cs-based oxides were reported to be very promising for lubricating Si-based ceramic components at high temperatures. At 600°C, friction coefficients of 0.02–0.1 have been reported for Cs_2O-lubricated Si_3N_4 ceramics [12]. During sliding at high

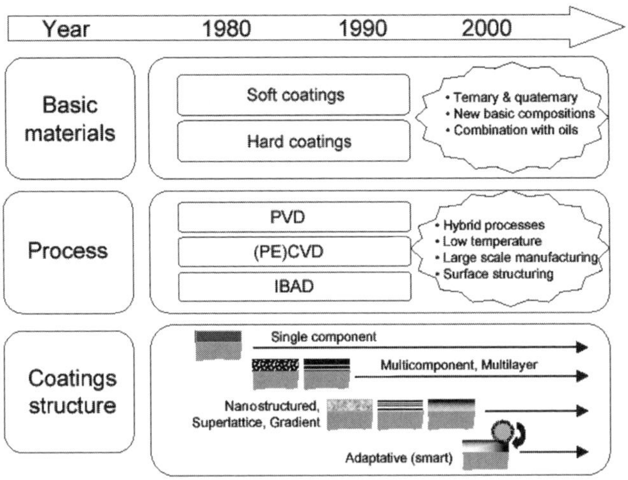

Figure 17.4: Historical development of tribologic coatings and solid lubricant films.

temperature, a mixed oxide layer consisting of Cs_2O and SiO_2 was found and believed to be responsible for low friction. As will be elaborated later (see Section 17.4), new formulations and exotic coating architectures (nanocomposites, alloying/doping, superlattice, gradient, etc.) have been used to produce various film structures to achieve better performance and durability in solid lubricant films. However, these developments did not prevent the emergence of new "single" structures or original concepts in solid lubrication. Let us first focus our attention on evolution of single and multicomponent solid lubricant films and then review the progress in other related products and processes.

17.3. Successive Generation of Solid Lubricant Coatings

Some of the major milestones in the historical development of solid lubricant coatings are graphically summarized in Fig. 17.4.

17.3.1. First Generation: Single-component Coatings

In recent years, enormous efforts have been made to refine the microstructure and the chemistry of solid lubricant coatings. Despite these efforts, only a few multicomponent and mostly "simple-structured" films prevailed on the commercial marketplace. With most commercial physical vapor deposition (PVD) and chemical vapor deposition (CVD) systems, it is much more practical and cost-effective to produce single layer coatings, often containing only one or two phases. The commercially successful solid lubricant coatings produced by these methods include single layers of MoS_2, WS_2 coatings (with or without some alloying elements, such as Ti, Ni, Au, Pb, and Sb); diamond and DLC coatings (with or without some H, N, B, Si, Ti, W, etc.), soft metals, and some polymers. To increase their lifetime, an adhesion interlayer may be necessary in most of these coatings. The transition-metal dichalcogenides (such as MoS_2, and WS_2) and other lamellar solids (such as graphite, hexagonal boron nitride, and boric acid) owe their lubricity primarily to their layered crystal structure (see Fig. 17.2). The lubrication mechanisms are not exactly the same from one lamellar solid to another; and the ambient temperature and environmental species (water vapor, oxygen, etc.) may play significant roles in their lubricity. For example, water vapor is necessary to lower the friction of graphite but rapidly inhibits the lubricating properties of MoS_2. The latter is widely used in vacuum lubrication, and considerable data have been accumulated on its tribologic behavior, leading to deep knowledge of its lubricating mechanisms in inert or vacuum environments [12,13], including the evidence of super low friction in the solid state [14,15]. Addition of such alloying elements as Ti, Ni, and Au reduces the environmental susceptibility of MoS_2, but the improvements achieved in its lubrication performance by such strategies have not yet lived up to expectations.

In recent years, considerable attention has been devoted to carbon-based solid lubricant films, such as DLC. These films are made of sp^2- and sp^3-hybridized carbon atoms and may contain some hydrogen (between <1 and ≈ 50 at.%), which affects not only the structure but also the properties [16]. Among all other solid lubricants, DLC probably exhibits the widest variations in friction and wear coefficients. Some DLC films exhibit friction coefficients below 0.01 in inert or vacuum environments, while others show friction coefficients of 0.6 or more. Fig. 17.5 is a diagram that classifies some of the carbon-based coatings with respect to their sp^2 vs. sp^3 characters as well as hydrogen content [16]. DLC films are generally synthesized by the PVD and CVD methods using a variety of hydrocarbon gases or solid carbon targets. Films derived from hydrocarbon gases contain large amounts of hydrogen in their microstructures and they are often referred to as hydrogenated DLC films. Films derived from solid carbon materials may contain very little or no hydrogen, hence they are often referred to as hydrogen-free DLC. Some of these films are made mostly of sp^3-bonded carbon atoms and they are often referred to as tetrahedral amorphous carbon (ta-C).

As mentioned above, the friction and wear behavior of DLC films are very sensitive to test environments and conditions. However, at atomic levels, the extent of tribochemical and adhesive interactions at sliding interfaces can play a major role. The main sources of adhesive interactions in DLC films are mainly due to covalent

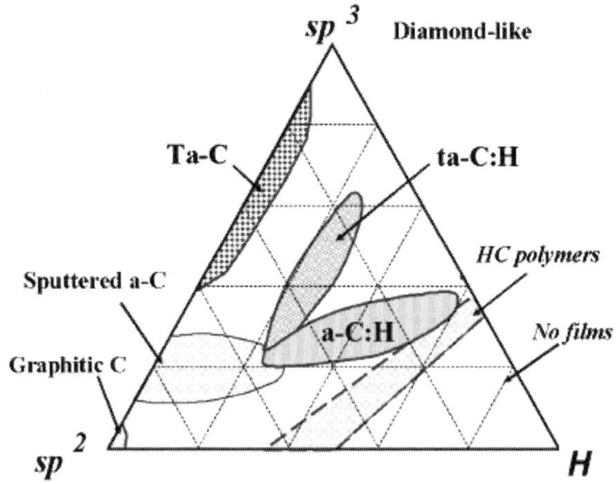

Figure 17.5: Ternary phase diagram of DLC films [16].

bonding between unoccupied or dangling σ-bonds at these interfaces. These bonds are the strongest and if they do exist between sliding carbon surfaces, they then cause very high friction as explained in Ref. [1]. Strong adhesive interactions can readily occur between sliding diamond surfaces in ultrahigh vacuum and/or at high temperatures and hence cause very high friction. Van der Waals forces, π–π^* interactions, capillary forces, and electrostatic attractions may also be present between sliding DLC surfaces and may lead to additional friction forces. The carbon-based materials and coatings are chemically inert and hence do not enter into strong reactions with other solids that are in static and/or sliding contacts. However, they may still interact with the gaseous species (such as water molecules, oxygen, hydrocarbons, etc.) in their surroundings and form a layer of physisorbed or even chemisorbed film on the sliding surfaces. Among others, both oxygen and water molecules can strongly interact with diamond and DLC surfaces and hence may cause higher friction. In the case of diamond films, the opposite is true. It gives very high friction coefficients in ultrahigh vacuum or at high temperatures, but less than 0.1 in open or moist air at room temperature. The more specific references related to the tribologic behavior of DLC films can be found in Refs. [12,17].

Diamond is another low friction material which is also extremely hard and offers several other outstanding properties, such as high mechanical strength and chemical inertness. At present, several CVD techniques can produce diamond as a thin hard coating on certain ceramic and metallic substrates (such as WC, SiC, Si, and W). The high-quality diamond coatings produced by CVD exhibit most of the desired mechanical and tribologic properties of natural diamond. Recently, nanocrystalline diamond (NCD) films with a very smooth surface finish have also been deposited by microwave CVD in the near absence of hydrogen and their superior friction and wear properties have been demonstrated [18]. Fig. 17.6 compares the surface morphology of a NCD film with that of a conventional microcrystalline diamond film. By polishing of rough diamond films, one can also achieve very low friction [19].

Polymers in various forms are used widely in tribology. They are lightweight, relatively inexpensive, and easy to fabricate. They can easily be blended with other solids to make self-lubricating composite structures. Certain polymers (polytetrafluorethylene (PTFE), polyimide, nylon, ultrahigh-molecular-weight polyethylene (UHMWPE), etc.) are inherently self-lubricating when used in both bulk and thin-film form, or as binders for other solid lubricants [20]. Except when they contain a self-lubricating inorganic filler [21], they generally cannot be used at temperatures higher than 250°C.

One of the most popular polymers is probably UHMWPE which is used widely in total joint replacements, even if its wear resistance against harder counterfaces must be improved to enhance the wear life of prostheses [17,22]. In total joint replacements, the synovial fluid largely provides the lubrication and the formation of UHMWPE wear particles is highly undesirable. Such particles may eventually migrate out of the contact interface and can lead to severe inflammatory reactions. In the long run, they can also trigger bone loosening which is a serious problem.

Soft metallic lubricants have crystal structures with multiple slip planes and do not work-harden appreciably during sliding contacts. Dislocations and point defects generated during shear deformation are rapidly nullified by the frictional heat

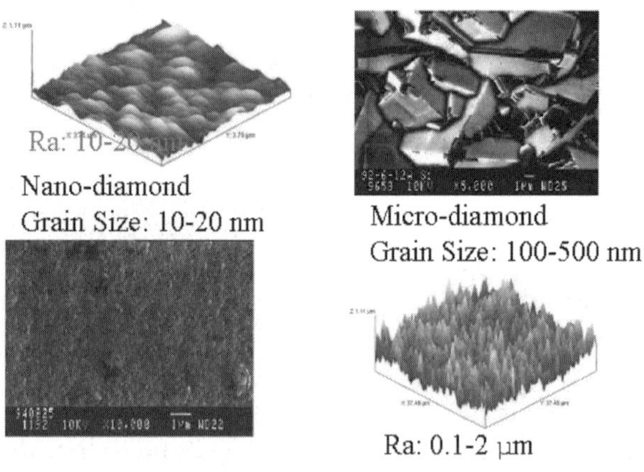

Figure 17.6: Scanning electron microscopy (SEM) and atomic force microscopy (AFM) images of nanocrystalline and microcrystalline diamond films grown in microwave plasmas. The friction coefficient of NCD was ~0.05, while that of microcrystalline diamond was more than 0.5 in dry nitrogen [18].

produced during sliding contact. Reported friction coefficients of soft metals range from 0.1 to 0.4, depending on the metal and test conditions [1]. Film-to-substrate adhesion is extremely critical for achieving long wear life or durability with soft metals, especially on the surfaces of ceramic tribomaterials.

Other solid lubricants include boron-containing materials: hexagonal boron nitride and boric acid, which can be formed *in situ* by reaction between B_2O_3 and water vapor during sliding contacts. The formation of such self-replenishing films on boron-based materials and coatings seems to provide very low friction and wear to sliding surfaces. Fig. 17.7 illustrates the chemical formation mechanism of boric acid on a boron carbide substrate as well as its frictional performance in open air. Observation of such low friction coefficients on boric acid containing surfaces is primarily attributed to the layered lattice structure of boric acid. Detailed microscopic examination of sliding surfaces that contained such boric acid films revealed sheets of boric acid on these sliding surfaces. They also provide some evidence for inter crystallite shear or slip in support of the solid lubrication mechanism (see Fig. 17.8).

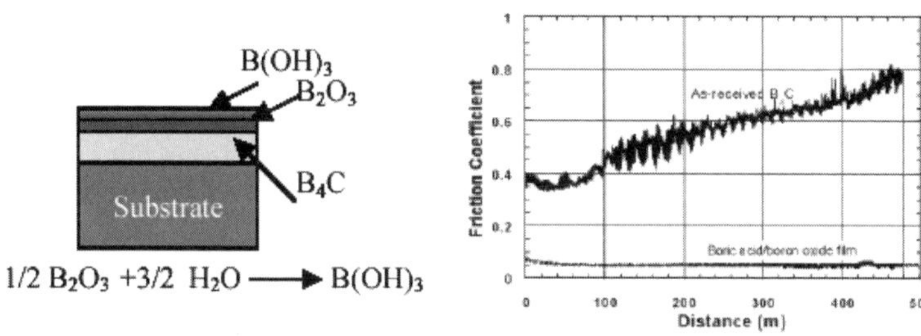

Figure 17.7: (a) Self-replenishing mechanism of boric acid on a boron oxide/boron carbide film and (b) friction coefficient of such a film in open air.

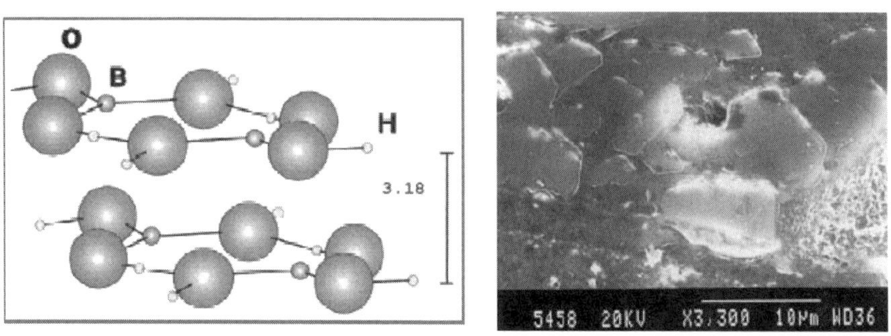

Figure 17.8: (a) Layered crystal structure of boric acid and (b) evidence of intercrystallite slip on a boric acid lubricated sliding surface.

As discussed in Ref. [1], (oxy-)thiomolybdates of Cs or Zn and a few other alkali metals, certain complex oxides and oxide-fluorides [1,11], and sulfate-based coatings ($CaSO_4$, $BaSO_4$, and $SrSO_4$) [23] were shown to be lubricious at elevated temperatures (300–600°C). Under that condition, a few complex materials can achieve both low friction and high wear resistance, and this is consistent with the use of "simple" coating structures, because more complex ones (such as multilayer and superlattice) may be altered by the combination of friction and high temperature during operation.

17.3.2. Second Generation: Duplex, Multicomponent Coatings

Modern coatings with multifunctional capabilities consist of hybrid, duplex, or multiplex coating architectures, as widely illustrated by Matthews *et al.* (see Ref. [24]). The main reasons behind such architectures are to improve not only wear but also some other application-specific properties of the coated system. A simple example would be deposition of an electroless coating on a surface, then a hard coating on the top. The intermediate layer can increase toughness and resistance to corrosion and oxidation, while the top layer can provide low friction and wear [25,26]. If the substrate material is very soft to start with, the use of an intermediate layer with higher hardness and toughness may be imperative. Such a layer can be selected from relatively hard materials that can more effectively support the top layer when subjected to severe loading conditions in practical applications [27,28]. In a similar fashion, the so-called duplex treatment processes that involve a prior thermal diffusion treatment (such as nitriding) followed by a top PVD coating (such as TiN) are carried out in sequence in the same deposition chamber. Such duplex surface layers are effective in extending the wear life of treated components. The same approach may be used to produce coatings with three or more layers. Such a multiplex coating architecture may permit additional functionality in practical applications. For example, some layers may provide high resistance to corrosion and oxidation, while others may provide superior friction and wear properties, while yet others may increase electrical and/or thermal conductivity of the coated systems. Such multiplex coating architectures may be very useful for achieving and maintaining smooth operations over broad application conditions.

The versatility of most vacuum deposition techniques such as PVD, CVD, and ion-beam-assisted deposition (IBAD) allows the production of complex or multicomponent coating systems mentioned above. Due to these techniques, great strides have been made in improving the wear resistance of hard tribologic coatings, such as transition-metal carbides, nitrides, and borides. For example, Knotek *et al.* [29] have shown that with the optimization of Ti–Al–V(C, N) compositions, the wear life of tools can be significantly improved. The incorporation of vanadium into the coating systems was found to be particularly beneficial since the oxides of vanadium were found to provide reasonably low friction and wear coefficients at elevated temperatures [30]. There are several excellent reviews on such multicomponent coatings that can be referred to for more information [31–37].

As with hard coating systems, multicomponent solid lubricant coatings are also made up of two or more constituents in the form of layers, grains, particles, or fibers. Multilayered coatings consist of a periodically repeated structure or lamellae of two or more materials, with thicknesses up to a few tens of a micrometer (if the thickness of each lamellae is in the nanometer range, these films are generally classified as superlattice, as discussed in Section 17.3.3). Generally, multilayered solid lubricant coatings are distinguished from sandwich layers, which include several superimposed films that possess complementary mechanical and/or tribologic properties. This second generation is generally fabricated by combining one of the solid lubricants described in Section 17.3.1 with another soft or hard layer. For example, a duplex coating system consisting of a TiN and a thin MoS_2 film was proven to have superior tribologic properties, especially in metal cutting and forming operations as compared to TiN or MoS_2 alone. Details on the diverse structures of such multicomponent coatings can be found in Refs. [2,38]. The other advantages of combining several structures and compositions within one coating include achievement of various individual physical properties (i.e. diffusion barrier + low friction), reduction of the mismatch in mechanical and chemical properties between the substrate and the coating (mainly to enhance adhesion), control of the residual strain and therefore the stress within the coatings, the prevention of crack formation under severe operating conditions, and enhancement of hardness and/or toughness by allowing layers or phases to slide over each other when they deflect under load. The versatility of vacuum technology (see Section 17.4) and the emergence of multiplex/hybrid processes have led to considerable advances in this second generation of coatings, which have recently found major industrial applications in transportation and manufacturing sectors. An excellent up-to-date overview of multilayered and multicomponent coatings can be found in Ref. [31].

17.3.3. Third Generation: Gradient, Superlattice and Nanostructured Coatings

Control of the structure and composition of coatings at the nanoscale is an exciting new scientific development. Despite many challenges, it now represents one of the hottest research topics in the field. There are three main approaches to controlling the structure and composition of coatings. The first involves the production of a functionally graded coating architecture, which is a logic progression from the multilayered coatings discussed above. In the past, grading of the composition of hard coatings (such as TiAlN) with a soft, solid lubricant film (such as MoS_2) has been successfully realized by controlling the PVD parameters [39]. The amount of MoS_2 has been increased in volume toward the end so that the top layer is essentially made of MoS_2 (for friction control) while the layer near the substrate is TiAlN (for wear control). Other benefits of grading the composition of tribologic coatings are that they can also improve adhesion and modify contact mechanics during sliding. In the case of DLC-based coatings, a bond layer is used in a graded fashion to increase film-to-substrate adhesion. Such improvements have been obtained by Ti and TiC(H) graded underlayers to increase the wear resistance of DLC films [40,41]. Other nanostructured coatings include nanocrystalline films (with grain sizes in the nanometer range) and nanocomposite films, as a logic progression of multicomponent films toward the nanoscale. Nanocomposites include structures that combine amorphous phases with crystallized ones, as widely emphasized by Veprek [42] in the case of superhard and wear-resistant coatings and by Zabinski *et al.* [43] in the case of solid lubricant films.

Another major development in solid lubricant architectures involves the production of nanolayered superlattice coatings, which is essentially an outgrowth of the multilayer concept mentioned above. The thickness of individual layers is in 1–50 nm. For example, nanolayering of DLC with W and Cr has been explored quite extensively in recent years. Fig. 17.9 shows the morphology of a nanolayered Cr/DLC film [44].

The major function of these structures is to significantly enhance hardness, fracture toughness, and adhesion of the coatings, and thus to improve their friction and wear properties. In recent years, the literature and the theoretic understanding of this new generation of films have grown substantially, in particular, with regard to the compositionally modulated superlattice films [45] and the nanocrystalline [42] and nanocomposite [46] coatings. On the macroscopic scale, the concept associated with this generation of coatings has resulted in the maximization of hardness (H) while ensuring an adequately low elastic modulus (E), to provide an appropriate "elastic stain to failure," as determined by the H/E ratio [47]. On the nanometer scale, the mechanism corresponds to a higher resistance to the creation and movement of stable dislocations. This effect is achieved by decreasing the grain size, in accordance with the well-known Hall–Petch relationship, or by controlling the presence of interfaces between nanocrystalline (nc) metal nitride/metal systems [48], or between amorphous (a) and nanocrystallized phases as with nc-M_nN/a-Si_3N_4 (where M is Ti, W, V, or other transition metal) [42]. The most recent efforts to extend the use of nanometer-scale coatings have been largely devoted to improving their stability at high temperature to mainly avoid phase transformation, grain

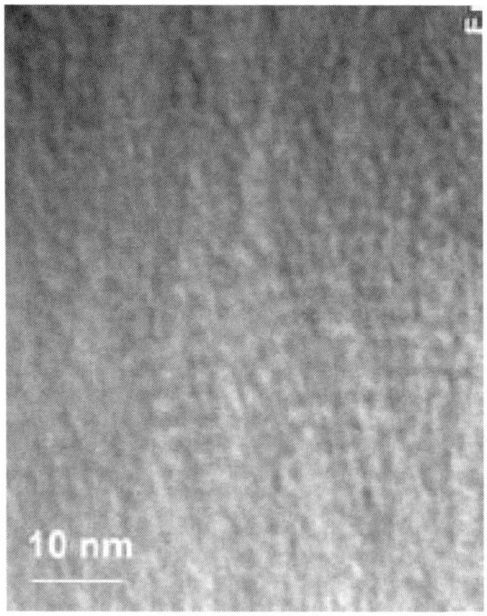

Figure 17.9: Z-contrast high-resolution transmission electron microscopy (HRTEM) image of a nanoscale multilayer C/Cr coating. Layer thickness is about 2 nm [44].

growth, and diffusion. This improvement seems to be possible if the layers are thermodynamically stable with respect to each other and are stable enough to form low-energy coherent interfaces [49].

17.3.4. Fourth Generation: Smart Adaptative or Chameleon Coatings

This fourth generation comprises coatings that change their properties as needed to meet the specific requirements of the operating conditions of a given application. More specifically, these coatings adapt to the changing conditions of a tribologic application. Design and production of such coatings require a deep knowledge of the film microstructure and chemistry as well as their interaction with or response to rapid or cyclic changes in temperature, contact pressure, and operating environments. As an example of this fourth generation, the most recently developed cutting-tool coatings (e.g. based on TiAlN with the addition of yttrium or chromium) produce a stable oxide during cutting at high temperature and enhance the wear resistance [39]. Another example is the composite coating produced within the W-C-S system, which consists of 1–2 nm WC and 5–10 nm WS_2 grains embedded in an amorphous DLC matrix [50,51]. The WC/DLC/WS_2 nanocomposite exhibits self-adaptation to tribologic conditions that occur in aerospace systems. This adaptation was found in crystallization and reorientation of initially nanocrystalline and randomly oriented

WS_2 grains, graphitization of the initially amorphous DLC matrix, reversible regulation of the composition of the transfer film between WS_2 and graphite with environmental cycling from dry to humid air, and a possible DLC/WS_2 synergistic effect, providing friction reduction in oxidizing environments. Friction lower than 0.05 was recorded during a space simulation test of two million cycles, and recovery of low friction in tests that simulate ambient/space environmental cycling was demonstrated. Another challenge for smart solid lubricants is achievement of lubrication over a wide range of temperatures. No single material is known to be lubricious from ambient temperature to 800°C. Thus, the way to produce a lubricant coating that can operate over this temperature range is to combine low- and high-temperature lubricants into a composite or layered structure, such as CaF_2 and WS_2, which interact during friction to form $CaSO_4$ [52]. A similar concept has been demonstrated by combining certain transition-metal dichalcogenides (MoS_2 or WS_2) with oxides (ZnO or PbO) to form $PbMoO_4$ or $ZnWO_4$, which are lubricious at high temperature [53,54]. One of the greatest challenges of future activities may be the design of better and much smarter coatings whose desirable properties are unaffected by large fluctuations in temperature or environment.

17.3.5. New Coating Composition: the Case of CN_x

It is unusual for a material to find large-scale applications only a few years after its discovery. One exception is the recently developed carbon nitride films, which are already used as protective coatings on hard disks and recorder heads [55]. The wear rates of carbon nitride-coated disks can be 10 times lower than those of disks coated with a commercial DLC film of the same thickness [56]. A flurry of attempts to produce the elusive C_3N_4 material that is theoretically harder than diamond has been reported. Most of the films deposited are, in fact, amorphous CN_x, with $x \approx 0.1$–0.3. Primarily because of a large variation in their microstructure and chemical stoichiometry, amorphous CN_x films produced by various research groups exhibit large variations in mechanical properties and tribologic performance (see Ref. [12] for a summary on CN_x films). Reported friction values range from 0.05 to $>$0.5 in air, and friction tends to increase as the film's nitrogen content is increased. In a dry environment or vacuum, much higher friction coefficients were observed, however, when sliding against Si_3N_4 balls in a stream of dry N_2, Kato *et al.* have reported friction coefficients of $\sim$0.01 for an ion-beam deposited CN_x film [57]. Such a large scatter in friction has been attributed to the difference in N content, C–N bondings, carbon hybridization ratios, and deposition methods and conditions. Noteworthy, however, is the extreme surface smoothing of the coating that takes place during sliding wear with root mean square (rms) surface roughness of <0.5 nm, and this smoothing has a large impact on applications that involve gliding and sliding contacts such as in magnetic hard disks. Normally, most hard materials are very brittle, but CN_x films are very elastic, which seems to be a contradiction. In a recent chapter [58], Hultman *et al.* present evidence that CN_x may exhibit a fullerene-like structure, which would give an elastic behavior consistent with a resilient material. The observed structure consists of sp^2-coordinated basal planes that are buckled by the incorporation of pentagons and cross-linked at sp^3-hybridized C sites, both of which are caused by structural incorporation of nitrogen. Carbon nitride, thus, deforms elastically due to bending of the structural units.

17.4. Recent Advances in Deposition Methodology

Solid lubricant films can be produced on a given surface by a wide range of methods. Most are based on vacuum technologies such as sputtering, ion plating, arc-PVD, pulse laser deposition, IBAD, or plasma-enhanced CVD (PECVD). These methods result in strong adhesion between solid lubricant films and substrate materials. Quite recently, these advanced methods were combined with smart surface engineering practices, such as micro patterning or texturing, to achieve much improved tribologic performance in practical applications. In this chapter, we cannot provide a comprehensive overview of all of the available methods, but several key book chapters, journal articles, and conference proceedings are currently available. For example Ref. [2] provides some basic considerations in deposition methodology.

17.4.1. Deposition Temperature

The temperature of thin film deposition is extremely important not only for achieving the optimum film microstructure and property in resultant films, but also for retaining the most desired mechanical properties of substrate materials. In reality, most solid lubricant coatings are produced by a PVD or CVD process, and the range of deposition temperatures available in these processes may vary from room temperature to more than 1000°C. The CVD process may not be used to deposit solid lubricant and tribologic coatings on most metallic substrates. High deposition temperatures can cause irreversible phase transformation, thermal softening, or shape changes in most metallic substrates during deposition. In particular, certain Al, Cu, and Mg alloys cannot be coated in a CVD system. The preferred deposition temperature for these non-ferrous metals and alloys should be less than 100°C. For steels, deposition temperatures may vary between 100°C and 500°C, which are still much lower than the typical deposition temperature of the conventional CVD process. For PECVD or plasma-assisted CVD (PACVD) processes, the deposition temperature may be less than 500°C, and these processes may be suitable for coating steel substrates with solid lubricant coatings. Conventional CVD processes are mainly used to produce hard coatings on ceramics, cemented carbides, and superalloy materials that are much more heat resistant than most steels and alloys. PVD processes are mainly used to deposit solid lubricant films (such as MoS_2) and hard coatings (such as TiN) on steels and other metallic substrates. Some heat-resistant steels, such as high-speed steel (HSS), can be treated by all types of PVD and some low-temperature PECVD processes. But the emerging need to deposit solid lubricant films on temperature-sensitive materials, such as polymers, is still very difficult to attain by PVD or CVD methods. Certain polymers (such as PEEK, polyimides, etc.) can be coated by these methods provided that the deposition temperature is kept low during deposition. The development of less conventional processes, such as sol-gel techniques, may be more appropriate for such heat-sensitive materials.

17.4.2. Large-scale Production of Solid Lubricant Coatings

Published data on most solid lubricant coatings are often based on work done on small scale, R&D-type PVD/CVD systems. When such coatings were applied to industrial parts and components with complex and intricate shapes and sizes in large production volumes, several difficulties were encountered. This was particularly true for industrial production of exotic coating architectures that include gradient, superlattice, and nanostructured films (Section 17.3.3). One of the difficulties in large-scale productions is to achieve and maintain a very uniform coating thickness over the exposed surface areas of industrial components. Also, most laboratory-scale coatings may not provide the same level of performance or desired property when produced in larger deposition systems. Moreover, in many cases, complicated installations, including shielding and shuttering, are not feasible in large-scale deposition systems. Finally, the coating supplier must be watchful of the reproducibility of the process vs. time. Some chapters present deposition runs performed on industrial vacuum devices. A combined cathodic arc/unbalanced magnetron deposition system with four vertical targets is detailed in Ref. [59]. These linear cathodes, in combination with up to threefold substrate rotation, must be used to achieve homogeneous rate distribution over the whole height of the vacuum chamber, with a distance of 1 m between two opposing targets and an outer diameter of the rotating substrate table of 750 mm. The magnets and the polarization of the electromagnetic coils are arranged in a close magnetic field configuration to achieve high plasma density and to control the temperature to as low as 200°C under specific conditions. The industrial production of nanolayered TiAlCrN/TiAlYN coatings has been shown to be feasible, and the process has proved to be stable and reproducible. Gradient Ti/TiC(H)/a-C:H films have been deposited in an industrial duplex reactor that combines magnetron sputtering for Ti deposition and PECVD for deposition of hydrogenated DLC [40]. The films exhibit friction in the 10^{-2} range with high wear resistance under high-vacuum conditions. Due to confidentiality requirements, less work in that field is published, and most data are kept within companies for their own use.

17.4.3. Hybridization of Deposition Methods with Surface Texturing and/or Patterning

In an effort to achieve better performance and durability, researchers have lately combined tribologic coatings with high-precision texturing and/or patterning, which require high-energy plasmas, and chemical or laser etching techniques. These techniques have existed for quite some time and been used extensively by the microelectronic industry for many years. They are suitable for producing specific geometric dimples and/or grooves on a surface. Production of special grooves and dimples was demonstrated by a special method developed by Petterson and Jacobson [60], but lasers have advantages due to their very robust and quick nature. In the past, excimer lasers have been used to induce patterns on solid lubricant DLC films [61], but emerging femtosecond (fs) laser facilities will probably introduce new perspectives in the field of laser patterning. Indeed, the ultrashort laser pulses (in the 100 fs range, with power density as high as $10^{13}\,W\,cm^{-2}$) can ablate all kinds of materials without any collateral effects (negligible thermal affected zone), contrary to nanosecond pulses [62]. Recent experimental studies have demonstrated that the lifetime of TiN [63] and TiCN [64] coatings can be increased by factors of up to 10 by patterning with a femtosecond laser. Dumitru *et al.* [65] have studied laser patterning of wear-resistant DLC films by coating of already patterned substrates (indirect processing) and by direct laser processing of deposited DLC films. Pore depths that yield positive tribologic improvements ($>10\,\mu m$) are larger than the film thickness ($<5\,\mu m$). Dumitru *et al.* indicate that debris particles were trapped in the surface pores obtained by indirect laser processing, thus preventing the breakdown of the tribologic system. According to Voevodin *et al.* [66], the three-dimensional design considerably improved the tribologic characteristics of hard coatings by permitting solid lubricant replenishment inside the friction contacts. Such patterning may also have a positive effect by carrying away heat from sliding interfaces and by trapping abrasive wear debris. In Ref. [66], a functionally gradient Ti–TiC–TiC/DLC coating with an upper layer of tough nanocrystalline/amorphous composite was used for load support, crack prevention, and stress equalization. This coating was processed by laser irradiation to form grooved tracks along wear paths, which were then filled with MoS_2 to provide a solid lubricant reservoir in the lateral dimension of the coating. The three-dimensional coating was tested in long-duration sliding tests at fixed and cycling humidity. The coating exhibited environmental adaptation, with friction coefficients of 0.15 in humid air and 0.02 in dry nitrogen. The wear life was increased by at least one order of magnitude when compared with that for a hard-gradient coating with a top layer of MoS_2 without three-dimensional laser processing. In the near future, coating patterning will probably be a major emphasis of development effort.

17.5. Concluding Remarks

Solid lubricants are critically important for safe and smooth operations of numerous tribologic systems. Lamellar solids, soft metals, diamond and DLC films, lubricious oxides, and certain polymers are well-known solid lubricants. Their uses are expected to further increase in coming years, mainly because the operating conditions of future tribosystems are becoming more and more demanding and liquid and grease-type lubricants are undesirable due to environmental concerns. In modern applications, most solid lubricants are produced as thin solid films on sliding surfaces. Several PVD and CVD techniques are now available for the deposition of strongly bonded solid lubricant films on various kinds of substrates, including metallic, ceramic, and polymeric types. These techniques produce solid lubricant films in gradient, duplex, multiplex, and nanostructured or nanocomposite forms, resulting in better performance and durability under severe applications. In recent years, researchers have also prepared coatings that are designed to adapt to the changing conditions of tribologic applications. These are called "adaptive" or "chameleon" solid lubricant coatings. As a novel approach, researchers have recently coupled solid lubricant films with smart surface engineering strategies (such as micro-texturing and/or -patterning) and thus achieved even higher levels of performance and durability under severe tribologic conditions. For applications involving high temperatures, most layered solid lubricants appear ineffective. Certain lubricious oxides and fluorides may be used to combat friction and wear at high temperatures. A new generation of adaptive solid lubricant films was also shown to be effective in achieving lubrication at broader temperature ranges. Certain polymers are also used as solid lubricants because the attractive properties they combine are unavailable in other solid lubricants. Polymers are particularly favored for applications where cost, weight, corrosion, and biocompatibility are the major considerations. In short, solid lubricants have been around for a long time, and they have been meeting some very important and critical tribologic needs. They are expected to be in high demand for many more years to come.

Acknowledgments

This work is supported by the US Department of Energy, Office of Transportation Technologies, under Contract W-31-109-Eng-38.

References

[1] A. Erdemir, in *Modern Tribology Handbook*, Ed. B. Bhushan, Vol. II, CRC Press, Boca Raton, FL, 2001, pp. 787–825.
[2] K. Holmberg and A. Matthews, in *Coatings Tribology*, Ed. D. Dowson, Elsevier, Netherlands, 1994, pp. 1–6.
[3] C. Donnet, *Condens. Matter News*, 4 (6) (1995) 9.
[4] B.A. Cook, J.L. Harringa, T.L. Lewis and A.M. Russell, *J. Adv. Mater.*, 36 (2004) 56.
[5] B. Carroll, Y. Gogotsi, A. Kovalchenko, A. Erdemir and M.J. McNallan, *Tribol. Lett.*, 15 (2003) 51–55.
[6] C. Mancinelli, C.J. Jenks, P.A. Thiel and A.J. Gelman, *J. Mater. Res.*, 18 (2003) 1447.
[7] R. McGrath, J. Ledieu, E. Cox and R.D. Diehl, *J. Phys. Condens. Matter.*, 14 (2002) 119.
[8] P. Sutor, *MRS Bull.*, 16 (1991) 24.
[9] K. Miyoshi, NASA Technical Memorandum No:1072249 (1991).
[10] I.L. Singer, *MRS Bull.*, 23 (6) (1998) 37.
[11] A. Erdemir, *Tribol. Lett.*, 8 (2000) 97.
[12] A. Erdemir and C. Donnet, in *Modern Tribology Handbook*, Ed. B. Bhushan, Vol. II, CRC Press, Boca Raton, FL, 2001, pp. 871–908.
[13] I.L. Singer, in *Fundamentals of Friction: Macroscopic and Microscopic Processes*, Eds. I.L. Singer and H.M. Pollock, *NATO-ASI Series*, Vol. 220, Kluwer Academic, Dordrecht, 1992, p. 237.
[14] J.M. Martin, C. Donnet, T. Le Mogne and T. Epicier, *Phys. Rev. B*, 48 (14) (1993) 10583.
[15] C. Donnet, J.M. Martin, T. Le Mogne and M. Belin, *Tribol. Int.*, 29 (2) (1996) 123.
[16] T.J. Robertson, *Mater. Sci. Eng. R*, 37 (4–6) (2002) 129.
[17] R. Hauert and U. Müller, *Diam. Relat. Mater.*, 12 (2003) 171.
[18] A. Erdemir, C. Bindal, G.R. Fenske, C. Zuiker, A.R. Krauss and D.M. Gruen, *Diam. Relat. Mater.*, 5 (1996) 923.
[19] K. Miyoshi, R.L.C. Wu, A. Garscadden, P.N. Barnes and H.E. Jackson, *J. Appl. Phys.*, 74 (1993) 4446.

[20] W.E. Jamison, in *CRC Handbook of Tribology and Lubrication*, Vol. 3, CRC Press, Boca Raton, FL, 1994, p. 121.
[21] T.A. Blanchet and F.E. Kennedy, *Wear*, 153 (1992) 229.
[22] A. Wang, D.C. Sun, C. Stark and J.H. Dumbleton, *Wear*, 181/183 (1995) 241.
[23] P.J. John and J.S. Zabinski, *Tribol. Lett.*, 7 (1999) 31.
[24] A. Matthews, *Vacuum*, 65 (2002) 237.
[25] M. Bin-Sudin, A. Leyland, A.S. James, A. Matthews, J. Housden and B. Garside, *Surf. Coat. Technol.*, 81 (1996) 215.
[26] A.L. Yerokhin, X. Nie, A. Leyland, A. Matthews and S.J. Dowey, *Surf. Coat. Technol.*, 122 (1999) 73.
[27] A. Matthews, in *Protective Coatings and Thin Films*, Eds. Y. Pauleau and P.B. Berna, Kluwer Academic, Dordrecht, 1997, pp. 1–12.
[28] T. Bell, in *New Directions in Tribology*, Ed. I.M. Hutching, Mechanical Engineering Publications, London, 1997, pp. 121–133.
[29] O. Knotek, M. Atzor and H.G. Prengel, *Surf. Coat. Technol.*, 36 (1988) 265.
[30] E. Lugscheider, O. Knotek, S. Barwulf and K. Bobzin, *Surf. Coat. Technol.*, 142–144 (2001) 137–142.
[31] H.A. Jehn, *Surf. Coat. Technol.*, 131 (2000) 433.
[32] R. Cremer, in *Value-Addition Metallurgy*, Eds. W.D. Cho and H.Y. Sohn, TMS, 1998, p. 249.
[33] A. Cavaleiro and C. Louro, *Vacuum*, 64 (3/4) (2002) 211.
[34] M. Balzer and H. Kappl, *Surf. Coat. Technol.*, 116–119 (1999) 766.
[35] B. Rother and H. Kappl, *Surf. Coat. Technol.*, 73 (1995) 14.
[36] J.C. Rebholtz, H. Ziegele, A. Leyland and A. Matthews, *J. Vac. Sci. Technol.*, A16 (1998) 2851.
[37] S. PalDey and S.C. Deevi, *Mater. Sci. Eng. A*, 342 (2003) 58.
[38] K. Holmberg and A. Matthews, in *Modern Tribology Handbook*, Ed. B. Bhushan, Vol. II, CRC Press, 2001, pp. 827–870.
[39] A. Savan, E. Pfluger, R. Goller and W. Gissler, *Surf. Coat. Technol.*, 126 (1999) 159.
[40] C. Donnet, J. Fontaine, T. Le Mogne, M. Belin, C. Héau, J.P. Térrat, F. Vaux and G. Pont, *Surf. Coat. Technol.*, 120/121 (1999) 548.
[41] A.A. Voevodin, S.D. Walck and J.S. Zabinski, *Wear*, 203/204 (1997) 516.
[42] S. Veprek, *J. Vac. Sci. Technol.*, A17 (5) (1999) 2401.
[43] J.S. Zabinski, J.H. Sanders, J. Nainaparampil and S.V. Prasad, *Tribol. Lett.*, 8 (2000) 103.
[44] P. Eh. Hovsepian, Y.N. Kok, A.P. Ehiasarian, A. Erdemir, J.-G. Wen and I. Petrov, *Thin Solid Films*, 447–448 (2004) 7.
[45] S.A. Barnett and M. Shinn, *Annu. Rev. Mater. Sci.*, 24 (1994) 481.
[46] J. Patscheider, *MRS Bull.*, 28 (3) (2003) 180.
[47] A. Leyland and A. Matthews, *Wear*, 246 (2000) 1.
[48] J. Musil, P. Zeman, H. Hruby and P. Mayrhofer, *Surf. Coat. Technol.*, 115 (1999) 32.
[49] S.A. Barnett, A. Madan, I. Kim and K. Martin, *MRS Bull.*, 28 (3) (2003) 169.
[50] A.A. Voevodin, J.P. O'Neull and J.S. Zabinski, *Surf. Coat. Technol.*, 116–119 (1999) 36.
[51] A.A. Voevodin and J.S. Zabinski, *Thin Solid Films*, 370 (2000) 223.

[52] P.J. John, S.V. Prasad, A.A. Voevodin and J.S. Zabinski, *Wear*, 219 (2) (1998) 155.
[53] S.D. Walck, M.S. Donley, J.S. Zabinski and V.J. Dyhouse, *J. Mater. Res.*, 9 (1994) 236.
[54] S.D. Walck, J.S. Zabinski, N.T. McDevitt and J.E. Bultman, *Thin Solid Films*, 305 (1997) 130.
[55] E.C. Cutiongco, D. Li, Y.W. Chung and C.S. Bhatia, *J. Tribol.*, 118 (1996) 543.
[56] A.D. Khurshudov and K. Kato, *Surf. Coat. Technol.*, 86/87 (1996) 664.
[57] K. Kato, N. Umehara and K. Adachi, *Wear*, 254 (2003) 1062.
[58] L. Hultman, J. Neidhardt, N. Hellgren, H. Sjöström and J.E. Sundgren, *MRS Bull.*, 28 (3) (2003) 194.
[59] W.D. Münz, *MRS Bull.*, 28 (3) (2003) 173.
[60] U. Pettersson and S. Jacobson, *Tribol. Int.*, 36 (2003) 857.
[61] V.P. Agreev, T.N. Glushko, V.F. Dorfman, A.V. Kuzmichev and B.N. Pypkin, *SPIE Proc.*, 1503 (1991) 453.
[62] R. Le Harzic, N. Huot, E. Audouard, C. Jonin, P. Laporte, S. Valette, A. Fraczkiewicz and R. Fortunier, *Appl. Phys. Lett.*, 80 (21) (2002) 3886.
[63] T.M. Kononenko, S.V. Garnov, S.M. Pimenov, V.I. Konov, V. Romano and B. Borsos, *Appl. Phys. A.*, 71 (2000) 627.
[64] G. Dumitru, V. Romano, H.P. Weber, H. Haefke and Y. Gerbig, *Proc. WLT Laser 2001*, (2001) 351.
[65] G. Dumitru, V. Romano, H.P. Weber, S.M. Pimenov, T.M. Kononenko, J. Hermann, S. Bruneau, Y. Gerbig and M. Shupegin, *Diam. Relat. Mater.*, 12 (3/7) (2003) 1034.
[66] A.A. Voevodin, J. Bultman and J.S. Zabinski, *Surf. Coat. Technol.*, 107 (1998) 12.

Chapter 18

Corrosion- and Wear-Resistant Coatings Formed by Ion Beam Techniques

Wolfgang Ensinger

18.1. Introduction

This chapter deals with ion-beam-based methods for protecting metals. It is divided into two parts, corresponding to the kind of protection. One is protection against *mechanical* attack, or tribological protection. The attack takes place by one or many hard bodies in contact with the sample to be protected. Wear of the material induced by these counter bodies can be decreased when the material is treated by ion beam techniques. The other one is *chemical* protection, that is against corrosion. The corrosive medium is an aggressive aqueous environment such as an acid or a salt brine. The specimen is chemically changed or is consumed by the chemical reaction. Again, ion beam techniques can be used for protecting the material and increasing its lifetime in the corrosive environment. When a material is in contact with another one or another phase, it might be damaged by this contact by either mechanical or chemical effects. In most cases, contact and damage take place at the surface. The material communicates with its surrounding environment via its surface. Therefore, surface modification is a method to influence this communication and the performance of the material in case of an attack. Ion beam methods are, to first order, surface or near-surface modification methods. As such they have a potential to serve as a materials modification method for wear and corrosion protection. The techniques considered here are *ion implantation* and *ion-beam-assisted deposition* (IBAD) of thin films. Both work with a directed beam of energetic ions directed towards the surface of the material to be modified. In ion implantation, ions penetrate into a material and are incorporated. Apart from their presence there, they transfer energy and momentum into the material and may change physical structure and chemical composition. Ion implantation can be divided with respect to the energy of the ions. Roughly, three energy domains can be distinguished: low energy (several keV kinetic ion energy), medium energy (several 10–100 keV), and high energy (MeV). While medium- and high-energy ions are mainly used for ion implantation, the low-energy range is more relevant

Materials Surface Processing by Directed Energy Techniques
Y. Pauleau (Editor)

for IBAD. Here, a growing film is irradiated with ions. Again, they transfer energy into it and change its structure and phase, and elemental composition. Technically, ion implantation in the medium-energy range is most important. It is commercially being used. As mentioned above, the chapter is divided in two parts, each dealing with the protection to be exerted, chemical or mechanical. Each part is divided in sub-chapters according to the ion beam technique or ion energy regime used. Before discussing results from the literature, the measurement techniques for evaluating the effect of the ion beam treatment are described. Eventually, the last sub-chapter deals with industrial application of ion implantation.

18.2. Protection Against Chemical Attack: Aqueous Corrosion of Ion-Beam-Treated Metals

Before the different ion beam techniques and results for protection of metals against corrosion are treated, the corrosion measurement techniques are briefly introduced so that the results can be better understood.

18.2.1. Electrochemical Corrosion Measurement Methods

Mostly, the corrosion performance of ion-beam-modified metals is evaluated by electrochemical techniques [1,2]. Either, corrosion potential or corrosion currents are recorded as a function of exposure time, or polarization measurements with a three-electrode set-up are carried out. The sample is given a defined potential vs. a reference electrode by means of a potentiostat. The potentiostat measures the potential between the working and the reference electrode. In order to maintain a particular potential, it establishes a current flow through the working electrode and the counter electrode. The latter has to carry the current, as the reference electrode must remain without current flow in order to be able to measure the potential. The potential is either kept constant (potentiostatic technique) for a certain time while the current response is recorded, or it is scanned over a certain potential range (potentiodynamic technique). In cyclic voltammetry, the potential is repeatedly scanned to a certain potential and back. In the potentiostatic technique the potential can be altered step-by-step to gain informations on a potential range of interest. Fig. 18.1 shows schematically a polarization or current density/potential curve of an actively corroding metal such as iron in acid. In the literature, the curves often are inversed, with the y-axis showing the potential.

At the left branch of the curve, the metal is cathodic. It develops hydrogen. At the anodic branch, it is dissolved. The potential were the curve changes the sign is the

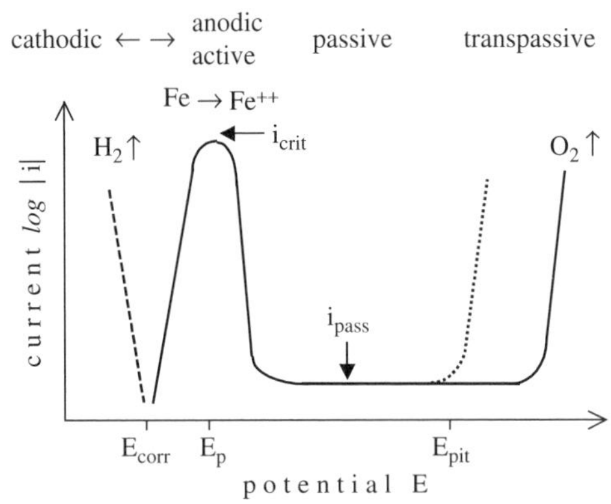

Figure 18.1: Polarization (current vs. potential) curve, E_{corr} corrosion potential, E_p passivation potential, E_{pit} pitting potential, i_{crit} critical current, i_{pass} passive current.

open-circuit, or rest, or corrosion potential E_{corr}. It is adopted by the sample in a corrosive environment. E_p is the potential where the active/passive transition takes place. The metal is oxidized. The oxide film protects it from further dissolution. The current at this position is the maximum current of metal dissolution, the so-called critical current i_{crit}. It is often used as a criterion of corrosion. When localized corrosion is investigated, the breakthrough or pitting potential E_{pit} can be determined, where the natural protective film of a material locally breaks down owing to attack of a corrosive species. An example is aluminum. It is protected by a stable oxide film. However, in the presence of aggressive anions, the film may locally be dissolved, leading to formation of deep pits (pitting corrosion). The onset of pitting is given by E_{pit}.

Electrochemical corrosion measurement methods have the advantage that results can be obtained in a comparatively short time. In contrast, field-type tests with immersion of the samples take longer periods of time. Another advantage is that they can give insight into corrosion mechanisms. The drawback is that they are artificial tests; the long-term behavior of the samples can usually not be predicted.

18.2.2. Corrosion of Medium-Energy Ion Implanted Metals

The first systematic studies on ion implantation for corrosion protection were carried out in the late 1970s. Ashworth *et al.* implanted 20 keV chromium ions with doses up to $2 \cdot 10^{17}\,\mathrm{cm}^{-2}$ into iron [3]. They studied the corrosion in neutral aqueous solution. Alloying iron with more than 12 at.% Cr yields excellent corrosion resistance. The formation of a Fe–Cr surface alloy was supposed to change the behavior of iron or low-alloy steel to the one of stainless steel. The authors carried out polarization measurements. Fig. 18.2 compares current/potential curves of untreated steel, metallurgically alloyed steel, and Cr-ion-implanted steel. Cr reduced the anodic Fe dissolution currents. The ion beam surface alloy compares to the metallurgically prepared bulk alloy. Ion implantation improves the corrosion resistance considerably.

One of the advantages of ion implantation is that virtually any element can be incorporated into any solid. A metal can be alloyed with any other metal even if they are not soluble in each other. Metastable alloys can be formed. This idea was used for corrosion protection. Transition metals other than Cr are even more corrosion resistant. Examples are titanium or tantalum. However, the latter is not soluble in iron. By means of ion implantation, a metastable alloy of Ta and Fe can be formed. Ashworth *et al.* implanted tantalum into iron at 20 keV ion energy, again with doses up to $2 \cdot 10^{17}\,cm^{-2}$ [4]. The result showed that the protection was more efficient than the one by Cr implantation. During corrosion, the iron matrix was dissolved, while the insoluble tantalum remained at the sample surface.

After these encouraging first results, several groups in the world started to work on the influence of ion implantation and IBAD on corrosion. There is a number of reviews on this topic in the literature with a large number of examples of ion implantation into and IBAD onto various metals and alloys, including alloys based on iron, aluminum, titanium, and nickel [2,5–10]. The implanted element is usually one which is more corrosion resistant than the base metal. All kinds of metals and non-metals including noble metals such as platinum and metalloids such as

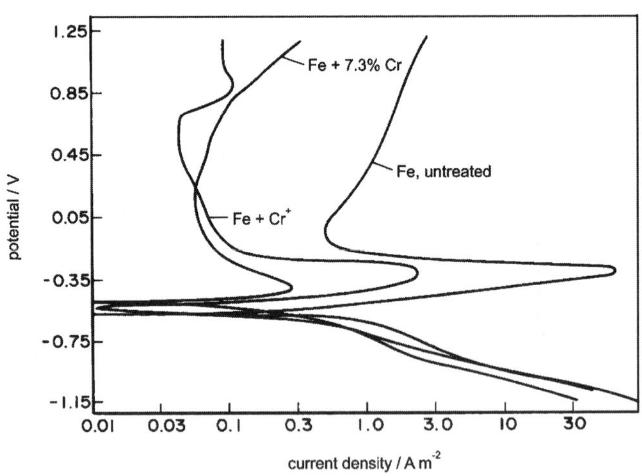

Figure 18.2: Polarization curves of untreated Fe, Fe alloyed with 7.3% Cr, and Fe ion implanted with Cr^{+} (adapted from Ref. [3]).

phosphorus and boron were used. The implantation was used to protect the metal against different types of corrosion, including uniform corrosion, localized corrosion such as pitting and crevice corrosion, which may occur when metals are exposed to water, salt brines, or acids.

Several aspects were considered, from solid solutions to metal/metalloid compounds, including metastable alloys and glassy alloys. Ferber and Wolf implanted noble metals such as Hg, Au, and Pt into iron [11,12]. They are not or only to a quite limited content soluble in iron. Hence, the formed surface alloys were metastable. The metal ion energies were 200–400 keV, the doses reached up to $10^{17}\,cm^{-2}$. The authors carried out cyclic voltammetry measurements in acetate buffer at pH 5.6 and in 1 N sulfuric acid. The results were quite different, depending on the metal. While mercury and lead reduced corrosion, gold and platinum enhanced it considerably. Another approach is amorphization. Corrosion often is promoted by grain boundaries. Amorphous metals are therefore in general often more corrosion resistant. Cooney *et al.* implanted stainless steel AISI 304L with 175 keV phosphorus ions with doses from $0.2 \cdot 10^{17}$ to $4 \cdot 10^{17}\,cm^{-2}$ [13]. At the dose of $3.7 \cdot 10^{17}\,cm^{-2}$, the implanted zone became amorphous. Austenitic stainless steels are protected by a dense inert oxide film. The steels do not corrode in water. However, if aggressive anions such as chloride are present, the oxide may locally fail. Deep pits may be formed. For testing the *P* ion-treated steel, anodic polarization measurements were carried out in 0.1 M H_2SO_4 + 0.1 M NaCl. These are pitting conditions. It turned out that the implant influences the pitting behavior strongly. In the polarization curves in Fig. 18.3, the dissolution currents of the implanted samples

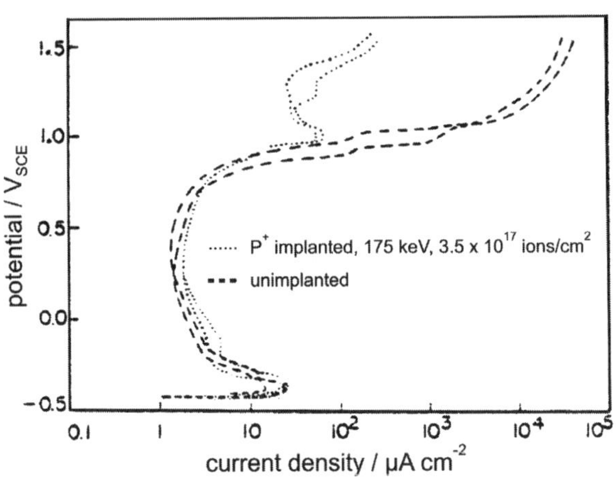

Figure 18.3: Anodic polarization curves of stainless steel samples in chloride-containing sulfuric acid (adapted from Ref. [13]).

remained at a low level, indicating resistance against pitting. In contrast, the unimplanted steel started severe pitting at around 1 V.

Nielsen *et al.* treated M50 steel with 180 keV tantalum and chromium ions with maximum doses of $5 \cdot 10^{16}$ and $4 \cdot 10^{17} cm^{-2}$ [14]. The objective of the study was to increase the lifetime of ball-bearing steel for gas turbine engine bearings. The samples were tested by cyclic polarization measurements in carbonate/bicarbonate (0.5 M/0.05M) solution containing 0.5 M chloride ions. Fig. 18.4 shows the peak currents extracted from polarization curves of Cr-implanted samples as a function of the potential scan rate. The steel dissolution currents decrease with increasing implantation dose, indicating a beneficial effect on corrosion protection with increasing Cr concentration.

In these examples, the corrosion protection effect has been achieved by alloying more corrosion-resistant elements into the surface-near regions of the sample. These reduce the dissolution rates. The next example is different. Here, the protective effect is given by an electrocatalytical effect. Due to a very stable Ta_2O_5 film tantalum is a particularly corrosion-resistant metal. It is used as a construction material in chemical industry when concentrated mineral acids have to be handled

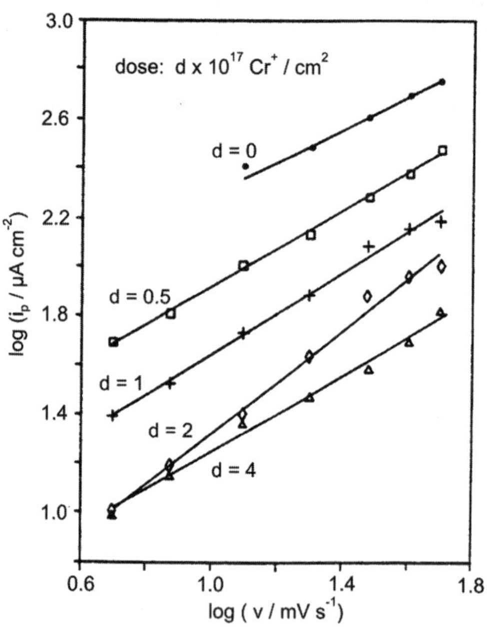

Figure 18.4: Critical current densities i_p of anodic ion dissolution as a function of the potential scan rate v of steel M50, implanted with 180 keV Cr ions (adapted from Ref. [14]).

at elevated temperature. The corrosion which takes place under these conditions proceeds with very low rates. The loss of tantalum can be tolerated. However, the anodic dissolution reaction is accompanied by a cathodic counter reaction. It, too, takes place at a low rate, however, in this case even low currents may be detrimental. The cathodic reaction is hydrogen evolution. When the hydrogen enters the metal lattice, it converts it to hydride. Tantalum hydride is brittle. Additionally, it grows in the metal under stress. The metal looses its ductility and may even break. A possible way of protection is to use a catalyst which increases the probability of reaction of hydrogen atoms to molecules at the metal surface. The molecules leave it as gas bubbles rather then entering the metal. Platinum is such a catalyst. However, it is a very expensive metal. Bulk alloying is not possible. Ion implantation offers a possibility to introduce the catalyst exactly in the place where it is needed, without affecting the Ta bulk and without the need of large quantities. Ensinger and Wolf studied the embrittlement of Pt-implanted Ta [15]. The Pt ions had an energy of 80 keV and were implanted to doses up to $5 \cdot 10^{16} cm^{-2}$. Under industrial service conditions, tantalum is in contact with concentrated H_2SO_4 at temperatures below 200°C. Under these conditions, corrosion takes place at acceptably low rates. For an enforced test, the samples were immersed in 230°C hot acid. This leads to rapid embrittlement. The loss in ductility was tested by a T-bend test. The samples were bend around an edge of 90° until fracture. The ductility was normalized with B_0 being the number of bends until fracture of pristine Ta. $B/B_0 = 1$ means that the metal is fully ductile. The result of the long-term immersion test is shown in Fig. 18.5. Pristine tantalum looses its mechanical stability within a few days. In contrast, the ion beam treated one embrittles at a much lower rate.

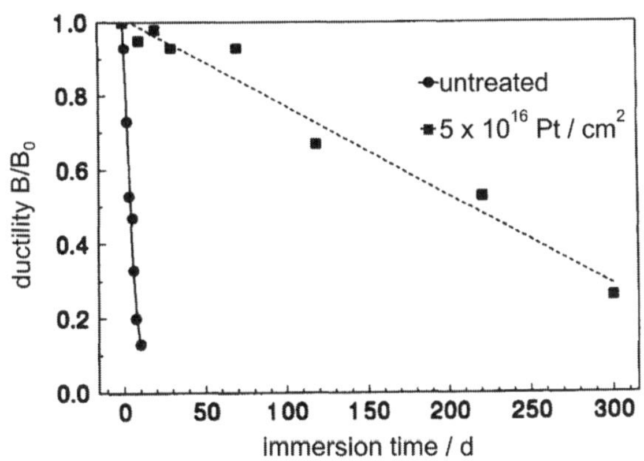

Figure 18.5: Embrittlement (reduction of ductility B) of tantalum as a function of corrosion time in concentrated sulfuric acid at 230°C (adapted from Ref. [15]).

The result shows that a very small amount of Pt is sufficient for this beneficial effect. Pt is not ionically dissolved, but remains for a long time at the sample surface and stays catalytically active.

Ion implantation has not only been used for transition metals and their alloys such as steels and Ti or Ta, but also for main group metals, particularly the light metals Al and Mg. After Al, now Mg and its alloys are increasingly being used as construction material for aeronautic and marine applications. Its advantage is the low density and high strength-to-weight ratio. A drawback of Mg is its minor corrosion resistance. In order to improve the situation, ion implantation can be used. Vilarigues *et al.* implanted Cr into Mg in order to form a Cr–Mg surface alloy with enhanced corrosion resistance [16]. They used an energy of 140 keV and doses up to $5 \cdot 10^{17}\,cm^{-2}$. The samples were immersed in a 3 wt.% NaCl solution saturated with $Mg(OH)_2$ at pH 10.2. For the unimplanted Mg, formation of crusts and their dissolution was observed. The Cr-implanted Mg corroded slower by a factor of 10. A possible mechanism is that Cr slows down the oxide-hydroxide conversion and dissolution of Mg.

Apart from implanting a single species, dual (or co-)implantations have been performed. In this case, one hopes to benefit from a ternary phase formed by the implanted species and the base material. Sugizaki *et al.* implanted Ta, Mo, and Ni as single ions, and mixtures of Mo/Ta and Ta/Ni into titanium [17]. The implantation doses ranged from 10^{16} to $10^{17}\,cm^{-2}$. The corrosion investigations were carried out by polarization measurements in boiling sulfuric acid. The results showed that tantalum reduced the anodic dissolution of titanium. Nickel and molybdenum shifted the corrosion potentials in noble direction, thus promoting the passivation of titanium. Co-implantation of Ta and Ni or Mo combined the advantages of both species, a stable passivation behavior with low-passive current densities. The authors emphasize that appropriate implantation doses of the respective species are required for a beneficial effect. Zhang *et al.* co-implanted C and W into H13 steel [18]. They used a MEVVA ion source with ion energies of 40, 84, and 136 keV. The doses ranged between 3 and $8 \cdot 10^{17}\,cm^{-2}$. They found a supersaturated solid solution. Nanometer sized precipitates of Fe_2W, FeW, WC, Fe_5C_3 and Fe_7C_3 were identified. Multi-scan cyclic voltammetry in acetate buffer at pH 5 showed that all implanted samples gave lower dissolution currents. From the curves, the anodic peak currents i_c were extracted. In Fig. 18.6, i_c is shown as a function of the number of the potential scan cycles.

Samples C5W5 with each C and W implanted at a dose of $5 \cdot 10^{17}\,cm^{-2}$ is compared to pristine H13 steel. While the dissolution currents of the untreated steel increase rapidly and stay at a high level, the ion implanted specimen stays at low currents, even after 80 potential cycles. An increase in the C implantation dose to $8 \cdot 10^{17}\,cm^{-2}$ further improved the situation, as shown in Fig. 18.6(b). Similar results were found for the co-implantation of C and Ti [19].

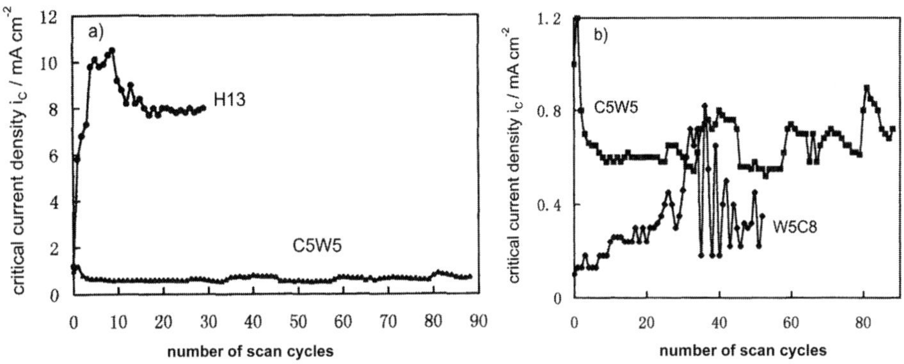

Figure 18.6: (a) Anodic dissolution currents of pristine H13 steel and C + W implanted steel as a function of the number of potential cycles. (b) Comparison of two samples implanted both with $5 \cdot 10^{17}\,\mathrm{W\,cm^{-2}}$, one additionally with $5 \cdot 10^{17}\,\mathrm{C\,cm^{-2}}$ (C5W5), the other one with $8 \cdot 10^{17}\,\mathrm{C\,cm^{-2}}$ (W5C8) (adapted from Ref. [18]).

A large number of studies in the literature have shown that ion implantation is an effective surface modification technique for protecting metals from aqueous corrosion. Implantation of corrosion-resistant metals such as Cr, Ti, or Ta or metalloids such as B or Si, forming alloys or compounds, either nanocrystalline or amorphous, increases the corrosion resistance. Metals and their alloys which had been protected by ion implantation are, for instance, iron, aluminum, titanium, and magnesium. The corrosion mechanisms which can be influenced are uniform corrosion, pitting corrosion, crevice corrosion, and others. As the modified zone is limited in thickness, the corrosion protection effect is limited in time. When the protective layer has been consumed, the material corrodes with the rate of the unprotected metal. In case of cathodic protection, when a noble metal which is not consumed during corrosion shifts the potential of the substrate metal in the passive region, a long-lasting effect can be expected. Examples are stainless steel, titanium and tantalum, implanted with Pt or Pd.

18.2.3. Ion-Beam-Assisted Coating of Metals for Corrosion Protection

In the previous section, ion implantation has been used to form a modified zone at the surface of the implanted metal. The substrate metal always was a part of the formed layer. Due to the limited projected range of the implanted ions, the thickness of the layer was quite small, usually below 0.2 μm. Ion irradiation/implantation can be combined with depositing a film of a third species. With this technique, the drawback of ion implantation, the shallowness of the implanted zone, can be overcome. IBAD allows the deposition of a number of highly corrosion-resistant coatings at thicknesses which are typically in the μm range. Typical corrosion-resistant coating elements from the different groups of elements of the periodic system are transition metals such as the noble metals gold, silver, and platinum or the reactive metals chromium, titanium, and tantalum which are protected by a native stable oxide film. Also, aluminum as a main group metal, and the non-metals boron, carbon, and silicon have been used as protective films. When reactive IBAD is used, with a nitrogen or oxygen ion beam, or in an atmosphere of backfilled nitrogen or hydrocarbon gas, or by co-evaporation of a metal and a non-metal, protective compound films can be deposited such as nitrides, oxides and carbides of titanium, aluminum, and silicon. Apart from their chemical stability, the protective power of the films strongly depends on their morphology and structure. Also, composition and structure of the interface region between film and substrate is important. Ion irradiation during film growth causes a number of physical and chemical effects which influence these features and may be advantageous for the corrosion stability.

In the literature, a few reviews on IBAD for depositing corrosion-resistant films can be found [2,8–10]. In the following, the effects of ion irradiation during thin film deposition will be discussed with a few examples. One will concentrate on structural effects, another one on chemical ones.

Ensinger *et al.* coated aluminum with carbon films of thicknesses between 120 and 150 nm by electron beam evaporation of graphite under simultaneous bombardment with argon ions at an energy of 10 keV [20,21]. The I/A ratio (arrival ratio of Ar^+ ions to C atoms on the sample) was varied from I/A = 0.025 to 0.2. For comparison, several samples were carbon coated by electron beam evaporation without ion bombardment during the deposition (I/A = 0). The corrosion performance was tested in 0.6% NaCl solution at 25°C by electrochemical polarization measurements. From the curves, Al dissolution currents were extracted. In Fig. 18.7 the Al dissolution currents at a potential of −500 mV are depicted as a function of the I/A ratios. The highest currents are observed by the sample grown without ion bombardment at I/A = 0. Al is being dissolved through micropores. The comparison shows that the corrosion protection ability of the film depends strongly on the relative ion irradiation intensity, the I/A ratio. The best results are obtained for the samples grown under moderate conditions. An I/A ratio around 0.05 gives the lowest currents, more than two orders of magnitude lower then the film grown without ion irradiation.

The damage of the samples by corrosion, that is loss of coating and large pits in the Al, can directly be seen. In Fig. 18.8 the photographs of the samples after the electrochemical test are shown. The surface of the sample prepared at I/A = 0.05 shows the lowest amount of damage, see Fig. 18.8(c).

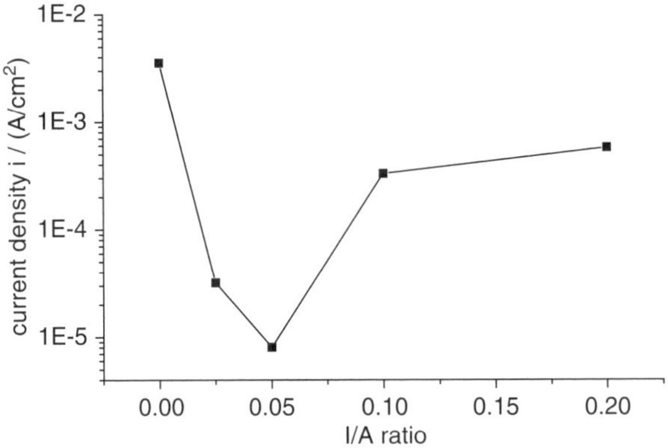

Figure 18.7: Current density at a potential of −500 mV vs. I/A-ratio (adapted from Ref. [21]).

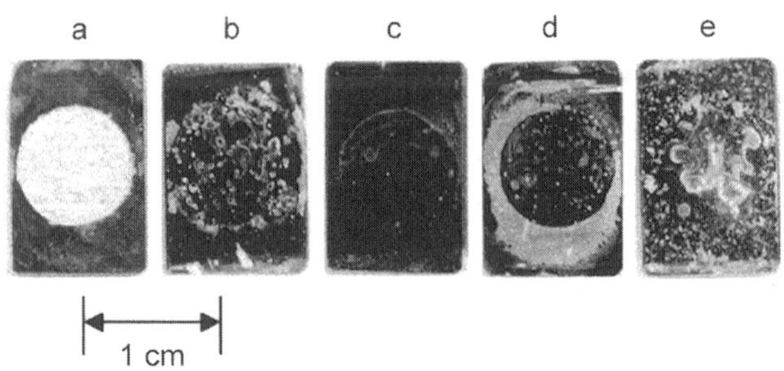

Figure 18.8: Photographs of samples after electrochemical testing in 0.6% NaCl solution under identical conditions; (a) uncoated Al, (b) I/A = 0.025, (c) I/A = 0.5, (d) I/A = 0.1, (e) I/A = 0.2. The central circular area was exposed to the corrosive medium.

This result is typical of corrosion of IBAD films [10]. When a film such as carbon is deposited by physical vapor deposition at a substrate at low temperature, it grows with micropores and with poor adhesion to the substrate. The weak adhesion is a result of a contamination layer on the substrate such as water or hydrocarbons. This layer hinders the formation of strong bonds. The microporosity is a consequence of the fact that the deposited atoms are not able to move over the substrate surface when the temperature is low. They stick close to the place where they have landed. This may lead to shadowing effects. Overhanging structures may be formed, leading to porosity. Through the pores, corrosive solution enters the film. It reaches the substrate and the interface. There, corrosion starts. The film is undermined and, eventually, is detached. When the substrate is irradiated prior to coating, its surface is cleaned from the contamination layers. As a consequence, adhesion is increased. The probability of local attack in the interface is reduced.

When the film grows under moderate ion bombardment, two major effects occur. In the initial phase of film deposition, ion-beam-mixing effects including forward sputtering, may lead to an increased bond strength between substrate and film. This happens when film and coating are chemically compatible. Often, a third phase, consisting of both film and coating material, the interphase, is formed. It is a transition between substrate and film, leading to improved adhesion. The second more important effect in terms of corrosion is that the films grow more densely with a lower density of micropores. Fig. 18.9 shows schematically some of the general effects of irradiation on the film structure. When the energetic ion impinges onto the film surface, the mobility of the freshly deposited surface atoms is increased. The formation of overhanging structures such as those in the bulk of the film is reduced. Thus, pores are not formed to the same extent. As the ions are able to penetrate into the deposited film, they change its structure in a certain depth. The ion transfers its kinetic energy into the film by electronic and nuclear stopping. This leads to a motion of film atoms around the trajectories of the ions. Already existing voids and pores can be closed or reduced in size by the motion of the atoms in collision cascades (Fig. 18.9(c)) or by forward sputtering (Fig. 18.9(b)).

When ion irradiation during film growth is too intense, effects may occur which reduce corrosion protection again. First, net film growth will be reduced due to intense re-sputtering. Thinner films are less protective because their porosity is usually larger. Thickness loss may be counterbalanced by a longer process time. However, more important, negative structural and chemical effects may become effective. The film may become strongly radiation-damaged with a large number

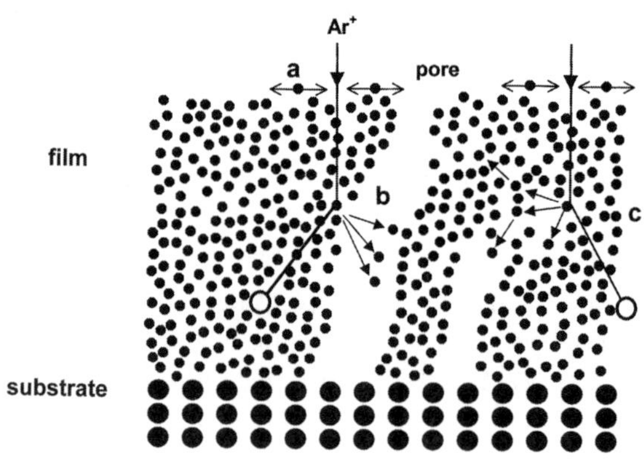

Figure 18.9: Schematic presentation of ion beam effects on an atomic level, (a) enhanced surface adatom mobility, (b) forward sputtering of film atoms into pores and voids, (c) collision cascades with structural rearrangement.

of accumulated damage clusters which reduces film density again. This may enhance microporosity. Another effect may be excessive rare gas ion incorporation, leading to stress in the film.

Ion irradiation during vapor deposition is not only used for controlling the structure of single element monolithic films, but also for designing multi-element films and multi-layered films. Wolf *et al.* coated large area low-alloy steel samples in an IBAD coil coater with several μm-thick coatings of Zn, partially with Ti, Cr, Zr, or Mn [22,23]. The latter were co-evaporated from a second crucible. As reference, a 8 μm-thick galvanically deposited Zn coating was used. This is the standard coating applied for automobiles. The samples were sheets of $100 \times 200\,mm^2$ size. The ion energy was mostly 1.5 keV. The I/A-ratios were below 0.001. Two deposition modes were used, dynamic, with the samples being moved, and static. The corrosion resistance was evaluated using a standard salt-spray test with 5% NaCl solution. The time until the appearance of red rust was recorded. In Table 18.1 the results are listed.

The pure Zn coatings stand 80 h or less. Samples with Cr and Ti included survive much longer, despite the thickness being lower. The 8 μm-thick coating with 3% Cr included shows a lifetime increase of five times. Similar results are found for Mn as alloying element. In the dynamic coating mode with moving steel substrate, multi-layers and gradient layers with Ti and Cr were formed. A 6–7 μm-thick Zn/Ti multi-layer reaches a maximum lifetime of 600 h, a Cr-containing one even 900 h. This example shows that IBAD with a slit ion source in combination with twin electron beam evaporation offers the possibility of designing coatings which are superior to the standard ones which are used in automotive industry.

The superior performance of IBAD thin films for corrosion protection has successfully been utilized in industry for coating blades of electric shavers. Miyano *et al.* developed a process line for TiN and, later, TiCN deposition onto foils of martensitic

Table 18.1: Results of salt-spray test of Zn-coated low-alloy steel

Sample: (Static)	8 μm Zn Reference	8 μm Zn	4 μm Zn/3% Cr	8 μm Zn/ 3% Cr	3 μm Zn/ 10% Ti
Time/h:	68	50	287	338	121
Sample: (Static)	8 μm Zn Reference	4 μm Zn + 1 μm Mn	4 μm Zn + 1 μm Zn/Mn alloy	5 μm Zn/Mn multi-layers	
Time/h:	80	131	110	55	
Sample: (Dynamic)	8 μm Zn Reference	6 μm Zn/Ti gradient	6–7 μm Zn/Ti multi-layer	5 μm Zn/Cr	6–7 μm Zn/Cr
Time/h:	80	140	600	350/500	900

Data from Ref. [23].

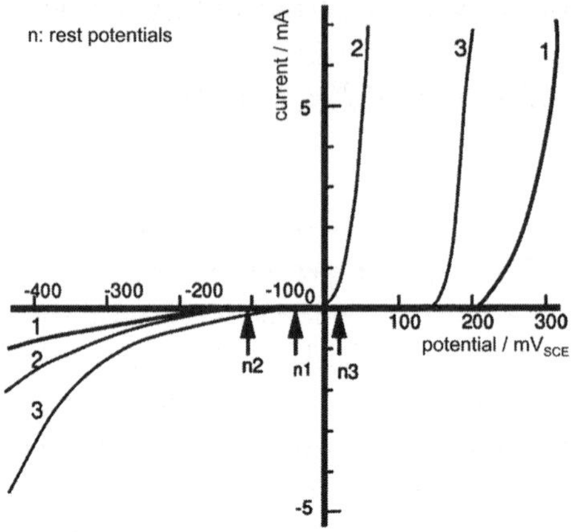

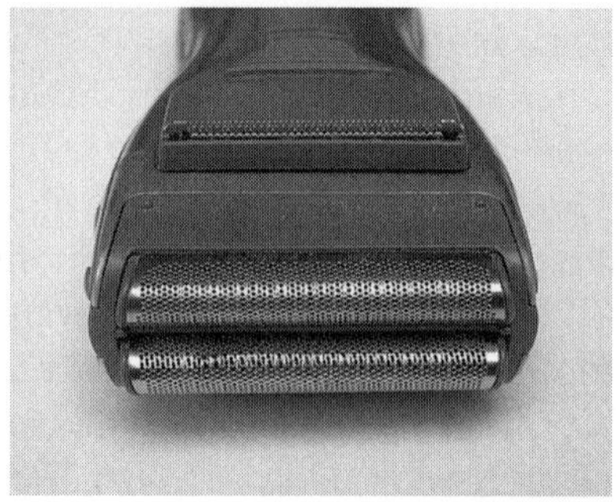

Figure 18.10: (a) Polarization curves of martensitic stainless steel in 3% NaCl solution at 25°C. 1. IBAD coated, 2. coated by conventional ion plating, 3. uncoated. n1, n2, and n3 are the rest potentials (adapted from Ref. [24]), (b) TiN-coated shaver blade.

stainless steel (AISI 410) [24]. The quality of the coating was tested for its usability in 3% neutral sodium chloride solution. In this environment, the steel may suffer from pitting corrosion. In Fig. 18.10 the potentiodynamic polarization curves are shown together with a coated shaver blade. At low anodic potential, both coated and uncoated steel are passive. The onset of pitting corrosion, indicated by the steep increase of dissolution current at the pitting potential, gives an indication of the pitting resistance.

While the coating deposited by ion plating showed pitting at low potentials, the pitting potential of the IBAD film was shifted to much more noble values which is an indication of stability against pitting. The electrochemical results was verified by a long-term immersion test. The IBAD-coated sample remained unstained for a long time in contrast to the others. The coating has several beneficial effects: it is decorative due to its golden color (gray-black in case of TiCN), it protects the blade against corrosion, that is the blade is not stained by sweat, it can be cleaned with water, and the blade stays longer sharp.

The examples both from basic and application-oriented studies have shown that the versatility of IBAD with its independently controllable parameters such as ion energy, ion-to-atom arrival ratio, and ion incidence angle, allows to design coatings with high chemical stability, low microporosity and excellent adhesion which are very well suitable for corrosion protection. The energy required for a favorable film growth is not delivered thermally but by the kinetic energy and momentum of the ions. The substrate can be kept at low temperature.

18.3. Protection Against Mechanical Attack: Friction and Wear

18.3.1. Methods for Measuring the Mechanical Properties: Hardness, Friction, Wear

With respect to mechanical properties of ion-beam-modified materials, mostly (micro)hardness, elastic modulus and tribological properties, including friction and wear, are measured. Hardness measurements are carried out to evaluate the strengthening obtained by the treatment. Often, a standard indenter such as Knoop or Vickers is used. The size of the indentation mark gives data on the hardness. A general problem of these tests is that, due to the shallowness of the implantation zone, mostly bulk properties affect the measurement. Therefore, often low loads with low indentation depths are used. However, lower loads lead to larger errors, as the indentation sizes become smaller. Additionally, elastic recovery plays a greater role. For these reasons, recently mostly nanoindentation is used. The indenter is usually of the triangular Berkovich type. This test gives a load-vs.-depth curve from which hardness and elastic modulus can be extracted [25,26].

For evaluating tribological properties, in most cases, pin-on-disk wear tests in combination with an inspection of the wear groove with light or electron microscopy and surface profilometry were used [27]. A pin with a certain shape, often a spherical tip, is pressed onto the sample with a defined normal force. Mostly, the pin is fixed and the sample rotates forming a circular wear track. Measurement of the tangential forces gives the friction coefficient. In some cases, the reciprocating wear test with an oscillating pin is used. It gives a linear wear track. The test is carried out without lubrication or under lubricated conditions. The atmosphere, gas, or air with a defined relative humidity, plays an important role.

18.3.2. Medium-Energy Ion Implantation of Metals for Wear Reduction

The pioneering work in the field of wear protection of metals by ion implantation was carried out in Harwell, UK, by Hartley, Dearnaley *et al.* [28–31]. Advantages of ion beams were seen in the non-equilibrium nature and in the low-process temperature. There were no limitations with conventional metallurgical solubilities, any kind of chemical element could be implanted into any solid. Thus, materials with improved properties could be created. Initially, mainly nitrogen was implanted. The implantation doses ranged between 10^{17} and $10^{18}\,cm^{-2}$, the ion energies were 50–100 keV. Fig. 18.11 shows the wear test results of mild steel implanted with ca. 10^{18} 35 keV N^+ cm^{-2}. The wear parameter on the steel pin gained from a pin-on-disk test is considerably reduced. With increasing load, the effect becomes more pronounced.

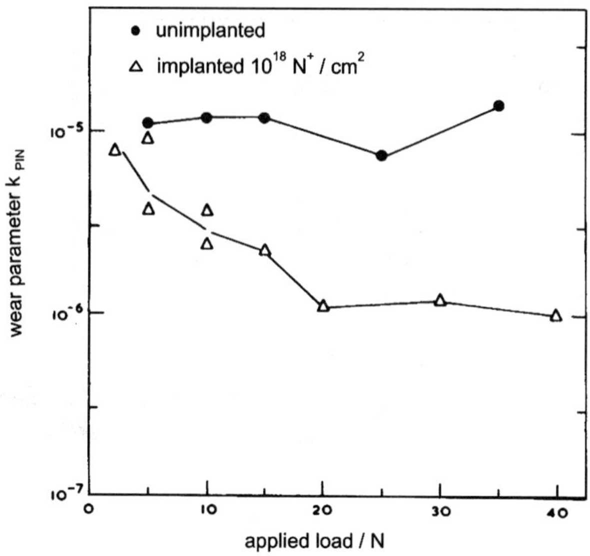

Figure 18.11: Wear parameter for nitrogen-implanted mild steel as a function of applied load (adapted from Ref. [29]).

The investigated materials in the beginning were steels and titanium alloys [31,32]. As hardening mechanism, solid solution hardening by interstitial nitrogen and dispersion strengthening by precipitation of hard nitrides such as iron nitride and chromium nitride in the case of steels, and titanium nitride in the case of Ti alloys were assumed.

Apart from nitrogen, B and metals such as Ti, Ta, Mo, Y, and Sn were used. Follstaedt, Singer and others found that high-dose ion implantation of Ti gave low friction coefficients [33–35]. It turned out that the combination of Ti with C, for example by surface carbonization induced by the carbide former Ti, leads to a surface strengthening by finely dispersed TiC which is particularly effective. Thin films of amorphous carbon and an amorphous ternary phase of Fe–Ti–C give the low friction coefficient. Iwaki implanted low-carbon steel with 10^{17} Cr^+ cm^{-2} at 150 keV [36]. He found a decrease in the friction coefficient in a pin-on-disk test against a steel pin under unlubricated conditions from 0.5–0.6 to 0.42. For H13 tool steel, N^+ implantation at 75 keV and a dose of $5 \cdot 10^{17}$ cm^{-2}, the friction coefficient was only reduced from 0.55 to 0.5, while B implantation with the same parameters reduced the coefficient considerably down to 0.17. The decrease in wear rate was a factor of 60–70 for N implantation, and almost 600 for B implantation. For N implanted TiAl6V4, Hutchings, Oliver *et al.* measured a reduction of the friction under dry conditions against a ruby ball from 0.48 to 0.15 [32,37]. The wear volume was reduced by 200–300 times. An example for metal implantations is the

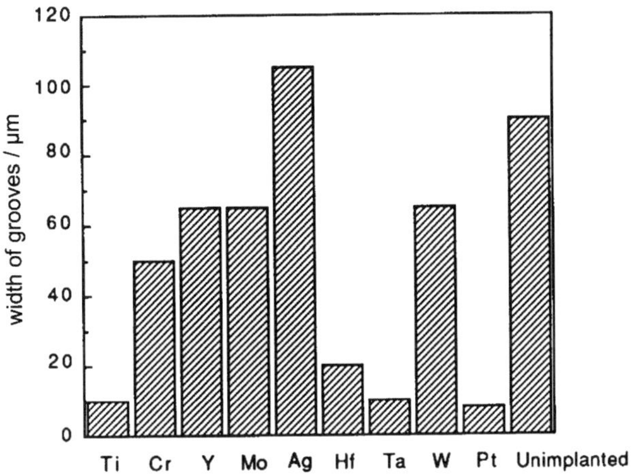

Figure 18.12: Widths of wear grooves after the wear test of stainless steel implanted with different metals (adapted from Ref. [38]).

work of Sasaki *et al.* [38]. They implanted nine different metals into 440C stainless steel and compared the friction and wear behavior against a 440C steel ball at a load of 0.245 N. Fig. 18.12 compares the depths of the wear grooves after the test. Ti, Hf, Ta, and Pt gave the best results. With the exception of the inert Pt, these metals had incorporated C during implantation. The strongest wear was found for Ag. Apparently, the soft metal is worn away.

Apart from implanting one species only, also dual implantation has been tested. Sánchez *et al.* implanted high-speed steel (HSS) (SP 30 powder metallurgy (PM)) and stainless steel AISI 304N with C and N at different implantation sequences [39]. The single doses were $3 \cdot 10^{18}\,cm^{-2}$, the ion energy was 100 keV. Table 18.2 shows the results of friction and wear tests of stainless steel at different relative humidities. The load was 100 gf, the counter body was a hardened 100Cr6 ball. The sample code gives the implanted species and the implantation sequence, for example C for C implantation, NCN for implantation of N, C, and N in this order.

It turns out that the tribological properties strongly depend on implanted species and order of implantation. Samples implanted with C, either pure or C as the last step, showed lowest friction and wear.

These are a few examples out of a large number of studies carried out all over the world on the effect of ion implantation on tribological features of metals. In [40], a compilation of results on a large number of different materials and implants can be found. These studies demonstrated the positive effect of ion implantation onto wear of metals and the feasibility of the technique. They led to industrial applications, as discussed later.

Table 18.2: Friction coefficient f and wear rates of stainless steel

Sample	f @ 15% RH	f @ 50% RH	f @ 85% RH	Wear rate ($10^{-13}\,m^2/N$)
Reference	0.70–0.90	0.70–0.90	0.60–0.80	1.34
C	0.20	0.20	0.20	0.00447
CN	0.45–0.52	0.45–0.52	0.45–0.50	0.627
NC	0.16	0.16	0.20	0.00482
NCN	0.50–0.55	0.50–0.55	0.38–0.50	0.0161

Data from Ref. [39].

18.3.3. Low-Energy High-Temperature Implantation

Ion implantation for wear reduction has mostly been performed at ambient or slightly elevated temperature. The thickness of the modified zone has been given by the projected range of the ions in the solid, based on ballistic laws. Diffusion into depth does not play a role. One of the drawbacks of this method was the shallowness of the implantation zone. This can be overcome when ion implantation is combined with thermal diffusion. In this case, low kinetic ion energies are used, usually around 1 keV. This technique can be applied when elevated temperatures are tolerable. With high-dose nitrogen ion implantation at elevated temperature nitrided layers of thicknesses of 10 μm can be obtained [41–44]. These comparatively thick films mostly show very good tribological features.

Fig. 18.13 shows composition depth profiles gained from Auger electron spectrometry of AISI 304 stainless steel after different nitriding treatments from a paper of Wei [44]. The treatment time was 30 min, the temperature was kept constant at 400°C. For comparison, the profile of 60 keV low-temperature ion implantation is included. It shows the usual shape of ballistic ion implantation, given by the ion energy. High-temperature ion implantation at the same energy yields a much thicker layer. When the ion energy is reduced by almost a factor of 100 down to 0.7 keV, but the ion current is increased to $2.5\,mA\,cm^{-2}$, a much higher dose is implanted and the layer becomes considerably thicker. Phase analysis showed that the nitrided zone consisted of expanded austenite γ_N. In Fig. 18.14 the Vickers hardnesses are depicted. By means of ion implantation at low energy, but elevated temperature and ion current density, the hardened zone expands much deeper into the substrate compared to the other methods. The ions do not need high kinetic energy for implantation, as they travel by diffusion at elevated temperature, however, they need some kinetic energy to overcome surface barrier layers, to sputter away native oxides and to form an implanted layer at the outer edge of the material for subsequent fast in-diffusion.

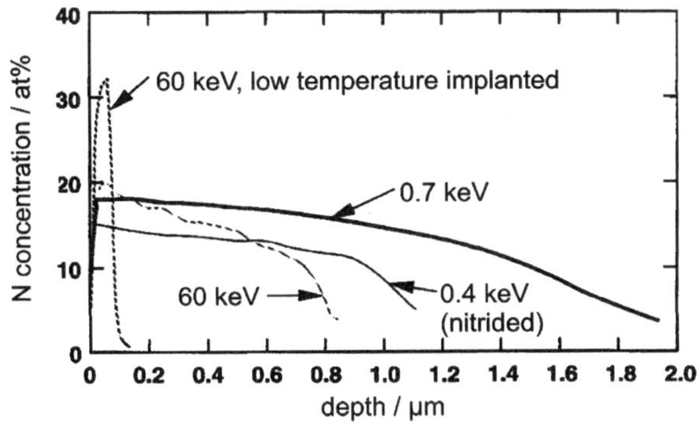

Figure 18.13: Auger depth profiles of nitrogen-implanted and ion-nitrided stainless steel AISI 304 at 400°C (adapted from Ref. [44]).

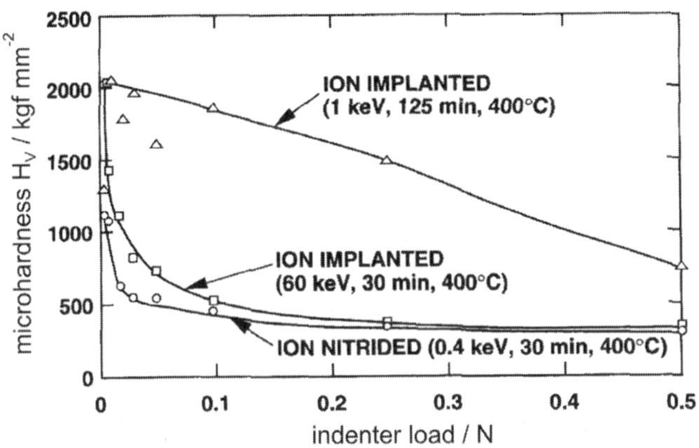

Figure 18.14: Vickers hardness of nitrogen-implanted and nitrided stainless steel AISI 304 (adapted from Ref. [44]).

Rivière *et al.* implanted stainless steel AISI 316 and INCONEL600 at 400°C for 1 h at an ion current density of 1 mA cm^{-2} at 1.2 kV acceleration voltage [45]. The hardness was measured at a load of 2 mN. Friction and wear tests were carried out against WC and 100Cr6 steel balls at a normal load of 1 N. Table 18.3 shows the results of the steel measurements. The hardness increases by a factor of 3.4.

While the friction coefficient is lower against a WC ball or the steel ball when the humidity is high, it is increased at low humidity. For the unimplanted sample, a wear coefficient could be measured; in case of the implanted sample the wear track after 20,000 cycles was practically invisible, and no wear rate could be determined.

Table 18.3: Mechanical test results on AISI 316 samples; the indentation depth was 0.042 μm, the nitrided layer thickness was 4 μm. For the wear test, the friction coefficients at a given counter body and relative humidity H_r are listed

Sample	H_v (GPa)	Elastic modulus (GPa)	WC ball 50% H_r	100Cr6 ball 15% H_r	100Cr6 ball 85% H_r	Wear coefficient ($\mu m^2 N^{-1}$)
Untreated	2.6 ± 0.2	187 ± 4	0.7	0.75	0.6	$2.3 \cdot 10^{-2}$
Implanted	8.8 ± 1.5	204 ± 5	0.3–0.7	1	0.45	0

Data from Ref. [45].

These examples demonstrate the effectiveness of low-energy high-current ion implantation at elevated temperature. This technique may be used for improving the tribological performance of metals, when an increased temperature can be tolerated.

18.3.4. High-Energy Ion Implantation

Another approach to extend the thickness of the implanted zone is to use higher kinetic energies of the ions. In this case, again the temperature can be kept low. In contrast to the just mentioned high-temperature implantation, a thick implant zone is not formed by diffusion but by ballistic implantation. With MeV ion energies, the thicknesses are larger than those of low-temperature medium-energy ion implantation, but smaller than those of high-temperature low-energy ion implantation. Another difference is that the implanted species does not reach to the surface of the sample but is buried in depth.

Taniguchi *et al.* performed dual ion implantation of silicon and carbon into copper and iron [46]. The doses were $10^{18}\,cm^{-2}$ each. The ion energies were 1.5 MeV for Si and 1.2 MeV for C. The hardness data of the samples were measured with nanoindentation with a Berkovich indenter. Fig. 18.15(a) shows the depth profiles. It turns out that hardness increases due to ion implantation, but the hardness peak is not located at the position of the implantation peak. The authors explain this effect as follows: At stage A, the indenter and its deformation zone do not reach the implantation layer. The hardness is still low, close to the one of the unimplanted material. At stage B, when the implanted layer is reached, the deformation is blocked. The yield strength increases. At stage C, the implanted layer begins to deform. Plastic deformation goes beyond the implanted layer. At stage D, the indenter penetrates through the implanted layer. Hardness decreases again.

This example demonstrates the effect of ion implantation on the mechanical response of the modified substrate. While this technique gives good results in the

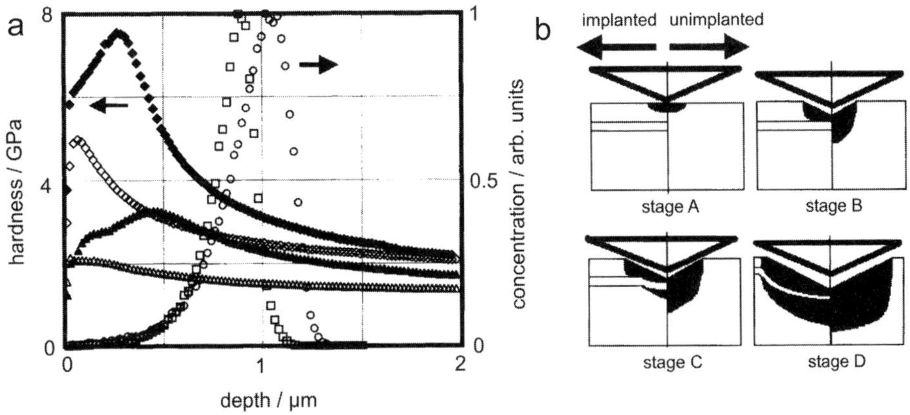

Figure 18.15: (a) Hardness/depth profiles of implanted Cu (▲), unimplanted Cu (Δ), implanted Fe (◆), and unimplanted Fe (◇). Concentration/depth profile of Si + C ions in Cu (○) and Fe (□); (b) Stages of the deformation process during indentation. The regions of plastic deformation are marked in black (adapted from Ref. [46]).

laboratory tests, it will certainly not be relevant for industrial applications, as the formation of high-energy ions is too difficult and expensive.

18.3.5. IBAD of Wear-Resistant Coatings

So far, the modified zone was shallow when low temperatures and medium ion energies were used. For a thicker modified zone either higher ion energies or higher temperatures were needed. By means of IBAD, comparatively thick wear-resistant coatings can be deposited at low sample temperature. Different approaches can be found in the literature. One is to deposit a hard- and wear-resistant coating which protects the substrate by its strength. This is the case with TiN. Another approach is to use a film with a low friction coefficient such as amorphous carbon.

In an early study, Erler *et al.* combined sputtering and ion irradiation to deposit nitride films [47]. Satou and Fujimoto formed hard boron nitride (BN) films by electron beam evaporation under nitrogen ion bombardment at medium ion energies [48]. The most widely investigated IBAD film for tribological protection is TiN. Kant and Sartwell deposited TiN films on AISI 52100 and on M50 steel by electron beam evaporation of Ti under 30 keV nitrogen ion bombardment in an atmosphere of backfilled nitrogen [49]. Scratch tests with a Rockwell indenter showed that a TiN film grown without ion bombardment failed adhesively, while the IBAD film deformed with the substrate and remained intact. A cavitation erosion test of M50 gave a considerable reduction in the rate of mass loss of the coated sample compared to the uncoated one. Smidt and Hubler reported on the wear performance of TiN films formed by titanium deposition in an atmosphere of backfilled nitrogen under irradiation with 0.5 keV argon ions [50]. The TiN phase is formed by reaction of the metal with backfilled nitrogen gas. Fig. 18.16 shows hardness and wear life for different Ar-to-Ti arrival ratios. While hardness constantly increases with ion irradiation intensity, for wear an optimum around an I/A-ratio of 0.3 is found. A certain energy and momentum transfer into the growing film results in an optimum structure and stress level of the film so that it is neither too soft nor too brittle.

Kluge *et al.* deposited 2.1 μm-thick TiN onto unhardened ferritic steel AISI 440 B by evaporation of Ti under irradiation with 50 keV nitrogen ions [51]. They compared the properties of the films with those prepared by ion plating by two companies. Fig. 18.17 shows the results of a wear test. An oscillating-ball-test without lubrication with a load of 2.2 N was used. The largest wear rate was observed for the uncoated steel. The IBAD-coated sample showed the lowest wear. For longer wear times, the ion plated samples exhibited increased wear, even higher than the one of the uncoated sample. This was due to formation of wear particles of the damaged film. For the IBAD film, this effect was not found.

Similar results were obtained by Wen and Zhang [52]. They compared ZrN films prepared by IBAD with 3 keV ion beam sputter deposition combined with 65 keV nitrogen ion implantation, that is the high-energy regime of IBAD, and by magnetron sputtering. A scratch test showed that the sputter deposited film failed adhesively at lower loads. Additionally, it tended to form cracks under the stress of the scratch indenter. This was not observed for the IBAD film. From these results, it can

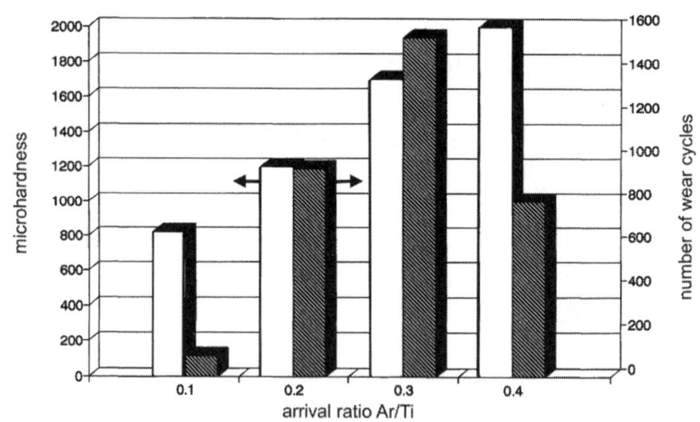

Figure 18.16: Microhardness and wear life of 0.5 μm-thick TiN films formed by IBAD as a function of Ar^+ to Ti arrival ratio (adapted from Ref. [50]).

be concluded that the IBAD films showed an improved adherence. Mitsuo *et al.* deposited ZrN by Zr evaporation under nitrogen bombardment with 0.5–2 keV ions [53]. The purpose was to form a coating which is less prone to oxidation than TiN when used at elevated temperatures. The substrate was Si (100). The authors tested the wear performance against a ball of stainless steel AISI 304 at 60–70% relative humidity without lubrication. Fig. 18.18 compares the friction behavior of both TiN, formed under similar conditions, and ZrN. Both show a high friction coefficient *f* due to adhesive wear, however, *f* of ZrN is lower than that of TiN.

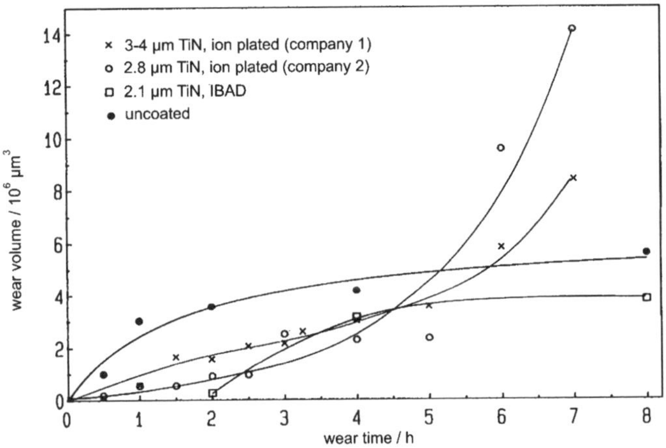

Figure 18.17: Wear volume vs. time of differently prepared TiN films (adapted from Ref. [51]).

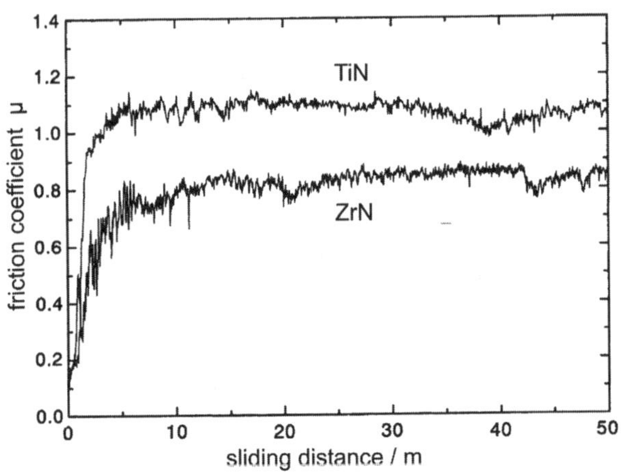

Figure 18.18: Comparison of friction behavior of TiN and ZrN films (adapted from Ref. [53]).

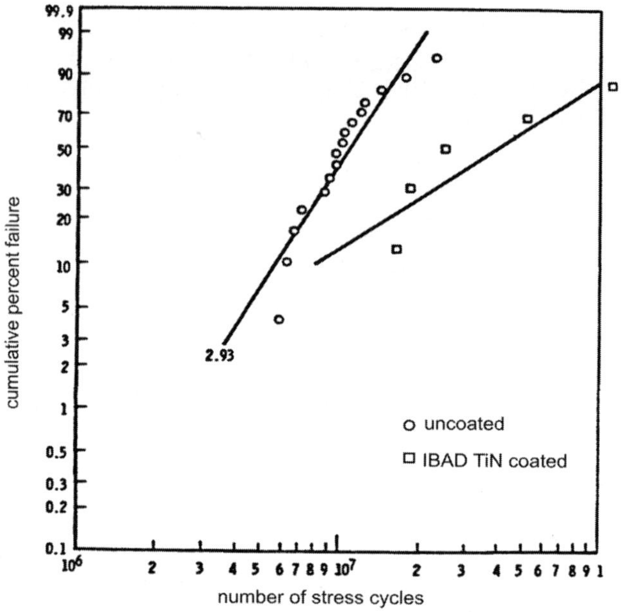

Figure 18.19: Weibull distribution of rolling contact fatigue test of uncoated M50 rod and rod coated with 1 μm TiN (adapted from Ref. [54]).

Apart from laboratory experiments on basic tribological properties, application-oriented tests on industrial use of IBAD films have been published. Hirvonen describes results on rolling contact fatigue testing of gear and bearing materials [54]. M50 rods were coated with TiN by IBAD. In the test, the rod revolved between three bearing balls at a load of 1282 MN m^{-2} at 3600 rev min^{-1}. The tests creates a wear track on the rod. The number of cycles until failure is measured. Fig. 18.19 shows a Weibull plot with the accumulated percentage of failed samples. For a 1 μm-thick IBAD TiN coating, a fourfold increase in lifetime at 50% accumulative failure was observed. Arps *et al.* deposited diamond-like carbon (DLC) coatings onto pins of hardened 52100 steel by evaporation of hydrocarbons (polyphenyl ether, trimethyl trisiloxane) under bombardment with 9 keV nitrogen ions [55]. Prior to forming the DLC coating, silicon was evaporated and deposited at 200°C under nitrogen irradiation onto some samples. Silicon formed a silicide bond coat between the steel substrate and the carbon film. The samples were tested by a reciprocating pin-on-flat setup. A comparison revealed that the sample with the Si interlayer showed around 40% lower wear. The results are relevant for diesel engine components such as roller pins and rocker shafts.

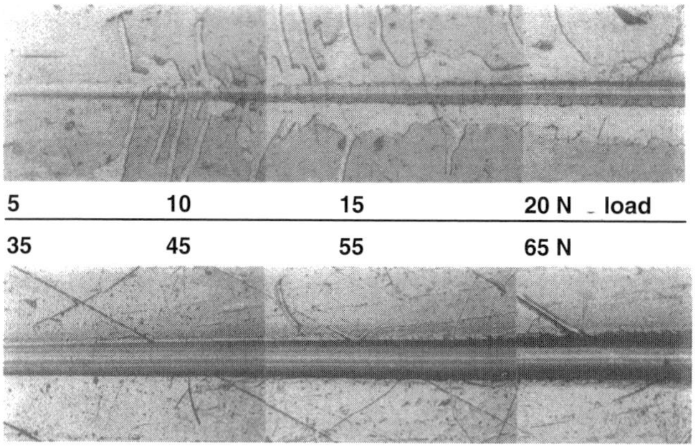

Figure 18.20: Grooves of scratch test of TiN films on 100Cr6 ball-bearing steel, upper: without ion irradiation, lower: with Ar^+ ion irradiation (from Ref. [57]).

Ensinger *et al.* compared TiN films deposited by sputtering and by IBAD in an industrial environment [56]. C35 tool steel cylinders with 50 μm hard chromium galvanic coating were additionally coated with 0.7 μm IBAD TiN under 12 keV Ar ion bombardment and with 1–2 μm-thick magnetron sputtered TiN films. The cylinders were used as fiber -guiding elements for polyester yarn. They were tested in a yarn friction tester. Yarn was drawn at a speed of 300 m min^{-1} around the cylinder at a tension of 0.75 N. The yarn was pigmented with TiO_2 (white) or with carbon (black). As the yarn is wet, the conditions are highly abrasive–corrosive. While the wear coefficients of sputtered and IBAD films for the white yarn were similar, the IBAD films were superior in case of the black yarn. In the latter case, the wear coefficients were around 10×10^{-9} mm^3 Nm^{-1} for the IBAD films, and 18 and 37×10^{-9} mm^3 Nm^{-1} for the sputtered films.

Kohlhof tested the wear performance of TiN films deposited onto heat-treated ball-bearing steel 100Cr6 for applications in automobile engines [57]. A problem with the hardened steel is that it must not exceed a temperature of 200°C. State-of-the-art ion plating at elevated temperature could not be used. This created problems with adhesion of the deposited films. In order to improve the situation, Kohlhof combined the deposition of TiN by magnetron sputter coating with irradiation with Ar ions at 30 keV from a radio frequency (RF) ion source. The ion beam was used for pre-sputtering the substrate surface and for ion beam mixing the TiN film with the substrate. The adhesion was tested by the scratch test with a Rockwell indenter. Fig. 18.20 shows the scratch tracks. The TiN film deposited without ion irradiation failed at a critical load of 8 N. In contrast, the ion beam mixed film stood the test up to 60 N.

The author states that due to the low doses required (10^{15}–$10^{16}\,cm^{-2}$) no cost problem arises with the extra ion beam treatment step.

These and other examples show that IBAD offers the possibility to deposit coatings at low-process temperature which are effective in wear protection. As the process parameters of IBAD can be selected over a wide range, an optimization of the coating with respect to friction, wear, and adhesion is possible.

18.4. Industrial Application of Ion Beam Methods for Metals Protection

Ion beam techniques have been industrially used over the last almost two decades. In the literature, reviews from different countries, including USA, UK, Denmark, Spain, France, Japan, China, and Thailand can be found [58–76].

The scientific work and field tests had shown that ion implantation for protecting metals from wear and corrosion is feasible and effective. Despite the shallowness of the treatment (typically $<0.2\,\mu m$) ion implantation works for sliding wear, adhesive wear and mild corrosive wear. With surface hardening and a strengthening of the material and sometimes lowering of friction and adherence, the implanted layer changes the wear condition to a less serious one.

The first industrial success in ion implantation of metals for wear and corrosion protection has been achieved by surface treatment of orthopedic prostheses [66]. Artificial joints, mostly knees and hips, made of TiAl6V4 were ion implanted. The hardening by formation of a hard nitride led to a reduction in wear and an increase in lifetime. In simulation studies, ion implanted knees or hip joints showed a considerably better performance than the untreated ones, even after several million of wear cycles. Since the late 1980s, the majority of titanium orthopedic bearings sold worldwide was treated by ion implantation. Since the early 1990s, also Co–Cr orthopedic prostheses were ion implanted.

While these examples showed that ion beams can be industrially successfully used for materials modification, the shallowness of the treatment and the invisibility remained a drawback for industrial acceptance in other sectors. With a thickness well below 1 μm, people from industry often believed that the effect would be too little. Additionally, the treatment did not show any visible change. The transfer from biomedical applications with its comparatively high prizes to applications in tooling industry took a number of years. Company representatives from Denmark say that it was difficult and time-consuming to convince the customers that ion implantation was beneficial [69].

For the process, there are two technical approaches: broad beam irradiation on one side and selected area implantation on the other [69]. Ion implantation developed from accelerators. They include a beam line and produce a beam with a small diameter. If a large work piece is to be treated, the work piece has to be manipulated and the beam has to be scanned. Such a technique is complicated and takes time. The desire came up to have a broad beam available which covers the whole work piece or large parts of it. This is possible with a broad beam ion source, emitting a high-current large area non-scanned beam. The work pieces are arranged in a close array which is fully irradiated by the beam. A disadvantage of such an ion source is that mass selection is not possible. All ions from the ion source are accelerated and implanted. This restricts this method to gaseous ions, mainly to nitrogen. The next step in development was to use a mass-analyzed beam with a larger diameter than the focused beam of early accelerators. Now, again all kinds of ions were available, including metal ions such as Cr. The final step of development was to use such a beam and combine it with beam steering and sample manipulation for selected area ion implantation [69]. With this technique, only those parts of a large work piece are irradiated which need to be such as the edges of a punch. Rather than irradiating the whole tool, only the critical areas are treated. Selected area implantation helped to reduce the cost of ion implantation. Both process time and cooling-down time are reduced.

Today, in industrial application mostly mass-analyzed ion beams are used. In 80–90% of all cases, nitrogen ion implantation which gives high currents is sufficient. The doses needed range between $4 \cdot 10^{17}$ and $8 \cdot 10^{17}\,\text{cm}^{-2}$. For practical applications, mainly HSS and other tool steels as well as stainless steel, both austenitic and martensitic, are treated. In Table 18.4 a number of examples from the literature are listed [67–75].

The beneficial effects will be discussed with some examples. In the paper industry, cutting and punching tools received lifetime increases of a factor up to four [75]. When paper is cut, the cellulose fibers create abrasive wear. As the fibers are not very hard and do not penetrate deeply into the metal, an ion implanted layer is sufficient for protection. There are similar problems in the food industry. The lifetime of an asparagus cutter was increased by a factor of 10. In the case of vegetables, there is not only the fiber problem, but sometimes sand particles are involved which create abrasive wear. Other examples from food industry are meat cutting and bread cutting. Here, ion implantation shows another advantage. In contrast to coated tools, there is no problems with contamination of processed food, because a coating might fail adhesively and create small particles, an implant not. Tools for cutting or shaping sheets of steel or aluminum such as punches and dies for can manufacturing suffer from adhesive and abrasive wear. This can be reduced by a factor of three to five. An important issue is not only thc increased lifetime which reduces the need of acquisition of tools, but the reduced downtime for tool changes. Often, the latter is the important cost factor.

Table 18.4: Examples of application of ion implantation for tools

Tool and application	Service lifetime increase
Cutting punches for thin metal sheets	5–10x
Knives for cutting aluminum	10x
Steel knives for cutting meat	4–5x
Precision knives for cutting plastic or paper	5x
Knives for cutting rubber	5x
CrN-coated punches for aluminum	10x
TiN-coated punches for metal sheet forming	5–7x
TiN-coated steel parts of plastic molds	12–13x
HSS milling cutters	3x
HSS phenolic honeycomb milling cutter	3x
Tool steel coining dies	3–4x
Hard chrome plated dies for medals	6x
Aluminum alloy injection molds for polyimide parts	100x
Tool steel nozzles for injection of polycarbonate	2–3x
Tool for perforating holes in stamps	7x
Cellophane cutter	9x
WC-Co drills for carbon fiber composites	3x
High-speed steel tools for perforating holes in foils (polyethyleneterephthalate (PET) for fruit packaging)	4x

In many cases, not only the lifetime is increased, but the performance of the tool is better and often the product shows a better quality. In aluminum processing, often the material pickup is the problem. Ion implanted tools show a much lower pickup and the time until repolishing is increased. Punches often show reduced adhesive wear which leads to a reduction in consumption of lubricants. Often, friction is reduced and the workpiece shows a better slide. And in many cases, the cutting edges of the products are smoother.

As mentioned before, mostly nitrogen ion implantation is sufficient. In some cases, carbon implantation is of interest, however, process times and cost is high. Another example of implantation of an ion of a solid species is chromium implantation. It helps in a particularly severe case of corrosion. This happens in plastic molds. During the process, the polymer is heated and releases corrosive gases such as hydrochloric acid and water in case of polyvinyl chloride (PVC). The gas attacks the steel molds at the air outlets. Chromium ion implantation increases the lifetime by three to five times [74].

All these applications are based on a cost-benefit calculation with respect to savings in tools, refurbishing, and downtime.

The features of ion implantation for industrial application are:

- Ion implantation is a low-temperature method. This is true in most cases, however, temperature control is required.
- It does not change dimensions or surface finish.
- Ion implanted layers have no adhesion problems; they are a part of the material itself.
- Ion implantation can be focused to the areas to be protected, that is treatment time and cost are lower.

Based on these, the niche sector for ion implantation and the most successful applications turned out to be tools with the following features:

- Tools and components that suffer from mild abrasive wear at not too high contact pressure, abrasive wear caused by fibers or small not too hard particles, corrosion or oxidation.
- Expensive tools where the risk of damage has to stay small.
- Tools where high-process temperatures, dimensional changes or risk of adhesive coating failure are unacceptable, for example tools which have been heat treated at low temperature, typically $<200°C$.
- Precision tools where shape or dimensions might change when they are heated.
- Highly polished tools.
- Large tools with small wear area.

Among these are large high-precision punches, tools for thin sheet forming, sharp knives for plastics, cardboards, paper, and food, and injection molds.

Ion implantation has proved to be a commercially feasible treatment. For a further growth, further niches have to be found where ion implantation is a better choice than competing techniques such as physical vapor deposition (PVD) or chemical vapor deposition (CVD).

18.5. Conclusion

The improvements in wear resistance in ion implanted materials is primarily based on the strengthening of a surface-near zone. This hardening is due to the formation of hard, secondary-phase precipitates. The precipitate particles are dispersed in a matrix of the substrate material. Their high hardness and cohesion strengthens the implanted zone. A further beneficial effect is the strengthening of the implanted matrix by the presence of strain fields around the precipitates. Thus, dislocation movements are hindered. As a consequence, wear is decreased. As an example, the hardening of chrome-containing steels is based on the precipitation of chromium nitride. Titanium alloys are hardened by titanium nitride. These nitrides are ceramic hard materials with large mechanical stability and hardness.

Only when a certain implantation dose is applied, an effect in wear reduction can be observed. This dose depends on the implanted ionic species and on the substrate material. Up to a dose of around 10^{16} ions cm^{-2} improvements in wear resistance are hardly observed. At such a low dose, second-phase precipitation plays a minor role. Hardening is mainly due to dislocation entanglements and solid solution strengthening. However, the effect is minor. Significant effects, based on second-phase strengthening, can be expected when the dose ranges between 10^{17} and 10^{18} ions cm^{-2}. Above these doses, usually no further improvement can be achieved. The retained doses are limited due to sputtering. Additionally, negative effects may appear. Sputter etching may lead to surface roughening with the consequence of enhanced friction and wear debris formation. When the implant concentration becomes very high, the precipitate phase may grow to such an extent that a kind of film is formed. Stress of this brittle film under load may lead to fracture. This effect may be particularly serious when the substrate under the film is comparatively soft. Experience has shown that doses between 3 and $8 \cdot 10^{17}$ ions cm^{-2} are optimal for wear reduction.

So far, ion implantation has been successfully industrially applied in several countries. The users are convinced that this technique has a potential of further growth.

The improvement of corrosion performance by ion implantation is based on chemical and structural modifications. When a species is implanted which is chemically more stable than the substrate, its stability may be transferred to it. This has been shown by the early example of Cr implantation into iron. An advantage is that metastable alloys can be formed. There are no metallurgical restrictions. Tantalum which is next to noble metals in its corrosion stability can be ion beam alloyed with steel, making the steel highly corrosion resistant. However, one has to take care about electrochemical effects. Noble metals such as gold implanted into iron will form a galvanic element thus enhancing iron corrosion. Another way of protecting materials from corrosion by means of a metastable phase is ion-beam-induced amorphization. When a species is implanted which is highly corrosion resistant and which stabilizes the glassy state, the amorphous surface alloy will show excellent corrosion resistance. This has been shown with metalloids such as boron or phosphorus. The glassy metal has no grain boundaries which often are a weak point with respect to corrosive attack.

A general advantage of ion implantation for formation of protective layers is that these layers do not suffer from adhesion problems. They have no sharp interface. There is no zone where shear stresses can lead to delamination or where the corrosive medium coming in through micropores can attack and also lead to delamination. However, a major disadvantage is that the protective layers are very shallow. It has been shown that this does not necessarily mean that the gained lifetime enhancements in application are not significant. In contrast, in application of nitrogen ion implantation for tribological applications, the achieved lifetimes are remarkable.

In case of IBAD, there is, basically, no such limitation. The coatings can be made much thicker. However, for very thick coatings the process times will become too long. Mostly, IBAD coatings are limited to several μm in thickness. A large number of results in the literature have shown that IBAD coatings are well suited for both wear and corrosion protection. The advantages of IBAD are that the process parameters, in contrast to other PVD techniques, can be chosen independently from each other, with the possibility of better optimization, and that the energy needed for growing a high-quality film does not need to be introduced thermally; the ions will do this effectively.

There are industrial applications of IBAD for formation of protective coatings, however, presumably most of them are unknown. Known examples are coatings for optical and biomedical applications and corrosion protection of shaver blades.

References

[1] ASTM Standard G5: Standard Reference Test Method for Making Potentiostatic and Potentiodynamic Anodic Polarization Measurements, *American Society for Testing and Materials*, West Conshohocken, PA, 1994.

[2] W. Ensinger, *Protective Coatings and Thin Films: Synthesis, Characterization and Applications*, Eds. Y. Pauleau and P.B. Barna, Kluwer Academic Publishers, Dordrecht, 1997, p. 585.

[3] V. Ashworth, D. Baxter, W.A. Grant and R.P.M. Procter, *Corros. Sci.*, 16 (1976) 775.

[4] V. Ashworth, D. Baxter, W.A. Grant and R.P.M. Procter, *Corros. Sci.*, 17 (1977) 947.

[5] C.R. Clayton, *Nucl. Instr. Meth.*, 182–183 (1981) 865.

[6] C.R. Clayton, *Surface Alloying by Ion, Electron, and Laser Beams*, Eds. L.E. Rehn, S.T. Picraux and H. Wiedersich, American Society for Metals, Metals Park, OH, 1987, p. 325.

[7] G.K. Wolf, *Surf. Coat. Technol.*, 83 (1996) 1.

[8] W. Ensinger and G.K. Wolf, *Mater. Sci. Eng.*, A116 (1989) 1.

[9] W. Ensinger, *Nucl. Instr. Meth. Phys. Res.*, B56/B57 (1991) 644.

[10] W. Ensinger, *Surf. Coat. Technol.*, 80 (1996) 35.

[11] H. Ferber and G.K. Wolf, *Ion Implantation into Metal*, Eds. V. Ashworth, W.A. Grant and R.P.M. Procter, Pergamon Press, Oxford, 1982, p. 1.

[12] G.K. Wolf and H. Ferber, *Nucl. Instr. Meth.*, 209/210 (1983) 197.

[13] E.C. Cooney, N.L. Lee, G.B. Fisher and D.I. Potter, *Mater. Sci. Eng.*, A116 (1989) 27.

[14] B.R. Nielsen, B. Torp, C.M. Rangel, M.H. Simplicio, A.C. Consiglieri, M.F. DaSilva, F. Paszti, J.C. Soares, A. Dodd, J. Kinder, M. Pitaval, P. Thevenard and R.G. Wing, *Nucl. Instr. Meth. Phys. Res.*, B59/B60 (1991) 772.
[15] W. Ensinger and G.K. Wolf, *Electrochimica Acta*, 39 (1994) 1159.
[16] V. Vilarigues, L.C. Alves, I.D. Nogueira, N. Franco, A.D. Sequeira and R.C. da Silva, *Surf. Coat. Technol.*, 158–159 (2002) 328.
[17] Y. Sugizaki, Y. Yasunaga and H. Tomari, *Surf. Coat. Technol.*, 83 (1996) 167.
[18] T. Zhang, Y. Wu, Z. Weng, F. Ma, H. Zhang, X. Wang, H. Lian and X. Zhang, *Surf. Coat. Technol.*, 128–129 (2000) 181.
[19] Y. Li, T. Zhang and X. Wang, *Surf. Coat. Technol.*, 128–129 (2000) 205.
[20] W. Ensinger, O. Lensch, J. Knecht, K. Volz, T. Matsutani and M. Kiuchi, *Surf. Coat. Technol.*, 158–159 (2002) 594.
[21] W. Ensinger, O. Lensch, F. Sittner, J. Knecht, K. Volz, T. Matsutani and M. Kiuchi, *Nucl. Instr. Meth. Phys. Res.*, B206 (2003) 334.
[22] G.K. Wolf, G. Preiß, R. Münz and L. Guzman, *Nucl. Instr. Meth. Phys. Res.*, B175–B177 (2001) 756.
[23] G.K. Wolf, G. Preiß, R. Münz and L. Guzman, *Surf. Coat. Technol.*, 158–159 (2002) 125.
[24] T. Miyano and H. Kitamura, *Surf. Coat. Technol.*, 65 (1994) 179.
[25] W.C. Oliver and G.M. Pharr, *J. Mater Res.*, 7 (1992) 1564.
[26] G.M. Pharr, *Mater. Sci. Eng.*, A253 (1998) 151.
[27] ASTM Standard G 99: Standard Test Method for Wear Testing with a Pin-on-disk Apparatus, *American Society for Testing and Materials*, West Conshohocken, PA, 1990.
[28] N.E.W. Hartley, W.E. Swindlehurst, G. Dearnaley and J.F. Turner, *J. Mater. Sci.*, 8 (1973) 900.
[29] N.E.W. Hartley, G. Dearnaley, J.F. Turner and J. Saunders, *Applications of Ion Beams to Metals*, Eds. S.T. Picraux, E.P. EerNisse and F.L. Vook, Plenum Press, London, 1974, p. 123.
[30] G. Dearnaley and N.E.W. Hartley, *Thin Solid Films*, 54 (1978) 215.
[31] G. Dearnaley, *Radiat. Eff.*, 63 (1982) 1.
[32] R. Hutchings and W.C. Oliver, *Wear*, 92 (1983) 143.
[33] D.M. Follstaedt, L.E. Pope, J.A. Knapp, S.T. Picraux and F.G. Yost, *Thin Solid Films*, 107 (1983) 259.
[34] I.L. Singer and R.A. Jeffries, *Appl. Phys. Lett.*, 43 (1983) 925.
[35] D.M. Follstaedt, F.G. Yost, L.E. Pope, S.T. Picraux and J.A. Knapp, *Appl. Phys. Lett.*, 43 (1983) 923.
[36] M. Iwaki, *Mater. Sci. Eng.*, 90 (1987) 263.
[37] W.C. Oliver, R. Hutchings and J.B. Pethica, *Metall. Trans.*, 15A (1984) 2221.
[38] J. Sasaki, K. Hayashi, K. Sugiyama, O. Ichiko and Y. Hashiguchi, *Surf. Coat. Technol.*, 51 (1992) 166.
[39] R. Sánchez, J.A. Garcia, A. Medrano, M. Rico, R. Martínez, R. Rodríguez, C. Fernández-Ramos and A. Fernández, *Surf. Coat. Technol.*, 158–159 (2002) 630.

[40] Chapter 12, *Surface Treatment by Ion Beams, Handbook of Tribology*, Eds. B. Bhushan and B.K. Gupta, McGraw-Hill, New York, 1991, p. 12.1.
[41] R. Wei, P.J. Wilbur, W.S. Sampah, D.L. Williamson and L. Wang, *Lubn. Eng.*, 47 (1991) 326.
[42] D.L. Williamson, O. Ozturk, S. Glick, R. Wei and P.J. Wilbur, *Nucl. Instr. Meth. Phys. Res.*, B59–B60 (1991) 737.
[43] R. Hutchings, *Mater. Sci. Eng.,* A184 (1994) 87.
[44] R. Wei, *Surf. Coat. Technol.*, 83 (1996) 218.
[45] J.P. Rivière, P. Méheust, J.A. García, R. Martínez, R. Sánchez and R. Rodríguez, *Surf. Coat. Technol.*, 158–159 (2002) 295.
[46] S. Taniguchi, A. Kitahara, S. Wakayama, E. Eriguchi and N. Suyama, *Nucl. Instr. Meth. Phys. Res.*, B175–B177 (2001) 647.
[47] H.-J. Erler, G. Reisse and C. Weissmantel, *Thin Solid Films*, 65 (1980) 233.
[48] M. Satou and F. Fujimoto, *Jpn. J. Appl. Phys.*, 22 (1983) L171.
[49] R.A. Kant and B.D. Sartwell, *Mater. Sci. Eng.*, 90 (1987) 357.
[50] F.A. Smidt and G.K. Hubler, *Nucl. Instr. Meth. Phys. Res.*, B80/B81 (1993) 207.
[51] A. Kluge, B. Haji-Sayed Javadi, H. Ruoff, R. Öchsner and H. Ryssel, *Surf. Coat. Technol.*, 51 (1992) 237.
[52] X. Wen and Z. Zhang, *Surf. Coat. Technol.*, 51 (1992) 252.
[53] A. Mitsuo, T. Mori, Y. Setsuhara, S. Miyake and T. Aizawa, *Nucl. Instr. Meth. Phys. Res.*, B206 (2003) 366.
[54] J.K. Hirvonen, *Surf. Coat. Technol.*, 65 (1994) 84.
[55] J.H. Arps, R.A. Page and G. Dearnaley, *Surf. Coat. Technol.*, 84 (1996) 579.
[56] W. Ensinger, A. Schröer and G.K. Wolf, *Nucl. Instr. Meth. Phys. Res.*, B80/B81 (1993) 445.
[57] K. Kohlhof, *Nucl. Instr. Meth. Phys. Res.*, B106 (1995) 662.
[58] P. Sioshansi, *Mater. Sci. Eng.*, 90 (1987) 373.
[59] P. Sioshansi, *Nucl. Instr. Meth. Phys. Res.*, B24–B25 (1987) 767.
[60] J.K. Hirvonen, *Mater. Sci. Eng.*, A116 (1989) 167.
[61] C.A. Straede, *Wear*, 130 (1989) 113.
[62] M. Iwaki, *Mater. Sci. Eng.*, A115 (1989) 369.
[63] G.I. Zhang, *Mater. Sci. Eng.*, A115 (1989) 377.
[64] W.L. Lin, X.J. Ding, H.X. Zhang, J.M. Sang, J. Xu and Z.Y. Wang, *Surf. Coat. Technol.*, 51 (1992) 534.
[65] H. Li and C. Ji, *Surf. Coat. Technol.*, 65 (1994) 189.
[66] P. Sioshansi and E.J. Tobin, *Surf. Coat. Technol.*, 83 (1996) 175.
[67] Ch. Straede, *Nucl. Instr. Meth. Phys. Res.*, B113 (1996) 161.
[68] R.J. Rodriguez, A.L. Sanz and A.M. Medrano, *Surf. Coat. Technol.*, 84 (1996) 594.
[69] B. Torp, B.R. Nielsen, N.J. Mikkelsen and C.A. Straede, *Surf. Coat. Technol.*, 84 (1996) 575.
[70] C.A. Straede and N.J. Mikkelsen, *Surf. Coat. Technol.*, 84 (1996) 567.
[71] T. Vilaithong, L.D. Yu, D. Suwannakachorn, S. Davydov, S. Thontem, B. Yotsabat and S. Intanasiri, *Surf. Coat. Technol.*, 83 (1996) 322.

[72] B. Torp, P. Abrahamsen, K.I. Blomqist, S. Eriksen, P.L. Hoeg, B.R. Nielsen, N.J. Mikkelsen, J. Pedersen, C.A. Straede and M. Dossing, *Nucl. Instr. Meth. Phys. Res.*, B127/B128 (1997) 940.
[73] J.I. Onate, F. Alonso, I. Braceras, A.L. Sanz and R.J. Rodriguez, *Surf. Coat. Technol.*, 103–104 (1998) 185.
[74] C.A. Straede and N.J. Mikkelsen, *Surf. Coat. Technol.*, 103–104 (1998) 191.
[75] R.J. Rodriguez, A. Medrano, M. Rico, R. Sanchez, R. Martinez and J.A. Garcia, *Surf. Coat. Technol.*, 158–159 (2002) 48.
[76] M. Klingenberg, J. Arps, R. Wei, J. Demaree and J. Hirvonen, *Surf. Coat. Technol.*, 158–159 (2002) 164.

Chapter 19

High-Temperature Behaviour of Thermal Barrier and Bond Coatings in Oxidizing and Corrosive Atmospheres

Robert Vaßen

19.1. Introduction

Thermal barrier coating systems (TBCs) are used in gas turbines or diesel engines to improve their performance. The isolative layer can provide a reduction of the temperature of the metallic substrate which results in an improved component durability. Alternatively, an increase of efficiency can be achieved by allowing an increase of the turbine inlet temperatures [1].

TBCs consist typically of two layers, a so-called bond-coat layer and an isolative, ceramic topcoat. The bond-coat is often a metal and has two major functions. It improves the bonding between the substrate and the topcoat and it protects the substrate from corrosion and oxidation. Two types of bond-coats are frequently used, a (platinum-) aluminide-based one and a so-called MCrAlY with M being Ni or Co. The choice of the adequate bond-coat depends on the used deposition technique for the topcoat. Electron beam physical vapor deposition (EB-PVD) and atmospheric plasma spraying (APS) are the most frequently used techniques (details are given below). The development of ceramic TBC for components of area gas turbines started already in the forties and fifties of the last century [2]. Already in the sixties zirconia was identified as a promising candidate material. Pure zirconia is not suitable for the application as it undergoes different phase transition [3]. The martensitic monoclinic-tetragonal at about 1000°C is accompanied by a large volume change (3–9%, [4]), and hence by the risk of cracking of the coating. Therefore, different doping additions, as MgO or CaO, were used to stabilize the tetragonal or also the cubic structure. At the end of the seventies 6–8 wt.% Y_2O_3 was established as a nearly ideal material for a TBC application [5] and frequently used in aero and stationary gas turbines from the beginning of the eighties [6,7].

Materials Surface Processing by Directed Energy Techniques
Y. Pauleau (Editor)

Since many decades the development of TBC remains an attracting research area. This can be illustrated by the large number of articles related to this topic. A list of review articles is given in the references [2,7–17]. The large number of articles also reflects the high complexity of the subject. All the different components, substrate, bond-coat, and topcoat, interact with each other or the environment to a more or less extent and/or they undergo detrimental changes due to thermo-mechanical treatments during operation.

In the present chapter, first a description of the main manufacturing routes for the deposition of the different coatings is given. Here also several new processes will be mentioned. In the next section thermo-physical properties and high-temperature behaviour of the different types of coatings are discussed and their main shortcomings and failure mechanisms will be given. Alternative materials will be discussed. Finally, monitoring tools, a research area of increasing interest, will be described.

19.2. Manufacture of Thermal Barrier and Bond Coatings

19.2.1. Thermally Sprayed MCrAlYs (M = Ni, Co) Bond Coatings

The TBCs developed in the fifties and sixties of the last century used simple alloys like NiCr, Mo or NiAl, deposited with flame spray techniques [6,18]. These alloys have been further developed for improved oxidation and corrosion resistance towards the nowadays used NiCoCrAlY bond-coats. They contain about 10 wt.% or more aluminium which leads to the formation of a dense aluminium oxide coating (thermally grown oxide: TGO). Of special importance are the so-called reactive elements as yttrium and others. Yttrium can reduce the formation of sulfur precipitations at the interface TGO/bond-coat and avoid the weakening of the interface. Furthermore, it is build into the grain boundaries and can reduce grain boundary diffusion, and hence the oxidation rate [19]. However, too high amounts of yttrium can lead to so-called internal oxidation and increased oxidation rates. Suitable amounts are in the range of 0.1–0.5 wt.%. In addition, elements like Hf, Si or Re can improve the oxidation resistance of bond-coats [20,21].

In most cases, MCrAlYs are deposited using thermal spray processes. In these deposition techniques different heat sources as combustion flames or plasma gases are used to melt injected particles and to accelerate the droplets of liquid material which are then deposited on the substrate [22]. At the beginning of the TBC development deposition techniques as flame spray or APS have been used, which lead to a considerable oxidation of the particles. Oxide layers within the coating might lead to a reduced bonding of the spray lamellae and to a failure within the bond coating ("black failure", [6]). Several techniques are nowadays frequently used to

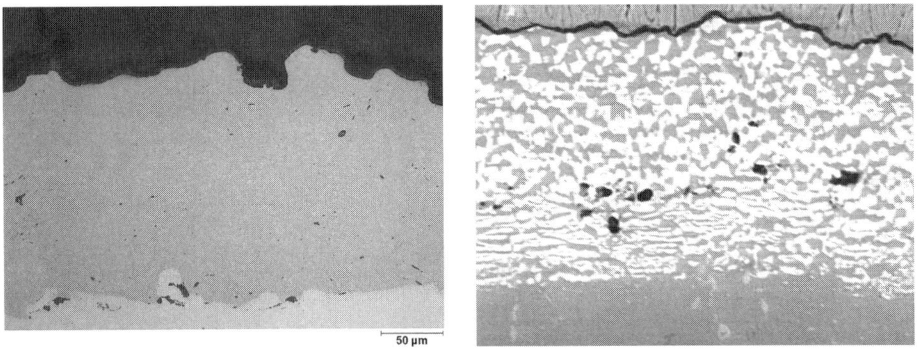

Figure 19.1: Micrographs of a vacuum plasma-sprayed MCrAlY (left) and a platinum aluminide bond-coat (bright phase (Ni/Pt)Al, dark β-NiAl [31]).

reduce the oxidation of particles during processing, as shrouded plasma-spraying which uses inert shroud gases, vacuum plasma-spraying or recently, high-velocity oxygen fuel (HVOF, [23]) spraying. A micrograph of a vacuum plasma-sprayed coating is shown in Fig. 19.1 left. It should be mentioned here that for gas turbine parts, which are not highly thermally and mechanically loaded as combustion chamber parts often, APS bond-coats have shown a sufficient performance and are therefore frequently used.

In MCrAlY bond coatings the major phases are the so-called β-phase (corresponding to NiAl) and a matrix of γ-(Ni) and γ′-(corresponding to Ni_3Al) phases. The β-phase appears dark in Fig. 19.1. In addition, according to composition and heat treatment other phases as α-Cr and σ-phases are observed [24,25].

Bond coatings for atmospheric-sprayed topcoats need a certain surface roughness to guarantee sufficient bonding. Typical values of bond-coat roughness are in the range of $R_a \approx 6$–15 μm. The surface roughness has also an influence on the stresses build up in the topcoat during operation (Section 19.5) and it can influence the oxide scale formation [26]. The obtained roughness values of thermally sprayed bond coatings depend strongly on the processing conditions, the used spraying technique, and the diameters of the used particles. Typically, HVOF coatings give rather low surface roughness values.

If thermally sprayed bond coatings are used for EB-PVD topcoats, the roughness values have to be reduced to 1–2 μm by a finishing process. In this case, the yttria-stabilized zirconia (YSZ) layer is bonded to the alumina scale on the bond coating formed during heat treatment. Such a heat treatment at about 1100°C is typically applied for MCrAlY bond-coats to improve bonding to the substrate.

Sometimes, also additional platinum coatings are applied to improve the oxidation behaviour [27].

19.2.2. (Platinum) Aluminide Bond Coatings

For EB-PVD TBCs in many cases aluminide bond coatings are used instead of MCrAlYs. These coatings often contain platinum to improve the oxidation resistance [28]. Typically these platinum containing coatings are produced by a galvanic deposition of a 5–10 μm thick platinum layer followed by a diffusion treatment. In a second step, an aluminium coating is produced by a chemical vapour deposition (CVD) process. An often used variant is the so-called pack cementation. The part is put into a granular mixture of an aluminium alloy (pure aluminium would melt at the coating temperature of 1000–1100°C), a so-called activator (in many cases AlF_3) and a filler material (e.g. Al_2O_3). During the coating process gaseous AlF_3 is decomposed at the substrate surface, producing NiAl by the reaction of Al with Ni from the substrate. The produced fluorine reacts with the granular aluminium materials to form again AlF_3. So the coating process can run till the aluminide is used up.

During the aluminizing process platinum is dissolved in the NiAl or $PtAl_2$ is formed in the outer part of the coating [29,30].

19.2.3. APS YSZ Topcoats

For the deposition of YSZ TBC by thermal spray processes in most cases the APS process is used. The process allows high gas temperature which are necessary to melt zirconia with its high melting temperature of about 2700°C during the spraying process. During the deposition of the YSZ topcoats typical microstructural features are generated. Spraying conditions with rather low substrate temperatures below about 300°C form coatings with two types of pores. Globular pores with diameters in the micrometer range are a result of the non-ideal arranging and deformation of the spray splats during deposition. Secondly, microcracks are formed during the cooling of the splats from the deposition temperature to the substrate temperature. In Fig. 19.2 these features are clearly visible. The microcracks are essential for the performance of TBC as they give the coating the necessary strain tolerance even at room temperature [32,33]. In combination with the globular porosity they also reduce the Young's modulus of the coating which is favorable for the stress state in the coating. Further details on the stress state and failures modes are given in Section 19.5. The total porosity level in APS TBCs is typically in the range between about 9% and 20% [34].

For coatings with thickness above about 500 μm often a different coating microstructure is used as the microcracked TBC which does not give sufficiently good performance. These coatings contain so-called segmentation cracks which can effectively reduce the stress state in the thick coatings in cyclic tests. The segmentation cracks are formed during the deposition process mainly by applying high

Figure 19.2: Micrographs of APS coatings (left) optical micrograph of a polished section (right) scanning electron micrograph of a fracture surface.

Figure 19.3: APS system (left) and plasma gun in operation (right).

substrate temperatures above 500°C in combination with high particle temperatures [35–37].

In industrial coating facilities the plasma guns are mounted on robotic systems which allow the homogenous coating of complex shaped parts, an example of a modern APS chamber is shown in Fig. 19.3 (left). In Fig. 19.3 (right) a plasma gun is shown in operation.

The complexity of the plasma spraying process with its large number of process parameters often leads to a poor reproducibility of the coatings. As a consequence several diagnostic tools have been developed to control directly particle parameters as velocity and temperature or plume characteristics during operation [38,39]. Meanwhile several industrial companies have introduced these systems for process control [40].

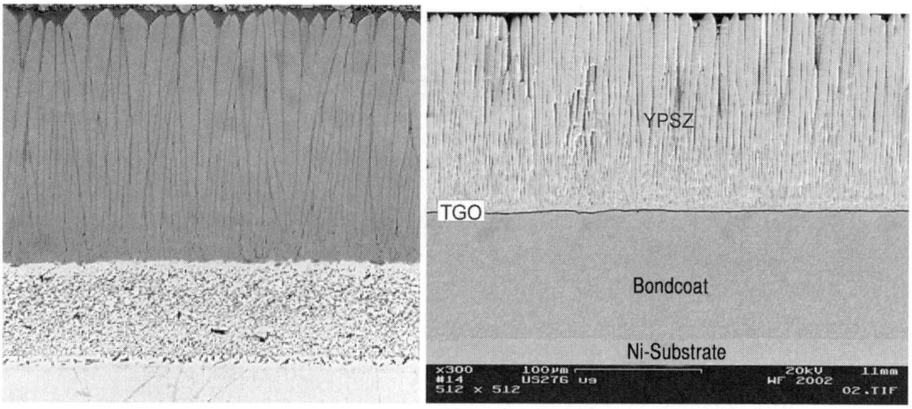

Figure 19.4: Microstructure of as-deposited EB-PVD coatings (left: courtesy Rodney Wing, Chromalloy, UK, right: courtesy Uwe Schulz, DLR, Germany).

19.2.4. EB-PVD YSZ Topcoats

During the EB-PVD process a porous ingot of YSZ is heated in a vacuum chamber with an electron beam above the melting temperature up to about 3500°C [41]. Material is vaporized from the ingot and condensates in a line-of sight process on the typically preheated substrates. Typically first a globular microstructure is formed which will transfer to a columnar structure with an often <100> texture during further coating growth (see Fig. 19.4, [42]). The columns show a feather-like outer structure and their microstructure can be adjusted by the proper choice of process conditions as substrate temperature, evaporation rate or geometrical conditions (e.g. rotation) during deposition. Special care has to be taken to control the evaporation rate as it determines the homogeneity of the coatings [43]. Also the substrate temperature has to be adjusted in a certain high-temperature range (about 950–1100°C) to achieve the desired columnar microstructure with the most suitable column diameters of about several micrometers.

As the individual columns can separate from each other during tensile loading, the PVD coatings are often considered to have a high strain tolerance. This behaviour in combination with the different used bond-coats have made these coatings widely used on thermally and mechanically highly loaded parts of gas turbines (especially blades). However, due to the high investment costs and the relatively low deposition rates (about 200 $\mu m\, h^{-1}$) the costs of this process are about a factor of

Figure 19.5: Industrial EB-PVD coater at Chromalloy, UK (top), enlarged view of ingots and parts (bottom, courtesy Rodney Wing, Chromalloy, UK).

10 higher than these of the APS process. An industrial coating system is shown in Fig. 19.5.

19.2.5. New Processes

An important target of many process developments is, besides property improvement, cost reduction. Efforts are made to improve the performance of plasma-sprayed TBCs so that they can substitute nowadays expensive EB-PVD coatings also for demanding applications. A certain cost reduction potential is also expected from chemical vapour deposition (CVD) and certain PVD processes [44,45], as the investment costs are remarkably lower than for EB-PVD equipment. Especially new CVD processes as plasma-enhanced or hollow cathode PVD seem to be promising for future applications [46].

An additional, major advantage of these processes is their capability to coat out of sight surfaces. This becomes increasingly important for the coating of complex shaped parts as vane clusters (Fig. 19.6).

Figure 19.6: PVD coating applied on turbine vane cluster (courtesy Dan Roth-Fagaraseanu, Rolls Royce Deutschland).

19.3. Thermo-physical Properties

19.3.1. Thermal and Radiative Properties

Zirconia with its low thermal conductivity (about 2–3 $W\,m^{-1}K^{-1}$ at 1000°C for bulk ceramics, [47]) and its high thermal expansion coefficient ($10–11 \times 10^{-6}K^{-1}$) has been identified quite early as an interesting TBC material. Depending on the microstructure thermal conductivity of the coatings can vary significantly. EB-PVD 7–8 wt.% YSZ coatings have typically relatively high thermal conductivities of about 1.5–1.9 $W\,m^{-1}K^{-1}$ due to the dense, columnar microstructure [48]. In contrast, the microcracked structure of plasma-sprayed TBC lead to lower as-sprayed thermal conductivities of about 0.8–1.1 $W\,m^{-1}K^{-1}$ [48,49]. During heat treatment the thermal conductivity of plasma-sprayed coatings tends to increase fast due to the sintering of the microcracks. Values of 1.5 $W\,m^{-1}K^{-1}$ are already observed after treatment for 50 h at 1000°C [49]. In contrast, EB-PVD show much more stable values after heat treatment.

It should be mentioned that heat transfer in YSZ based TBCs is not limited to transport via phonons as also radiation can play especially at higher temperature an important role. This is due to the fact that YSZ is highly transparent below about 6 μm [50].

19.3.2. Mechanical Properties

Elastic properties of TBCs are directly related to their performance, and hence of special importance. It is simply not possible to characterize the response of a TBC on an applied load in an adequate way by only giving a single Young's modulus. This has several reasons which are all related to the complex microstructure of both APS and EB-PVD coatings. Typically the elastic properties are highly non-isotropic, giving for example higher values perpendicular to the spray direction as in spray direction for APS coatings [51]. In addition, properties depend on the loading conditions, so PVD coatings have extremely low values for tensile, however, high values for compressive loading (about 150 GPa in Ref. [52]). This is explained by the fact that the columns can easily separate during tensile loadings. A similar trend is also found in APS TBCs in a less pronounced way [53]. Typical values of the Young's modulus of as-sprayed YSZ TBCs can vary between about 10 and 80 GPa.

The behaviour becomes even more complex as the TBCs often show pronounced viscous behaviour at elevated temperatures, to a certain extent even at room temperature [33].

All these lined out properties are not stable during heat treatment, instead extensive changes are observed after annealing above 1000°C [54]. An increase of stiffness by nearly a factor of 2 is observed in Ref. [55] after 50 h at 1100°C (Section 19.4.1).

19.4. Behaviour at High Temperatures

19.4.1. Stability of YSZ Coatings and Alternative Materials

Pure zirconia undergoes a phase transformation from monoclinic to tetragonal at about 1170°C during heating which is accompanied by a volume change and leads to crack formation in the coating. Stabilizers as yttria can suppress this. The most frequently used TBC material is 6–8 wt.% yttria partially stabilized zirconia (YSZ). During both APS and EB-PVD deposition this material will not form the equilibrium phase consisting of monoclinic and cubic zirconia but a so-called t'-phase. During the fast cooling of the sprayed splats and also during the deposition from the gas phase the time at elevated temperatures is not sufficiently long to obtain the equilibrium phase. The t'-phase contains a high amount of yttria in the tetragonal phase and does not transform to cubic and monoclinic phase even at longer operation times at elevated temperatures of up to about 1200°C. However, with about 1200°C an upper temperature limit is reached for YSZ based TBCs as the material undergoes a diffusion induced transformation from t'-phase to tetragonal and cubic phases at higher temperatures. This is true for both APS [56,57] and EB-PVD coatings [58].

One reason for the frequent use of YSZ TBCs with 6–8 wt.% yttria content is the relatively good fracture toughness of this material due to transformation toughening [59]. A higher amount of stabilizer would lead to a fully stabilized, cubic zirconia. This materials does not show any transformation-induced toughening processes.

A lower amount of stabilizing additives leads to a higher toughness at room temperature, but also, to detrimental phase transformations during heating.

Besides the phase transformation also significant sintering has a detrimental effect on the performance of YSZ TBCs. It leads to the formation of a kind of "sintering necks" and hence to a reduction of the strain tolerance in combination with an increase of the Young's modulus [60]. Higher stresses will be originated in the coating, which lead to a reduced life under thermal cyclic loading.

These disadvantageous properties of YSZ at high temperatures led to an intense search for new TBC materials in the past [16,61–66].

Among the interesting candidates for TBC, especially those materials with pyrochlore (e.g. $La_2Zr_2O_7$, $Gd_2Zr_2O_7$), spinel ($MgAl_2O_4$), perovskite (e.g. $SrZrO_3$) or magnetoplumbite (e.g. $LaMgAl_{11}O_{19}$) structures and high melting points show promising thermo-physical properties. Previous investigations showed good physical properties of these materials; that is, thermal conductivity comparable or lower than YSZ and high thermal stability [63,67]. However, the thermal expansion coefficient is typically lower than that of YSZ, which leads to higher thermal stresses in the TBCs as both substrate and bond-coat have higher thermal expansion coefficients (about $15 \times 10^{-6} K^{-1}$). In addition, relatively low toughness values are observed in these materials [68].

As a result, the thermal cycling properties are worse than for YSZ coatings. This problem is probably relevant for most of the new TBC materials, as the need for thermal stability seems to contradict the ability of transformation toughening (Section 19.4.1). A way to overcome this shortcoming is the use of layered topcoats (Fig. 19.7). Failure of TBCs often occur within the TBC close to the bond-coat/topcoat interface. At this location, YSZ is used as a TBC material with a relatively high thermal expansion coefficient and high toughness. The YSZ layer is then coated with the new TBC material (e.g. $La_2Zr_2O_7$) which is able to withstand the typically higher temperatures at this location. It was shown in the past years in several publications on $La_2Zr_2O_7$/YSZ double layer systems that this concept really works [69–73]. Recently, a temperature increase compared to YSZ of 100 K could be demonstrated [74].

19.4.2. Static and Cyclic Oxidation Behaviour

Most static (isothermal) and cyclic oxidation tests on TBCs are performed at temperatures between 1000°C and about 1150°C in furnaces under static or flowing air atmosphere [75,76]. The higher end of the temperature ranges are typically higher than the normal operation temperatures at least for stationary gas turbines, however, as the degradation processes are temperature activated, the higher temperatures are used to reduce the testing times to affordable lengths.

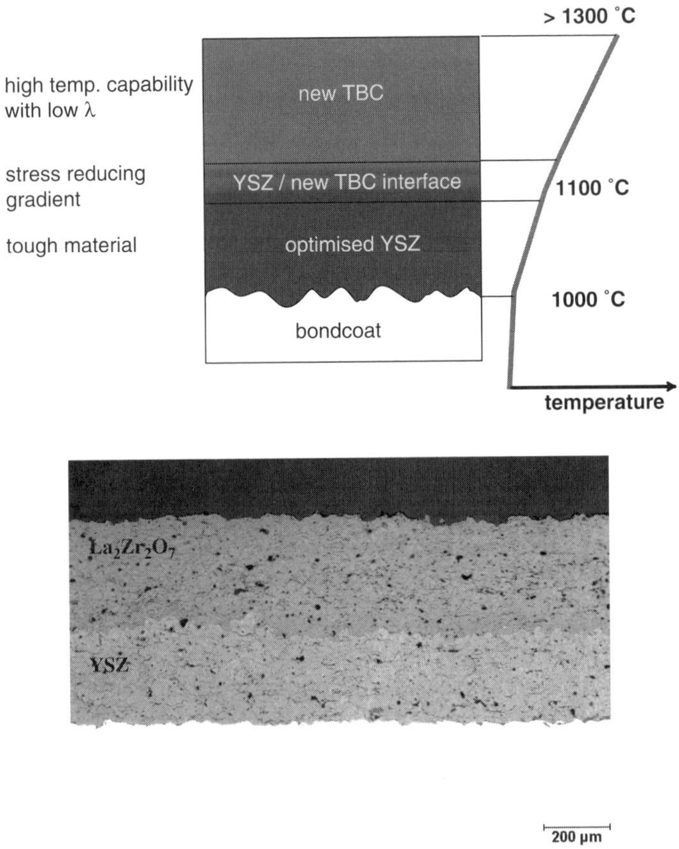

Figure 19.7: Principle of double-layer coatings (top) and plasma-sprayed version consisting of $La_2Zr_2O_7$/YSZ (bottom).

In the cyclic oxidation experiments, the specimens are typically removed out of the furnace and cooled to near room temperature. Cooling is often promoted by blowing the surface with compressed air.

In the isothermal experiments, which are discussed first, frequently thermo-gravimetric analysis (TGA) is used. In these types of experiments weight changes can be precisely determined as a function of time. The observed weight gains are a result of the growth of an oxide layer on the bond-coat. As this so-called TGO is generally regarded as a key factor which determines lifetime of the coatings (Section 19.5), these weight gain measurements are expected to be correlated to the performance of the TBCs. In the thermo-gravimetric experiments the growth constant k_p is determined from the steady-state portion of the weight gain curves:

$$\frac{\Delta W}{A} = (k_p t)^{1/n} \tag{19.1}$$

Often $n = 2$ fits the experimental data quite well. The determined constants k_p differ considerably depending on the exact composition and the processing of the used bond-coat materials. Literature values of the parabolic ($n = 2$) rate constants k_p for VPS – NiCrAlY bond-coats at 1083°C and 1150°C – are determined to be about $4 \times 10^{-12}\,g^2cm^{-4}s^{-1}$ and $1.5 \times 10^{-12}\,g^2cm^{-4}s^{-1}$, respectively [76,77]. Results for NiAl(Pt) layers at 1100°C appear to be somewhat lower (about $7 \times 10^{-13}\,g^2cm^{-4}s^{-1}$, [75]). The corresponding oxide thickness values after 100 h at 1100°C for this given rate constant is about 3 μm.

Studies on the influence of the TBC top layer on the oxidation rate of bond coatings have not found a distinct effect in most cases [76]. The reason for this is on the one hand the porosity of plasma-sprayed and partially, also EB-PVD coatings. On the other hand, YSZ is a material with a high oxygen ion conductivity which allows considerable oxygen transfer even through dense coatings.

In addition to the oxide growth rate also the scale adhesion is an important factor relevant for lifetime as early oxide scale spallation will also lead to the spallation of the TBC. Typically a sharp reduction of the sample mass occurs when scale loss occurs ("break away", [79]).

For MCrAlYs, the formed oxide layer does in many cases not only consist of a pure α-Al_2O_3 layer. It contains different oxides depending on the chemical composition of the bond-coat and sometimes even the substrate alloy [78]. For MCrAlY bond-coats $\theta - Al_2O_3$ can be found at the beginning of the oxidation which transforms into stable α-Al_2O_3 after longer oxidation times [79]. $\theta - Al_2O_3$ with a needle-like morphology formed before the EB-PVD coating with YSZ can support the formation of the so-called mixed oxide phase (TGO + YSZ, [80]).

Often spinels ($Ni(Cr, Al)_2O_4$) or also YAG ($Al_5Y_3O_{12}$) can be detected in the scales formed on MCrAlYs [76]. These oxides might reduce the performance of the TBCs as they influence the adhesion of the scale or lead to crack formation or to faster oxidation rates.

PtAl bond-coats form frequently pure alumina scales, however, also here the TGO morphologies and growth rates can vary considerably [81] depending on surface treatment or composition.

At 1150°C test temperature lifetimes between 200 h (24-h cycles) and 700 h (isothermal) have been found for PtAl bond-coats with EB-PVD topcoats [78,81].

Systems with MCrAlY bond-coats and EB-PVD topcoats can even perform better than those with PtAl bond-coats [78].

Failure of the PVD systems often occurs between oxide scale and bond-coat, while for the APS coating failure within the topcoat close to the bond-coat is also found in many cases.

In cyclic oxidation experiments, the growth rate of the oxide scale does not differ significantly from isothermal experiments. In addition, also the decohesion appears in PVD coatings at the metal–oxide interface as often found in isothermal oxidation. It was also stated that for both oxidation experiments the spallation of the topcoat occurred during cooling. However, a significant difference is that due to the thermal cycling the formation of cavities [75] or the roughening of the scale/bond-coat interface is promoted. Especially, the roughening is considered as an important mechanism which leads to failure in some TBCs and will be discussed in Section 19.5.

Typical number of 1-h cycles to failure at 1150°C are for plasma-sprayed systems between 150 and 350 [82]. For EB-PVD systems with MCrAlY bond-coats lifetimes between 500 h and more than 3000 h, and between 216 and 2000 h have been observed at 1000°C and 1050°C, respectively (cycles of 2 h at high temperatures) [80].

Certainly, the TGO growth has a significant influence on the failure of TBCs. On the other hand, it is meanwhile accepted that no single critical TGO thickness exists which leads to a spallation of the TBC when it is reached. The TBC system is too complex to be described with such a single parameter, and hence each TBCs in combination with the given loading conditions leads to different critical thickness values. Correspondingly, the TGO thickness at failure varies considerably between about 4 and 10 μm.

19.4.3. Cyclic Burner Testing

During cyclic burner testing the coated surface of the specimen is heated by a gas burner. Depending on the cooling conditions at the backside more or less steep temperature gradients can be established. In general, the heat flows are much larger than those realized in furnaces. For rigs operating with cylindrical specimens, which are not cooled from inside, gradients exist mainly during the transient heating and cooling phases. Similar conditions are found for strip-like samples which are cooled mainly by radiation from the backside [83]. Rigs actively cooled, for example, by compressed air can establish well-defined temperature gradients throughout the sample [84].

Many companies active in the field of TBC development have their own thermal cycling rigs. As the thermal conditions in these rigs strongly depend on the used conditions as geometry, type of gas burner and cooling technology, the obtained results are hardly comparable and often not accessible to the public. Sometimes even complete sections of combustion chambers are used to generate hot combustion gases for

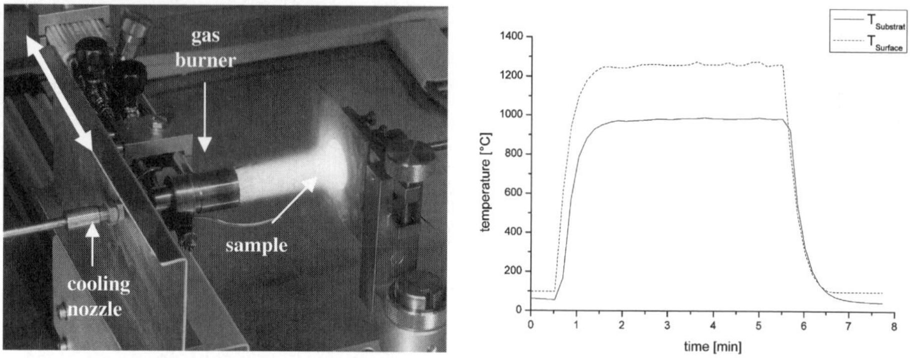

Figure 19.8: Photo of one of the thermal cycling rigs at IWV1, Forschungszentrum Jülich (left) and a typical temperature profile of one cycle (right).

testing. In general, the costs of this type of testing are considerably higher than for furnace testing. On the other hand, the test conditions are certainly more comparable to the actual conditions in a gas turbine as isothermal treatment. Recently it becomes evident that the cooling and heating rates seem to have a major influence on the lifetime of the coatings. Hence, efforts should be made to adjust these as close to the once observed in real gas turbines as possible.

A general finding of thermal cycling tests compared to cyclic furnace tests is the reduced lifetime in the first ones. In Ref. [83], for example, the lifetime in furnace tests with 50-min cycles at 1137°C was about 300, whereas thermal rig tests with approximately the same bond-coat temperature gave only a lifetime of less than 100 cycles.

In the following our test rigs at IWV1, Forschungszentrum Jülich will be described in more detail (Fig. 19.8). Disk-shaped samples (thickness 3 mm, diameter 30 mm) coated from one side with a TBCs are heated with a natural gas/oxygen burner. From the backside the specimens are cooled by compressed air. During cooling the gas burner is moved aside and an additional nozzle cools the surface by compressed air with an initial cooling rate of about $100\,\mathrm{K\,s^{-1}}$. All gas flows are controlled by mass flow controllers giving stable and reproducible conditions. The surface temperature is measured by a pyrometer operating at a long wavelength of about 10 μm as YSZ is highly transparent at wavelengths below 6 μm. The substrate temperature is measured with a thermocouple located in the center of the substrates.

In Fig. 19.9 the results of thermal cycling tests performed on IN738 samples with about 300–400 μm thick YSZ APS TBCs on a 150-μm VPS NiCoCrAlY bond-coat are shown. A typical linear correlation between the logarithm of the lifetime and the bond-coat temperature is found (more correctly the inverse temperature should be plotted). This relation indicates that a thermally activated process is

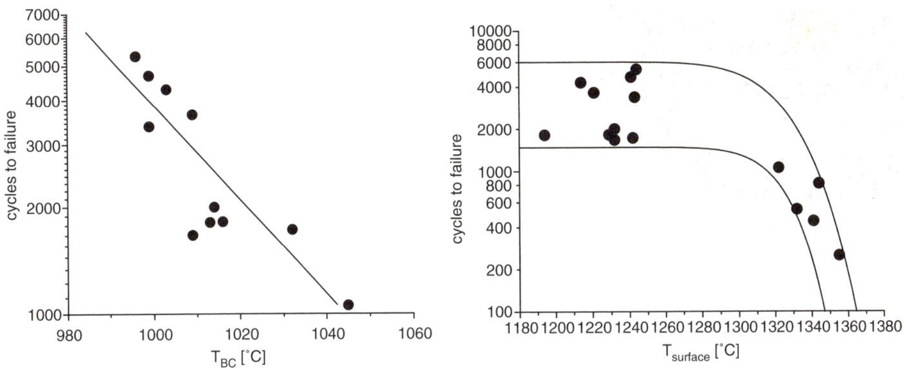

Figure 19.9: Lifetime of thermally cycled plasma-sprayed TBCs as a function of bond-coat temperature (left) and surface temperature (right). Duration of cycles: heating 5 min, cooling 2 min.

responsible for failure. Typically TGO growth is considered as most relevant. A detailed discussion will be given in Section 19.5.

The dependence on surface temperature (Fig. 19.9 right) reveals a sharp drop of lifetime at about 1350°C which is caused by the limited temperature capability of YSZ. The reasons for this behaviour were already discussed in the Section 19.4.1.

19.4.4. Corrosion Tests

The materials applied in hot sections of gas turbine are in contact with aggressive combustion atmospheres. Three potentially harmful impurities are vanadium, sulfur and sodium. The degree of degradation of the gas turbine materials by corrosion is related to the impurity levels of the used fuels. While aviation gas turbines are typically operated with high-purity fuels with sulfur contents below 0.05 wt.%, industrial gas turbine fuel can contain sulfur levels of up to 4 wt.% [85]. In addition to the impurities from the fuel also airborne impurities, sea-salt containing aerosols or dust and salts can accelerate the corrosive degradation. Due to the high air-to-fuel ratio of more than 10 the impurity level in the air is amplified by this factor compared to the impurity level of the fuel [86].

Although the focus of this chapter will be on the corrosion of TBCs, it should be mentioned that the hot corrosion behaviour of the ceramic topcoat may often be better than the one of the bond-coat [86–88]. Often sulfur plays a crucial role in high-temperature (type I) corrosion of the bond-coat. Depending on SO_3 and partial pressure of oxygen, the protective oxide scale (both Al_2O_3 and Cr_2O_3) can be dissolved in, for example, Na_2SO_3 which penetrates through the TBC to the interface of bond-coat and TBC. Typically a higher Cr content is advantageous for a good corrosion behaviour as Cr bonds sulfur by forming thermodynamically stable CrS [89].

Corrosion tests of TBCs are often made in furnaces. Samples are coated by salt solutions of, for example, Na_2SO_4, K_2SO_4, Mg_2SO_4 or mixtures of them or directly with V_2O_5 powder. The used atmosphere is often air which can contain typically up to 1% SO_2 gas. In some cases burner rigs have been used in which aqueous solutions of the corrosive materials have been injected in the combustion chamber [90]. Typical applied temperatures during the tests are in the range of 800–1000°C.

The major reactions taking place between the corrosive media and YSZ have been summarized in the overview article by Jones [87]. Reactions which contain V- and S-species are:

$$ZrO_2(Y_2O_3) + V_2O_5 \rightarrow ZrO_2\,(\text{monoclinic}) + 2YVO_4 \tag{19.2}$$

$$ZrO_2(Y_2O_3) + 3SO_3\,(+Na_2SO_4) \rightarrow ZrO_2\,(\text{monoclinic}) + Y_2(SO_4)_3 \quad (\text{in } Na_2SO_4 \text{ solution}) \tag{19.3}$$

Both reactions lead to the destabilization of YSZ which is accompanied by phase transitions and volume changes during thermal cycling (Section 19.4.1). The reaction with V_2O_5 is more pronounced than the one with SO_3 which needs relatively high SO_3 partial pressures. As a result degradation due to sulfur is less significant than degradation by V-containing species. These facts are underlined by the photos shown in Fig. 19.10. The YSZ sample in contact with V_2O_5 underwent severe degradation while sulfur had only a minor effect on the coating.

An opposite trend was recently observed in the new TBC material $La_2Zr_2O_7$. This coating showed a better corrosion resistance against V_2O_5 than YSZ, however, a worse behaviour under sulfur corrosion [91].

Figure 19.10: YSZ coatings after corrosion tests for 3 h at 1000°C under contact with V_2O_5 (left) and for 360 h at 900°C under exposure with sulfur-containing salts (right, courtesy B.R. Marple, National Research Council of Canada, Canada, [85]).

Recently, a new type of degradation process called CMAS (calcium–magnesium–alumino-silicate) has attracted considerable attention especially in the USA [92,93]. Debris with chemical composition close to CMAS are deposited as layer on the surface of airfoils. As the melting point is about 1240°C CMAS can penetrate at the highest surface temperatures into the coating. During cooling additional stresses are build up which lead to the spallation of parts of the coatings.

19.5. Failure Mechanisms and Lifetime Modeling of TBCs

19.5.1. Overview

TBCs undergo detrimental changes during operation due to different factors:

- Stresses build up due to:
  - thermal expansion mismatch between TBC (10–11 × $10^{-6}K^{-1}$) and substrate/bond-coat (about 15 × $10^{-6}K^{-1}$),
  - growth of a TGO layer,
  - phase transformations within TBC, bond-coat or TGO,
  - rough interface between bond-coat and TBC,
  - creep and sintering effects within TBC, bond-coat or TGO.
- Additional mechanisms like:
  - erosion,
  - corrosion,
  - foreign object damage.

An example of a spalled coating in an industrial gas turbine is shown in Fig. 19.11.

Figure 19.11: Heavy industrial vane showing loss at engine strip (courtesy Rodney Wing, Chromalloy, UK).

Nowadays TBCs are used in most industrial applications not fully design integrated parts of the gas turbines, that is the gas turbine can survive a complete spallation of the TBCs without catastrophic failure till the next shutdown. This automatically implies that the design does not take full advantage of the potential of TBCs. At present two different strategies are followed to overcome this problem. One method is to further improve lifetime modeling to obtain more reliable predictions. This will be outlined in the following section. Another possibility is the use of monitoring tools to determine the remaining lifetime of the TBCs directly in the gas turbine. This will be discussed in Section 19.6.

Intense work in the area of lifetime prediction of TBCs started at least 20 years ago [94,95]. Especially within the NASA a lot of work was carried out leading to a strain-controlled model with a relatively large number of parameters [96, 97].

The difficulties in accurate lifetime prediction are a result of the complexity of the TBCs and their failure mechanisms (see above). In addition, the often relatively poor reproducibility of the coating processes lead to a large scatter in lifetime data.

Corresponding to the different microstructures of APS and EB-PVD TBCs the developed models for both systems differ considerably. Recent results will be given in the next chapters focusing on mechanisms related to the growth of a TGO scale on the bond-coat.

Other failure mechanisms as erosion, foreign object damage (FOD), or corrosion are in most cases not included in the lifetime models although they might play an important role depending on the used fuel or the environment in which the gas turbine is operated.

19.5.2. EB-PVD TBCs

In EB-PVD systems failure often occurs between the interface TGO and bond-coat (see Fig. 19.12 left). A problem for the explanation of such kind of failure is the fact that for an ideal smooth interface no stress perpendicular to the interface exists. Therefore, in most cases imperfections at the interface are considered as critical. Especially for PtAl bond-coats work has been concentrated on the growth of imperfections during isothermal and cyclic oxidation ("morphological instabilities" [98,99]).

Ratcheting or morphological instabilities are characterized by a local imperfection in the TGO that grows on a cyclic basis, eventually causing crack propagation in, and spallation of, the topcoat (Fig. 19.13 left). This failure mechanism is driven by a combination of three non-linear constitutive behaviours in the coating: (1) high-temperature inelasticity in the TGO, (2) growth strain in the TGO and (3) cyclic yielding in the bond-coat. The high-temperature inelastic strength of the TGO is often referred to as "growth stress." The growth strain is induced due to the oxidation process when the

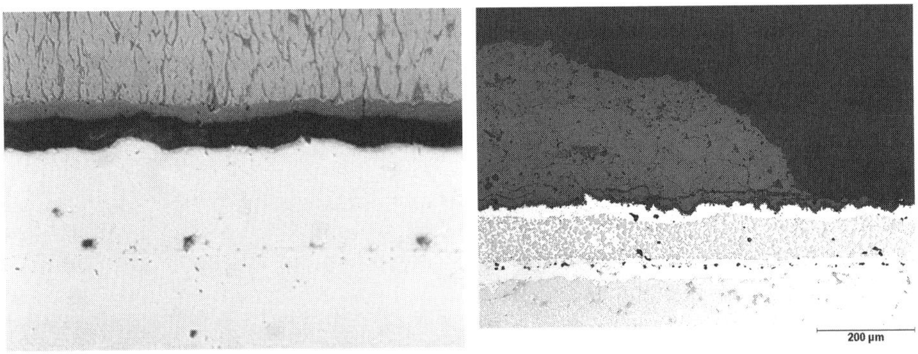

Figure 19.12: De-bonding at bond-coat interface for a PVD system (left, courtesy Rodney Wing, Chromalloy, UK) and within TBC close to bond-coat for an APS system (right).

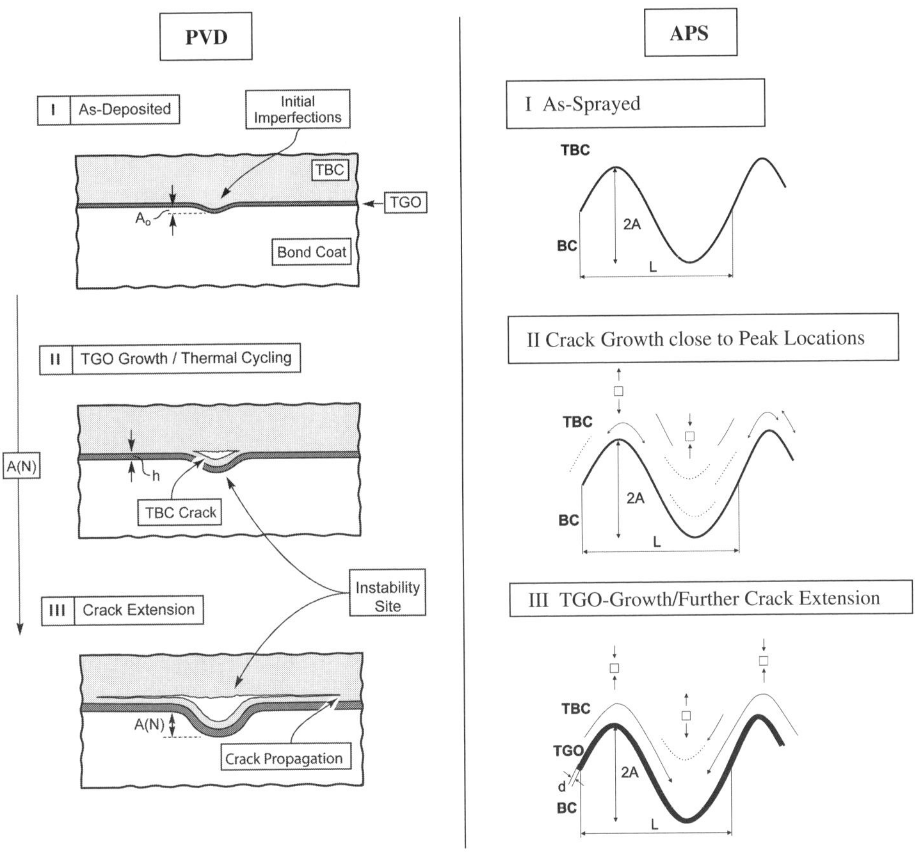

Figure 19.13: Proposed failure mechanisms for PVD (left, [98]) and APS TBCs (right, see text).

new alumina is formed. Most of the TGO growth occurs as thickening, but a small part is distributed in the grains of the TGO, leading to a lengthening component. The growth strain is limited by the growth stress, and once the TGO stress reached the level of the growth stress, the lengthening strain is reallocated into thickening strain. In the isothermal scenario, the changes in the displacement essentially stop, while under cyclic conditions, displacements and stresses continue to change once the point of TGO inelasticity is reached. This is so, since the stresses in the TGO are relaxed during each cycle, allowing for additional accumulation of lengthening strain in the TGO when high temperature is reached. Ratcheting is observed for systems that have low bond-coat yield strength at elevated temperatures, such as Pt-modified alumina, and is subjected to frequent and short cycles, since this allow for relaxation of the TGO during cycling.

The outcome of the modeling can in principle explain experimental findings as enhanced durability of TBCs with a nearly perfect, smooth interface or with bond-coats having enhanced high-temperatures strength [98].

For EB-PVD TBCs with MCrAlY bond-coats the clear evidence of the growth of instabilities are still missing. In these systems the imperfections might be the result of pores and cracks within the TGO. These imperfections can reduce the bonding between TBC and bond-coat.

Fracture mechanical approaches have been developed which compare the stored elastic energy in the coating during thermal cycling with a critical energy release rate related to the energy necessary to drive a crack at the interface [100]. The first quantity increases, for example, as a result of sintering and the resulting rise of Young's modulus. The second one typically decreases during time due to the formed defects within the TGO. Failure occurs if both are equal.

19.5.3. APS TBCs

In plasma-sprayed TBCs a rough interface between bond-coat and topcoat is essential for the performance as the bonding mechanism of plasma-sprayed coatings is mechanical interlocking. On the other side, this wavy structure of the interface leads to off-plane stresses which can promote crack growth in this region (see Fig. 19.14 left).

In principle these mechanisms are known since many years [101], however more detailed finite element calculations have been made during the last decade [102,103], partially also considering cracks within the coating [104]. A result of such calculations is shown in Fig. 19.14. The stress relaxes at high temperatures, cooling to room temperature will generate tensile stress in the TBC perpendicular to the bond-coat/TBC interface (see Fig. 19.14 left). Cracks can start propagating here, however

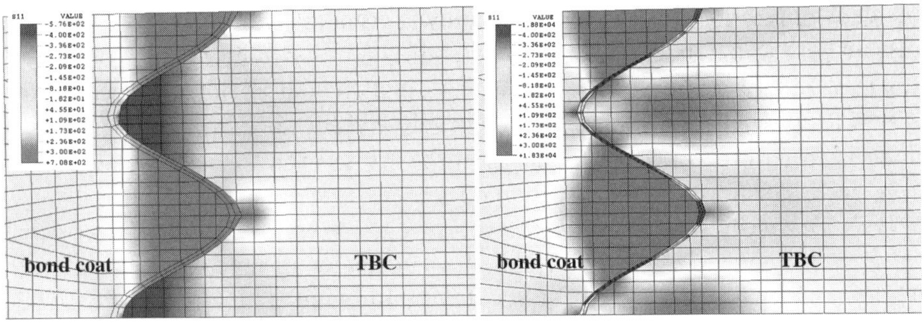

Figure 19.14: Stress distribution in TBC and bond-coat after cooling from a stress-free state at 1000°C to room temperature, left: as-sprayed condition and right: after growth of a TGO.

they will stop close to the valley locations as there compressive stress states are found. If now TGO growth sets in, the stress state in the TBC will be converted after a certain thickness is build up (see Fig. 19.14 right). The reason is the very low thermal expansion coefficient (about $8 \times 10^{-6} K^{-1}$) of the TGO. The valley locations can be approximated by a cylindrical geometry with the TBC in the center, a bond-coat as outer layer and a TGO in the middle. Due to its high thermal expansion coefficient the bond-coat will generate compressive stresses on the ceramics during cooling. However, as the thermal expansion of the TGO is below the one of the TBC, the intermediate TGO layer is able to shield the pressure. After a certain TGO thickness is reached even tensile stresses perpendicular to the interface are build up in the TBC. The described stress states in the TBC can promote the growth of pre-existing microcracks in the TBC. A simplified outline of such crack extension is shown in Fig. 19.13 right. In principle, the combination with sub-critical crack growth laws allow the calculation of lifetimes [105], however, the complex geometrical and thermo-mechanical situation makes a precise calculation still difficult.

In parallel to the outlined model also fracture mechanical approaches as discussed in Section 19.5.2 can be applied [100].

19.6. Non-destructive Testing and Remaining Lifetime Monitoring

Different methods have been investigated in the last decade to characterize TBCs by non-destructive testing [106,107] and several of the used techniques deliver results which can be related to the remaining lifetime of the coatings. As already mentioned in Section 19.5.1 development of such remaining lifetime monitoring tools is one way to take better advantage of the potential of TBCs.

Five techniques will be outlined in the following:

19.6.1. Acoustic Emission

Acoustic emission (AE) analysis is a common technique to investigate the fracture behaviour, for example, of plasma-sprayed coatings. Crack initiation or growth leads to emission of acoustic signals with characteristic duration, energy, amplitude and frequency spectrum. In addition, the location of the AE, and hence the location of the crack can be determined by the use of several sensors at different positions. AE analysis has also been combined with fracture mechanical measurements as bending tests to correlate the emission directly with crack initiation and growth [108]. In thermal cycling tests intense AE signals have been observed during cooling. This indicates that crack growth during cooling plays an important role in accordance with the discussion in Section 19.5.3 [109,110]. Reliable results of the AE analysis can only be expected if the sensors have a good mechanical contact to the material under investigation. As the sensors have typically only a low-temperature capability an application of this technique directly in a gas turbine is probably hardly possible.

19.6.2. Impedance Spectroscopy

In impedance spectroscopy (IS) the complex resistance of a system is measured as a function of frequency. By applying this well-established techniques to TBCs it is possible to distinguish between different layers and so to follow, for example, the growth of the TGO layer [111,112]. However, interpretation of the complex impedance spectra is rather difficult and makes use of models in which the different layers are described by specific capacitance and resistance values.

For IS the TBCs has to be connected by electrodes (typically Pt) to the impedance analyzer. More easily removable electrodes would certainly be an advantage for the investigation of TBCs, for example, during shutdown off gas turbines.

19.6.3. Piezospectroscopy

Cr^{3+} photoluminescence piezospectroscopy (PS) is capable of measuring stresses in the α-Al_2O_3 (TGO) layer of TBCs [113]. Impurity Cr^{3+} ions are excited by a laser beam and subsequently relax resulting in photon emission. Due to the so-called piezospectroscopic effect the frequency of the photons is shifted when the crystal is strained. The resolution limit of this technique is in the order of 100 MPa, which is sufficiently low to measure the stress state in the TGO. The technique is well-established for EB-PVD coatings, however, the use in APS systems is limited to coatings below 200 μm thickness as the microstructural defects in the APS TBC greatly attenuate the optical signal [114].

Investigations have shown that the compressive stress level in the TGO is changed during thermal treatment, and hence during the degradation of the TBCs [115,116]. This correlation indicates the potential of PS for remaining lifetime monitoring of EB-PVD TBC and thin APS systems.

19.6.4. Phosphor Thermometry

In phosphor thermometry (PT) the luminescence of thermographic phosphors consisting of rare earth ions (e.g. Dy) in a ceramic host (YSZ) is used to determine temperature. After excitation with a short pulse of UV light, the phosphor exhibits exponentially decaying luminescence at a longer wavelength. Decay time decreases with temperature and also intensity ratio of certain emission lines are temperature dependent. Both the properties are used to measure temperatures [117,118].

Using designed TBCs consisting of rare earth doped and pure YSZ layers it is also possible to get additional information about the system, for example, the loss of surface layers by an erosion process.

19.6.5. Infrared Cameras

Infrared cameras can be used to measure the temperature distribution on hot surfaces. Especially for YSZ, which is transparent below about 6 μm, longer wavelengths are essential to measure the surface temperature as otherwise mainly bond-coat temperatures will be measured. Of special interest is the detection of delamination cracks close to the interface TBC/bond-coat. Depending on crack size and crack opening a more or less pronounced temperature increase will be observed in the TBC above the crack. On-line monitoring of failures in TBCs might be possible in operating gas turbines by the use of fast infrared cameras [119].

During overhaul an inspection of TBC-coated gas turbine components is possible by the use of pulsed thermography. In this analysis the surface is heated by a pulsed energy source, such as a flash lamp array, and the surface temperature response is measured with an infrared camera [120]. Sub-surface structures as delaminations and pores will affect the flow of heat and can therefore be detected. Finally, the thermal wave interferometry is mentioned. In this technique a periodic heating of the surface by, for example, a laser diode is used. The thermal response allows the determination of different physical properties of the coatings as thermal diffusivity and conductivity [121].

19.7. Conclusions

Different types of TBCs, EB-PVD and APS YSZ topcoats with (platinum) aluminide and MCrAlY bond-coats, are applied in both aero and stationary gas turbines. Under operation conditions all TBCs show a degradation which is in many cases promoted by the growth of a TGO on the bond-coat. The lifetime of the coatings varies significantly with the applied testing conditions and the used coating systems. With respect to the testing conditions lifetime is typically reduced successively in the following sequence: isothermal, cyclic, in thermal gradients, under corrosive conditions.

At high surface temperatures above 1200°C the limited thermal capability of the YSZ becomes increasingly important. As a consequence new TBC materials with improved temperature capability are under development.

During the last decade a detailed understanding of several major failure mechanisms has been developed for both EB-PVD and APS systems. However, still open questions exist due to the complexity of TBCs and applied thermal and mechanical loading.

As an accurate prediction of the lifetime of TBCs is not yet possible, new tools for the monitoring of remaining lifetime are of increasing interest. Both improved lifetime prediction and new monitoring tools will contribute to a better integration of TBCs into future gas turbine design.

Acknowledgements

The author would like to thank Annette Carlson (for the contribution on failure mechanisms of PVD coatings). Also the support especially with pictures from Uwe Schulz (DLR, Germany), Rodney Wing, (Chromalloy, UK) and Dan Roth-Fagaraseanu (Rolls-Royce, D) is gratefully acknowledged.

Special thanks also to the members of the IWV1 at the Forschungszentrum Jülich GmbH, Germany, who made contributions to the present paper.

References

[1] P. Hancock and M. Malik, *Materials for Advanced Power Engineering Part 1*, Eds. D. Coutsouradis *et al.*, Kluwer Academic Publishers, Dordrecht, 1994, pp. 658–704.

[2] R.A. Miller, *Surf. Coat. Technol.*, 30 (1987) 1–11.

[3] E.C. Subbarao, *Science and Technology of Zirconia, Advances in Ceramics*, Vol. 3, Eds. A.H. Heuer and L.W. Hobbs, The American Ceramic Society, Columbus, OH, USA, 1981, pp. 1–24.

[4] R. Bürgel and I. Kvernes, *High Temperature Alloys for Gas Turbine and Other Applications*, Eds. W. Betz *et al.*, D. Reidel Publishing Company, Dordrecht, NL, 1986, pp. 327–356.

[5] S. Stecura, *Adv. Ceram. Mater.*, 1 (1) (1986) 68–76.

[6] S. Bose and J. DeMasi-Marcin, *J. Therm. Spray Technol.*, 6 (1) (1997) 99–104.

[7] W.A. Nelson and R.M. Orenstein, *J. Therm. Spray Technol.*, 6 (2) (1997) 176–180.

[8] M.B. Beardsley, *J. Therm. Spray Technol.*, 6 (2) (1997) 181–186.

[9] T.M. Yonushonis, *J. Therm. Spray Technol.*, 6 (1) (1997) 50–56.

[10] J. Wigren and L. Pejryd, *Proceedings of the 15th International Thermal Spray Conference*, ASM International, OH, USA, 1998, pp. 1531–1541.

[11] J. Thornton, Materials Forum 22 (1998) 159–181, V.Arnault, R. Mévrel, S. Alpérine, and Y.Jaslier, La Revue Métallurgie – CIT/Science et Génie des Matériaux 96 (5) (1999) 585–597.

[12] M.J. Stiger, N.M Yanar, M.G. Topping, F.S. Pettit and G.H. Meier, *Z. Metallkd.*, 90 (1999) 1069–1078.

[13] D. Stöver and C. Funke, *Mater. Process. Technol.*, 92–93 (1999) 195–202.

[14] U. Schulz, C. Leyens, K. Fritscher, M. Peters, B. Saruhan-Brings, O. Lavigne, J.-M. Dorvaux, M. Poulain, R. Mevrel and M. Caliez, *Aerosp. Sci. Technol.*, 7 (2003) 73–80.

[15] J.R. Nicholls, *MRS Bull.*, September (2003) 659–670.

[16] D.R. Clarke and C.G. Levi, *Annu. Rev. Mater. Res.*, 33 (2003) 383–417.

[17] C.G. Levi, *Curr. Opin. Solid State Mat. Sci.*, 8 (2004) 77–91.

[18] R.A. Miller, *J. Therm. Spray Technol.*, 6 (1) (1997) 35–42.

[19] H. Nickel, D. Clemens, W.J. Quadakkers and L. Singheiser, *Trans. ASME*, 121 (1999) 384–387.

[20] B.H. Pilsner, *NASA-CR-174851*, Lewis Research Centre, Cleveland, OH, (1985).

[21] W. Beele, N. Czech, W.J. Quadakkers and W. Stamm, *Surf. Coat. Technol.*, 94–95 (1997) 41–45.

[22] P. Fauchais, A. Vardelle and B. Dussoubs, *Proceedings of the International Thermal Spray Conference, Thermal Spray 2001*, Eds. C.C. Berndt, K.A. Khor and E.F. Lugscheider, ASM International, Materials Park, OH, USA, 2001, 1–32.

[23] W. Brandl, D. Toma, J. Krüger, H.J. Grabke and G. Matthäus, *Surf. Coat. Technol.*, 94–95 (1997) 21–26.

[24] H.J. Penkalla, J. Wosik, D. Clemens, W.J. Quadakkers and L. Singheiser, *Proceedings of the XV Physical Metallurgy and Materials Science Conference on Advanced Materials and Technologies* (AMT'98), Eds. J. Kusinski, I. Suliga and S. Kac, Krakow-Krynica, Poland, 17–21 May, 1998, 982–985.

[25] G. Schumacher, T. Rehfeldt, R. Vaßen and R.P. Wahi, *Scripta Mater.*, 43 (11) (2000) 963–968.

[26] N. Czech, M. Juez-Lorenzo, V. Kolarik and W. Stamm, *Surf. Coat. Technol.*, 108–109 (1998) 36–42.

[27] K.E. Rosteck, D. Löhe and O. Vöhringer, *Mater. Corrosion*, 50 (1999) 417–421.

[28] H.M. Tawancy, N. Sridhar, N.M. Abbas and D.S. Rickerby, *J. Mat. Sci.*, 25 (2000) 315–329.

[29] R.S. Parzuchowski, *Thin Solid Films*, 45 (1997) 349–355.

[30] M. Gell, K. Vaidyanathan, B. Barber, J. Cheng and E. Jordan, *Metall. Mater. Trans. A*, 30A (1999) 427–435.

[31] H. Lau, C. Leyens, U. Schulz and C. Friedrich, *Surf. Coat. Technol.*, 165 (3) (2003) 217–223.

[32] S.R. Choi, *Ceram. Eng. Sci. Proc.*, 19 (4) (1998) 293–301.

[33] M. Ahrens, S. Lampenscherf, R. Vaßen and D. Stöver, *J. Therm. Spray Technol.*, 13 (3) (2004) 432–442.

[34] N. Czech, *cfi/Ber. DKG*, 77 (9) (2000) 18–21.

[35] D.M. Gray, Y.-C. Lau, C. A. Johnson, M.P. Borom and W.A. Nelson, *US Patent* 6,180,184 , filed October 24, 1997.

[36] P. Bengston and J. Wigren, *Gas Turbine Materials Technology: Conference Proceedings from ASM Materials Solutions*, Eds. P.J. Maziasz and I.G. Wright, Rosemont, IL, USA, 1999, 92–101.

[37] H. Guo, R. Vaßen and D. Stöver, *Surf. Coat. Technol.*, 186 (2004) 353–363.

[38] P. Fauchais and M. Vardelle, *Thermal Spray 2003*, Eds. C. Moreau and B. Marple, ASM International, Materials Park, OH, USA, 2003, pp. 1165–1173.

[39] E. Hämälinen, N. Kriikka and G. Barbezat, *Thermal Spray 2003*, Eds. C. Moreau and B. Marple, ASM International, Materials Park, OH, USA, 2003, pp. 1249–1253.

[40] J. Wigren and P. Nylén, *Thermal Spray 2003*, Eds. C. Moreau and B. Marple, ASM International, Materials Park, OH, USA, 2003, pp. 1237–1242.

[41] O. Unal, T.E. Mitchell and A.H. Heuer, *J. Am. Ceram. Soc.*, 77 (4) (1984) 984–992.

[42] U. Schulz and M. Schmücker, *Mater. Sci. Eng. A*, 276 (2000) 1–8.

[43] P. Seserko, J. Hotz, J. Lemke, P.R. Smith and M. Mede, *ebeam 2002*, October 27–29 (2002) Paper 6, 1–6.

[44] J.R.V. Garcia and T. Goto, *Sci. Technol. Advan. Mater.*, 4 (2003) 397–402.

[45] G. Wahl, W. Nemetz, M. Giannozzi, S. Rushworth, D. Baxter, N. Archer, F. Cernuschi and N. Boyle, *Trans. ASME*, 123 (2001) 520–524.

[46] B. Preauchat and S. Drawin, *Surf. Coat. Technol.*, 142–144 (2001) 835–842.

[47] S. Raghavan, H. Wang, R.B. Dindiddie, W.D. Porter and M.J. Mayo, *Scripta Mater.*, 39 (8) (1998) 1119–1125.

[48] J.R. Nicholls, K.J. Lawson, A. Johnstone and D.S. Rickerby, *Mater. Sci. Forum*, 369–372 (1991) 595–606.

[49] R.W. Trice, Y. Jennifer Su, J.R. Mawdsley, K.T. Faber, A.R. Arellano-López, Hsin Wang and W.D. Porter, *J. Mater. Sci.*, 37 (2002) 2359–2365.

[50] J.I. Eldridge, C.M. Spuckler, K.W. Street and J.R. Markham, *Ceram. Eng. Sci. Proc.*, 23 (4) (2002) 417–430.

[51] J. Malzbender and R.W. Steinbrech, *J. Mater. Res.*, 18 (8) (2003) 1975–1984.

[52] A.G. Evans, D.R. Mumm, J.W. Hutchinson, G.H. Meier and F.S. Pettit, *Prog. Mater. Sci.*, 46 (2001) 505–553.
[53] T. Wakui, J. Malzbender and R.W. Steinbrech, *J. Therm. Spray Technol.*, 13 (3) (2004) 390–395.
[54] D. Basu, C. Funke and R.W. Steinbrech, *J. Mater. Res.*, 14 (12) (1999) 4643–4650.
[55] J.A. Thompson and T.W. Clyne, *Acta Mater.*, 49 (2001) 1565–1575.
[56] R.A. Miller, J.L Smialek and R.G Garlick, *Science and Technology of Zirconia, Advances in Ceramics*, Vol. 3, Eds. A.H. Heuer and L.W. Hobbs, The American Ceramic Society, Columbus, OH, USA, 1981, pp. 241–251.
[57] J. Ilavsky and J.K. Stalick, *Surf. Coat. Technol.*, 127 (2000) 120–129.
[58] U. Schulz, *J. Am. Ceram. Soc.*, 83 (4) (2000) 904–910.
[59] P.D. Harmsworth and R. Stevens, *J. Mater. Sci.*, 27 (1992) 611–615.
[60] C. Funke, B. Siebert, R. Vaßen and D. Stöver, *Proceedings of the United Thermal Spray Conference* (15 –19, September 1997, Indianapolis, IN), Ed. C.C. Berndt, ASM International, Materials Park, OH, USA, 1998, pp. 277–284.
[61] R.L. Jones, R.F. Reidy and D. Mess, *Surf. Coat. Technol.*, 82 (1996) 70–76.
[62] R. Vaßen, F. Tietz, G. Kerkhoff, R. Wilkenhöner and D. Stöver, *Proceedings of the 6th Liège Conference, Part III, Materials for Advanced Power Engineering*, Eds. J. Lecomte-Beckers, F. Schubert and P.J. Ennis, Forschungszentrum Jülich GmbH, Jülich, Germany, 1998, pp. 1627–1635.
[63] R. Vassen, X. Cao, F. Tietz, D. Basu and D. Stöver, *J. Am. Ceram. Soc.*, 83 (8) (1999) 2023–2028.
[64] G.W. Schäfer and R. Gadow, *Ceram. Eng. Sci. Proc.*, 20 (4) (1999) 291–297.
[65] M. Dietrich, V. Verlotski, R. Vaßen and D. Stöver, *Materialwiss. Werkst.*, 8 (2001) 669–672.
[66] D.R. Clarke and C.G. Levi, *Annu. Rev. Mater. Res.*, 33 (2003) 383–417.
[67] H. Lehmann, D. Pitzer, G. Pracht, R. Vaßen and D. Stöver, *J. Am. Ceram. Soc.*, 86 (8) (2003) 1338–1344.
[68] U. Bast and E. Schumann, *Ceram. Eng. Sci. Proc.*, 23 (4) (2002) 525–532.
[69] R. Vaßen, G. Pracht and D. Stöver, *Proceedings of the International Thermal Spray Conference 2002*, Verlag für Schweißen und verwandte Verfahren DVS-Verlag GmbH, Düsseldorf, 2001, pp. 202–207.
[70] R. Vaßen, G. Barbezat and D. Stöver, in *Materials for Advanced Power Engineering*, Eds. J. Lecomte-Becker, M. Carton, F. Schubert and P.J. Ennis, Schriften des Forschungszentrum Jülich, Reihe Energietechnik, Vol. 21, Part 1, 2002, pp. 511–521.
[71] R. Vaßen, X. Cao and D. Stöver, *Ceram. Eng. Sci. Proc.*, 22 (4) (2001) 435–442.
[72] R. Vaßen, M. Dietrich, H. Lehmann, X. Cao, G. Pracht, F. Tietz, D. Pitzer, D. Stöver, *Materialwiss. Werkst.*, 8 (2001) 673–677.
[73] U. Schulz, B. Saint-Ramond, O. Lavigne, P. Moretto, A. van Lieshout, A. Borger and J. Wigren, *Ceram. Eng. Sci. Proc.*, to be published.
[74] R. Vaßen, F. Traeger and D. Stöver, *Int. J. Appl. Ceram. Technol.*, 1 (4) (2004), 351–61.

[75] F. Ibegazene-Quali, R. Mevrel, C. Rio and Y. Renollet, *Mater. High Temp.*, 17 (2) (2000) 205–218.

[76] J.A. Haynes, E.D. Rigney, M.K. Ferber and W.D. Porter, *Surf. Coat. Technol.*, 86–87 (1996) 102–108.

[77] W.J. Brindley and R.A. Miller, *Surf. Coat. Technol.*, 43/44 (1990) 446–457.

[78] H.M. Tawancy, N. Sridhar, N.M. Abbas and D.S. Rickerby, *J. Mater. Sci.*, 35 (2000) 3615–3629.

[79] D. Clemens, V.R. Vosberg, H.J. Penkalla, U. Breuer, W.J. Quadakkers, H. Nickel and Fresenius, *J. Anal. Chem.*, 358 (1997) 122–126.

[80] R. Anton and W.J. Quadakkers, Thesis, RWTH Aachen, 2003.

[81] V.K. Tolpygo, D.R. Clarke and K.S. Murphy, *Surf. Coat. Technol.*, 146–147 (2001) 124–131.

[82] J.A. Haynes, M.K. Ferber and W.D. Porter, *J. Therm. Spray Technol.*, 9 (1) (2000) 38–48.

[83] K.F.J. Koolloos and J.M. Houben, *J. Therm. Spray Technol.*, 9 (1) (2000) 49–58.

[84] F. Traeger, R. Vaßen, K.-H. Rauwald and D. Stöver, *Adv. Eng. Mater.*, 5 (6) (2003) 429–432.

[85] B.R. Marple, J. Voyer, C. Moreau and D.R. Nagy, *Mater. High Temp.*, 17 (3) (2000) 397–412.

[86] C. Leyens, I.G. Wright and B.A. Pint, *Oxid. Met.*, 54 (5/6) (2000) 401–424.

[87] R.L. Jones, *J. Therm. Spray Technol.*, 6 (1) (1997) 77–84.

[88] Y. Yoshiba, K. Abe, T. Aranami and Y. Harada, *J. Therm. Spray Technol.*, 5 (3) (1996) 259–268.

[89] R. Bürgel, *Handbuch Hochtemperatur-Werkstofftechnik*, Vieweg Technik, Wiesbaden, 1998.

[90] P.E. Hodge, R.A. Miller and M.A. Gedwil, *Thin Solid Films*, 73 (1980) 447–453.

[91] B.R. Marple, J. Voyer, M. Thibodeau, D.R. Nagy and R. Vaßen, *J. Eng. Gas Turb. Power*, 2004, to be published.

[92] M.P. Borom, C.A. Johnson and L.A. Peluso, *Surf. Coat. Technol.*, 86–87 (1996) 116–126.

[93] C. Mercer, S. Faulhaber, A.G. Evans and R. Darolia, *Acta Mater.*, submitted.

[94] R.A. Miller, *J. Am. Ceram. Soc.*, 67 (8) (1984) 517–522.

[95] R.A. Miller, *J. Eng. Gas Turb. Power*, 111 (1989) 301–305.

[96] J.T. DeMasi, K.D. Sheffler and M. Ortiz, *Thermal Barrier Coating Life Prediction Model Development*, Phase I Final Report, Contract NAS3-23944, NASA CR 182230 (1989).

[97] S.M. Meier, D.M. Nissley and K.D. Sheffler, *Thermal Barrier Coating Life Prediction Model Development*, Phase II Final Report, Contract NAS3-23944, NASA CR 189111, (1991).

[98] A.M. Karlsson, J.W. Hutchinson and A.G. Evans, *J. Mech. Phy. Solid.*, 50 (2002) 1565–1589.

[99] A.M. Karlsson, J.W. Hutchinson and A.G. Evans, *Mater. Sci. Eng.*, A351 (2003) 244–257.

[100] M. Öchsner, Thesis, in Fortschr.-Ber. VDI Reihe 18 Nr. 263, VDI Verlag Düsseldorf, Germany, 2001.
[101] G.C. Chang, W. Phucharoen and R.A. Miller, Finite element thermal stress solutions for thermal barrier coatings, *Surf. Coat. Technol.*, 32 (1987) 307–325.
[102] M.Y. He, J.W. Hutchinson and A.G. Evans, *Mater. Sci. Eng.*, A345 (2003) 172–178.
[103] R. Vaßen, G. Kerkhoff and D. Stöver, *Mater. Sci. Eng.*, A303 (2001) 100–109.
[104] K. Sfar, J. Aktaa and D. Munz, *Mater. Sci. Eng.*, A333 (2002) 351–360.
[105] F. Traeger, M. Ahrens, R. Vaßen and D. Stöver, *Mater. Sci. Eng.*, A358 (2003) 255–265.
[106] W.A. Ellingson, J.G. Sun, C. Deemer, S. Erdman and C. Prested, *International Conference On Advances in Life Assessment and Optimization of Fossil Power Plants*, March 11–13, Orlando, FL, May 2002, https://www.osti.gov/servlets/purl/795035-gaT1hz/native/
[107] A. Fahr, B. Rogé, J.S.R. Giguère, K.I. McRae and J. Thornton, *Nondestructive Characterization of Materials XI, Proceedings of the 11th International Symposium*, Berlin, Germany, June 24–28, 2002, Eds. R.E. Green Jr., B.B. Djordjevic and M.P. Hentschel, Springer-Verlag, Berlin, Germany, 2002, pp. 565–573.
[108] X.Q. Ma, S. Cho and M. Takemoto, *Surf. Coat. Technol.*, 139 (2001) 55–62.
[109] J. Voyer, F. Girthofer and M.I. Boulos, *J. Therm. Spray Technol.*, 7 (2) (1998) 181–190.
[110] F. Traeger, private communication.
[111] K. Ogawa, N. Gotoh, D. Minkov and T. Shoji, *Review of Progress in Quantitative Nondestructive Evaluation*, Vol. 20, Eds. D.O. Thompson and D.E. Chimenti, American Institute of Physics, 2001, pp. 1140–1147.
[112] T. Hilpert, E. Ivers-Tiffee, S. Wagner and R. Oberacker, *Nondestructive Characterization of Materials XI, Proceedings of the 11th International Symposium*, Berlin, Germany, June 24–28, 2002, Eds. R.E. Green Jr., B.B. Djordjevic and M.P. Hentschel, Springer-Verlag, Berlin, Germany, 2002, pp. 725–731.
[113] J.R. Christensen, D.M. Lipkin, D.R. Clarke and K. Murphy, *Appl. Phys. Lett.*, 69 (1996) 3754.
[114] K.W. Schlichting, K. Vaidyanathan, Y.H. Sohn, E.H. Jordan, M. Gell and N.P. Padture, *Mater. Sci. Eng.*, A291 (2000) 68–77.
[115] J.P. Singh, B. Nair, D. Renusch, M. Sutaria and M. Grimsditch, *J. Am. Ceram. Soc.*, 84 (2001) 2385.
[116] J.A. Nychka and D.R. Clarke, *Surf. Coat. Technol.*, 146–147 (2001) 110–116.
[117] J.P. Feist, A.L. Heyes and J.R. Nicholls, *Proc. Inst. Mech. Eng.*, 215, G (2001) 333–341.
[118] M.M. Gentleman, D.R. Clarke, *Surface and Coatings Technology*, 188–189 (2004), 93–100.
[119] D.H. Lemieux, On-line thermal barrier coating monitoring for real-time failure protection and life maximization, *Semi-Annual Reports*, October 2001–March 2002, April 2003–September 2003, Prepared for US DOE, Cooperative Agreement No. DE-FC26-01NT41232.
[120] S.W. Shepard, T. Ahmed, B.A. Rubadeux, D. Wang and J.R. Lhota, *Insight*, 43 (9) (2001) 587–589.
[121] A. Bendada, *Meas. Sci. Technol.*, 13 (2002) 1946–1951.

Chapter 20

Polymer Films Produced by Plasma Polymerization

Norihiro Inagaki

20.1. Introduction

Plasma, which means low-temperature plasma, is chemically active toward organic materials such as monomers and polymers to initiate some chemical reactions. There are three types of the chemical reactions initiated by plasma: plasma polymerization, plasma treatment, and plasma graft copolymerization. When plasma interacts with monomer molecules, the monomer molecules are activated to initiate polymerization reactions. Polymers deposit in a shape of thin film on surface of substrates, and then all of the substrate surfaces are covered with the deposited thin films. As a result, surface properties of the substrates are drastically modified. The polymerization reaction initiated by plasma is called as "plasma polymerization", and is a useful technique for creation of new surfaces with functionalities. On the other hand, when plasma irradiates surfaces of polymeric materials, polymer chains existing on the surfaces are activated by interactions of the plasmas to initiate modification reactions. The modification reactions contain oxidation, nitration, fluorination, etc. As a result, some functional groups such as oxygen-, nitrogen-, and fluorine-containing groups are introduced into the polymer chains. The introduction reactions are restricted to thin layer of only a few hundred nanometers from the topmost layer of the surface, and the other deeper layers (bulk of the materials) are never modified by plasma irradiation. This process is usually called as "plasma treatment", and is a useful technique for modification of surface properties, especially polymeric materials. When surfaces of polymer materials are exposed to chemically inactive plasmas such as argon or helium plasmas, ion bombardment by the plasmas occurs to form radicals in polymer chains on the surfaces. The radicals can initiate polymerization reactions of monomers, when the surfaces are exposed to monomer solution or monomer vapor, and new polymers grow up from the radical sites on the surfaces. As a result, some polymers deposit on the surfaces, and surface properties of the polymeric materials are changed noticeably.

Materials Surface Processing by Directed Energy Techniques
Y. Pauleau (Editor)

This process is called "plasma graft copolymerization reactions", and is a useful technique for surface modification.

As described above, plasmas have capabilities to initiate polymerization reactions of organic molecules (plasma polymerization and plasma graft copolymerization), and also to introduce functional groups on polymer surfaces (plasma treatment). Plasma polymerization and plasma graft copolymerization deposit polymers in a shape of thin film (from a few nanometers to a few micrometers) on the polymeric substrate surfaces. The polymer deposition leads to remarkable changes in surface properties. Although plasma treatment does not deposit polymer on the substrate surface, functional groups can be introduced on polymer surfaces. The introduction reactions of functional groups are restricted to a thin layer as deep as a few hundred nanometers from the surface. As a result, the chemical composition of the surface layer is drastically changed as if new polymer deposits.

In this chapter, plasma polymerization, plasma treatment, and plasma graft copolymerization are picked out as surface modification processes using plasmas. The basic mechanism of the modification processes is interpreted, and also recent developments of the modification processes are discussed.

20.2. Plasma Polymerization

Plasma polymerization is a thin film-forming process, where thin films deposit directly on surfaces of substrates without fabrication. In the plasma polymerization process, monomers (low-molecular-weight molecules) are converted into polymers (high-molecular-weight molecules) with the assistance of plasma (growing of molecules), and the polymers deposit in a shape of thin films on substrate surfaces (fabrication of polymers). Therefore, plasma polymerization process contains two fundamental technologies, "growing of molecules" and "fabrication of polymers". In the chemical meaning, plasma polymerization is different from conventional polymerizations such as radical and ionic. The term "radical polymerization" means that propagating reactions of monomers in the polymerization step are initiated by radical species at polymer chain end. "Ionic polymerization" means chemical reactions propagated by ionic species in the polymerization step. Therefore, the adjectives "radical" and "ionic" signify a kind of propagating species in the polymerization step. However, the term "plasma polymerization" means that the propagating species in the polymerization step are not plasma, but the energy source for initiation of polymerization reactions is plasma. Radical species created from monomer molecules by plasma play an important role in plasma polymerization process. The main topics covered here are (1) mechanism of chemical reactions in plasma polymerization process, (2) chemical structure of plasma polymers and controlling factors in plasma polymerization reactions, and (3) new technology of plasma polymerization.

20.2.1. Mechanism of Chemical Reactions in Plasma Polymerization Process

Table 20.1 shows the atomic composition of plasma polymers when typical monomers used for radical and ionic polymerizations (conventional polymerization) are plasma polymerized [1,2]. There are some differences in atomic composition between the plasma polymers and conventionally polymerized polymers even though the same monomers are used for the plasma polymerization and conventional polymerization. For example, we watch the plasma polymerization of ethylene (C_2H_2). Polymer conventionally polymerized is polyethylene of which atomic composition is C:H=1:2. The atomic composition is identical to that of the starting monomer, ethylene, C_2H_2. On the other hand, plasma polymer of ethylene is clear and transparent thin film of which properties are similar to those of conventionally polymerized polymer, polyethylene. However, the atomic composition for the plasma polymer, as shown in Table 20.1, is $C_2H_{2.6}O_{0.4}$, which is different from that

Table 20.1: Atomic composition of plasma-polymerized polymers

Monomers used for plasma polymerization	Atomic composition of monomers	Atomic composition of plasma polymers
Ethylene	C_2H_2	$C_2H_{2.6}O_{0.4}$
Ethylene/N_2	C_2H_2/N_2	$C_2H_3N_{0.6}O_{0.8}$
Acetylene	C_2H_2	$C_2H_{1.6}O_{0.3}$
Acetylene/N_2	C_2H_2/N_2	$C_2H_{2.2}N_{0.5}O_{0.3}$
Acetylene/H_2O	C_2H_2/H_2O	$C_2H_{2.7}O_{0.6}$
Acetylene/N_2/H_2O	$C_2H_2/N_2/H_2O$	$C_2H_{2.9}N_{0.5}O_{0.7}$
Allene	C_3H_4	$C_3H_{3.7}O_{0.4}$
Allene/N_2	C_3H_4/N_2	$C_3H_{3.8}N_{0.7}O_{0.5}$
Allene/H_2O	C_3H_4/H_2O	$C_3H_{4.2}O_{0.6}$
Allene/N_2/H_2O	$C_3H_4/N_2/H_2O$	$C_3H_{24.4}N_{0.45}O_{0.6}$
Acrylonitrile	C_3H_3N	$C_3H_3NO_{0.4}$
Propyonitrile	C_3H_5N	$C_3H_{4.7}NO_{0.8}$
Propylamine	C_3H_9N	$C_3H_5NO_{0.4}$
Allylamine	C_3H_6N	$C_3H_{4.7}NO_{0.4}$
Ethyleneoxide	C_2H_4O	$C_2H_{2.9}O_{0.4}$

of conventionally polymerized polymer, polyethylene, C_2H_2. The differences are (1) lower hydrogen content (1.6) than polyethylene and (2) incorporation of oxygen atom (oxygen content of 0.4), although the monomer used, ethylene, possesses never oxygen atom. Similar differences in atomic composition can be observed in olefin monomers such as acetylene and allene, and vinyl monomers such as acrylonitrile and propyonitrile in Table 20.1. Furthermore, when ethylene is mixed with nitrogen, and served as a monomer for plasma polymerization, the plasma polymer formed from the mixture shows large difference in atomic composition with the plasma polymer from only ethylene. The atomic composition for the plasma polymer from the ethylene/N_2 mixture is $C_2H_3N_{0.6}O_{0.8}$, and that for the plasma polymer from only ethylene is $C_2H_{2.6}O_{0.4}$. Similar formation of nitrogen-containing plasma polymers can be observed in the acetylene/N_2 and allene/N_2 systems. Furthermore, when water vapor instead of nitrogen is mixed with acetylene or allene, the deposited plasma polymers show higher hydrogen and oxygen concentrations than the plasma polymers from acetylene or allene alone: The atomic composition for the plasma polymers from the ethylene/H_2O and acetylene/H_2O mixtures is $C_2H_{2.7}O_{0.6}$, and $C_3H_{4.2}O_{0.6}$, respectively; while, the composition for the plasma polymers from ethylene and acetylene alone is $C_2H_{1.6}O_{0.3}$ and $C_3H_{3.7}O_{0.4}$, respectively. These experimental results indicate that plasma polymers cannot be interpreted by a concept of the repeating unit of monomers used for plasma polymerization, because there is a large difference in atomic composition between the plasma polymer and the monomer used. Furthermore, molecules existing in the plasma zone used for the plasma polymerization, even nitrogen and water vapor, are incorporated and become constituents of the plasma polymers. Consequently, plasma polymerization reaction is not chain reaction of monomer molecules.

Yasuda [2,3] has proposed "atomic polymerization" as a concept of plasma polymerization reactions. The propagation reaction in plasma polymerization is not chain reactions through double bonds, but stepwise reactions of recombination between biradicals formed by plasma. When monomer molecules are injected into plasma, the molecules are bombarded by active species such as electrons and ions in the plasma to be broken down into small fragments with biradicals (radical formation). Then, the fragments are stepwise recombined each other, and grow up larger molecules (recombination). Such radical formation and recombination are repeated in plasma, and finally plasma polymers deposit on substrate surfaces. In an extreme case, starting molecules are fragmented into atoms, and as a result the sequence of the formed polymer chains is not identical to that of the starting molecules. Plasma polymerization reactions are schematically illustrated in Fig. 20.1. Question whether such reaction mechanism is reasonable has been discussed from X-ray photoelectron spectroscope (XPS) analysis of plasma polymers deposited from fluorine-containing monomers.

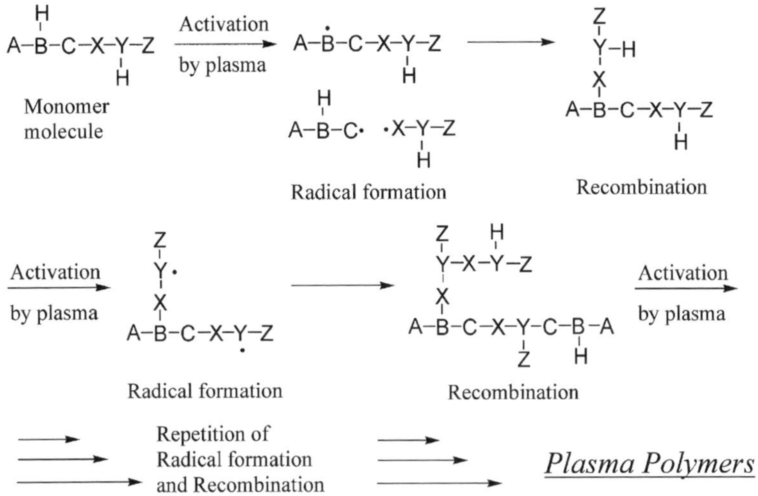

Figure 20.1: Overall plasma polymerization mechanism.

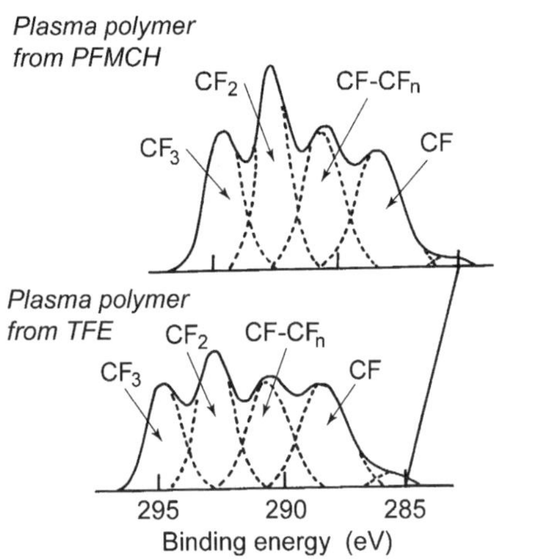

Figure 20.2: XPS (C1s) spectra of plasma polymers deposited from TFE and PFMCH.

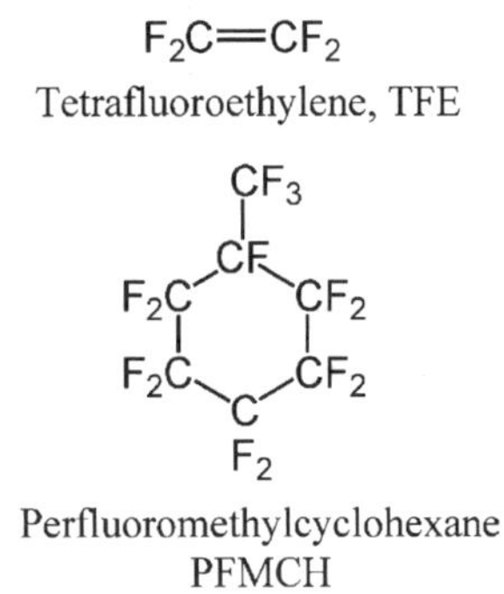

Figure 20.3: Chemical structure of TFE and PFMCH.

Fig. 20.2 shows XPS spectra for polymers plasma polymerized from fluorocarbons, tetrafluoroethylene (TFE) and perfluoromethylcyclohexane (PFMCH), of which chemical structure is shown in Fig. 20.3 [4]. If TFE and PFMCH are plasma polymerized according to chain reactions through double bond and ring opening of cyclohexane, the formed polymers should be composed of a sequence of CF_2—CF_2

Table 20.2: XPS spectra of plasma polymers from TFE and PFMCH

Monomers	W/FM parameter ($MJ\,kg^{-1}$)	C1s components (%)			
		Component #1	Component #2	Component #3	Component #4
TFE	220	23	26	23	25
PFMCH	190	19	32	24	23
	270	22	27	27	22

Component #1: CF at 288.0–288.4 eV; component #2: CF—CF_n at 290.3–290.8 eV; component #3: CF_2 at 292.4–292.8 eV; component #4: CF_3 at 294.4–294.8 eV.

for TFE and CF_2—CF_2—CF_2—CF_2—CF_2—CF(CF_3) for PFMCH. We expect XPS spectrum of CF_2 group (100%) for TFE and that of CF (14.3%), CF—CF_n (14.3%), and CF_2 (71.4%) for PFMCH. The two plasma polymers, as shown in Fig. 20.2, show four C1s components, CF, CF—CF_n, CF_2, and CF_3 groups. The concentration of their groups is tabulated in Table 20.2. The plasma polymer from TFE is not composed of CF_2—CF_2 alone but of the four components, CF, CF—CF_n, CF_2, and CF_3 groups. Similarly, the plasma polymers from PFMCH are composed of the four components, CF, CF—CF_n, CF_2, and CF_3 groups. Such disparity from the prediction of the monomer structure indicates that the propagation reaction in plasma polymerization process is not chain reaction. A concept of "atomic polymerization" is a reasonable mechanism to understand plasma polymerization reactions. "Atomic polymerization" can reasonably interpret a question on why nitrogen-containing groups are incorporated in the deposited polymers when a mixture of ethylene and nitrogen is plasma polymerized.

20.2.2. Controlling Factors for Plasma Polymerization Reactions

The concept of "atomic polymerization" tells us new ideas that how to control the plasma polymerization process. Monomer molecules are fragmented by action of plasma to form small fragments with radicals (radical formation), and then the fragments are recombined to grow up the size of molecules (recombination of radicals). Repetition of the radical formation and the radical recombination leads to the formation of plasma polymers. Therefore, the chemical composition of plasma polymers depends mainly on how monomer molecules are fragmented recombined in plasma. Fig. 20.4 shows XPS spectra for plasma polymers deposited from TFE as a function of the W/FM parameter [5].

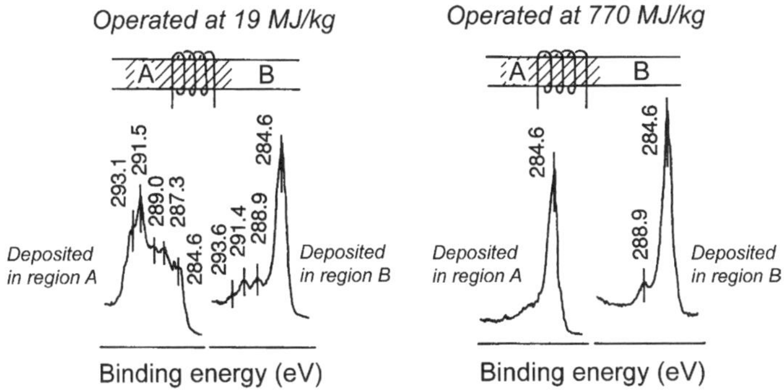

Figure 20.4: Effects of the W/FM parameter on plasma polymers deposited from TFE. The plasma polymerization processes were operated at W/FM parameters of 19 and 770 MJ kg^{-1}.

The W/FM parameter is defined as an apparent input energy of the plasma per the unit of monomer molecule in plasma polymerization, where W, F, and M are plasma power, the monomer flow rate, and the molecular weight of the monomer, respectively. In other words, the W/FM parameter means that how much energy is invested in the fragmentation process of monomer molecules. This conception regarding the W/FM parameter is firstly proposed by Yasuda [5]. We believe that heavy fragmentation of monomer molecules will occur at a high W/FM parameter. Fig. 20.4 compares polymers plasma polymerized at W/FM parameters of 19 and 770 MJ kg^{-1}. In the plasma polymerization process at a W/FM parameter of 19 MJ kg^{-1}, the deposited plasma polymer shows a complex XPS spectrum on which three peaks appear at 290.0, 291.5, and 293.1 eV. The three peaks are assigned to CF unit at 290.0 eV, CF_2 unit at 291.5 eV, and CF_3 unit at 293.1 eV. This spectrum indicates that the plasma polymer is composed of fluorocarbon chains of CF, CF_2, and CF_3 units. Furthermore, appearance of CF and CF_3 peaks in the XPS spectrum suggests that the detachment (formation of CF units) and rearrangement of fluorine atoms (formation of CF_3 units) occurred in the plasma polymerization process. While, the polymer plasma polymerized at a W/FM parameter of 770 MJ kg^{-1} shows a simple spectrum in which a peak alone appears at 284.6 eV (CH_2 and CH unit). This spectrum indicates that the polymer is not composed of fluorocarbons but hydrocarbons. Complete detachment of fluorine atoms from TFE molecules occurred in the plasma polymerization process to form polymers. This comparison between the plasma polymerization processes at W/FM parameters of 19 and 770 MJ kg^{-1} shows that even when the same monomer molecule is used for plasma polymerization, the chemical composition of the deposited plasma polymers depends strongly on how to operate the plasma polymerization process. In other words, how plasma fragments monomer molecules is an important factor to determine the chemical composition of deposited plasma polymers.

Conclusively, plasma polymerization is the stepwise process of radical formation and radical recombination. Fragmentation and rearrangement of monomer molecules also occur in the plasma polymerization process. As a result, the chemical composition of the deposited polymers is not analogous to that of the monomer used. Theoretically, if we can control the fragmentation and rearrangement reactions, plasma polymers whose chemical composition is tailored will be formed. However, anybody has not yet succeeded in controlling the fragmentation and rearrangement reactions. Many investigators have sought a new technique that can tailor the chemical composition of plasma polymers. The new technique sought by them is "pulsed plasma polymerization".

20.2.3. Pulsed Plasma Polymerization

Plasma polymers are entirely different in the chemical composition from those polymerized by conventional polymerizations such as radical and ionic, even when the same monomer molecules are used for plasma and conventional polymerizations. Plasma polymers in many cases are cross-linked, and their chemical structure cannot be represented in the reiteration of the used monomer unit. While, polymers polymerized conventionally are composed of the reiteration of the monomer unit. This discrepancy arises from different propagation reactions in plasma and conventional polymerizations. In conventional polymerization, monomer molecules are added to active sites at polymer chain ends to grow polymer chains. This is the propagation process in conventional polymerization, and the process is chain reactions. On the other hand, in the plasma polymerization process, monomer molecules are interacted with energetic species such as electrons and ions in plasma, and are fragmented into small fragments with radicals (radical formation). The fragments recombine with other fragments to grow up a molecular size of the fragments (recombination of radicals). The repetition of the radical formation and recombination of radicals yield plasma polymers. Therefore, the chemical composition of plasma polymers is not analogous to that of the monomer used, but is resultant of the recombination of the radicals. As a result, it is very difficult to predict and control the chemical composition of deposited plasma polymers from monomer used for plasma polymerization. Theoretically, if we can control the radical formation (including fragmentation and rearrangement reactions) from monomer molecules, we will succeed in tailoring the plasma polymers. However, we cannot yet succeed in controlling the radical formation.

If propagation reactions in plasma polymerization process proceed without fragmentation of monomer molecules, the formed polymers will retain completely the chemical composition of the monomer used, and structure-tailored plasma polymers will be formed. How to initiate such propagation reactions? Pulsed plasma polymerization may be a possible technique for formation of structure-tailored plasma polymers. The pulsed plasma polymerization means that electric discharge for plasma is pulsed on the ms–μs time scale, and plasma polymerization reactions are operated in such a state of pulsing the discharge. During on-period of the electric discharge, plasma polymerization occurs to deposit polymers on substrate surfaces, while during off-period of the electric discharge, not plasma polymerization but post-discharge reactions occur on surfaces of the deposited polymers. The post-discharge reactions are propagation reactions of monomers.

Inagaki and Yasuda [6] firstly showed that when plasma polymerization was operated in a state of pulsing electric discharge, acetylene and styrene could be polymerized even during an off-period of electric discharge. Polymerization reactions occurring in the off-period was due to the initiation by radicals trapped on surfaces of the deposited plasma polymers. This experimental work is the first illustration of pulsed plasma polymerization. After this experiment, especially for the last 5 years, many investigators have focused on pulsed plasma polymerization to form structure-tailored plasma polymers, and challenged the formation [7–18] of polymer surfaces having special groups such as amino, carboxyl, and hydroxyl groups. They expect that such surfaces will be superior in surface properties such as hydrophilicity, hydrophobicity, adhesion, immobilization of biomolecules and bioconjugates, because these properties are closely related to amino, carboxyl, and hydroxyl groups on polymer surfaces. We interpret here some of their challenges briefly.

Han and co-workers [7] applied pulsed plasma polymerization technique for thin film formation from 3-(pentafluorophenyl)pentafluoropropene-1, of which chemical structure is shown in Fig. 20.5. Of course, many people know that when

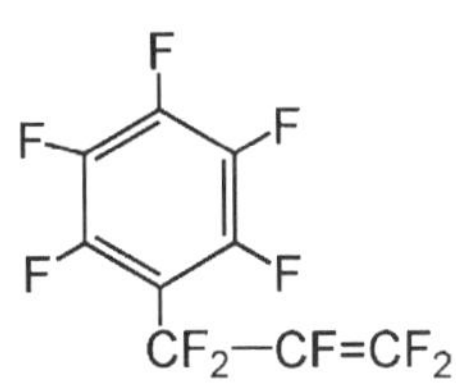

3-(pentafluorophenyl)
pentafluoropropene-1

Figure 20.5: Chemical structure of 3-(pentafluorophenyl)pentafluoropropene-1.

aromatic compounds are exposed to plasma, the aromatic molecules undergo the rupture of aromatic rings to form plasma polymers without aromatic groups. How much concentration of aromatic groups is retained in the formed plasma polymer is an important factor in pulsed plasma polymerization. Plasma polymerization of 3-(pentafluorophenyl)pentafluoropropene-1 were plasma polymerized in a state of pulsing electric discharge. Fig. 20.6 shows typical XPS (C1s) spectra for polymers formed by pulsed plasmas polymerization as well as continuous plasma polymerization. Polymers formed by continuous plasma polymerization, as shown in Fig. 20.6 (spectrum A), are composed of five peaks that are assigned to C—CF_n (component #1 at 286.1 eV), CF (aromatic) (component #2 at 288.2 eV), CF (non-aromatic) (component #3 at 289.3 eV), CF_2 (component #4 at 291.1 eV), and CF_3 groups (component #5 at 293.2 eV). In the XPS spectrum, there is no peak due to aromatic rings (π-bond shake-up satellite, which has the expected the peak appearance at 295.5 eV, component #6). On the other hand, polymers formed by pulsed plasma polymerization (on-time/off-time = 10/300 ms) show different C1s spectrum from that by continuous plasma polymerization. The spectrum is composed of five peaks, which are assigned to C—CF_n (component #1 at 286.1 eV), CF (aromatic) (component #2 at 288.2 eV), CF (non-aromatic) (component #3 at 289.3 eV), CF_2 (component #4 at 291.1 eV), CF_3 groups (component #5 at 293.2 eV), and π-bond shake-up satellite (component #6 at 295.5 eV). Surely, the XPS spectrum shows the peak appearance due to π-bond shake-up satellite.

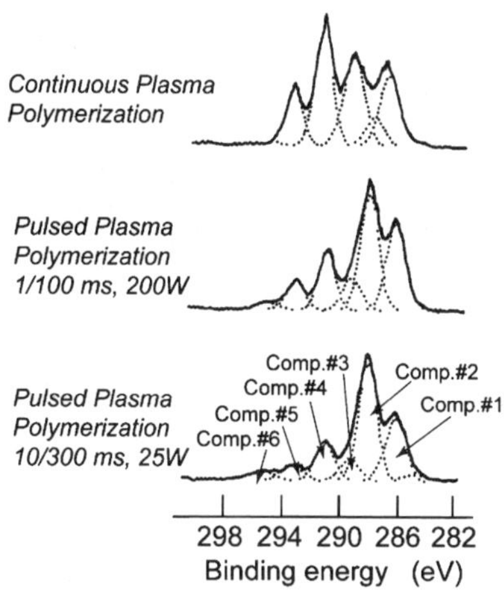

Figure 20.6: XPS (C1s) spectra of plasma polymers deposited from 3-(pentafluorophenyl) pentafluoropropene-1. The continuous plasma polymerization was operated at 200 W at a frequency of 13.56 MHz. The pulsed plasma polymerizations were operated at an on-time/off-time interval of 1/100 ms and at 25 W at 10/300 ms.

Infrared (IR) spectra also show characteristics of the plasma polymers (Fig. 20.7). Absorption peaks due to aromatic rings appear at 1509 (due to C═C ring stretching) and 999 cm^{-1} (due to C—F ring stretching), and the intensity of these peaks becomes strong with increasing the off-time period from 1/5 to 1/300 ms. From these experimental results, it is obvious that pulsing electric charge in the plasma polymerization process can contribute to formation of plasma polymers with aromatic rings.

The next example of the structure-tailored plasma polymers is thin film formation having cyano groups. Cyano compounds such as acrylonitrile, fumaronitrile, tetracyanoethylene, acetonitrile, etc. have been already plasma polymerized by many investigators to deposit thin films [19–21]. They expected that the thin films would show special capability of the cyano groups to make complex with metal ions. In practice, continuous plasma polymerization led to extensive modification reactions of cyano groups into amide, amino, and keten-imine groups, and could not succeed in the formation of thin films having cyano groups. Tarducci and co-workers [11] tried the pulsed plasma polymerization technique for the formation of cyano group-containing films. They used 2-cyanoethyl acrylate, of which the chemical composition was shown in Fig. 20.8, as a monomer for the pulsed plasma polymerization. Fig. 20.8 compares in the chemical composition between thin films formed by pulsed plasma polymerization and continuous plasma polymerization techniques. The pulsed plasma polymerization was operated at discharge

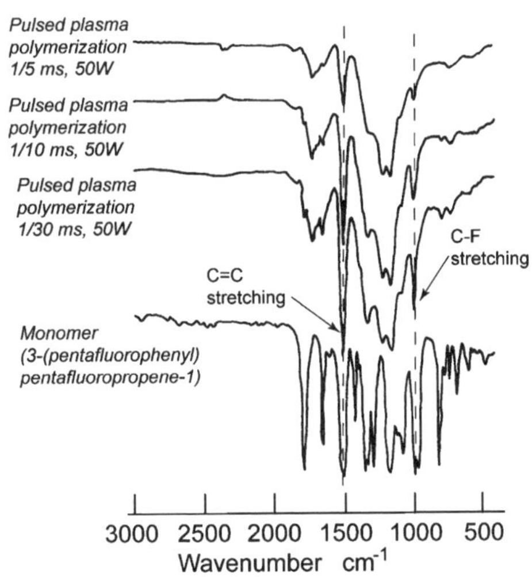

Figure 20.7: IR spectra of plasma polymers deposited from 3-(pentafluorophenyl) pentafluoropropene-1. The pulsed plasma polymerizations were operated at 50 W at on-time/off-time intervals of 1/5, 1/10, and 1/30 ms.

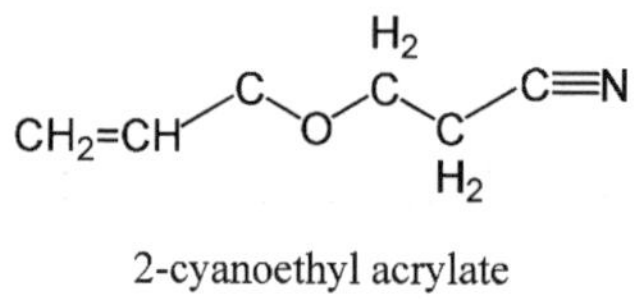

Figure 20.8: Chemical structure of 2-cyanoethyl acrylate.

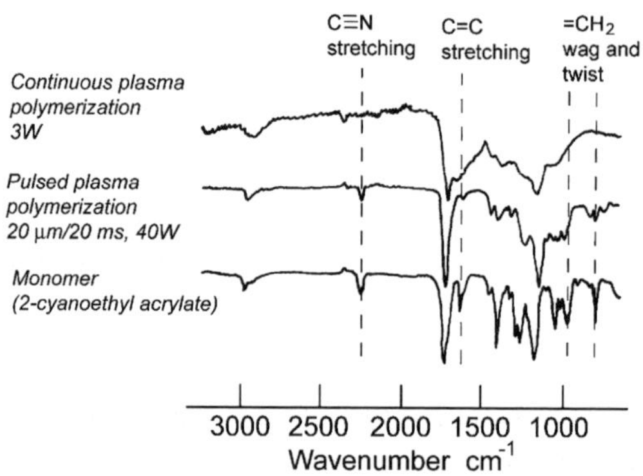

Figure 20.9: IR spectra of plasma polymers deposited from 2-cyanoethyl acrylate. The continuous plasma polymerization was operated at 3 W at a frequency of 13.56 MHz. The pulsed plasma polymerization was operated at 40 W at an on-time/off-time interval of 20 μm 20 ms^{-1}.

power of 40 W at 13.56-MHz frequency at a pulsed cycle of on-time period of 20 μs and off-time period of 20 ms. The IR spectrum for the formed film, as shown in Fig. 20.9 spectrum A, shows strong absorptions due to C═N stretching vibration (at 2254 cm^{-1}), C═O stretching vibration (at 1726 cm^{-1}), and C—O stretching vibration (at 1273 and 1185 cm^{-1}), indicating that cyano groups as well as ester groups are the main components of the film. On the other hand, the continuous plasma polymerization of 2-cyanoethyl acrylate shows dull IR spectrum shown in Fig. 20.9 spectrum B, even when the polymerization was operated at an extremely low electric discharge power of 3 W. The absorption peak due to C═N stretching vibration at 2254 cm^{-1} does not appear on the spectrum. From this comparison, they concluded that the pulsed plasma polymerization technique was effective in protecting cyano compounds from rearrangement reactions by plasma, and led to cyano group-containing films.

Acrylic acid and methacrylic acid are interesting monomers because of hydrophilic and biomedical functionalities of carboxylic acid groups. Many people have tried thin film formation from acrylic acid and methacrylic acid, but they do not

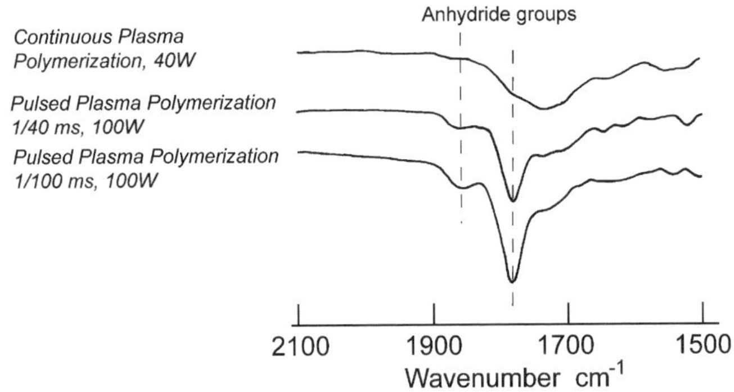

Figure 20.10: IR spectra of plasma polymers deposited from 2-cyanoethyl acrylate. The continuous plasma polymerization was operated at 3 W at a frequency of 13.56 MHz. The pulsed plasma polymerizations were operated at 100 W at on-time/off-time intervals of 1/40 and 1/100 ms.

have succeed in the formation of thin films having such functionalities. Carboxyl groups in acrylic acid and methacrylic acid are easy to be undergone rearrangement reactions when exposed to plasma, and carbon dioxides are eliminated from the molecules. Therefore, continuous plasma polymerization is a difficult way to form thin films containing carboxyl groups. Inagaki and co-workers [22] modified the continuous plasma polymerization process in order to inhibit the elimination of carbon dioxide from acrylic acid molecules during the plasma polymerization process. The modified process was that acrylic acid molecules were plasma polymerized under carbon dioxide atmosphere. Such modification of continuous plasma polymerization process was effective in retaining carboxyl groups, but did not lead to the complete retention (formation of plasma polymers without rearrangement of carboxyl groups). In place of the modified plasma polymerization process, a pulsed plasma polymerization technique has been applied for the formation of plasma polymers having carboxyl groups. Fraser and co-workers [12], and Hu *et al.* [13] separately have done pulsed plasma polymerization of acrylic acid and maleic anhydride, respectively. Plasma polymers formed from acrylic acid and maleic anhydride contain much concentration of carboxyl groups. The concentration of carboxyl groups in the plasma polymers showed strong effects of the electric discharge power as well as the duty cycle ([on-time period]/[sum of on-time and off-time periods]). Fig. 20.10 compares the plasma polymers formed from maleic anhydride by pulsed and continuous plasma polymerization. There appear characteristic absorption peaks due to anhydride groups appear at 1860 and 1780 cm^{-1}. Plasma polymers formed by a continuous plasma polymerization at an electric discharge power (rf power) of 40 W, as shown in Fig. 20.10, show no characteristic

Table 20.3: Concentration of carboxyl groups formed from maleic anhydride on PET film surfaces by pulses plasma polymerization

On-time/off-time (ms/ms)	rf power (W)	Carboxyl concentration (nmol cm^{-2})
1/100	1.0	16.2
1/40	2.4	14.7
5/40	11.1	2.9

peaks due to anhydride groups, but those formed at an rf power of 5 W barely shows the characteristic peaks. On the other hand, the pulsed plasma polymerization operated even at rf power of 100 W, when the duty cycle is 1/100 in ms, leads to plasma polymers, which show strong absorption peaks due to carboxyl groups at 1860 and 1780 cm^{-1} (Fig. 20.10). Finally, Table 20.3 shows improved effects of the duty cycle on the retention of carboxyl groups. The carboxyl concentration (COOH concentration) is 5 times higher at a duty cycle of 1/100 than at 5/40.

Amino group-containing polymers are attractive materials in biomedical application. The application involves the immobilization of biomolecules such as protein, enzymes, deoxyribonucleic acid; and the adhesion of cells on the polymer surface. Many investigators have been interested in such applications, and have tried the formation of amino group-containing polymer surfaces. Plasma polymerization technique is one of important processes for the formation of such polymer surfaces. The continuous plasma polymerizations of amine compounds such as allyl amine never led to successful formation of amino groups-containing polymers, because heavy modification reactions of amino groups into amide and cyano groups occur in plasma. Choukourov and co-workers [15] have showed important effects of pulsing electric discharge on formation of amino group-containing polymers. They used diaminocyclohexane (DACH) as a starting monomer, and the chemical composition of DACH is shown in Fig. 20.11. The pulsed plasma polymerization was performed at rf power of 5–30 W and duty cycle of 1–0.1. Amino groups formed on the plasma polymer surfaces were determined using a combination of chemical derivatization of nitrogen groups and XPS analysis of the nitrogen derivatives. Practically, primary and secondary amino groups in the plasma polymers were modified into fluorine-containing groups, and analyzed by XPS. The concentration of the primary and secondary amino groups was calculated from fluorine intensity in the XPS spectra. Fig. 20.12 shows typical effects of pulsing on formation of primary and secondary amino groups in plasma polymerization of DACH. The formation of primary and secondary amino groups is strongly influenced by the on-time period even if the duty cycle is a constant of 0.1. The maximum concentrations of the primary and secondary amino groups are only 8% and

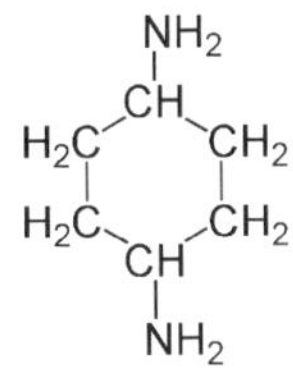

Diaminocyclohexane
DACH

Figure 20.11: Chemical structure of DACH.

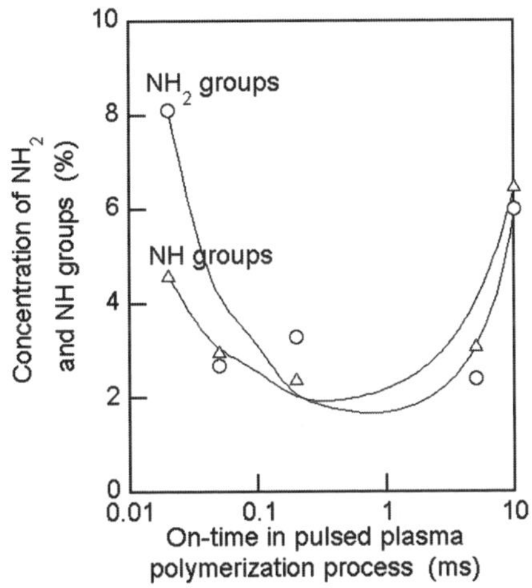

Figure 20.12: Concentration of primary (NH_2) and secondary amino (NH) groups in the plasma polymers formed from DACH as a function of the on-time in pulsed plasma polymerization processes.

6% of all the nitrogen groups in plasma polymers formed from DACH, respectively. Most of nitrogen groups in plasma polymers formed from DACH is derivatives such as amide and cyano groups. The retention of amino groups from DACH molecules to plasma polymers during the pulsed plasma polymerization process is extremely small (6–8%). From this viewpoint, we are skeptical about superiority of the pulsed plasma polymerization.

20.3. Plasma Treatment

Polymeric materials are exposed to non-polymer-forming plasmas such as argon, oxygen, and nitrogen plasmas, active species in the plasmas interact with the surfaces of the polymeric materials to modify the chemical composition of the surfaces. As a result, surface properties such as hydrophilicity and adhesion are drastically changed by the plasma exposure in itself. This process is called as "plasma treatment", and has been commercially used as one of effective surface modification techniques of polymeric materials. Plasma treatment leads to formation of functional groups such as oxygen- and nitrogen-containing groups on surfaces of polymeric materials. The formation reactions of functional groups are restricted to a thin layer as deep as a few hundred nanometers from the surface. As result, a topmost layer of polymeric surfaces is modified, and a new polymeric layer, whose chemical composition is different from the original, appears, although plasma treatment does not deposit new polymers on substrate surfaces. From easy operating of plasma treatment process, plasma treatment is one of the most useful techniques for the surface modification of polymeric materials. Main topics in this section are (1) mechanism of surface modification process by plasmas, (2) influences of plasmas on surface modification of polymeric materials, (3) influences of chemical composition of polymeric materials on modification reactions, and (4) remote plasma treatment for effective surface modification.

20.3.1. Mechanism of Surface Modification Process by Plasmas

When polymeric materials are exposed to plasmas for a few seconds or minutes, surfaces of the materials show large changes in surface properties such as hydrophilicity and hydrophobicity. Table 20.4 shows changes in contact angle of water on polyethylene and polypropylene films when these films are exposed to plasmas [23]. By exposure to argon and oxygen plasmas, polyethylene film surfaces show large decrease in contact angle of water from 94 to 36 degrees (by exposure to argon plasma) and 32 degrees (by exposure to oxygen plasma). Polypropylene film surfaces also show similar decrease in contact angle of water from 98 to 50 degrees (by exposure to argon plasma) and 60 degrees (by exposure to oxygen plasma). On the other hand, when polyethylene and polypropylene film surfaces are exposed to CF_4, C_2F_6, and SF_6 plasmas instead of argon and oxygen plasmas, the film surfaces show large increases in contact angle of water. The contact angle on the polyethylene film surfaces increases from 94 to 121 degrees (by exposure to CF_4 plasma), 114 degrees (by exposure to C_2F_6 plasma), and 117 degrees (by exposure to SF_6 plasma) (Table 20.4). Similar decreases in contact angle occur in exposing polypropylene film surfaces to CF_4, C_2F_6, and SF_6 plasmas. The contact angle decreases from 98 to 123 degrees (by exposure to CF_4 plasma), 116 degrees (C_2F_6 plasma), and 117 degrees (SF_6 plasma) (Table 20.4). These changes in contact

Table 20.4: Contact angle of water on polyethylene and polypropylene film surfaces modified by plasmas

	Contact angle of water (degree)	
Plasma	Polyethylene film	Polypropylene film
Ar	36	50
O_2	32	60
CF_4	121	123
C_2F_6	114	116
SF_6	117	117
None	94	98

angle indicate that plasma exposure can modify surface properties of polymeric materials. What species in the plasmas make such modification?

Plasma contains many kinds of active species that can initiate chemical reactions against polymeric materials. These active species existing in plasmas are electrons, negative and positive ions, and radicals, which can interact with polymeric materials to initiate modification reactions. Which species contribute to the surface modification of polymeric materials, electrons, negative and positive ions, or radicals? Electrons and ions are active in electric properties, but radicals are neutral in electric properties. Therefore, radicals are distinguished from electrons and ions by electric properties. Radicals in plasma could be isolated from the plasma using two electrodes with electrically negative and positive potentials against the electrode for the initiation of plasma discharge. Fig. 20.13 shows a special reactor for separation of radicals alone from plasma to modify surfaces of polymeric materials [24]. Nitrogen gas is injected at an end of the reactor, and nitrogen plasma is initiated by electric power at 20-kHz frequency. Active species involving electrons, nitrogen ions, and nitrogen radicals in the nitrogen plasma as well as nitrogen molecules travel in the direction of downstream, and reach two mesh electrodes, which are placed at the downstream of nitrogen plasma and which are electrically biased against the electrode for the plasma initiation. Charged species such as electrons and ions are trapped with the mesh electrodes, and radicals and nitrogen molecules that are neutral in electric properties pass through the mesh electrodes. Using the special reactor, nitrogen radicals can be isolated from nitrogen plasma, and interacted with polyethylene films. Polyethylene films were exposed to nitrogen radicals or nitrogen plasma using the special reactor shown in Fig. 20.13, and the contact angle on the exposed film surfaces were measured [24]. Table 20.5 compares the surface energy between the modification reactions with nitrogen radicals and those

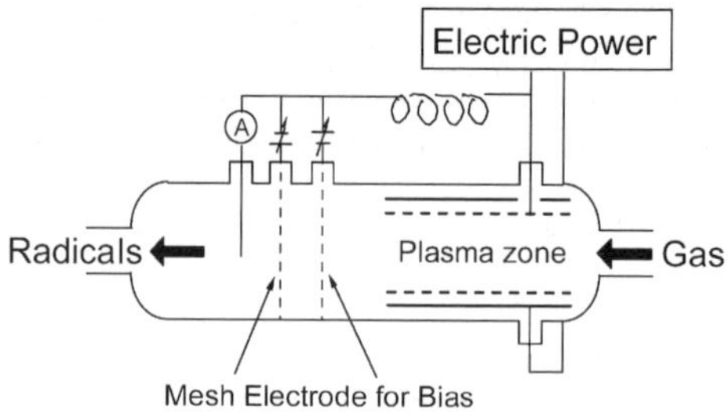

Figure 20.13: Schematic presentation of radical beam generator.

Table 20.5: Surface energy of polyethylene film surfaces modified by nitrogen radical and nitrogen plasma

		Surface energy (mJ m^{-2})		
Active species	Modification time (min)	γ_s	γ_s^d	γ_s^p
Nitrogen radicals	5	49.6	29.7	19.9
	10	59.5	22.5	37.0
	15	57.1	23.6	33.5
	30	54.9	23.8	31.1
Nitrogen plasma	10	58.2	23.2	35.0
	30	54.3	24.3	30.0
None	–	32.9	30.7	2.2

with nitrogen plasma [24]. The surface energy, γ_s, increases from 32.9 to 49.6–57.1 Jm m^{-2} for the polyethylene film surfaces exposed to nitrogen radical. On the other hand, nitrogen plasma also leads to a large decrease in the surface energy, γ_s, from 32.9 to 54.3–58.2 mJ m^{-2}. This comparison indicates that the surface energy, γ_s, for the polyethylene film surfaces exposed to the nitrogen radicals alone is the same magnitude as that for polyethylene films exposed to the nitrogen plasma. For example, the surface energy, γ_s, for polyethylene films exposed to the nitrogen radicals for 10 and 30 min is 59.5 and 54.9 mJ m^{-2}, respectively, which correspond to 58.2 and 54.3 mJ m^{-2} for polyethylene films exposed to the nitrogen plasma for 10 and 30 min, respectively. This comparison emphasizes that the nitrogen radicals rather than nitrogen ions play an important role in the modification of polyethylene film surfaces. Furthermore, this assumption is supported by the chemical analysis of the polyethylene films exposed to oxygen radical and oxygen

Table 20.6: Concentration of oxygen functional groups formed on polyethylene film surfaces by oxygen radical and oxygen plasma

	Concentration of oxygen functional groups (number/100 carbons)		
Active species	C=O groups	COOH groups	OH groups
Oxygen radical	4.5	1.6	0
Oxygen plasma	5.3	1.6	0.02

plasma. Table 20.6 compares the chemical composition of oxygen functional groups formed on the polyethylene film surfaces by the oxygen radical and oxygen plasma [24]. The oxygen functional groups that were formed by exposure to the oxygen radicals and the oxygen plasma were analyzed by the chemical derivatization method. The polyethylene films exposed to the oxygen radicals, as shown in Table 20.6, contain hydroxyl, carbonyl, and carboxyl groups as oxygen functional groups. The composition of these groups and the relative concentration are almost same as the polyethylene films exposed to the oxygen plasma (Table 20.6).

Conclusively, these correspondences between the polyethylene films exposed to the nitrogen and oxygen radicals and the films exposed to nitrogen and oxygen plasmas suggest that radical species rather than ion species in plasmas may contribute to the modification reactions.

20.3.2. Influences of Plasmas on Surface Modification of Polymeric Materials

Plasma contains activated species such as electrons, ions, radicals, and photons, which are able to initiate chemical reactions on polymer surfaces. Main reactions initiated by plasmas are etching reactions of polymer surfaces as well as introduction reactions of new functional groups onto polymer surfaces. The introduction reactions are due to recombination of carbon radicals formed on the polymer surfaces with activated species in the plasma such as oxygen, nitrogen atoms, etc. On the other hand, the etching reactions are due to degradation of polymer chains on the surfaces. As a result, topology of the surfaces is changed. Argon, oxygen, nitrogen, and ammonia plasmas are frequently used for hydrophilic modification of polymeric surfaces. However, details of the modification reactions by the plasmas are not yet completely interpreted. Such fundamental matters to be interpreted are (1) Is there any difference in the hydrophilic modification among these plasmas? (2) Is there any difference in the etching process among these plasmas? From above

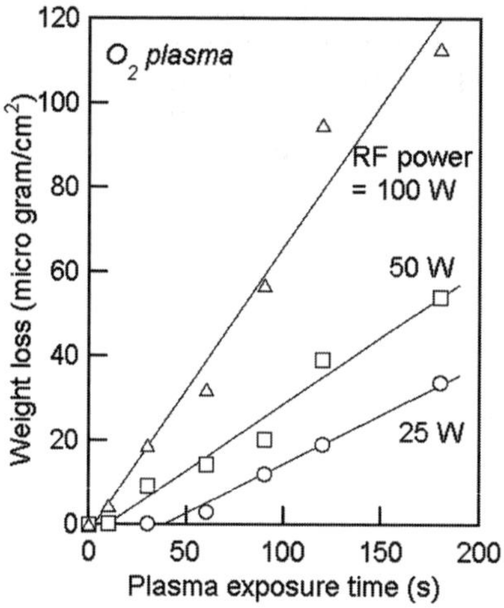

Figure 20.14: Weight loss from PET film surfaces by exposing to the O_2 plasma as functions of the rf power and plasma exposure time.

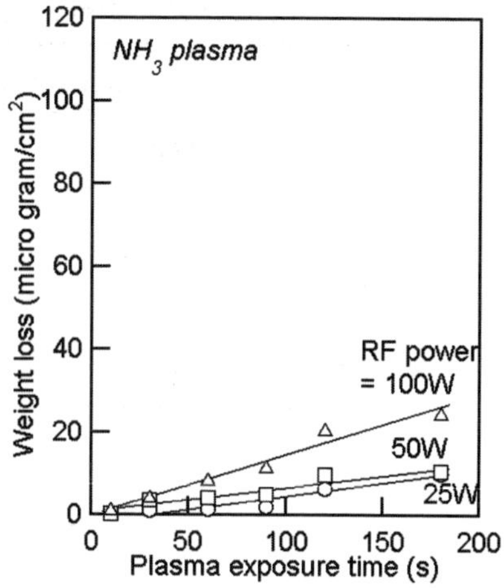

Figure 20.15: Weight loss from PET film surfaces by exposing to the NH_3 plasma as functions of the rf power and plasma exposure time.

viewpoint, this section focuses on modification reactions initiated on polyethylene terephthalate (PET) film surfaces by five plasmas containing argon, oxygen, hydrogen, nitrogen, and ammonia plasmas. What was different in the introduction reactions and etching reactions among the five plasmas was investigated using contact angle meter, XPS, and a scanning probe microscope.

PET was used as a polymeric material for plasma modification, and was exposed to five plasmas, argon, oxygen, hydrogen, nitrogen, and ammonia plasmas. Figs. 20.14 and 20.15 show typical results of weight loss caused by exposing to the O_2 and NH_3 plasmas as functions of the plasma exposure time and the rf power, respectively [25]. A magnitude of weight loss, as shown in Figs. 20.14 and 20.15, is linear with the plasma exposure time. From a slop of the linear relationship between the weight loss and the plasma exposure time, an etching rate (in nm/s) was calculated. The etching rate by the five plasmas, Ar, O_2, H_2, N_2, and NH_3 plasmas, is summarized in Table 20.7. There was large difference in the etching rate among the five plasmas. The O_2 plasma (2.3 nm s^{-1} at 50 W) is the highest rate of the five plasmas, and Ar, N_2, and NH_3 plasmas (0.79, 0.93, and 0.58 nm s^{-1} at 50 W, respectively) show slow rate. The H_2 plasma (1.15 nm s^{-1} at 50 W) is in middle among the five plasmas. Fig. 20.16 shows the etching rate as functions of the rf power and the plasmas. The etching rate, as shown in Fig. 20.16, is linear with a magnitude of the rf power, indicating that degradation reactions are accelerated at high rf power.

Table 20.7: Etching rate of PET film surfaces by plasmas

Plasma	Average of etching rate from PET film surfaces ($nm\,s^{-1}$)		
	At 25 W	At 50 W	At 100 W
Ar	0.72	0.79	1.15
O_2	1.15	2.30	5.68
H_2	0.43	1.15	2.09
N2	0.22	0.93	1.22
NH_3	0.36	0.58	1.22

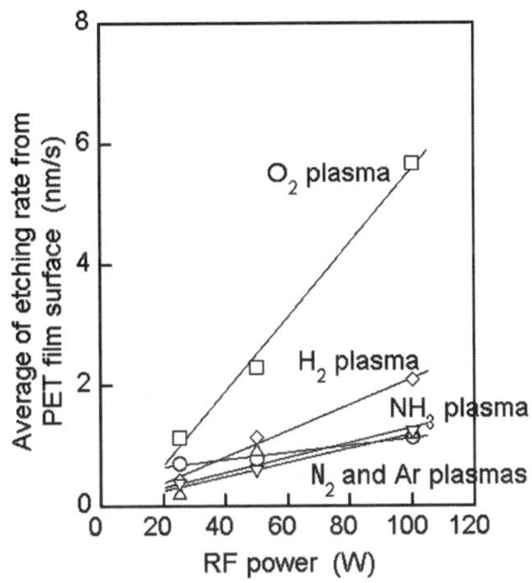

Figure 20.16: Average etching rate from PET film surfaces as functions of the plasmas and rf power.

Topographic figures for the plasma-modified PET film surfaces were scanned with a scanning probe microscope. Figs. 20.17–20.19 show typical pictures for the original and plasma-modified PET film surfaces [25]. The original PET film in Fig. 20.17 shows surfaces inlaid wholly with conical protuberances, and the surface roughness, Ra, for the original PET film was 1.20 nm. The O_2 plasma-modified PET film in Fig. 20.18 also shows similar topographic picture to that of the original PET film. The surface roughness, Ra, is 1.75 nm. There is less difference in surface roughness between the O_2 plasma-modified PET film and the original PET film, although heavy etching occurs on the PET film surfaces. On the other hand, the NH_3 plasma-modified PET film shows largely different surface from that of the original PET film and of the O_2 plasma-modified PET films (Fig. 20.18).

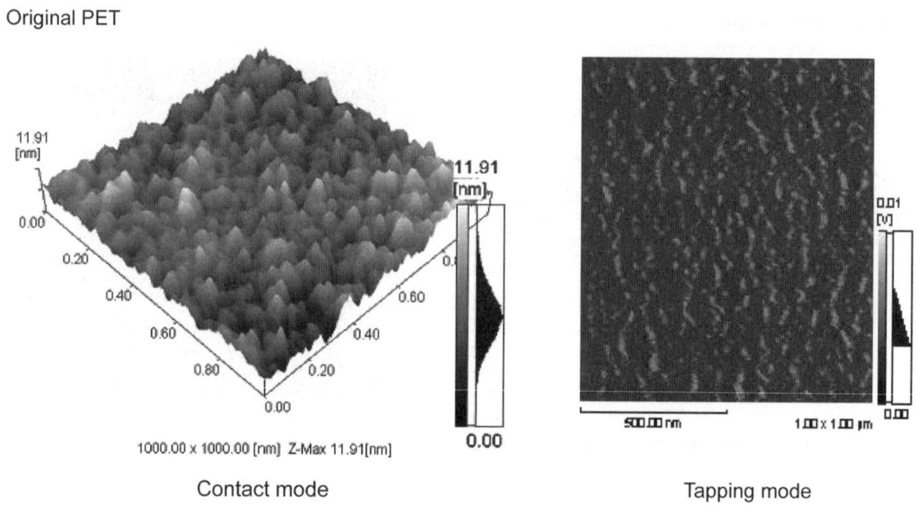

Figure 20.17: Topographical figures for the original PET film surface.

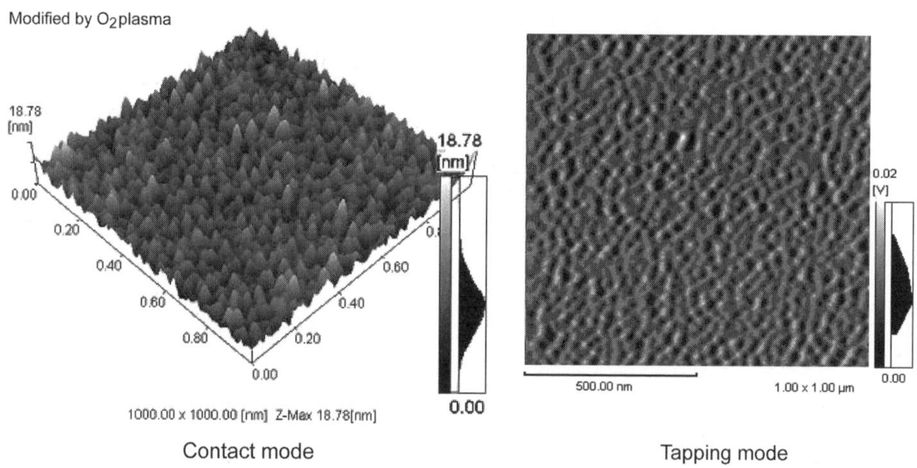

Figure 20.18: Topographical figures for the original PET film surface modified by O_2 plasma at 50 W for 60 s.

The surface roughness, Ra (7.56 nm) is 4 times larger than that of the O_2 plasma-modified PET film (1.75 nm), although the NH_3 plasma is very slower in etching rate than the O_2 plasma. The comparison between the O_2 plasma and NH_3 plasma modifications suggests that the topographic figures may be closely related to what kind of the plasmas was used for the modification. Table 20.8 summarizes the

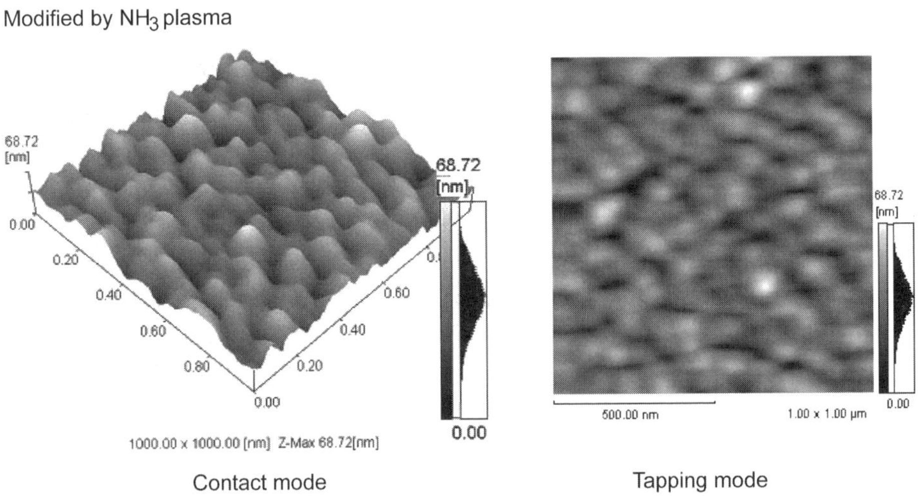

Figure 20.19: Topographical figures for the original PET film surface modified by NH_3 plasma at 50 W for 60 s.

Table 20.8: Surface roughness of plasma-modified PET film surfaces

	Plasma modification			Surface roughness	
Plasma	rf power (W)	Treatment time (s)	Film thickness etched by plasma (nm)	Ra (nm)	Ry(nm)
Ar	50	60	47.5	2.26	25.6
O_2	50	60	138	1.75	18.8
H_2	50	60	69	3.31	41.5
N_2	50	60	56	6.81	48.6
NH_3	50	60	34.5	7.56	68.7
None	–	–	–	1.20	11.9

Ra: arithmetical average of surface roughness.
Ry: Maximum distance between the top of mountains and the bottom of valleys.

surface roughness for the PET film surfaces modified with the five different plasmas: Ar, O_2, H_2, N_2, and NH_3 plasmas. The five plasma modifications never show similar Ra but different ones each other. Ra is in the order of NH_3 plasma > N_2 plasma > H_2 plasma > Ar plasma > O_2 plasma. Therefore, heavy etching reactions do not always lead to large increase in surface roughness. There are two categories of the etching process by the plasmas: one is high etching rate but less change in topographic figure before and after the plasma modification. The O_2

plasma belongs to this category. We believe that the O_2 plasma will etch homogeneously whole surface, and as a result there is less change in topographic figure. The other category is low etching rate but large changes in topographic figure. The NH_3 and N_2 plasmas belong to this category. We believe that the NH_3 and N_2 plasmas will not homogenously etch whole surface of the PET film, but will etch special parts of the PET surface. As a result, the etching rate is slow but surface roughness become extremely large after the plasma modification.

Conclusively, what plasma is used for surface modification of polymeric materials is an important factor to control surface topography.

The PET film surfaces were exposed to the five plasmas, Ar, O_2, H_2, N_2, and NH_3 plasmas. The surfaces were evaluated in water contact angle as functions of the plasma exposure time and rf power. Figs. 20.20 and 20.21 show typical results for the O_2 plasma-modified and NH_3 plasma-modified PET film surfaces [25]. In Fig. 20.20, when the PET film surface is exposed to the O_2 plasma for a short time of 10 s, the contact angle of water on the surface decreases largely from 78 to 47 degrees (at an rf power of 25 W), 41 degrees (at 50 W), and 39 degrees (at 100 W), thereafter, decreases gradually with increasing the plasma exposure time; and leveled off at a constant angle of 36 degrees (at 25 W), 25 degrees (at 50 W), and 21 degrees (at 100 W) after plasma exposure time more than 90 s. The NH_3 plasma exposure also leads to similar decrease in contact angle (Fig. 20.21): A contact

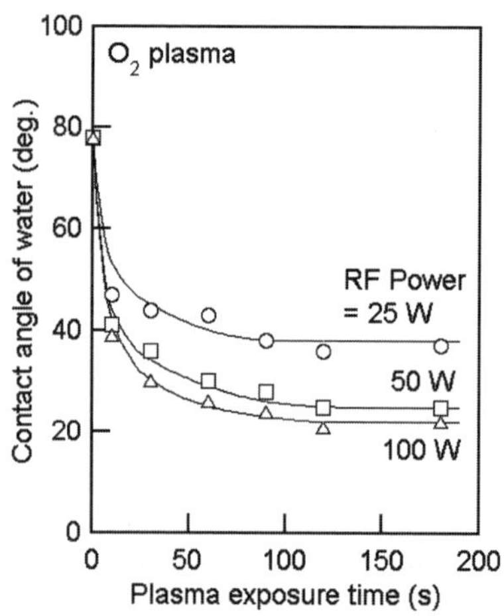

Figure 20.20: Contact angle of water on the O_2 plasma-modified PET film surfaces as functions of the rf power and plasma exposure time.

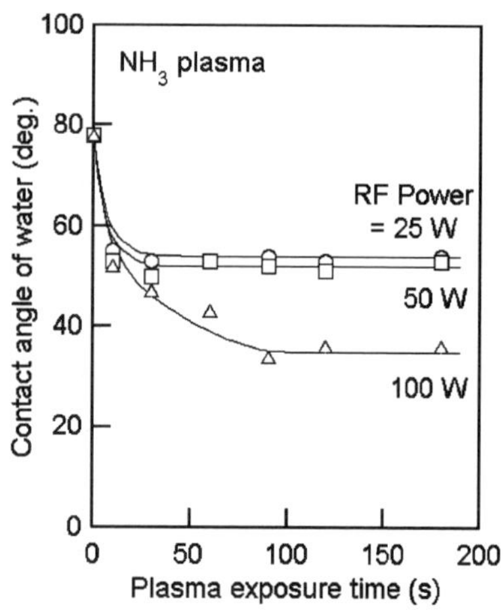

Figure 20.21: Contact angle of water on the NH_3 plasma-modified PET film surfaces as functions of the rf power and plasma exposure time.

angle in exposing to the NH_3 plasma for a short time of 10 s is to 55 degrees at 25 W, to 53 degrees at 50 W, and to 52 degrees at 100 W. Thereafter, small but unceasing decrease continues up to a plasma exposure time of 90 s, and the contact angle reaches a constant degree of 53 degrees at 25 W, 51 degrees at 50 W, and 36 degrees at 100 W. The two figures show that the modification of the PET film surfaces is closely related to what kind of plasmas was used, how much the rf power was used, and how long the PET surfaces were exposed to the plasma. And the surface modification reactions are accomplished already at a plasma exposure time of 120 s. The contact angle at exposure time of 120 s is summarized in Table 20.9 as functions of plasmas and rf power. The effectiveness is closely related to what kind of the plasmas was used, and is in the order of O_2 plasma > N_2 plasma > Ar plasma > H_2 plasma > NH_3 plasma. It is as a matter of course that the plasmas at higher rf powers are effective in decreasing the contact angle.

What functional groups were formed by the plasma exposure was determined using XPS at a take-off angle of 20 degrees. Table 20.9 shows the O/C and N/C atomic ratios for the PET film surfaces modified by the five plasmas [25]. The specimens provided for the XPS measurements are the same as those used for the

Table 20.9: Atomic composition for plasma-modified PET film surfaces

	Plasma modification		Contact angle	Atomic composition	
Plasma	rf power (W)	Modification time (s)	(degree)	O/C ratio	N/C ratio
Ar	25	120	47	–	–
	50	120	45	0.37	0.00
	100	120	32	–	–
O_2	25	120	36	–	–
	50	120	25	0.62	0.00
	100	120	21	–	–
H_2	25	120	51	–	–
	50	120	47	0.39	0.00
	100	120	45	–	–
N_2	25	120	38	–	–
	50	120	28	0.58	0.10
	100	120	21	–	–
NH_3	25	120	53	–	–
	50	120	51	0.35	0.03
	100	120	36	–	–
None	–	–	78	0.36	0.00

contact angle measurements. The four plasma modifications except for the NH_3 plasma modification leads to increase in the O/C atomic ratio from 0.36 to 0.35–0.62. This increase suggests that new oxygen-containing groups are formed on the film surfaces. Furthermore, the N_2 and NH_3 plasma modifications show the formation of some nitrogen-containing groups on the surfaces. The N/C atomic ratios for the N_2 plasma-modified and NH_3 plasma-modified PET film surfaces are 0.10 and 0.03, respectively. These N/C atomic ratios for the N_2 plasma-modified and NH_3 plasma-modified PET film surfaces are fairly lower than their O/C atomic ratios. The nitrogen-containing groups may not be predominant products formed by the N_2 and NH_3 plasmas.

XPS (C1s, O1s, and N1s) spectra were scanned at a take-off angle of 15 degrees to emphasize components at the topmost layer. Figs. 20.22 and 20.23 show typical C1s and O1s spectra for the O_2 and NH_3 plasma-modified PET film surfaces [25]. The C1s spectrum for the original PET film is decomposed into groups, at 286.7 eV (component #C2) due to —$\underline{C}H_2$—O— groups in ester linkages, at 289.1 eV (component #C3) due to —$\underline{C}$(O)—O— groups in ester linkages, and at 291.5 eV (component #C4) due to π–π^* shake-up satellite in phenyl groups. Their concentration of these components without the component #C4 is 62%, 19%, and 19%, respectively.

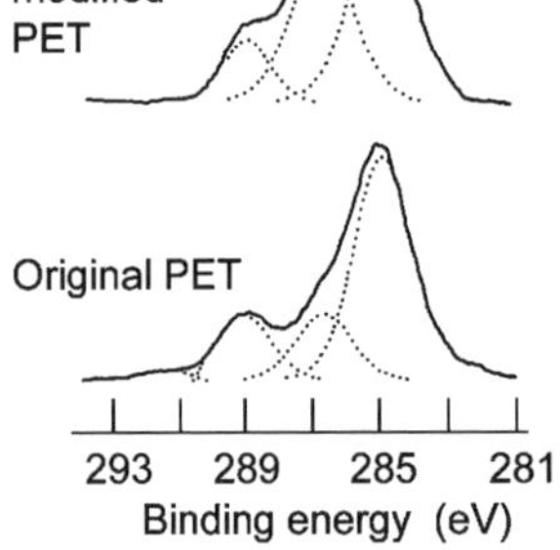

Figure 20.22: XPS (C1s) spectra of the O_2 plasma-modified and NH_3 plasma-modified PET film surfaces. The spectra were measured at a take-off angle of 15 degrees.

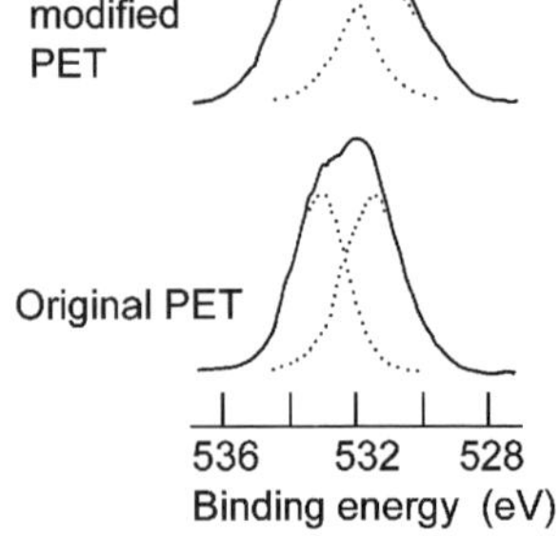

Figure 20.23: XPS (O1s) spectra of the O_2 plasma-modified and NH_3 plasma-modified PET film surfaces. The spectra were measured at a take-off angle of 15 degrees.

Also the O1s spectrum for the original PET film was decomposed into two components, which are at 531.7 eV (component #O1) due to O═C—O— in ester linkages and at 532.2 eV (component #O2) due to —O—C(O)— groups in ester linkages. Their concentrations are 50% and 50%, respectively. The plasma modification, as shown in Fig. 20.22 and 20.23, leads to large changes in C1s and O1s spectra. In the O_2 plasma modification, the component #C2 at 286.7 eV increases from 19% to 40% and the components #C1 at 285.0 eV and #C3 decreases from 62% to 47% and from 19% to 13%, respectively. Furthermore, the component #C4 due to π–π^* shake-up satellite becomes a negligible intensity. The component #O2 increases from 50% to 58%, and the component #O1 decreases from 50% to 42%. These changes in the C1s and O1s spectra indicate that the O_2 plasma attacked mainly CH_2 or phenyl rings rather than ester groups in the PET polymer chains to form C—O groups. As a result, the O/C atomic ratio for the modified PET film surfaces increased, and the contact angle also decreased. On the other hand, in the NH_3 plasma modification also, the component #C2 increases from 19% to 35%, and the component #C3 decreases from 19% to 3%. The component #O2 increases from 50% to 66%, and the component #O1 decreases from 50% to 34%. These changes also indicate that the NH_3 plasma reacted predominantly with ester (C(O)O) groups rather than phenyl rings in the PET polymer chains, and modified (C(O)O) groups into C—O groups. In consequence, the O/C atomic ratio for the modified PET film surfaces does not increase, but the contact angle decreases because of increased concentration of the C—O groups. The other plasmas, Ar, H_2, and N_2 plasmas, lead to similar effects on the XPS spectra. Results of the XPS analyses are summarized in Table 20.10. From the results, we can

Table 20.10: XPS (C1s and O1s) spectra for plasma-modified PET film surfaces

	Plasma modification		C1s components (%)			O1s components (%)	
Plasma	rf power (W)	Modification time (s)	Comp. #1	Comp. #2	Comp. #3	Comp. #1	Comp. #2
Ar	50	120	52	35	13	39	61
O_2	50	120	47	40	13	42	58
H_2	50	120	72	19	9	43	57
N_2	50	120	56	34	10	47	53
NH_3	50	120	62	35	3	34	66
None	–	–	62	19	19	50	50

C1s components – component #1: —$\underline{C}H_2$— and —$\underline{C}$H— groups; component #2: —$\underline{C}H_2$—O— groups in linkages; component #3: —$\underline{C}$(O)—O— groups in ester linkages.

O1s components – component #1: $\underline{O}$═C—O— in ester linkages; component #2: —$\underline{O}$—C(O)— groups in ester linkages.

conclude that the Ar and N_2 plasma modify mainly CH_2 or phenyl rings rather than ester groups in the PET polymer chains to form C—O groups. In the H_2 plasma modification, ester groups in the PET polymer chains are mainly modified.

Conclusively, what plasma is used for surface modification is an important factor to control modification reactions occurring on polymeric surfaces.

20.3.3. Influences of Chemical Composition of Polymeric Materials on Modification Reactions

Plasma exposure makes two essentially different modifications against polymer surfaces, physical and chemical modifications. The physical modification is sometimes called as etching, and means degradation reactions of the polymer chain. In the physical modification process, bond scission of the polymer chains will occur, and some degradation products will deposit on the polymer surfaces or the topmost layer of the polymer specimens will be tripped off. When the plasma-treated surface was rinsed with organic solvents, weight loss of the specimens will be observed. The chemical modification means introduction reactions of functional groups on the polymer surfaces, and as a result the chemical composition of the polymer surfaces will be modified. Some decrease in water contact angle will occur on the polymer surfaces. How chemical composition of polymer chains influences the two modifications? Three polyesters, poly(lactic acid) (PLA), poly(oxybenzoate-co-oxynaphthoate) (PBN), and PET films, were used as specimens for the investigation. PLA film is composed of aliphatic carbon chains and ester groups, PBN film consists of aromatic carbon chains and ester groups, and PET film is composed of a mixture of aliphatic and aromatic carbon chains and ester groups.

Three polyester films, PLA, PET, and PBN films, were exposed to argon and oxygen plasmas at 25 to 100 W as a function of the exposure time. After finishing the plasma exposure, the plasma-treated films were rinsed with ethanol or acetone to remove deposited degradation products from the film surfaces, and then dried in vacuum. Weight changes for the polyester films before and after the plasma exposure were evaluated as a function of the plasma exposure. Figs. 20.24–20.27 are typical results of the weight loss (in $\mu g\,cm^{-2}$) for PLA and PBN films as a function of the argon or oxygen plasma exposure time [26]. The weight loss, as shown in Figs. 20.24–20.27, is a linear relationship with the plasma exposure time. From the linear relationship, the weight loss rate (in $\mu g\,cm^{-2}s$) for PLA, PET, and PBN films in the argon or oxygen plasma-exposing processes is calculated in a dimension $\mu g\,cm^{-2}s$, and is summarized in Table 20.11 as functions of a kind of the plasmas and the rf power. The weight loss rate is an index of susceptibility to plasma.

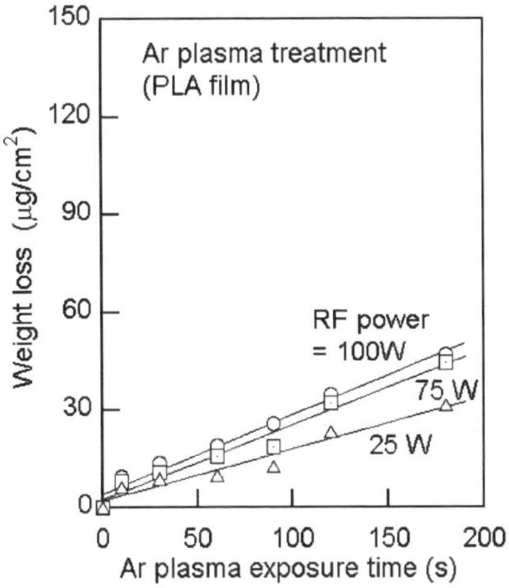

Figure 20.24: Weight loss from PLA film surfaces by exposing to the Ar plasma as functions of the rf power and plasma exposure time.

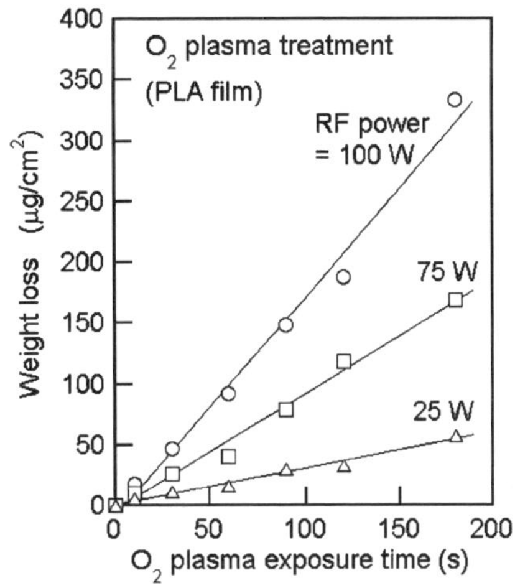

Figure 20.25: Weight loss from PLA film surfaces by exposing to the O_2 plasma as functions of the rf power and plasma exposure time.

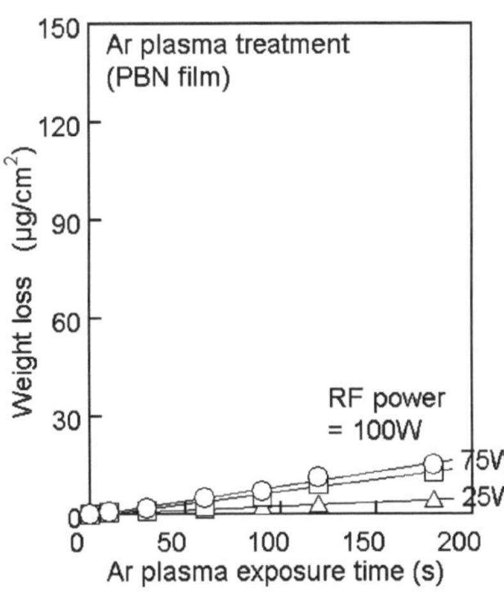

Figure 20.26: Weight loss from PBN film surfaces by exposing to the Ar plasma as functions of the rf power and plasma exposure time.

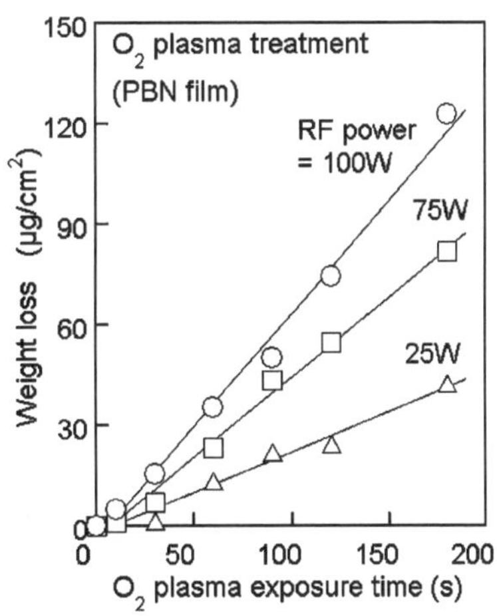

Figure 20.27: Weight loss from PBN film surfaces by exposing to the O_2 plasma as functions of the rf power and plasma exposure time.

Table 20.11: Weight loss rate of PLA, PET, and PBN films in exposing to Ar and O_2 plasmas

Plasma exposure		Weight loss rate ($\mu g\,cm^{-2}\,s$)			Weight loss rate ($10^{-9}\,mol\,cm^{-2}\,s$)		
Plasma	rf power (W)	PBN	PET	PLA	PBN	PET	PLA
Ar	25	0.025	0.085	0.161	0.080	0.443	2.78
	75	0.076	0.130	0.229	0.242	0.677	3.94
	100	0.088	0.123	0.242	0.280	0.641	4.17
O_2	25	0.241	0.287	0.303	0.768	1.50	5.22
	75	0.478	0.527	0.950	1.52	2.75	16.4
	100	0.674	0.670	1.79	2.17	3.49	30.9

From this viewpoint, the weight loss is compared among the three polyester films, PLA, PET, and PBN films.

Although the weight loss rate is conventionally estimated in a dimension of $\mu g\,cm^{-2}\,s$, the rate in $\mu g\,cm^{-2}\,s$ is not an adequate index for evaluation of the plasma susceptibility. The plasma susceptibility does not mean how much weight degradation products were formed during the plasma exposure, but how many numbers of C—C bond scission occurred during the plasma exposure. When one molecule (degradation product) was split out from a polymer chain by the bond scission, the molecular weight of the product from PBN film being composing of large repeating unit ($290\,g\,mol^{-1}$) may be larger than that from PLA being composing of small repeating unit ($78\,g\,mol^{-1}$). To compensate large differences in the repeating unit among the three polyester films, PLA, PET, and PBN films whose molecular weight of the repeating units is 78, 192, and $290\,g\,mol^{-1}$, respectively; the weight loss rate in $\mu g\,cm^{-2}\,s$ is divided with the molecular weight of the repeating unit for each polyester film. This compensation never solves completely large difference in the repeating unit, and does not lead to accurate concentration of the degradation products formed on the polyester film surfaces by the plasma exposure. The compensated rates for PLA, PET, and PBN films are appended to Table 20.11 as functions of the kind of plasmas and rf power. There are large differences in the compensated rate among the three polyester films. PLA film shows the highest rate for weight loss process of the three films, even when either of argon or oxygen plasmas was used at rf powers of 25–100 W. PLA film shows higher compensated rate by about 15 times than PBN film. PET film also shows higher rate by about 2 times than PBN film. On the other hand, PBN film shows the lowest rate for weight loss process of the three films. From this comparison, we conclude that PLA film is easy to be subjected to degradation reactions, and that PBN film is difficult to be subject to degradation

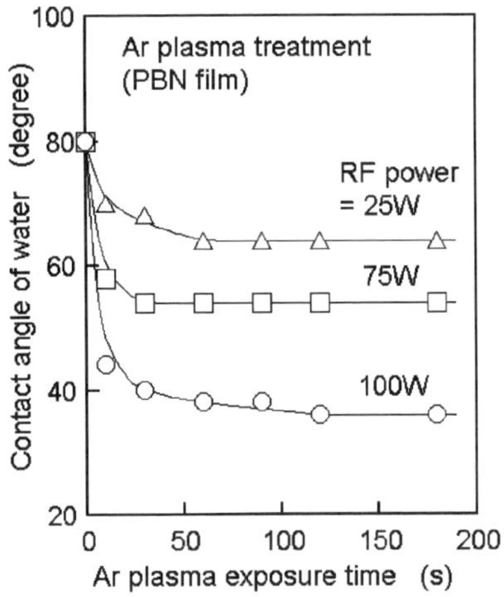

Figure 20.28: Contact angle of water on the Ar plasma-modified PBN film surfaces as functions of the rf power and plasma exposure time.

reactions. This conclusion means that the plasma susceptibility is related to the chemical composition of the polymer chains. Polyester chains containing aliphatic groups are susceptible to plasma and are easy of weight loss reactions. On the other hand, polyester chains containing aromatic groups are not so susceptible to plasmas and are not easy of weight loss reactions.

Contact angle of water on the plasma-treated polyester film surfaces was measured in order to investigate how to be modified by the plasmas. Three polyester films, PLA, PET, and PBN films, were exposed to argon or oxygen plasma at rf powers of 25–100 W for 10–180 s, and then the film surfaces were rinsed with ethanol or acetone to remove degradation products deposited on the surface. Afterwards, against the plasma-treated and rinsed film surfaces, the water contact angle was measured as functions of the plasma exposure time and rf power. Fig. 20.28 shows typical results of water contact angle on the argon plasma-treated PBN film surfaces as functions of the argon plasma exposure time and rf power [26]. The argon plasma exposure for a short time of 10 s leads to large decrease in contact angle. The contact angle decreases from 80 degrees (not exposed film) to 58 degrees (exposed for 10 s). Afterwards, the contact angle continues small decrease with increasing the exposure time, and reaches a constant (54 degrees) at an exposure time of 120 s. To evaluate effects of the plasma exposure on surface modification reactions, the contact angle for the three polyester (PLA, PET, and PBN) films treated with argon and oxygen plasmas for 120 s are summarized in Table 20.12 as functions of kind of plasma and rf power. The contact angle for the original (not yet plasma-treated) PLA, PET, and PBN films is

Table 20.12: Contact angle of water on PLA, PET, and PBN film surfaces modified with Ar and O_2 plasmas for 120 s

Plasma modification conditions			Contact angle of water (degree)		
Plasma	rf power (W)	Modification time (s)	PLA	PET	PBN
No treated	–	–	74	78	80
Ar	25	120	74	32	64
	75	120	74	28	54
	100	120	74	24	36
O_2	25	120	74	26	56
	75	120	74	16	52
	100	120	74	16	44

74, 78, and 80 degrees, respectively. There is not so large difference in contact angle among the three original films. Once these films are exposed to argon or oxygen plasmas, there is large difference in contact angle among the plasma-exposed films. PLA showed no change in contact angle by the argon and oxygen plasma exposures, even at an rf power of 100 W. On the other hand, PET films shows large decrease in contact angle by the argon and oxygen plasma exposures. For example, the contact angle for PET films decreases from 78 degrees (for the original PET film) to 28 and 16 degrees in exposure to the argon and oxygen plasmas at 75 W. PBN films also show some decrease in contact angle by the argon and oxygen plasma exposures, but the magnitude of the decrease is not so large as the PET films. The contact angle for PBN films treated with argon and oxygen plasma exposures is 54 and 52 degrees, respectively. These comparisons show that PET and PBN film surfaces are effectively modified into hydrophilic by either of argon and oxygen plasmas. However, PLA film surface is never modified even by argon and oxygen plasmas. Why is PLA film never modified by the plasma exposure, although PLA films are easy of weight loss reactions?

The C1s and O1s spectra for the original PLA film surface, as shown in Figs. 20.29 and 20.30, are composed into three and two main components, respectively [26]. The three decomposed components are illustrated as dotted lines in the figure. The three components for the C1s spectrum are assigned CH groups, C—O groups in ester linkages, and C=O groups in ester linkages at 285.0, 287.0, and 289.0 eV, respectively. The concentration of these CH, C—O, and C=O components are 36%, 31%, and 33%, respectively, which correspond well to the values (CH, C—O, and C=O groups are 33.3% each) calculated from the repeat unit of the PLA film (Table 20.12). On the other hand, the two components for the O1s

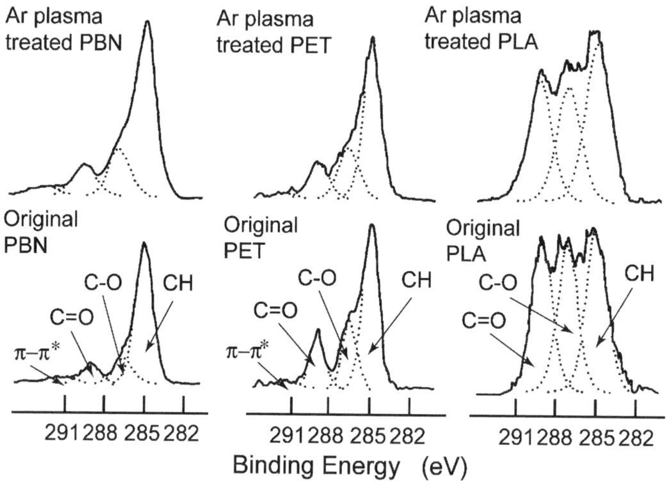

Figure 20.29: XPS (C1s) spectra of the Ar plasma-modified PLA, PET, and PBN film surfaces.

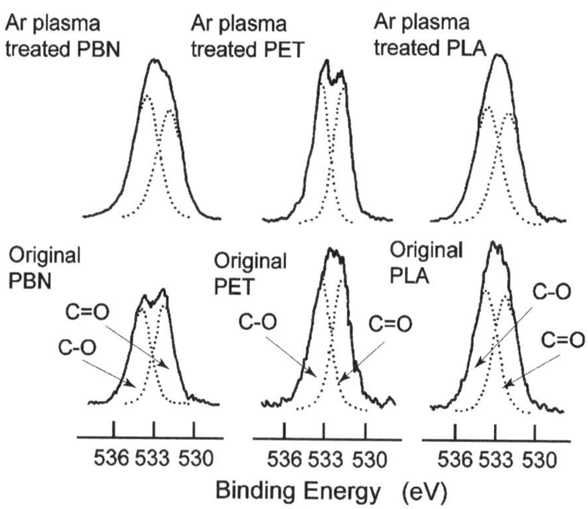

Figure 20.30: XPS (O1s) spectra of the Ar plasma-modified PLA, PET, and PBN film surfaces.

spectrum are due to O═C and O—C groups at 532.3 and 533.9 eV, respectively. The concentrations of the O═C and O—C groups are 49% and 51%, respectively, which correspond well to the values (50% and 50%) calculated from the repeat unit of the PLA film (Table 20.13). The argon plasma-treated PLA film surface shows

Table 20.13: Chemical composition of PLA, PET, and PBN film surfaces treated with argon and oxygen plasmas

Polymers	O/C atomic ratio	C1s components (%)			O1s components (%)	
		Comp. #1	Comp. #2	Comp. #3	Comp. #1	Comp. #2
Original PLA	0.59	36	31	33	49	51
Ar plasma-modified PLA	0.61	41	30	29	49	51
Original PET	0.34	58	22	20	50	50
Ar plasma-modified PET	0.31	59	24	17	47	53
O_2 plasma-modified PET	0.35	62	19	19	52	48
Original PBN	0.24	75	13	12	51	49
Ar plasma-modified PBN	0.30	69	19	14	47	53
O_2 plasma-modified PBN	0.34	69	21	10	45	55

C1s component #1 at 285.0 eV due to $\underline{C}H_3$, $\underline{C}H_2$, and $\underline{C}H$ components.
C1s component #2 at 286.6–287.0 eV due to $\underline{C}$—O component.
C1s component #3 at 288.9–289.1 eV due to $\underline{C}$═O component.
C1s component #4 at 291.4–291.8 eV due to π–π^* component.
O1s component #1 at 531.9–532.3 eV due to $\underline{O}$═C component.
O1s component #2 at 533.5–533.9 eV due to $\underline{O}$—C component.

similar C1s and O1s spectra, which are decomposed into three and two components, respectively (Figs. 20.29 and 20.30). The O/C atom ratio for the argon plasma-treated PLA film is 0.61, which was changeless in comparison with that for the original PLA (O/C atom ratio = 0.59). The concentrations of CH, C—O, and C═O groups are 41%, 30%, and 29%, respectively; and that of O═C and O—C groups are 49% and 51%, respectively (Table 20.13). These spectral results indicate that the Ar plasma-treated PLA film surface may be almost same in chemical composition as the original PLA film surface, even although the film surface was exposed to argon plasma for 120 s. These results indicate that the argon plasma exposure leads always to fresh surface of the PLA film surface, and as a result the argon plasma-treated film surface shows the same chemical composition as the original PLA film surface. The PBN film surfaces treated with the argon and oxygen plasmas show increase in O/C atom ratio from 0.24 for the original PBN film to 0.30 and 0.34, respectively (Table 20.13). This increase indicates the formation of some oxygen functional groups on the PBN film surface by the argon and oxygen plasma exposures. The C1s spectrum for the original PBN film is decomposed into four components due to CH (at 285.0 eV), C—O (at 286.6 eV), C=O groups (at 289.1 eV), and π–π^* shake-up satellite (at 291.4 eV) (Fig. 20.30). The O1s spectrum also is decomposed into two components due to O=C (at 532.2 eV) and O—C groups (at 533.9 eV). The argon and oxygen plasma-treated PBN film surfaces show similar C1s and O1s spectra to those for the original PBN film surface. These spectra are decomposed into four components for the C1s spectra and two components for the O1s spectra. Their concentrations are summarized in Table 20.11. The C—O component (at 286.6–287.0 eV) for the plasma-treated PBN films is larger concentration than that for the original PBN film, indicating that new C—O groups are formed on the PBN film surfaces by the plasma exposure. And the relative concentration of π–π^* shake-up satellite (at 291.4–291.8 eV) is changeless by the plasma exposure. The concentration of π–π^* shake-up satellite is 4.9% for the original PBN film, and 4.5% and 4.3% for the argon and oxygen plasma-treated PBN films, respectively. This indicates that aromatic groups in the PBN polymer chains are never broken down by the plasma exposure, and that C—O groups are formed in the PBN polymer chains. Conclusively, we can conclude that the PLA film surface is not modified in chemical composition. On the other hand, the PBN film surfaces are modified in chemical composition. C—O groups are formed in the PBN polymer chains. This is the main difference in modification reactions occurring on the PLA and PBN film surfaces. This difference may be due to chemical composition of the polymer chains: PLA film is composed of aliphatic carbon chains and ester groups. PBN film consists of aromatic carbon chains and ester groups.

Why the PLA film surface was not modified but the PBN film surface was modified to form C—O groups on film surface is discussed here. It is well known that in exposing polyesters to high-energy radiation such as gamma and ionizing rays, C—O bonds in the ester groups are scissored off to form radicals at the ends of polymer chains. Degradation reactions are initiated from the radicals, and degradation products with low-molecular-weight deposit on the surface. A similar bond scission will occur in the plasma exposure, and radicals will be formed at the end of the polymer chains. Typical degradation reactions are shown in Fig. 20.31 for the PLA film and in Fig. 20.32 for the PBN film. In degradation reactions for the PLA film (Fig. 20.31), the C—O bond scission will lead to the formation of two radicals, A1 and A2 at polymer chain ends. Subsequently, A1 and A2 will transform into A3 and A4 radicals, respectively, with elimination of small molecules such as CO_2 and CH_2=CH_2. A3 and A4 radicals will transform into A5 and A2 radicals, respectively. A5 radical will transform into A3 radical to eliminate CO. Once the C—O bond scission in PLA film occurs, chain reactions of A1–A3–A5–A1 and A2–A4–A2 are

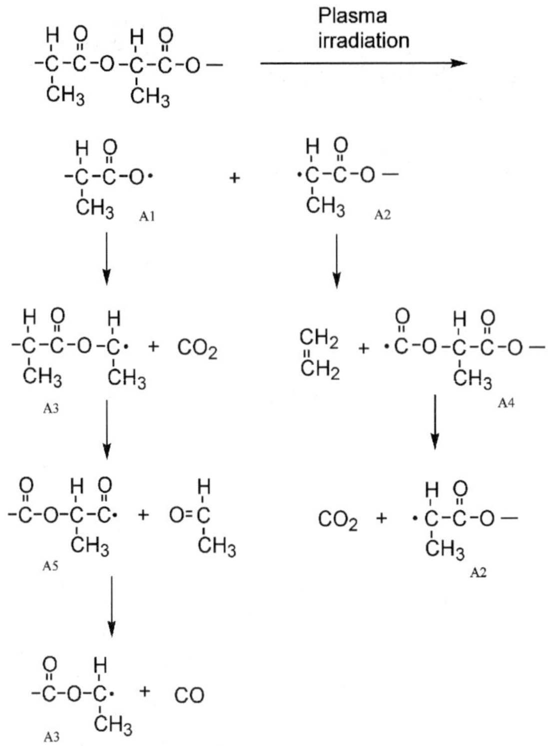

Figure 20.31: Overall degradation reactions initiated from PLA by plasma.

initiated from the polymer chain ends. The chain reactions continue as long as the radicals exist at the polymer chain ends. Consequently, no radical remains on the PLA film surface any more after finishing the plasma exposure; and thus no oxygen functional group forms on the PLA film surface. The oxygen functional groups formed by the plasma exposure may be due to a post-oxidation of the radicals remaining on film surfaces. The plasma-treated PLA film surface shows no improvement in hydrophilic properties compared with the original film surface. This is a story of the modification of PLA film surface by plasma irradiation. Instability of radicals such as A2 and A3 may be a possible answer regarding why the PLA film surfaces were not modified into hydrophilic by the plasma exposure. On the other hand, PBN films also undergo scission of ester groups to form two radicals, B1 and B2 (Fig. 20.32). Each of these radicals will subsequently convert into B3 radicals with elimination of small molecules such as CO_2. Some of B2 and B3 radicals will remain on the PBN film surface because of their high stability due to delocalization. The B2 and B3 radicals will be oxidized to form oxygen functional groups. The PBN

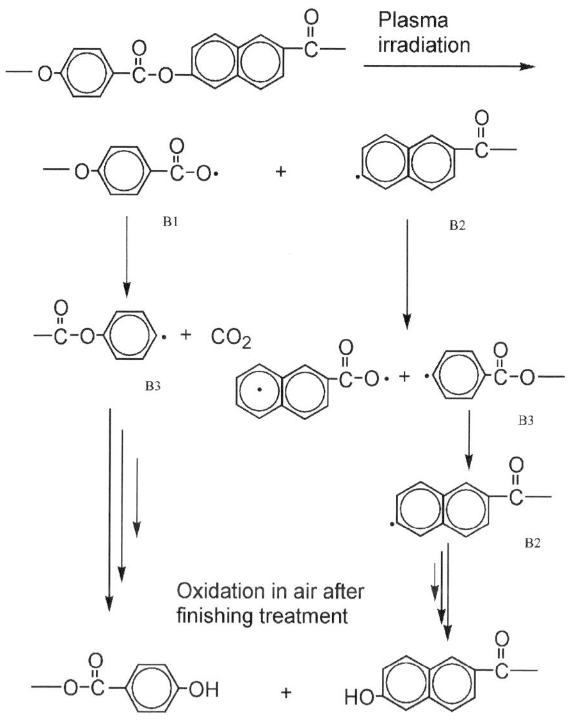

Figure 20.32: Overall degradation reactions initiated from PBN by plasma.

film surface is modified into hydrophilic (low contact angle of water). This is a story of the modification of PBN film surface by the plasma irradiation.

20.3.4. Remote Plasma Treatment for Effective Surface Modification

Surface modification of polymeric materials, as described in 20.3.1, is due to radical species rather than ion species in plasmas. If radicals alone could be isolated from plasmas and could be interacted predominantly with polymeric surfaces, surface modification will succeed effectively. Plasma contains electrons, positive and negative ions, and radicals. These ions and radicals are generated from the collision between accelerated electrons and gas molecules in plasma reactor. These species disappear in reactions of the electron–positive ion recombination, and the radical recombination. The electron–positive ion recombination reaction is very faster than the radical recombination reaction. The rate constant of these recombination

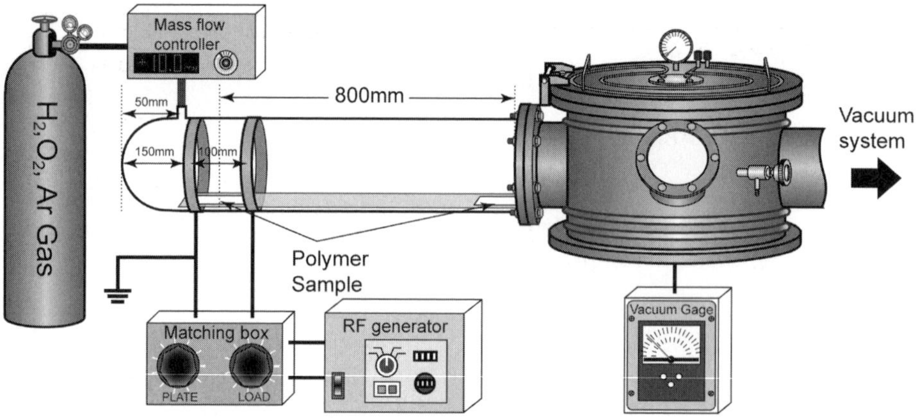

Figure 20.33: Schematic presentation of remote plasma reactor.

reactions is in the order of $10^{-7}\,cm^3\,s^{-1}$ and $10^{-33}\,cm^6\,s^{-1}$, respectively. Therefore, radicals possess longer lifetime than electrons and ions. The radical concentration at the position away from the plasma zone may be predominantly higher than the electron and ion concentration, because of lifetime of difference between radicals and ions. This is an essential concept of the remote plasma treatment. In practice of the remote plasma treatment process, surfaces of polymeric materials to be modified are placed in the position far away from the plasma zone. In the conventional plasma treatment process, the surfaces are placed in the plasma zone. We call the conventional plasma treatment the "direct plasma treatment" for distinguishing it from the "remote plasma treatment".

To evaluate effectiveness of the remote plasma treatment, surface modification of poly(tetrafluoroethylene) (PTFE) sheets by remote and direct hydrogen plasmas was compared [27]. A special reactor for the remote hydrogen plasma treatment of the PTFE sheet was used. The reactor consists of a cylindrical Pyrex glass tube (45-mm diameter, 1000-mm long) and a columnar stainless steel chamber (300-mm diameter, 300-mm height). The Pyrex glass tube has two gas inlets for the injection of hydrogen and argon gases and a copper coil of nine turns for the energy input of rf power (13.56-MHz frequency). The gas inlets and the copper coil are attached to the Pyrex glass tube at a distance of 50 and of 200 mm from the glass tube end, respectively. The schematic diagram of the reactor is shown in Fig. 20.33.

The PTFE sheets were placed at a distance of 0, 25, 50, 75, and 80 cm from the center of the rf coil in the Pyrex glass tube of the reactor, and were exposed to the hydrogen plasma at rf power levels of 25, 50, 75, and 100 W. The modification of the PTFE sheet by the remote hydrogen plasma was evaluated from the advancing

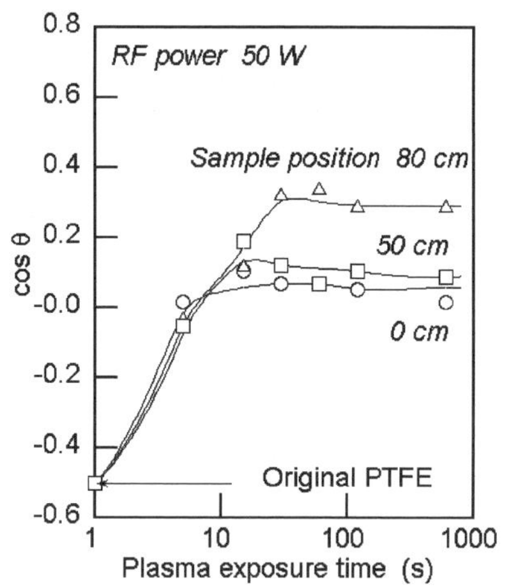

Figure 20.34: Cosine of the contact angle of water on the H_2 plasma-modified PTFE film surfaces as functions of the plasma exposure time.

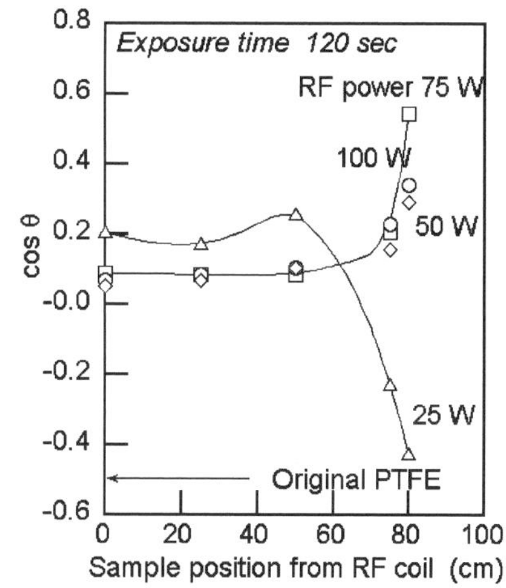

Figure 20.35: Cosine of the contact angle of water on the H_2 plasma-modified PTFE film surfaces as functions of the sample position from RF coil.

contact angle of water as functions of the sample position as well as the plasma exposure time and the rf power. Typical results are shown in Fig. 20.34. Regardless the sample position, $\cos\theta$ increases with increasing plasma exposure time, and reaches a constant value after exposure time of more than 100 s. The sample position leads to large effects on the contact angle of water. For example, the contact angle for the original PTFE sheet is 120 degree. The contact angle at a sample position of 0, 50, and 80 cm is 89, 84, and 73 degrees, respectively. Fig. 20.35 shows $\cos\theta$ for the PTFE sheet modified with the remote hydrogen plasma for 120 s as a function of the sample position. Cos θ at sample positions of 0 to 50 cm distance is almost constant. However, $\cos\theta$ at sample positions of more than 75-cm distance except that the treatments at an rf power of 25 W increases with increasing the sample position. Cos θ at a sample position of 80-cm distance reaches 0.29 for 50 W, 0.55 for 75 W, and 0.39 for 100 W. These values are fairly higher than that at 0-cm distance, indicating that the remote hydrogen plasma is superior in hydrophilic modification to the direct plasma.

The atomic composition at the surface of the PTFE sheets treated with the remote and direct hydrogen plasmas was analyzed by XPS. Table 20.14 shows the atomic composition (F/C and O/C atom ratio) for the PTFE sheets treated with the remote and direct hydrogen plasmas [27]. The F/C atom ratio for the PTFE sheets modified by the remote and direct hydrogen plasmas is 0.41 and 0.60, respectively. It is

Table 20.14: Atomic composition of PTFE film surfaces modified by remote and direct H_2 plasmas

Surface modification			Atomic composition	
Plasma	rf power (W)	Modification time (s)	F/C	O/C
None	–	–	1.92	0.05
Remote H_2 plasma	75	120	0.41	0.12
Direct H_2 plasma	75	120	0.60	0.07

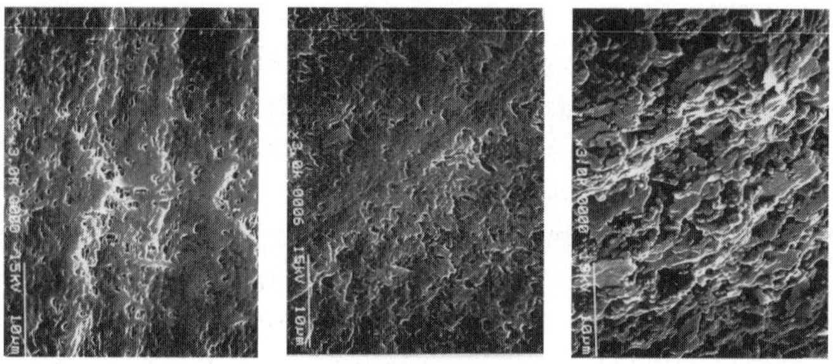

Original PTFE

PTFE treated with remote hydrogen-plasma for 120 s at 100W

PTFE treated with direct hydrogen-plasma for 30 s at 100W

Figure 20.36: Topographical figures for the PTFE film surfaces modified by the remote H_2 plasma at 100 W for 120 s and by the direct H_2 plasma at 100 W for 30 s.

obvious that defluorination occurred in the remote and direct hydrogen plasma treatment processes. The remote hydrogen plasma treatment leads to lower F/C atom ratio (F/C = 0.41) than the direct hydrogen plasma treatment (F/C = 0.60), indicating effective modification by the remote hydrogen plasma rather than the direct hydrogen plasma.

PTFE sheet surfaces modified by the remote and direct hydrogen plasmas are compared in Fig. 20.36 [27]. The specimens for XPS measurement are the PTFE sheets treated with the remote and direct hydrogen plasma at 75 W for 120 s, and their $\cos\theta$ is 0.55 and 0.09, respectively. The original PTFE sheet shows a symmetrical spectrum appearing at 292.5 eV due to CF_2—CF_2 unit. Both remote and direct hydrogen plasma-modified PTFE sheets show complex C1s spectra having two peaks, which are decomposed into five components appearing at 285.9–296.0, 287.6–287.9, 289.6–289.8, 292.5, and 293.7–294.1 eV. These five components are assigned $\underline{C}H_2$—CHF groups (at 285.9–296.0 eV), $\underline{C}$HF—CH_2 and C═O groups (at 287.6–287.9 eV), $\underline{C}H(OR)$—CHF groups (at 289.6–289.8 eV), $\underline{C}F_2$—CF_2 groups (at 292.5 eV), and $\underline{C}F_3$—CF_2 and/or $\underline{C}F(OR)_2$—CF_2 groups (at 293.7–294.1 eV). The underlined carbon atom is the objective carbon atom for the assignment. The fifth component at 293.7–294.1 eV is assigned a mixture of the $\underline{C}F_3$—CF_2 and $\underline{C}F(OR)_2$—CF_2 groups. These spectra show that both remote and direct hydrogen plasmas make the substitution of hydrogen atom for fluorine atom of the PTFE surface to form CHF and CH_2 units. The relative concentration of the five components is estimated from the relative peak area of the decomposition (Table 20.15). Table 20.15 shows that a main product is dihydrogen-substituted carbon atom (CH_2 groups), which accounts for 63% of the total carbon atoms for the remote hydrogen plasma-treated PTFE sheet and for 55% for the direct hydrogen plasma treated.

The mono-hydrogen-substituted carbon atom (CHF groups) is 12% for the two plasma-treated PTFE sheets. The unsubstituted carbon atom (CF_2 groups) is only 9% and 26% for the remote and direct hydrogen plasma-treated PTFE sheets, respectively. Therefore, the reactivity of CF_2 groups is 91% and 74% for the remote and direct hydrogen plasma treatments, respectively. This indicates that the remote hydrogen plasma has higher capability of the substitution than the direct hydrogen plasma.

Finally, the remote hydrogen-modified PTFE surface is compared in scanning electron microscope (SEM) with the direct hydrogen plasma-modified surface (Fig. 20.36) [27]. The original PTFE surface is not smooth but rough, and covered with may scales. Similar surface is observed even when PTFE surface is treated with the remote hydrogen plasma for 120 s at 100 W. It is clear that there is less difference in topographic picture between the remote hydrogen plasma-treated and original PTFE surfaces. However, when treated with the direct hydrogen plasma only for 30 s at 100 W, the PTFE surface shows a different picture from the original. The SEM picture shows that edges of the scales are etched off and valleys become deeper. This comparison indicates that the remote hydrogen plasma leads to surface modification without topographic changes, while the direct hydrogen plasma leads to surface modification with topographic changes.

20.4. Plasma Graft Copolymerization

When polymeric surfaces are exposed to plasma, radicals in plasma attack mainly on the surfaces and functional groups such as oxygen-containing groups are formed on the surfaces. This process is called "plasma treatment", which is used as a useful technique for surface modification. At the same time, electrons and ions existing in the plasma also attack on the polymeric surfaces to generate carbon radicals. The carbon radicals are chemically active to initiate polymerization reactions when monomers contact with the radicals. New polymers grow up from the radicals and

Table 20.15: Chemical composition of PTFE film surfaces modified by remote and direct H_2 plasmas

Plasma	rf power (W)	Modification time (s)	C1s components (%)					O1s components (%)	
			$\underline{C}H_2$—CHF	$\underline{C}$HF—CH_2 and $\underline{C}$=O	$\underline{C}$H(OR)—CHF	$\underline{C}F_2$—CF_2	$\underline{C}$F(OR)—CF_2 and $\underline{C}F_2$—CF_2	$\underline{O}$=C	$\underline{O}$—C
None	–	–	0	0	0	100	0	–	–
Remote H_2 plasma	75	120	63	12	3	9	13	27	73
Direct H_2 plasma	75	120	55	12	5	26	2	26	74

deposit over the surfaces of the polymeric materials. This polymerization process is called "plasma graft copolymerization". Therefore, plasma in the plasma graft copolymerization process contributes only to formation of radicals on polymeric surfaces, and never contributes to the polymerization process. Recently, many investigators focus on their interest in plasma graft copolymerization because of a new technique for surface modification with well-defined functional groups.

Plasma polymerization is a process of the formation of a new polymer layer on polymeric surfaces by plasma. On the other hand, plasma graft copolymerization is also a process of a new polymer formation on the polymeric surfaces. There is large difference in chemical composition between polymers formed by the two processes even when the same monomer molecules are used for plasma polymerization and plasma graft copolymerization. In the plasma polymerization process, monomer molecules are polymerized with assistance of plasma. As a result, a part of monomer molecules is fragmented or rearranged, and the chemical composition of the formed polymers is not repetition of the used monomer. The plasma polymerization process is difficult to form polymers with well-defined functional groups. While, in the plasma graft copolymerization, once radicals are formed on the polymeric surfaces with assistance of plasma, monomer molecules polymerize spontaneously to deposit polymers on the surfaces. The chemical composition of the graft-polymerized polymers is repetition of the monomer molecules used. Functional groups in the used monomer molecules are completely retained in the formed plasma polymers. From the viewpoint, plasma graft copolymerization is applied for new technique of surface modification. Main topics in this section are (1) radical formation on polymer surfaces by plasma, (2) graft copolymerization process as a new technique for selective surface modification with special functional groups.

20.4.1. Radical Formation on Polymer Surfaces

When polymeric materials are exposed to plasma, active species such as electrons and ions interacted with the surfaces of the materials to make bond scission of C—C, C—H, C—O, and C—N bonds on the surfaces. As a result, radicals are generated on the surfaces of the polymeric materials [29]. Fig. 20.37 shows typical results of the peroxide concentration formed on PET film surfaces as functions of the argon plasma exposure time and rf power for initiation of the argon plasma. In Fig. 20.37, the peroxide concentration instead of the radical concentration is used, because as soon as the films are taken out from the plasma reactor to air atmosphere after finishing the argon plasma exposure process, radicals existing on the PET film surfaces are oxidized into peroxides. Therefore, the peroxide concentration is equivalent to the radical concentration. A large amount of peroxides is formed within a

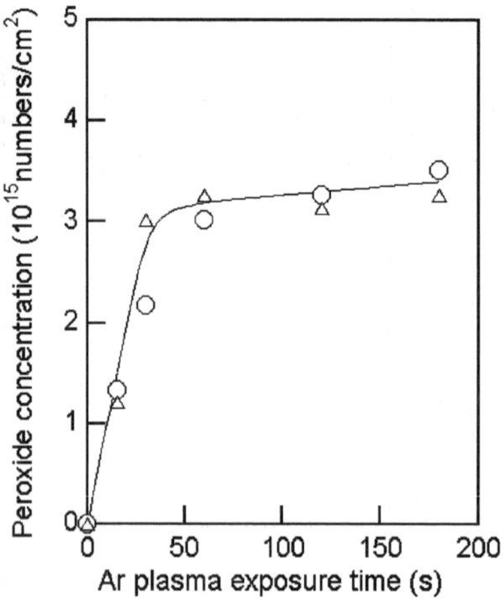

Figure 20.37: Peroxide concentration formed on PET film surfaces by exposing to Ar plasma as functions of the rf power (○: rf power of 25 W; △: 50 W) and plasma exposure time.

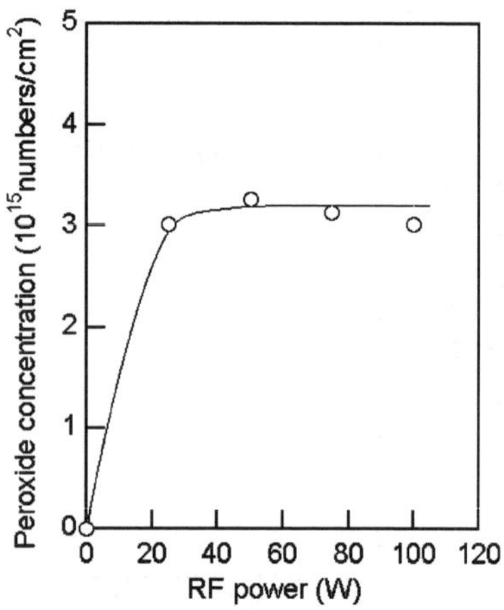

Figure 20.38: Peroxide concentration formed on PET film surfaces by exposing to Ar plasma for 60 s as a function of the rf power.

short period of 60 s, as soon as PET films are exposed to argon plasma, and after this time the peroxide formation is small (Fig. 20.37). The peroxide concentration at an exposure time of 60 s is 3.3×10^{15} numbers cm^{-2}. Fig. 20.38 shows effects of the rf power for initiating the argon plasma on the peroxide formation. A magnitude of the rf power gives less influence on the peroxide formation. The peroxide concentration formed on PET surfaces is a constant of 3.2–3.5×10^{15} numbers cm^{-2} as high as the rf power is from 25 to 100 W.

For other polymers also besides PET, peroxides are formed on their surfaces by exposing to argon plasma. Polyimide films are exposed to argon plasma at 25 W, and peroxides formed on the film surfaces are analyzed. Fig. 20.39 shows the peroxide concentration formed on polyimide surfaces by argon plasma at 25 W as a function of argon plasma exposure time [30]. The concentration of peroxides rapidly increases with increasing the exposure time, and large increase in the peroxide formation is accomplishes within a short time of 60 s. After this time, the peroxide concentration is leveled off at about 4.6×10^{14} numbers cm^{-2}. Conclusively, the peroxide formation may occur within a short period of 60 s after exposing to argon plasma. A magnitude of the rf power for the argon plasma may make less influence on the peroxide formation.

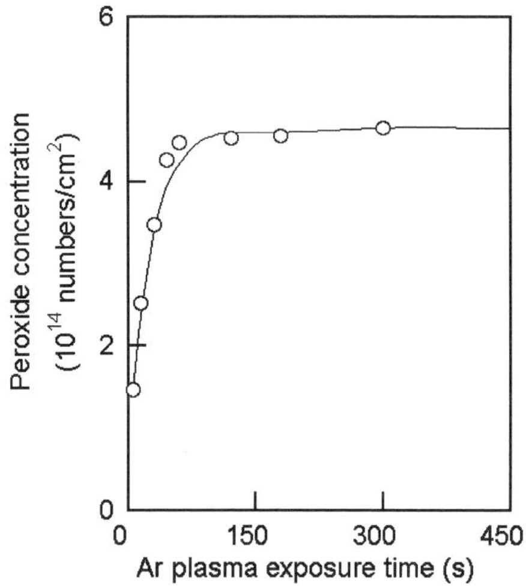

Figure 20.39: Peroxide concentration formed on Kapton film surfaces by exposing to Ar plasma at 50 W.

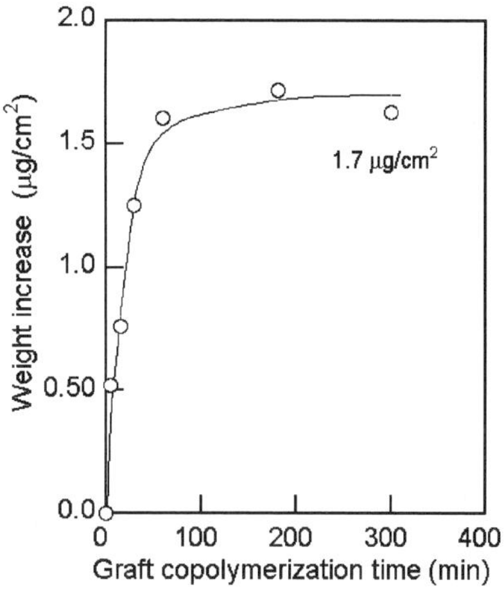

Figure 20.40: Weight increase by graft copolymerization of acrylonitrile on PET film surfaces as a function of the graft copolymerization time.

20.4.2. Plasma Graft Copolymerization Process as a New Technique for Selective Surface Modification with Special Functional Groups

When radicals (peroxides) were formed on polymer surfaces by argon plasma contact with monomer molecules, polymerization reactions of the monomer molecules initiate from the radicals, and new polymers deposit on the surfaces. The polymer deposition rate is very slow, and the complete covering with new polymers requests sometimes reaction time as long as overnight time. Figs. 20.40 and 20.41 show graft copolymerization of acrylonitrile on PET film surfaces and that of vinylimidazole on polyimide film surfaces as a function of the graft copolymerization time, respectively [29,30]. The rate of weight increase due to the graft copolymerization, as shown in Figs. 20.40 and 20.41, is very slow. Polymer deposition of 17–18-nm thickness (17–18 $\mu g\,cm^{-2}$) requests reaction times as long as 5–24 h. The graft copolymerization is a time-consumed process, which is a disadvantage of this technique. The disadvantage obstructs commercial application of the graft copolymerization technique.

Recently, graft copolymerization is reperceived as a new functionalization technique rather than the polymer deposition technique [31–56]. The functionalization proceeds in two steps. In the first step, radicals are formed on the polymer surfaces by plasma. In the second step, selected and stable precursor molecules are reacted

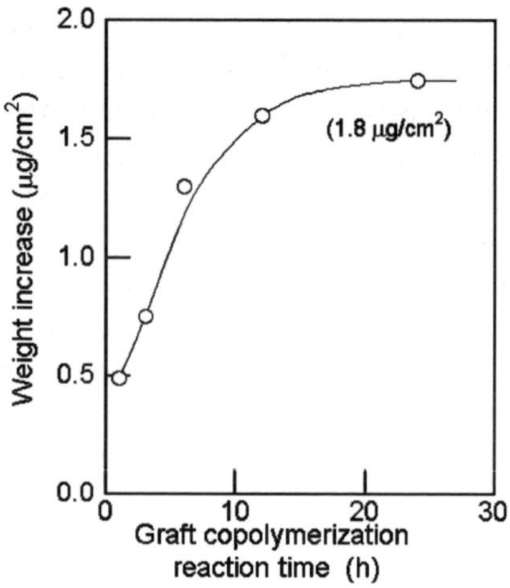

Figure 20.41: Weight increase by graft copolymerization of vinylimidazole on Kapton film surfaces as a function of the graft copolymerization time.

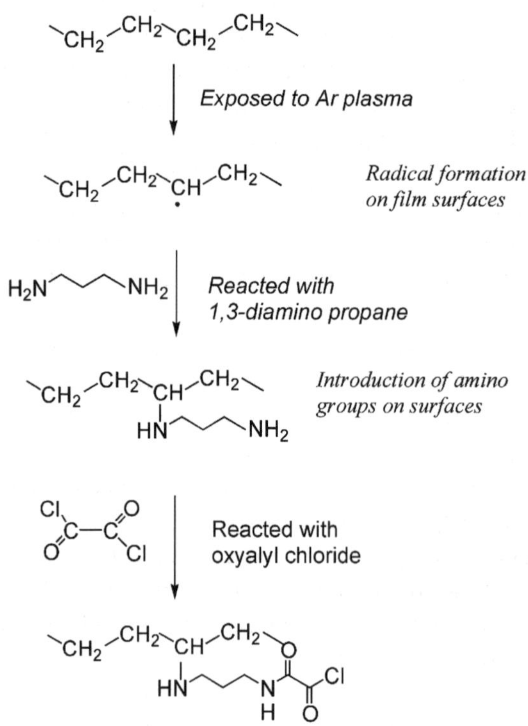

Figure 20.42: Surface modification reactions on polyethylene film surfaces by 1,3-DP and OC.

with the radicals in the absence of plasma. König and co-workers [57] selected acrylic acid as a precursor molecule. On PTFE surfaces, carboxylic acid groups were introduced. Oehr and co-workers [58] selected many precursor molecules containing allyl alcohol, glycidyl methacrylate, acrylic acid, 4-pentene acid, and 6-heptene acid to introduce hydroxyl, epoxy, and carboxylic groups on poly(propylene), PTFE, and poly(vinylidene fluoride) (PVDF) film surfaces. Alvarez-Blanco and co-workers [59] selected 1,3-diamino propane (1,3-DP) as a precursor molecule. Polyethylene surfaces were exposed to argon plasma, and then exposed to DP vapor *in situ* to combine DP with polyethylene surface. The amino functionalities introduced into polyethylene surfaces are further modified with oxyalyl chloride (OC). Series reactions are shown in Fig. 20.42.

20.5 Conclusion

Plasma is a good tool for surface modification of polymeric surfaces. When organic materials are exposed to plasma, many chemical reactions are initiated on polymeric surfaces by the plasma, and the polymeric surfaces are modified. These chemical reactions are plasma polymerization, plasma treatment, and plasma graft copolymerization. Plasma polymerization is the polymer deposition process by plasma. Monomer molecules are polymerized by plasma, and the polymerization mechanism is completely different from conventional polymerization such as radical and ions. The plasma polymerization is the stepwise process of radical formation and radical recombination. Plasma treatment is the introduction process of functional groups into polymer surfaces by plasma. Oxygen- and nitrogen-containing groups are introduced into polymer surfaces, and the surface properties are drastically changed. The introduction reactions are due to radicals rather than ions in plasma. Plasma graft copolymerization is the polymer deposition process. The process by the plasma graft copolymerization is completely different from that by the plasma polymerization. On polymer surface, radicals are formed by plasma exposure, and then from the radicals on the polymer surfaces, monomer molecules polymerize spontaneously to deposit polymers on the surface. The three processes are useful techniques for surface modification of polymeric materials. Recently, many investigators have sought new techniques that can tailor the chemical composition and special functional groups on polymer surfaces in order to operate effective functionalities on polymer surfaces. These techniques may be pulse plasma polymerization, remote plasma treatment, and plasma graft copolymerization. Recent progresses of these techniques have been summarized in Sections 20.2.3, 20.3.4, and 20.4.2.

References

[1] H. Yasuda and C.E. Lamaze, *J. Appl. Polym. Sci.*, 17 (1973) 1519.
[2] H. Yasuda, *Plasma Polymerization*, Academic Press, New York, 1985.
[3] H. Yasuda and C.R. Wang, *J. Polym. Sci., Polym. Chem. Ed.*, 23 (1985) 87.
[4] N. Inagaki and Nakanishi, *Polym. Bull.*, 9 (1983) 502.
[5] H. Yasuda, *J. Polym. Sci., Macromol. Rev.*, 16 (1981) 199.
[6] N. Inagaki and H. Yasuda, *J. Appl. Polym. Sci.*, 26 (1981) 3557.
[7] L.M. Han, R.B. Timmons and W.W. Lee, *J. Vac. Sci. Technol.*, B18 (2000) 799.
[8] L.M.H. Groenewoud, G.H.M. Engbers and J.G.A. Terlingen, *Langmuir*, 16 (2000) 6278.
[9] C. Tarducci, W.C.E. Schofield and J.P.S. Badyal, *Chem. Mater.*, 13 (2001) 1800.
[10] Y.J. Wu, A.J. Griggs, J.S. Jen, S. Manolache, F.S. Denes and R.B. Timmons, *Plasmas Polym.*, 6 (2001) 123.
[11] C. Tarducci, W.C.E. Schofield and J.P.S. Badyal, *Chem. Mater.*, 14 (2002) 2541.
[12] S. Fraser, R.D. Short, D. Barton and J.W. Bradley, *J. Phys. Chem.*, B106 (2002) 5596.
[13] J. Hu, C. Yin, H.-Q. Mao, K. Tamada and W. Knoll, *Avd. Funct. Mater.*, 13 (2003) 692.
[14] J. Zhang, X. Feng, H. Xie, Y. Shi, T. Pu and Y. Guo, *Thin Solid Films*, 435 (2003) 108.
[15] A. Choukourov, H. Biederman, D. Slavinska, M. Trchova and A. Hollander, *Surf. Coatings Technol.*, 174–175 (2003) 863.
[16] I. Keen, G.A. George and P.M. Fredricks, *J. Appl. Polym. Sci.*, 88 (2003) 1643.
[17] L. Zajícková, S. Rudakowski, H.-W. Becker, D. Meyer, M. Valtr and K. Wiesemann, *Thin Solid Films*, 425 (2003) 72.
[18] P.A. Tamirisa, K.C. Liddell, P.D. Pedrow and M.A. Osman, *J. Appl. Polym. Sci.*, 93 (2004) 1327.
[19] A.E. Lefohn, N.M. Mackie and E.R. Fisher, *Plasmas Polym.*, 3 (1998) 197.
[20] N. Inagaki, S. Tasaka, H. Ohmori and S. Mibu, *J. Adhesion Sci. Technol.*, 10 (1996) 243.
[21] N. Inagaki, S. Tasaka and Y. Yamada, *J. Polym. Sci., Polym. Chem. Ed.*, 30 (1992) 2003.
[22] N. Inagaki, S. Tasaka and Z. Chengfei, *Polym. Bull.*, 26 (1991) 187.
[23] M. Strobel, C.S. Lyons, P.A. Thomas and M.C. Morgen, *J. Appl. Polym. Sci., Appl. Polym. Symp.*, 42 (1988) 73.
[24] N. Inagaki, *Plasma Surface Modification and Plasma Polymerization*, Technomic Pub., Lancaster Pen, 1996.
[25] N. Inagaki, K. Narushima, N. Tsuchida and K. Miyazaki, *J. Polym. Sci., Polym. Phys. Ed.*, 42 (2004) 3727.
[26] N. Inagaki, K. Narushima and S.K. Lim, *J. Appl. Polym. Sci.*, 89 (2003) 96.
[27] Y. Yamada, T. Yamada, S. Tasaka and N. Inagaki, *Macromolecules*, 29 (1996) 4331.
[28] N. Inagaki, S. Tasaka and T. Umehara, *J. Appl. Polym. Sci.*, 71 (1999) 2191.
[29] N. Inagaki, unpublished data.
[30] N. Inagaki, S. Tasaka and S. Masumoto, *Macromolecules*, 29 (1996) 1642.
[31] V.N. Vasilets, G. Hermel, U. König, C. Werner, M. Müller, F. Simon, K. Grundke, Y. Ikada and H.-J. Jacobasch, *Biomaterials*, 18 (1997) 1139.
[32] N. Inagaki, S. Tasaka and Y. Goto, *J. Appl. Polym. Sci.*, 66 (1997) 77.

[33] N. Inagaki, S. Tasaka and T. Inoue, *J. Appl. Polym. Sci.*, 69 (1998) 1179.
[34] S. Mishima and T. Nakagawa, *J. Appl. Polym. Sci.*, 73 (1999) 1835.
[35] K. Yamada, T. Haraguchi and T. Kajiyama, *J. Appl. Polym. Sci.*, 75 (2000) 284.
[36] L. Liang, M. Shi, V.V. Viswanathan, L.M. Peurrung and J.S. Youngm, *J. Membr. Sci.*, 177 (2000) 97.
[37] B. Gupta, J.G. Hilborn, I. Bisson and P. Frey, *J. Appl. Polym. Sci.*, 81 (2001) 2993.
[38] B. Gupta, C. Plummer, I. Bisson, P. Frey and J. Hilborn, *Biomaterials*, 23 (2001) 863.
[39] S.Y. Kim, T. Kanamori and T. Shinbo, *J. Appl. Polym. Sci.*, 84 (2002) 1168.
[40] X.P. Zou, E.T. Kang and K.G. Neoh, *Plamas Polym.*, 7 (2002) 151.
[41] D.S. Wavhal, E.R. Fisher and R. Ellen, *J. Membr. Sci.*, 209 (2002) 255.
[42] X.-L. Wang, J. Huang, X.-Z. Chen and X.-H. Yu, *Desalination*, 146 (2002) 337.
[43] W.C. Wang, E.T. Kang and K.G. Neoh, *Appl. Surf. Sci.*, 199 (2002) 52.
[44] Z.J. Yu, E.T. Kang and K.G. Neoh, *Langmuir*, 18 (2002) 10221.
[45] D.S. Wavhal and E.R. Fisher, *Langmuir*, 19 (2003) 79.
[46] L. Ying, C. Yin, R.X. Zhuo, K.W. Leong, H.Q. Mao, E.T. Kang and K.G. Neoh, *Biomacromolecules*, 4 (2003) 157.
[47] F. Hochart, J.R. De and J. Levalois-Grutzmacher, *Surf. Coatings Technol.*, 165 (2003) 201.
[48] I. Keen, G.A. George and P.M. Fredericks, *J. Appl. Polym. Sci.*, 88 (2003) 1643.
[49] Y.-J. Choi, S.-H. Moon, T. Yamaguchi and S. Nakao, *J. Polym. Sci., Polym. Chem. Ed.*, 41 (2003) 1216.
[50] R.-Q. Kou, Z.-K. Xu, H.-T. Deng, Z.-M. Liu, P. Seta and Y. Xu, *Langmuir*, 19 (2003) 6869.
[51] C.-L. Hsueh, Y-J. Peng, C.-C. Wang and C.-Y. Chen, *J. Membr. Sci.*, 219 (2003) 1.
[52] V. Sciarratta, U. Vohrer, D. Hegemann, M. Müller and C. Oehr, *Surf. Coatings Technol.*, 174–175 (2003) 805.
[53] H.-S. Choi, Y-S. Kim, Y. Zhang, S. Tang, S.-W. Myung and B.-C. Shin, *Surf. Coatings Technol.*, 182 (2004) 55.
[54] N. Abidi and E. Hequet, *J. Appl. Polym. Sci.*, 93 (2004) 145.
[55] C.-Y. Tu, C.-P. Chen, Y.-C. Wang, C.-L. Li, H-A. Tsai, K.-R. Lee and J.-Y. Lai, *Eurp. Polym. J.*, 40 (2004) 1541.
[56] Y. Chen and P. Liu, *J. Appl. Polym. Sci.*, 93 (2004) 2014.
[57] U. König, M. Nitschke, A. Menning, C. Sperling, F. Simon, C. Arnhold, C. Wener, H.-J. Jacobasch, *Surf. Coatings Technol.*, 116–119 (1999) 1011.
[58] C. Oehr, M. Müller, B. Elkin, D. Hegemann and U. Vohrer, *Surf. Coatings Technol.*, 116–119 (1999) 25.
[59] S. Alvarez-Blanco, S. Mnolache and F. Denes, *Polym. Bull.*, 47 (2001) 329.

Index